Petrophysics

Djebbar Tiab

This book is dedicated to my late parents, children, brothers, and sister. Last but not least, to my ex-wives: brainy Teresa, beautiful Twylah, and crazy Salima for giving me the best years (20) of their lives.

Erle C. Donaldson

This book is dedicated to my children; to the late Robert T. Johansen, for his encouragement, inspiration, and contributions to the field of enhanced oil recovery; and to the most important woman in my life: my wife Grace.

Petrophysics

Theory and Practice of Measuring Reservoir Rock and Fluid Transport Properties

Third edition

Djebbar Tiab

Erle C. Donaldson

ELSEVIER

AMSTERDAM • BOSTON • HEIDELBERG
LONDON • NEW YORK • OXFORD • PARIS
SAN DIEGO • SAN FRANCISCO • SINGAPORE
SYDNEY • TOKYO
Gulf Professional Publishing is an imprint of Elsevier

GP
PP

Gulf Professional Publishing is an imprint of Elsevier
225 Wyman Street, Waltham, MA 02451, USA
The Boulevard, Langford Lane, Kidlington, Oxford OX5 1GB, UK

First edition 1996
Second edition 2003
Third edition 2012

Notice
Knowledge and best practice in this field are constantly changing. As new research and experience broaden our understanding, changes in research methods, professional practices, or medical treatment may become necessary.

Practitioners and researchers must always rely on their own experience and knowledge in evaluating and using any information, methods, compounds, or experiments described herein. In using such information or methods they should be mindful of their own safety of others, including parties for whom they have a professional responsibility.

To the fullest extent of the law, neither the Publisher nor the authors, contributors, or editors, assume any liability for any injury and/or damage to persons or property as a matter of products liability, negligence or otherwise, or from any use operation of any methods, products, instructions, or ideas contained in the material herein.

Library of Congress Cataloging-in-Publication Data
A catalog record for this book is available from the Library of Congress

British Library Cataloguing-in-Publication Data
A catalogue record for this book is available from the British Library

ISBN: 978-0-12-383848-3

For information on all Elsevier publications
visit our website at elsevierdirect.com

Printed and bound in United States of America

12 11 10 9 8 7 6 5 4 3 2 1

Contents

Preface

Petrophysics was revised with the addition of two new chapters: Chapter 10: Reservoir Characterization and Chapter 12, Basic Well-log Interpretation (including FORTRAN programs). The addition of these chapters extends the scope of the book with the basics of two topics which are intended to furnish lucid introductions leading to more extensive study of the topics.

The other chapters have remained with some additions suggested by readers who generously conveyed their support and advice. A more extensive discussion of the concept of flow units is included in Chapter 3. The art of hydraulic fracturing which is currently being modified and adapted to use in extended horizontal wells in shale beds is addressed in two chapters related to rock mechanics: Chapter 8 on naturally fractured reservoirs and Chapter 9 on the effect of stress on reservoir rock properties. Rock mechanics and hydraulic fracturing has become important as greater reserves of natural gas is discovered in shale beds world wide. The experimental study of petrophysics has not changed and hence there were no modifications to this section.

The intent of this book is to present the developed concepts, theories, and laboratory procedures related to the porous rock properties and their interactions with fluids (gases, hydrocarbon liquids, and aqueous solutions). The properties of porous subsurface rocks and the fluids they contain govern the rates of fluid flow and the amounts of residual fluids that remain in the rocks after all economical means of hydrocarbon production have been exhausted. It is estimated that the residual hydrocarbons locked in place after primary and secondary production, on a worldwide scale are about 40% of the original volume in place. This is a huge hydrocarbon resource target for refined reservoir characterization (using the theories and procedures of petrophysics) to enhance the secondary recovery or implement tertiary (enhanced oil recovery) technology. The use of modern methods for reservoir characterization with a combination of petrophysics and mathematical modeling is bringing new life into many old reservoir that are near the point of abandonment. This book brings together the theories and procedures from the scattered sources in the literature.

Djebbar Tiab
Erle C. Donaldson

About the Authors

Djebbar Tiab is a senior professor of petroleum engineering at the University of Oklahoma, and a petroleum engineering consultant. He received his B.Sc. (May 1974) and M.Sc. (May 1975) degrees from the New Mexico Institute of Mining and Technology and his Ph.D. degree (July 1976) from the University of Oklahoma, all in petroleum engineering. He is the director of the University of Oklahoma Graduate Program in Petroleum Engineering in Algeria.

At the University of Oklahoma, he taught 15 different petroleum and general engineering courses including well-test analysis, petrophysics, oil reservoir engineering, natural gas engineering, and properties of reservoir fluids. Dr. Tiab has consulted for a number of oil companies and offered training programs in petroleum engineering in the United States and overseas. He worked for over 2 years in the oil fields of Algeria for Alcore, S.A., an association of Sonatrach and Core Laboratories. He has also worked and consulted for Core Laboratories and Western Atlas in Houston, Texas, for 4 years as a senior reservoir engineer advisor.

As a researcher at the University of Oklahoma, he received several research grants and contracts from oil companies and various U.S. agencies. He supervised 23 Ph.D. and 94 M.S. students at the University of Oklahoma. He is the author of more than 150 conference and journal technical papers. In 1975 (M.S. thesis) and 1976 (Ph.D. dissertation), he introduced the pressure derivative technique, which revolutionized the interpretation of pressure transient tests. He developed two patents in the area of reservoir characterization (identification of flow units). Dr. Tiab is a member of the U.S. Research Council, Society of Petroleum Engineers, the Society of Core Analysts, Pi Epsilon Tau, Who is Who, and American Men and Women of Science. He served as a technical editor of various SPE, Egyptian, Kuwaiti, and U.A.E. journals, and as a member of the SPE Pressure Analysis Transaction Committee. He is a member of the SPE Twenty-Five Year Club.

He has received the Outstanding Young Men of America Award, the SUN Award for Educational Achievement, the Kerr–McGee Distinguished Lecturer Award, the College of Engineering Faculty Fellowship of Excellence, the Halliburton Lectureship Award, the UNOCAL Centennial Professorship, and the P&GE Distinguished Professorship. In October 2002, Dr. Tiab was elected to the Russian Academy of Natural Sciences as a foreign member because of his "outstanding work in petroleum engineering." In October 2002, he was also awarded the Kapista Gold Medal of Honor for his "outstanding contributions to the field of engineering." He received the prestigious 1995 SPE Distinguished Achievement Award for Petroleum Engineering Faculty. The citation read, "He is recognized for his role in student development and his excellence in classroom instruction. He pioneered the pressure derivative technique of well testing and has contributed considerable understanding to petrophysics and reservoir engineering through his research and writing." He is also the recipient of the 2003 SPE Formation Evaluation Award for "distinguished contributions to petroleum engineering in the area of formation evaluation."

Erle C. Donaldson began his career as a pilot plant project manager for Signal Oil and Gas Research in Houston, Texas. Later he joined the U.S. Bureau of Mines Petroleum Research Center in Bartlesville, Oklahoma, as a project manager of subsurface disposal and industrial wastes and reservoir characterization; when the laboratory transferred to the U.S. Department of Energy, Dr. Donaldson continued as the chief of petroleum reservoir characterization. When the laboratory shifted to private industry for operations, he joined the Faculty of the School of Petroleum and Geological Engineering at the University of Oklahoma as an associate professor. Since retiring from the university in 1990, he has consulted for various oil companies, universities, and U.S. agencies including the Environmental Protection Agency, the U.S. Naval Ordnance Center, King Fahd Research Institute of Saudi Arabia, and companies in the United States, Brazil, Venezuela, Bolivia, and Thailand. He is currently the senior consulting engineer for Tetrahedron, Inc.

Dr. Donaldson has earned four degrees: a B.Sc. in chemistry from The Citadel, an M.S. in organic chemistry from the University of South Carolina, a B.Sc. in chemical engineering from the University of Houston, and a Ph.D. in chemical engineering from the University of Tulsa. He has served as the chairman of committees and sessions for the Society of Petroleum Engineers and the American Chemical Society, as well as other national and international conferences. He was the managing editor of the *Journal of Petroleum Science and Engineering* for 20 years. He is a member of the SPE Twenty-Five Year Club.

Acknowledgment

The authors are especially indebted to academician George V. Chilingar, professor of civil and petroleum engineering at the University of Southern California, Los Angeles, who acted as the technical, scientific, and consulting editor.

They can never thank him enough for his prompt and systematic editing of this book. He is forever their friend.

Units

Units of Area

$$\text{Acre} = 43,540\ \text{ft}^2 = 4046.9\ \text{m}^2$$
$$\text{ft}^2 = 0.0929\ \text{m}^2$$
$$\text{Hectare} = 10,000\ \text{m}^2$$

Constants

$$\text{Darcy} = 0.9869\ \ \text{mm}^2$$
$$\text{Gas constant} = 82.05\ (\text{atm} \times \text{cm}^3)/(\text{g mol} \times \text{K})$$
$$= 10.732\ (\text{psi} \times \text{ft}^3)/(\text{lb mol} \times {}^\circ\text{R})$$
$$= 0.729\ (\text{atm} \times \text{ft}^3)/(\text{lb mol} \times {}^\circ\text{R})$$
$$\text{Mol. wt. of air} = 28.97$$

Units of Length

$$\text{Angstrom} = 1 \times 10^{-8}\ \text{cm}$$
$$\text{cm} = 0.3937\ \text{in.}$$
$$\text{ft} = 30.481\ \text{cm}$$
$$\text{in.} = 2.540\ \text{cm}$$
$$\text{km} = 0.6214\ \text{mile}$$
$$\text{m} = 39.370\ \text{in.} = 3.2808\ \ \text{ft}$$

Units of Pressure

$$\text{atm} = 760\ \text{mmHg}(0^\circ\text{C}) = 29.921\ \text{in. Hg} = 14.696\ \text{psi}$$
$$\text{atm} = 33.899\ \text{ft water at}\ 4^\circ\text{C}$$
$$\text{bar} = 14.5033\ \text{psi} = 0.987\ \text{atm} = 0.1\ \text{MPa}$$
$$\text{dyne}/\text{cm}^2 = 6.895\ \text{kPa(kilopascal)}$$
$$\text{ft water} = 0.4912\ \text{psi}$$
$$\text{kg(force)}/\text{cm}^2 = 14.223\ \text{psi}$$
$$\text{psi} = 2.036\ \text{in.Hg}\ (0^\circ C) = 6.895\ \text{kPa}$$

Units of Temperature

$$\text{Degrees Fahrenheit}(^\circ\text{F}) = 1.8\ ^\circ\text{C} + 32$$
$$\text{Degrees Rankine}(^\circ\text{R}) = 459.7 + {}^\circ\text{F}$$
$$\text{Degrees Kelvin}(\text{K}) = 273.16 + {}^\circ\text{C}$$

Units of Volume

$$\text{acre-ft} = 43{,}560\ \text{ft}^3 = 7{,}758.4\ \text{bbl} = 1.2335 \times 10^3\text{m}^3$$
$$\text{bbl} = 42\ \text{US gal} = 5.6145\ \text{ft}^3 = 0.1590\ \text{m}^3$$
$$\text{cu ft}(\text{ft}^3) = 7.4805\ \text{gal} = 0.1781\ \text{bbl} = 0.028317\ \text{m}^3$$
$$\text{cu in}(\text{in}^3) = 16.387\ \text{cm}^3$$
$$\text{cu m}(\text{m}^3) = 6.2898\ \text{bbl}$$
$$\text{gal} = 231\ \text{in}^3 = 3{,}785.43\ \text{cm}^3$$

Molarity = mass of solute equal to the molecular weight per 1,000 g of solvent

Normality = equivalent weight of solute per 1,000 g of solvent (mass of solute equal to the molecular weight divided by the valence per 1,000 g of solvent)

Chapter 1

Introduction to Mineralogy

Petrophysics is the study of rock properties and their interactions with fluids (gases, liquid hydrocarbons, and aqueous solutions). The geologic material forming a reservoir for the accumulation of hydrocarbons in the subsurface must contain a three-dimensional network of interconnected pores in order to store the fluids and allow for their movement within the reservoir. Thus, the porosity of the reservoir rocks and their permeability are the most fundamental physical properties with respect to the storage and transmission of fluids. Accurate knowledge of these two properties for any hydrocarbon reservoir, together with the fluid properties, is required for efficient development, management, and prediction of future performance of the oilfield.

The purpose of this book is to provide a basic understanding of the physical properties of porous geologic materials, and the interactions of various fluids with the interstitial surfaces and the distribution of pores of various sizes within the porous medium. Procedures for the measurement of petrophysical properties are included as a necessary part of this text. Applications of the fundamental properties to subsurface geologic strata must be made by analyses of the variations of petrophysical properties in the subsurface reservoir.

Emphasis is placed on the testing of small samples of rocks to uncover their physical properties and their interactions with various fluids. A considerable body of knowledge of rocks and their fluid flow properties has been obtained from studies of artificial systems such as networks of pores etched on glass plates, packed columns of glass beads, and from outcrop samples of unconsolidated sands, sandstones, and limestones. These studies have been used to develop an understanding of the petrophysical and fluid transport properties of the more complex subsurface samples of rocks associated with petroleum reservoirs. This body of experimental data and production analyses of artificial systems, surface rocks, and subsurface rocks makes up the accumulated knowledge of petrophysics. Although the emphasis of this text is placed on the analyses of small samples, the data are correlated to the macroscopic performance of the petroleum reservoirs whenever applicable. In considering a reservoir as a whole, one is confronted with the problem of the distribution of these properties within the reservoir and its stratigraphy. The directional distribution of thickness, porosity, permeability, and geologic features that contribute to heterogeneity governs the natural pattern of fluid flow. Knowledge of this natural pattern is sought to design the most efficient injection–production system for economy of energy and maximization of hydrocarbon production [1].

Petrophysics is intrinsically bound to mineralogy and geology because the majority of the world's petroleum occurs in porous sedimentary rocks. The sedimentary rocks are composed of fragments of other rocks derived from mechanical and chemical deterioration of igneous, metamorphic, and other sedimentary rocks, which is constantly occurring. The particles of erosion are frequently transported to other locations by winds and surface streams and deposited to form new sedimentary rock structures. Petrophysical properties of the rocks depend largely on the depositional environmental conditions that controlled the mineral composition, grain size, orientation or packing, amount of cementation, and compaction.

MINERAL CONSTITUENTS OF ROCKS—A REVIEW

The physical properties of rocks are the consequence of their mineral composition. Minerals are defined here as naturally occurring chemical elements or compounds formed as a result of inorganic processes. The chemical analysis of six sandstones by emission spectrography and X-ray dispersive scanning electron microscopy [2] showed that the rocks are composed of just a few chemical elements. Analysis of the rocks by emission spectroscopy yielded the matrix chemical composition since the rocks were fused with lithium to make all of the elements soluble in water, and then the total emission spectrograph was analyzed. The scanning electron microscope X-ray, however, could only analyze microscopic spots on the broken surface of the rocks. The difference between the chemical analysis of the total sample and the spot surface analysis is significant for consideration of the rock–fluid interactions. The presence of the transition metals on the surface of the rocks induces preferential wetting of the surface by oil through Lewis acid–base type reactions between the polar organic compounds in crude oils and the transition metals exposed in the pores [3]. The high surface concentration of aluminum reported in Table 1.1 is probably due to the ubiquitous presence of clay minerals in sandstones.

The list of elements that are the major constituents of sedimentary rocks (Table 1.1) is confirmed by the averages of thousands of samples of the crust reported by Foster [4] (Table 1.2). Just eight elements make up 99% (by weight) of the minerals that form the solid crust of the Earth; these are the elements, including oxygen, listed in the first seven rows of Table 1.1 from analysis of six sandstones. Although the crust appears to be very heterogeneous

TABLE 1.1 Average of the Compositions of Six Sandstone Rocks (Reported as Oxides of Cations) Obtained by Emission Spectroscopy and the Scanning Electron Microscope [2]

Mineral Oxides	Total Analysis (Emission Spectrograph)	Surface Analysis (Scanning Electron Microscope)
Silicon oxide (SiO_2)	84.1	69.6
Aluminum oxide (Al_2O_3)	5.8	13.6
Sodium oxide (NaO)	2.0	0.00
Iron oxide (Fe_2O_3)	1.9	10.9
Potassium oxide (K_2O)	1.1	3.0
Calcium oxide (CaO)	0.70	2.1
Magnesium oxide (MgO)	0.50	0.00
Titanium oxide (TiO)	0.43	1.9
Strontium oxide (SrO)	0.15	0.00
Manganese oxide (MnO)	0.08	2.0

TABLE 1.2 Weight and Volume of the Principal Elements in the Earth's Crust

Element	Weight Percent	Volume Percent
Oxygen	46.40	94.05
Silicon	28.15	0.88
Aluminum	8.23	0.48
Iron	5.63	0.48
Calcium	4.15	1.19
Sodium	2.36	1.11
Magnesium	2.33	0.32
Potassium	2.09	1.49

Source: *Courtesy C.E. Merrill Publishing Co., Columbus, OH.*

TABLE 1.3 List of the Principal Sedimentary Rocks

		Sedimentary Rocks
Mechanism	**Formation**	**Composition**
Mechanical weathering	Sandstone	Quartzose—Quartz grains, deltaic in origin
		Arkosic—more than 20% feldspar grains
		Graywacke—Poorly sorted grains of other rocks with feldspar and clay
		Calcareous—Fragments of limestone
	Friable sand	Clastics—Loosely cemented grains of other rocks
	Unconsolidated sand	Clastics—Loose sand grains from other rocks
	Siltstone	Clastics—Compacted, cemented, fine-grained clastics with grain size less than 1/16 mm
	Conglomerate	Gravel and boulders cemented with mud and fine sand
Chemical weathering	Shale	Clay—Compacted fine-grained particles with grain size less than 1/256 mm. Usually laminated in definite horizontal bedding planes. As oil shale it contains organic matter (kerogen)
	Evaporites	Salts and some limestone
		Gypsum ($CaSO_4 \cdot 2H_2O$)
		Anhydrite ($CaSO_4$)
		Chert (SiO_2)
		Halite (NaCl)
		Limestone ($CaCO_3$)
	Dolomite	Carbonate—Chemical reaction with limestone ($CaMg(CO_3)_2$)
	Limestone	Carbonate—Biological extraction of calcium and precipitation of $CaCO_3$
Biological origin	Reefs	Carbonate—Fossil remains of marine organisms
	Diatomite	Silicates—Silicate remains of microscopic plants

with respect to minerals and types of rocks, most of the rock-forming minerals are composed of silicon and oxygen together with aluminum and one or more of the other elements listed in Table 1.2.

The chemical compositions and quantitative descriptions of some minerals are listed in Tables 1.3 and 1.4. Some of the minerals are very

TABLE 1.1 Mineral Compositions and Descriptions. Parentheses in the Formulas Mean That the Elements Enclosed May Be Present in Varying Amounts

Agate (Chalcedony)—SiO_2: silicon dioxide; variable colors; waxy luster; H = 7*.
Anhydrite—$CaSO_4$: calcium sulfate; white-gray; H = 2.
Apatite—$Ca_5(PO_4)_3F$: fluorapatite; H = 4.
Asbestos (serpentine)—$Mg_6Si_4O_{10}(OH)_8$: hydrous magnesium silicate; light green to dark gray; greasy or waxy; H = 3.
Augite (pyroxene group)—(Ca, Na)(Al, Fe^{2+}, Fe^{3+}, Mg)(Si, Al)$_2O_6$: Alkali, ferro-magnesium, aluminum silicates; dark green to black; exhibits cleavage; large, complex group of minerals; H = 5.
Barite—$BaSO_4$: barium sulfate; white, light blue, yellow, or red; pearly luster; H = 3.
Beryl—$Be_3Al_2Si_6O_{18}$: clear beryl forms the blue-green aquamarine and green emerald gems; exhibits cleavage. Beryl is an ore of the element beryllium; H = 7 to 8.
Biotite (mica)—$K(Fe, Mg)_3(AlSi_3O_{10})(OH)_2$: hydrous potassium, ferro-magnesium, aluminum silicate; dark green to black (black mica); vitreous; exhibits cleavage; rock-forming mineral; H = 3.
Calcite—$CaCO_3$: calcium carbonate; colorless or white to light brown, vitreous; effervesces in dilute HCl; H = 3.
Celestite—$SrSO_4$: strontium sulfate; colorless; H = 3.
Chalk—$CaCO_3$: calcite; white; soft fine-grained limestone formed from microscopic shells; effervesces with dilute HCl; H = 2 to 3.
Chlorite—$(Al, Fe, Mg)_6(Al, Si)_4O_{10}(OH)_8$: hydrous ferro-magnesium, aluminum silicate; shades of green (green mica); exhibits cleavage; rock-forming mineral; H = 3.
Cinnabar—HgS: mercury sulfide; red to brownish-red; luster is dull; only important ore of mercury; H = 2.5.
Cordierite—$Al_4(Fe, Mg)_2Si_5O_{18}$: ferro-magnesium, aluminum silicate; blue; vitreous; H = 7.
Corundum—Al_2O_3: red varieties are rubies and other colors are known as sapphire; H = 9.
Diatomite—SiO_2: silica; white; formed from microscopic shells composed of silica; distinguished from chalk by lack of effervescence with dilute HCl; H = 1 to 2.
Dolomite—$CaMg(CO_3)_2$: calcium-magnesium carbonate; pink or light brown, vitreous-pearly; effervesces in HCl if powdered; H = 3.
Feldspar (orthoclase, potassium feldspar)—$KAlSi_3O_8$: white to pink; vitreous; large crystals with irregular veins; exhibits cleavage; rock-forming mineral; H = 6.
Feldspar (plagioclase)—$CaAl_2Si_2O_8$ and $NaAlSi_3O_8$(albite): calcium and sodium aluminum silicate; white to green; vitreous; exhibits cleavage; rock-forming mineral; H = 6.
Fluorite—CaF_2: calcium fluorite; H = 4.
Galena—PbS: lead sulfide; lead-gray; bright metallic luster; lead ore; H = 2.5.
Graphite—C: carbon; gray to black; metallic luster; H = 2.

(Continued)

TABLE 1.4 (Continued)

Gypsum—$CaSO_4 \cdot 2H_2O$: hydrous calcium sulfate; transparent to white or gray; vitreous-pearly-silky; H = 2.
Halite—NaCl: sodium chloride; colorless to white; vitreous-pearly; H = 2.
Hematite—Fe_2O_3: iron oxide (the most important iron ore); reddish-brown to black or gray, H = 6.
Hornblende (amphibole group)—$Ca_2Na(Fe^2, Mg)_4(Al, Fe^3, Ti)(Al, Si)_8(O, OH)_2$: hydrous alkali, ferro-magnesium, aluminum silicates; dark green to black; exhibits cleavage. The iron and magnesium impart the dark color; H = 5.
Illite (muscovite)—$KAl_2(AlSi_3O_{10})(OH)_2$: hydrous potassium-aluminum silicate; clear to light green, vitreous; not chemically well defined but with the approximate composition of muscovite; H = 2.5.
Kaolinite (clay)—$Al_4(Si_4O_{10})(OH)_4$: hydrous aluminum silicate; light colored; H = 1 to 2.
Limonite (goethite)—$FeO(OH) \cdot H_2O$: hydrous iron oxide; yellow-brown to dark brown; H = 5.
Magnetite—Fe_3O_4: iron oxide; black; metallic luster; strongly magnetic iron ore; H = 6.
Montmorillonite (smectite clay)—$(CaNa)(Al, Fe, Mg)_4(Si, Al)_8(OH)_8$; generally light colored; H = 1.
Muscovite (mica)—$KAl_2(AlSi_3O_{10})(OH)_2$: hydrous potassium-aluminum silicate; clear to light green; vitreous; rock-forming mineral; H = 2.5.
Olivine—$(Fe, Mg)_2SiO_4$: ferro-magnesium silicate; clear to light green, various shades of green to yellow; vitreous (glassy) luster with crystals in the rock; H = 7.
Opal—$SiO_2 \ldots nH_2O$: hydrous silicon dioxide; variety of almost any color; glassy luster; H = 5.
Pyrite—FeS_2: iron sulfide; pale yellow; bright metallic luster; H = 6.
Quartz—SiO_2: silicon dioxide; clear (transparent) or with a variety of colors imparted by impurities (purple amethyst, yellow citrine, pink rose quartz, brown smoky quartz, snow-white chert, multiple-colored agate); glassy luster; H = 7.
Serpentine—$Mg_3Si_2O_5(OH)_4$: hydrous magnesium silicate; beige color; H = 3.
Siderite—$FeCO_3$: ferrous carbonate; light colored to brown; H = 3 to 4.
Sphalerite—ZnS: zinc sulfide; yellow to dark brown or black; resinous luster; exhibits cleavage; zinc ore; H = 3.
Sulfur—S: yellow; resinous; H = 1 to 2.
Sylvite—KCl: potassium chloride; colorless to white; H = 1 to 2.
Talc—$Mg_3(Si_4O_{10})(OH)_2$: hydrous magnesium silicate; green, gray, or white; soapy to touch; H = 1.
Topaz—$Al_2(SiO_4)(Fe,OH)_2$: yellow, pink, blue-green; exhibits cleavage; H = 8.
Turquoise—$CuAl_6(PO_4)_4(OH)_8 \cdot 2H_2O$: blue or green color; H = 5.
Vermiculite—$Mg_3Si_4O_{10}(OH)_2 \cdot nH_2O$: hydrous magnesium silicate; light colored; H = 1.

[*]H, hardness; defined in the Glossary.

complex and their chemical formulas differ in various publications; in such cases the most common formula reported in the list of references was selected.

IGNEOUS ROCKS

Igneous rocks (about 20% of all rocks) are the product of the cooling of molten magma intruding from below the mantle of the crust. Igneous (plutonic) rocks are divided into three easily recognizable rocks, which are subdivided by the rate of cooling (Figure 1.1). The granites are intrusive rocks that cooled slowly (at high temperature) below the surface, whereas gabbro is a rock resulting from more rapid (low temperature) cooling in the subsurface. Diorite is a rock that cooled below the surface at a temperature intermediate between that of granite and gabbro. The minerals differentiate during the slow cooling, forming large recognizable, silica-rich crystals with a rough (phaneritic) texture.

The second classification is extrusive (volcanic) rock that has undergone rapid cooling on or near the surface, forming silica-poor basaltic rocks. Rhyolite, or felsite, is light colored and estimated to be produced on the surface at a lower temperature than the darker andesite that formed at a temperature intermediate between that of rhyolite and the dark-colored basalt. As a result of rapid cooling on the surface, these rocks have a fine (alphanitic) texture with grains that are too small to be seen by the unaided eye [5].

Minerals precipitating from melted magma, or melt, do not crystallize simultaneously. Generally, a single mineral precipitates first and, as the melt cools slowly, this is joined by a second, third, and so on; thus the earlier-formed minerals react with the ever-changing melt composition. If the reactions are permitted to go to completion, the process is called equilibrium crystallization. If the crystals are completely or partially prevented from reacting with the melt (by settling to the bottom of the melt or by being removed), fractional crystallization takes place and the final melt composition will be different from that predicted by equilibrium crystallization. The mechanism by which crystallization takes place in a slowly cooling basaltic melt was summarized by Bowen [6] as two series of simultaneous reactions; after all of the ferro-magnesium minerals are formed, a third series of minerals begins to crystallize from the melt. From laboratory experiments, Bowen discovered that the first two series of reactions have two branches.

The Bowen series of specific crystallization occurs only for some basaltic magmas (a variety of different reaction series occur within different melts), but the processes discussed by Bowen are significant because they explain the occurrence of rocks with compositions different from that of the original melted magma.

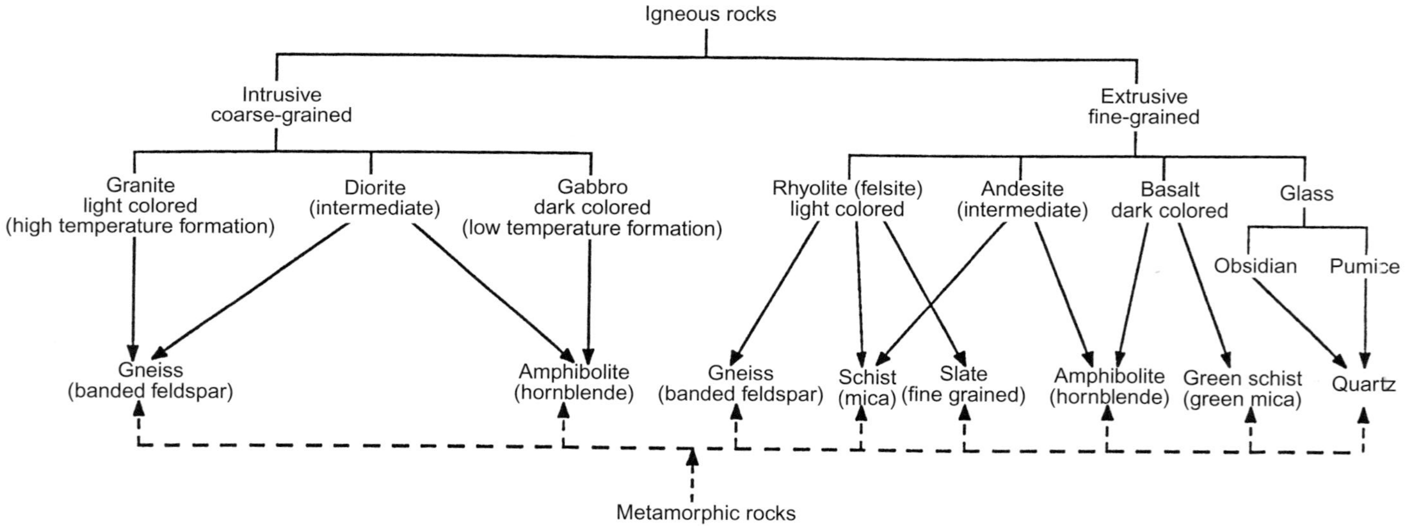

FIGURE 1.1 Origin of the principal metamorphic rocks. (a) The plagioclase grade into each other as they crystallize; the crystals react continually with the melt and change composition from an initial calcium plagioclase crystal to sodium plagioclase. (b) The other series of crystallization that is taking place simultaneously forms minerals that are compositionally distinct. The reaction series (olivine–pyroxene–amphibole–biotite) is discontinuous; thus the reaction between crystals and the melt occurs only during specific periods of the cooling sequence. (c) After all of the ferro-magnesium minerals and plagioclase are formed, the third series of minerals begins to crystallize as the melt continues to cool slowly. First potassium feldspar precipitates, followed by muscovite and finally quartz [7–9].

METAMORPHIC ROCKS

The metamorphic rocks (about 14% of all rocks) originate from mechanical, thermal, and chemical changes of igneous rocks [10]. Mechanical changes on or near the surface are due to the expansion of water in cracks and pores, tree roots, and burrowing animals. If the igneous rocks undergo deep burial due to subsidence and sedimentation, the pressure exerted by the overlying rocks, shear stress from tectonic events, and the increased temperature result in mechanical fracturing. When unequal shear stress is applied to the rocks as a result of continental motion of other force fields, cleavage of the rocks (fracturing) occurs; alternatively, slippage of a regional mass of rocks and sediments (faulting) occurs. The pressure produced by overlying rocks is approximately 1.0 psi per foot of depth (21 kPa per meter of depth). The changes induced by overburden pressure occur at great depth in conjunction with other agents of metamorphism.

Chemical metamorphosis of igneous intrusive rocks, aided by high pressure, temperature, and the presence of water, results in chemical rearrangement of the elements into new minerals. This produces foliated rocks with regularly oriented bands of mineral grains because the new crystals tend to grow laterally in the directions of least stress. This chemical metamorphism of granite yields gneiss: a foliated granite with large recognizable crystals of banded feldspars. Gabbro changes to amphibolite, whose main constituent is the complex mineral known as hornblende.

The chemical metamorphosis of the extrusive rocks, rhyolite, basalt, etc., produces changes to easily recognizable rocks. Rhyolite, light-colored volcanic rock, undergoes change principally to three types of metamorphic rocks, depending on the environmental conditions inducing the changes: (1) gneiss, which has foliated bands of feldspars; (2) schist or mica; and (3) slate, which is a fine-grained smooth-textured rock. Basalt, the dark-colored volcanic rock, produces two main types of metamorphic rocks: (1) amphibolite and (2) greenschist, or green mica, as illustrated in Figure 1.1.

On a regional scale, the distribution pattern of igneous and metamorphic rocks is belt-like and often parallel to the borders of the continents. For example, the granitic rocks that form the core of the Appalachian mountains in eastern United States are parallel to the east coast and those in the Sierra Nevada are parallel to the west coast.

Igneous and metamorphic rocks are not involved in the origin of petroleum as source rocks. In some cases, they do serve as reservoirs, or parts of reservoirs, where they are highly fractured or have acquired porosity by surface weathering prior to burial and formation into a trap for oil accumulated by tectonic events.

SEDIMENTARY ROCKS

All of the sedimentary rocks (about 66% of all rocks) are important to the study of petrophysics and petroleum reservoir engineering. It is possible to interpret them by considering the processes of rock degradation. The principal sedimentary rocks may be organized according to their origin (mechanical, chemical, and biological) and their composition, as illustrated in Table 1.3.

Mechanical weathering is responsible for breaking large preexisting rocks into small fragments. The most important mechanism is the expansion of water upon freezing, which results in a 9% increase of volume. The large forces produced by freezing of water in cracks and pores results in fragmentation of the rocks. Mechanical degradation of rocks also occurs when a buried rock is uplifted and the surrounding overburden is removed by erosion. The top layers of the rock expand when the overburden pressure is relieved, forming cracks and joints that are then further fragmented by water. Mechanical weathering produces boulder-size rocks, gravel, sand grains, silt, and clay from igneous and metamorphic rocks. These fragments remain in the local area, or they may be transported by winds and water to other sites to enter into the formation of conglomerates, sandstones, etc., as shown in Table 1.3.

Water is the principal contributor to chemical weathering, which occurs simultaneously with mechanical weathering. Mechanical weathering provides access to a large area for contact by water. Chemicals dissolved in the water, such as carbonic acid, enter into the chemical reactions that are responsible for rock degradation. One of the processes that takes place is leaching, which is the transfer of chemical constituents from the rock to the water solution. Some minerals react directly with the water molecules to form hydrates. Carbonic acid, formed from biogenic and atmospheric carbon dioxide dissolved in water, plays an important role in the chemical weathering process by reacting with the minerals to form carbonates and other minerals such as clays. The feldspars react with carbonic acid and water forming various clays, silica, and carbonates, as illustrated in the reaction below for potassium feldspar:

$$2KAlSi_3O_8 + H_2CO_3 + H_2O \rightarrow Al_2SiO_5(OH)_4 + 4SiO_2 + K_2CO_3$$

The sedimentary deposits that make up the large variety of rocks are continually altered by tectonic activity, resulting in deep burial of sediments in zones that are undergoing subsidence. Uplift of other areas forms mountains. The continual movement and collisions of continental plates cause folding and faulting of large blocks of sedimentary deposits. This activity forms natural traps that in many cases have accumulated hydrocarbons migrating from the source rocks in which they were formed. The geologic processes of sedimentation, subsidence, compaction, cementation, uplift, and other structural changes occur continuously on a gradual scale and are intrinsically associated with the physical properties of the rocks as well as the migration and accumulation of hydrocarbon reserves. The physical properties of rocks, such as density, rate of sound transmission, compressibility, and the wetting properties of fluids, are the consequence of the mineral composition of the rocks.

Thus the basic materials that make up the rocks and their chemistry are associated with the petrophysical characteristics of rocks.

Siltstones (Mudrocks)

Quartz grains (originating from weathering of igneous and metamorphic rocks) are very hard; they resist further breakdown, but are winnowed by currents of winds and water and distributed according to size. Larger grains accumulate as sandstones, and grains having an average size of 15 μm mix with clays and organic materials in turbulent aqueous suspensions that are transported and later deposited in quiet, low-energy, valleys from flooding rivers, lakes, and the continental shelves. Tidal currents on the continental shelves effectively sort the grains of sand, silt, and clay once more until they settle in quiet regions, forming very uniform thick beds. Bottom-dwelling organisms burrow through the mud, kneading and mixing it until the depth of burial is too great for this to happen. The material then undergoes compaction and diagenesis, with the clay minerals changing composition as they react with chemicals in the contacting water. The compacted mud forms the siltstones and beds of shale that are encountered throughout the stratigraphic column, making up two-thirds of the sedimentary deposits. Where they overlie hydrocarbon reservoirs, the compacted layers of mud provide seals for the petroleum traps.

Beds of mud containing organic materials that are deposited in anaerobic environments, such as swamps, form siltstones and shales that are gray to black in color. Many of these are the source rocks of petroleum hydrocarbons. Red deposits of mud were exposed to oxygen during burial and the organic material was lost to oxidation while iron compounds formed ferric oxide (Fe_2O_3) that produced the bright red coloration. Brown muds underwent partial oxidation with iron constituents, forming the hydroxide goethite [FeO(OH)]. If the mud does not contain iron, it will exhibit the coloration of the clays (biotite, chlorite, illite, etc.) that range in color from beige to green.

Sandstones

The quartz grains and mixed rock fragments resulting from mechanical and chemical degradation of igneous, metamorphic, and sedimentary rocks may be transported to other areas and later transformed into sandstones.

After the loose sediments of sand, clay, carbonates, etc., are accumulated in a basin area they undergo burial by other sediments forming on top. The vertical stress of the overlying sediments causes compaction of the grains. Transformation into sedimentary rocks occurs by lithification, or cementation, from minerals deposited between the grains by interstitial water. The main cementing materials are silica, calcite, oxides of iron, and clay. The composition of sandstones is dependent on the source of the minerals (igneous, metamorphic, and sedimentary) and the nature of the depositional environment.

Theodorovich [11] used the three most general constituents of sandstones to establish a scheme of classification, which is useful in petroleum engineering because it encompasses the majority of the clastic petroleum reservoirs (Figure 1.2). Only the three most important classifications are shown; many other subdivisions of these were developed by Theodorovich and other investigators, and are summarized by Chilingarian and Wolf [12].

A distinctive feature of sandstones is the bedding planes, which are visible as dark horizontal lines. The bedding planes are the consequence of layered deposition occurring during changing environmental conditions over long periods of deposition in the region. Layering introduces a considerable difference between the vertical (cross-bedding plane direction) and horizontal (parallel to the bedding planes) flow of fluids. The vertical permeability can be 50–75% less than the horizontal permeability; therefore, any fluid flow experiments, or numerical simulations, must account for the directional permeability.

Sandstones that originate from the cementation of wind-blown sand dunes have bedding planes that are oriented at various angles (cross-bedding). Cross-bedding also can be produced by ripples and swirling currents in water while it is transporting the grains.

Clastic sediments transported to continental shelves by rivers are subjected to wave action and currents that sort and transport the grains over large distances. The sediments tend to form rocks that are quite uniform in properties and texture over large regions. The deposits can be several

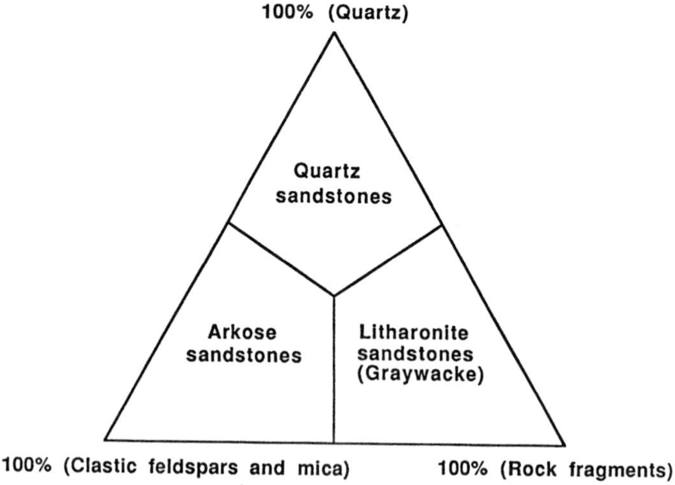

FIGURE 1.2 The major classifications of sandstones, based on composition [11].

kilometers in thickness due to contemporaneous subsidence of the zone during the period of deposition.

Carbonates

Carbonate rocks form in shallow marine environments. Many small lime (CaO)-secreting animals, plants, and bacteria live in the shallow water. Their secretions and shells form many of the carbonate rocks. In addition, calcite can precipitate chemically: calcite is soluble in water containing carbon dioxide; however, if the amount of dissolved carbon dioxide is decreased by changes of environmental conditions, or uplift, the dissolved calcite will precipitate because it is only slightly soluble in water free of carbon dioxide.

There are three major classifications of limestone (which is generally biogenic in origin): oolitic limestone is composed of small spherical grains of calcite (encapsulated fossils and shell fragments); chalk is composed of accumulated deposits of skeletal or shell remains of microscopic animals; and coquina is fossiliferous limestone composed almost entirely of fossil fragments cemented by a calcareous mud.

Dolomite forms in areas where seawater has been restricted, or trapped, by land enclosure where the concentration of salts increases due to evaporation. As the concentration of magnesium increases it reacts with the calcite that has already been deposited to form dolomite by the following reaction:

$$2CaCO_3 + Mg^{2+} \rightarrow CaMg(CO_3)_2 + Ca^{2+}$$

In some cases, the limestone formations are changed to dolomite by reaction with magnesium, which is dissolved in water percolating through pores and fractures in the limestone. Porous carbonate rocks derived from chemical and biogenic precipitation of calcium carbonate form a large portion of the petroleum reservoirs [13].

Evaporites

Evaporites are salts that are deposited in isolated marine basins by evaporation of the water and subsequent precipitation of salts from the concentrated solutions. Salt Lake in Utah, the United States, and the Dead Sea in the Middle East are examples of lakes that are gradually forming beds of evaporites as the water evaporates. Anhydrite ($CaSO_4$), sodium halite (NaCl), sylvite (KCl), and other salts are associated with evaporites.

Table 1.5 contains a general description of the rocks that have been discussed. The principal rock-forming minerals are feldspars, olivine, pyroxene, amphibole, mica, and quartz. Almost all coarse-grained rocks contain feldspars. There are three types of feldspars: calcium-, potassium-, and sodium-aluminum silicates. Other descriptive names that are used for them are placed in parentheses.

TABLE 1.5 General Descriptions of Rocks

Andesite: fine-grained extrusive igneous rock; intermediate color between rhyolite and basalt.
Basalt: fine-grained extrusive igneous rock; dark colored.
Coquina: a form of limestone that is composed of shells and shell fragments
Diorite: coarse-grained intrusive igneous rock intermediate in color between granite and gabbro; composed principally of potassium feldspar (~25%), sodium plagioclase (~35%), biotite (~20%), and hornblende (~20%).
Evaporite: sedimentary rock originating from the evaporation of water.
Gabbro: coarse-grained intrusive igneous rock; dark colored with an approximate composition of calcium plagioclase (~40%), augite (~50%), and olivine (~10%).
Gneiss: coarse-grained, foliated metamorphic rock; contains feldspar and is generally banded. The rock has recrystallized under pressure and temperature with growth of new crystals in bands.
Granite: coarse-grained intrusive, quartz-bearing rock. The coarse texture implies that it came from a large, slowly cooled, intrusive body and has been exposed by uplift and deep erosion (light colored to dark). Granite is generally composed of a mixture of quartz (~35%), potassium feldspar (~45%), biotite (~15%), and hornblende (~5%).
Marble: originates from metamorphosis of limestone or dolomite; the fine crystals of limestone grow bigger and develop an interlocking texture to yield marble.
Rhyolite (felsite): fine-grained, extrusive igneous rock; light colored.
Schist: coarse-grained, foliated, metamorphic rock containing mica; derived from high temperature and pressure metamorphosis of shale.
Shale: fine-grained sedimentary rock composed of clay and silt.
Slate: fine-grained metamorphic rock derived from shale.

PROPERTIES OF SEDIMENTARY PARTICLES

There are a large number of tests that can be made to obtain quantitative and qualitative data for characterization of sedimentary rocks. All of the methods listed in Table 1.6 are discussed at various locations in the book and can be found by reference to the Index. The loose particle analyses are made on disaggregated rock particles that are obtained using a crushing apparatus, or by carefully breaking the rock with a hammer. The other analyses are obtained from core samples of rock, which are oriented parallel to the bedding planes. Tests of the vertical fluid flow properties can be useful for analyses of gravity drainage of oil, vertical diffusion of gas released from solution, and transport properties using mathematical simulation. More recent microgeometry analyses are discussed by Ceripi et al. [14] and Talukdar and Torsaeter [15].

TABLE 1.6 List of Tests for Rock Characterization

Disaggregated Rock Particles
1. Particle size distribution by sieve analysis
2. Sphericity and roundness of the grains by microscopic analyses
3. Chemical composition of the fraction by instrumental analyses
4. Type of grains (quartz, feldspar, older rock fragments, etc.)
5. Clay mineral analyses
6. Organic content of the particle size fractions
Core Samples
1. Geologic setting and origin of the rock
2. Bedding plane orientation
3. Fluid content by retort analysis
4. Capillary pressure curves
5. Pore size distribution
6. Surface area
7. Porosity
8. Absolute permeability
9. Irreducible water saturation
10. Oil–water wettability
11. Residual oil saturation
12. Cation exchange capacity
13. Point-load strength
14. Surface mineral analyses by scanning electron microscope
15. Formation resistivity factor

A simplifying theme resulting from the analysis of the sources of sedimentary rocks is that they are composed of materials from two different sources: (1) detrital sediments are composed of discrete particles, having a wide range of sizes, that are derived from weathering of preexisting rocks; and (2) chemical sediments are inorganic compounds precipitated from aqueous solutions, and may be subdivided into carbonates and evaporites as shown in Figure 1.2. The detrital sediments form beds of unconsolidated sands, sandstones, and shales. In the process of being transported from the source to a depositional basin, the grains are reduced in size and rounded, and as a result they cannot pack together without having pore spaces between the grains.

Chemical sediments originate from soluble cations, particularly sodium, potassium, magnesium, calcium, and silicon. They form beds of evaporites with very low to zero porosity because they have a granular, interlocking texture. Chemical sediments also serve as most of the cementing agents for sandstones by forming thin deposits between the rock grains.

Sedimentary particles range in size from less than 1 μm to large boulders of several meters diameter (Table 1.7). The classification of sizes, from

TABLE 1.7 Standard Size Classes of Sediments

Limiting (mm)	Particle Diameter (ϕ units)	Size Class			
2,048	−11	V. large			
1,024	−10	Large	Boulders		Im
512	−9	Medium			
256	−8	Small			
128	−7	Large	Cobbles		10^{-1}
64	−6	Small		GRAVEL	
32	−5	V. coarse			
16	−4	Coarse			
8	−3	Medium	Pebbles		10^{-2}
4	−2	Fine			
2	−1	V. fine			
1	0 (microns, μm)	V. coarse			
1/2	+1–500	Coarse			10^{-3}
1/4	+2–250	Medium	Sand		
1/8	+3–125	Fine			
1/16	+4–62	V. fine			10^{-4}
1/32	+5–31	V. coarse			
1/64	+6–16	Coarse			
1/128	+7–8	Medium	Silt	MUD	
1/256	+8–4	Fine			10^{-3}
1/512	+9–2	V. fine			
		Clay			

boulders to clay, is indicative of their source, mode of transportation, and hardness. Angular particles remain close to their source of origin, whereas spherical, smooth particles indicate transportation by streams. Sand, silt, and clay may be transported long distances by water and winds. Soft carbonates will rapidly pulverize in the process of transport, eventually being dissolved and later precipitated from a concentrated solution.

The phi-size classification of Table 1.7 is based on a geometric scale in which the size of adjacent orders differs by a multiple of 2. The phi-scale is used as a convenient scale for graphical presentations of particle size distributions since it allows plotting on standard arithmetic graph paper. It is based on the negative base-2 logarithm of the particle diameter (d):

$$\text{phi} = -\log_2(d) = -3.322 \times \log_{10}(d) \tag{1.1}$$

The size distribution may be represented as the cumulative curve of grains that are retained on a given sieve size "percent larger," or the grains that pass through a given sieve, "percent finer." The cumulative curve is often represented as a histogram, which is more amenable to visual inspection. Figures 1.3 and 1.4 compare the cumulative curves and histograms of the Berea sandstone outcrop from Amherst, Ohio, to the coarse-grained Elgin sandstone outcrop from Cleveland, Oklahoma [16]. Although the porosities of these two sandstones are not very different (0.219 and 0.240, respectively) the permeability of the Elgin sandstone is about 10 times greater because it is composed of a relatively large amount of coarse grains, which produces a network of large pores.

The sphericity and roundness of particles are two important attributes that affect the petrophysical properties of the rocks and consequently may be used to explain differences between rocks and their properties. For example, these two attributes control the degree of compaction and thus can explain the differences between rocks that have the same sedimentary history but differ in porosity and permeability.

Sphericity is a measure of how closely a particle approximates the shape of a sphere. It is a measure of how nearly equal are the three mutually perpendicular diameters of the particle, and is expressed as the ratio of the surface area of the particle to the surface area of a sphere of equal volume [17, 18].

Roundness is a measure of the curvature, or sharpness, of the particle. The accepted method for computing the roundness of a particle is to view the particle as a two-dimensional object and obtain the ratio of the average radius of all the edges to the radius of the maximum inscribed circle.

Krumbein [19] established a set of images for visually estimating roundness, ranging from a roundness of 0.1 to 0.9. Later, Pettijohn [20] defined five grades of roundness as (1) angular, (2) subangular, (3) subrounded, (4) rounded, and (5) well rounded. The degree of roundness is a function of the maturity of the particle. The particles are more angular near their source

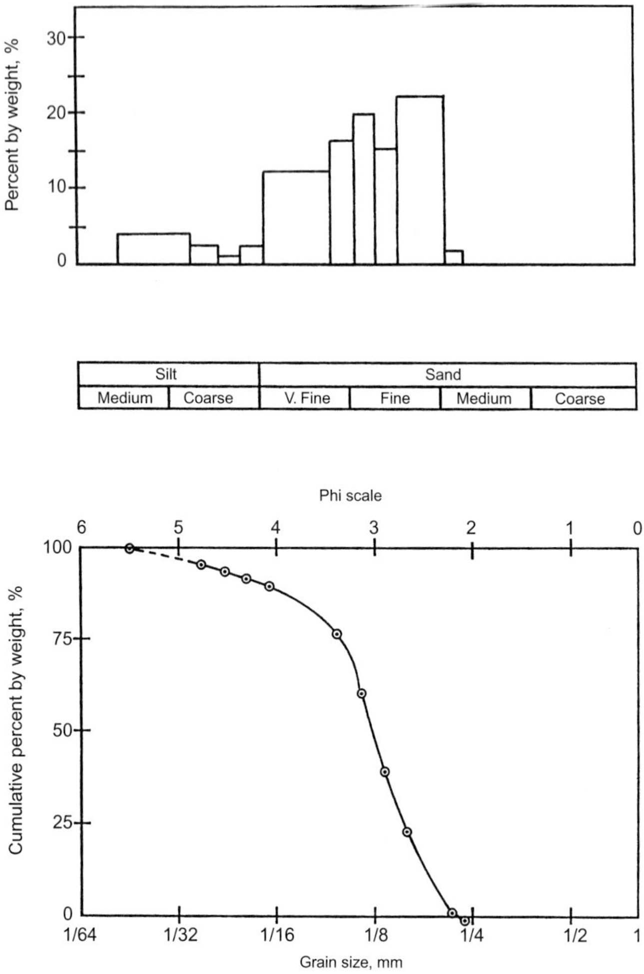

FIGURE 1.3 Histogram and cumulative size curves showing textural parameters for the Berea sandstone, Amherst, Ohio. Porosity = 0.219; permeability = 363 mD.

just after genesis and acquire greater roundness from abrasion during transportation to a depositional basin.

The texture of clastic rocks is determined by the sphericity, roundness, and sorting of the detrital sediments from which they are composed. The sphericity and roundness are functions of the transport energy, distance of transport from the source, and age of the particles. Young grains, or grains near the source, are angular in shape while those that have been transported long distances, or reworked from preexisting sedimentary rocks, have higher sphericity and roundness.

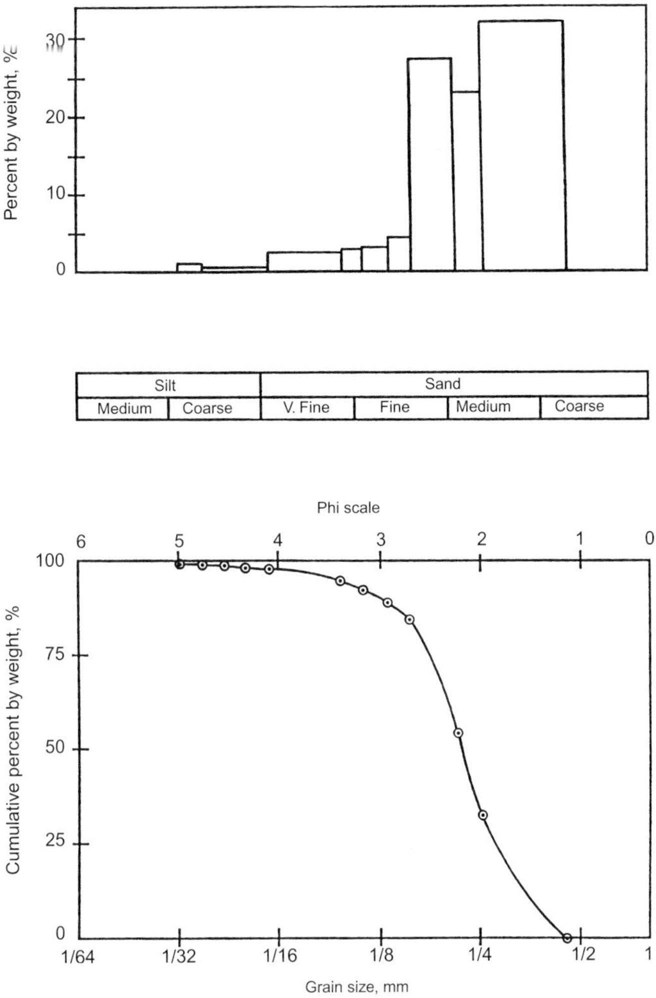

FIGURE 1.4 Histogram and cumulative size curves showing textural parameters for the Elgin sandstone, Cleveland, Oklahoma. Porosity = 0.240; permeability = 3,484 mD.

DEVELOPMENT AND USE OF PETROPHYSICS

The study of fluid flow in rocks and rock properties had its beginnings in 1927 when Kozeny [21] solved the Navier–Stokes equations for fluid flow by considering a porous medium as an assembly of pores of the same length. He obtained a relationship between permeability, porosity, and surface area.

At about the same time the Schlumberger brothers introduced the first well logs [22]. These early developments led to rapid improvements of equipment, production operations, formation evaluation, and recovery efficiency. In the decades following, the study of rock properties and fluid flow was intensified and became a part of the research endeavors of all major oil companies. In 1950, Archie [23] suggested that this specialized research effort should be recognized as a separate discipline under the name of petrophysics. Archie reviewed an earlier paper and discussed the relationships between the types of rocks, sedimentary environment, and petrophysical properties. Earlier, in 1942, Archie [24] discussed the relationships between electrical resistance of fluids in porous media and porosity. Archie proposed the equations that changed well-log interpretation from a qualitative analysis of subsurface formations to the quantitative determination of in situ fluid saturations. These and subsequent developments led to improvements in formation evaluation, subsurface mapping, and optimization of petroleum recovery.

The Hagen–Poiseuille equation [25], which applies to a single, straight capillary tube, is the simplest flow equation. By adding a tortuosity factor, however, Ewall [25] used pore size distributions to calculate the permeability of sandstone rocks. The calculated values matched the experimentally determined permeability within 10%. She was then able to show the relative amount of fluid flowing through pores of selected pore sizes. Thus the Hagen–Poiseuille equation, with modification to account for the tortuous flow path in a rock, may be used for non-rigorous analysis of fluid flow characteristics.

The general expression for fluid flow in porous media was developed by Darcy in 1856 from investigations of the flow of water through sand filter beds [26]. Darcy developed this expression from interpretation of the various parameters involved in the flow of water through sand filters, to yield the expression known as Darcy's law.

Although Darcy's law was developed for the single-phase flow of a fluid through a porous medium, it applied also to multiphase flow. In 1936, Hassler et al. [27] discussed procedures and apparatus for the determination of multiphase flow properties in rocks. Morse et al. [28] introduced a dynamic steady-state method for simultaneous flow of fluids in rocks, using a small piece of rock at the face of the core to evenly distribute the fluids entering the test sample. They showed that consistent values of the relative permeabilities of two flowing fluids could be obtained as a function of the wetting phase saturation. In 1952, Welge [29] developed a method for calculating the ratio of the relative permeabilities as a function of the wetting phase saturation for unsteady-state displacement of oil from rocks, using either gas or water as the displacing phase. Then in 1959, Johnson et al. [30] extended Welge's work, enabling the calculation of individual relative permeabilities for unsteady-state displacements. This method is the most consistently used method because it can be run in a short time and the results are consistent with those of other methods that require several days for complete analysis.

In 1978, Jones and Roszelle [31] presented a graphical method for the evaluation of relative permeabilities by the unsteady-state method.

Applications of the concept of relative permeability to analysis of reservoir performance and prediction of recovery were introduced by Buckley and Leveret [32], who developed two equations that are known as the fractional flow equation and the frontal advance equation. These two equations enabled the calculation of oil recovery resulting from displacement by an immiscible fluid (gas or water).

Research in petrophysics reached a plateau in the 1960s but received increased emphasis in the following decades with the advent of efforts to improve ultimate recovery by new chemical and thermal methods; this has generally been recognized as enhanced oil recovery [33]. Enhanced oil recovery techniques are new and developing technologies and only a few processes (thermal and miscible phase displacement) have been proven on a large scale. Research on the displacement mechanisms of chemical solutions, trapping of residual oil, measurement of residual oil saturation, phase relationships of multiple fluids in porous media, and other complex characteristics of fluid behavior in rocks have become the new areas of petrophysical research. More emphasis is now placed on the origin of rocks and petroleum, since the mineral composition of the rocks and the chemical characteristics of crude oils are involved in the fluid flow properties and the amount of residual oil saturation.

The evaluation of any petroleum reservoir, new or old, for maximum rate of production and maximum recovery of the hydrocarbons requires a thorough knowledge of the fluid transport properties of rocks and the fluid–rock interactions that influence the flow of the fluids. General knowledge of fluid flow phenomena can be gained through the study of quarried outcrop samples of rocks. The behavior of a specific reservoir, however, can only be predicted from analyses of the petrophysical properties of the reservoir and fluid–rock interactions obtained from core samples of the reservoir. Analyses of the cores only yield data at point locations within the reservoir; therefore, the petrophysical analyses must be examined with respect to the geological, mineralogical, and well-log correlations of the reservoir to develop a meaningful overall performance estimate.

OBJECTIVES AND ORGANIZATION

This text is a presentation of the theories and methods of analyses of rock properties, and of single, multiple, and miscible phase transport of fluids in porous geologic materials. The presentation is oriented to petroleum engineering and is designed to provide the engineer with the required theory, together with methods of analyses and testing, for measurement of petrophysical and fluid flow properties for application to reservoir evaluation, reservoir production engineering, and the diagnosis of formation damage.

The physical and fluid transport properties of rocks are a consequence of their pore structure, degrees of grain cementation, and electrolytic properties. Chapter 1 therefore begins with a brief review of mineralogy and the origin of sedimentary rocks. Chapter 2 is a general discussion of the origin and composition of hydrocarbons and water solutions of salts and gases that form an integral part of petroleum reservoirs. Chapter 3 follows naturally from this by describing specific rock properties, and procedures for measurement, that are important to petroleum engineering. Porosity, permeability, surface area, etc. are all considered in the development and prediction of the fluid transport behavior of petroleum reservoirs. Some of these properties are more important than others at various stages of petroleum production. During initial development porosity, permeability, and wettability, together with hydrocarbon saturation, are important; but at later stages of development, especially if enhanced recovery techniques (EOR) are being considered, pore size distribution, surface area, and capillary pressure become very important petrophysical properties in the planning and design of continued reservoir development.

Chapter 4 presents various fundamental theories establishing quantitative and qualitative relationships among porosity, electrical resistivity, and hydrocarbon saturation of reservoir rocks. A brief discussion of core analysis, well logging, and well testing is included. Laboratory techniques for measuring core properties are presented in the Appendix. Well logging techniques are presented solely for the purpose of explaining the applications of the Archie [24] and Waxman and Smits [34] equations. A discussion is included on how well logs provide data not directly accessible by means other than coring; and how well logs can be used to extend core analysis data to wells from which only logs are available. Several field examples are included in this chapter.

Capillary pressure and its measurement by several methods are presented in Chapter 5. Laboratory techniques (semipermeable disk, mercury injection, and centrifuge) are presented for measuring capillary pressure. Chapter 6 is in many ways an extension of the capillary phenomena to the measurement and determination of the influence of wettability on oil recovery, pore size distribution, and relative permeability. Methods for determining the wettability index are also included in this chapter.

The flow of fluids (oil and gas) through porous rocks is presented in Chapter 7. The analysis of linear, laminar flow is followed by a discussion of radial and turbulent flow. Equations for calculating the average permeability of naturally fractured rocks and stratified formations are derived in this chapter. This chapter concludes with a discussion of rocks of multiple porosity.

Chapter 8 is a discussion of naturally fractured rocks and their properties.

The effect of stress on reservoir rock properties, including permeability, porosity, compressibility, and resistivity, is the subject of Chapter 9.

Chapter 10 present criteria and methods for: determination of subsurface formation fluid quantities, and prediction of hydrocarbon rates of production and the practical ultimate recovery from a field.

Chapter 11 is principally an analysis of formation damage around the borehole as a result of production-injection operation and work-overs. Loss of permeability due to matrix structural compaction and fines migration followed by deposition of fines in the rock pores are defined, and empirical mathematical correalation describe the conditions.

Chapter 12 is a discussion of basic electrical well-log interpretation accompanied by FORTRAN computer programs.

PROBLEMS

1. What are the principal natural processes that affect the petrophysical properties of sedimentary rocks?
2. As shown in Table 1.1, the total bulk chemical analysis of rock samples is clearly different from the surface analysis. What effect does this have on the rock properties?
3. Since all rocks have a single source (molten magma from below the crust), what general processes produce the differentiation into many different recognizable rocks?
4. List three natural processes that are constantly operating to produce sedimentary rocks.
5. The average particle sizes from a sieve analysis are 2.00, 0.050, 0.10, and 0.06 mm. What are the respective phi-sizes?

GLOSSARY

Aphanitic refers to rock texture that contains minerals that are too small to see.

Arkose sandstone that contains a large amount of feldspar.

Batholith large intrusive body of rock, generally granite.

Breccia similar to tuff, but contains large angular fragments (>2 mm) within the fine matrix.

Cleavage a separation along a plane of weakness that produces a smooth plane that reflects light when broken. A fracture is an irregular break of the rock.

Conglomerate rock composed of fragments of preexisting rocks greater than 2 mm and inclusion of other rocks (pebbles, cobbles, and boulders; see Table 1.7)

Continental shelf the gently inclined, flat portions of the continent below sea level, extending from the shore to the continental slope where it slopes into the deep ocean platform. The shelf is generally covered with clastic sediments and the slope with fine sediments.

Diagenesis the chemical and physical changes that a sediment undergoes after deposition. Most of the diagenesis occurs after burial of the sediment. In deep burial (>3,000 m), the principal diagenetic changes are compaction and lithification.

Fissility the property of breaking along thinly spaced sheets, or planes, parallel to the depositional bedding orientation.

Foliation directional property of metamorphic rocks caused by layered deposition of minerals.

Hardness (H) an arbitrary scale of approximately equal steps between numerical hardness numbers, except for 9 and 10, which is a very large step (the hardness value is followed by a mineral that represents that value): 1—Talc, 2—Gypsum, 3—Calcite, 4—Fluorite, 5—Apatite, 6—Orthoclase, 7—Quartz, 8—Topaz, 9—Corundum, 10—Diamond. The minerals 1–3 can be scored by a fingernail, 4 and 5 by a copper penny, 5 and 6 by a knife or piece of glass, 6–8 by a piece of quartz, but 9 and 10 cannot be scored by any of the above.

Igneous rocks solidify from a melt, or magma. They are classified according to texture and mineralogy; however, they are not uniform in either composition or texture. A homogeneous magma produces a variety of chemically different rocks by the process of fractional crystallization, or differentiation. Igneous rocks that are rich in light-colored minerals are generally referred to as felsic because they contain a relatively large amount of feldspar. Composition and texture (grain size) are used for classification. The common groups of rocks fall into various steps in the differentiation of a basaltic magma according to the Bowen series. Igneous rocks occur in two ways: intrusive (below the surface) and extrusive (on the surface). The source is magma from the upper part of the mantle.

Lithification the process of changing accumulated unconsolidated sediments into a rock. The grains are compacted by the overburden sediments and cemented by deposition (from interstitial water) of silica, calcite, clays, iron oxide, and other minerals, between the grains.

Luster reflection of light by a clean surface.

Metamorphic rocks form as a result of a new set of physical and chemical conditions being imposed on preexisting rocks. Metamorphic rocks differ significantly in mineralogy and texture. Most are regional and related to orogenic events. The naming of metamorphic rock is based principally on textural features, but some names are based on composition. Most have distinct anisotropic features: foliation, lineation, and rock cleavage.

Obsidian a dark-colored, or black, essentially non-vesicular volcanic glass. It usually has the composition of rhyolite.

Pegmatic having crystals greater than 1 cm.

Porphyritic named for the texture of the matrix. Porphyritic basalt is fine-grained dark rock, with inclusions of large crystals. Porphyritic granite is coarse-grained granite with much larger crystals imbedded in it.

Porphyroblasts crystals created during metamorphism that are larger than the mineral grains in the rock.

Pyroclasts viscous magma containing gas erupting at the surface; the gas expands rapidly, blowing the plastic magma into fragments high in the air. Pyroclasts less than 2 mm in size are called ash, between 2 and 64 mm are called lapilli, and greater than 64 mm are known as blocks or bombs.

Pumice formed from a froth of small bubbles in magma, which has erupted suddenly. It is light, glassy, and floats on water.

Sedimentary rocks composed of the weathered fragments of older rocks that are deposited in layers near the earth's surface by water, wind, and ice.

Shale composed of clay particles less than 1/256 mm. Not gritty when tested by biting. Exhibits fissility.

Siltstone (mudstone) composed of particles between 1/256 and 1/16 mm in size. Noticeably gritty to the teeth.

Tuff a deposit of volcanic ash that may contain as much as 50% sedimentary material.

Vitreous (glassy) variously described as greasy, waxy, pearly, or silky.

REFERENCES

1. Allen TO, Roberts AP. *Production operations*, vol. I. Tulsa, OK: Oil & Gas Consultants International; 1982. 290 pp.
2. Crocker, ME, Donaldson, EC, Marchin, LM. *Comparison and analysis of reservoir rocks and related clays*. DEO/BETC/RI-83/7, October 1983. National Technical Information Service, US Department of Commerce, Springfield, VA, 25 pp.
3. Donaldson, EC, Crocker, ME. *Characterization of the crude oil polar compound extract*. DOE/BETC/RI-80/5, October 1980. National Technical Information Service, US Department of Commerce, Springfield, VA, 27 pp.
4. Foster RJ. *Physical geology*. 3rd ed. Columbus, OH: C. E. Merrill; 1971. 550 pp.
5. Ehlers EG, Blat H. *Petrology: igneous, sedimentary and metamorphic*. San Francisco: W. H. Freeman; 1982. 731 pp.
6. Bowen NL. *The evolution of igneous rocks*. Princeton, NJ: Princeton University Press; 1956. 333 pp.
7. George RD. *Minerals and rocks*. New York: D. Appleton-Century; 1943. 595 pp.
8. Correns CW. *Introduction to mineralogy*. New York: Springer-Verlag; 1969. 484 pp.
9. Dickey PA. *Petroleum development geology*. Tulsa, OK: Penn Well Books; 1986. 530 pp.
10. Chilingarian GV, Wolf KH. *Diagenesis*. Amsterdam: Elsevier Science; 1988. 591 pp.
11. Theodorovich GI. Expanded classification of sandstones based upon composition. *Izv Akad Nauk USSR Ser Geol* 1965;**6**:75–95.
12. Chilingarian GV, Wolf KH, editors. *Compaction of coarse-grained sediments, I*. New York: Elsevier Scientific; 1975. 808 pp.
13. Chilingarian, GV, Mazzullo, SJ, Rieke, HH. *Carbonate reservoir characterization, part I*. Amsterdam: Elsevier Science. 1992. p. 639; Part II, 1996. 994 pp.
14. Ceripi A, Durand C, Brosse E. Pore microgeometry analysis in low-resistivity sandstone reservoirs. *J Petrol Sci Eng* 2002;**35**:205–35.

15. Talukdar MS, Torsaeter O. Reconstruction of chalk pore networks from 2D backscatter electron micrographs using a simulated annealing technique. *J Petrol Sci Eng* 2002;**33**:265–82.
16. Nguyen, DD. *Capillary pressure and wettability phenomena in sandstones for a wide range of API gravity oils*. M.S. thesis, Mewbourne School of Petroleum & Geological Engineering, University of Oklahoma, 1985.
17. Carman PC. Some physical aspects of water flow in porous media. *Discussions Faraday Soc* 1948;**72**(3):72.
18. Donaldson EC, Kendall RF, Baker BA, Manning FS. Surface-Area Measurement of Geologic Materials. *Soc Petrol Eng J* 1975;**15**(2):111–16.
19. Krumbein WC. Measurement and geologic significance of shape and roundness of sedimentary particles. *J Sediment Petrol* 1941;**11**:64–72.
20. Pettijohn FJ. *Sedimentary rocks*. 3rd ed. New York: Harper & Row; 1975. 628 pp.
21. Kozeny, J. *Sitzungsber. Akad. Wiss. Wien. Math-Naturwiss. KL*, Abt, Vol. 136, 2A, 1927 pp. 271–306.
22. Schlumberger C, Schlumberger M, Leonardon E. Electric coring: a method of determining bottom-hole data by electrical measurements. *Trans AIME* 1936;**110**:237–72.
23. Archie GE. Introduction to petrophysics of reservoir rocks. *Am Assoc Pet Geol Bul* 1950;**34** (5):943–61.
24. Archie GE. The electrical resistivity log as an aid in determining some reservoir characteristics. *Trans AIME* 1942;**146**:54–62.
25. Ewall, NR. *Relationship of pore size distribution to fluid flow*. M.S. thesis, Mewbourne School of Petroleum & Geological Engineering, University of Oklahoma, 1985.
26. Darcy H. *Les fontaines publiques de la ville de Dijon*. Paris: Victor Dalmont; 1856.
27. Hassler GL, Rice RR, Leeman EH. Investigations on the recovery of oil from sandstones by gas drive. *Trans AIME* 1936;**118**:116–37.
28. Morse RA, Terwilliger PL, Yuster ST. Relative permeability measurements on small core samples. *Oil Gas J* 1947;**23**:109–25.
29. Welge HJ. A simplified method for computing oil recovery by gas or water drive. *Trans AIME* 1952;**195**:91–8.
30. Johnson EF, Bossler DP, Naumann VO. Calculation of relative permeability from displacement experiments. *Trans AIME* 1959;**216**:370–2.
31. Jones SC, Roszelle WO. Graphical Techniques for Determining Relative Permeability from Displacement Experiments. *J Petrol Technol* 1978;**30**(5):801–17.
32. Buckley SE, Leverett MC. Mechanism of fluid displacement in sands. *Trans AIME* 1941;**146**:107–16.
33. Donaldson EC, Chilingarian GV, Yen TF, editors. *Enhanced oil recovery, I—fundamentals and analyses*. New York: Elsevier; 1985.
34. Waxman MH, Smits LJH. Electrical conductivities in oil-bearing shaly sands. *Soc Petrol Eng* 1968;**8**(3):107–22.

Chapter 2

Introduction to Petroleum Geology

REVIEW OF COMPOSITION OF THE GLOBE

Geology is the study of the Earth, which is a dynamic system covered by crustal plates that are constantly moving and changing in structure. The crustal plates are driven by deep-lying forces that are not yet completely understood. New crustal plates are being formed by magma rising from molten regions deep in the Earth at mid-ocean rifts. Other crustal plates are being consumed as they are drawn downward into the mantle at subduction zones at the edges of some continents, such as the Pacific coasts of North and South America.

Detailed analyses of earthquake wave seismograms, waves that travel on the earth's surface, gravity and magnetic differences, heat flow from the interior, and electrical conductivity have been used to develop a composite picture of the globe. Four distinct zones have been identified:

1. The lithosphere, which includes the continental and ocean crusts
2. The mantle underlying the lithosphere, which is readily recognized because the seismic (earthquake) waves increase in velocity at the boundary known as the Mohorovicic discontinuity in honor of its discoverer (generally called the Moho discontinuity)
3. A liquid outer core composed principally of nickel and iron
4. The solid inner core

More than 100,000 detectable earthquakes occur each year around the globe, and most of these originate at specific focal points (a point of maximum intensity within the crust) [1–3]. Two types of waves emanate from the focal point of the earthquake: compression and shear waves. Compression waves travel through all materials by moving particles forward and backward. Shear waves, however, can propagate only through solids by moving the particles back and forth perpendicular to the direction of travel. A worldwide network of seismographs records the paths and velocities of these waves, making it possible to locate the focal point of any earthquake and to infer the composition of the interior of the Earth.

Compression waves (P waves) travel at a velocity approximately two times the velocity of the shear waves (S waves). The velocities are functions of the elastic properties and density of the materials through which they travel:

$$V_c = \left[\frac{(K + 4G/3)}{\rho}\right]^{1/2} \tag{2.1}$$

and

$$V_s = \left[\frac{G}{\rho}\right]^{1/2} \tag{2.2}$$

where

V_c = velocity of the compression wave, m/s
V_s = velocity of the shear wave, m/s
K = bulk modulus, Pa
G = shear modulus, Pa
ρ = density of material, kg/m^3

Example

Calculate the velocities of the compression and shear waves through limestone: $K = 7.0336 \times 10^{10}$ Pa, $G = 3.1026 \times 10^{10}$ Pa, $\rho = 2{,}710.6\ kg/m^3$.

Solution

$$V_c = \left\{\frac{[7.0336 + (4/3)(3.1026 \times 10^{10})]}{2710.6}\right\}^{1/2} = 6419.5\ m/s$$

$$V_c = \left\{\frac{3.1026 \times 10^{10}}{2710.6}\right\}^{1/2} = 3383.2\ m/s$$

In the crustal plates, the P-wave velocity ranges from about 6.4 to 7 km/s. At the Moho discontinuity, where the P waves enter the mantle, the velocity increases to about 8 km/s. The velocity ranges from 9 to 10 km/s in the upper mantle, 12 to 13 in the middle mantle, and peaks at 13.7 km/s at 2,800 km depth. When the P and S waves encounter the liquid core, the P-wave velocity decreases sharply to about 8 km/s and the S waves disappear, because a liquid cannot support a shear wave. At the inner solid core of the Earth, the P-wave velocity increases once more to about 11.3 km/s.

Crust is the term that originated for the outer solid shell of the Earth when it was generally believed that the interior was completely molten, and it is still used to designate the outer shell, which has different properties than the underlying mantle. The crust varies in thickness and composition. The continental masses are composed of a veneer of sediments over a layer of

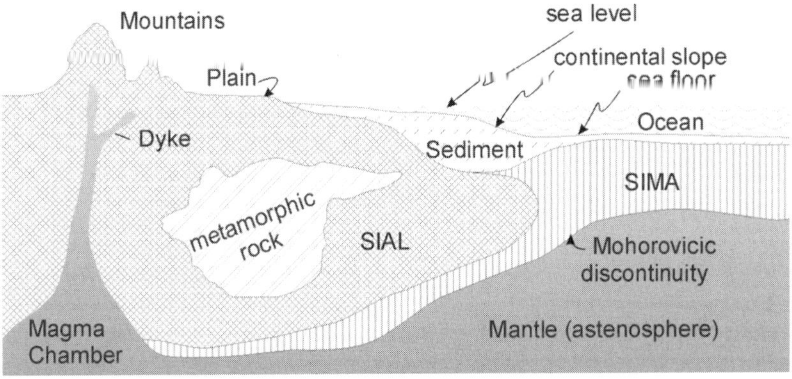

FIGURE 2.1 Cross section of the crust at a continental shelf showing the relationship between the SIAL (granite rocks) and SIMA (basalt) layers under the continents and oceans [2].

light-colored granitic rocks. The granite-type layer is called the SIAL layer because its most abundant components are silicon and aluminum, with an average density of 2.7 g/cm^3. Below the SIAL layer, there is a layer of dark rocks resembling basalt and gabbro, which is known as the SIMA layer because its principal constituents are silicon and magnesium. The density of SIMA is slightly higher than that of the SIAL layer, about 2.9 g/cm^3. Under the oceans, the SIMA layer is covered only by a thin layer of soft sediments (Figure 2.1).

The mantle is a shell, which is apparently a plastic-like solid that extends about 2,900 km deep from the Moho discontinuity to the outer liquid core. The movement of crustal plates and continents on top of the mantle is partially explained by the theory of convective currents within the mantle. Theoretically, the mantle responds to the continuous stress created by heat rising from the interior of the globe by developing current cells of very slowly ascending and descending material. Continental masses accumulate over the descending zones, and the ocean basins lie over the ascending zones. Thus, the slow movement of the mantle, as a plastic material, could be the mechanism causing the drift of the continental masses and the spreading of the ocean floor at mid-ocean rifts around the globe. Continuous drifting motion of the crustal plates also may be influenced by body forces generated by gravitational Earth tides and by the rotation of the Earth.

Rocks and magma at volcanic eruptions that have apparently come from the upper mantle are basic in composition, and are rich in magnesium and iron. The density of the mantle is greater than that of the lithosphere, approximately 3.3 g/cm^3.

The boundary at the base of the mantle, where the S waves disappear and the P-wave velocity decreases, marks the beginning of the outer liquid core. The fact that the P waves increase in velocity once more at a depth of

5,000 km suggests that the inner core is a solid. It is believed to be composed principally of nickel and iron with a density of about 10.7 g/cm^3, which is more than twice as dense as the mantle. The Earth's magnetic field is assumed to be created by an electric field resulting from circulation of currents within the liquid core [1–5].

PLATE TECTONICS

Theories of plate tectonics are based on spreading of the sea floor at mid-ocean rifts and the motion, or drift, of the continents. The Earth's lithosphere is composed of six major plates whose boundaries are outlined by zones of high seismic activity [4]. The continents appear to be moved by the convection currents within the mantle at rates of 2 to 3 in. (5.1–7.6 cm) per year. The convection cells apparently occur in pairs and thus provide the kinetic energy for movement of the continental masses.

Mid-ocean ridges form a network of about 65,000 km of steep mountains with branches circling the globe. Some of the mountains are as high as 5,500 m above the ocean floor, and some emerge above the ocean as islands.

The crustal plates are manufactured from magma rising to the surface through rifts at the sites of the mid-ocean ridges. Material from the mantle apparently liquefies as it nears the surface and is relieved of a great part of its pressure. The liquid, or magma, rises to the crust and adds to the mass of the plate. As the plate moves across the ocean floor, it accumulates a layer of sediments that was eroded from the continents. The sedimentary layer that accumulates on the ocean floor is thin in comparison to the sedimentary layers on the continents because the ocean floor is very young. Driven by convective, rotational, and gravitational forces, the plates move around until they are eventually drawn into the mantle at subduction zones before sedimentation has time to form thick layers [1, 2, 6].

If two ocean plates of equal density collide, they will slowly deform at the edges and become a range of mountains. If the colliding forces remain active long enough, the range of mountains will rise above sea level. The Alp Mountains in Switzerland constitute an example of this process due to a collision between Eurasia and Africa that began about 80 million years ago when the region was covered by a sea. Marine sediments can be found high in the Alpine regions.

India was once a separate continent riding on a plate moving in a northerly direction. The plate carrying the Indian continent was diving under the Asian continental plate. Eventually, India collided with Asia and pushed up the massive Himalayan mountain range [3, 4].

Island arcs, such as those that have developed in the Pacific Ocean east of Asia, also occur as a result of plate collisions. The Asian plate is more or less stationary with respect to the Pacific Ocean plate, which is slipping under the large land mass and forming a range of offshore islands. As the denser ocean plate returns to the high-temperature mantle, selective melting of some of its material takes place, and the lighter materials are squeezed upward as rising columns called diapirs. Diapirs are pushed through the overriding plate and form chains of offshore volcanoes that eventually rise above sea level to form islands. Lavas from the island arc volcanoes are generally intermediate in composition between granitic continental rocks and basaltic rocks. Deep-focused earthquakes occur along the arcs, indicating deep fracture zones between the continent and ocean plates.

The plates also may slip laterally with respect to each other, forming transform faults. These faults may be very long (hundreds of miles) such as the San Andreas Fault of California, where the Pacific plate abuts the North American continental plate. The Pacific plate is moving in a northwest direction with respect to the American plate, which is moving west. The difference in the relative motions of the plates produces a shear-type phenomena at the junction and results in a transform fault, many thrust faults parallel to the Earth's surface, and devastating earthquakes.

The ancient supercontinent known as Pangaea was formed by the union of a number of other continents. North America apparently moved east about 500 million years ago to collide into Pangaea, and the collision brought about the formation of the Appalachian Mountains. This movement of North America apparently crushed a chain of ancient island arcs and welded them onto the continent, because layers that appear to be crushed island arcs have been located east of the Appalachian Mountains. The junction between Pangaea and North America was apparently weak, leading to the development of a line of rising magma between them with the formation of spreading ocean plates on both sides that gradually pushed the two continental masses apart and formed the Atlantic Ocean [1, 2, 6, 7].

The convolutions of old crustal plates and sedimentary rocks at continental margins provide conditions for entrapment of hydrocarbons in porous sedimentary rocks under impermeable layers that seal the oil in place. Continental margins bordering a sea with restricted circulation permits the collection of sediments and salt deposits, which are associated with the genesis, migration, and trapping of oil. Margins that are separating from one another also are zones where oil is formed and trapped. Usually, if oil is formed on one side of a continental margin, it also will be found across the gulf, or ocean, on the margin of the other continent. Divergent, convergent, and transform continental margins provide the necessary conditions for sedimentation and accumulation of hydrocarbon deposits [1, 8–10].

GEOLOGIC TIME

Geologic timescales in use today were developed by numerous geologists working independently. Different methods for subdividing the records of flora, fauna, minerals, and radioactive decay found in sedimentary rocks were suggested; some were repeatedly used and have been generally accepted. Table 2.1 shows the subdivisions of geologic time, approximate dates in millions of years, and recognized physical events that took place during the long record of geologic history. The Earth's age is estimated to be about 4.6 billion years. The Paleozoic Era began 580 million years ago; therefore, approximately 87% of the Earth's history occurred during the Precambrian age. The approximate dates of most of the boundaries in the geologic time column are established from extensive analyses of radioactive isotopes and the flora and fauna records in sedimentary rocks. Isotopic dating also allows estimates of the rates of mountain building and sea-level changes [5].

TABLE 2.1 Subdivisions of the Three Geologic Eras and the Estimated Times of Major Events [5]

Subdivisions Based on Strata/Time			Radiometric Dates (Millions of Years Ago)	In Physical History
Era	Systems/ Periods	Series/ Epochs		
Cenozonic	Quaternary	Recent or Holocene	0	Several glacial ages
		Pleistocene	2	
	Tertiary	Pliocene	6	Colorado River begins
		Miocene	22	Mountains and basins in Nevada
		Oligocene	36	Yellowstone Park volcanism
		Eocene	58	
		Paleocene	63	Rocky Mountains begin
				Lower Mississippi River begins
Mesozonic	Cretaceous	(Many)	145	Atlantic Ocean begins
	Jurassic		210	Appalachian Mountains climax
	Triassic		255	

(*Continued*)

TABLE 2.1 (Continued)

Subdivisions Based on Strata/Time			Radiometric Dates (Millions of Years Ago)	In Physical History
Era	Systems/ Periods	Series/ Epochs		
Paleozoic	Permian		280	
	Pennsylvanian (Upper Carboniferous)		320	
	Mississippian (Lower Carboniferous)		360	
	Devonian		415	Appalachian Mountains begin
	Silurian		465	
	Ordovician		520	
	Cambrian		580	
Precambrian (mainly igneous and metamorphic rocks, no worldwide subdivisions)			1,000	
			2,000	
			3,000	Oldest dated rocks
Birth of Planet Earth			4,650	

Geologic age dating using radioisotopes is carried out by determining the amount of the specific daughter isotope present with the radioactive element and then multiplying by the rate of decay of the parent element (Table 2.2). The rate of radioactive element decay is exponential and is characterized by the following equations:

$$
\begin{aligned}
Ct &= \ln(N_o/N_t) \\
Ct_{1/2} &= \ln(1.0/0.5) = 0.693 \qquad (2.3) \\
t_{1/2} &= 0.693/C
\end{aligned}
$$

where

C = radioactive decay constant
N_o = original amount of parent element
N_t = amount of daughter isotope currently present

TABLE 2.2 Radioactive Elements, Their Half-lives and Radioactive Decay "Daughter" Elements [3]

Element	Half-life	Stable Daughter
Carbon-14	5,710 years	Nitrogen-14
Potassium-40	1.3 billion years	Argon-40
Thorium-232	13.9 billion years	Lead-208
Uranium-235	0.71 billion years	Lead-207
Uranium-238	4.5 billion years	Lead-206

$t_{1/2}$ = half-life of the parent element
t = age, years

Dating early events from the decay of carbon-14 is possible because the radiocarbon is formed in the atmosphere by collision of cosmic rays with nitrogen. The carbon dioxide in the atmosphere thus contains a small amount of radiocarbon and, therefore, all plants and animals contain carbon-14 along with the stable carbon-12. When the plant or animal dies, the accumulation of carbon-14 stops and its content of radiocarbon decays steadily. The carbon dating is then made possible by measuring the ratio of ^{14}C to ^{12}C in the remains of organism and comparing it to the ratio of these isotopes in current living plants or animals; for example, if the relative radiocarbon content of a specimen of bone $[(^{14}C/^{12}C)_{dead}/(^{14}C/^{12}C)_{living}]$ is one-fourth that of the modern specimen, the age of the specimen is 11,420 years. This is because $1/4 = 1/2 \times 1/2$ of two half-lives (2 half-lives $\times$ 5,710 years/half-life = 11,420 years).

Example

If 0.35 g of nitrogen-14 per 1.0 g of carbon-14 is found in a sediment, determine the age of the sediment.

Solution

$$C = \frac{0.693}{5710} = 1.2 \times 10^{-14}$$

$$t = \frac{1}{C} \ln\left(\frac{1.0}{0.35}\right) = (8.3 \times 10^{3})(1.051)$$

$$\text{Age} = 8,723 \text{ years}$$

(refer to Eq. (2.3) and U-238 in Table 2.2).

Several important events in the geologic history of the Earth already have been mentioned, and others are shown in the geologic column of Table 2.1. The Appalachian Mountains were formed by collision of North America with Pangaea about 500 million years ago, and the climax of their growth coincides with the birth of the Atlantic Ocean at the beginning of the Mesozoic Era about 255 million years ago. The Mississippi River and the Rocky Mountains originated at about the same time (63–65 million years ago), and Yellowstone Park volcanism is estimated to have begun about 40 million years ago. Several ice ages occurred in the Recent or Holocene Epoch that began about 2 million years ago [3, 5].

SEDIMENTARY GEOLOGY

Sedimentary geology is fundamental to the exploration and development of petroleum reservoirs. It establishes the criteria for petroleum exploration by providing the geologic evidence for prediction of the location of new petroleum provinces. Petroleum is found in many areas in a variety of sedimentary basins. Hydrocarbons may occur at shallow depths along the edges of the basin, the deep central areas, and in the far edges where tectonic motion may have provided sealed traps for oil and gas [1–10].

Basins

Sedimentary basins differ in origin and lithology. Each is individually unique, but all share several common characteristics. Basins represent accumulations of clastic and evaporite materials in a geologically depressed area (an area that has undergone subsidence with respect to the surrounding land mass) or an offshore slope. They have thick sedimentary layers in the center that thin toward the edges. The layers represent successive sedimentary episodes.

Dynamic sedimentary basins exist when sediment accumulation occurs simultaneously with subsidence of the basin area. The forces producing localized subsidence are not fully understood, but they have been related to isostatic adjustment of unbalanced gravitational forces. The theory of isostatic equilibrium is that the outer, lighter SIAL crust of the Earth is essentially floating on a plastic-type mantle in a state of equilibrium. Therefore, part of the Earth's crust can gradually subside into the plastic mantle while an adjacent area is slowly uplifted.

No earthquake foci have been recorded deeper than about 1,600 km, where the pressure and temperature are probably great enough to transform the mantle into a plastic-type material that can develop slow convective currents and gradually move to adjust for changing gravitational loads on the crust. The Great Lakes area of the United States, Canada, and the

Scandinavian Peninsula are still gradually rising in response to the melting of Pleistocene glaciers.

Continental masses have stable interiors known as cratons, or shields, which are composed of ancient metamorphosed rocks. Examples are the Canadian, Brazilian, Fennoscandian, and Indian shields that form the nuclei of their respective continents. Sedimentary deposits from the cratons have accumulated to form much of the dry land of the Earth's surface, filling depressions and accumulating on the shelves of continental margins.

Divergent Continental Margins

Sediments accumulated on the shelves at the margins of the continents form several types of geologic structures that are the result of the direction and stress imposed on them by motion of the drifting crustal plates. Divergent continental margins develop on the sides of continents that are moving away from the spreading ocean rifts. Examples are the east coasts of North and South America and the west coasts of Europe and Africa, which were originally joined together at the mid-ocean rift. The continents are extending, leaving wide, shallow subsea continental shelves where carbonate sediments originate from the reefs in shallow areas and clastic sediments result from the washing down of clastics from the land surface.

In considering sedimentation and the attributes of a sedimentary basin, one must include the entire region that has furnished the detrital materials that have accumulated in the basin as sediments, and the environmental conditions of the various episodes of sedimentation. Chapman defined this as the physiographic basin, an area undergoing erosion, which will furnish material for the sediments accumulating in a depositional basin or depression on the surface of the land or sea floor [9]. Thus, the nature of the sediments is determined by the geology of the peripheral areas of weathering and erosion, and by the physiography and climate of the entire interacting area.

Convergent Continental Margins

Convergent continental margins develop when two crustal plates collide. When an ocean plate collides with a less dense continental plate, a marginal basin forms between the island arc and the continent. This basin fills with carbonate deposits from marine animals and clastics from the land mass, forming large areas for accumulation of hydrocarbons such as the oilfields of Southeast Asia.

Continual movement of the plates against each other will result in the formation of a long, narrow trough (several hundreds of miles long) called a geosyncline. The resulting trough is filled with great thicknesses of sediments that may become uplifted and folded as mountain building (orogeny) begins, accompanied by volcanic activity. The Appalachian Mountains in the eastern United States and the Ural Mountains in Russia are the result of convergent continental margins where sediments accumulated. Subsequently, they were uplifted during the orogenic period to form the stable mountains that are eroding today and furnishing sediments to the lowland areas on both sides of the mountains.

Some of the petroleum that may have accumulated in the sediments is lost during the orogenic period, because the seals (caprocks) holding the oil in geologic traps are destroyed, allowing the hydrocarbons to migrate to the surface. Folding and faulting of the sediments, however, also produce structural traps in other areas of the region.

Transform Continental Margins

When two crustal plates slide past each other, they create a long transform fault with branches at 30° to the main fault, creating fault blocks at the edge of the transform fault. Numerous sealed reservoirs occur along such faults where clastic sediments have accumulated. An example is the San Andreas Fault in California and its associated oilfields. Transform faults on the ocean floor are sites of sea mounts, some of which project above the ocean floor and are accompanied by volcanic activity [9].

Transgressive-Regressive Cycles

A transgressive phase occurs when the sea level is rising or the basin is subsiding. During this period, the volume created by subsidence generally exceeds the volume of sediments entering the basin, and hence the depth of the sea increases. As the sea advances over the land surface, the depositional facies also migrate inland, creating a shallow, low-energy environment along the shore that tends to accumulate fine-grained particles. The fine-grained sediments have low permeability and are potential petroleum source rocks rather than reservoirs [9].

During a regressive phase in the formation of a basin, the basin becomes shallower and the depositional facies migrate seaward into a high-energy environment. A regressive sequence may develop because the supply of sediments is greater than the amount accumulating in the basin that can be removed by the available energy. This occurs in river deltas where the delta is growing because the supply of sediments to the delta is greater than the amount of sediments being removed from the area by sea currents and waves. Thus, one of two elements may be active: (1) the sea level may be decreasing, or (2) the sediment supply may exceed the capacity for removal and redistribution. The sediments accumulating during the regressive phase tend to be coarse-grained because of the higher energy level in the depositional basin during this period. The rocks of this sequence, therefore, have relatively high permeabilities and are potential reservoirs laid down on top of potential source rocks deposited during the transgressive phase.

The transgressive-regressive stages tend to accumulate sequences of sediments that are either shale/sand or shale/carbonate-evaporite. The carbonate-evaporite sequences are associated with some, but not all, of the transgressive phases resulting in periodic accumulations of carbonate-evaporite lithologies. The low-energy environment of the shallow shelves provides opportunities for development of abundant shellfish whose shells become beds of limestone. Calcium and magnesium tend to precipitate from the shallow seas resulting in depositions of limestone ($CaCO_3$) and dolomite [$CaMg(CO_3)_2$]. Porosity is developed by dolomitization, chemical leaching by percolating waters (solution porosity), and mechanical fissuring from structural movements leading to jointing and vertical cracks. Carbonates also are deposited as reefs at the edge of continental shelves and along the continental slope.

Accumulation of Sediments

The accumulation of sediments in a given area depends on equilibrium between the energy of the environment and the inertia of sedimentary particles. For example, sediments transported to the mouth of a river may be moved by waves and currents to another location where the environmental energy is not high enough to move the particles. This is the concept of base level [9]. Sediments of a given size and density will accumulate in an area at their base level of energy, but finer grades of the material cannot accumulate in that location and are carried in suspension to an area of lower energy equivalent to their base level. This is the process that leads to sorting with accumulation of sand grains in one area and silt and clay in another area. The base level of a given area fluctuates with time; thus, during one period of accumulation sand particles are deposited, whereas later, finer particles of silt and clay are deposited on top of the sand. This sequence may be repeated many times, leading to alternate deposition of sand and shale, and formation of sand-shale sequences.

Pirson identified three types of physiographic areas that lead to the accumulation of quartzose, graywacke, or arkose sands in basins [11]. Each depends on the relief of the land mass and thus the time available for chemical weathering of the rocks and particles prior to accumulation in the sedimentary basin. This is a simplification of the sedimentary process, which is a complex interplay of the numerous depositional situations including those idealized by Pirson. Nevertheless, the simplifications present a clear explanation of sedimentary accumulations that lead to different lithologies.

During periods of negligible orogenic activity in flat plains bordered by shallow seas, erosion of the land mass is at a minimum, whereas chemical weathering is occurring at a rapid rate because the residence time of interstitial fluids at and near the surface is relatively long. Under these conditions,

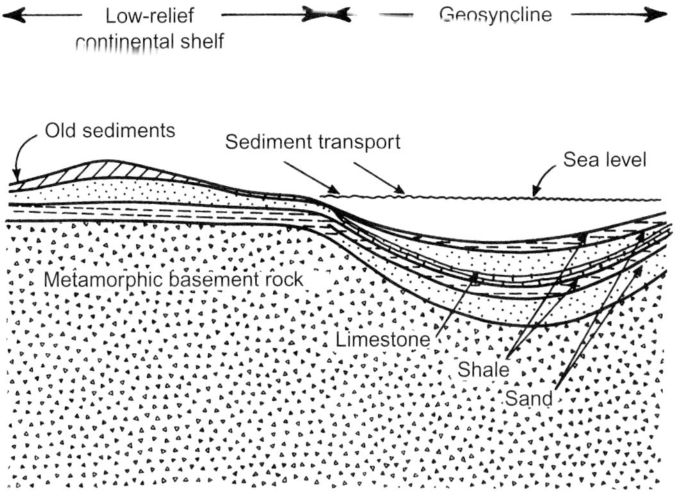

FIGURE 2.2 Accumulation of quartzose-type sediments in a basin from a low-relief continental shelf. On a low-relief land surface, erosion is at a minimum and chemical degradation of rocks to quartz is at a maximum [11].

weathering processes go to completion, furnishing stable components from igneous and metamorphic rocks, such as quartz and zircon, for clastic sediments. These materials are carried into the depression forming the sea and are accumulated as clean, well-sorted sediments with uniform composition and texture. The sediments may remain as unconsolidated sand formations, or the grains may be cemented by carbonate, siliceous, clayey, and ferruginous compounds precipitated from the seawater, interstitial solution, or ferruginous waters percolating gradually through the deposits at some later stage (Figure 2.2). Changes of the climatic conditions of the physiographic area can change the type of sediments accumulating in the basin, from clean granular material to mixtures of silt, clay, and organic materials. These become shale beds that can serve as source rocks for hydrocarbons as well as impermeable caprocks.

Well-sorted, granular, quartzose reservoirs exhibit relatively high vertical permeability (k_v) with respect to the horizontal permeability (k_h); however, k_h is still higher than k_v. Therefore, primary oil recovery will be relatively high, whereas secondary recovery will be very low due to severe fingering and early water breakthrough. Pirson lists the Oriskany Sandstone in Pennsylvania, St. Peter Sandstone in Illinois, Wilcox Sandstone in Oklahoma, and Tensleep Sandstone in Wyoming as examples of quartzose-type reservoirs [11].

In conditions where the uplifted land areas bordering seas are steep enough to prevent total chemical weathering of the exposed rocks to stable minerals such as quartz, the detrital material accumulating in the basin

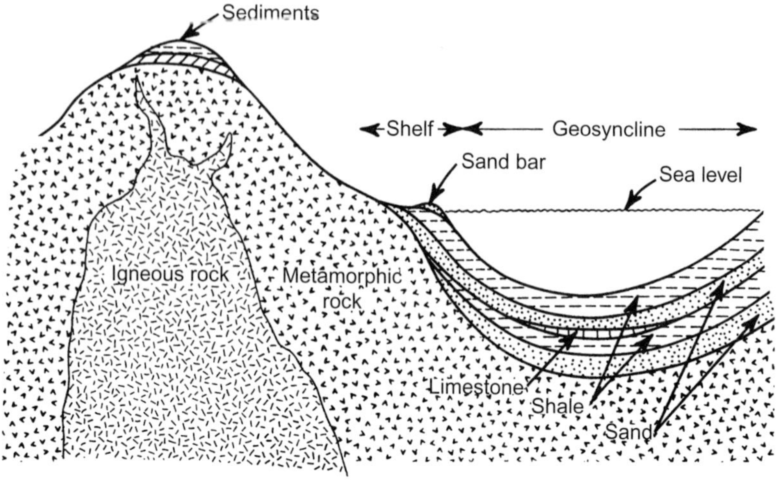

FIGURE 2.3 Accumulation of graywacke-type sediments in a geosyncline adjacent to a land mass of moderate relief [11].

will be composed of mixed rock fragments, or graywacke-type sediments. The sedimentary particles are irregular in shape and are poorly sorted, with variable amounts of intergranular clay particles. Changes of the climatic conditions of the physiographic area result in variable episodes of fine clastic deposition on top of the coarse particles forming the layers that become the caprocks of the reservoirs (Figure 2.3). The permeabilities of these reservoirs vary considerably over short distances, and the vertical permeability is usually much lower than the horizontal permeability. The permeability variation is one reason why graywacke-type reservoirs do not produce as well during primary production as the quartzose-type reservoirs, but exhibit excellent secondary recovery. Due to the mixed sediments containing clay minerals, the reservoirs are generally subject to water sensitivity problems (clay swelling and clay particle movement). The Bradford Sandstone in Pennsylvania and the Bartlesville Sandstone in Oklahoma are examples of graywacke sandstone formations.

A third general class of clastics, arkose-type sediments, will accumulate in basins or dendritic canyons adjacent to land areas of steep relief. Due to the steep relief, chemical weathering of the sediments is incomplete, resulting in deposition of angular grains with considerable size variation. Reactive clays and unstable minerals such as feldspars are mixed with other grains and also make up a large portion of the cementing agents. Variable climatic conditions of the physiographic area result in periods of deposition of coarse clastics followed by fine sediments that eventually become the caprocks of reservoirs (Figure 2.4). Thick reservoirs are formed, but the permeability is

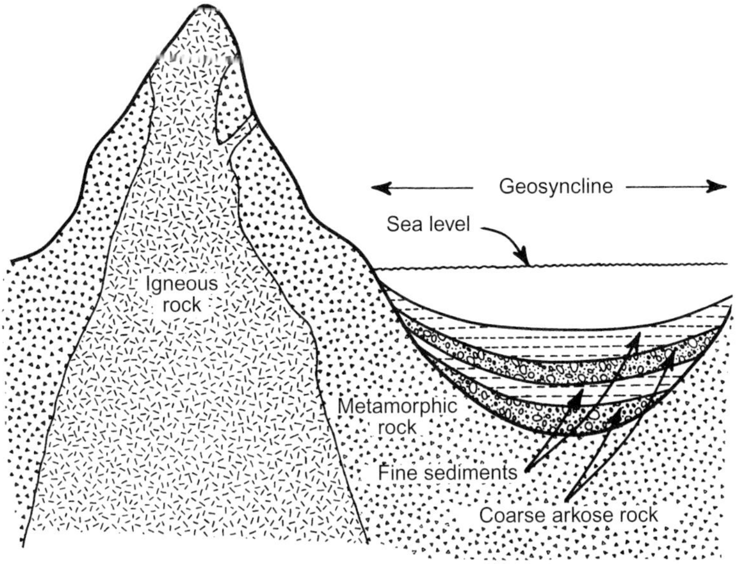

FIGURE 2.4 Idealized conditions that lead to deposition of arkose-type sediments. Steep land relief results in incomplete chemical weathering that yields arkose-type sediments [11].

extremely variable, both vertically and horizontally. Consequently, both primary and secondary production may be poor and the reactive clays produce severe water sensitivity. Examples of arkose-type formations are the Kern River formation in California and the Granite Wash in the Oklahoma-Texas panhandle area [11].

HYDROCARBON TRAPS

Hydrocarbon traps may be illustrated by considering a porous, permeable formation that has been folded into an anticlinal trap by diastrophism and is enclosed between impermeable rocks (Figure 2.5). The closure of the trap is the distance between the crest and the spill point (lowest point of the trap that can contain hydrocarbons). In most cases, the hydrocarbon trap is not filled to the spill point. It may contain a gas cap if the oil contains light hydrocarbons and the pressure–temperature relationship of the zone permits the existence of a distinct gas zone at the top of the reservoir. If a gas cap exists, the gas–oil contact is the deepest level of producible gas. Likewise, the oil–water contact is the lowest level of producible oil. Transition zones exist between various zones grading from a high oil saturation to hydrocarbon-free water. For example, the water zone immediately below the oil–water contact is the bottom water, whereas the edge water is laterally

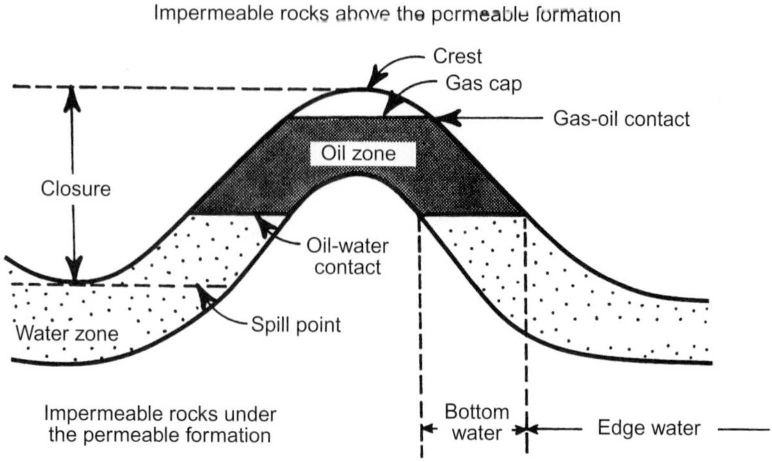

FIGURE 2.5 Idealized cross section through an anticlinal trap formed by a porous, permeable formation surrounded by impermeable rocks. Oil and gas are trapped at the top of the anticline.

adjacent to the oil zone. The gas–oil and water–oil contacts are generally planar, but they may be tilted due to hydrodynamic flow of fluids, a large permeability contrast between opposite sides of the reservoir, or unequal production of the reservoir.

An anticline structure may contain several oil traps, one on top of the other, separated by impermeable rocks. Furthermore, the lithology of the individual traps may vary from sands to limestone and dolomite [9, 11].

Hydrocarbon traps are generally classified as either structural or stratigraphic, depending on their origin. Structural traps were formed by tectonic processes acting on sedimentary beds after their deposition. They may generally be considered as distinct geological structures formed by folding and faulting of sedimentary beds. Structural traps may be classified as (1) fold traps formed by either compressional or compactional anticlines, (2) fault traps formed by displacement of blocks of rocks due to unequal tectonic pressure, or (3) diapiric traps produced by intrusion of salt or mud diapirs (Figure 2.6).

Stratigraphic traps are produced by facies changes around the porous, permeable formation such as pinch-outs and lenticular sand bodies surrounded by impermeable shales. Stratigraphic traps may develop from offshore sand bars, reefs, or river channels. The processes of formation are more complex than those of structural traps because they involve changes of the depositional environment that lead to isolation of permeable zones by different lithologies. Distinctions are made between those that are associated with unconformities and those that are not [6].

Many hydrocarbon accumulations are associated with unconformities. An unconformity forms when a site of sedimentation is uplifted, eroded, and

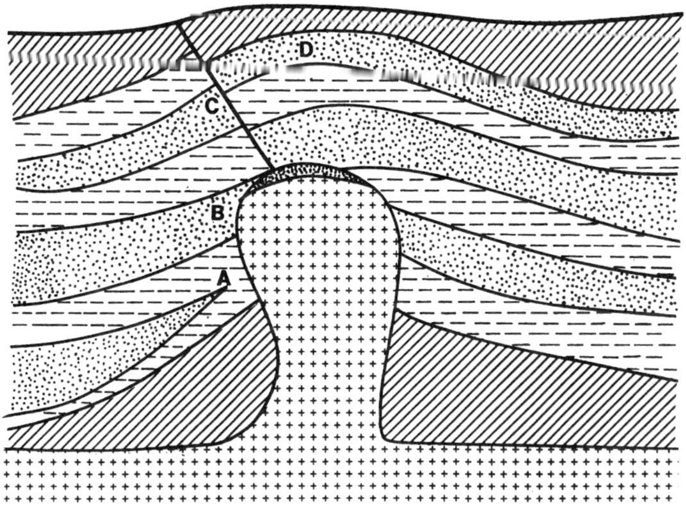

FIGURE 2.6 Illustration of several types of traps: (A) stratigraphic pinch-out trap, (B) trap sealed by a salt dome, (C) trap formed by a normal fault, and (D) domal trap.

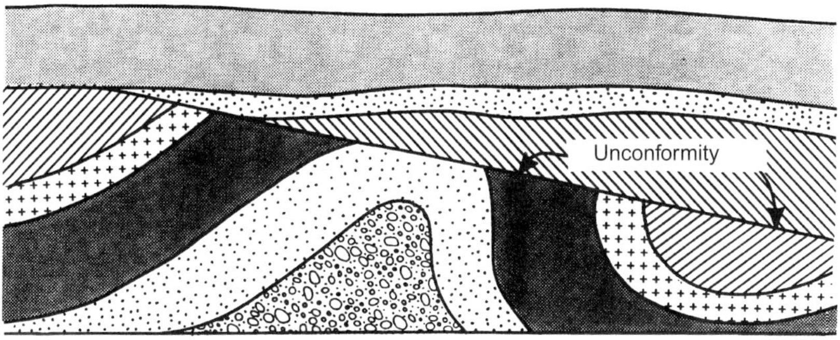

FIGURE 2.7 Unconformity, showing the uplifted, eroded strata overlain in an unconforming pattern by younger sediments.

buried again under a new layer of sediments that may delineate the boundaries of an oil trap, because unconformities generally separate formations that developed under very different environmental conditions (Figure 2.7). The rocks immediately below an unconformity are likely to be porous and permeable because an unconformity is a zone of erosion that is on top of a weathering zone where water is percolating through the rocks causing solution of some minerals and precipitation of others as cementing agents. This is especially true of carbonate formations underlying unconformities.

In addition, the mixed debris deposited on top of an unconformity can form permeable conduits for migration of oil from source rocks to geologic traps [12].

ORIGIN OF PETROLEUM

The biogenic origin of petroleum is widely accepted on the basis of geochemical studies. Petroleum contains compounds that have characteristic chemical structures related to plants and animals such as porphyrins, isoprenoids, steranes, and many others. In addition, the source rocks where the precursors of petroleum were originally deposited are the fine-grained sediments that are deposited in shallow marine environments during the low-energy transgressive phases of geologic basin formation. Particulate organic matter is not much denser than water and, therefore, sedimentation along with clay and fine carbonate precipitates will take place slowly in a low-energy environment. Depletion of oxygen takes place in quiet water leading to an anaerobic condition and preservation of organic matter. Anaerobic bacteria tend to reduce organic compounds by removal of oxygen from the molecules in some cases, but they do not attack the carbon-to-carbon bond of hydrocarbons. The evidence for the origin of petroleum in low-energy, anaerobic environments is supported by the fact that in the opposite condition (high-energy, aerobic environments) aerobic bacteria decompose organic matter to carbon dioxide and water [9, 13, 14].

Transformation of Organics into Kerogen

Organic materials from dead plants and animals are either consumed by living organisms or left to be decomposed by bacteria. If the organic material remains in an oxygen-rich, aerobic environment, aerobic bacteria will decompose it to carbon dioxide and water. If the environment is anaerobic, the products of decomposition will be essentially compounds of carbon, hydrogen, and oxygen. The hydrocarbons of crude oils can originate from the fundamental biological molecules: proteins (amino acids), lipids (fats, waxes, and oils), carbohydrates (sugars and starches), and lignins (polymeric hydrocarbons related to cellulose) of plants. If these are preserved in a low-energy environment free of oxygen, they can be mixed with the clays and precipitates that are forming the fine-grained sediments characteristic of the low-energy transgressive phase of basin formation. Therefore, to be preserved, this organic matter must be buried as it is supplied with fine-grained sediments. The source rocks of petroleum are, therefore, those rocks formed from fine-grained sediments mixed with organic materials. Not all fine-grained sediments are source rocks for petroleum, which implies that a necessary criterion is the availability of abundant organic matter in an area of fine-grained deposition. This implies a sedimentary basin along a gentle continental slope and the presence of aquatic life (plankton, algae, etc.), in addition to copious terrestrial plant life. Land vertebrates are not a very likely source for organic matter in shallow marine sediments.

Higher-order land plants contain abundant quantities of cellulose and lignin, yielding aromatic-type compounds with a low hydrogen-to-carbon ratio (1.0–1.5). Marine algae contain proteins, lipids, and carbohydrates; these are aliphatic in character with a high hydrogen-to-carbon ratio of 1.7–1.9. (The hydrogen-to-carbon ratios of specific compounds are benzene-1.0, cyclohexane-2.0, and *n*-pentane-2.4.)

The organic materials, fine-grained sediments, and bacteria that are mixed together and deposited in the quiet, low-energy environments are not in thermodynamic equilibrium. The system approaches thermodynamic equilibrium during initial burial while it is undergoing diagenetic transformations. Inasmuch as burial is shallow during this stage, the temperature of the environment is low, and the sediment undergoes diagenetic changes slowly under mild conditions. The first 10 ft or so of sediment represents an interface where the biosphere passes into the geosphere. The residence time in this shallow sediment, before deeper burial, may range from 1,000 to 10,000 years. During this time, the organic matter is subjected to both microbial and chemical actions that transform it from the biopolymers (e.g., proteins) to more stable polycondensed compounds that are the precursors of kerogen. In time the sediments are buried deeper, where the anaerobic environment prevails and where the organic matter continues to transform to more insoluble high-molecular-weight polymers due principally to the increase of pore fluid pressure and temperature.

Anaerobic bacteria reduce sulfates to hydrogen sulfide and may remove oxygen from some low-molecular-weight organic compounds, but otherwise they add to the total biomass rather than depleting it, which occurs in the aerobic regions. Some organically produced compounds of calcium and silica dissolve in the water and later are precipitated with the mixture of clay minerals and organics as they reach saturation in the aqueous layer. The organic matter is gradually transformed into new polymeric organic compounds that eventually become kerogen. Considerable methane is formed and released—mixed with hydrogen sulfide—as marsh gas. Low-molecular-weight water-soluble compounds formed during diagenesis probably are lost to the interstitial water percolating upward, leaving behind a solid organic mass compacted into fine kerogen particles.

Transformation of Kerogen into Oil and Gas

Consecutive deposition of sediments in the basin leads to deeper burial reaching several thousand feet deep, which imposes an increase of temperature and pressure on the kerogen mixed with the fine-grained sediments. The increase of temperature with burial places the materials once more out of thermodynamic equilibrium, which induces further reactions and transformations (catagenetic stage). During catagenesis, the reactions are catalyzed to some extent by the inorganic matrix. While the organic material is undergoing major transformations, the sediments are being compacted with expulsion of water and decrease of porosity and permeability. The kerogen evolves through a liquid bitumen to liquid petroleum. If the petroleum remains in the compacted source rock undergoing deeper burial with continued heating, the kerogen is ultimately reduced to graphite and methane.

The thermodynamic stability of the organic matter is never reached because of the gradual increase in temperature as burial proceeds. Chilingarian and Yen describe the approximate depths for the various diagenetic and catagenetic changes: 10–20 ft is the zone of change to humic materials; 20–1,500 ft is where the diagenetic changes take place; 1,500–6,000 ft is the zone of catagenetic changes and formation of oil from kerogen; and below 6,000 ft there is a zone of metagenesis where petroleum changes to graphite and methane (Figure 2.8) [15].

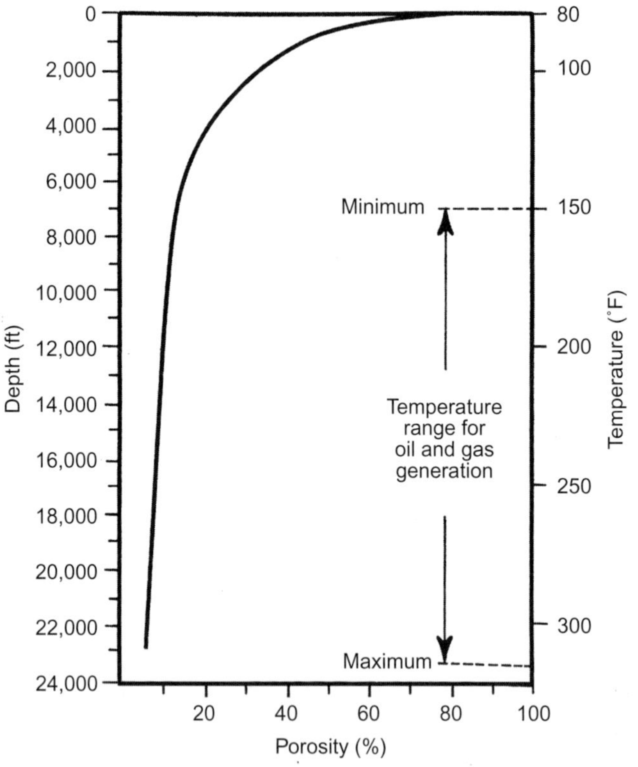

FIGURE 2.8 Average relationship between porosity and depth of burial for shales, and the temperatures and depths for the genesis of oil and gas [15].

MIGRATION AND ACCUMULATION OF PETROLEUM

The genesis of petroleum occurs in compacted clay and shale beds, which are essentially impermeable to fluid flow. Therefore, the processes by which hydrocarbons migrate from the source rock to a porous, permeable reservoir (called primary migration) are not completely understood. Numerous theories have been advanced to explain the processes. Possibly, several different mechanisms may be operative under different environmental and geological conditions. Some of these are as follows:

1. Transport in colloidal solutions as micelles
2. Transport as a continuous hydrocarbon phase
3. Buoyant movement of individual droplets
4. Solution of hydrocarbons in water moving out of the source rock
5. Transport by mechanical forces during clay diagenesis
6. Movement through microfractures in the source rock

After leaving the source rock, the hydrocarbons migrate upward through permeable beds until they reach a sealed hydrocarbon trap where accumulation occurs forming a hydrocarbon reservoir. This process has been called the secondary migration, which is governed principally by buoyancy and hydrodynamic flow [9].

Primary Migration

The geochemical evidence of the generation for petroleum shows that hydrocarbons do not generally originate in the structural and stratigraphic traps in which they are found. The petroleum reservoirs are porous, permeable geologic structures, whereas the source rocks have been identified as compacted, impermeable shales. Inasmuch as the source rocks are impermeable, the method of expulsion of oil from the shales where it is generated is not obvious. Considerable data on the expulsion of water from shale during compaction show that most of the pore water is squeezed out during burial before the temperature required for the generation of petroleum is attained (Figure 2.8).

Compaction of sediments begins as soon as the sediments begin to accumulate. During original accumulation, the loose, fine-grained sediments contain more than 50% water. As they are buried deeper, due to subsidence and continued deposition of sediments on top, the interstitial water from the deeper sediments is expelled, resulting in a decrease in porosity and an increase in density. The material acquires cohesive strength as the grains are pressed together tightly. Chemical changes occurring in the interstitial fluids produce precipitates that cement the grains into an even more cohesive formation.

The major oil generation occurs well below the depth at which compaction of the shale is almost complete. Consequently, the displacement of oil from most source rocks could not have taken place when the shales were being compacted [6]. Expulsion of oil during compaction may have taken place in a few isolated cases where rapid burial resulted in the development of abnormally high pore pressures or zones of abnormally high temperatures at shallow depths. Barker contends that petroleum may be expelled from the top and bottom of source rocks due to the pressure gradient that develops during deep burial [16]. After expulsion of the pore water, petroleum formed in the organically rich shale may form a continuous phase and move along a network of fine, thread-like channels under the applied physical stress [13].

Some clay minerals (smectites in general) contain bound water within the lattice structure of the clay particles. This bound water is expelled when the smectites are transformed to illite, which begins at a temperature of about 200°F. This temperature is well within the temperature range for generation of petroleum and thus may assist in the primary migration of oil when smectites are present in the shale body [6].

Secondary Migration

Inasmuch as petroleum reservoirs exist in a water environment, the migration of hydrocarbons from the point of release from a source rock to the top of the trap is intimately associated with capillary pressure phenomena and hydrology. The pore size distributions, tortuosity of continuous channels, porosity, permeability, and chemical characteristics of reservoir rocks and their interstitial fluids differ widely. Nevertheless, because of the ubiquitous presence of water, capillarity, buoyancy, and hydrology apply in all cases [14].

The migration of oil as distinct droplets in water-saturated rock is opposed by the capillary forces, which are functionally related to pore size, interfacial tension between oil and water, and adhesion of oil to mineral surfaces (wettability). This is expressed through a contact angle for a capillary of uniform size as:

$$P_c = \frac{2\sigma\cos\theta}{r_c} \tag{2.4}$$

where

P_c = capillary pressure, Pa
σ = interfacial tension, $(N \times 10^{-3})/m$
θ = contact angle
r_c = radius of the capillary, m

The more usual case is one in which the oil droplet exists within the confines of a large pore containing several smaller-sized pore throat exits (Figure 2.9). Under these conditions, the pressure required to displace the

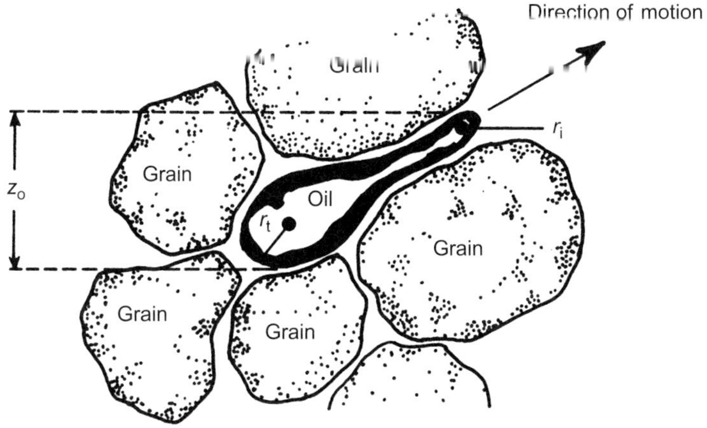

FIGURE 2.9 Displacement of an oil droplet through a pore throat in a water-wet rock.

droplet from the large pore through the constriction of a pore throat (the displacement pressure) is the difference between the capillary pressures of the leading (l) and trailing (t) pores [6]:

$$P_d = 2\sigma\left(\frac{\cos\theta_l}{r_l} - \frac{\cos\theta_t}{r_t}\right) \tag{2.5}$$

where

P_d = displacement pressure, Pa
θ_l = contact angle of the leading edge
θ_t = contact angle of the trailing edge
r_l = radius of the leading pore, m
r_t = radius of the trailing pore, m

The two forces in a reservoir that are most likely to be operating on the droplet are buoyancy and hydrodynamic pressure, neither of which is normally sufficient to dislodge an isolated droplet of oil.

The displacement pressure due to buoyancy is expressed as:

$$P_d = Z_o g_c(\rho_w - \rho_o) \tag{2.6}$$

where

Z_o = height of the oil column
g_c = gravitational constant, 9.81 m/s^2
ρ_w = water density, kg/m^3
ρ_o = oil density, kg/m^3
P_d = displacement pressure, Pa

Since the combined buoyant and hydrodynamic pressures acting on an isolated droplet are insufficient to exceed the displacement pressure required by the capillary forces, isolated drops of oil cannot migrate under the influence of these forces alone [14].

As the oil leaves the source rock under the forces of compaction, large saturations develop at the entry to the reservoir rock. The oil then begins to migrate upward as a continuous phase in long filaments within the pores. Under these circumstances, sufficient buoyant and hydrodynamic forces can develop to cause migration of the oil.

It also has been suggested that oil migration may occur by molecular solution of oil in water that is in motion, or by colloidal solution brought about by surfactants present in petroleum. Both theories have been challenged because the solubility of oil molecules in water is extremely low and the actual concentration of surfactant-type molecules in crude oils is very small [9, 17]. Leaching of sand containing discrete droplets of oil is possible, however, if the sand is flushed with large quantities of hot water. This process may help account for the oil-free sand found below many hydrocarbon-saturated reservoirs, given the enormous amount of geologic time accompanied by changes of temperature and diastrophism.

Secondary migration of petroleum ends in the accumulation in a structural or stratigraphic trap, and sometimes in a trap that is a complex combination of the two. Levorsen observed that oil has been found in traps that were not developed until the Pleistocene Epoch, which implies that the minimum time for migration and accumulation is about 1 million years [18]. The hydrocarbons accumulate at the highest point of the trap and the fluids are stratified in accordance with their densities, which show that individual hydrocarbon molecules are free to move within the reservoir. Inasmuch as the sedimentary rocks may have formed during the Cretaceous Period or earlier, it is entirely possible that the oil accumulation may have been disturbed by diastrophism, and many changes of temperature and pressure. The petroleum accumulation may (1) become exposed by an outcrop and develop an oil seep, or (2) become uplifted and eroded to form a tar pit. In addition, petroleum may be transported to another sedimentary sequence as a result of rapid erosion and clastic transport. Levorsen identifies this type of secondary accumulation as recycled oil, which should be low in paraffins because of attack by aerobic bacteria [18]. Thus, the geologic history of an oil reservoir may have been quite varied, and knowledge of the sedimentary history, origin, migration, and accumulation is valuable for the overall understanding of oil recovery processes and formation damage that may develop during production of the oil.

The caprock, or oil trap seal, may not be absolutely impermeable to light hydrocarbons. The capillary pressure relationship of the rocks overlying the oil trap may form an effective vertical seal for liquid petroleum constituents (C_{5+} hydrocarbons), but the seal may not be completely effective in retaining lighter hydrocarbons.

PROPERTIES OF SUBSURFACE FLUIDS

A basic knowledge of the physics and chemistry of subsurface waters and petroleum is essential for petroleum engineers because many problems associated with exploration, formation damage or production problems, enhanced oil recovery, wettability, and others are directly associated with the physical and chemical behavior of subsurface waters and petroleum as a whole, or as groups of constituents such as paraffins and asphaltenes.

Hydrostatic Pressure Gradient

An important physical property of reservoir fluids is the density and its relationship to the hydrostatic gradient (the increase of the fluid pressure with increasing depth due to the increasing weight of the overlying fluid). Density measurements are made relative to the maximum density of water, which is 1.0 g/cm^3 at 15°C (60°F) and 1 atm pressure. When the specific weight (or mass) of any substance is divided by the specific weight (or mass) of an equal volume of water at 15°C and 1 atm pressure, the resulting dimensionless value is described as the specific gravity (SG) relative to water. The pressure gradient (G_p) of any fluid is determined from the specific gravity as follows:

$$\begin{aligned} G_p &= 1000\ \text{kg}/m^3 \times 9.81\ m/s^2 \times \gamma_w \\ &= 9810\gamma_w\ \text{Pa}/m\ (0.00694\gamma_w\ \text{psi/ft}) \end{aligned} \tag{2.7}$$

where γ_w = specific weight of water in kg/m^3 (lb/ft^3)

The hydrostatic gradient of subsurface waters is greater than 9.81 kPa per meter of depth (0.433 psi/ft) because the brines contain dissolved solids that increase the density of the fluids. The gradient also is affected by the temperature and in some areas by dissolved gas, both of which decrease the hydrostatic pressure gradient. An average hydrostatic gradient of 10.53 kPa/m (0.465 psi/ft) generally is used in the literature for subsurface brines [28]. This value corresponds to about 80,000 ppm of dissolved solids at 25°C (SG = 1.074).

Lithostatic Pressure Gradient

The lithostatic pressure gradient is caused by the density of the rocks and is transmitted through the grain-to-grain contacts of successive layers of rocks. The lithostatic weight is, however, supported by the pressure of the subsurface fluids in the pore spaces. Thus, the overburden pressure is equal to the grain-to-grain lithostatic pressure plus the fluid pressure of the porous formation, yielding an average overburden pressure gradient of 22.7 kPa per meter

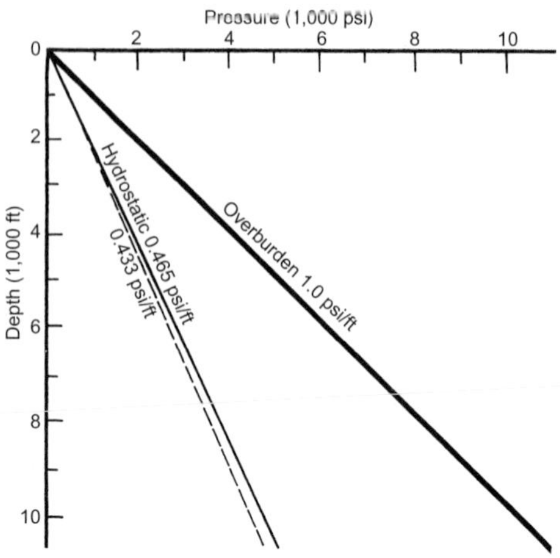

FIGURE 2.10 Subsurface pressure gradients.

of depth (1.0 psi/ft), which corresponds to an overall bulk specific gravity of the rocks plus the interstitial fluids equal to 2.31 (Figure 2.10):

$$p_{ob} = p_l + p_f \tag{2.8}$$

where

p_{ob} = overburden pressure
p_l = lithostatic pressure
p_f = fluid pressure

When the hydrostatic pressure gradient for any region is approximately 10.53 kPa/m, it is known as the normal pressure gradient. Abnormal pressure gradients may be either abnormally low or abnormally high. Abnormally high hydrostatic pressure gradients of 21.5 kPa/m (0.95 psi/ft) have been encountered, for example, in the geopressured/geothermal zones (1) along the Gulf Coast of the United States extending from New Orleans into Mexico, (2) the Niger Delta, and (3) the North Sea [6, 18]. Abnormally low pressures have been encountered, for example, in some gas fields of Pennsylvania and the Morrow Formation in Northwest Oklahoma.

Geothermal Gradient

Heat rising from the mantle produces a heat flux in midcontinent regions ranging from 0.8 to 1.2 μcal/cm$^2 \cdot$s (30–4.4 μBTU/ft^2-s) measured at the surface, which results in a geothermal gradient, G_t [5]. The geothermal gradient varies at different areas on the globe depending on the annual mean surface temperature and the thermal conductivity of the subsurface formations, but an overall average temperature gradient G_t of 18.2°C/km (1.0°F/100 ft) of depth has been recorded around the world. Using this average value and the region's mean annual surface temperature T_s, an estimate of subsurface formation temperatures T_f can be obtained as follows:

$$T_f = T_s + G_t D \tag{2.9}$$

When the bottomhole temperature T_f of a well is accurately measured, the local geothermal gradient G_t may be obtained from Eq. (2.9) and used to estimate the temperature of formations at any other depth D.

Example

The bottomhole temperature at 2.2 km was found to be 70°C. The mean surface temperature for the region is 24°C. Determine the geothermal gradient G_t and a temperature of the formation at a depth of 1,700 m.

Solution

Solving for G_t from Eq. (2.9):

$$G_t = \frac{T_f - T_s}{D} = \frac{70 - 24}{2.2} = 20.9\,°\text{C/km}$$

The formation temperature at $D = 1.7$ km is obtained from Eq. (2.9):

$$T_f = 24 + 20.9 \times 1.7 = 59.5\,°\text{C}$$

There are zones at various locations on the globe where the geothermal and geopressure gradients are abnormally high. Some areas in the United States where abnormally high pressures and temperatures have been reported are Gulf Coast Basin post-Cretaceous sediments, Pennsylvanian Period sediments in the Anadarko Basin in Oklahoma, Devonian zone in the Williston Basin in North Dakota, and the Ventura area of California. Outside of the United States, geopressure/geothermal zones have been reported in many areas, for example, the Arctic Islands, Africa (Algeria, Morocco, Mozambique, and Nigeria), Europe (Austria, the Carpathians, the Ural Mountains, Azerbaijan, and Russia), Far East (Burma, China, India, Indonesia, Japan, Malaysia, and New Guinea), Middle East (Iran, Iraq, and Pakistan), and South America (Argentina, Colombia, Trinidad, and Venezuela) [19, 20]. The pressure and temperature gradients range up to 20 kPa/m (0.9 psi/ft) and 30°C/km (1.7°F/100 ft), respectively, as shown in Figures 2.11 and 2.12.

Many possible causes for the geopressured zones are presented in the literature. Fertl and Timko discussed 17 causes [21]. Among these are rapid sedimentation accompanied by contemporary faulting, which is apparently the greatest contributing cause of the abnormally high pressures found in the

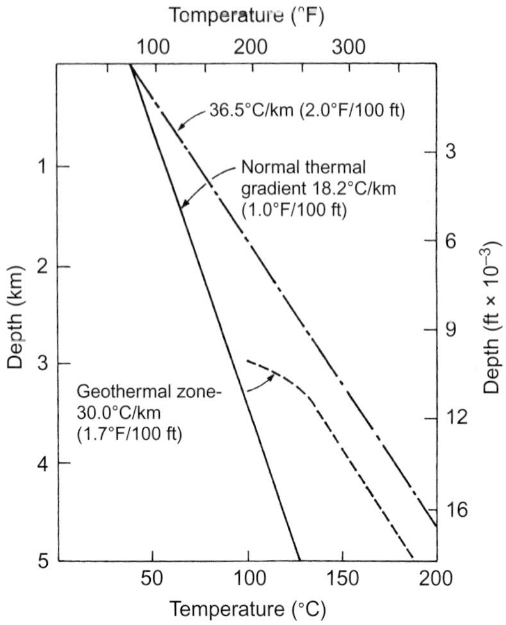

FIGURE 2.11 Subsurface temperature gradient showing the change within the geopressured zone. The 36.5°C/km gradient was included for reference only.

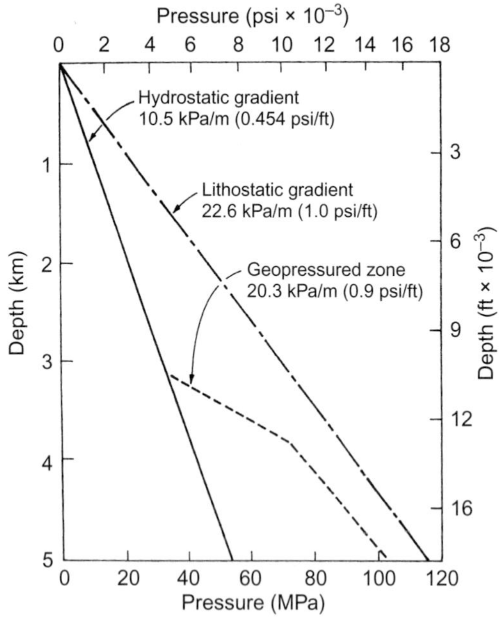

FIGURE 2.12 Subsurface pressure gradients showing the change in hydrostatic pressure gradient within the geopressured zone.

Gulf Coast Basin of the United States. Undercompaction of the sediments can occur during rapid sedimentation and burial of sediments containing a large quantity of clay minerals. The complete expulsion of water does not occur, leaving the sediments as a loosely bound system of swollen clay particles with interlayer water. Continued sedimentary deposition caused a shear zone to develop by overloading the undercompacted shale. Expulsion of the water was accompanied by subsidence of blocks of sediments. Thus, the contemporaneous fault zone of the Gulf Basin is characterized by the cycle of deposition, temperature increase, expulsion of water, and subsidence of blocks of sediments. As the depth of burial continued, the increase in temperature induced dehydration of the clays within the buried zone and contributed to the shearing stresses. The transformation of montmorillonite to illite during diagenesis and catagenesis occurs between 150° and 250°F, releasing an amount of water equal to about half of the original volume, leading to undercompaction in the geopressured zone. When the fluid pressure exceeds the total overburden pressure, the faults act as "valves" for discharge of water upward into the hydropressured aquifers overlying the zone. As the pressure declines, the "valves" close until the pressure once more exceeds the total overburden pressure [22, 23].

Another contributor to the fluid overpressure is the temperature increase that occurs within the geopressured zone. The overlying, normally pressured sediments that are well compacted possess a lower thermal conductivity and act as a "blanket," decreasing the transfer of heat from the mantle. The heat trapped by the blanket above the geopressured zone produces an abnormally high temperature in the formation, which contributes another incremental pressure increase to the fluid [24].

Geopressured zones along the Gulf Coast generally occur at depths below 8,000 ft and require careful and expensive drilling technology whenever the zones are penetrated. The zones usually contain about 3.6 cm^3 of methane per meter square of brine (20 SCF/bbl).

Oilfield Waters

The genesis of petroleum is intimately associated with shallow marine environments; hence, it is not surprising that water found associated with oil generally contains dissolved salts, especially sodium and calcium chlorides. Petroleum source rocks originally formed in lakes or streams, and the porous sediments that became today's petroleum reservoirs could have acquired saline waters by later exposure to marine waters. Thus, the original waters present in the sediments when they were developed may have been either fresh water or saline marine water. After the original deposition, however, the oilfield sedimentary formations have had histories of subsidence, uplift, reburial, erosion, etc. Therefore, the chemistry of the original water may have been altered by meteoric water, marine water infiltration at a later time, changes of salt types and concentrations due to solution of minerals as subsurface waters moved in response to tectonic events, and precipitation of some salts that may have exceeded equilibrium concentration limits [25].

The origin of deep saline subsurface waters has not been completely explained. The most plausible explanation is that they were originally derived from seawater. If seawater is trapped in an enclosed basin, it will undergo evaporation, resulting in precipitation of the dissolved salts. The least soluble salts will precipitate first, leaving a concentrated brine deficient in some cations and anions when compared to seawater. The common order of evaporative deposition from seawater in a closed basin is calcium carbonate (limestone)>calcium-magnesium carbonate (dolomite)>calcium sulfate (gypsum)>sodium chloride (halite)>potassium chloride (sylvite). Dolomite begins to precipitate when the removal of calcium from solution increases the Mg/Ca ratio. The residual brines (containing unprecipitated salts at any period) may migrate away from the basin and leave the evaporites behind, or they may become the interstitial water of sediments that are rapidly filling the basin [19]. In accumulating marine clastic sediments, aerobic bacteria consume the free oxygen in the interstitial waters and create an anaerobic environment in which the anaerobes become active and attack the sulfate ion, which is the second-most important anion in seawater. The sulfate is reduced by the bacteria to sulfide, which is liberated as hydrogen sulfide (marsh gas) [19]. Thus, the composition of saline oilfield waters, or brines, is quite different from the average composition of seawater (Table 2.3). With the exception of sulfate, all of the ions in the Smackover Formation (carbonate) brine are enriched with respect to seawater. Several mechanisms of enrichment are possible: (1) the original seawater may have evaporated if it was trapped in a closed basin; (2) movement of the waters through beds of clay may have concentrated cations by acting like a semipermeable membrane allowing water to pass through, but excluding or retarding the passage of dissolved salts; and (3) mixing with other subsurface waters containing high salt concentrations may have occurred. The content of alkali cations is many times greater in the oilfield brines than in the water that owes its salinity to the dissolution of salts from the Earth or to the infiltration of high-salinity waters from other sources.

There are many reactions between the ions that can occur as the environmental conditions change with burial. Consequently, the composition of oilfield waters varies greatly from one reservoir to another. Commonly, the salinity (total amount of dissolved salts, or TDS) of petroleum-associated waters increases with depth (there are a few exceptions to this). The principal anions change in a characteristic manner as depth increases: (1) sulfate is the major anion in the near-surface waters; (2) below about 500 m, bicarbonate may become the principal anion; and (3) in brines from deeper formations, chloride is the principal anion. The ratios of the cations also change with

TABLE 2.7 Average Composition of Seawater Compared to Smackover, Arkansas, Oilfield Brine [19]

Constituent	Seawater (mg/L)	Smackover Brine (mg/L)
Lithium	0.2	174
Sodium	10,600	67,000
Potassium	380	2,800
Calcium	400	35,000
Magnesium	1,300	3,500
Strontium	8	1,900
Barium	0.03	23
Boron	5	130
Copper	0.003	1
Iron	0.01	41
Manganese	0.002	30
Chloride	19,000	172,000
Bromide	65	3,100
Iodide	0.05	25
Sulfate	2,690	45

respect to depth. The Ca/Na ratio increases, whereas the Mg/Na ratio decreases [19].

The concentrations of salts in formation waters are expressed as weight percent (wt%): milligrams per liter (mg/L) or parts per million (ppm). The quantities are related as follows: $1\% = 10{,}000\ \text{ppm}$ and $\text{mg/L} = (\text{ppm}) \times (\text{density})$.

Where ionic reactions are involved, the contents of ions are expressed as milliequivalents per liter (meq/L). One milliequivalent of a cation reacts quantitatively with exactly 1 meq of an anion:

$$\text{meq/L} = (\text{mg/L}) \times \left(\frac{\text{valence}}{\text{molecular weight}}\right) \tag{2.10}$$

The calcium and magnesium cation concentrations of subsurface waters are probably functions of the origin of the specific oilfield water as well as its history of contact with infiltrating waters. These cations undergo reactions forming dolomite and enter into ion exchange reactions; consequently, they are normally found in lower concentrations than sodium cations. Other cations are present in concentrations less than 100 mg/L [13].

Oilfield waters are frequently referred to as connate or interstitial water, which is found in small pores and between fine grains in water-wet rocks. As defined by Collins, the two terms are synonymous and they are indistinguishable as used in the petroleum literature [26]. "Connate" implies that the water is the original fossil water present in the rocks from the time of original deposition. One cannot be certain of this because the original water may have been displaced or mixed with other waters during the geologic history of the sedimentary formation. Collins considers connate water as fossil water that has not been in contact with water from other sources for a large part of its geologic history.

Compressibility

Compressibility of water is a function of the environmental pressure and temperature as shown in Figure 2.13 [27]. At any given pressure, the compressibility decreases as the temperature is increased from ambient, reaching a minimum compressibility at about 55°C. Then, the compressibility increases continuously with temperature increase. At any given temperature, the compressibility decreases as the pressure is increased. The isothermal compressibility (c_w) is expressed as follows:

$$c_w = -\frac{1}{V_1}\left(\frac{dV}{dp}\right)_T = \left(1 - \frac{V_2}{V_1}\right)\frac{1}{p_2 - p_1} \tag{2.11}$$

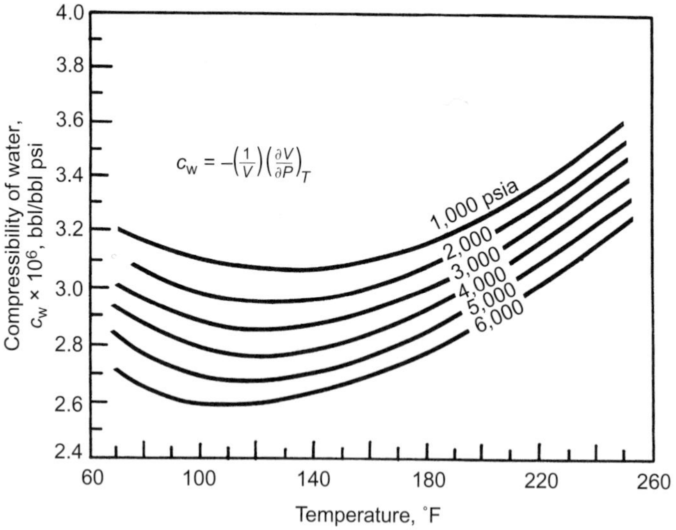

FIGURE 2.13 Compressibility of water as a function of temperature and pressure [27].

where V_1 and V_2 are the volumes at pressures p_1 and p_2. The ratio V_2/V_1 is equivalent to the amount of water expansion as the pressure drops from p_2 to p_1.

Example

The bottomhole temperature of a gas reservoir is 140°F. Calculate the amount of water expansion, per unit volume, that will occur when the pressure is decreased from 4,000 to 3,270 psi. From Figure 2.13, the estimated compressibility of water at the given reservoir conditions (i.e., at 4,000 psi) is 2.8×10^{-6} psi^{-1}.

Solution

$$\frac{V_2}{V_1} = \left[1 - (2.8 \times 10^{-6})(3,270 - 4,000)\right] = 1.02$$

Water compressibility decreases when the water contains hydrocarbon gases in solution according to the following empirical equation [26, 27]:

$$c_{sw} = c_w(1.0 + 0.0088 \times R_{sw}) \tag{2.12}$$

Where

c_{sw} = compressibility of water containing solution gas (1/kPa or 1/psi)
c_w = compressibility of water
R_{sw} = solubility of gas in water, m^3 gas/m^3 water (ft^3/bbl)

Gas Solubility

The solubility of hydrocarbon gases in water at any given pressure does not change very much as the temperature is increased. The behavior is similar to compressibility because the solubility decreases slightly as the temperature is increased from ambient temperature reaching a minimum solubility at about (66°C) 150°F and then increasing continuously as the temperature is increased (Figure 2.14). On the other hand, pressure has a large influence. According to Figure 2.14, the solubility of natural gas in water at 500 psi and 150°F is about 4.1 ft^3/bbl and at 2,000 psi and 150°F the solubility increases to about 11.9 ft^3/bbl (2.1 m^3 gas/m^3). The solubility of gas in water also is influenced by the amount of dissolved salts. Increasing salinity decreases the solubility of hydrocarbon gases in water according to the following empirical relationship:

$$R_B = R_{wp}[1 - X_c \times (\text{salts, ppm})(10^{-7})] \tag{2.13}$$

where

R_{WP} = solubility of gas in pure water, m^3/m^3 (SCF/bbl)
R_B = solubility of gas in brine, m^3/m^3 (SCF/bbl)
X_c = salinity correction factor (Table 2.4)

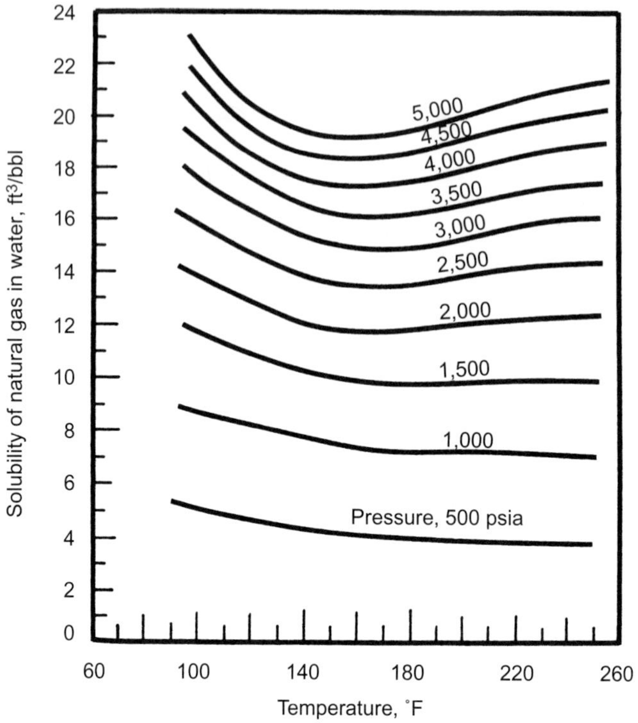

FIGURE 2.14 Solubility of natural gas in water as a function of temperature and pressure [27].

TABLE 2.4 Salinity Correction Factor for Estimation of the Solubility of Hydrocarbon Gases in Brine [22]

X_c (Salinity Correction Factor)	T (°F)
75	100
50	150
44	200
33	250

Example

Brine from a 7,000 ft deep reservoir in Kansas where the mean annual surface temperature is 70°F contains 80,000 ppm of total dissolved solids (TDS). If the reservoir pressure is 3,300 psi, estimate the solubility of hydrocarbon gas in the oilfield brine at reservoir conditions. Assume a geothermal gradient of 1°F/100 ft of depth and use Eq. (2.11) to estimate the reservoir temperature (T_f):

Solution

$$T_f = 70 + 1.0(7{,}000/100) = 140°F$$

Use Figure 2.14 to obtain the solubility of gas in pure water ($R_{wp} = 16$ ft^3/bbl). Then, extrapolate the salinity correction factor (X) to 140°F using Table 2.4 ($X = 55$).

$$R_B = 16[1 - 55(80{,}000 \times 10^{-7})] = 8.96 \text{ SCF/bbl}$$

Viscosity

All fluids resist a change of form, and many solids exhibit a gradual yield in response to an applied force. The force acting on a fluid between two surfaces is called a shearing force because it tends to deform the fluid. The shearing force per unit area is the shear stress (τ). The absolute viscosity is defined by:

$$\tau = -\mu(dv/dx) \tag{2.14}$$

where

τ = shear stress
μ = absolute viscosity
v = fluid velocity
x = distance

Viscosity is reported in terms of several different units: poise (CGS unit of absolute viscosity) = g/cm · s = 14.88 lbm/ft · s; centipoise = 0.01 poise; stoke (CGS unit of kinematic viscosity) = g/[(cm · s)(g/cm^3)]; centistoke = 0.01 stoke; and pascal seconds (SI units) = 0.1 poise.

Figure 2.15 may be used to estimate the viscosity of oilfield waters as a function of salinity, temperature, and pressure. A separate chart (insert on Figure 2.15) is used to obtain a factor relating the viscosity to pressure.

Example

Estimate the viscosity of brine containing 12% salts that was obtained from a reservoir with a fluid pressure of 6,000 psi and temperature of 180°F.

Solution

Obtain the pressure correction factor from the chart in Figure 2.15 (pressure correction factor = 1.018).

Viscosity of 12% brine at 180°F and 14.7 psia = 0.48 cP.

Viscosity at 180°F and 6,000 psia = (0.48)(1.018) = 0.49 cP.

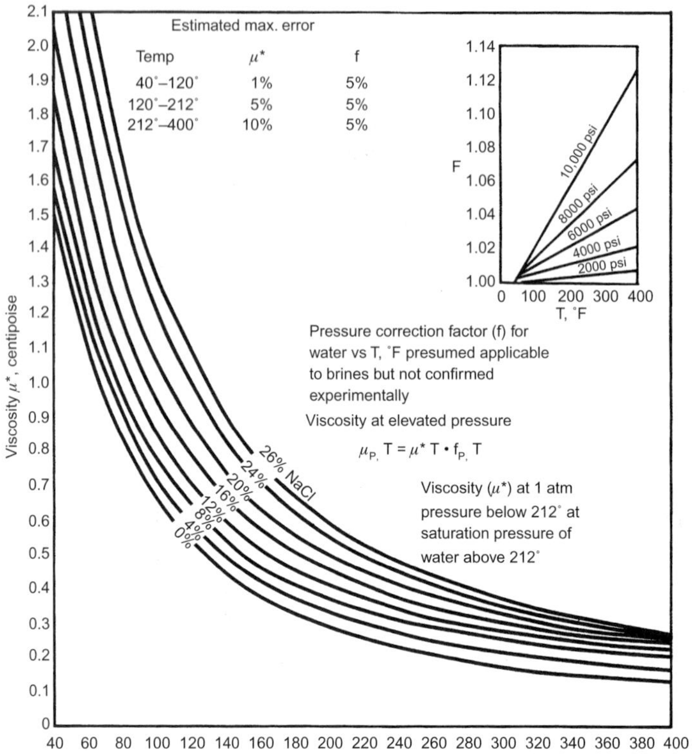

FIGURE 2.15 Viscosity of water as a function of temperature, salinity, and pressure [22].

PETROLEUM

Petroleum is a complex mixture containing thousands of different compounds, most of which are composed exclusively of hydrogen and carbon (hydrocarbons). Included in the mixture are compounds containing nitrogen, sulfur, oxygen, and metals compounds. In 1927, the American Petroleum Institute (API) initiated Research Project 6, "The Separation, Identification, and Determination of the Chemical Constituents of Commercial Petroleum Fractions," which was designed to elucidate the structure of compounds in crude oil from the Ponca City oilfield, Oklahoma. By 1953, 130 hydrocarbons had been identified. The number of compounds clearly identified has increased greatly since then after introduction of gas chromatography and mass spectroscopy [13].

The density and viscosity of hydrocarbon gases and liquids are very important physical quantities. They are used to characterize pure and mixed hydrocarbons and to evaluate their fluid flow behavior in the reservoir.

Gas Density

The density of gases may be calculated from the equation of state for real gases (Eq. (2.15)), which is corrected for nonideal behavior by a compressibility factor Z. The factor Z is the ratio of the actual volume occupied by a real gas to the volume it would occupy if it behaved like an ideal gas where $Z = 1.0$ [29]:

$$pV = ZmRT/M \tag{2.15}$$

or

$$\rho = m/V = pM/ZRT \tag{2.16}$$

where

p = pressure, psi
V = volume, ft^3
Z = real gas deviation factor
m = mass of gas, lb
R = gas constant (10.73 psi-ft^3/lbmol-°R)
T = temperature, °R
M = molecular weight of the gas

Gravitational units are used because, to date, engineering charts in the United States have not been converted to SI units.

The compressibility factor, or real gas deviation factor, is obtained from the reduced temperatures and pressures and the compressibility charts for pure and mixed gases (Figure 2.16). The reduced temperature and pressure are calculated from the gas pseudocritical temperatures and pressures as follows:

$$T_{pr} = \frac{T}{T_{pc}}; \quad p_{pr} = \frac{p}{p_{pc}} \tag{2.17}$$

where

T_{pr} and p_{pr} = pseudoreduced temperature and pressure, respectively.
T_{pc} and p_{pc} = critical temperature and pressure, respectively (Table 2.5).

Viscosity of Gases

Gas viscosity varies with respect to temperature, pressure, and molecular weight. The exact mathematical relationships have not been developed; however, Carr et al. developed two charts that may be used to estimate gas viscosities at various temperatures and pressures (Figures 2.17 and 2.18) [30].

Oil Density

The most commonly measured physical property of crude oils and its fractions is the API gravity. It is an arbitrary scale adopted for simplified

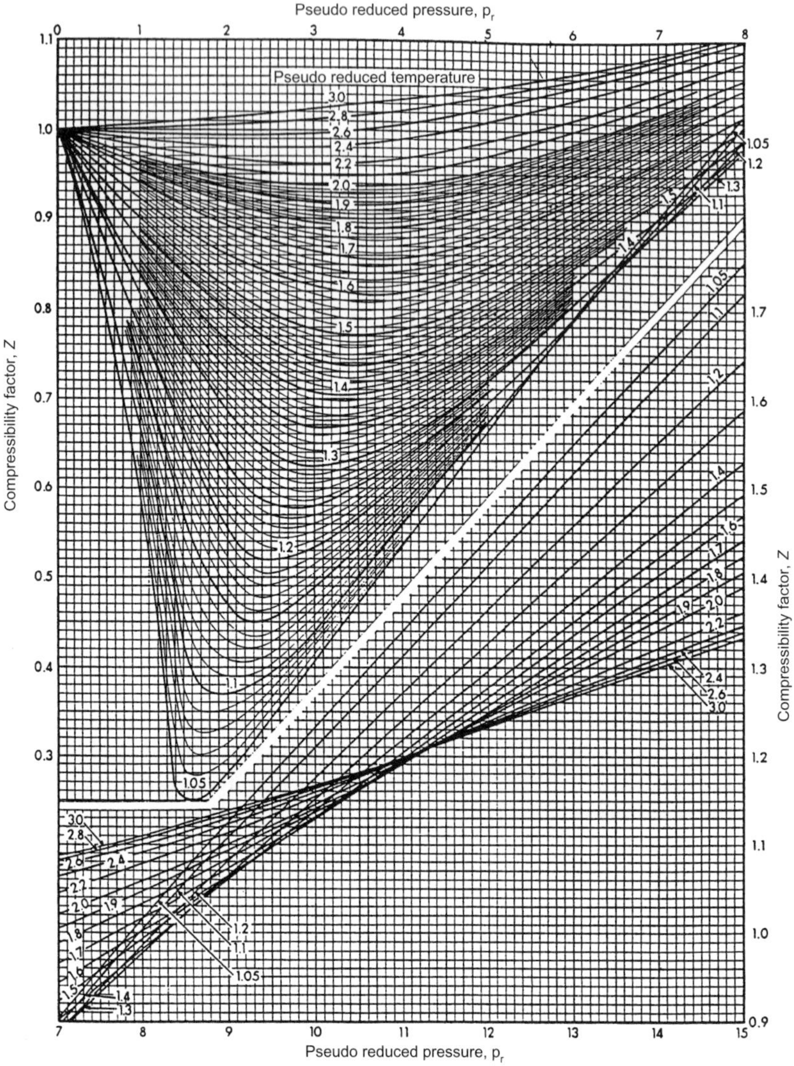

FIGURE 2.16 Real gas deviation factor as a function of p_{pr} and T_{pr}.

measurement by hydrometers, because it enables a linear scale for gravity measurement. The °API gravity is directly related to the specific gravity as follows:

$$^{\circ}\text{API} = \left(\frac{141.5}{\text{SG}_{60^{\circ}\text{F}}}\right) - 131.5 \tag{2.18}$$

TABLE 2.5 Physical Properties of Various Hydrocarbons and Associated Compounds [31]

Constituent	Molecular Weight	Normal Boiling Point °F	Normal Boiling Point °R	Liquid Density (lbm/ft³)	Gas Density at 60°F, 1 atm (lbm/ft³)	Critical Temperature (°R)	Critical Pressure (psia)
Methane, CH_4	16.04	−258.7	201	18.72*	0.04235	344	673
Ethane, C_1H_6	30.07	−127.5	332	23.34*	0.07986	550	712
Propane, C_3H_8	44.09	−43.8	416	31.68**	0.1180	666	617
iso-butane, C_4H_{10}	58.12	10.9	471	35.14**	0.1577	735	528
n-butane, C_4H_{10}	58.12	31.1	491	36.47**	0.1581	766	551
iso-pentane, C_3H_{10}	72.15	82.1	542	38.99	—	830	483
n-pentane, C_3H_{12}	72.15	96.9	557	39.39	—	847	485
n-hexane, C_3H_{14}	86.17	155.7	615	41.43	—	914	435
n-heptane, C_7H_{16}	100.20	209.2	669	42.94	—	972	397
n-octane, C_8H_{18}	114.22	258.1	718	44.10	—	1,025	362
n-nonane, C_9H_{20}	128.25	303.3	763	45.03	—	1,073	335
n-decane, $C_{10}H_{22}$	142.28	345.2	805	45.81	—	1,115	313
Nitrogen, N_2	28.02	−320.4	140	—	0.0739	227	492
Air ($O_2 + N_2$)	29	−317.7	142	—	0.0764	239	547
Carbon dioxide, CO_2	44.01	−109.3	351	68.70	0.117	548	1,073
Hydrogen sulfide, H_2S	34.08	−76.5	383	87.73	0.0904	673	1,306
Water	18.02	212	672	62.40	—	1,365	3,206

*Apparent density in liquid phase.
**Density at saturation pressure.

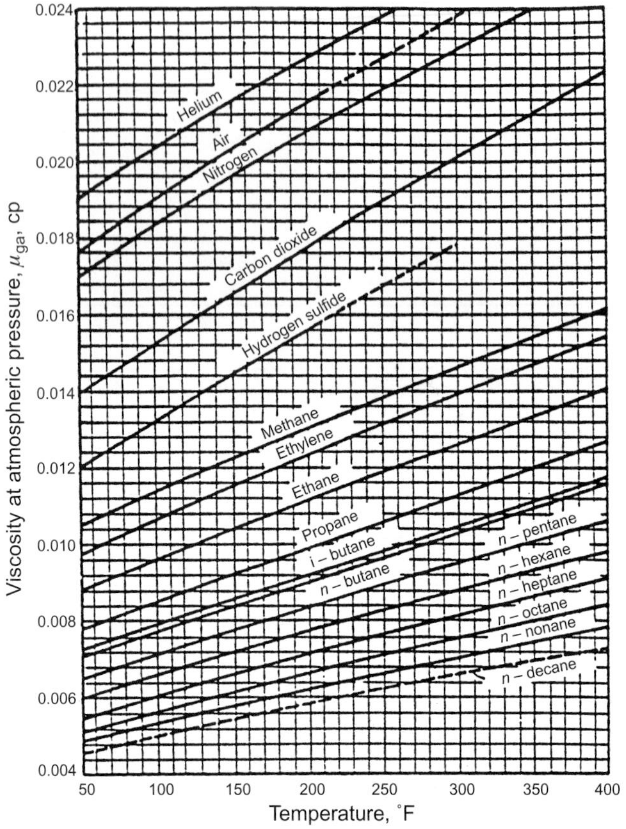

FIGURE 2.17 Viscosity of gases at 1 atm as a function of temperature [30].

The °API gravity does not have a linear relationship to the physical properties of petroleum or its fractions; therefore, it is not a measure of the quality of petroleum. The measurements are important, however, because the °API gravity is used with other parameters for correlation of physical properties. Also, the price of petroleum is commonly based on its API gravity.

A comparison of API gravity and specific gravity is shown in Table 2.6. Specific gravity (SG) is the density of the fluid at any temperature and pressure divided by the density of water at 60°F and 14.7 psia (62.34 lbm/ft^3; where lbm = pounds mass). Note that the API gravity is inversely proportional to the specific gravity and an °API gravity of 10° corresponds to the specific gravity of water at 60°F (SG = 1.0).

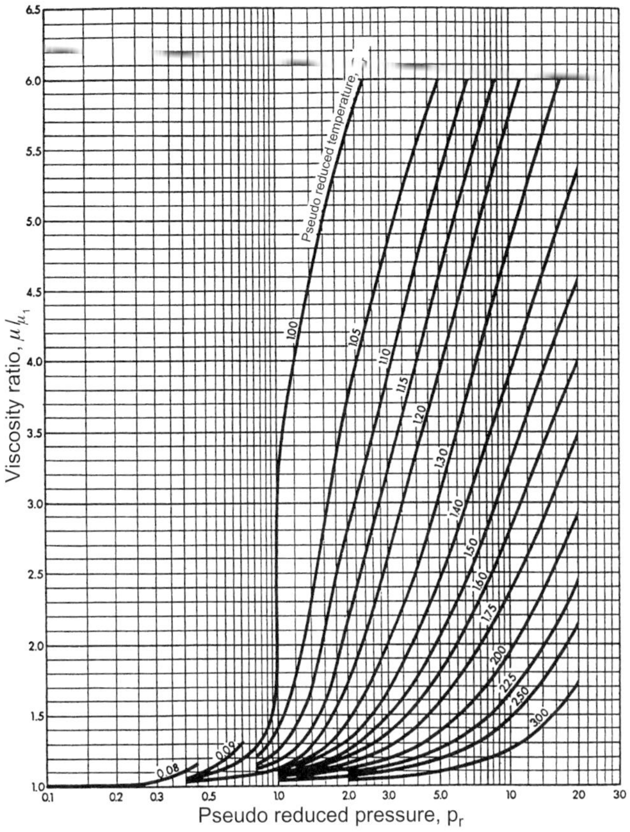

FIGURE 2.18 Viscosity ratio as a function of pseudoreduced pressure [30].

Oil Viscosity

Two methods for measuring the viscosity of crude oils and their fractions that have received universal acceptance are (1) the kinematic viscosity measurement, which is obtained by timing the flow of a measured quantity of oil through a glass capillary, yielding the viscosity in centistokes, and (2) the Saybolt viscosity measurement, which is the time (seconds) required for a standard sample of oil to flow through a standard orifice (ASTM Test D-88). The Saybolt Universal viscometer is used for refined oil fractions and lubricating oils, and the Saybolt Furol ("fuel and road oil") viscometer is used for high-viscosity crude oils and fractions. (The Furol viscometer has a larger diameter orifice.) Results of the test are expressed in Saybolt or Furol seconds at a specified temperature.

TABLE 2.6 Comparison of API Gravity and Specific Gravity at 60°F and 1 atm Pressure [13]

API Gravity	Fluid Type	Specific Gravity
−8	Heavy oil and brine	1.1460
−4	Heavy oil and brine	1.1098
0	Heavy oil and brine	1.0760
5	Heavy oil and brine	1.0366
10	Heavy oil and fresh water	1.000
15	Heavy oil	0.9659
20	Heavy oil	0.9340
30	Light oil	0.8762
40	Light oil	0.8251
50	Condensate fluids	0.7796

Tables 2.7 and 2.8 are used to convert from Saybolt seconds to centistokes. Absolute viscosity (centipoises) is obtained by multiplying centistokes by the density of the oil [32].

PETROLEUM CHEMISTRY

Petroleums are frequently characterized by the relative amounts of four series of compounds. The members of each series are similar in chemical structure and properties. The four series (or classes of compounds) that are found in petroleums are (1) the normal and branched alkane series (paraffins), (2) cycloalkanes (naphthenes), (3) the aromatic series, and (4) asphalts, asphaltenes, and resins (complex, high-molecular-weight polycyclic compounds containing nitrogen, sulfur, and oxygen atoms in their structures: the NSO compounds). The petroleums are generally classified as paraffinic, naphthenic, aromatic, and asphaltic according to the relative amounts of any of the series [14].

Tissot and Welte refined this classification further into six groups by adding intermediate types of oils using a ternary diagram (Figure 2.19) [14]. According to this classification, an oil is considered aromatic if the total content of aromatics, asphaltenes, and resins is 50% or greater. Paraffinic oils contain at least 50% of saturated compounds, 40% of which are paraffins. Likewise, naphthenic oils are those composed of 50% or more saturated compounds, of which 40% or more are naphthenes. The gases and low-boiling

TABLE 2.7 Conversion of Viscosity Measured as Saybolt Universal Seconds at Two Temperatures to Centistokes [32]

Centistokes	Saybolt 100°F	Seconds at 210°F	Centistokes	Saybolt 100°F	Seconds at 210°F
2	32.6	32.8	28	132.1	133.0
3	36.0	36.3	30	140.9	141.9
4	39.1	39.4	32	149.7	150.8
5	42.3	42.6	34	158.7	159.8
6	45.5	45.8	36	167.7	168.9
7	48.7	49.0	38	176.7	177.9
8	52.0	52.4	40	185.7	187.0
9	55.4	55.8	42	194.7	196.1
10	58.8	59.2	44	203.8	205.2
12	65.9	66.4	46	213.0	214.5
14	73.4	73.9	48	222.2	223.8
16	81.1	81.7	50	231.4	233.0
18	89.2	89.8	60	277.4	279.3
20	97.5	98.2	70	323.4	325.7
22	106.0	106.7	80	369.6	372.2
24	114.6	115.4	90	415.8	418.7
26	123.3	124.2	100	462.9	465.2

point fractions of petroleum contain greater amounts of the low-molecular-weight alkanes. Intermediate boiling fractions contain greater amounts of the cyclic alkanes and aromatics, where the higher boiling point fractions (>750°F–399°C) are composed predominantly of the naphthenic aromatics. Hunt presented the composition of a crude oil, which is classified as naphthenic according to Figure 2.19, because the oil contains 49% naphthenes and the total amount of saturated hydrocarbons (paraffins and naphthenes) is 79% (Table 2.9) [33].

Also listed in the table are the molecular size ranges (number of carbon atoms per molecule) of average refinery fractions of this crude oil and the approximate weight percentages of each fraction that can be obtained from the naphthenic crude oil described above.

TABLE 2.8 Conversion of Viscosity Measured as Furol Seconds at 122°F to Centistokes [32]

Centistokes	Furol Seconds at 122°F	Centistokes	Furol Seconds at 122°F
48	25.3	140	67.0
50	26.1	145	69.4
52	27.0	150	71.7
54	27.9	155	74.0
56	28.8	160	76.3
58	29.7	165	78.7
60	30.6	170	81.0
62	31.5	175	83.3
64	32.4	180	85.6
66	33.3	185	88.0
68	34.2	190	90.3
70	35.1	195	92.6
72	36.0	200	95.0
74	36.9	210	99.7
76	37.8	220	104.3
78	38.7	230	109.0
80	39.6	240	113.7
82	40.5	250	118.4
84	41.4	260	123.0
86	42.3	270	127.7
88	43.2	280	132.4
90	44.1	290	137.1
92	45.0	300	141.8
94	45.9	310	146.5
96	46.8	320	151.2
98	47.7	330	155.9
100	48.6	340	160.6
105	50.9	350	165.3
110	53.2	360	170.0
115	55.5	370	174.7

(Continued)

TABLE 2.8 (Continued)

Centistokes	Furol Seconds at 122°F	Centistokes	Furol Seconds at 122°F
120	57.8	380	179.4
125	60.1	390	184.1
130	62.4	400	188.8
135	64.7		

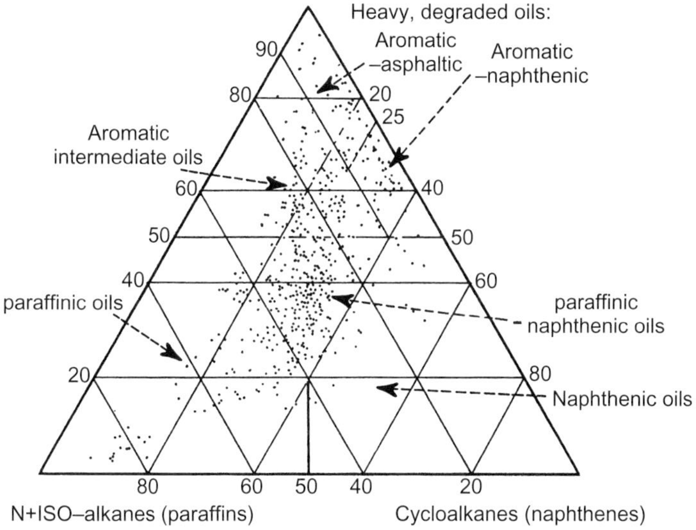

FIGURE 2.19 Ternary diagram for classification of crude oils as paraffinic, naphthenic, or aromatic [14].

TABLE 2.9 Composition and Refinery Fractions of a Naphthenic Crude Oil [33]

Molecular Type	wt%	Molecular Size	wt%
Naphthenes	49	Gasoline (C_4–C_{10})	31
Paraffins	30	Kerosene (C_{11}–C_{12})	10
Aromatics	15	Gas oil (C_{13}–C_{20})	15
Asphalts/resins	6	Lubricating oil (C_{20}–C_{49})	20
		Residum (C_{40+})	

The U.S. Bureau of Mines Research Center at Bartlesville, Oklahoma, standardized the classification of crude oils by distillation and characterized a large number of oils from oilfields around the world. The distillation of a crude oil from the Oklahoma City oilfield is shown in Table 2.10. A liter of oil is placed in the flask and the temperature is raised gradually while the volume percents of condensed vapors collected at specific temperatures are recorded. After reaching 275°C, the flask is placed under a vacuum of 40 mm Hg and the distillation is continued as shown in Table 2.10.

Alkanes also are referred to as saturated hydrocarbons because the valence (or bonding capacity) of all of the carbon atoms is satisfied by hydrogen atoms (Figure 2.20). Each carbon atom is connected to another carbon atom by a single covalent bond, and the remaining bonding capacity is occupied by hydrogen atoms as illustrated for ethane, propane, butane, and pentane in Figure 2.20. Isomers are compounds that have the same atomic composition but differ in molecular structure and properties. There are three structurally different pentanes although they all have the same number of carbon and hydrogen atoms—*n*-pentane, iso-pentane, and 2,2-dimethyl propane (Figure 2.21). The structural difference results in slight differences in chemical reactivity and physical properties as indicated by the difference of the boiling points of the three pentanes. As the number of carbon atoms increases in a homologous series, the number of possible isomers also increases; for example, there are 18 isomers of octane (eight carbon atoms) and 75 isomers of decane (10 carbon atoms). Thus, a single homologous series of compounds exhibits enormous complexity. Even though crude oils from different locations may have the same °API gravity and viscosity, they can vary widely with respect to chemical composition.

The alkanes with 25 or more carbon atoms are solids at room temperature and are extracted from the crude oils to make industrial paraffin waxes. Crude oils containing these alkanes become cloudy when cooled. The temperature at which this occurs is called the cloud point and is used in refineries as a general indication of the abundance of paraffin waxes. The formation around the wellbore and production tubing must be cleaned periodically to remove precipitated high-molecular-weight alkanes, which reduce the rate of production [34].

Crude oils derived principally from terrestrial plant organic material contain high amounts of alkanes, whereas the oils generated from marine organic materials generally contain greater amounts of cyclic saturated and unsaturated compounds. If, after it has migrated from the source rock to an oil trap, a paraffinic oil is exposed to the percolation of meteoric water due to diastrophism, aerobic bacteria will remove the paraffins by gradual degradation to carboxylic acids and carbon dioxide [14]. A crude oil that has been exposed to aerobic bacterial degradation will be chiefly composed of aromatics, asphalts, and resins.

TABLE 2.10 U.S. Bureau of Mines Distillation Method for Analysis of Crude Oil, Paul-Kune No. 1

Oklahoma City Field		Oklahoma
Prue Sand		Oklahoma County
6,511–6,646 ft	Sample 38005	11N-3W-Indian
	General Characteristics	
Specific gravity, 0.844		API gravity, 36.2°
Sulfur, %, 0.16		Color: brownish green
Saybolt Universal viscosity at 77°F, 62 s; at 100°F, 50 s		
	Distillation, Bureau of Mines Hempel Method	
Distillation at atmospheric pressure, 752 mm		First drop 86°F

Fraction No.	At °F	Percent	Sum Percent	SP. Gr. 60/60°F	API 60°F	CI	SU Visc. 100°F	Cloud Test °F
1	122	—	—	—	—	—		
2	167	1.7	1.7	0.672	79.1	—		
3	212	3.0	4.7	0.702	70.1	13		
4	257	4.9	9.6	0.734	61.3	19		
5	302	4.7	14.3	0.755	55.9	21		
6	347	4.7	19.0	0.772	51.8	23		
7	392	4.7	23.7	0.787	48.3	23		
8	437	5.0	28.7	0.801	45.2	24		
9	482	5.3	34.0	0.815	42.1	26		
10	527	6.7	40.7	0.829	39.2	28		
			Distillation Continued at 40 mm					
11	392	3.6	44.3	0.844	36.2	31	41	10
12	437	6.7	51.0	0.851	34.8	30	47	25
13	482	5.9	56.9	0.866	31.9	34	61	45
14	527	6.3	63.2	0.876	30.0	36	87	65
15	572	5.6	68.8	0.884	28.6	37	150	80
Residum		28.6	97.4	0.925	21.5			

Carbon residue of residum—4.2%; carbon residue of crude—1.2%.

(*Continued*)

TABLE 2.10 (Continued)

	Approximate Study			
Light Gasoline	**Percent 4.7**	**Sp. Gr. 0.691**	**API 73.3**	**Viscosity**
Total gasoline and naphtha	23.7	0.748	57.7	
Kerosene distillate	10.3	0.808	43.6	
Gas oil	15.0	0.838	37.4	
Nonviscous lubricating distillate	12.4	0.854–0.878	34.2–29.7	50–100
Medium lubricating distillate	7.4	0.878–0.888	29.7–27.9	100–200
Viscous lubricating distillate	—	—	—	+100
Residum	28.6	0.925	215	
Distillation loss	2.6			

The cycloalkanes (naphthenes) are composed of carbon atoms bonded in a cyclic chain with the remaining valence satisfied by hydrogen atoms. Figure 2.22 shows the structure of cyclohexane and decalin, which, along with the methyl derivatives, are important constituents of petroleum. Tri-, tetra-, and pentacycloalkanes are present in crude oils in smaller quantities than are the mono- and dicycloalkanes. The naphthenes are important constituents of petroleum-derived commercial solvents.

The series of compounds known as aromatics are composed of multiples of benzene, a six-membered carbon ring linked with alternate double and single bonds (Figure 2.23). Aromatic compounds occurring in petroleum contain side chains of various lengths. The asphalts and resins are composed of high-molecular-weight condensed ring structures containing aromatics, saturated ring compounds, and alkane side chains and are interspersed with nitrogen, sulfur, and oxygen compounds [15].

The various homologous series discussed above may be readily separated by first diluting a sample of crude oil with pentane and then filtering. The asphaltenes are insoluble in pentane and can thus be removed and weighed. The diluted sample may then be percolated through a double column of active clay mineral on top of a column of silica gel, as shown in Figure 2.24, and eluted with pentane. The resins are adsorbed by the clay, whereas the paraffins and aromatics pass through the clay column. The aromatics are adsorbed by the silica gel column and the non-adsorbing paraffins are collected in the bottom flask. The resins and aromatics are removed from the

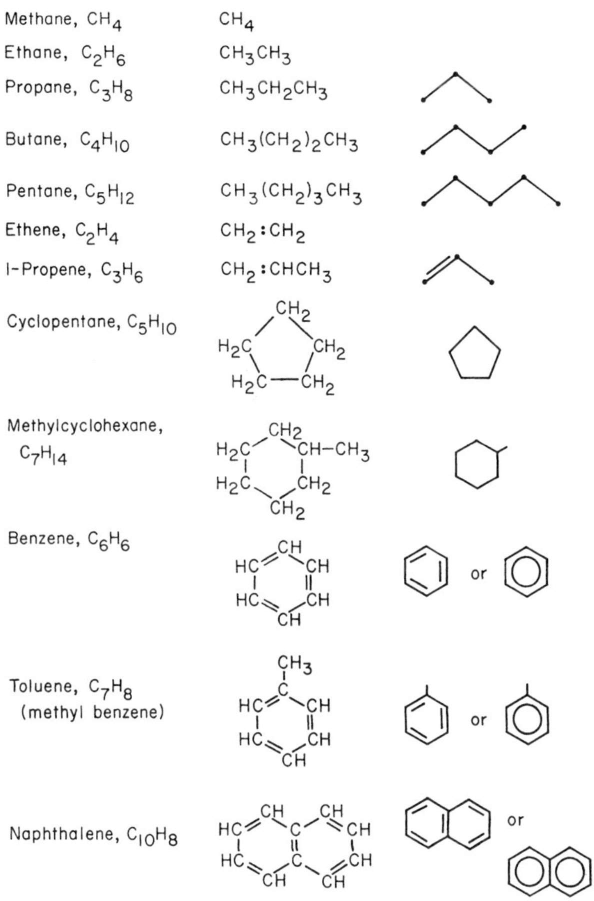

FIGURE 2.20 Chemical structure of a few important hydrocarbons found in many crude oils.

clay and silica gel with a mixture of equal parts benzene and acetone, and can be obtained quantitatively by evaporation of the solvent [35] (Figure 2.25). A high-pressure liquid chromatograph (HPLC) also may be used to obtain the same fractions using less than 1 mL of sample. The response from the HPLC is shown in Figure 2.26.

Organometallic compounds are usually associated with the resins because of their polar characteristics. Alkyl derivatives of nickel and vanadium porphyrins have been isolated from crude oils, especially the Boscan heavy oil from Venezuela. The porphyrins are characterized by a tetrapyrrolic nucleus, which also is the base structure of chlorophyll in plants and hemin in blood (Figure 2.27). The transformation of the natural compounds to porphyrins probably takes place only during sedimentation with replacement of the

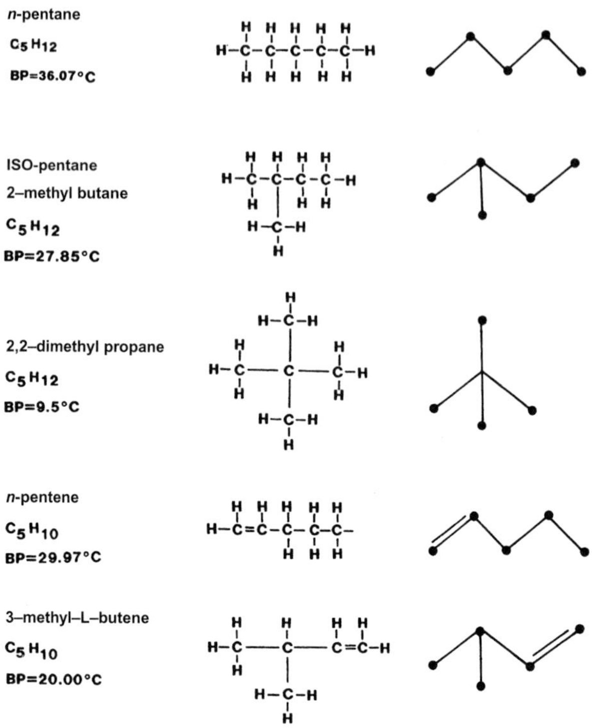

FIGURE 2.21 Chemical structure of hydrocarbons found in crude oils.

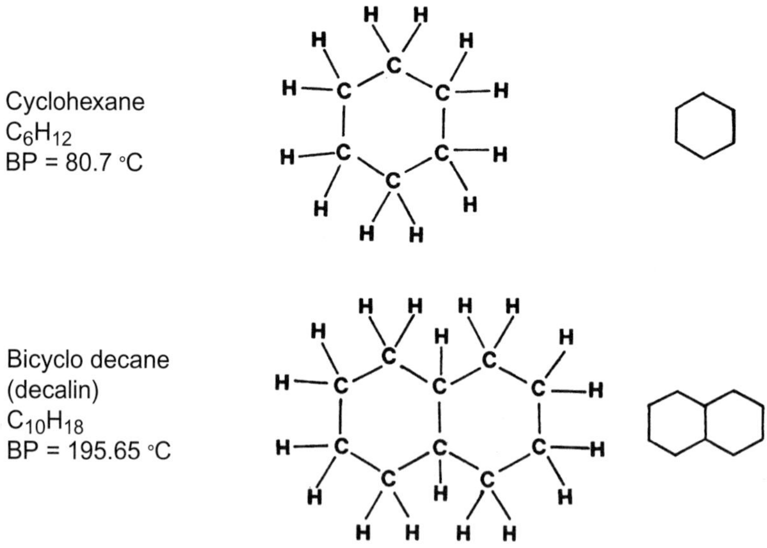

FIGURE 2.22 Chemical structure of cyclohexane and decalin.

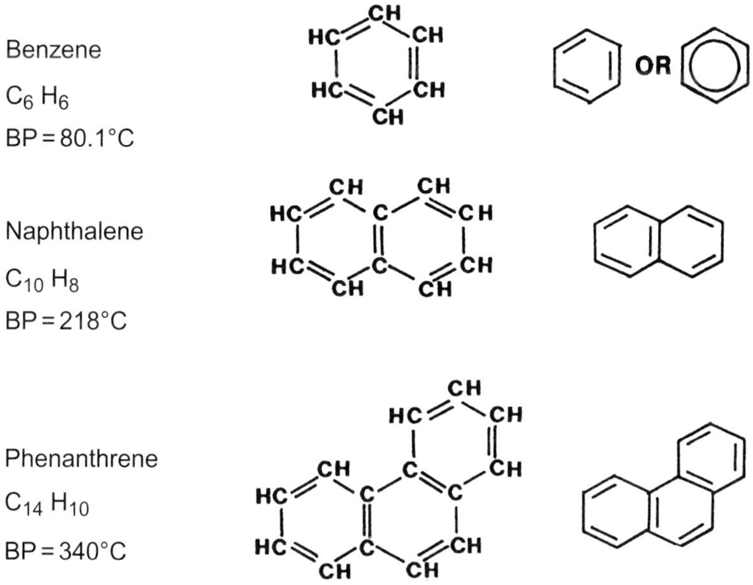

FIGURE 2.23 Chemical structure of several aromatic compounds found in crude oils.

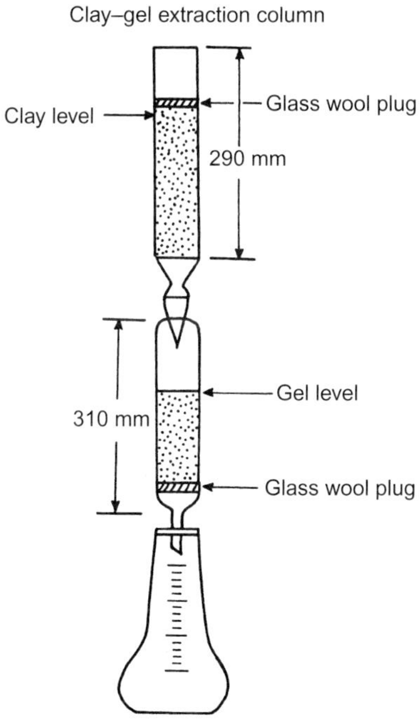

FIGURE 2.24 Clay–gel column used for separation of resins, aromatics, and paraffins from crude oils [35].

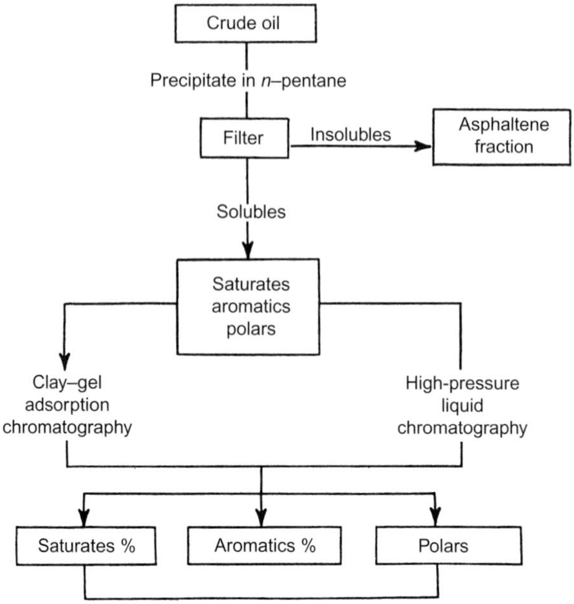

FIGURE 2.25 Schematic diagram of the chromatographic separation procedures used for extraction of resins, aromatics, and paraffins from crude oils [35].

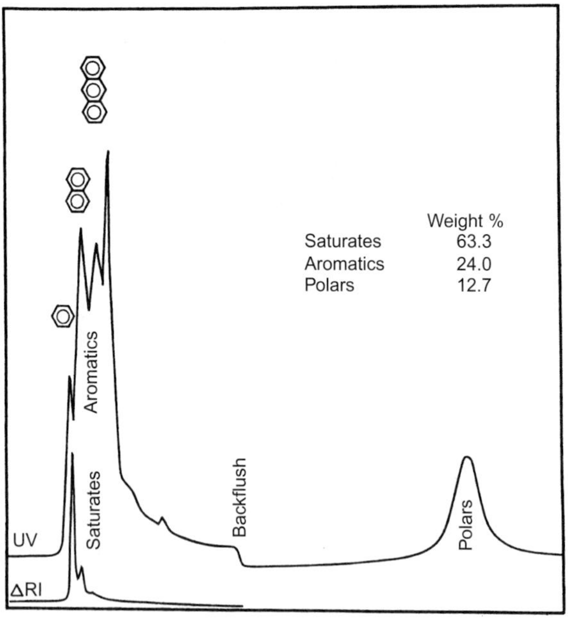

FIGURE 2.26 Typical response peaks for high-pressure liquid chromatographic separation of paraffins, aromatics, and polar organic compounds from crude oils [35].

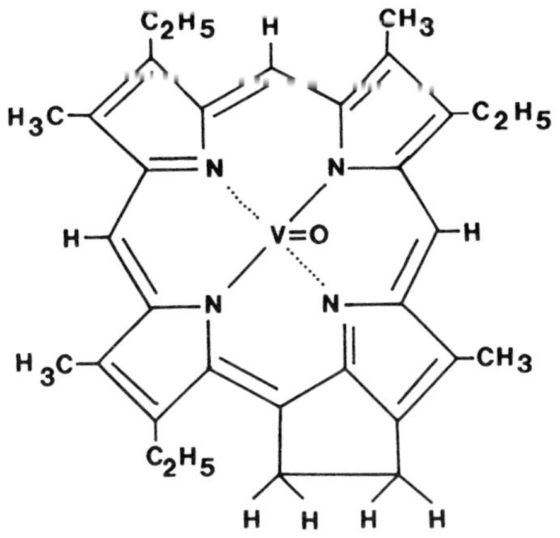

FIGURE 2.27 Chemical structure of desoxophylloery throetioporphyrin (vanadyl porphyrin).

magnesium in chlorophyll and iron in hemin by vanadium or nickel. This stabilizes the molecule, insuring its preservation. Blumer and Snyder suggested that the precursors of the porphyrins are incorporated into kerogen and are later transformed to porphyrins during the various changes that take place as the kerogen-type organics change to crude oil [17].

The alkenes are unsaturated hydrocarbons that contain double bonds between the carbons. Thus, the balance of the carbon atoms is not completely satisfied with hydrogen atoms. An example of this is 1-propene, in which the second and third carbon atoms are joined by a double bond and the rest are single bonded (Figure 2.20).

Asphalt is a black colloidal solution composed of high-molecular-weight polynuclear aromatic compounds, high-molecular-weight unsaturated compounds, and heterogeneous hydrocarbons containing nitrogen, sulfur, oxygen, and metals in their structure. The heavy oils and bitumens generally contain more nitrogen, sulfur, oxygen, and metal compounds than do the light oils, and many oils contain free hydrogen sulfide gas.

PROBLEMS

1. Convection currents in the mantle are apparently responsible for the movements of continents. Explain the location (accumulation) of continents and basins in response to rising and descending convection currents in the mantle.
2. Calculate the seismic velocities through sandstone from the following data and compare them to the velocities in limestone. Why are the velocities different?

$$B = 3.4 \times 10^{10}\ \text{Pa};\ S = 3.1 \times 10^{10}\ \text{Pa};\ \rho = 2.64\ \text{g/cm}^3$$

3. Explain the initial formation of the Appalachian mountain range. What were the geologic periods and estimated time when this began and reached its climax?
4. If the relative radiocarbon content of the remains of a plant is 1/7, how long ago did the plant live? What geologic period and epoch was this?
5. Define "craton." Where are cratons located?
6. Discuss transgressive and regressive periods of sedimentary deposition. Which period leads principally to the formation of hydrocarbon source rocks? Why?
7. What are "clastics," "granite wash," "arkose," and "graywacke"? What are some general locations of these types of rocks?
8. Well logs of an area show that the temperature at the bottom of a 3,140-m deep well is 92°C. If the mean surface temperature is 27°C, what is the geothermal gradient?
9. The composition of a brine sample from a geopressured zone 2,929 m deep is listed below. Compare the brine sample analysis to that of sea-water (Table 2.3) and give a reasonable explanation for the differences. What is the TDS of the brine?

Ion	Concentration (ppm)
Na^+	29,400
Ca^{2+}	2,662
Mg^{2+}	1,011
K^+	172
Ba^{2+}	5
Cl^-	46,618
HCO^-	714
SO^-	60
Br^-	40
I^-	23

10. The Saybolt viscosity of an oil is 117 s at 100°C. What is the viscosity in centipoises if the oil density is 0.885 g/cm^3?
11. Show the chemical structures of the following compounds: iso-propane, 1-methyl-2-ethyl cyclohexane, para-xylene, and anthracene.

NOMENCLATURE

B_w water formation volume factor (FVF)
c_w water compressibility
c_{sw} compressibility of water with solution gas

C radioactive decay constant
D depth
F_c salinity correction factor
g_c gravitational constant
G_t geothermal gradient
G_p pressure gradient
G shear modulus
h_o height of oil column
k permeability
K bulk modulus
m mass of gas (grams or lbm)
M molecular weight
N moles
N_o original amount of parent element
N_t amount of daughter isotope currently present
NSO nitrogen, sulfur, and oxygen
p pressure
p_d displacement pressure
p_f fluid pressure
p_{ob} overburden pressure
p_l lithostatic pressure
p_{pc} pseudocritical pressure
p_{pr} pseudoreduced pressure
P_c capillary pressure
r radius
r_c radius of a capillary
R universal gas constant
R_b solubility of gas in brine
R_{wp} solubility of gas in pure water
R_{sw} solubility of gas in water
SG specific gravity
SCF standard cubic feet
t time
$t_{1/2}$ half-life of parent element
T temperature
T_f formation temperature
T_{pc} pseudocritical temperature
T_{pr} pseudoreduced temperature
T_R reservoir temperature
T_s surface temperature
TDS total dissolved solids
v velocity
V volume
x Cartesian distance coordinate
z valence
Z real gas deviation factor
Z_o height of a column of oil

GREEK SYMBOLS

γ specific weight, lb/ft^3
θ contact angle
μ viscosity
μ_{ga} gas viscosity at atmospheric pressure
μ_T^* viscosity at reservoir temperature and atmospheric pressure
ρ density
ρ_o oil density
ρ_w water density
σ interfacial tension
τ shear stress

SUBSCRIPTS

c compressional wave
d displacement
f fluid
h horizontal
l leading pore or edge
o oil
ob overburden
s shear wave
t trailing pore or edge
v vertical
w water
1,2 reservoir zones

REFERENCES

1. Link PK. *Basic petroleum geology*. Tulsa, OK: Oil & Gas Consultants International;1982. 235 pp.
2. Stokes WL. *Essentials of Earth history*. Englewood Cliffs, NJ: Prentice-Hall;1966. 468 pp.
3. Dott Jr RH, Batten RL. *Evolution of the Earth*. New York: McGraw-Hill;1976. 504 pp.
4. LePichon x. Sea-floor spreading and continental drift. *J Geophys Res* 1968;**73**(12):3661–97.
5. Flint RF, Skinner FJ. *Physical geology*. New York: Wiley;1974. 407 pp.
6. Selley RC. *Elements of petroleum geology*. Chap. 4. New York: W. H. Freeman;1985. 449 pp.
7. Lowell JD. *Structural styles in petroleum exploration*. Tulsa, OK: Oil & Gas Consultants International;1985. 460 pp.
8. Hobson GD. *Developments in petroleum geology—I*. London: Applied Science Publishers;1977. 335 pp.
9. Chapman RE. *Petroleum geology—I*. New York: Elsevier Science;1986. 328 pp.
10. Magara K. *Geological models of petroleum entrapment*. New York: Elsevier Science;1986. 328 pp.
11. Pirson SJ. *Elements of oil reservoir engineering*, 2nd ed. New York: McGraw-Hill;1958. 441 pp.
12. Hobson GD, Tiratsoo EN. *Introduction to petroleum geology*. Houston, TX: Gulf Publishing Company;1985. 352 pp.
13. Dickey PE. *Petroleum development geology*, 2nd ed. Tulsa, OK: PennWell Books;1979. 424 pp.

14. Tissot BP, Welte DH. *Petroleum formation and occurrence*. Heidelberg: Springer-Verlag;1978. 538 pp.
15. Chilingarian GV, Yen TF, editors. *Bitumens, asphalts and tar sands*. New York: Elsevier Science;1978. 331 pp.
16. Barker C. Origin, composition and properties of petroleum. In: Donaldson EC, Chilingarian GV, Yen TF, editors. *Enhanced oil recovery*. New York: Elsevier Science;1985. pp. 11–42. Chap. 2.
17. Blumer M, Snyder WD. Porphyrins of high molecular weight in a Triassic oil shale. *Chem Geol* 1967;**2**:35–45.
18. Levorsen AI. *Geology of petroleum,* 2nd ed. San Francisco: W.H. Freeman;1967. 724 pp.
19. Collins AG. *Geochemistry of some petroleum-associated waters from Louisiana*. U.S. Bureau of Mines RI 7326, National Technical Information Service, Department Of Commerce, Springfield, VA, 1970, 31 pp.
20. Rieke III HH, Chilingarian GV, editors. *Argillaceous sediments*. New York: Elsevier Science;1974. 424 pp.
21. Fertl WH, Timko DJ. Prepressured formations. *Oil Gas J* 1970;**68**(1):97–105.
22. Bebout DG. Subsurface techniques for locating and evaluating geopressured/geothermal reservoirs along the Texas Gulf Coast. In: Dorfman MH, Beller RW, editors, *Proceedings of the 2nd Geopressured/Geothermal Energy Conference*, vol. II. Austin, TX: University of Texas, February 23–25; 1976. pp. 1–12.
23. Jones PH. Geothermal and hydrocarbon regimes, Northern Gulf of Mexico Basin. In: Dorfman MH, Beller RW, editors, *Proceedings of the 1st Geopressured/Geothermal Energy Conference*. Austin, TX: University of Texas, June 2–4; 1975. pp. 15–22.
24. Kreitler CW, Gustavson TC. Geothermal resources of the Texas Gulf Coast—environmental concerns arising from the production and disposal of geothermal waters. In: Dorfman MH, Beller RW, editors, *Proceedings of the 2nd Geopressured/Geothermal Energy Conference*, vol. V, Part 3. Austin, TX: University of Texas, February 23–25; 1976. pp. 1–9.
25. Collins AG, Wright CC. Enhanced oil recovery injection waters. In: Donaldson EC, Chilingarian GV, Yen TF, editors. *Enhanced oil recovery*. New York: Elsevier Science;1985. pp. 151–217. Chap. 6.
26. Collins AG. *Geochemistry of oilfield waters*. New York: Elsevier Science;1975. 496 pp.
27. Dodson CR, Standing MB. Pressure-volume temperature and solubility relations for natural-gas-water mixtures. *API Drill Prod Prac* 1944;173–9.
28. Mathews CS, Russell DG. *Pressure buildup and flow tests in wells*, Vol. I. Richardson, TX: Soc Petrol Engr Monogram;1967. 167 pp.
29. Standing MB, Katz DL. Density of natural gases. *Trans AIME* 1942;**146**:140–9.
30. Carr NL, Kobayashi R, Burrows DB. Viscosity of hydrocarbon gases under pressure. *Trans AIME* 1954;**201**:264–72.
31. Amyx JW, Bass Jr. DM, Whiting RL. *Petroleum reservoir engineering*. New York: McGraw-Hill;1960. 610 pp.
32. Fisher Scientific Co. *Fisher/tag manual for inspectors of petroleum,* 28th ed. New York: Fisher Scientific Co.;1954. 218 pp.
33. Hunt JM. *Petroleum geochemistry and geology*. San Francisco: W. H. Freeman;1979. 617 pp.
34. Donaldson EC, Chilingarian GV, Yen TF, editors. *Enhanced oil recovery, I—fundamentals and analyses*. New York: Elsevier Science;1985. 357 pp.
35. Donaldson EC, Crocker ME. Characterization of the crude oil polar compound extract. *DOE/BETC/RI-80/5, National. Technical Information Service*, Springfield, VA, 1980, 27 pp.

Chapter 3

Porosity and Permeability

The nature of reservoir rocks containing oil and gas dictates the quantities of fluids trapped within the void space of these rocks, the ability of these fluids to flow through the rocks, and other related physical properties. The measure of the void space is defined as the porosity of the rock, and the measure of the ability of the rock to transmit fluids is called the permeability. Knowledge of these two properties is essential before questions concerning types of fluids, amount of fluids, rates of fluid flow, and fluid recovery estimates can be answered. Methods of measuring porosity and permeability comprise much of the technical literature of the oil industry. Other reservoir properties of importance include the texture, the resistivity of the rock and its contained fluids to electrical current, the water content as a function of capillary pressure, and the tortuous nature of the interstices or pore channels.

The texture of sedimentary rocks is determined largely by grain shape and roundness, grain size and sorting, grain orientation and packing, and chemical composition. A specific combination of these variables may reveal information about diagenetic and catagenetic processes and mechanisms operating during transportation, deposition, and compaction and deformation of sedimentary materials. In some cases, texture may yield some information about formation permeability and porosity. For example, fine-grained sandstones with poorly sorted angular grains will generally have lower porosity than sandstones composed of coarse, well-sorted grains. Variation in permeability may be predicted from variation in grain size and shape, and from distribution of pore channels in the rock.

The resistivity of any formation to the electrical current flow is a function of the amount of water in that formation and the resistivity of the water itself. The rock grains and hydrocarbons are normally insulators. Changes in water saturation combined with changes in the resistivity of the fluids filling the pores create resistivity profiles in well logs. These profiles help locate hydrocarbon-bearing formations.

POROSITY

Sand grains and particles of carbonate materials that make up sandstone and limestone reservoirs usually never fit together perfectly due to the high degree of irregularity in shape. The void space created throughout the beds between grains, called pore space or interstice, is occupied by fluids (liquids and/or gases). The porosity of a reservoir rock is defined as that fraction of the bulk volume of the reservoir that is not occupied by the solid framework of the reservoir. This can be expressed in mathematical form as:

$$\phi = \frac{V_b - V_{gr}}{V_b} = \frac{V_p}{V_b} \tag{3.1}$$

where

ϕ = porosity, fraction
V_b = bulk volume of the reservoir rock
V_{gr} = grain volume
V_p = pore volume

According to this definition, the porosity of porous materials could have any value, but the porosity of most sedimentary rocks is generally lower than 50%.

Example

A clean and dry core sample weighing 425 g was 100% saturated with 1.07 specific gravity (γ) brine. The new weight is 453 g. The core sample is 12 cm in length and 4 cm in diameter. Calculate the porosity of the rock sample.

Solution

The bulk volume of the core sample is:

$$V_b = \pi(2)^2(12) = 150.80 \text{ cm}^3$$

The pore volume is:

$$V_p = \frac{1}{\gamma}(V_{wet} - V_{dry}) = \frac{453 - 425}{1.07} = 26.17 \text{ cm}^3$$

Using Eq. (3.1), the porosity of the core is:

$$\phi = \frac{V_p}{V_b} = \frac{26.17}{150.80} = 0.173 \text{ or } 17.3\%$$

FACTORS GOVERNING THE MAGNITUDE OF POROSITY

In an effort to determine approximate limits of porosity values, Fraser and Graton determined the porosity of various packing arrangements of uniform spheres as shown in Figure 3.1 [1].

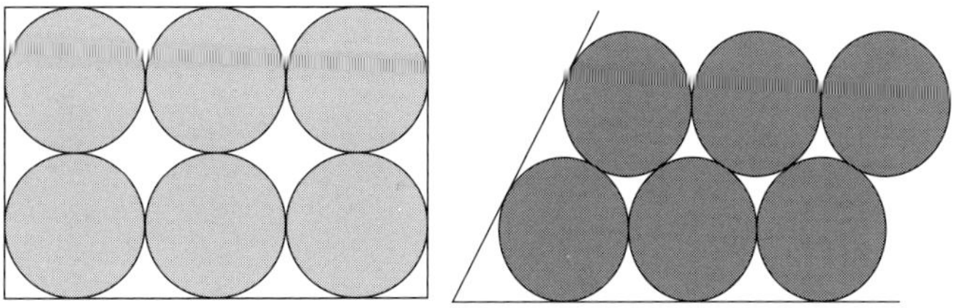

FIGURE 3.1 Cubic (left) and rhombohedral (right) packing of spherical grains.

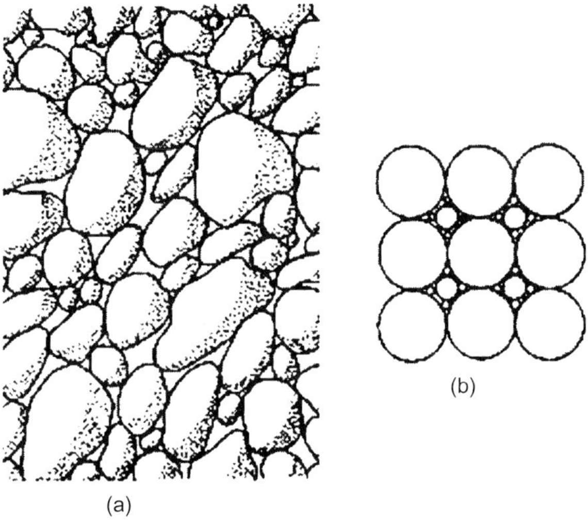

FIGURE 3.2 Collection of (a) different sized and shaped sand grains and (b) spheres illustrating a cubic packing of three grain sizes. Cubic (left) and rhombohedral (right) packing of spherical grains.

They have shown that the cubic, or wide-packed system, has a porosity of 47.6% and the rhombohedral, or close-packed system, has a porosity of 25.9%. The porosity for such a system is independent of the grain size (sphere diameter). However, if smaller spheres are mixed among the spheres of either system, the ratio of pore space to the solid framework becomes lower and porosity is reduced [2]. Figure 3.2(b) shows a three-grain-size cubic packing. The porosity of this cubic packing is now approximately 26.5%.

The porosities of petroleum reservoirs range from 5% to 30%, but most frequently are between 10% and 20% (Table 3.1). Any porosity less than 5% is very seldom commercial, and any porosity more than 35% is extremely unusual. The following table defines what typically constitutes poor, good,

TABLE 3.1 Range of Porosity and Practical Cutoff

What is good porosity
• 0–5%—Negligible
• 5–10%—Poor
• 10–15%—Fair
• 15–20%—Good
• >20%Very good
• Practical cutoff for oil sandstone ~8%
• Cut off for limestone ~5%
• For gas reservoirs, the cutoff is lower

and very good porosity. The factors governing the magnitude of porosity in clastic sediment are as follows:

(a) *Uniformity of grain size:* Uniformity or sorting is the gradation of grains. If small particles of silt or clay are mixed with larger sand grains, the effective (intercommunicating) porosity will be considerably reduced as shown in Figure 3.2. These reservoirs are referred to as dirty or shaly. Sorting depends on at least four major factors: size range of material, type of deposition, current characteristics, and the duration of the sedimentary process.

(b) *Degree of cementation or consolidation:* The highly cemented sand stones have low porosities, whereas the soft, unconsolidated rocks have high porosities. Cementation takes place both at the time of lithification and during rock alteration by circulating groundwater. The process is essentially that of filling void spaces with mineral material, which reduce porosity. Cementing materials include calcium carbonate, magnesium carbonate, iron carbonate, iron sulfides, limonite, hematite, dolomite calcium sulfate, clays, and many other materials including any combination of these materials.

(c) *Amount of compaction during and after deposition:* Compaction tends to lose voids and squeeze fluid out to bring the mineral particles close together, especially the finer-grained sedimentary rocks. This expulsion of fluids by compaction at an increased temperature is the basic mechanism for primary migration of petroleum from the source to reservoir rocks. Whereas compaction is an important lithifying process in claystones, shales, and fine-grained carbonate rocks, it is negligible in closely packed sandstones or conglomerates. Generally, porosity is lower in deeper, older rocks, but exceptions to this basic trend are common. Many carbonate rocks show little evidence of physical compaction.

(d) *Methods of packing:* With increasing overburden pressure, poorly sorted angular sand grains show a progressive change from random packing to a closer packing. Some crushing and plastic deformation of the sand particles occurs.

ENGINEERING CLASSIFICATION OF POROSITY

During sedimentation and lithification, some of the pore spaces initially developed became isolated from the other pore spaces by various diagenetic and catagenetic processes such as cementation and compaction. Thus, many of the pores will be interconnected, whereas others will be completely isolated. This leads to two distinct categories of porosity, namely, total (absolute) and effective, depending upon which pore spaces are measured in determining the volume of these pore spaces. The difference between the total and effective porosities is the isolated or noneffective porosity. Absolute porosity is the ratio of the total void space in the sample to the bulk volume of that sample, regardless of whether or not those void spaces are interconnected. A rock may have considerable absolute porosity and yet have no fluid conductivity for lack of pore interconnections. Examples of this are lava, pumice stone, and other rocks with vesicular porosity.

Effective porosity is affected by a number of lithological factors including the type, content, and hydration of the clays present in the rock; the heterogeneity of grain sizes; the packing and cementation of the grains; and any weathering and leaching that may affect the rock. Many of the pores may be dead-ends with only one entry to the main pore channel system. Depending on wettability, these dead-end pores may be filled with water or oil, which are irreducible fluids. Experimental techniques for measuring porosity must take these facts into consideration.

In order to recover oil and gas from reservoirs, the hydrocarbons must flow several hundred feet through the pore channels in the rock before they reach the producing wellbore. If the petroleum occupies nonconnected void spaces, it cannot be produced and is of little interest to the petroleum engineer. Therefore, effective porosity is the value used in all reservoir engineering calculations.

GEOLOGICAL CLASSIFICATION OF POROSITY

As sediments were deposited in geologically ancient seas, the first fluid that filled pore spaces in sand beds was seawater, generally referred to as connate water. A common method of classifying porosity of petroleum reservoirs is based on whether pore spaces in which oil and gas are found originated when the sand beds were laid down (primary or matrix porosity), or if they were formed through subsequent diagenesis (e.g., dolomitization in carbonate rocks), catagenesis, earth stresses, and solution by water flowing through the rock (secondary or induced porosity). The following general classification of porosity, adapted from Ellison, is based on the time of origin, mode of origin, and distribution relationships of pores spaces [3].

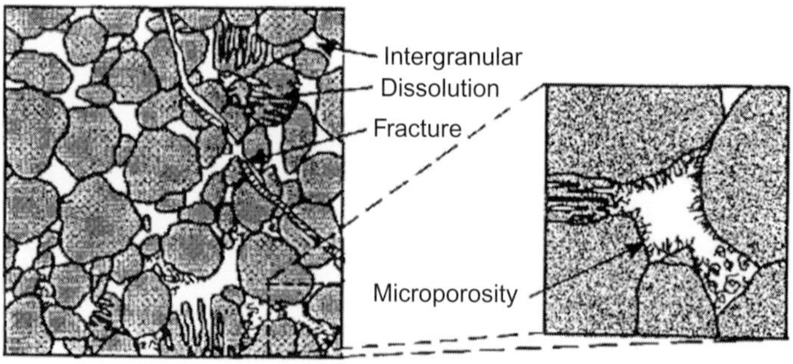

FIGURE 3.3 Types of porosity found in sandstone reservoirs (courtesy of Core Laboratories).

Primary Porosity

1. *Intercrystalline voids:* Voids between cleavage planes of crystals, voids between individual crystals, and voids in crystal lattices. Many of these voids are sub-capillary, i.e., pores less than 0.002 mm in diameter. The porosity found in crystal lattices and between mud-sized particles is called "micro-porosity" by Pittman, as shown in Figure 3.3 [4]. Unusually, high recovery of water in some productive carbonate reservoirs may be due to the presence of large quantities of microporosity.
2. *Intergranular or interparticle voids:* Voids between grains, i.e., interstitial voids of all kinds in all types of rocks. These openings range from sub-capillary to super-capillary size (voids greater than 0.5 mm in diameter).
3. *Bedding planes voids:* Voids of many varieties are concentrated parallel to bedding planes. The larger geometry of many petroleum reservoirs is controlled by such bedding planes. Differences of sediments deposited, of particle sizes and arrangements, and of the environments of deposition are causes of bedding plane voids.
4. *Miscellaneous sedimentary voids:* (1) Voids resulting from the accumulation of detrital fragments of fossils, (2) voids resulting from the packing of oolites, (3) vuggy and cavernous voids of irregular and variable sizes formed at the time of deposition, and (4) voids created by living organisms at the time of deposition.

Secondary Porosity

Secondary porosity is the result of geological processes (diagenesis and catagenesis) after the deposition of sediment. The magnitude, shape, size, and interconnection of the pores may have no direct relation to the form of original sedimentary particles. Induced porosity can be subdivided into three groups based on the most dominant geological process:

1. *Solution porosity:* Channels due to the solution of rocks by circulating warm or hot solutions; openings caused by weathering, such as enlarged

joints and solution caverns; and voids caused by organisms and later enlarged by solution.

2. *Dolomitization:* A process by which limestone is transformed into dolomite according to the following chemical reaction:

$$\underset{\text{Limestone}}{2CaCO_3} + Mg^{2+} \rightarrow \underset{\text{Dolomite}}{CaMg(CO_3)} + Ca^{2+} \tag{3.2}$$

Some carbonates are almost pure limestones, and if the circulating pore water contains significant amounts of magnesium cation, the calcium in the rock can be exchanged for magnesium in the solution. Because the ionic volume of magnesium is considerably smaller than that of the calcium, which it replaces, the resulting dolomite will have greater porosity. Complete replacement of calcium by magnesium can result in a 12–13% increase in porosity [5,6].

3. *Fracture porosity:* Openings created by structural failure of the reservoir rocks under tension caused by tectonic activities such as folding and faulting. These openings include joints, fissures, and fractures. In some reservoir rocks, such as the Ellenburger carbonate fields of West Texas, fracture porosity is important. Porosity due to fractures alone in the carbonates usually does not exceed 1% [7].
4. *Miscellaneous secondary voids:* (1) Saddle reefs, which are openings at the crests of closely folded narrow anticlines; (2) pitches and flats, which are openings formed by the parting of beds under gentle slumping; and (3) voids caused by submarine slide breccias and conglomerates resulting from gravity movement of seafloor material after partial lithification.

In carbonate reservoirs, secondary porosity is much more important than primary porosity: dolomites comprise nearly 80% of North American hydrocarbon reservoirs [6]. Primary porosity is dominant in clastic – also called detrital or fragmental – sedimentary rocks such as sandstones, conglomerates, and certain oolitic limestones [7]. However, it is important to emphasize that both types of porosity often occur in the same reservoir rock.

VISUAL DESCRIPTION OF POROSITY IN CARBONATE ROCKS

The role played by the visual description of pore space in carbonate rocks has changed considerably since the development of a method for classifying carbonate reservoir rocks in 1952 by Archie [8]. The development of well logging technology has provided the petroleum industry with effective and direct methods to measure the *in situ* porosity of a formation. The visual description of the pore geometry, however, is still needed to estimate the effects of (1) the grain size; (2) the amount of interparticle porosity; (3) the amount of unconnected vugs; (4) the presence of fractures and cavities; and (5) the presence or absence of connected vugs on the porosity–permeability relationship and other petrophysical parameters of naturally fractured

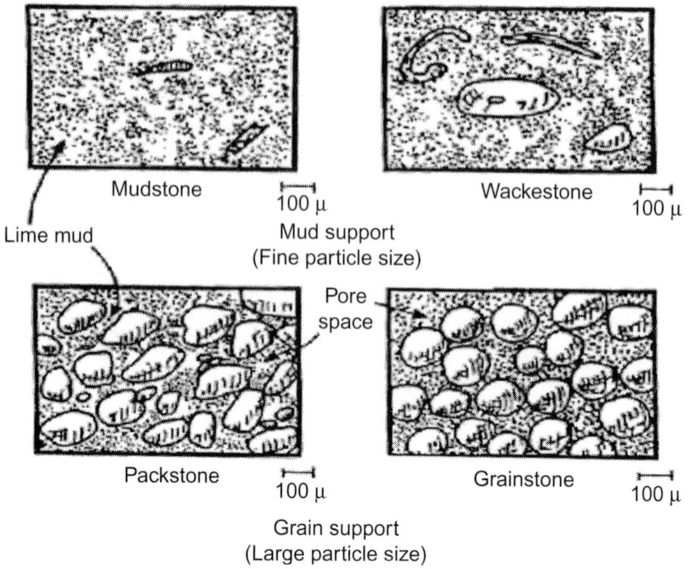

FIGURE 3.4 Mud and grain support in depositional fabric [9].

reservoirs. Lucia presented field classification of carbonate rock pore space based on the visual description of petrophysical parameters of a large number of samples [9]. He also discussed basic geological characteristics necessary for the visual estimation of particle size and recognition of interparticle pore space, and connected and unconnected vugs.

Figure 3.4 shows two common types of particle sizes based on artificially prepared samples containing various kinds of carbonate particles: large sand-sized particles such as those found in packstone or grainstone deposits, small silt-to-clay-sized particles such as mudstone or wackestone [9]. The particle size of primary interest is that of the supporting framework because interparticle porosity of the matrix rock is controlled by the size of the particles.

The concept of support in defining particle size in dolomites is illustrated in Figure 3.5 [9]. If the dolomite crystals form a continuous, supporting network, their size controls the connected pore size. The dolomite crystal size is of primary interest when it is the same or larger than the sediment particle size, such as observed in dolomitized limestone or wackestone rocks. However, the sediment particle size becomes of primary interest if the sediment particle size is larger than the dolomite crystal size, as is usually the case in dolomitized grainstones or packstones [9].

Recognition of intergranular porosity depends on the size and shape of grains in the rock matrix. In coarsely grained rocks, the intergranular pore space may be identified with the naked eye. In finely grained limestones or dolomites, e.g., the intergranular pores are more difficult to identify, and

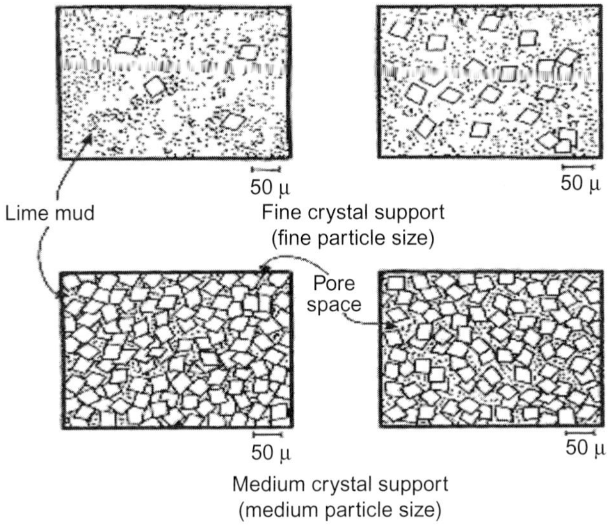

FIGURE 3.5 Fine and medium crystal support in dolomites [9].

scanning electron microscopy (SEM) and petrographic techniques are necessary to observe this porosity.

Visual recognition of unconnected vug porosity depends on the granular texture of the rock and origin of the vugs. Intrafossil, shelter, and fenestral porosity, as well as leached grains and leached anhydrite crystals, are unconnected vug types. Vugs and cavities can be connected by intergranular pore channels or by fractures. Visual evaluation of fracture-connected porosities in core samples is complicated by the possibility of fractures induced by the coring operations [10]. Based on these observations, Lucia proposed a field classification of carbonate porosity as follows [9]: (1) for fine particle size ($d_{gr} < 20\ \mu m$), the displacement pressure, P_D, is greater than 70 psia; (2) for medium particle size ($20 < d_{gr} < 100\ \mu m$), the P_D is in the range of 15–70 psia; (3) for large grains ($d_{gr} > 100\ \mu m$), the displacement pressure is less than 15 psia. The term P_D is the extrapolated displacement pressure, which is determined from the mercury capillary-pressure curves discussed in Chapter 5. Figure 3.6 shows the relationship between P_D and the average grain size as a function of the intergranular porosity for nonvuggy rocks with permeability greater than 0.1 mD. This relationship is the basis for dividing particle size into the three groups.

FLUID SATURATION

The porosity of a reservoir rock is very important because it is a measure of the ability of that rock to store fluids (oil, gas, and water). Equally important

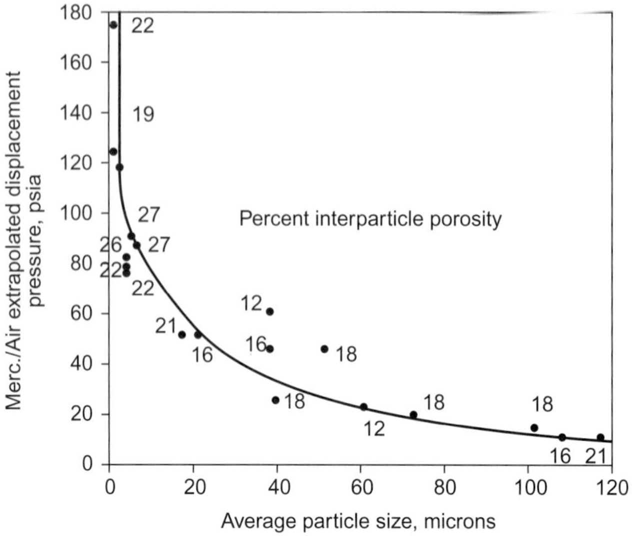

FIGURE 3.6 Relationship between displacement pressure and particle size for nonvuggy rock, with $k > 0.1$ mD [9].

is the relative extent to which the pores of the rock are filled with specific fluids. This property is called fluid saturation and is expressed as the fraction, or percent, of the total pore volume occupied by the oil, gas, or water. Thus, for instance, the oil saturation S_o is equal to:

$$S_o = \frac{\text{Volume of oil in the rock, } V_o}{\text{Total pore volume of the rock, } V_p}$$

Similar expressions can be written for gas and water. It is evident that:

$$S_o + S_g + S_w = 1 \tag{3.3}$$

and

$$V_o + V_g + V_w = V_p \tag{3.4}$$

Ideally, because of the difference in fluid densities, a petroleum reservoir is formed in such a way that, from top to bottom of the sand bed there will be gas, oil, and water, as shown in Figure 3.7.

Connate water, however, is nearly always found throughout the petroleum reservoir. Connate water is the seawater trapped in porous spaces of the sediments during their deposition and lithification, long before the oil migrated into the reservoir rock. In addition to density, wettability and interfacial tension combine to alter the manner in which the three fluids are distributed in the reservoir.

The amount of connate water present in the porous space varies from 100% below the oil zone to theoretically zero at heights above the free water

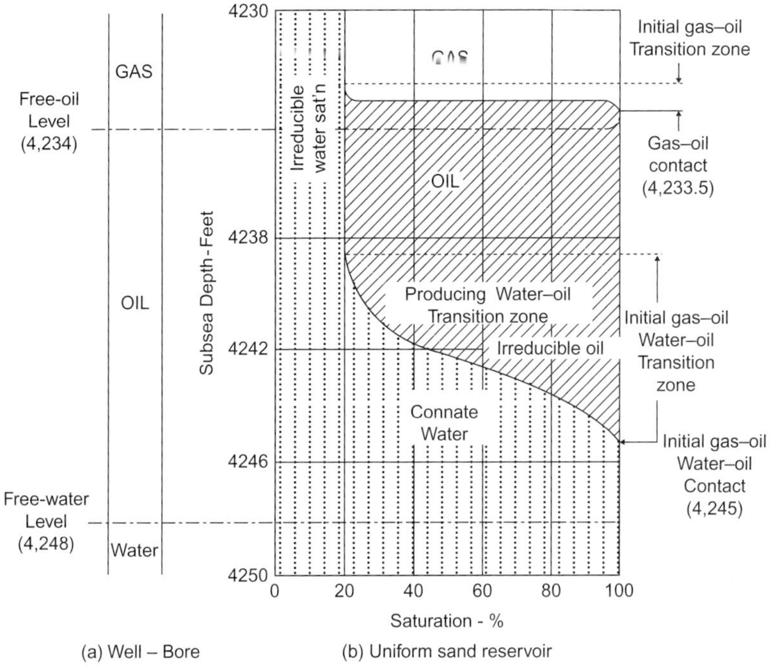

FIGURE 3.7 Distribution of fluids in reservoirs.

level. However, in practical cases a nearly constant content of irreducible connate water (S_{iw}) exists above the transition zone. The magnitude of S_{iw} and height of the transition zone depend on the pore size and texture. High S_{iw} values are indicative of small pore sizes. The transition zone corresponds to the zone of varying water saturation. Wells completed within this zone will produce hydrocarbons and water, and wells completed above this zone, i.e., within the zone of irreducible water saturation, will produce only hydrocarbons [11].

QUANTITATIVE USE OF POROSITY

One of the simplest methods of calculating reservoir oil content is called the volumetric method. The mathematical expression for the initial oil-in-place (N, in bbl) by this method is:

$$N = 7,758 A_s h \phi S_{oi} \tag{3.5}$$

where

A_s = surface area of the reservoir, acres
h = thickness of the formation, ft
ϕ = porosity, fraction
S_{oi} = initial oil saturation, fraction

Equation (3.5) gives the volume of oil contained in the porous rock at reservoir conditions of pressure and temperature. However, the surface or "stock tank" oil as finally sold by the producer is different from the liquid volume that existed underground. The difference is due to the changes in the oil properties as the pressure is decreased from high underground pressure and temperature to surface pressure and temperature. This reduction in p and T causes some of the volatile components to come out of solution (evaporate), causing the liquid volume to shrink. This reduction in volume is expressed by the oil formation volume factor, B_{oi}. Thus, the stock tank oil initially in place is:

$$N = 7,758 \frac{A_s h \phi (1 - S_{iw})}{B_{oi}} \tag{3.6}$$

where B_{oi} is in reservoir barrels per stock tank barrel or bbl/STB. In this equation, S_{oi} is replaced by $(1 - S_{iw})$, where S_{iw} is the irreducible or connate water saturation. This implies that no free gas is present in the pore space. Because no petroleum reservoir is homogenous, the factors A_s, h, ϕ, and S_{iw}, must be averaged. The constant 7,758 becomes 10,000 if A_s, h are expressed in hectares (ha) (1 hectare = 10,000 m^2) and m, respectively, and N in m^3.

Example

Calculate the initial oil-in-place (N) of an oil reservoir if $A = 1,600$ acres, $h = 32$ ft, $\phi = 22\%$, $S_{iw} = 20\%$, and $B_{oi} = 1.23$ bb1/STB.

Solution

Using Eq. (3.6), we have:

$$N = \frac{7,758(1,600)(32)(0.22)(1-0.20)}{1.23} = 56.8 \times 10^6 \text{ STB}$$

An expression similar to Eq. (3.6) may be derived for estimating the initial gas-in-place. In this case, it is convenient to express the gas volume in cubic feet. At standard conditions, i.e., $P_{sc} = 14.7$ psia and $T_{sc} = 60$ °F, the initial gas-in-place in a volumetric reservoir is given by:

$$G = 43,560 \frac{Ah\phi(1 - S_{iw})}{B_{gi}} \tag{3.7}$$

where B_{gi}, the initial gas formation volume factor in ft^3/SCF, is calculated as:

$$B_{gi} = 0.02829 \left(\frac{z_i T}{p_i} \right) \tag{3.8}$$

The initial gas deviation (also called compressibility) factor, z_i, is calculated at the initial pressure, p_i, of the gas reservoir. This factor accounts for the difference between the actual and ideal gas volumes. The reservoir temperature, T, is in degree Rankin (°R).

Example

A volumetric gas reservoir has the following characteristics:

$A = 1{,}320$ acres	$T = 200°F$
$h = 45$ ft	$P_i = 4{,}000$ psia
$\phi = 0.175$	$z_i = 0.916$
$S_{iw} = 0.23$	

Solution

The initial gas formation volume factor is:

$$B_{gi} = 0.02829 \frac{0.916(460 + 200)}{4{,}000} = 0.004276\ \text{ft}^3/\text{SCF}$$

The initial gas in place is:

$$G = 43{,}560 \frac{(1{,}320)(45)(0.175)(1 - 0.23)}{0.004276} = 81.539 \times 10^9\ \text{SCF}$$

PERMEABILITY

In addition to being porous, a reservoir rock must have the ability to allow petroleum fluids to flow through its interconnected pores. The rock's ability to conduct fluids is termed as permeability. This indicates that nonporous rocks have no permeability. The permeability of a rock depends on its effective porosity, consequently, it is affected by the rock grain size, grain shape, grain size distribution (sorting), grain packing, and the degree of consolidation and cementation. The type of clay or cementing material between sand grains also affects permeability, especially where fresh water is present. Some clays, particularly smectites (bentonites) and montmorillonites swell in fresh water and have tendency to partially or completely block the pore spaces.

In 1856, French engineer Henry Darcy developed a fluid flow equation that since has become one of the standard mathematical tools of the petroleum engineer [12]. This equation, which is used to measure the permeability of a core sample as shown in Figure 3.8, is expressed in differential form as follows:

$$v = \frac{q}{A_c} = -\frac{k}{\mu}\frac{dp}{dl} \tag{3.9}$$

where

v = fluid velocity, cm/s
q = flow rate, cm^3/s
k = permeability of the porous rock, Darcy ($0.986923\ \mu m^2$)
A_c = cross-sectional area of the core sample, cm^2

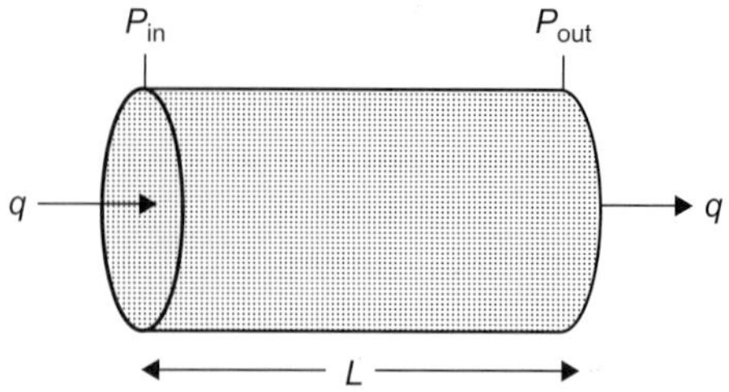

FIGURE 3.8 Core sample.

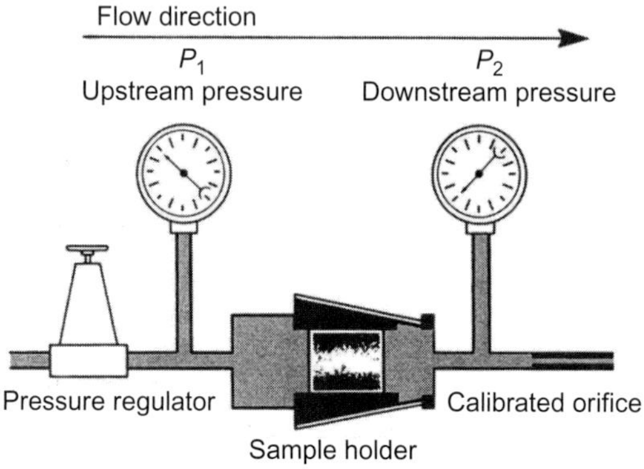

FIGURE 3.9 Schematic flow diagram of permeameter.

μ = viscosity of the fluid, centipoises (cP)
l = length of the core sample, cm
dp/dl = pressure gradient in the direction of the flow, atm/cm

Equation (3.9) can be expressed as:

$$q = -\frac{A_c k}{\mu}\frac{dp}{dx} \tag{3.9a}$$

Separating the variable and integrating between 0 and L and inlet pressure P_1 and outlet pressure P_2, and solving for k gives:

$$k = \frac{q\mu L}{(p_1 - p_2)A_c} \tag{3.9b}$$

where k is measured by causing a fluid to flow through a clean and dry core sample (plug) of measured dimensions (A_c and L). A schematic of the principle involved in permeability measurements is shown in Figure 3.9.

A clean dry plug is placed in a holder. Upstream and downstream pressures are measured to determine the pressure differential across the core (see Experiment 10 in Appendix for details). Flow rate, in cm^3/s, is measured at atmospheric pressure. This steady-state method is acceptable for high permeability rocks. In low permeability samples this method may take several hours. For low-k rocks, unsteady-state methods are preferable, as they allow determination of k to be made in minutes.

Example

A 10-cm long cylindrical core sample was subjected to a laboratory linear flow test under a pressure differential of 3.4 atm using a fluid of viscosity 2.5 cp. The diameter of the core is 4 cm. A flow rate of 0.35 cm^3/s was obtained. Calculate the permeability of this core sample.

Solution

The cross-sectional area of the core is $A_c = \pi r^2 = 3.1415 \times 2^2 = 12.57\ \text{cm}^2$. Using Eq. (3.9b) the permeability of the core samples is:

$$k = \frac{q\mu L}{A_c\,\Delta P} = \frac{0.35 \times 2.5 \times 10}{12.57 \times 3.4} = 0.204\ \text{Darcy} = 204\ \text{mD}$$

Dry gas (air) has been selected as the standard fluid for use in permeability determination because it minimizes fluid–rock reaction and is easy to use.

Equation (3.9b) is valid for noncompressible or slightly compressible fluids (liquid). For compressible fluids (gas) k is obtained from:

$$k = \frac{2q\mu_g L}{(P_1^2 - P_2^2)A} \tag{3.9c}$$

where μ_g is the gas viscosity in cP.

Air permeability measured in a routine core analysis laboratory on a (nonfractured) core sample will give higher values than the actual reservoir permeability, especially with a liquid as the flowing fluid. The difference is due to gas slippage (or Klinkenberg) effect and overburden pressure effects. Klinkerberg [13] showed that at low mean pressure P_m (e.g., 1 atm) the gas molecules are so far apart that they "slip" through the pore spaces with little friction loss, and yield a higher permeability. At high P_m (e.g., 1000 psia or greater) the gas molecules are close together and experience a friction drag at the side of the pore walls. This increases as P_m increases, with the gas acting more and more like liquid.

If a plot of measured permeability versus $1/P_m$ was extrapolated to the point where $1/P_m = 0$ (i.e., P_m = infinity), as shown in Figure 3.10, this permeability would be approximately equal to the liquid permeability k_L.

The relationship between air permeability and liquid permeability is:

$$k_a = c\left(\frac{1}{P_m}\right) + k_L \tag{3.9d}$$

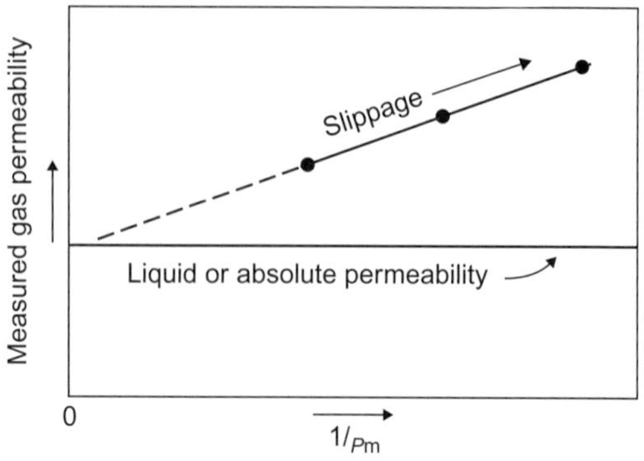

FIGURE 3.10 Klinkenberg effect.

where

P_m = mean pressure, $(P_1 + P_2)/2$
c = slope of the line
k_L = equivalent liquid permeability, i.e., absolute permeability k

Klinkenberg expressed the slope c as:

$$c = bk_L \tag{3.9e}$$

where b is the pore geometry factor. It depends on the size of the pore opening, and is inversely proportional to the radius of capillaries. b may be determined from unsteady state measurements. Jones experimentally showed that:

$$b = 6.9k_L^{-0.36} \tag{3.9f}$$

For high permeability cores, testing laboratories will often make their tests with one high mean pressure and neglect the Klinkenberg correction. If only one measurement of air permeability (k_a) is made, the correct k_L can be obtained from the following expression:

$$6.9k_L^{0.64} + P_m k_L - P_m k_g = 0 \tag{3.9g}$$

Equation (3.9g) is nonlinear and can only be solved by trial-and-error using the Newton Raphson technique.

One darcy is relatively high permeability. The permeability of most petroleum reservoir rocks is less than one darcy. Thus, a smaller unit of permeability, the millidarcy (mD), is widely used in the oil and gas industry. In SI units, the square micrometer (μm^2) is used instead of m^2.

The permeability, k, in Eq. (3.9) is termed the "absolute" permeability if the rock is 100% saturated with a single fluid (or phase), such as oil, gas, or water. In presence of more than one fluid, permeability is called the "effective" permeability (k_o, k_g, or k_w being oil, gas, or water effective permeability, respectively). Reservoir fluids interface with each other during their movement through the porous channels of the rock; consequently, the sum of the effective permeabilities of all the phases will always be less than the absolute permeability.

In presence of more than one fluid in the rock, the ratio of effective permeability of any phase to the absolute permeability of the rock is known as the "relative" permeability (k_r) of that phase. For example, the relative permeability of the oil, gas, and water would be $k_{ro} = k_o/k$, $k_{rg} = k_g/k$, $k_{rw} = k_w/k$, respectively.

CLASSIFICATION OF PERMEABILITY

Petroleum reservoirs can have primary permeability, which is also known as the matrix permeability, and secondary permeability. Matrix permeability originated at the time of deposition and lithification (hardening) of sedimentary rocks. Secondary permeability resulted from the alteration of the rock matrix by compaction, cementation, fracturing, and solution.

Whereas compaction and cementation generally reduce the permeability, as shown in Figure 3.11, fracturing and solution tend to increase it [14]. In some reservoir rocks, particularly low-porosity carbonates, secondary permeability provides the main flow conduit for fluid migration, e.g., in the Ellenburger Field, TX.

FACTORS AFFECTING THE MAGNITUDE OF PERMEABILITY

Permeability of petroleum reservoir rocks may range from 0.1 to 1,000 or more millidarcies, as shown in Table 3.2. The quality of a reservoir as determined by permeability, in mD, may be judged as: poor if $k < 1$, fair if $1 < k < 10$, moderate if $10 < k < 50$, good if $50 < k < 250$, and very good

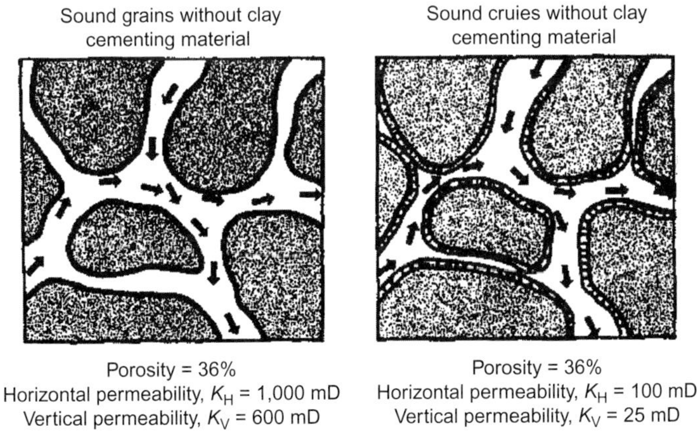

FIGURE 3.11 Effects of clay cementing material on porosity and permeability [14].

TABLE 3.2 Permeability and Porosity of Selected Oil Sands

Name of Sand	Porosity (%)	Permeability (mD)
"Second Wilcox" (Ordovician) Oklahoma Co., OK	12.0	100.0
Clinch (Silirian) Lee Co., VA	9.6	0.9
Strawn (Pennsylvanian) Cook Co., TX	22.0	81.5
Bartlesville (Pennsylvanian) Anderson Co., KS	17.5	25
Olympic (Pennsylvanian) Hughes Co., OK	20.5	35.0
Nugget (Jurassic) Fremont Co., WY	24.9	147.5
Cut Bank (Cretaceous) Glacier Co., MT	15.4	111.5
Woodbine (Cretaceous) Tyler Co., TX	22.1	3,390.0
Eutaw (Cretaceous) Choctaw Co., AL	30.0	100.0
O'Hern (Eocene) Dual Co., TX	28.4	130.0

if $k>250$ mD. In East Texas fields, permeability may be as high as 4,600 mD. Reservoirs having permeability below 1 mD are considered "tight." Such low permeability values are found generally in limestone matrices and in tight gas sands of the western United States. Stimulation techniques such as hydraulic fracturing and acidizing increase the permeability of such rocks and allow the exploitation of such low permeability reservoirs, which were once considered uneconomical. Only 50 years ago rocks with permeability of 50 mD or less were considered tight.

Factors affecting the magnitude of permeability in sediments are as follows:

(a) *Shape and size of sand grains:* If the rock is composed of large and flat grains uniformly arranged with the longest dimension horizontal, as illustrated in Figure 3.12, its horizontal permeability (k_H) will be very high, whereas vertical permeability (k_V) will be medium-to-large.

If the rock is composed of mostly large and rounded grains, its permeability will be considerably high and of same magnitude in both directions, as shown in Figure 3.13.

Permeability of reservoir rocks is generally low, especially in the vertical direction, if the sand grains are small and of irregular shape (Figure 3.14). Most petroleum reservoirs fall in this category.

Reservoirs with directional permeability are called anisotropic. Anisotropy greatly affects fluid flow characteristics of the rock. The

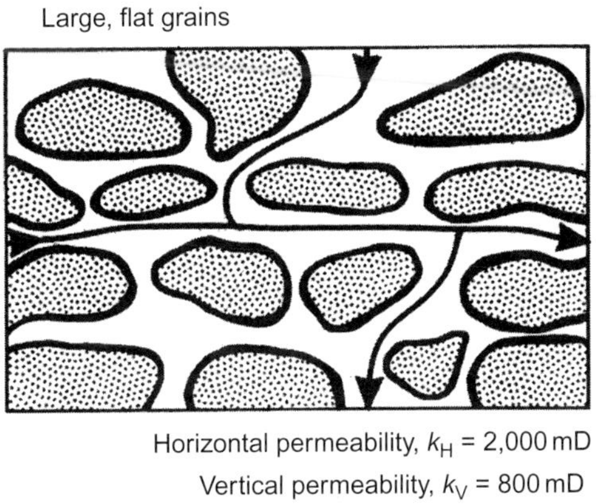

FIGURE 3.12 Effects of large flat grains on permeability [14].

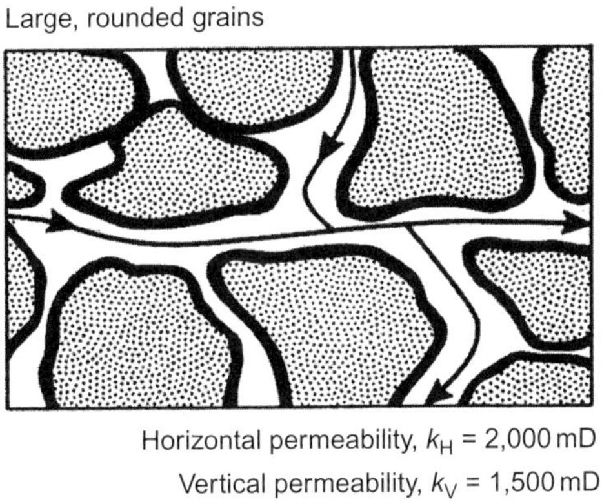

FIGURE 3.13 Effects of large rounded grains on permeability [14].

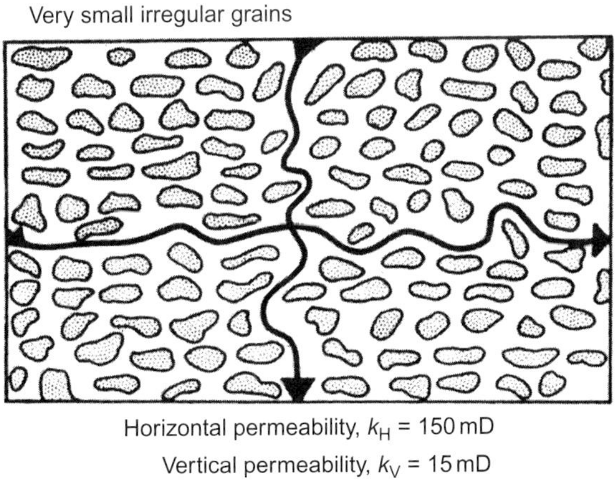

FIGURE 3.14 Effects of small, irregular grains on permeability [14].

difference in permeability measured parallel and vertical to the bedding plane is the consequence of the origin of the sediment, because grains settle in the water with their longest and flattest sides in a horizontal position. Subsequent compaction of the sediment increases the ordering of the sand grains so that they generally lie in the same direction [14].

(b) *Lamination:* Platy minerals such as muscovite, and shale laminations, act as barriers to vertical permeability. In this case, the k_H/k_V ratio generally ranges from 1.5 to 3 and may exceed 10 for some reservoir rocks. Sometimes, however, k_V is higher than k_H due to fractures or vertical jointing and vertical solution channels. Joints act as barriers to horizontal permeability only if they are filled with clay or other minerals. The importance of the clay minerals as a determinant of permeability is often related not only to their abundance but also to their mineralogy and composition of the pore fluids. Should the clay minerals, which coat the grain surfaces, expand and/or become dislodged due to changes in the chemistry of the pore fluids or mud filtrate invasion, as explained in Chapter 10, the permeability will be considerably reduced.

(c) *Cementation:* Figure 3.11 shows that both permeability and porosity of sedimentary rocks are influenced by the extent of the cementation and the location of the cementing material within the pore space.

(d) *Fracturing and solution:* In sandstone rocks, fracturing is not an important cause of the secondary permeability, except where sandstones are interbedded with shales, limestones, and dolomites. In carbonates, the solution of minerals by percolating surface and subsurface acidic waters as they pass along the primary pores, fissures, fractures, and bedding planes, increase the permeability of the reservoir rock. As shown by Chilingarian et al. [15], horizontal and vertical permeabilities are equal in many carbonate reservoirs.

PERMEABILITY–POROSITY RELATIONSHIPS

Figure 3.15 shows a plot of permeability versus porosity data obtained from a large number of samples of a sandstone formation. Even though this formation is generally considered very uniform and homogeneous, there is not a specifically defined trendline between permeability and porosity values. In this case, the relationship between permeability and porosity is qualitative and is not directly or indirectly quantitative in any way. It is possible to have very high porosity without having any permeability at all, as in the case of pumice stone (where the effective porosity is nearly zero), clays, and shales. The reverse of high permeability with a low porosity might also be true, such as in micro-fractured carbonates.

In spite of this fundamental lack of correspondence between these two properties, there often can be found a very useful correlation between them within one formation, as shown in Figure 3.16.

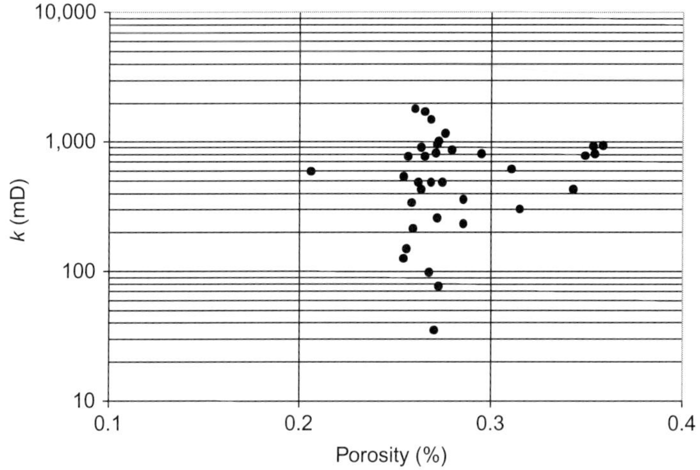

FIGURE 3.15 Permeability–porosity relationship in sandstone reservoir.

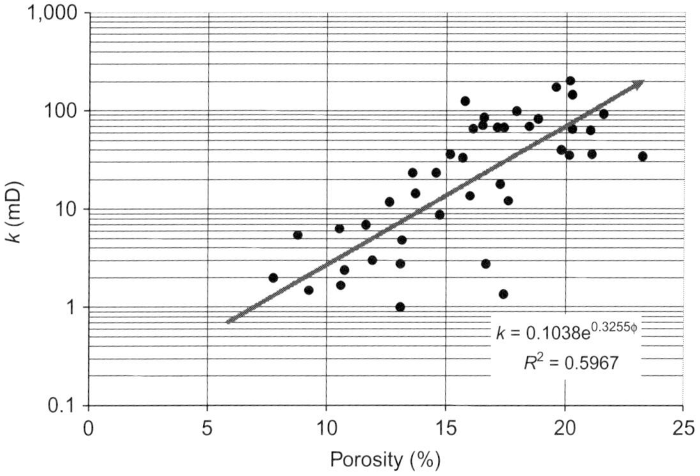

FIGURE 3.16 Permeability–porosity correlation.

Chilingarian showed that the granulometric composition of sandstones influences the relationship between permeability and porosity [15]. Figure 3.17 is a semilog plot of permeability versus porosity for (1) very coarse-grained, (2) coarse and medium-grained, (3) fine-grained, (4) silty, and (5) clayey sandstones.

Figure 3.18 shows typical permeability and porosity trends for various rock types. Such a relationship is very useful in the understanding of fluid flow through porous media. Many correlations relating permeability,

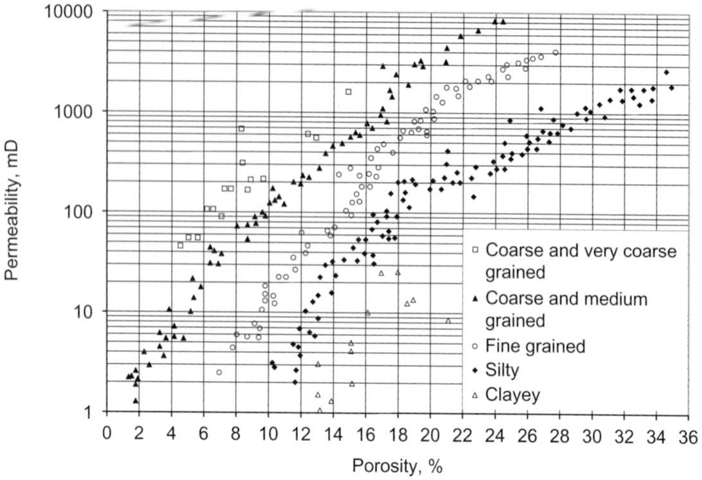

FIGURE 3.17 Influence of grain size on the relationship between permeability and porosity [16].

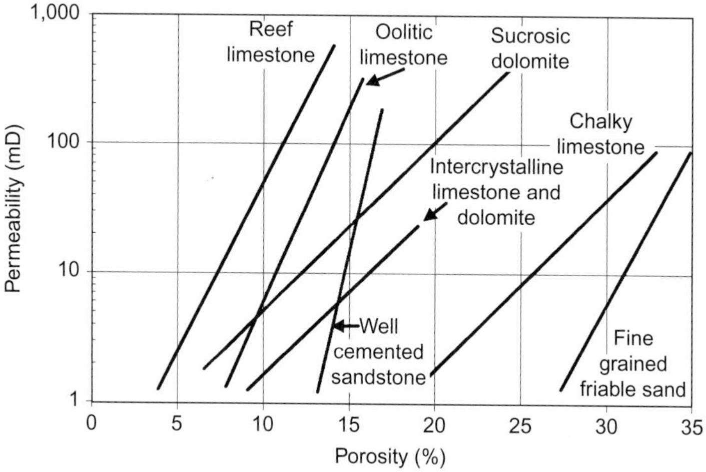

FIGURE 3.18 As typical permeability–porosity relationship for various rock types (courtesy of Core Laboratories).

porosity, pore size, specific surface area, irreducible fluid saturation, and other variables have been made. Some of these relationships are presented here for the sole purpose of enabling the reader to form a reasonable understanding of the interrelation of the rock properties in petroleum reservoirs.

KOZENY CORRELATION

Kozeny derived one of the most fundamental and popular correlations expressing permeability as a function of porosity and specific surface area [17]. Consider a porous rock sample of cross-sectional area A and length L as being made up of a number, n, of straight capillary tubes in parallel, with the spaces between the tubes sealed by a cementing material. If the capillary tubes are all of the same radius r (cm) and length L (cm), the flow rate q (cm^3/s) through this bundle of tubes, according to Poiseuille's equation, is:

$$q = \left(\frac{n\pi r^4}{8\mu}\right)\frac{\Delta p}{L} \tag{3.10}$$

where the pressure loss ΔP over length L is expressed in dynes/cm^2. Darcy's law can also approximate the flow of fluids through these n capillaries:

$$q = \left(\frac{kA_c}{\mu}\right)\frac{\Delta P}{L} \tag{3.11}$$

where A_c is the total cross-sectional area, including cemented zones, of this bundle of capillary tubes.

Equating (3.10) and (3.11) and solving for k gives:

$$k = \left(\frac{n\pi r^4}{8A_c}\right) \tag{3.12}$$

By definition, porosity is

$$\phi = \frac{V_p}{V_b} = \frac{n\pi r^2}{A_c} \tag{3.13}$$

Substituting $A_c = n\pi r^2/\phi$ from Eq. (3.13) into Eq. (3.12), one obtains a simpler relationship between permeability and porosity for pores of the same size and radii equal to r:

$$k = \frac{\phi r^2}{8} \tag{3.14}$$

where k is in cm^2 (1 cm^2 = 1.013×10^8 darcy) or in μm^2 (1 mD = 9.871×10^{-4} μm^2) and ϕ is a fraction.

Let s_{V_p} be the internal surface area per unit of pore volume; here, the surface area A_s for n capillary tubes is $n(2\pi rL)$ and the pore volume V_p is n $(\pi r^2 L)$:

$$s_{V_p} = \frac{A_s}{V_p} = \frac{n(2\pi rL)}{n(\pi r^2 L)} = \frac{2}{r} \tag{3.15}$$

Let $s_{V_{gr}}$ be the specific surface area of a porous material or the total area exposed within the pore space per unit of grain volume. For a bundle of capillary tubes, the total area exposed, A_t, is equivalent to the internal surface area A_s, and the grain volume, V_{gr}, is equal to $A_cL(1-\phi)$. Thus

$$s_{V_{gr}} = \frac{n(2\pi rL)}{A_cL(1-\phi)} = \frac{2\pi nr}{A_c(1-\phi)} = \frac{\pi nr^2}{A_c}\left(\frac{2}{r}\right)\frac{1}{(1-\phi)} \tag{3.16}$$

Combining Eqs. (3.13), (3.15) and (3.16) gives:

$$s_{V_{gr}} = s_{V_p}\left(\frac{\phi}{1-\phi}\right) \tag{3.17}$$

Equation (3.14) can be written as:

$$k = \left(\frac{\phi}{2}\right)\frac{1}{(2/r)^2} = \left(\frac{1}{2s_{V_p}^2}\right)\phi \tag{3.18}$$

Substituting for s_{V_p} from Eq. (3.18) yields:

$$k = \left(\frac{1}{2s_{V_{gr}}^2}\right)\frac{\phi^3}{(1-\phi)^2} \tag{3.19}$$

After the specific surface area per unit of pore volume, s_{V_p}, is determined from capillary data or petrographic image analysis (PIA), then Eq. (3.17) is used to obtain $s_{V_{gr}}$.

Example

A core sample from a uniform sandstone formation has a permeability of 480 mD and a porosity of 0.17. Estimate:

(a) The average pore throat radius of the core.

(b) Specific surface areas s_{V_p} and $s_{V_{gr}}$.

Solution

(a) Assuming the flow channels in the core sample may be represented by a bundle of capillary tubes, the pore throat radius can be estimated from Eq. (3.14). First, the permeability is converted from mD to μm^2:

$$k = (480)(9.8717 \times 10^{-4}) = 0.4738\ \mu m^2$$

Solving Eq. (3.14) for the pore throat radius r and substituting for permeability and porosity gives:

$$r = \sqrt{\frac{8k}{\varphi}} = \sqrt{\frac{8 \times 0.4738}{0.17}} = 4.72\ \mu m \text{ or } 4.72 \times 10^{-4}\ cm$$

(b) The specific surface area per unit pore volume is given by Eq. (3.15):

$$s_{V_p} = \frac{2}{r}$$

$$s_{V_p} = \frac{2}{4.72 \times 10^{-4}} = 4,237\ cm^{-1}$$

The specific surface area per unit grain volume can be estimated using Eq. (3.17):

$$s_{V_{gr}} = s_{Vp}\left(\frac{\phi}{1-\phi}\right)$$

$$s_{V_{gr}} = 4{,}237\left(\frac{0.17}{1-0.17}\right) = 868\ \text{cm}^{-1}$$

All the above equations used in deriving the relationship between the permeability and porosity (Eq. (3.19)) are based on the assumption that the porous rock can be represented by a bundle of straight capillary tubes. However, the average path length that a fluid particle must travel is actually greater than the length L of the core sample. The departure of a porous medium from being made up by a bundle of straight capillary tubes can be measured by the tortuosity coefficient, τ, which is expressed as [18,19]:

$$\tau = \left(\frac{L_a}{L}\right)^2 \tag{3.20}$$

where L_a is the actual flow path and L is the core length. Note that in the literature tortuosity is sometimes defined as L_a/L. Equation (3.20) is preferred here because in most laboratory experiments, the product of the formation resistivity factor (F) and porosity is related to the ratio L_a/L by the following correlation [19]:

$$F\phi = \left(\frac{L_a}{L}\right)^C \tag{3.21}$$

The exponent C is the correlation constant, which ranges from 1.7 to 2. Note that this range is rather similar to that of the cementation factor m. Thus, for a bundle of tortuous capillary tubes, Poiseuille's law becomes:

$$q = \left(\frac{n\tau r^4}{2\mu}\right)\frac{\Delta p}{L\sqrt{\tau}} \tag{3.22}$$

Combining Eq. (3.22) with Eq. (3.11) and using the same approach as above, one can show that Eqs. (3.14), (3.18), and (3.19), respectively, become:

$$k = \left(\frac{r^2}{8\tau}\right)\phi \tag{3.23a}$$

$$k = \left(\frac{1}{2\tau s_{V_p}^2}\right)\phi \tag{3.23b}$$

$$k = \left(\frac{1}{2\tau s_{V_{gr}}^2}\right)\frac{\phi^3}{(1-\phi)^2} \tag{3.23c}$$

Wylie and Spangler suggested that the factor 2 be replaced by a more general parameter, namely, the pore shape factor K_{ps} [20]. Carman reported that the product $K_{ps}\tau$ may be approximated by 5 for most porous materials [21]. Equation (3.25) for porous rocks can then be written as follows:

$$k = \left(\frac{1}{5 s_{V_{gr}}^2}\right)\frac{\phi^3}{(1-\phi)^2} \tag{3.24}$$

Equation (3.26) is the most popular form of the Kozeny equation, even though in actual porous rock $K_{ps}\tau$ is variable and much greater than 5.

Example

A sandpack of uniform fine grains has an effective porosity of 0.2. The average diameter of these spherical grains is approximately 1/8 mm. Calculate:

(a) The permeability (in cm^2 and mD) of this sandpack.

(b) The specific surface area per unit pore volume.

Solution

(a) Permeability of sandpack

The specific surface area of the grains can be estimated, assuming that the grains are spherical, as follows:

$$S_{V_{gr}} = \frac{6}{d_{gr}}$$

where d_{gr} is in cm; 1/8 mm = 0.0125 cm.

$$S_{V_{gr}} = \frac{6}{(0.0125)} = 480\ \text{cm}^{-1}$$

$$\frac{\phi^3}{(1-\phi)^2} = \frac{(0.2)^3}{(1-0.2)^2} = 0.0125$$

Now, using the Kozeny Eq. (3.19), the permeability can be estimated:

$$k = \left(\frac{1}{2 s_{V_{gr}}^2}\right)\frac{\phi^3}{(1-\phi)^2}$$

$$k = \left(\frac{1}{2\times(408)^2}\right)(0.0125) = 2.71\times10^{-8}\ \text{cm} = 2{,}750\ \text{mD}.$$

The Carman–Kozeny equation can also be used to estimate the permeability:

$$k = \left(\frac{1}{5 s_{V_{gr}}^2}\right)\frac{\phi^3}{(1-\phi)^2}$$

$$k = \left(\frac{1}{5\times(408)^2}\right)(0.0125) = 1.085\times10^{-8}\ \text{cm}^2 = 1{,}071\ \text{mD}.$$

Changing the constant from 2 to 5 yields a 40% change in the value of k.

(b) The specific surface area per unit pore volume S_{V_p} is obtained from Eq. (3.17):

$$S_{V_p} = S_{V_{gr}}\left(\frac{1-\phi}{\phi}\right) = 480 \times \frac{1-0.2}{0.2} = 1920\ \text{cm}^{-1}$$

Mapping the permeability distribution in reservoirs is one of the most crucial parts of geologic model preparation for performance–prediction studies. In the absence of field-wide *in situ* measurements, permeability is typically estimated indirectly using petrophysical properties acquired through well-log measurement. Because of the abundance of reservoir data from well-log measurements obtainable for every foot or so, a large number of statistical correlations relating these rock properties to permeability have been made. Babadagli and Alsalmi [22] provided an extensive review of permeability–prediction correlations. Many of the correlations are based on the Kozeny–Carman model, which is primarily applicable to homogeneous clastics and nonvuggy, well-sorted carbonate rocks.

CONCEPT OF FLOW UNITS

Petroleum geologists, engineers, and hydrologists have long recognized the need of defining quasi geological/engineering units to shape the description of reservoir zones as storage containers and reservoir conduits for fluid flow. Several authors have various definitions of flow units, which are resultant of the depositional environment and diagenitic process. Bear defined the hydraulic (pore geometrical) unit as the representative elementary volume of the total reservoir rock within which the geological and petrophysical properties of the rock volume are the same [23]. Ebanks defined hydraulic flow units as a mappable portion of the reservoir within which the geological and petrophysical properties that affect the flow of fluid are consistent and predictably different from the properties of other reservoir rock volume [24]. Hear et al. defined flow unit as a reservoir zone that is laterally and vertically continuous, and has similar permeability, porosity, and bedding characteristic [25]. Gunter et al. defined flow unit as a stratigraphically continuous interval of similar reservoir process that honors the geologic framework and maintains the characteristic of the rock type [26].

From these definitions, the flow units have the following characteristics:

1. A flow unit is a specific volume of reservoir, composed of one or more reservoir quality lithologies.
2. A flow unit is correlative and mapable at the interval scale.
3. A flow unit zonation is recognizable on wire-line log.
4. A flow unit may be in communication with other flow units.

Winland [27] developed the following empirical equation that has proved most valuable as a cutoff criterion to delineate commercial hydrocarbon reservoirs in stratigraphic traps:

$$R_{35} = 5.395 \left(\frac{k^{0.588}}{\phi} \right)^{0.864} \tag{3.25}$$

where R_{35} is the calculated pore throat radius (μm) at 35% mercury saturation from a mercury-injection capillary pressure test. Core samples of a given rock type or flow unit will have similar R_{35} values. Winland determined the correlation constants using data from 56 sandstone and 26 carbonate samples. Winland's equation is primarily for predicting pore throat radius for use in defining rock types (flow units) in wells without capillary pressure data. From Jennings and Lucia's work [28], it is clear that Winland's equation can be used to estimate permeability if the pore-throat radius has been independently determined, e.g., capillary pressure measurements, petrographic image analysis (PIA).

Aguilera [29] developed the following equation for calculating pore-throat radius at 35% mercury saturation:

$$R_{35} = 2.665 \left(\frac{k}{\phi} \right)^{0.45} \tag{3.26}$$

where permeability is in mD and porosity is a percentage in both correlations. Aguilera's correlation was developed with data from more than 2,500 sandstone and carbonate samples [30].

Gunter et al. introduced a graphical method for quantifying reservoir flow units based on geological framework, petrophysical rock/pores types, storage capacity, flow capacity, and reservoir process speed. According to them, the five steps for identifying and characterizing flow units are [26]:

1. Identify rock type and illustrate the Withland porosity–permeability cross plot (Figure 3.19).
2. Construct the stratigraphic modified Lorenz plot (SMLP) by computing on a foot–foot basis the percent flow capacity (permeability thickness) and percent flow storage (porosity thickness) (Figure 3.20).
3. Select flow unit intervals based on inflection points from SMLP. These preliminary flow units must be verified using the SFP geologic framework *R*35 (calculated pore throat radius (μm) at 35% mercury saturation) curve and k/ϕ ratio.
4. Prepare final stratigraphic flow profile (SFP) with correlation curve, porosity–permeability, k/ϕ, ratio, R35, percent storage, and percent capacity.
5. Construct a modified Lorenz plot (MLP) by ordering final flow units in decreasing unit speed.

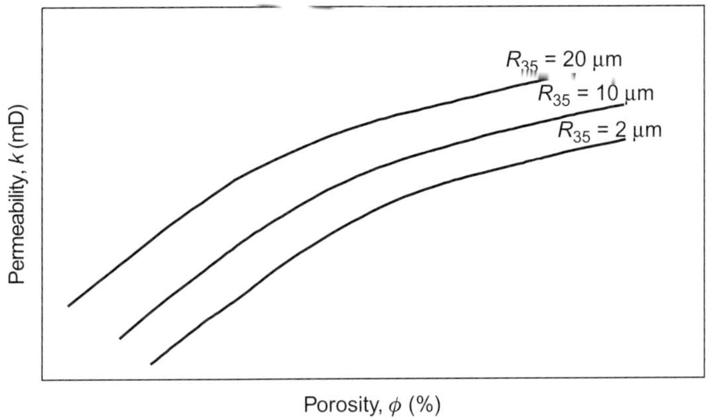

FIGURE 3.19 Sketch of $k-\phi$ Winland plot [26].

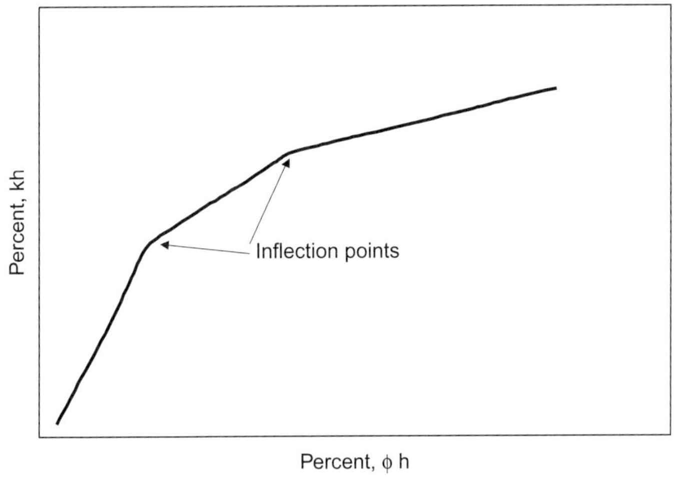

FIGURE 3.20 Sketch of stratigraphic modified Lorenz plot (SMLP) [26].

MATHEMATICAL THEORY OF FLOW UNITS

According to Tiab [31], a hydraulic flow unit is a continuous body over a specific reservoir volume that practically possesses consistent petrophysical and fluid properties, which uniquely characterize its static and dynamic communication with the wellbore. Tiab, Tiab et al., and Amaefule et al. developed a technique for identifying and characterizing a formation having similar hydraulic characteristics, or flow units, based on the microscopic measurements of rock core samples [32–34]. This technique is based on a modified Kozeny–Carman equation and the concept of mean hydraulic radius. The general form of Eq. (3.24) is:

$$k = \left(\frac{1}{K_T s_{V_{gr}}^2}\right)\left(\frac{\phi_e^3}{(1-\phi_e)^2}\right) \tag{3.27}$$

where

k = permeability, μm^2
ϕ_e = effective porosity
$s_{V_{gr}}$ = specific surface area per unit grain volume
τ = tortuosity of the flow path
$K_T = K_{ps}\tau$ = effective zoning factor

Equation (3.27) may be written as:

$$k = \frac{\phi_R}{K_T s_{V_{gr}}^2} \tag{3.28}$$

where ϕ_R is:

$$\phi_R = \frac{\phi_e^3}{(1-\phi_e)^2} \tag{3.29}$$

The parameter K_T, called here the pore-level effective zoning factor, is a function of pore size and shape, grain size and shape, pore and grain distribution, tortuosity, cementation, and type of pore system, e.g., intergranular, intercrystalline, vuggy, or fractured. This parameter varies between flow units, but is constant within a given unit.

The parameter K_T for a homogeneous sandstone formation can be estimated from [32]:

$$K_T = \frac{1}{J_1^2} \tag{3.30}$$

The lithology index J_1 is determined from capillary pressure data. Experimental data show that the plot of the Leverett J-function, $J(S_w^*)$, against the normalized water saturation S_w^* on a log–log graph yields a straight line according to the following equation:

$$\log J(S_w^*) = -\lambda \log(S_w^*) + \log(J_1) \tag{3.31}$$

where J_1 is the intercept of the straight line (extrapolated IF necessary) at $S_w^* = 1$, as shown in Figure 3.21. The normalized water saturation is defined as:

$$S_w^* = \frac{S_w - S_{wi}}{1 - S_{wi}} \tag{3.32}$$

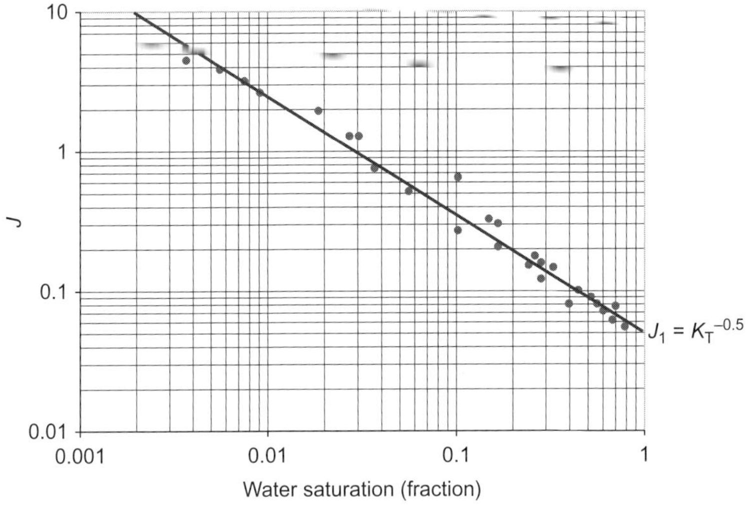

FIGURE 3.21 Determining K_T from a plot of J versus S_w^*.

The pore size distribution index λ is the slope of the line. The lithology index J_1 ranges from 0.44 for an unconsolidated spherical grain to 0.20 for a consolidated clean sandstone formation with homogeneous pore size distribution. Figure 3.21 shows typical values of J_1 and K_T for different formations. High values of J_1 are usually found in high permeability reservoirs, while low values of J_1 correspond to low permeability reservoirs (Table 3.3).

Low values of λ (<1) and J_1 (<0.10) typically indicate that the formation has a heterogeneous pore size distribution and poorly connected pores, which is the case of the reservoir depicted in Figures 3.21 and 3.22, where $J_1 = 0.05$ and, therefore, $K_T = 400$.

SPECIFIC SURFACE AREA

The specific surface area can be estimated by at least three techniques: the gas adsorption method, petrographic image analysis (PIA), and nuclear magnetic resonance (NMR). The basic method for measuring surface area from the gas adsorption technique involves determining the quantity of inert gas, typically nitrogen, argon, or krypton, required to form a layer one molecule thick on the surface of a sample at cryogenic temperature. The area of the sample is then calculated by using the area known, from other considerations, to be occupied by each gas molecule under these conditions. The gas adsorption method is widely used in the determination of specific surface area of porous materials. It should be, however, limited to porous media that do not have large specific

TABLE 3.3 Typical Values of J_1 and K_T for Several Formations

Reservoir	Formation	J_1	K_T
Hawkins	Woodbine	0.347	8.3
Rangely	Weber	0.151	43.9
El Robie	Moreno	0.18	30.9
Kinselia shale	Viking	0.315	10.1
Katia	Deese	0.116	74.3
Leduc	Devonian	0.114	76.9

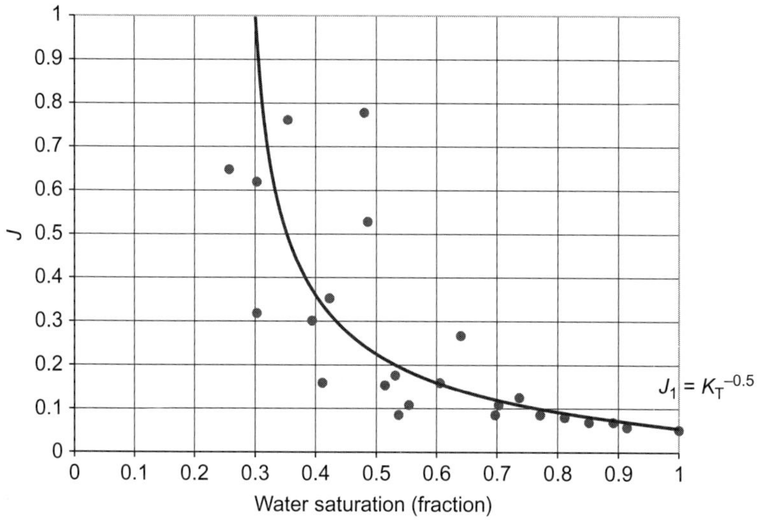

FIGURE 3.22 Determining K_T from a plot of J versus S_w^*.

surfaces, and where the grains of the matrix are singularly smooth and regular, i.e., sphericity >0.7 and roundness >0.5, as shown in Figure 3.23.

The adsorption method, as currently practiced, does not measure the same surface area as that involved in fluid flow experiments of most porous rocks, especially when the rock samples are crushed. However, for unconsolidated porous systems, the specific surface area obtained by this technique is very adequate. A log–log plot of $s_{V_{gr}}$ (cm^{-1}) versus the mean grain diameter d_{gr} (cm) yielded the following correlation:

$$S_{V_{gr}} = \frac{4.27}{d_{gr}} \tag{3.33}$$

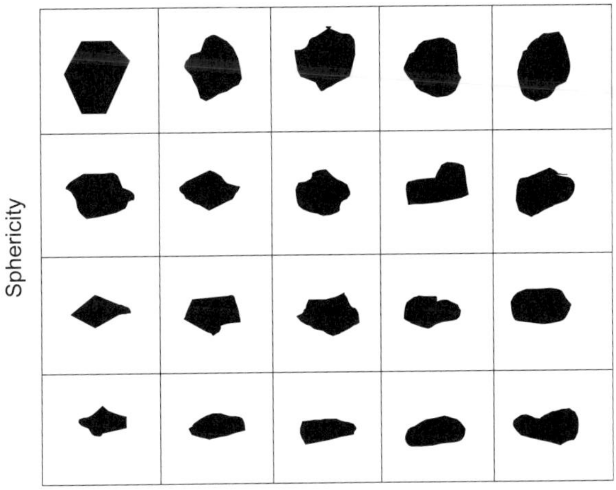

FIGURE 3.23 Roundness and sphericity of different shapes.

The numerator is actually the grain-shaped factor K_{gs}, as indicated in Eq. (3.34). The mean grain diameter can be obtained from several methods: sieve analysis, PIA, a compactor and micrometer. This correlation is applicable to grains with sphericity ≥0.7 and roundness ≥0.5. The general form of Eq. (3.33) is:

$$S_{V_{gr}} = \frac{K_{gs}}{d_{gr}} \tag{3.34}$$

As the sphericity and roundness approach unity, the grain shape factor approaches 6 that is valid only for the case of perfectly spherical grains. As the sphericity and roundness approach 0.1 or zero, the grain shape factor approaches π, thus $\pi < K_{gs} < 6$. The petrographic image analysis or PIA method may be used to characterize the porous rock if well-prepared samples are available, i.e., samples with good optical contrast between the pores and grains, and the thin sections are obtained at overburden conditions. The specific pore surface can be determined from:

$$s_{pv} = \frac{4L_p}{\pi A_p} \tag{3.35}$$

where L_p and A_P are the pore perimeter and the pore cross-section, respectively. Using PIA, a planar pore shape factor f_{ps} can be determined as:

$$f_{ps} = \frac{L_p^2}{4\pi A_p} \tag{3.36}$$

The factor f_{ps}, shown in Table 3.4, indicates what the perimeter L_p would be if the planar or 2D feature were a circle. The range of f_{ps} is 3.75

TABLE 3.4 2D PIA Pore Shape Factors

Pore Shape	f_{ps} (2D)
○	1
□	1.27
△	1.65
◇	3.75
⬭	5.84

(sphericity of 0.5 or less and roundness of 0.3 or less) to 5.84 (sphericity of 0.5 or less and roundness greater than 0.5). There is no practical relation between the 2D pore-shaped factor f_{ps} and the 3D pore-shaped factor K_{ps}. For an ideal spherical pore $f_{ps} = 1$ and $K_{ps} = 6$.

The nuclear magnetic resonance or NMR method appears to be currently the most accurate technique for estimating the specific surface area. In this case, the specific surface areas $s_{V_{gr}}$ and s_{pv} are obtained from:

$$s_{V_{gr}} = A_{NMR}\,\rho_m \tag{3.37}$$

$$s_{pv} = \left(\frac{1-\phi}{\phi}\right) A_{NMR}\,\rho_m \tag{3.38}$$

where

A_{NMR} = NMR surface area of dry material, m^2/g
ρ_m = grain–matrix density, g/cm^3
$s_{V_{gr}}$ = specific surface area per unit grain volume, m^2/cm^3

Values of s_{pv} and $s_{V_{gr}}$ obtained from NMR are generally higher than values obtained by PIA or the gas adsorption technique. Several studies have found that the specific surface area, measured with any of these three methods, is related to the irreducible water saturation or simply water saturation by a relationship of the general form:

$$s_{pv} = a\,e^{bS_w} \tag{3.39}$$

where a and b are constants of correlation. Zemanek investigated low-resistivity sandstone reservoirs and found surface areas measured with the NMR technique were quantitatively consistent with the irreducible water saturations from the capillary pressure curve data [35]. Figures 3.19 and 3.20 demonstrate a good correlation between s_{pv} and S_w and S_{wi}, yielding, respectively:

$$s_{pv} = 66.894\,e^{0.0224 S_{wi}} \tag{3.40}$$

$$s_{pv} = 66.493\,e^{0.0339 S_w} \tag{3.41}$$

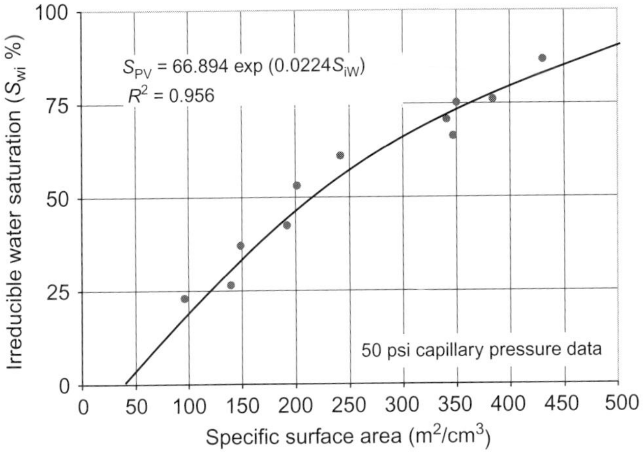

FIGURE 3.24 Irreducible water saturation as a function of surface area (S_{pv}) in sandstone formation [35].

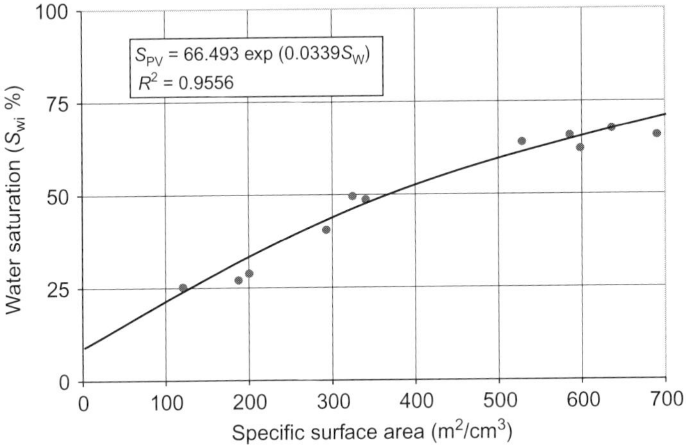

FIGURE 3.25 Water saturation as a function of surface area (S_{pv}) in sandstone formation [35].

where S_w and S_{wi} are expressed in percent and s_{pv} in m^2/cm^3. The values of s_{pv} used in Figures 3.24 and 3.25 were obtained from Eq. (3.38).

FLOW UNIT CHARACTERIZATION FACTORS

(a) *Reservoir quality index (RQI):* Amaefule et al. also introduced the concept of reservoir quality index (RQI), $(k/\phi)^{1/2}$, considering the pore-throat, pore and grain distribution, and other macroscopic parameters

[34]. Dividing both sides of Eq. (3.27) by porosity and taking the square root of both sides yields:

$$\sqrt{\frac{k}{\phi}} = \frac{1}{S_{V_{gr}}\sqrt{K_T}}\left(\frac{\phi}{1-\phi}\right) \tag{3.42}$$

If permeability is expressed in millidarcies and porosity as a fraction, the left-hand side of Eq. (3.42) becomes:

$$\text{RQI} = 0.0314\sqrt{\frac{k}{\phi}} \tag{3.43}$$

where RQI is expressed in micrometers or μm (1 μm = 10^{-6} m).

(b) *Flow zone indicator (FZI):* The flow zone indicator is defined from Eq. (3.42) as:

$$\text{FZI} = \frac{1}{S_{V_{gr}}\sqrt{K_T}} \tag{3.44a}$$

Combining Eqs. (3.30), (3.37), and (3.44a) yields:

$$\text{FZI} = \frac{J_1}{\rho_m A_{NMR}} \tag{3.44b}$$

This equation gives the most accurate value of FZI.
Thus, Eq. (3.42) can be written as:

$$\text{RQI} = \text{FZI}(\phi_z) \tag{3.45}$$

where ϕ_z is the ratio of pore volume to grain volume:

$$\phi_z = \frac{\phi}{1-\phi} \tag{3.46}$$

Taking the logarithm of Eq. (3.45) on both sides yields:

$$\log(\text{RQI}) = \log(\phi_z) + \log(\text{FZI}) \tag{3.47}$$

Equation (3.47) yields a straight line on a log–log plot of RQI versus ϕ_z with a unit slope. The intercept of this straight line at $\phi_z = 1$ is the flow zone indicator. Samples with different FZI values will lie on other parallel lines. Samples that lie on the same straight line have similar pore throat characteristics and, therefore, constitute a flow unit. Straight lines of slopes equal to unity should be expected primarily in clean sandstone formations. Slopes greater than one indicate a shaly formation.

The flow zone indicator (FZI) is a unique parameter that includes the geological attributes of the texture and mineralogy in the structure of distinct pore geometrical facies. In general, rocks containing authogenic pore lining, pore filling, and pore bridging clay as well as fine grained, poorly sorted sands tend to exhibit high surface area and high tortuosity, hence low FZI. In contrast, less shaly, coarse-grained, and well-sorted sand exhibit a lower surface area, low shape factor, lower tortuosity, and higher FZI. Different depositional environments and diagenetic processes control the geometry of the reservoir and consequently the flow zone index.

(c) *Tiab flow unit characterization factor (H_T):* Sneider and King showed that most of the petrophysical properties of sandstones and conglomerates can be related to grain size and sorting, degree of rock consolidation, cementation, sizes of pores, and pore interconnections [36]. They also showed that there are a finite number of rock types and corresponding pore geometries that characterize geologic units. However, geologic units may or may not coincide with hydraulic flow units. It is also possible that a geologic unit may contain several flow units. Equation (3.27) can be written as:

$$H_{\mathrm{T}} = K_{\mathrm{T}} s_{V_{\mathrm{gr}}}^2 = \frac{1}{k}\left(\frac{\phi^3}{(1-\phi)^2}\right) \tag{3.48}$$

where H_{T} is called the Tiab flow unit characterization factor. Substituting for $K_{\mathrm{T}} = \tau K p_{\mathrm{s}}$ and $S_{V_{\mathrm{gr}}}$ (Eq. (3.34)), H_{T} becomes:

$$H_{\mathrm{T}} = K_{\mathrm{ps}} \tau \left(\frac{K_{\mathrm{gs}}}{d_{\mathrm{gr}}}\right)^2 \tag{3.49}$$

Tortuosity can be estimated from:

$$\tau = \phi^{1-m} \tag{3.50}$$

Substituting for t into Eq. (3.49) yields a general expression for the Tiab flow unit characterization factor:

$$H_{\mathrm{T}} = K_{\mathrm{ps}} \phi^{1-m} \left(\frac{K_{\mathrm{gs}}}{d_{\mathrm{gr}}}\right)^2 \tag{3.51}$$

The Tiab flow unit characterization factor H_{T} clearly combines all the petrophysical and geological properties mentioned above by Snyder and King [36]. Note that H_{T} and FZI are related by the following equation:

$$H_{\mathrm{T}} = \frac{1}{\mathrm{FZI}^2} \tag{3.52}$$

The right-hand side of Eq. (3.48) is also, of course, H_{T}, i.e.:

$$H_{\mathrm{T}} = \frac{1}{k} \frac{\varphi^3}{(1-\varphi)^2} \tag{3.53}$$

However, H_T obtained from Eq. (3.51) reflects microscopic petrophysical properties, whereas H_T calculated using Eq. (3.53) reflects the flow unit at the macroscopic scale. If the petrophysical parameters in Eq. (3.51) can be accurately measured, then a log–log plot of the two H_T parameters may be used to normalize the data. Substituting for H_T (Eq. (3.51)) into Eq. (3.48) and solving for permeability gives:

$$k = \frac{1}{K_{ps}\phi^{1-m}(K_{gs}/d_{gr})^2} \frac{\phi^3}{(1-\phi)^2} \tag{3.54}$$

This is the generalized permeability–porosity equation, where the mean grain diameter d_{gr} and the permeability k are expressed in cm and cm^2, respectively. The porosity term ϕ is a fraction.

(d) *Free fluid index (FFI)*: The bulk volume water is commonly used to indicate whether or not a reservoir is at its irreducible water saturation, S_{wir}. It is equal to the product of total porosity and water saturation, S_w:

$$\text{BVW} = \phi S_w \tag{3.55}$$

Reservoirs with water saturation equal to irreducible or connate water saturation produce water-free hydrocarbons since water occupies small pores and is held by surface tension and high capillary pressure. In such a case bulk volume water is termed bulk water volume irreducible (BVI) and is estimated as:

$$\text{BVI} = \phi S_{wir} \tag{3.56}$$

One may consider the bulk volume of irreducible water saturation (BVI) to be represented by the consistent minimum value of the BVW curve. The BVW concept generally provides a good estimate of the irreducible water saturation if the porosity is intergranular, not secondary, and if the rock contains little clay in pore throats. BVW will remain constant in zones of irreducible water saturation and will increase toward the free-water level. S_{wir} cannot be determined confidently from resistivity logs when the reservoir is not at irreducible conditions and when the pay zone produces water. In this case, local experience is considered the best guide to the percentage of water saturation likely to be irreducible.

The free fluid index (FFI) is defined as the product of hydrocarbon saturation and porosity. It is a measure of movable liquids, oil and/or water, and, therefore, it is connected to the flow unit. It is obtained from the MNL tool. Mathematically, it is expressed as:

$$\text{FFI} = \phi(1 - S_{wir}) \tag{3.57}$$

Coates and Denoo [37] related permeability to FFI as follows:

$$k = 10^4\phi^4\left(\frac{\text{FFI}}{\phi - \text{FFI}}\right)^2 \tag{3.58}$$

The correlation constant 10 limits this equation to reservoirs in which (a) the irreducible water saturation is well defined, (b) the porosity is intergranular, and (c) the rock contains little clay in pore throats. Combining this equation and the definition of RQI yields a useful relationship between RQI and FFI:

$$\mathrm{RQI} = 3.14\left(\frac{\mathrm{FFI}}{\phi - \mathrm{FFI}}\right)\sqrt{\phi^3} \tag{3.59}$$

where FFI and porosity are expressed as a fraction, permeability in mD, and RQI in μm.

Taking the logarithm of both sides of Eq. (3.59) yields:

$$\log \mathrm{RQI} = \log(\sqrt{\varphi^3}) + \log\left(\frac{3.14\ \mathrm{FFI}}{\varphi - \mathrm{FFI}}\right) = \log(\sqrt{\varphi^3}) + \log(I_\mathrm{F}) \tag{3.59a}$$

where

$$I_\mathrm{F} = \frac{3.14\ \mathrm{FFI}}{\phi - \mathrm{FFI}} \tag{3.59b}$$

Thus, a log–log plot of the reservoir quality index versus $\sqrt{\phi^3}$ should yield a straight line of slope unity, assuming the flow unit is in a clean homogeneous sandstone formation. The intercept I_F at $\sqrt{\phi^3} = 1$, as shown in Figure 3.26, may be used similarly to FZI, i.e., core samples that lie on the same straight line have similar pore-throat characteristics and irreducible water saturation and, therefore, constitute a flow unit.

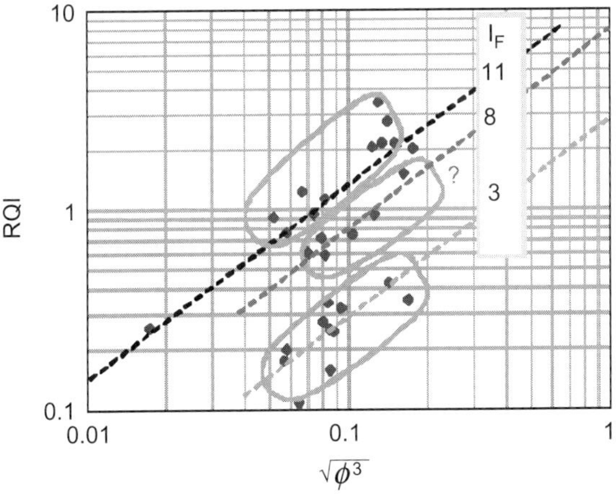

FIGURE 3.26 RQI versus $\sqrt{\phi^3}$ shows three flow units.

I_F can be used to estimate the free-fluid index and the average value of irreducible water saturation for individual flow units as follows:

$$\text{FFI} = \frac{\overline{\phi} I_F}{3.14 + I_F} \quad (3.59c)$$

$$S_{wir} = 1 - \frac{\text{FFI}}{\overline{\phi}} \quad (3.59d)$$

where $\overline{\phi}$ is the average porosity of the flow unit. Equation (3.59c) is valid only if the correlation constant in Eq. (3.58), i.e., 10^4, is applicable. The exponent 4 may be quite different in shaly formation and is a function of shale fraction.

It is highly recommended to generate a correlation similar to Eq. (3.58), but with a correlation constant specific to the subject reservoir or even flow unit. Replacing the constant 3.14 with C_{PP} in Eq. (3.59c) gives:

$$S_{wir} = 1 - \frac{\overline{\phi} I_F}{C_{PP} + I_F} \quad (3.59e)$$

where C_{PP} is the permeability-porosity correlation constant. If S_{wir} is known from capillary pressure measurements or well logs, the intercept I_F is determined as discussed above, then C_{PP} can be calculated from:

$$C_{PP} = I_F \left(\frac{\overline{\phi}}{\text{FFI}} - 1 \right) \quad (3.59f)$$

Coates and Denoo Eq. (3.59) can be generalized as follows:

$$k = 10^{C_{cd}} \phi^4 \left(\frac{\text{FFI}}{\phi - \text{FFI}} \right)^2 \quad (3.59g)$$

where C_{cd} is a correlation constant unique to the subject reservoir. In formations containing high clay contents, expect different values of C_{cd} than in clean formations. There is not enough field data available to establish a specific constant for shaly formations. It may be useful to generate a relationship between this parameter and shale fraction V_{sh}.

The exponent C_{cd} can be calculated from the following equation, once C_{pp} is determined from the intercept I_F:

$$C_{cd} = \log \left(\frac{C_{PP}}{0.0314} \right)^2 \quad (3.59h)$$

Combining Eqs. (3.45) and (3.59) yields the following relationship between the intercepts FZI and I_F:

$$I_F = \left(\frac{1}{(1 - \overline{\phi}) \sqrt{\overline{\phi}}} \right) \text{FZI} \quad (3.59i)$$

Therefore, it is not necessary to plot both RQI versus ϕ_Z and RQI versus $\sqrt{\phi^3}$. $\overline{\phi}$ is the arithmetic average porosity of the flow unit.

Example

(a) Estimate the permeability (mD) of a reservoir rock that has a porosity of 15% and an irreducible water saturation of 24%.

(b) Assuming this permeability is representative of the flow unit, calculate the reservoir quality index RQI (μm).

(c) Calculate the flow zone indicator FZI.

(d) Calculate the Tiab flow unit characterization factor H_T.

Solution

(a) The free fluid index is calculated from Eq. (3.57):

$$\text{FFI} = (1 - S_{wi})\phi = (1 - 0.24)(0.15) = 0.111$$

The permeability is estimated from Eq. (3.58):

$$k = (10\phi)^4 \left(\frac{\text{FFI}}{\phi - \text{FFI}}\right)^2$$

$$= (10 \times 0.15)^4 \left(\frac{0.111}{0.15 - 0.111}\right)^2 = 41 \text{ mD}$$

(b) The reservoir quality index RQI is calculated from Eq. (3.59):

$$\text{RQI} = 3.14\left(\frac{\text{FFI}}{\phi - \text{FFI}}\right)\sqrt{\phi^3}$$

$$= 3.14\left(\frac{0.111}{0.15 - 0.111}\right)\sqrt{0.15^3} = 0.519 \ \mu\text{m}$$

Equation (3.43) also yields the same value:

$$\text{RQI} = 0.0314\sqrt{\frac{k}{\phi}} = 0.0314\sqrt{\frac{41}{0.15}} = 0.519 \ \mu\text{m}$$

(c) Calculate the flow zone indicator (FZI) from Eq. (3.45):

$$\phi_z = \frac{\phi_e}{1 - \phi_e} = \frac{0.15}{1 - 0.15} = 0.176$$

$$\text{FZI} = \frac{\text{RQI}}{\phi_z} = \frac{0.519}{0.176} = 2.95 \ \mu\text{m}$$

(d) The Tiab flow unit characterization factor H_T is:

$$H_T = \frac{1}{\text{FZI}^2} = \frac{1}{2.95^2} = 0.115 \ \mu\text{m}^{-2}$$

Accurate estimates of reservoir rock parameters should not be made from log data alone. A judicious combination of core analysis and log data is required to link these parameters in order to achieve a more global applicability of the equations and relationships presented here. A consistent and systematic approach is required to integrate such petrophysical data in order to develop meaningful relationships between microscopic and macroscopic measurements. The flow chart in Figure 3.27 provides such an approach.

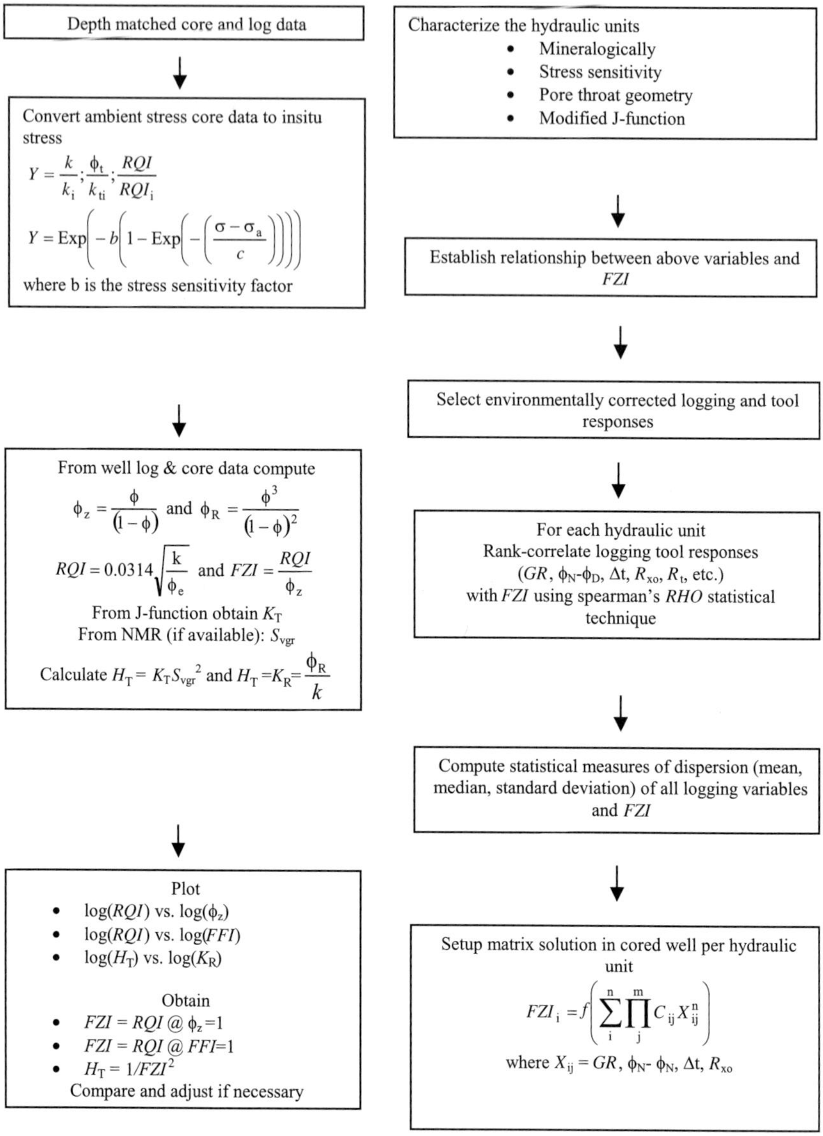
Depth matched core and log data
Convert ambient stress core data to insitu stress
$Y = \frac{k}{k_i}; \frac{\phi_t}{k_{ti}}; \frac{RQI}{RQI_i}$
$Y = \mathrm{Exp}\left(-b\left(1 - \mathrm{Exp}\left(-\left(\frac{\sigma - \sigma_a}{c}\right)\right)\right)\right)$
where b is the stress sensitivity factor
From well log & core data compute
$\phi_z = \frac{\phi}{(1-\phi)}$ and $\phi_R = \frac{\phi^3}{(1-\phi)^2}$
$RQI = 0.0314\sqrt{\frac{k}{\phi_e}}$ and $FZI = \frac{RQI}{\phi_z}$
From J-function obtain K_T
From NMR (if available): S_{vgr}
Calculate $H_T = K_T S_{vgr}^2$ and $H_T = K_R = \frac{\phi_R}{k}$
Plot
• $\log(RQI)$ vs. $\log(\phi_z)$
• $\log(RQI)$ vs. $\log(FFI)$
• $\log(H_T)$ vs. $\log(K_R)$
Obtain
• $FZI = RQI$ @ $\phi_z = 1$
• $FZI = RQI$ @ $FFI = 1$
• $H_T = 1/FZI^2$
Compare and adjust if necessary
Characterize the hydraulic units
• Mineralogically
• Stress sensitivity
• Pore throat geometry
• Modified J-function
Establish relationship between above variables and FZI
Select environmentally corrected logging and tool responses
For each hydraulic unit
Rank-correlate logging tool responses
(GR, ϕ_N-ϕ_D, Δt, R_{xo}, R_t, etc.)
with FZI using spearman's RHO statistical technique
Compute statistical measures of dispersion (mean, median, standard deviation) of all logging variables and FZI
Setup matrix solution in cored well per hydraulic unit
$FZI_i = f\left(\sum_i^n \prod_j^m C_{ij} X_{ij}^n\right)$
where $X_{ij} = GR$, ϕ_N- ϕ_N, Δt, R_{xo}

FIGURE 3.27 A generalized flow chart for characterizing flow units using core and well log data, modified after Amaefule et al. [34].

The chart indicates the different steps for identifying and characterizing flow units in clastic reservoirs. This zoning process is best suited for reservoirs in which intergranular porosity is dominant. Because of the similarity in distribution and movement of fluids within clastic and carbonate rock having

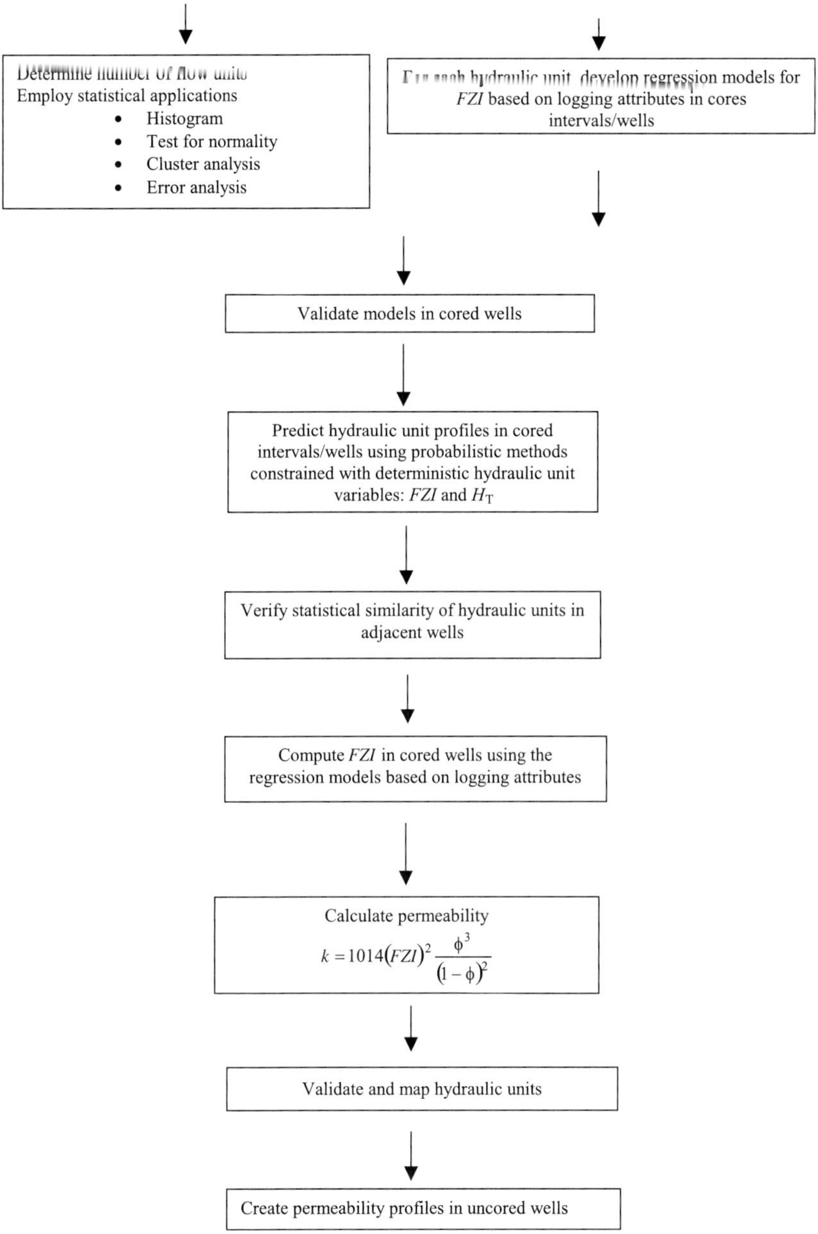

FIGURE 3.27 (*Continued*)

intercrystalline–intergranular porosity, this zoning process can be directly applied to these reservoir systems. This process is, however, not applicable to carbonate reservoirs with vugular solution channels and/or fractures.

Example

Assuming that the permeability and porosity data shown in Table 3.5 are representative of several hundred data points taken from a shaly oil reservoir. The average irreducible water saturations is 21%.

(a) Identify the number of flow units and their corresponding values of flow zone indicator, FZI, and the Tiab flow unit characterization factor, H_T.

(b) Calculate tortuosity and plot versus reservoir quality index, RQI, on a log–log graph. Does this plot confirm the number of flow units?

(c) Calculate the arithmetic average permeability, porosity, FFI, and FZI of each flow unit.

(d) Calculate the correlation constant C_{PP} and modify Eqs. (3.58) and (3.59a) for each flow unit.

Solution

Calculation of different parameters is only presented for sample #1. All other values are given in Table 3.6.

(a) Calculate RQI using Eq. (3.43), i.e.:

$$\text{RQI} = 0.0314\sqrt{\frac{k}{\phi_e}} = 0.0314\sqrt{\frac{22}{0.08}} = 0.52$$

The ratio ϕ_z is calculated from Eq. (3.46):

$$\phi_z = \frac{\phi}{1-\phi} = \frac{0.08}{1-0.08} = 0.087$$

Calculate the formation resistivity factor using the Humble equation:

$$F = \frac{0.81}{\phi^2} = \frac{0.81}{0.08^2} = 126.56$$

The plot of RQI versus $\phi/(1-\phi)$ shows two straight lines of slope unity, indicating two hydraulic units. The corresponding FZI values are 11 and 6.3, respectively (Figure 3.28).

TABLE 3.5 Permeability and Porosity

k (mD)	ϕ (fraction)	k (mD)	ϕ (fraction)
22	0.08	112	0.09
51	0.1	430	0.19
315	0.12	250	0.16
344	0.13	490	0.14
90	0.11		

TABLE 3.6 Example Results

k (mD)	ϕ	RQI (μm)	$\phi/(1-\phi)$	F	[illegible]	$\sqrt{\phi^3}$
22	0.08	0.521	0.087	126.56	12.5	0.0226
51	0.1	0.709	0.111	81	10	0.0316
315	0.12	1.609	0.136	56.25	8.3	0.0416
344	0.13	1.615	0.149	47.92	7.7	0.0469
90	0.11	0.898	0.124	66.94	9.1	0.0365
112	0.09	1.108	0.099	100	11.11	0.0270
430	0.19	1.535	0.22	25	5.5	0.0828
250	0.16	1.241	0.19	31.64	6.2	0.0640
490	0.14	1.858	0.163	41.33	7.1	0.0524

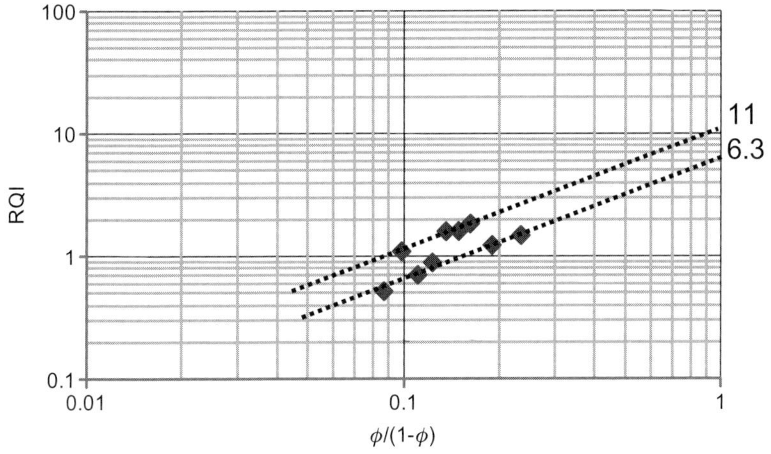

FIGURE 3.28 RQI versus ϕ_z indicating the presence of two flow units.

The corresponding values of H_T are obtained from Eq. (3.52):
Hydraulic unit #1, FZI = 11:

$$H_T = \frac{1}{\mathrm{FZI}^2} = \frac{1}{11^2} = 8.26 \times 10^{-3}\ \mu\mathrm{m}^{-2}$$

Hydraulic unit #2, FZI = 6.3:

$$H_T = \frac{1}{\mathrm{FZI}^2} = \frac{1}{6.3^2} = 0.0252\ \mu\mathrm{m}^{-2}$$

(b) Calculate tortuosity using Eq. (3.50), and assuming $m = 2$:

$$\tau = \phi^{1-m} = 0.08^{1-2} = 12.5$$

Figure 3.29 verifies that there are two hydraulic flow units in the reservoir.

(c) From Figure 3.30, the intercepts I_F at $\sqrt{\phi^3} = 1$ are 36 and 20.

Flow unit 1 (IF1 = 36):

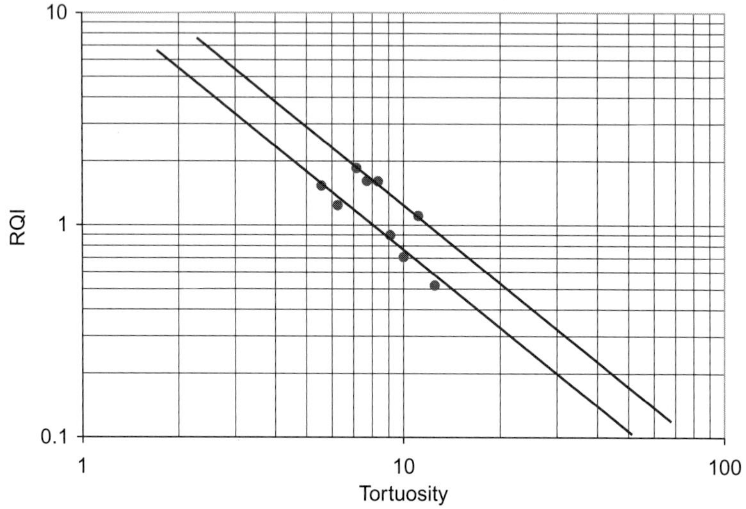

FIGURE 3.29 RQI versus tortuosity confirming the presence of two flow units.

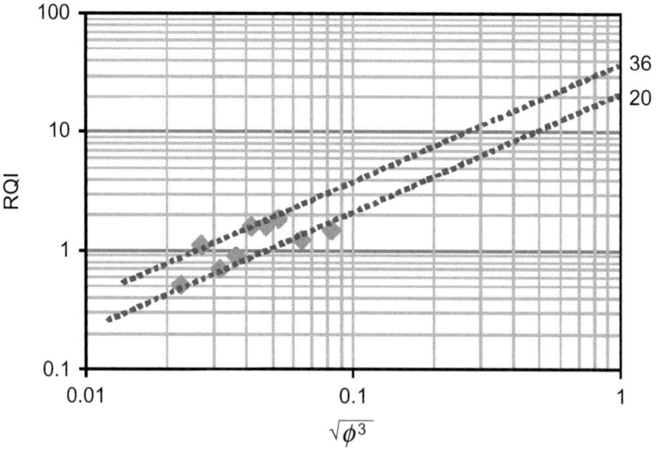

FIGURE 3.30 RQI versus free fluid index.

The arithmetic average porosity for this flow unit is:

$$\overline{\phi}_1 = \frac{0.12 + 0.13 + 0.09 + 0.14}{4} = 0.120$$

The average values of FFI and FZI of flow unit 1 are:

$$\text{FFI}_1 = \overline{\phi}(1 - S_{wir}) = 0.12(1 - 0.21) = 0.095$$

$$\text{FZI}_1 = ((1 - \overline{\phi})\sqrt{\overline{\phi}})I_{F1} = ((1 - 0.12)\sqrt{0.12}) \times 36 = 11$$

Flow unit 2 ($I_{F2} = 20$):

Similarly, the average values of porosity, FFI and FZI of the second flow unit are:

$$\overline{\phi}_2 = \frac{0.108 + 0.10 + 0.11 + 0.19 + 0.16}{5} = 0.128$$

$$\text{FFI}_2 = \overline{\phi}(1 - S_{wir}) = 0.128(1 - 0.21) = 0.1011$$

$$\text{FZI}_2 = ((1 - \overline{\phi})\sqrt{\overline{\phi}})I_{F2} = ((1 - 0.128)\sqrt{0.128}) \times 20 = 6.3$$

(d) The correlation constant C_{PP1} for flow unit 1 is calculated from Eq. (3.59f):

$$C_{PP1} = I_{F1}\left(\frac{\overline{\phi}}{\text{FFI}} - 1\right) = 36\left(\frac{0.12}{0.1011} - 1\right) = 5.31$$

The correlation constant 10 in Coates and Denoo equation (i.e., Eq. (3.58)) for flow unit 1 becomes:

$$C_{cd1} = \log\left(\frac{C_{PP1}}{0.0314}\right)^2 = \log\left(\frac{9.6}{0.0314}\right)^2 = 4.968$$

Therefore, a representative permeability–porosity equation for this flow unit are:

$$k = 10^{4.968}\phi^4\left(\frac{\text{FFI}}{\phi - \text{FFI}}\right)^2$$

Similarly, C_{PP2}, C_{cd2}, and permeability–porosity equation of flow unit 2 are:

$$C_{PP2} = I_{F2}\left(\frac{\overline{\phi}}{\text{FFI}} - 1\right) = 20\left(\frac{0.128}{6.3} - 1\right) = 5.3$$

$$C_{cd2} = \log\left(\frac{C_{PP2}}{0.0314}\right)^2 = \log\left(\frac{5.3}{0.0314}\right)^2 = 4.457$$

$$k = 10^{4.457}\phi^4\left(\frac{\text{FFI}}{\phi - \text{FFI}}\right)^2$$

Note that (a) the permeability–porosity correlation constant C_{cd} is slightly greater than 4 in this formation, and (b) the FZI obtained from Eq. (3.59e) matches the value obtained from the RQI versus ϕ_Z plot (Figure 3.22(a)).

FLOW UNITS FUNCTIONS

Capillary pressure concepts are used to evaluate reservoir rock quality, pay versus nonpay, fluid saturations and contacts, thickness of transition zones, as well as approximate recovery efficiency during primary or secondary recovery. Water saturation versus height can be estimated from the *J*-function for each well using (a) foot-by-foot porosity and permeability data, and (b) measured or estimated reservoir fluid densities and interfacial tension. Core data are used where available. Porosity logs and permeability–porosity correlations are used if a well was not cored. For each $k-\phi$ data point, a value of *J* is calculated for that foot of the well by substituting into the *J*-function the value of capillary pressure. The *J*-function, which was proposed by Leverett [38], is expressed as:

$$J(S_W) = \frac{0.2166\,P_C}{\sigma \cos\theta}\sqrt{\frac{k}{\phi}} \tag{3.60a}$$

where

k = permeability, mD
ϕ = porosity, fraction
σ = interfacial tension between oil and water, dyne/cm
θ = contact angle, degree (equal to 0 for a strongly water wet reservoir)
$J(S_W)$ = Leverett *J*-function, dimensionless

Capillary pressure data, especially when used to estimate connate water saturations, are often correlated using the *J* function. This function averages capillary pressure data by accounting for sample-to-sample variations in permeability and porosity, and for differences in interfacial tension if tests were performed using a pair of fluids, e.g., oil and water. Capillary pressure data are obtained from the following equation:

$$P_C = \frac{(\rho_W - \rho_O)H_{fw}}{144} \tag{3.60b}$$

where

ρ_O = density of oil, lb/ft^3
ρ_W = density of water, lb/ft^3
H_{fw} = height above free water level or level where $P_C = 0$, ft
P_C = capillary pressure, psia

If density is expressed in g/cm^3, P_C in psi, and H_{fw} in ft, Eq. (3.60a) becomes:

$$P_C = 0.433(\rho_W - \rho_O)H_{fw} \tag{3.60c}$$

Capillary pressure data are affected by the pore throat sizes and, hence, by the permeability of the rock. The smaller the pores, the lower the permeability, and the higher the water saturation. Water wet coarse grain sand and oolitic and vuggy carbonates with large pores have low capillary pressure and low interstitial water content. Often the water–oil contact elevation is substituted as an approximation for the free water level. Depending upon density differences and rock pore sizes, the water–oil contact may be very close or tens of feet higher than the free water level.

Theoretically, a single universal *J*-curve will result from a set of capillary pressure curves for cores covering a wide permeability range. This implies that the irreducible water saturation for all samples will be the same. Amyx et al. [39] and Rose and Bruce [40] and Brown [41] showed that the irreducible water saturation can cover a wide range depending on the permeability of individual core samples, so that no universal curve can be obtained. There exists however a unique *J*-curve defining a flow unit in which K_T is constant (Figure 3.21). Combining Eqs. (3.43), (3.45), and (3.60a) yields the following relationships between the *J*-function and the reservoir quality index and the flow zone index:

$$J(S_w) = \left(\frac{6.9P_C}{\sigma \cos\theta}\right)\text{RQI} \quad (3.60d)$$

$$J(S_w) = \left(\frac{6.9P_C}{\sigma \cos\theta}\right)\phi_Z\text{FZI} \quad (3.60e)$$

Taking the logarithm of both sides of Eq. (3.60e) gives:

$$\begin{aligned}\log(J(S_w)) &= \log\left(\frac{6.9P_C\phi_Z}{\sigma \cos\theta}\right) + \log(\text{FZI}) \\ &= \log\left(\frac{3(\rho_w - \rho_O)H_{fw}\phi_Z}{\sigma \cos\theta}\right) + \log(\text{FZI})\end{aligned} \quad (3.60f)$$

This equation provides an independent method to determine the flow units. The log–log plot of water saturation versus the *J*-function is a straight line of the general form:

$$S_w = \frac{C_1}{J(S_w)^{C_2}} \quad (3.61a)$$

Combining Eqs. (3.60a), (3.60c), and (3.61a) gives:

$$S_w = \frac{C_1}{((0.09379(\rho_W - \rho_O)H_{fw}/\sigma\cos\theta)\sqrt{k/\phi})^{C_2}} \quad (3.61b)$$

Substituting RQI into Eq. (3.61b) yields:

$$S_w = \frac{C_1}{((2.987(\rho_W - \rho_O)H_{fw}/\sigma\cos\theta)\text{RQI})^{C_2}} \quad (3.61c)$$

Thus, the flow-unit (saturation-based) function is:

$$S_w = \frac{C_1}{((2.987(\rho_W - \rho_O)H_{fw}/\sigma \cos\theta)\phi_Z \mathrm{FZI})^{C_2}} \tag{3.61d}$$

where FZI is determined from a plot of RQI versus $\varnothing_Z$.

The saturation–height relationship for each flow unit can be derived by modifying Cuddy's function [42] as follows:

$$S_w = \frac{C_3}{H_{fw}^{C_4}\phi^{C_5}} \tag{3.61e}$$

where $C_1 - C_5$ are regression coefficients unique to the flow unit.

Using Eqs. (3.61a), (3.61b), and (3.61e), Amabeoku et al. [43] developed several saturation–height functions on the basis of unique flow units as a part of an integrated petrophysical analysis of a gas field in Saudi Arabia. These functions were developed by (1) linking depositional and diagenetic rock fabric to flow units, (2) linking the flow units to zones with similar capillary pressure relationships, and (3) determining saturation–height functions for each flow unit. It was found that the modified Cuddy's function, which is also referred to as the modified FOIL function, is the most direct and the easiest to use. Amabeoku et al. suggested that, because the modified FOIL function does not require permeability in field application, it can be used in uncored wells. Application of the *J*-function is, however, constrained by the requirement that good permeability models be available. They also suggested that improved fluid-saturation profiles derived from the modified FOIL function can be used to validate log models and petrophysical parameters.

Example

Table 3.7 show core data obtained from a gas field in Saudi Arabia consisting of clastics of Permian Carboniferous age. The field is divided into two reservoirs: (1) the upper Reservoir A, which consists of highly laminated and fine-to-coarse grained sands, and (2) the lower Reservoir B, which is viewed as the principal reservoir in this field, is dominantly sandstone of fluvial origin with very low clay content (less than 5%). Table 3.8 shows ambient conditions high-speed centrifuge air/brine capillary pressure data obtained by use of core samples from three wells in this field. Table 3.9 shows the conversion factors and fluid densities used to generate the capillary pressure curves shown in Figure 3.31. Using this data Amabeoku et al. [43] identified five flow units on the basis of the FZI values of the samples, as shown in Figure 3.32.

1. Confirm the number of flow units by using Eq. (3.59a)
2. Calculate Swi, FFI and BVI for each flow unit.

Solution

1. Figure 3.32 clearly indicates many more points were used to identify the five flow units than those reported in Table 3.7.

   However, using the data in Table 3.7, a log–log plot of RQI versus $\sqrt{\phi^3}$ was made (Figure 3.33 and 3.34). Using straight lines of slope = 1 across the

TABLE 3.7 Permeability and Porosity Data for Saudi Gas Field [43]

Sample No.	Porosity (fraction)	Kilnkenberg Permeability (mD)	ϕ_z	RQI (m)	FZI (m)	HU
X718.2	0.126	58.900	0.144	0.679	4.709	1
X713.8	0.120	44.000	0.136	0.601	4.409	1
X724.2	0.119	41.300	0.135	0.585	4.331	1
X742.8	0.091	16.900	0.100	0.428	4.274	1
X719.8	0.119	30.600	0.135	0.504	3.728	1
X711a	0.079	7.260	0.086	0.301	3.506	2
X704a	0.096	12.800	0.106	0.363	3.431	2
X748.8	0.096	11.200	0.106	0.339	3.194	2
X704b	0.077	3.760	0.083	0.219	2.630	2
X725.2	0.071	2.370	0.076	0.181	2.374	2
X734.4	0.080	2.590	0.087	0.179	2.055	3
X741.8	0.082	2.610	0.089	0.177	1.983	3
X739.5	0.143	15.800	0.167	0.330	1.983	3
X745.5	0.134	10.200	0.155	0.274	1.770	3
X740.4	0.086	2.240	0.094	0.160	1.703	3
X711b	0.069	0.988	0.074	0.119	1.603	3
X747.8	0.078	1.380	0.085	0.132	1.591	3
X727.3	0.085	0.787	0.093	0.096	1.029	4
X748.7	0.089	0.628	0.098	0.083	0.854	4
X733.3	0.085	0.391	0.093	0.067	0.725	4
X717a	0.068	0.074	0.073	0.033	0.448	5
X693a	0.075	0.070	0.081	0.030	0.375	5
X717b	0.050	0.003	0.053	0.008	0.146	5
X577.6	0.096	0.013	0.106	0.012	0.109	5
X693b	0.063	0.003	0.067	0.007	0.120	5

same points identified as belonging to the five flow units, the intercepts I_F at $\sqrt{\phi^3} = 1$ are: 14.3 (HU_1), 11.3 (HU_2), 6.4 (HU_3), 3.2 (HU_4), and 1.0 (HU_5).

2. The FFI and irreducible water saturation for each flow unit are shown in Table 3.10. S_{wi} is calculated from Eq. (3.59c), assuming the constant is 3.14 (clean sandstone formation, clay content is less than 3% in Reservoir B). Equations (3.57) and (3.56) are used to calculate S_{wi}, FFI and BVI.

TABLE 3.8 Capillary Pressure data for Saudi Gas Field [43]

	HU-1			HU-2		HU-3			HU-4	HU-5	
Sample No.	X718.2	X713.8	X742.8	X711a	X704a	X734.4	X741.8	X739.5	X727.3	X717a	X693a
Permeability (mD)	58.900	44.000	16.900	7.260	12.800	2.590	2.610	15.800	0.787	0.074	0.070
Porosity (fraction)	0.126	0.120	0.091	0.079	0.096	0.080	0.082	0.143	0.085	0.068	0.075
ϕ_z	0.144	0.136	0.100	0.086	0.106	0.087	0.089	0.167	0.093	0.073	0.081
RQI (m)	0.679	0.601	0.428	0.301	0.363	0.179	0.177	0.330	0.096	0.033	0.030
FZI (m)	4.709	4.409	4.274	3.506	3.431	2.055	1.983	1.978	1.029	0.448	0.375
P_c (psi)	Sw	Sw	Sw	Sw	Sw	Sw	Sw	Sw	Sw	Sw	Sw
0	1.000	1.000	1.000	1.000	1.000	1.000	1.000	1.000	1.000	1.000	1.000
2	0.685	0.825	0.654	1.000	1.000	0.788	0.920	1.000	0.935	1.000	1.000
5	0.442	0.577	0.434	0.875	0.901	0.695	0.800	0.783	0.845	1.000	1.000
10	0.265	0.350	0.312	0.695	0.471	0.495	0.588	0.504	0.647	1.000	1.000
25	0.146	0.185	0.200	0.379	0.309	0.293	0.335	0.290	0.397	1.000	1.000
50	0.089	0.111	0.139	0.325	0.223	0.178	0.196	0.189	0.255	1.000	1.000
100	0.059	0.064	0.103	0.278	0.159	0.105	0.125	0.130	0.160	0.963	0.958
200	0.039	0.044	0.078	0.246	0.133	0.060	0.074	0.089	0.094	0.861	0.829
300	0.030	0.036	0.066	0.226	0.118	0.040	0.057	0.070	0.065	0.799	0.686
400	0.025	0.033	0.059	0.211	0.107	0.034	0.050	0.061	0.058	0.741	0.541

TABLE 3.9 Conversion Constants and Fluid Densities Used in Saudi Gas Field [43]

Laboratory θ (degrees)	0
Laboratory σ (dyne/cm)	72
Reservoir θ (degrees)	0
Reservoir σ (dyne/cm)	50
Laboratory $\sigma\cos\theta$ (dyne/cm)	72
Reservoir $\sigma\cos\theta$ (dyne/cm)	50
Fluid density (g/cc)	
Water	1.039
Gas	0.392

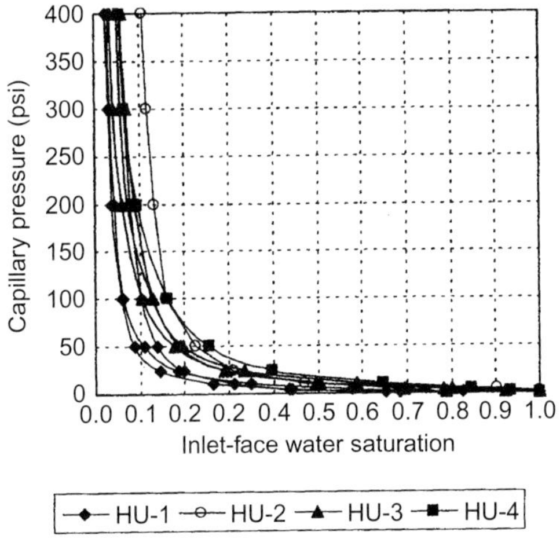

FIGURE 3.31 Correlation of the *J*-function with water saturation [43].

Amabeoku et al. identified five flow units, but they only retained the top four hydraulic units. The fifth hydraulic unit was not considered net-reservoir rock ($k = 10^{-3}$ mD). They excluded the samples in this unit from further analysis (Table 3.11).

The values of FFI, BVI, and S_{wi} in Table 3.9 will probably be slightly different if it were possible to generate a new correlation constant (different from 4) in Eq. (3.58), which is relevant to this field case.

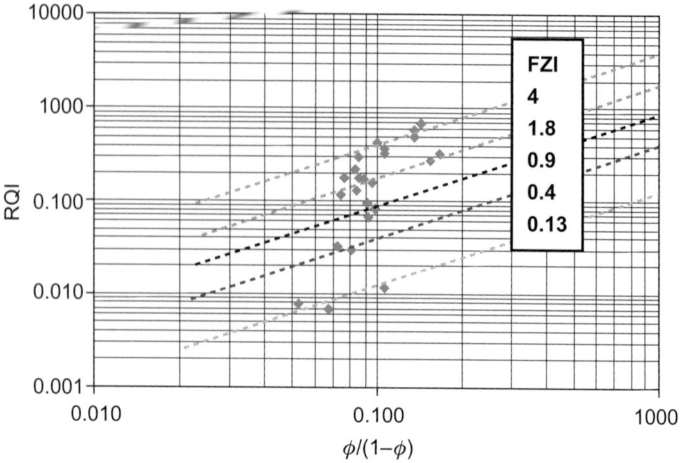

FIGURE 3.32 Hydraulic unit zonation [43].

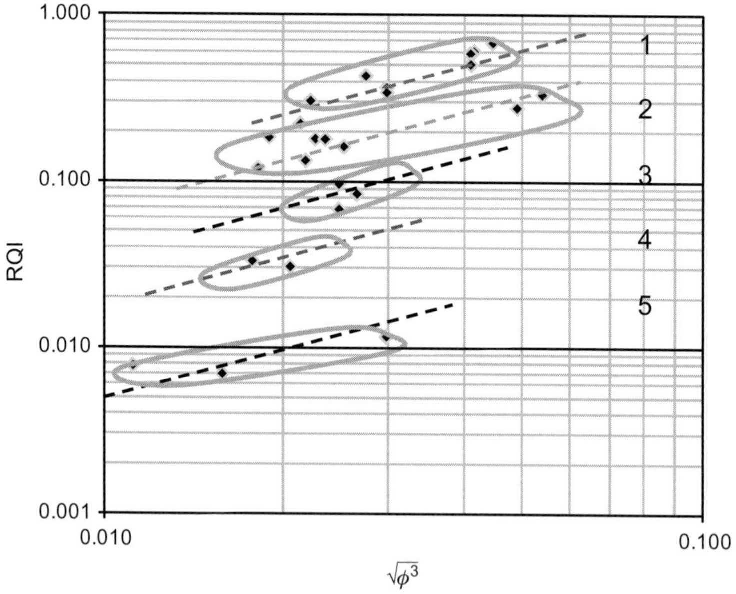

FIGURE 3.33 RQI versus $\sqrt{\phi^3}$ for Saudi Gas Field (identifying flow units).

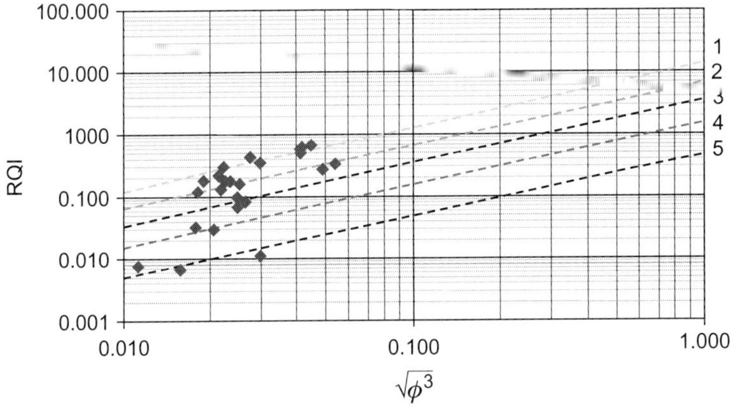

FIGURE 3.34 RQI versus $\sqrt{\phi^3}$ for Saudi Gas Field.

TABLE 3.10 S_{wi} **in Hydraulic Units 1 and 4**

Property	HU-1		HU-4	
Sample	34	41	42	44
Permeability (mD)	25.500	52.100	2.870	0.860
Porosity (fraction)	0.089	0.120	0.071	0.089
ϕ_z	0.098	0.136	0.076	0.098
RQI (μm)	0.532	0.654	0.200	0.098
FZI (μm)	5.440	4.798	2.612	0.999
S_{wi} (fraction)	0.116	0.238	0.107	0.094

TABLE 3.11 S_{wi}, **FFI, and BVI for Saudi Gas Field**

HU	K	Porosity	S_{wir}	Cdc
1	34	0.089	0.116	3.97
2	41	0.12	0.238	4.29
3	42	0.071	0.107	4.38
4	44	0.089	0.094	3.88

The values of FZI in Table 3.9 were obtained from Eq. (3.59e). Note that the calculated values of FZI approximately match the FZI values observed in Figure 3.32.

RESERVOIR ZONATION USING NORMALIZED RQI

Siddiqui et al. [44] proposed a practical technique for selecting representative samples for special core analysis (SCAL) tests. The technique requires accurate identification of intervals with similar porosity/permeability relationships. They used a combination of wireline logs, gamma scan, quantitative CT, and preserved state brine permeability data to calculate appropriate depth-shifted reservoir quality index (RQI) and flow zone indicator (FZI) data, which are then used to select representative plug samples from each reservoir compartment. According to Siddiqui et al. the combination of ϕ and k data, as discussed above, in terms of reservoir quality index provides a convenient starting point to address the differences between samples and between reservoir zones. If, however, the total productivity of a well is assumed to be a linear combination of individual flow zones, then a simple summation and normalization of permeability, RQI, or FZI starting at the bottom of the well provides a convenient comparison with the normalized cumulative plot of an open-hole flow meter test [44]. In such a plot, consistent zones are characterized by straight lines with the slope of the line indicating the overall reservoir quality within a particular depth interval. The lower the slope the better the reservoir quality. In general, these lines will coincide with Testerman's layers [45,46]. The equation used for calculating normalized cumulative *reservoir quality index* is:

$$\mathrm{RQI_{nc}} = \frac{\sum_{x=1}^{i} \sqrt{k_i/\phi_i}}{\sum_{x=1}^{n} \sqrt{k_i/\phi_i}} \tag{3.61f}$$

Sididiqui et al. [44] showed that by plotting the normalized cumulative reservoir quality index ($\mathrm{RQI_{nc}}$) versus depth it is possible to divide the reservoir into several zones by observing changes in the slope, as shown in Figure 3.35.

EFFECT OF PACKING ON PERMEABILITY

The lack of global applicability of the Kozeny model has led researchers to generate empirical correlations on a formation-by-formation level. Since no plausible physical models exist, these correlations have not been accurate enough to gain wide acceptance. This lack of accuracy may, however, be acceptable when gauging the relative difference of permeability of different zones, but not for obtaining an accurate permeability.

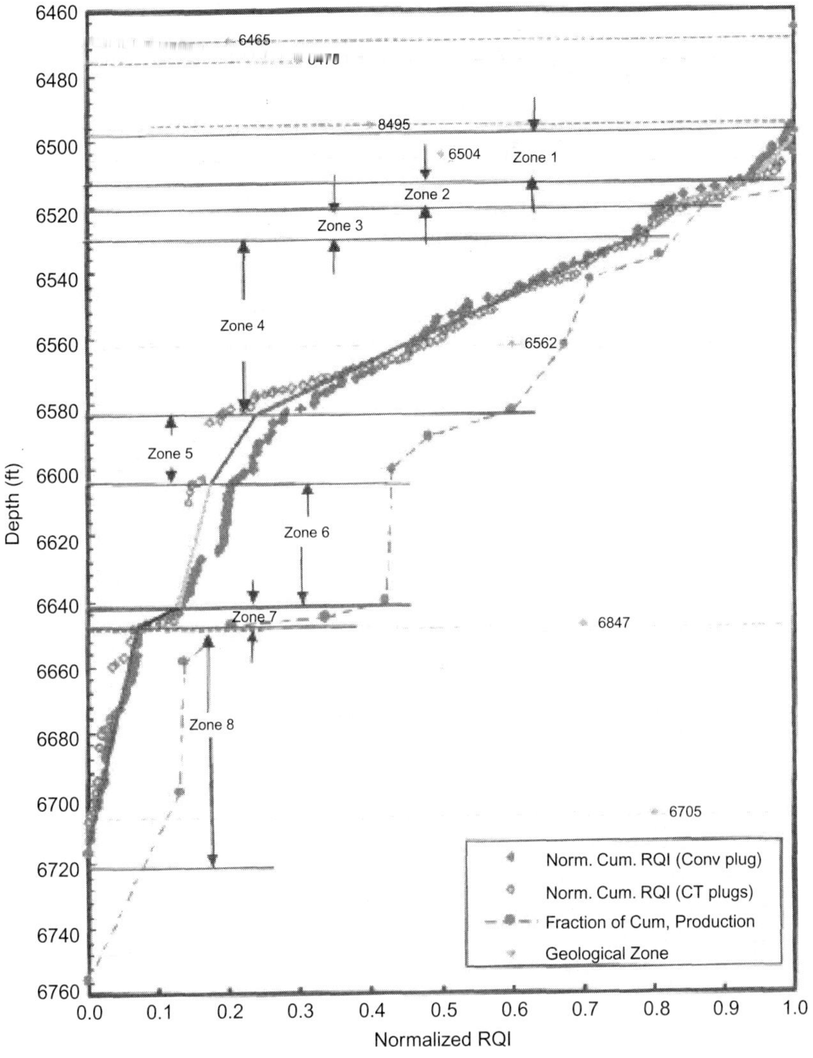

FIGURE 3.35 Normalized cumulative RQI versus depth showing the existence of eight distinct zones within the cored section (shown by solid line segments). The cumulative production data and the geological zone markers (dotted horizontal lines) are also shown for reference [44].

Slichter was the first to demonstrate mathematically the influence of packing and grain size on permeability [47]. His semi-empirical equation is:

$$k = 10.2 \frac{d_{gr}}{a_p} \tag{3.61}$$

where k is the permeability in Darcy, d_{gr} is the diameter of spherical grains in mm, and a_p is the packing constant, which may be estimated from:

$$a_p = 0.97\phi^{-3.3} \quad (3.62a)$$

Substituting Eq. (3.62a) into Eq. (3.61) gives:

$$k = 10.5 d_{gr}\phi^{3.3} \quad (3.62b)$$

This correlation is valid primarily for sandstone formations.

EFFECT OF WATER SATURATION ON PERMEABILITY

Wyllie and Rose investigated the effect of irreducible water saturation S_{wi} and porosity on the absolute permeability, and developed the following empirical correlation [48]:

$$k = \left(\frac{a_{wr}\phi^3}{S_{wi}}\right)^2 \quad (3.63)$$

where a_{wr} is a constant depending on the hydrocarbon density. For a medium gravity oil $a_{wr} = 250$ and for dry gas $a_{wr} = 79$, k is in mD, and ϕ and S_{wi} are fractions [49]. Inasmuch as $250^2 \approx 10 \times 79^2$, Eq. (3.63) indicates that, for the same ϕ and S_{wi}, $k_o \approx 10k_g$, which is not always the case. Equation (3.63) should be used only in the presence of clastic sediments. A similar expression was derived by Timur [50]:

$$k = 0.136\frac{\phi^{4.4}}{S_{wi}^2} \quad (3.64)$$

where permeability is in mD and S_{wi} and ϕ are expressed in percentages. Equation (3.64) is independent of the type of hydrocarbon present in the porous medium.

It is important to emphasize that Eqs. (3.60)–(3.64) are empirical. They are commonly used to obtain an estimate of permeability distribution from well log data. If porosity and irreducible water saturation are used in fractional form, Eq. (3.64) has the form:

$$k = (93^2)\frac{\phi^{4.4}}{S_{wi}^2} \quad (3.65)$$

Langnes et al. presented another empirical equation that was used successfully for sandstones [51]. It relates the specific surface area per unit of pore volume, s_{V_p}, to the porosity ϕ (fractional), permeability k (in millidarcies), and formation resistivity factor F_R (R_o/R_w, where R_o is equal to the electric resistivity of a formation 100% saturated with formation water and R_w is equal to the formation water resistivity):

$$S_{Vp} = \frac{2.11 \times 10^5}{\sqrt{F_R^{2.2} \phi^{1.2} k}} \tag{3.66}$$

The formation resistivity factor captures the effects of grain size, grain shape, grain distribution, and grain packing.

Example

An oil-bearing core sample recovered from a clean sandstone formation has a porosity of 24% and an irreducible water saturation of 30%. Estimate:

(a) The permeability of the core sample using the Wyllie and Rose correlation (Eq. (3.63)) and compare the result with that obtained from the Timur correlation (Eq. (3.64)).

(b) The average grain size.

Solution

(a) The permeability of an oil-bearing core sample according to the Wyllie and Rose correlation (Eq. (3.63)) is [48]:

$$k = \left(\frac{a_{wr}\phi^3}{S_{wi}}\right)^2$$

$$k = \left(250\frac{(0.24)^3}{0.30}\right)^2 = 133 \text{ mD}$$

The Timur correlation (Eq. (3.64)) gives:

$$k = 0.136\frac{\phi^{4.4}}{S_{wi}^2}$$

$$k = 0.136\frac{(0.24)^{4.4}}{(30)^2} = 179 \text{ mD}$$

The permeability obtained from the Timur equation is 25.7% higher than that obtained from the Wyllie and Rose equation. One of the reasons is that the Timur permeability–porosity correlation was obtained from core samples with high permeability.

(b) The average grain size can be estimated from Eq. (3.62). Solving for the grain diameter d_{gr} we have, for $k = 133$ mD:

$$k = 10.5 d_{gr}\phi^{3.3}$$

$$d_{gr} = \frac{k}{10.5\phi^{3.3}}$$

$$d_{gr} = \frac{133}{10.5(0.24)^{3.3}} = 1.4 \text{ mm}$$

For $k = 179$ mD, the grain diameter is $d_{gr} = 1.9$ mm. Thus, according to Table 1.7, the particles of the sandstone are very coarse and range in diameter from 1 mm to 2 mm.

Permeability from NMR and GR Logs

The NMR log uses a permanent magnet, a radio frequency (RF) transmitter, and an RF receiver. The tool responds to the fluids in the pore space and is used to measure lithology-independent effective porosity, pore size distribution, bound and moveable fluid saturation, and permeability on a foot-by-foot basis. Mathematical models, which include pore-size distribution, predict permeability more accurately than those that include effective porosity, since permeability is controlled by the pore throat size.

A small relaxation time from an NMR tool corresponds to small pores and a large relaxation time reflects the large pores. The distribution of the time constant T_2 in clastic rocks tends to be approximately log-normal. A good single representation of the T_2 is therefore obtained from the geometric or logarithmic mean value. Schlumberger-Doll Research (SDR) developed the following model for permeability [52]:

$$k = C_N T_{2ML}^2 \phi^4 \quad (3.67)$$

where

k = Permeability, mD
T_{2ML} = logarithm mean of relaxation (NMR) time T_2 distribution, also called the geometric mean of T_2, milliseconds. For brine: $1 < T_2 < 500$, for oil: $300 < T_2 < 1000$, and for gas: $30 < T_2 < 60$.
ϕ = NMR porosity, fraction
C_N = correlation constant, 4 for sandstones, 0.1 for carbonates.

Experience has shown that the mean T_2 model works very well in zones containing only water. However, if oil is present, the mean T_2 is skewed toward the bulk liquid T_2, and permeability estimates are erroneous. In unflushed gas zone, T_2 values are too low relative to the flushed gas zone, and permeability is consequently underestimated. Because hydrocarbon effects on T_{2ML} are not correctable, the mean T_2 model fails for hydrocarbon-bearing formations. Most NMR permeability–porosity correlations for carbonates assume that vugs do not contribute to permeability. Hidajat et al. [53] studied vuggy carbonate samples from a West Texas field using NMR and conventional core analysis. They concluded that in fractured carbonate formations, permeability estimates from the SDR model are too low because this model can represent only the matrix permeability. The SDR correlation is generally very useful for high permeability formations, but is not applicable in low permeability rocks.

The SDR model is sensitive to the presence of a hydrocarbon phase in the pores. T_2 response appears to be bimodal in water-wet rocks due to the partial presence of hydrocarbons (see Figures 3.36 and 3.37).

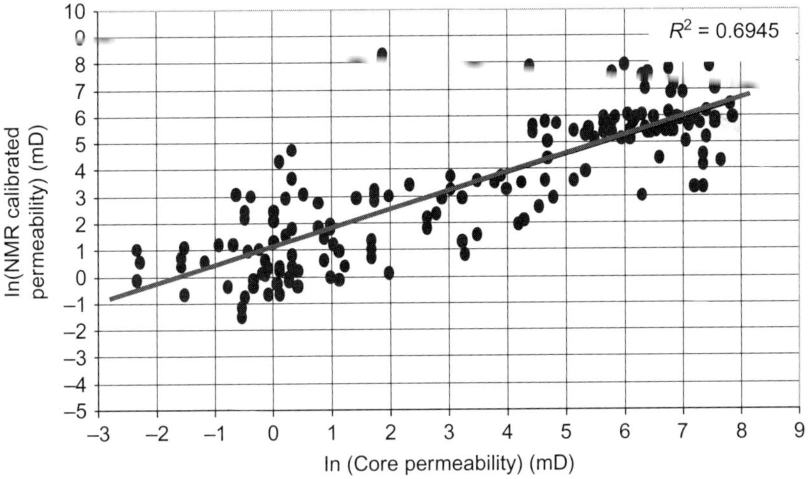

FIGURE 3.36 Core-measured permeability against calibrated NMR log-derived permeability, reproduced after Al-Ajmi and Holditch [52].

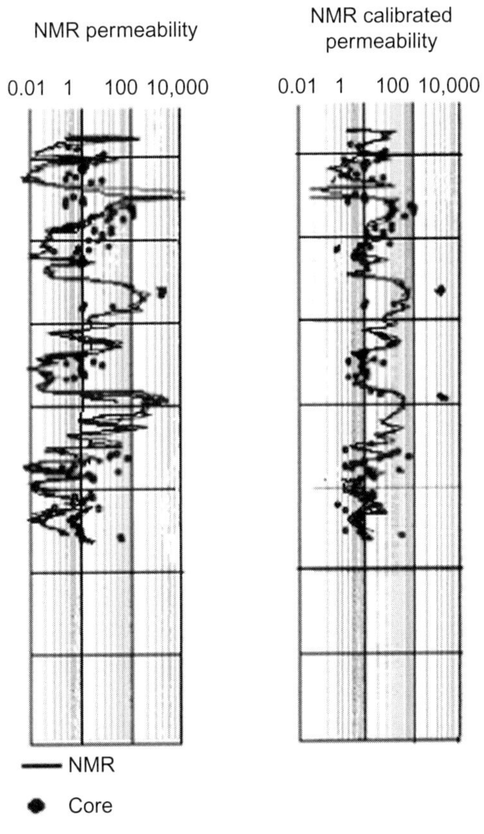

FIGURE 3.37 NMR permeability before and after calibration [52].

Example

An NMR log was run in an oil well located in a sandstone formation. The porosity is 0.10 and the log mean of relaxation time is 350 at the depth of 6,000 feet. Estimate the permeability at this depth for this well, using SDR correlations.

Solution

Using Eq. (3.67) a yields:

$$k = 4 \times 350^2 \times 0.10^4 = 49 \text{ mD}$$

PERMEABILITY FROM GR LOGS

The permeability of a sandstone formation depends upon its effective porosity, type, and amount of shale and clay content. Shokir [54] used the TSK fuzzy logic model [55] to estimate permeability in heterogeneous sandstone oil reservoirs using effective core porosity and gamma ray (GR) logs. According to Shokir "Fuzzy logic uses the benefits of approximate reasoning. Under this type of reasoning, decisions are made on the basis of fuzzy linguistic variables such as "low," "good," and "high," with fuzzy set operators such as "and" or "or." This process simulates the human expert's reasoning process much more realistically than do conventional expert systems. Fuzzy-set theory is an "efficient tool for modeling the kind of uncertainty associated with vagueness, imprecision, and/or a lack of information regarding a particular element of the problem at hand" [54]. He used a large core/log calibrated data set from a sandstone reservoir in the Middle East to develop the following permeability–porosity relationship:

$$\log(k) = C_{G1}\phi + C_{G2}\,\mathrm{GR} + C_{G3} \tag{3.67b}$$

where

k = permeability, mD
GR = gamma ray, API units, 20 (clean sands) $< \mathrm{GR} <$ 140 (very shaly formation)
ϕ = effective porosity, %
C_G = Correlation constants (Table 3.12)

The effective porosity is determined with sonic, density, and/or neutron logs.

Equation (3.67b) indicates that the permeability decreases as the GR value increases (as an indication of higher shale content), which is consistent with what is physically observed in cores. This correlation is strictly valid within the maximum and minimum values of the permeability, porosity, and GR values shown in Table 3.13.

TABLE 3.12 Correlation Constants for Eq. (3.67b)

	C_{G1}	C_{G2}	C_{G3}
High GR and low porosity	0.1769	−0.0165	−1.3986
Above medium GR and medium porosity	0.1486	−0.016	−0.9148
Medium GR and above medium porosity	0.1299	−0.0196	−0.2487
Low GR and high porosity	0.099	−0.0221	0.7877

TABLE 3.13 Range of Applicability of Equation 3.67b

	Minimum	Maximum	Average
GR, API units	7.64	83.56	29.5
Core Porosity (%)	5.1	36.9	26.84
Core Permeability (mD)	0.1	9,358	287.6

Example

Knowing GR = 39 API units and porosity = 20%, estimate permeability.

Solution

A GR reading of 39 API units is considered above medium. Using Eq. (3.67b) and the correlation constants in line of Table 3.12 yields:

$$\log(k) = 0.1486 \times 20 - 0.016 \times 39 - 0.9148 = 1.4332$$

The permeability is:

$$k = 10^{1.4332} = 27.1 \text{ mD}$$

Mohaghegh et al. [56] developed a robust model that could predict the permeability with only well log data for wells from which core data is not available. Using gamma ray log (GR, in API units), bulk density log (ρ_B, g/cc), and deep-induction log (I_D, Ohm-m) data from the Granny Creek field in West Virginia, which is a highly heterogeneous formation, they obtained the following multi-variable equation:

$$k = C_1 \mathrm{GR}^{C_2} \rho_B^{C_3} I_D^{C_4} \tag{3.67c}$$

where $C_1 = 38.254$, $C_2 = -0.5874$, $C_3 = -0.409438$, and $C_4 = 0.4066$.

This equation was used in other heterogeneous anisotropic sandstone fields and approximately matched the permeability profile obtained with core analysis.

Example

Estimate the permeability of a clean formation where GR = 22 API units, $I_D = 100$ Ohm m, $\rho_B = 2.55$ g/cm^3, and porosity = 18%.

Solution

Using Eq. (3.67c), the permeability is:

$$k = 38.2542 \times 22^{-0.5874} \times 2.55^{-0.409438} \times 100^{0.4066} = 27.6 \text{ mD}$$

Using Eq. (3.67b) gives:

$$\log(k) = 0.1486 \times 18 - 0.016 \times 22 - 0.9148 = 1.408$$

$$k = 10^{1.408} = 25.6 \text{ mD}$$

PERMEABILITY–POROSITY RELATIONSHIPS IN CARBONATE ROCKS

The relationship between permeability and porosity in carbonate rock formations is related to the grain size of the rock matrix, the size of intergranular pore space, the amount of unconsolidated vugs (fractures and solution cavities), and the presence or absence of connected vugs [57]. Figure 3.38 is a log–log plot of the permeability–porosity relationship for various particle size groups in the uniformly cemented nonvuggy rocks.

This plot indicates that there is a reasonably good relation between three petrophysical parameters and, therefore, if the particle size and the matrix porosity are known, the permeability (in millidarcies) of the nonvuggy portion of the carbonate rock can be estimated from:

$$k_{ma} = A_{gr} \phi_{ma}^{A_{mcp}} \tag{3.68a}$$

where

ϕ_{ma} = matrix porosity, fraction
A_{gr} = grain size coefficient, dimensionless
A_{mcp} = cementation-compaction coefficient, dimensionless.

The values of these coefficients are related to the average particle diameter d_{gr} as follows:

(1) For $d_{gr} < 20\ \mu$m, the values of A_{gr} and A_{mcp} average 1.5×10^3 and 4.18, respectively;
(2) If d_{gr} is in the range of 20–100, $A_{gr} = 2.60 \times 10^5$ and $A_{mcp} = 5.68$; and
(3) For $d_{gr} > 100\ \mu$m, the values of A_{gr} and A_{mcp} are 8.25×10^8 and 8.18, respectively.

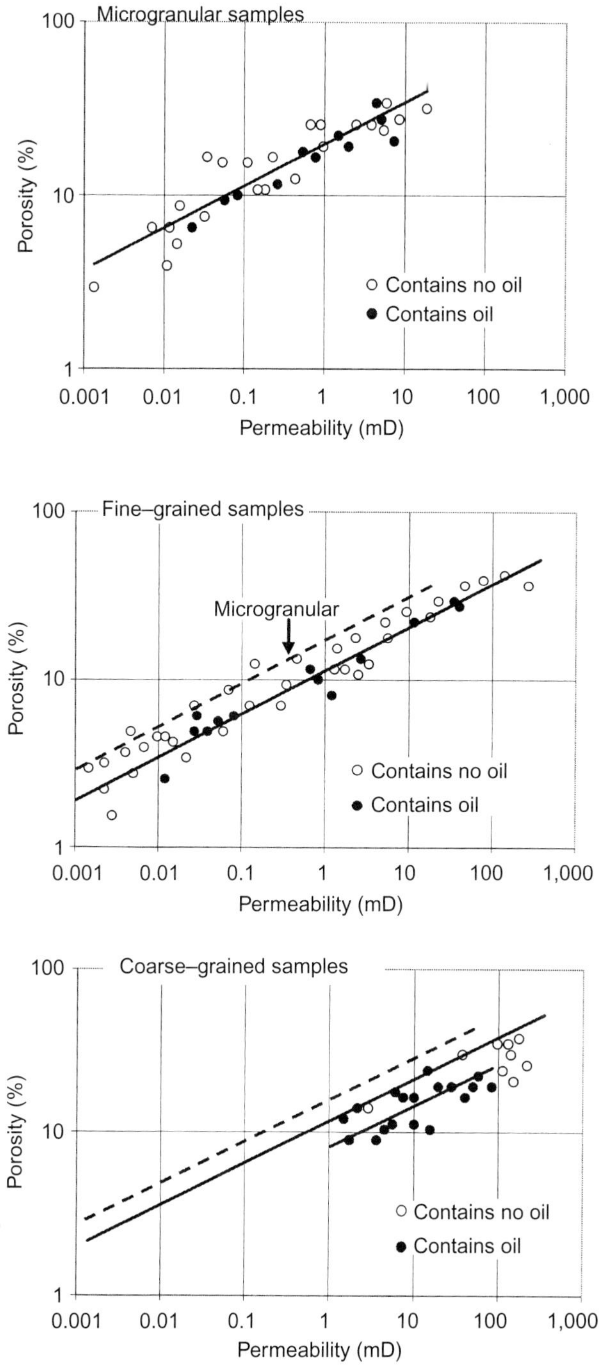

FIGURE 3.38 Effects of particle size on the permeability–porosity relationship in uniformly cemented, nonvuggy carbonate rocks [57].

If the distribution of compaction or cementation is not uniform, the constant A_{mcp} will be affected. Patchy cementation tends to yield higher values of A_{mcp}, thus reducing the permeability.

To quantify the effect of unconsolidated fractures and cavities on the inter-particle porosity, Lucia examined a large number of carbonate rocks and measured visually the fraction of the total matrix porosity due to these types of vugs [9]. It was found that their effect is to increase the interparticle matrix porosity with little or no increase in the permeability of the matrix. The following procedure is suggested for estimating the permeability in carbonate rocks containing unconnected vugs:

1. Measure the total porosity (interparticle and unconnected vugs), ϕ_t, from well logs or core analysis.
2. Estimate visually unconnected vug porosity, ϕ_u.
3. Calculate the intergranular porosity of the matrix (ϕ_{ma}) as:

$$\phi_{ma} = \frac{\phi_t - \phi_u}{1 - \phi_u} \tag{3.68b}$$

4. Estimate the average particle size, d_{gr}, using a compactor or micrometer.
5. Calculate the permeability of the nonvuggy matrix, k_{ma}, using Eq. (3.68).

A negligibly small increase in the permeability of the matrix (with unconsolidated vugs) will be observed if the total porosity, ϕ_t, is used in Eq. (3.68b) instead of ϕ_{ma}. Craze and Bagrintseva demonstrated the influence of lithology on the relationship between porosity and permeability [57,58]. On the basis of core data from cretaceous Edward limestone (Figure 3.39), Craze noted that as the texture changes from microgranular to coarse-grained, the permeability increases for a given porosity [57].

Bagrintseva investigated the interrelationships among various rock properties of several carbonate reservoirs in the former Soviet Union [58]. Chilingarian et al. used Bagrintseva's data and derived several useful correlations between permeability and porosity by considering two additional variables: irreducible fluid saturation and specific surface area [6]. The general form of the correlation is as follows:

$$\log k = a_1 + a_2\phi_o + a_3 S_{V_p} + a_4 S_{wr} + a_5 S_{V_p} S_{wr} \tag{3.69}$$

where

k = permeability to air, mD
S_{wr} = residual water saturation, %
S_{V_p} = specific surface area
ϕ_o = open porosity, %

a_1, a_2, a_3, a_4, and a_5 are constants for a given formation, determined empirically.

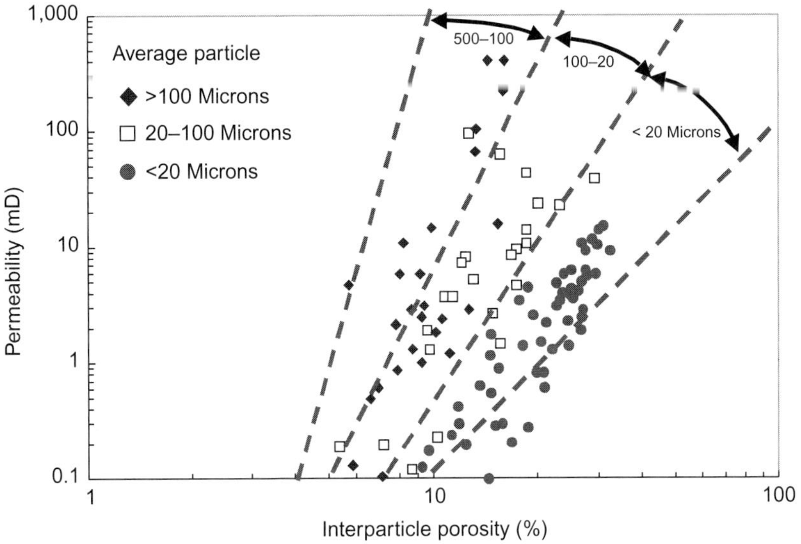

FIGURE 3.39 Relationship between porosity and permeability for various types of cretaceous Edwards limestone [57].

Mortensen et al. [59] showed analytically that Kozeny Eq. (3.37) is applicable in homogeneous reservoir chalk and the effective zoning factor K_T varies from 4.76 to 4.16 for porosity between 20% and 40%, respectively, where k = permeability, mD, ϕ = porosity, fraction, and $S_{V_{gr}}$ = specific surface per unit pore volume, μm.

Practically, all permeability–porosity correlations should be used only for qualitative purposes. To obtain an accurate correlation between the porosity and permeability, one must include a large number of physical factors that characterize a porous medium, including irreducible fluid saturation, specific surface area, grain size distribution, grainshape, packing and layering, lithology and mineralogy, degree and type of cementing, etc. Although some formations may show a correlation between permeability and porosity, a large number of physical factors influencing these two parameters differ widely in different formations.

Estimating Permeability in Carbonate Rocks

Although the absolute porosity provided by natural fractures is negligible ($<3\%$), the effective porosity is considerably enhanced because fractures connect the available pore volume. Consequently, the reservoir permeability and the petroleum recovery are greatly enhanced. The net impact of fracture connectivity may be a decisive factor in exploiting a particular reservoir.

Many methods have been proposed for estimating fracture permeability, including parallel-plate models, electric analog systems, core analysis, well logging, and pressure transient testing.

The equation for volumetric flow rate between the two smooth plates, combined with Darcy's law, provides the basic approach for estimating fracture permeability and its influence on fluid flow in naturally fractured rocks. Parsons used this approach to express the total permeability of the fracture–matrix system in which vertical fractures occur in sets of specified spacing and orientation relative to overall pressure gradient [60]. Murray used a parallel-plate model and a geometric approach applicable to folded rocks to demonstrate that, in folded beds with extension fractures normal to the bedding and parallel to the fold axis, the fracture porosity and permeability are functions of bed thickness and curvatures [61]. He assumed that extension fractures form primarily in the outer layers of curved beds. Murray applied this approach to the Spanish pool in McKenzie County, North Dakota, and demonstrated a good coincidence between areas of maximum curvature and areas of best productivity.

The flow of fluid through porous media is directly analogous to the flow of electricity. McGuire and Sikora used this analogy and showed that the width of artificial fractures is much more important than their length in affecting communication among natural fractures [62]. Steams and Friedman summarized that the permeability of a naturally fractured formation can be expected to be greatest where the reservoir bed contains wide, closely spaced, smooth fractures oriented parallel to the fluid pressure gradient [63].

Fracture permeability cannot be estimated directly from well logs. The modern trend is to combine core-derived parameters with computer-processed log data to establish a statistical relationship between the permeability of the matrix–fracture system and various parameters, such as porosity and irreducible water saturation. With such a relationship established, the formation's petrophysical parameters, including permeability distribution, can be deduced from log data alone in wells or zones without core data. In carbonate formations, however, where structural heterogeneity and textural changes are common, and only a small number of wells are cored because of the difficulty and cost of the coring, the application of statistically derived correlations is extremely limited. Watfa and Youssef developed a sound theoretical model that relates directly to the flow of path length (tortuosity), pore radius changes, porosity, and cementation factor, *m* [64]. This model assumes that:

(1) a porous medium can be represented by a bundle of tubes, as shown in Figure 3.40;
(2) the cross-sectional area of each tube, A_a, is constant; and
(3) the fluid path and the electric current path are same and the true conductivity, i.e., the reciprocal of resistivity, of the bundle of tubes is:

$$C_{tr} = C_w \phi^m \tag{3.70}$$

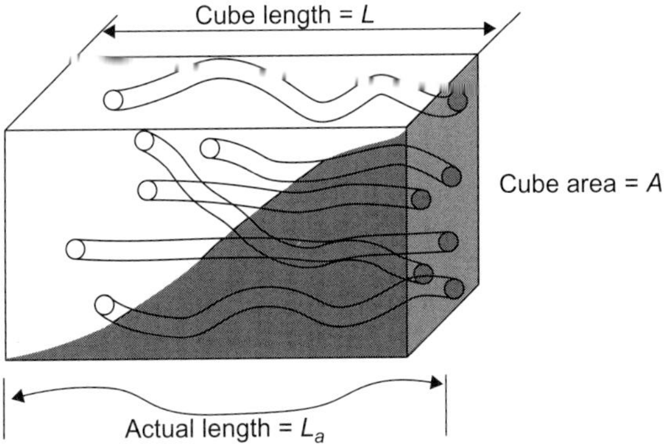

FIGURE 3.40 A bundle-of-tubes model [64].

where C_w is the water conductivity and m is the cementation factor. Because the apparent conductivity C_a of a block having a cross-sectional area A and length L is related to the true conductivity C_{tr} by the following expression:

$$\frac{C_a}{C_{tr}} = \frac{A}{L} \tag{3.71}$$

where A is the cross-sectional area of all tubes, then

$$C_a = C_w \phi^m \frac{A}{L} \tag{3.72}$$

Assuming that the bundle-of-tubes model contains n tubes, the conductivity of the ith tube (C_i) can be defined as:

$$C_i = C_w \frac{A_i}{L_i} \tag{3.73}$$

where L_i is the length of the ith tube.

The porosity of the tube is unity. The apparent conductivity of the block is the sum of the individual conductivities of all the tubes. Thus:

$$C_a = C_w \sum_{i=1}^{n} \frac{A_i}{L_i} \tag{3.74}$$

From the assumption that A_2 is constant:

$$C_a = C_w \frac{A_i}{L_i} \tag{3.75}$$

By definition:

$$\phi = n\left(\frac{L_a A_a}{LA}\right) \tag{3.76}$$

Combining Eqs. (3.70) and (3.75) yields:

$$\phi^{m-1}\left(\frac{L_a}{L}\right)^2 = \phi^{m-1}\tau = 1 \tag{3.77}$$

and the tortuosity is:

$$\tau = \phi^{1-m} \tag{3.78}$$

Using the same approach, the effects of flow path on the permeability can be evaluated. Applying Poiseiulle's equation to the ith tube, the flow rate in the ith tube, q_i, is equal to:

$$q_i = \left(\frac{\pi r_{\text{pai}}^4}{8\mu}\right)\frac{\Delta p}{L_i} \tag{3.79}$$

where r_{pai}, μ, and ΔP are, respectively, the apparent radius of the ith tube, fluid viscosity, and pressure differential across the unit block. For n tubes, the total flow rate q is:

$$q = \left(\pi\frac{\Delta p}{8\mu}\right)\sum_{i=1}^{n}\frac{r_{\text{pai}}^4}{L_i} \tag{3.80}$$

and, assuming A_2 is constant, the flow rate is:

$$q = n\left(\frac{\pi r_{\text{pa}}^4}{\mu}\right)\left(\frac{\Delta p}{8L_a}\right) \tag{3.81}$$

Applying Darcy's law to the unit block, the flow rate is equal to:

$$q = k_a A\left(\frac{\Delta p}{\mu L_a}\right) \tag{3.82}$$

Combining Eqs. (3.76), (3.77), (3.79)–(3.81), and solving for the apparent permeability of the block, k_a:

$$k_a = \left(\frac{r_{\text{pa}}^2}{8}\right)\phi^m \tag{3.83}$$

This equation is similar to Eq. (3.14) for $m = 1$. Combining Eqs. (3.83) and (3.78) gives:

$$k_a = \left(\frac{r_{\text{pa}}^2}{8}\right)\frac{\phi}{\tau} \tag{3.84}$$

Assuming $\tau = \phi F_R$, where F_R is the formation resistivity factor, Eq. (3.59) becomes:

$$k_a = \left(\frac{r_{\text{pa}}^2}{8}\right)\frac{1}{F_R} \tag{3.85}$$

Expressing tortuosity as $\tau = (\phi F_R)^2$, Eq. (3.84) results in:

$$k_a = \left(\frac{r_{pa}^2}{8}\right)\frac{1}{\phi F_R^2} \tag{3.86}$$

and, for $\tau = \phi(F_R)^2$, Eq. (3.84) gives:

$$k_a = \left(\frac{r_{pa}^2}{8}\right)\frac{1}{F_R^2} \tag{3.87}$$

These equations clearly indicate that no single correlation can be used to determine the formation permeability from logs alone.

If k_a is expressed in mD, r_{pa} in μm, Eq. (3.83) becomes:

$$k_a = 126.7 r_{pa}^2 \phi^m \tag{3.88}$$

Figure 3.41 is a semilog plot of this relationship. The Cartesian axis on this plot is ϕ^m instead of the conventional ϕ. The importance of including dimensions of the flow channels in developing $k - \phi$ relationships for carbonates is clearly demonstrated by this plot. Equation (3.88), which is also applicable to sandstones, is derived on the basis that the average pore radius of the flow channels remains constant along the length of the unit block. As shown in Figure 3.42(a), however, the true pore radius changes along the flow path length. The effect of changing cross-sectional area along the flow path can be evaluated by considering the system of Figure 3.42(b) as two resistors in series. The total conductivity C of this system is related to the two conductivities C_1 and C_2 by the parallel-conductivity equation:

$$\frac{1}{C} = \frac{1}{C_1} + \frac{1}{C_2} \tag{3.89}$$

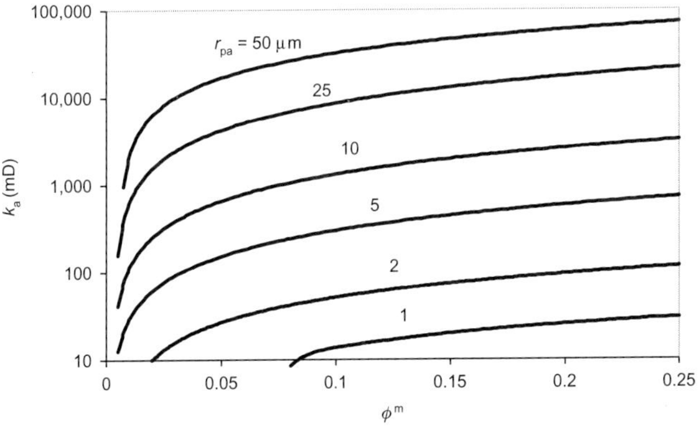

FIGURE 3.41 Variations of k_a, ϕ^m, and r_{pa} for an ideal system of tube bundles [64].

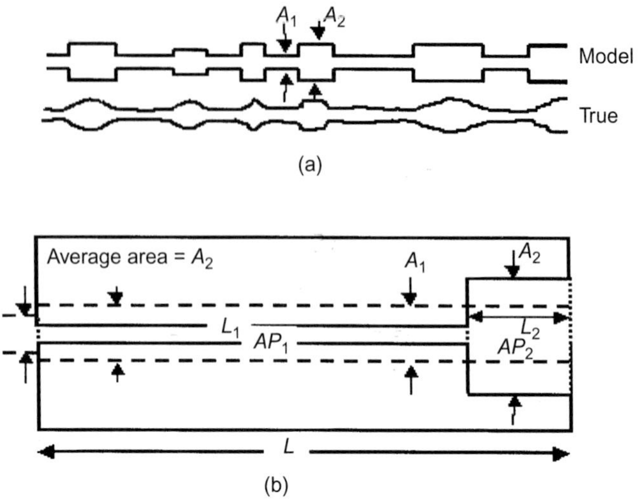

FIGURE 3.42 (a) and (b) Variations in flow path length and pore radius with variations in grain size [64].

Substituting Eq. (3.73) into the above expression, it can be shown that the change in conductivity caused by the change in pore radius is as follows:

$$C_p = \frac{C_a}{C} = \frac{L_r}{1 - A_r} + \frac{(1 - L_r)^2}{1 - L_r(1 - A_r)} \tag{3.90}$$

where $L_r = L_1/L$ (Figure 3.31(b)), $A_r = 1 - A_1/A_a$, and C_a is the apparent conductivity of the block such that Eq. (3.52) is true. Combining the Darcy and Poiseuille equations, it can be shown that the effect of pore-radius changes on the true permeability k is:

$$k_R = \frac{k_a}{k} = \frac{L_r}{1 - A_r} + \frac{(1 - L_r)^2}{1 - L_r(1 - A_r)} \tag{3.91}$$

where k_R is the ratio of apparent permeability to absolute permeability. Assuming $\Delta P = \Delta P_1 + \Delta P_2$ and $A_a L = A_1 L_1 + A_2 L_2$, Eq. (3.88) becomes:

$$k = 126.7 \frac{r_{pa}^2}{k_R} \phi^m \tag{3.92}$$

Letting r_{pe}, the effective pore radius, be equal to $r_{pa}/\sqrt{k_R}$, Eq. (3.92) becomes similar to Eq. (3.88):

$$k = 126.7 r_{pe}^2 \phi^m \tag{3.93}$$

The value of r_{pe} can vary considerably from the average radius value $\bar{r}$, depending on the texture and heterogeneity present in the system. Consider

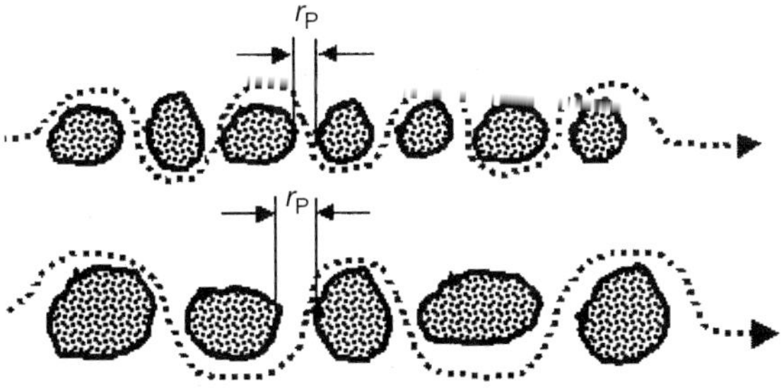

FIGURE 3.43 Two systems with different grain sizes and pore radii [64].

two systems with different grain sizes and with no vugs or fractures (Figure 3.43).

Because:

(a) the path of the current represents the true inter-matrix tortuosity,
(b) the tortuosity is a function of grain size and usually decreases with decrease in the grain size, and
(c) the value of r_{pe} varies with the variation in grain size,

a relationship between τ and r_{pe} must exist for a particular formation. Using experimental data of r_{pa}, m, ϕ, and k, Wafta and Youssef showed that r_{pe} and τ are related as follows [64]:

$$\log r_{pe} = a_1\sqrt{\tau} + a_2 \tag{3.94a}$$

where the coefficients a_1 and a_2 can be determined according to the following procedure:

(a) Obtain values of m and 0 from well logs and k from core analysis.
(b) Determine the cementation factor of the matrix m_m from:

$$m_m = \frac{m \log \phi}{\log(\phi - I_{S2})} \tag{3.94b}$$

where I_{S2} is the secondary porosity index, SPI, i.e., $\phi_t - \phi_{SL}$, where ϕ_t and ϕ_{SL} are, respectively, the total porosity and the sonic log porosity.

Figure 3.44 shows how to obtain m_m from a plot of the cementation factor m versus SPI. To compensate for the effects of fractures, data points for $I_{S2} < 1\%$ are not used to obtain m_m.

Inasmuch as the curve is not linear, one needs to be careful when extrapolating the curve to $I_{S2} = 1$ to obtain m_m on the m axis.

(a) Determine the value of the effective pore radius r_{pe} from Eq. (3.93).
(b) Calculate the tortuosity from Eq. (3.78).
(c) Establish a data bank for r_{pe} and τ, and plot $\log r_{pe}$ versus $\sqrt{\tau}$.

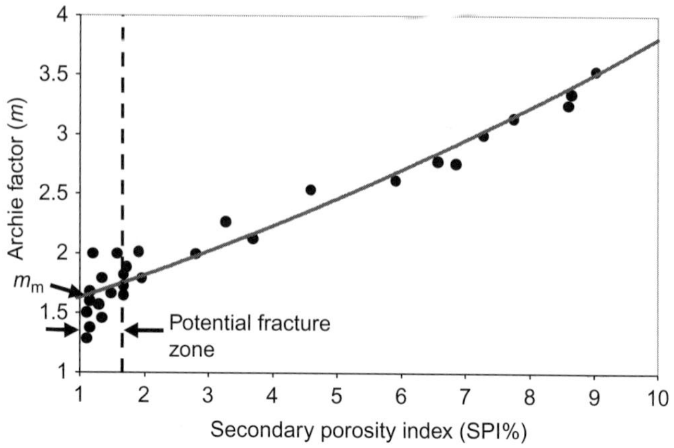

FIGURE 3.44 Estimation of the value of the inter-matrix Archie factor (m_m) from Archie factor (m) and secondary porosity index (I_{S2}) cross-plot [64].

(d) Draw the best-fit straight line. The general form of this line is given by Eq. (3.94b).

(e) Determine the correlation coefficient a_2 from the r_{pe} log-axis at $\sqrt{\tau} = 0$ and a_1 from the $\sqrt{\tau}$ axis at $r_{pe} = 1$.

Few permeability–porosity correlations generated in one reservoir can be extended to a different reservoir, particularly in naturally fractured carbonate reservoirs. These correlations are, however, very useful when it is necessary to populate matrix permeability for the purpose of constructing a geological model and reservoir simulation. The Mauddud reservoir in the Greater Burgan field, Kuwait, is a thin carbonate reservoir. Matrix permeability is low and natural fracture density is variable in this reservoir. Using data from 304 wells in the Mauddud reservoir Ambastha et al. [65] obtained the following matrix permeability–porosity transform:

$$k = 10^{C_1\phi - C_2} \tag{3.95}$$

where $k = \text{mD}$, $\phi = \text{fraction}$, $C_1 = 14.882$ and $C_2 = 2.863$. This correlation was tested in another field of similar characteristics as in Burgan. It was found that the factor C_1 reflects the fracture density, whereas C_2 is influenced by the degree to which the fractures are filled with minerals. It is important to emphasize that this correlation is only useful in carbonate reservoirs with low fracture density.

Because rock fabric changes tend to be systematically organized in a predictable manner within a sequence stratigraphic framework, Lucia [66] and Jennings and Lucia [28] showed that permeability for carbonates can best be estimated from well logs on the basis of rock-fabric.

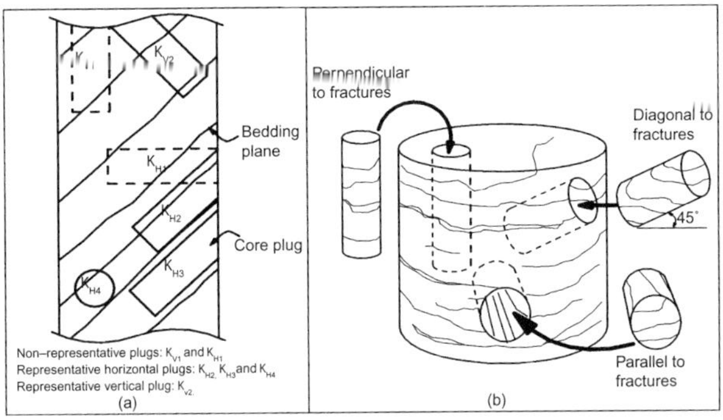

FIGURE 3.45 Orientation or core plugs used for measuring horizontal and vertical permeability.

DIRECTIONAL PERMEABILITY

In homogeneous reservoirs, permeability is assumed to be the same in all directions. However, in heterogeneous reservoirs, permeability in the x-direction may be considerably different than the permeability in the y and z directions. The net impact of such changing permeability in different directions on the natural recovery of a reservoir and the efficiency of a waterflood project can be of significant importance. Horizontal well test analysis and selective zonal well testing techniques provide the estimates of directional permeability. Discussion of horizontal well test analysis is beyond the scope of this book. Core samples are also analyzed for directional permeability in the laboratory. Usually, core plugs used for permeability measurement in the laboratory are cut perpendicularly from the main large core taken from the wellbore. However, to measure the vertical permeability, a core plug has to be cut in the direction of the main core taken from the wellbore, i.e., perpendicular to the bedding plane, as shown in Figure 3.45. The latest technological developments in well logging also provide the estimates of directional permeability.

Anisotropy

Directional permeability is frequently used to express the degree of heterogeneity in the formation. From the engineering point of view, the net effect of anisotropy is the loss or gain in effective permeability of a reservoir rock. Such loss or gain in effective permeability may be due to increased permeability in one direction and reduced permeability in other direction; thereby, the resulting average permeability is always less than the highest permeability in any direction in the reservoir. For example, reservoirs with vertical fractures have higher permeability in the vertical direction and low matrix permeability in the horizontal direction. Such variation in permeability is termed anisotropy.

$$I_A = \frac{k_V}{k_H} \tag{3.96}$$

Horizontal (k_H) and vertical (k_V) permeabilities are determined from core analysis on a regular basis. k_H and k_V can more accurately be determined from interference testing. Selective zonal well test analysis in the same wellbore is typically used to estimate vertical permeability. Partially penetrated or perforated wells (Figure 3.46), for instance, may develop a spherical flow

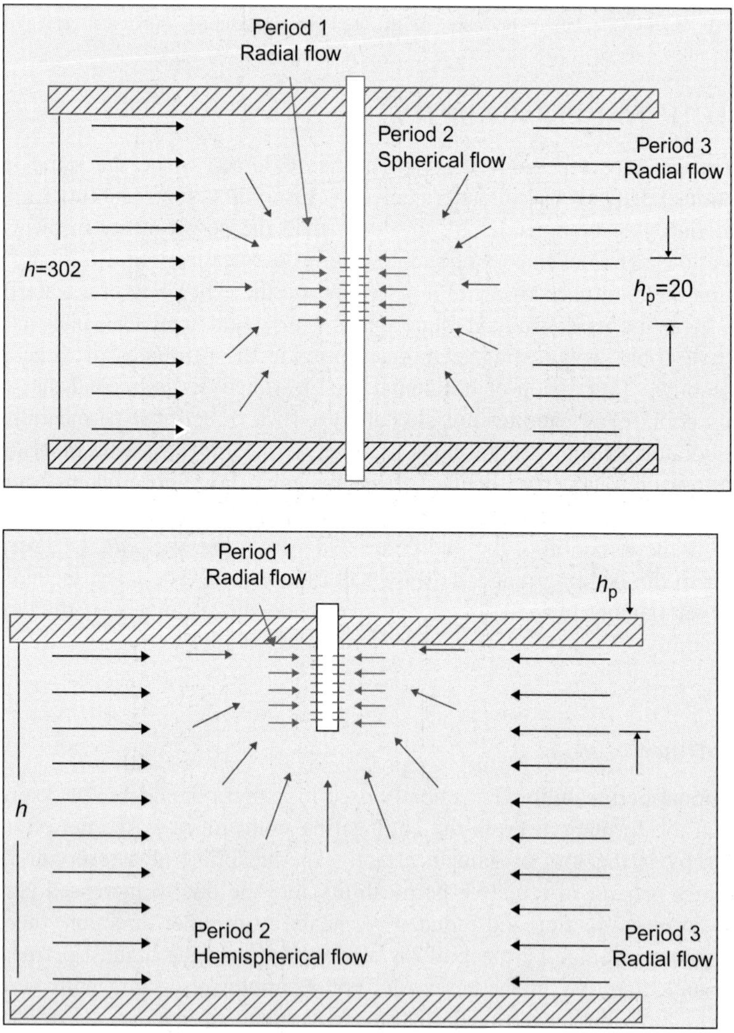

FIGURE 3.46 Partially penetrated/perforated well.

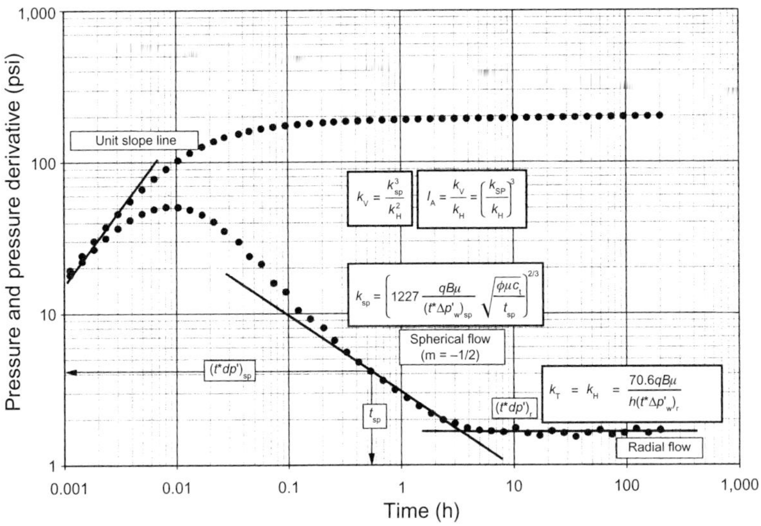

FIGURE 3.47 Pressure and pressure derivative curve indicating spherical flow from which vertical permeability is determined [67].

regime, which can be analyzed to estimate vertical and horizontal permeability as shown in Figure 3.47 [67].

Three flow periods can be identified from this plot:

Period 1: It corresponds to the initial radial flow over the completion interval.

During this period the reservoir behaves as if the formation thickness were equal to the length of the open zone. Period 1 is generally masked by wellbore storage effect.

Period 2: It corresponds to a transition period during which spherical flow (or hemispherical flow) may be identified.

Period 3: It corresponds to a second radial flow, but this time over the total formation thickness.

The following procedure, which is based on the Tiab's Direct Synthesis (TDS) Technique, can be used to calculate the values of k_H and k_V from a log–log plot of pressure and pressure derivative versus time.

Step 1. Calculate Δp and $(t \times \Delta p')$ versus test time.

It is important to remember that:

For a drawdown test:

$$\Delta p = p_i - p_{wf}$$

For a buildup test:

$$\Delta p = p_{ws} - p_{wf(at\ \Delta t = 0)}.$$

The derivative can be determined from:

$$t\left(\frac{\partial \Delta P}{\partial t}\right)_i = \left(\frac{\partial \Delta P}{\partial \ln t}\right)_i = \left\{\frac{\ln(t_i/t_{i-1})\,\Delta P_{i+1}}{\ln(t_{i+1}/t_i)\ln(t_{i+1}/t_{i-1})} + \frac{\ln(t_{i+1}t_{i-1}/t_i^2)\,\Delta P_i}{\ln(t_{i+1}/t_i)\ln(t_i/t_{i-1})} - \frac{\ln(t_{i+1}/t_i)\,\Delta P_{i-1}}{\ln(t_i/t_{i-1})\ln(t_{i+1}/t_{i-1})}\right\} \quad (3.97)$$

Step 2. Plot Δp and $(t \times \Delta p')$ versus test time
Step 2. Draw straight line with − ½ slope, which is the characteristic of the *spherical flow* (*and hemispherical flow*) *regime*.

Select *any convenient time*, t_{sp}, during spherical flow (or hemispherical flow) and read the corresponding value of $(t \times \Delta p')_{sp}$ from the pressure derivative curve.

Spherical permeability:

$$k_{sp} = \left(1227\frac{qB\mu}{(t \times \Delta p'_w)_{sp}}\sqrt{\frac{\phi\mu c_t}{t_{sp}}}\right)^{2/3} \quad (3.98a)$$

Hemispherical permeability:

$$k_{hs} = \left(2453\frac{qB\mu}{(t \times \Delta p'_w)_{hs}}\sqrt{\frac{\phi\mu c_t}{t_{hs}}}\right)^{2/3} \quad (3.98b)$$

Step 3. Read the value of the pressure derivative corresponding to the infinite acting radial (horizontal) flow line: $(t \times \Delta p')_R$
Step 4. Calculate the horizontal permeability:

$$k_H = \frac{70.6q\mu B}{h(t \times \Delta P')_r} \quad (3.99)$$

Step 5. Calculate the vertical permeability from:

$$k_V = \frac{k_{sp}^3}{k_H^2} \quad (3.100a)$$

or

$$k_V = \frac{k_{hs}^3}{k_H^2} \quad (3.100b)$$

Relationship Between k_H and k_V

The relationship between different petrophysical properties and fluid saturation is well established for clean sandstone rocks. Several empirical models have been developed to calculate water saturation, and other reqired parameters, for the evaluation of clean reservoirs. Vertical permeability in the formation is normally different from horizontal permeability, even when the system is homogenous. Such vertical anisotropy effects are generally the result of depositional environment and post-depositional compaction history of the formation.

Clean Sandstone Formations

In shale-free sandstone formations grain size, shape factor, and particle orientation are the most important factors in the $k_V - k_H$ relationship (see Figures 3.12–3.14). Tiab and co-workers and Zahaf and Tiab correlated horizontal permeability and vertical permeability for the lower Devonian sandstone from Illizi Basin, Algeria, as shown in Figure 3.48, and obtained the following correlation [68,69,89]:

$$k_v = 0.0429\left(\sqrt{\frac{k_H}{\phi_e}}\right)^{2.4855} \tag{3.101a}$$

where

k_v = vertical permeability, mD
k_H = horizontal permeability, mD
ϕ_e = effective porosity, fraction

This equation indicates a strong relationship between the mean hydraulic radius and the vertical permeability.

Figure 3.48 indicates an excellent correlation between the calculated values of k_V, using Eq. (3.97), and core-measured k_V. Figure 3.49 shows a plot of core vertical permeability values versus the product of the mean grain diameter and mean hydraulic radius. The curve fit between these two parameters is as follows:

$$k_v = 13.336\left(d_{gr}\sqrt{\frac{k_H}{\phi_e}}\right)^{1.333} \tag{3.101b}$$

The Coates and Denoo model takes into account porosity and irreducible water saturation in the estimation of horizontal permeability [37]. Their correlation for estimating horizontal permeability from porosity and irreducible water saturation is:

$$k_H = (10\phi_e)^4\left(\frac{1 - S_{wi}}{S_{wi}}\right)^2 \tag{3.102a}$$

Substituting Eq. (3.99) into Eq. (3.101a) and simplifying yields:

$$k_v = 4.012 \times 10^3 \phi_e^{3.728}\left(\frac{1 - S_{wi}}{S_{wi}}\right)^{2.4855} \tag{3.102b}$$

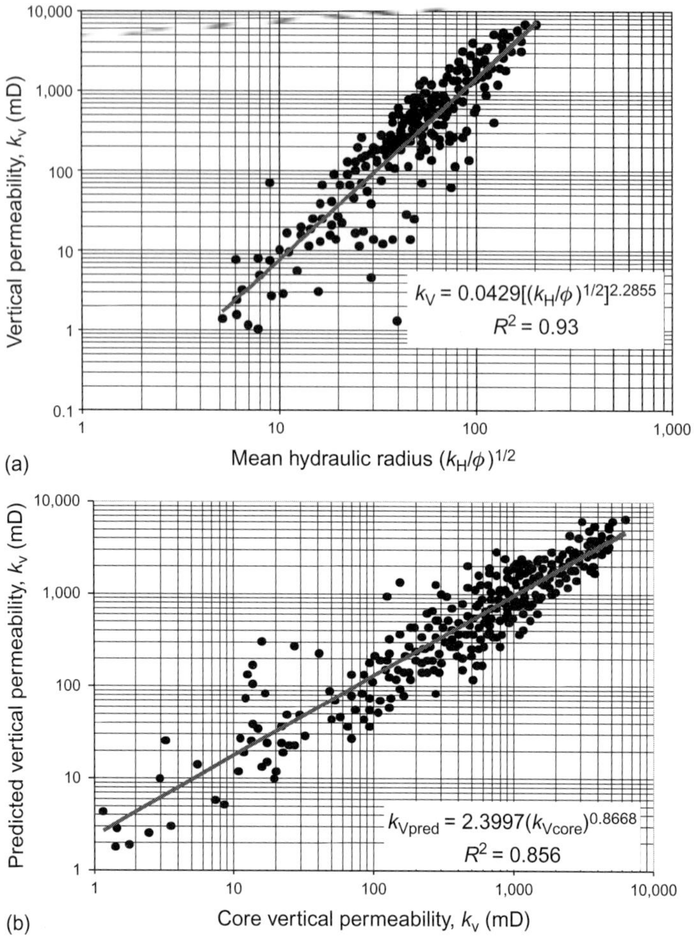

FIGURE 3.48 (a) Relationship between mean hydraulic radius $\sqrt{(k_H/\phi)}$ and vertical permeability in lower Devonian sandstone from Illizi Basin, Algeria. (b) Relationship between vertical permeability measured on core samples and vertical permeability predicted by Eq. (3.97) [68].

Equation (3.100) can be used to develop a vertical permeability profile in the well, using the S_{wi} and ϕ from logs in clean sandstone rocks.

Shaly Sandstone Formations

Permeability in shaly heterogeneous formations is extremely influenced by the nature of shale distribution in the rock. Shale exists in dispersed and laminated forms. Overall reservoir quality in heterogeneous sandstones is controlled by diagenesis, dissolution of feldspars and carbonate, crystal feeding, mineralogical redistribution of clay, and various cementation processes. Tiab

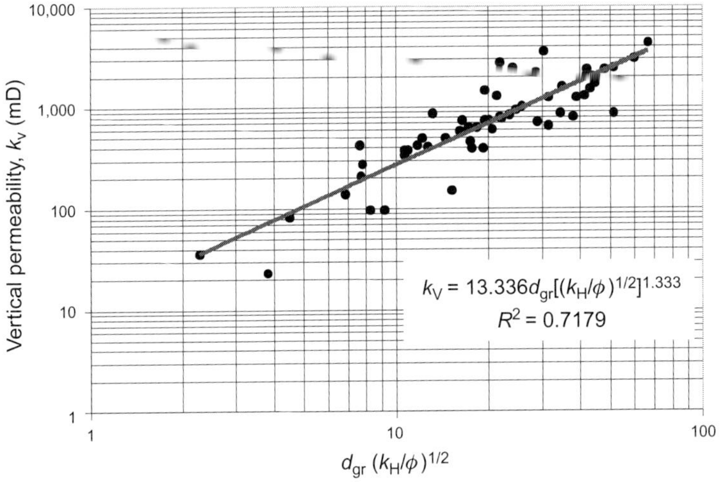

FIGURE 3.49 Relationship between vertical permeability and the product of average grain diameter and mean hydraulic diameter in lower Devonian sandstone from Illizi Basin, Algeria.

and co-workers related vertical permeability to the mean hydraulic radius [69]. They defined three general forms of correlation between vertical and horizontal permeability as follows:

(a) Vertical permeability as a function of hydraulic mean radius:

$$k_v = A_1\left(\sqrt{\frac{k_H}{\phi_e}}\right)^{B_1} \tag{3.103a}$$

(b) Vertical permeability as a function of clay content:

$$k_v = A_2(1 - V_{sh})\left(\sqrt{\frac{k_H}{\phi_e}}\right)^{B_2} \tag{3.103b}$$

(c) Vertical permeability as a function of mean grain size:

$$k_v = A_3 d_{gr}\left(\sqrt{\frac{k_H}{\phi_e}}\right)^{B_3} \tag{3.103c}$$

where A_1, A_2, A_3, B_1, B_2, and B_3 are coefficients and have to be determined for specific formation.

Field Example

Trias Argileux Greseux Inferior (TAGI) sandstone is a fluvial formation located in Algeria. The Triassic depositional environment involves facies changes as well as reservoir extension. Sandstone units of TAGI formation are multilayered producing zones, isolated by clay intercalation from flood plain deposition. TAGI has long been producing in Algeria in various basins with porosities ranging from 10% to 21% and often exceeding these values. Horizontal permeability ranges from 10 mD to 100 mD.

On the basis of a radio-crystallography study of the clay fraction, TAGI may be subdivided into two parts:

(a) The first part is characterized by the presence of relatively equal Kaolinite–Illite content.
(b) The second part is characterized by a high content of Illite (80–90%) and only traces of Kaolinite.

Figure 3.50 shows the results of a mineralogical study of TAGI. It is clear that TAGI is composed of very fine sandstone and has three types of porosity: inter-granular, dissolution, and fissured.

A log–log plot of vertical permeability versus horizontal permeability values measured on cores obtained from the TAGI formation yielded the following correlations (Figure 3.51):

(a) Equal Kaolinite–Illite content:

$$k_v = 0.598 k_H^{0.9707} \tag{3.104}$$

(b) High Illite content:

$$k_v = 0.159 k_H^{0.6675} \tag{3.105}$$

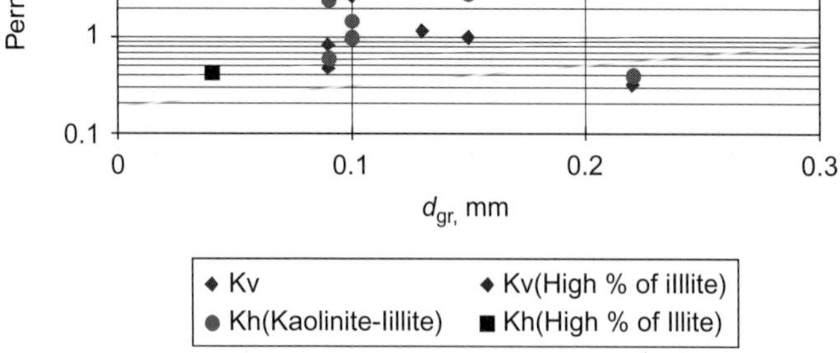

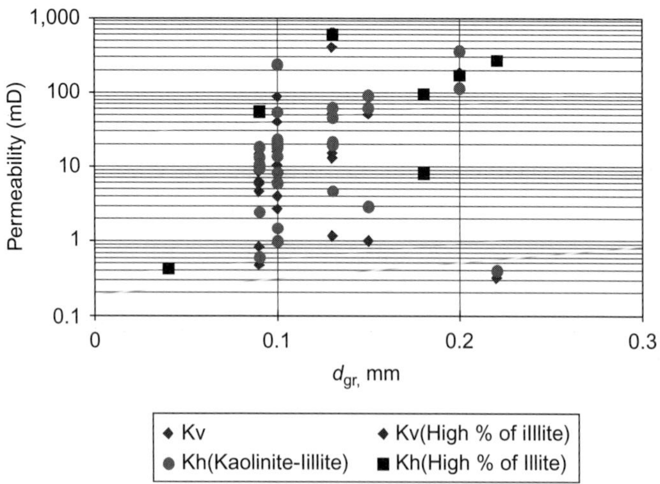

FIGURE 3.50 Permeability-mean grain size relationship in TAGI formation (kaolinite–illite).

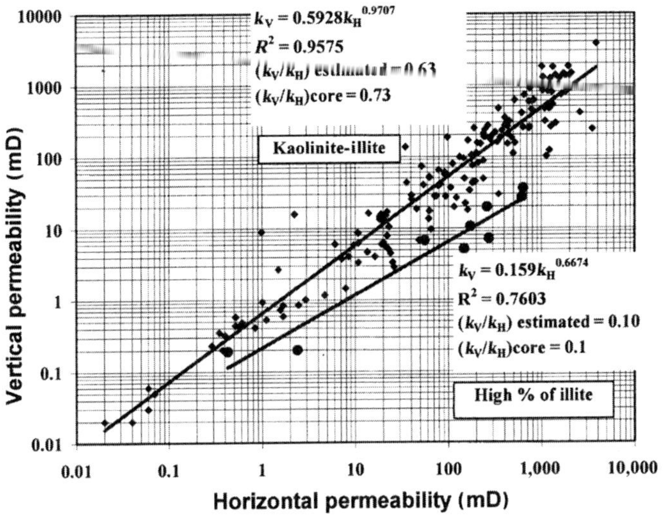

FIGURE 3.51 Vertical–horizontal permeability relationship in the TAGI formation (kaolinite–illite).

Both k_H and k_v are in millidarcies. The samples that have equal proportions of Kaolinite and Illite have a higher slope (0.97) than those with a higher content of Illite (0.66). From the regression lines, the ratios of anisotropy k_v/k_H are, respectively, 0.63 and 0.10. This implies that the k_v/k_H ratio decreases with increase in Illite content.

Peffer and O'Callagan published some values of k_v/k_H obtained from a formation dynamic tester (FDM) in the TAGI formation [70]. A good agreement was observed from the comparison of results of the formation tester and core analysis for both laminated and massively bedded sandstone rock types. Additionally, plotting vertical permeability versus the mean hydraulic radius yields two trends depending on Illite content in the formation, as indicated by Figure 3.52 which shows two trendlines represented by the following equations:

(a) Equal Kaolinite and Illite content:

$$k_v = 0.0535\left(\sqrt{\frac{k_H}{\phi}}\right)^{2.1675} \tag{3.106}$$

(b) High percentage of Illite:

$$k_v = 0.049\left(\sqrt{\frac{k_H}{\phi}}\right)^{1.3939} \tag{3.107}$$

Permeability in Eqs. (3.106) and (3.107) is in millidarcy. The slope of the straight line decreases with increase in Illite content. A good fit ($R^2 = 0.77$)

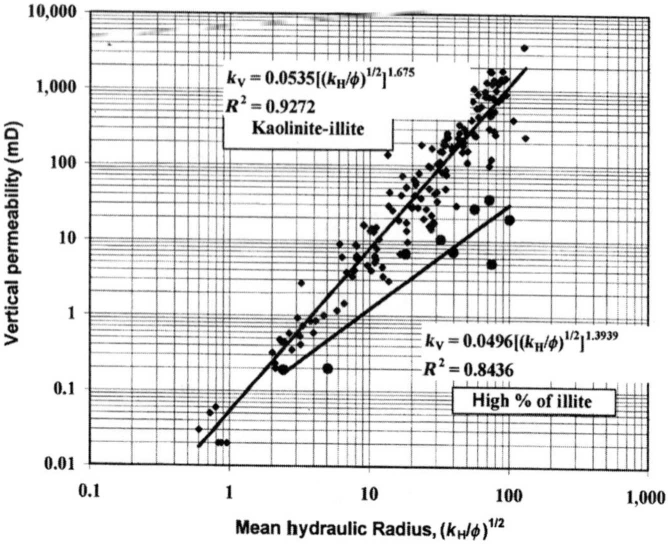

FIGURE 3.52 Relationship between vertical permeability and mean hydraulic radius in TAGI formation (kaolinite–illite).

for both samples having equal Kaolinite and Illite and the sample having a high percentage of Illite is also evident.

Other correlations have also been established by plotting vertical permeability versus $(1 - V_{sh})\sqrt{k_H/\phi_e}$ (Figure 3.53) and $d_{gr}\sqrt{k_H/\phi_e}$ (Figure 3.54) to show the impact of shale and grain size on the prediction of vertical permeability.

The following k_V–k_H correlations are applicable in formations containing Kaolinite and Illite.

(a) Equal contents of Kaolinite and Illite:

$$k_v = 0.1283\left((1 - V_{sh})\left(\sqrt{\frac{k_H}{\phi}}\right)\right)^{1.9658} \tag{3.108}$$

$$k_v = 7.7445\left(d_{gr}\left(\sqrt{\frac{k_H}{\phi}}\right)\right)^{1.8009} \tag{3.109}$$

(b) High percentage of Illite:

$$k_v = 0.0461\left((1 - V_{sh})\left(\sqrt{\frac{k_H}{\phi}}\right)\right)^{1.58} \tag{3.110}$$

$$k_v = 2.5054\left(d_{gr}\left(\sqrt{\frac{k_H}{\phi}}\right)\right)^{0.9383} \tag{3.111}$$

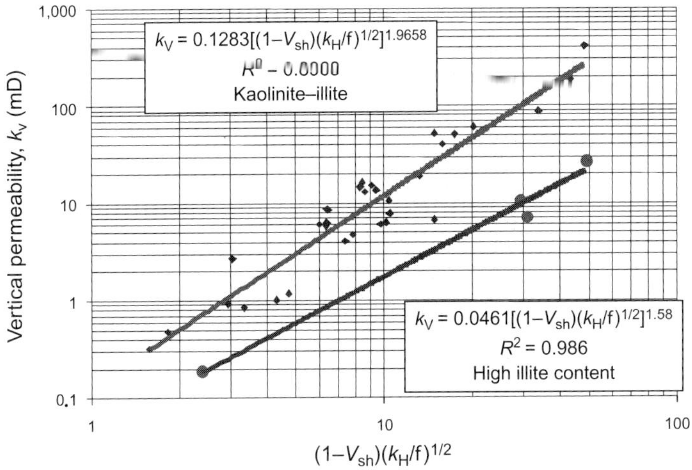

FIGURE 3.53 Relationship between vertical permeability and mean hydraulic radius in the TAGI formation (kaolinite–illite) with V-shale correlation.

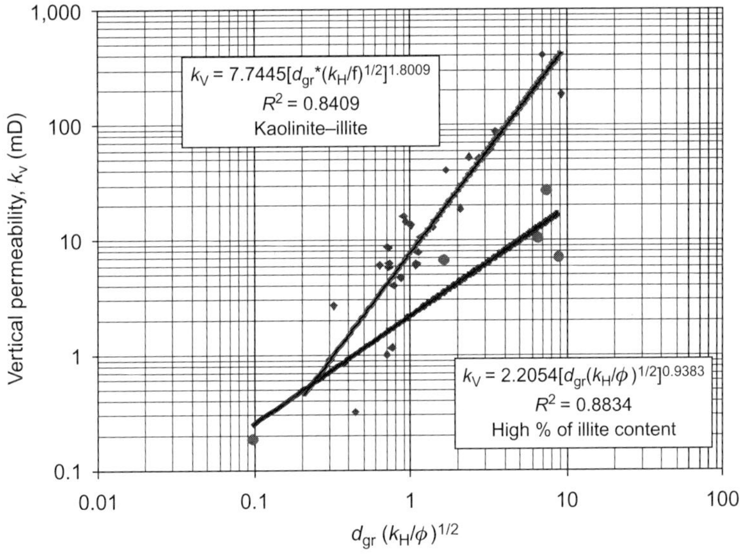

FIGURE 3.54 Relationship between vertical permeability and mean grain size in the TAGI formation (kaolinite–illite).

where k_H and k_v are in millidarcy, ϕ is the effective porosity (fraction), and d_{gr} is in centimetre. Figure 3.55 shows an acceptable match of k_v and k_H values obtained from well logs and core measurements.

All the above correlations relating vertical permeability to horizontal permeability have a strong physical meaning in the sense that they depict the

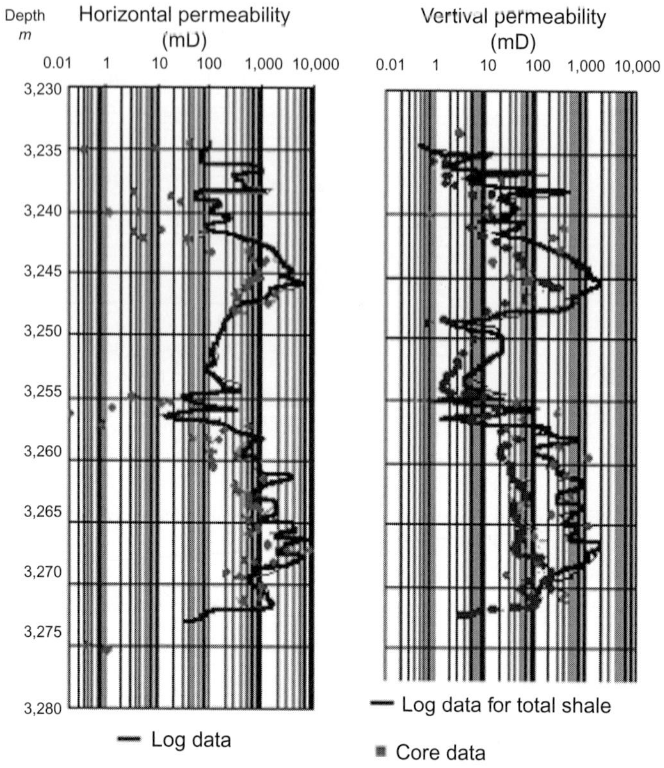

FIGURE 3.55 Core- and log-derived permeability for the TAGI formation [69].

degree of anisotropy, expressed by Eq. (3.96), in a formation containing various types of shale. This is very critical to the efficient field development and productivity of the formation.

Example

Core analysis in a new well in the TAGI formation revealed a horizontal permeability of 0.58 mD. Assuming that the TAGI is an equal Kaolinite–Illite formation, estimate the vertical permeability and anisotropy ratio for this well using the models developed for this formation for the following cases:

(a) Only horizontal permeability is known.

(b) Only horizontal permeability and porosity data are available (porosity = 11%).

(c) Horizontal permeability, porosity, and average grain diameter data are available (grain diameter = 0.085 cm).

Repeat the above example assuming that the TAGI formation contains a high percentage of Illite clay.

Solution

(1) Equal Kaolinite–Illite model

(a) Using Eq. (3.104):

$$k_V = 0.598 k_H^{0.9707} = 0.598 \times 0.58^{0.9707} = 0.35 \text{ mD}$$

Anisotropy ratio, Eq. (3.96):

$$I_A = \frac{k_V}{k_H} = \frac{0.35}{0.58} = 0.60$$

(b) Using Eq. (3.106):

$$k_V = 0.0535\left(\sqrt{\frac{k_H}{\phi}}\right)^{2.1675} = 0.0535\left(\sqrt{\frac{0.58}{0.11}}\right)^{2.1675} = 0.35 \text{ mD}$$

$$I_A = \frac{k_V}{k_H} = \frac{0.35}{0.58} = 0.60$$

(c) Using Eq. (3.109):

$$k_V = 7.7445\left(d_{gr}\sqrt{\frac{k_H}{\phi}}\right)^{1.8009} = 7.7445\left(0.085\sqrt{\frac{0.58}{0.11}}\right)^{1.8009} = 0.40 \text{ mD}$$

$$I_A = \frac{k_V}{k_H} = \frac{0.40}{0.58} = 0.70$$

(2) Assuming high percentage Illite model

(a) Using Eq. (3.105):

$$k_v = 0.159 k_H^{0.6675}$$

$$k_v = 0.159(0.58)^{0.6675} = 0.11 \text{ mD}$$

$$I_A = \frac{I_V}{I_H} = \frac{0.11}{0.58} = 0.19$$

(b) Using Eq. (3.107):

$$k_V = 0.049\left(\sqrt{\frac{k_H}{\phi}}\right)^{1.3939} = 0.049\left(\sqrt{\frac{k_H}{\phi}}\right)^{1.3939} = 0.15 \text{ mD}$$

$$I_A = \frac{k_V}{k_H} = \frac{0.15}{0.58} = 0.26$$

(c) Using Eq. (3.111):

$$k_V = 2.5054\left(d_{gr}\sqrt{\frac{k_H}{\phi}}\right)^{0.9383} = 2.5054\left(0.058\sqrt{\frac{0.58}{0.11}}\right)^{0.9383} = 0.37 \text{ mD}$$

$$I_A = \frac{k_V}{k_H} = \frac{0.37}{0.58} = 0.64$$

It is clear from the anisotropy values that high Illite content formations are more anisotropic. Incorporating the porosity and grain diameter reduces the scatter and predicts anisotropy more reliably, as confirmed from transient pressure test analysis.

RESERVOIR HETEROGENEITY

Heterogeneity is viewed on a broader scale than anisotropy. The degree of variation in the petrophysical properties of petroleum-bearing rocks varies from pore level to field level. Consequently, petrophysical properties are better understood by using the scales of heterogeneity.

Microscopic Heterogeneity

The microscopic scale of heterogeneity represents the scale volume at which the rock properties such as porosity and permeability are determined by (1) grain size and shape; (2) pore size and shape; (3) grain, pore size, and pore throat distribution; (4) packing arrangements; (5) pore wall roughness; and (6) clay lining of pore throats. The major controls on these parameters are the deposition of sediments and subsequent processes of compaction, cementation, and dissolution. Microscopic scale parameters are measured using scanning electron microscope (SEM), pore image analysis (PIA), magnetic resonance imaging (MRI), and NMR.

Macroscopic Heterogeneity

Core analysis represents the domain scale of macroscopic heterogeneity. Laboratory measurement of porosity, permeability, fluid saturation, capillary pressure, and wettability are physically investigated at the macroscopic level. Rock and fluid properties are determined to calibrate logs and well tests for input into reservoir simulation models.

Mesoscopic Heterogeneity

Information on this scale of heterogeneity is collected from well logs. Well logs are represented at grid cell scale in the reservoir simulation where variation in rock and fluid properties, along with small-scale geological features, is averaged to be assigned single values for the whole grid block. Core calibrated well logs are used to (1) establish the correlations and compatibility between the measured parameters; (2) integrate downhole measurements with data from pore studies, core analysis, and geophysical surveys through interscale reconciliation (using upscaling functions); (3) identify lithofacies; (4) relate and integrate petrophysical interpretation with geochemical, sedimentological, stratigraphic, and structural information; and (5) contour different reservoir parameters such as porosity, permeability, net thickness, tops and bottoms, fluid saturation, and fluid contact.

Megascopic Heterogeneity

This scale of heterogeneity represents the flow units, usually investigated through reservoir simulation. In fact, reservoirs are engineered and managed at this scale of interwell spacing, which is commonly inferred from transient pressure well test analysis, tracer tests, well logs correlations, and high resolution seismic (3-D seismic, conventional and reverse VSP (vertical seismic profile), cross-well seismic, and 3D AVC (advanced video coding)).

Megascopic heterogeneity determines well-to-well recovery variation and is the result of primary stratification and internal permeability trends within reservoir units. It is at this scale that internal architecture and heterogeneity become critical for identifying the spatial distribution of reservoir flow units. Examples of megascopic heterogeneities include (1) lateral discontinuity of individual strata; (2) porosity pinch-outs; (3) reservoir fluid contacts; (4) vertical and lateral permeability trends; (5) shale and sand intercalation; and (6) reservoir compartmentalization (see Figure 3.56).

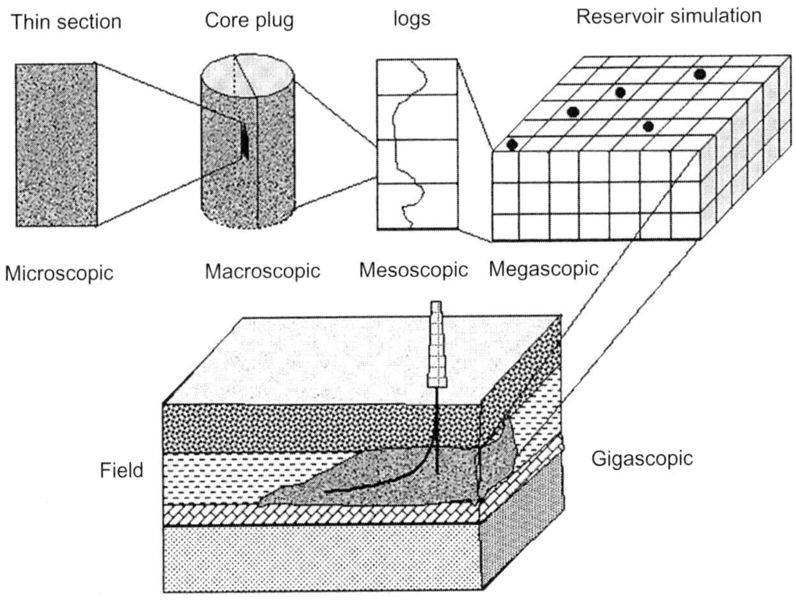

FIGURE 3.56 Scales of reservoir heterogeneity.

Gigascopic Heterogeneity

The whole field (depositional basin) is encompassed in this largest scale of heterogeneities. Reservoirs are explored for, discovered, and delineated at this level. This gigascopic field-wide scale, utilized to define the reservoir outline, is the domain of structural and stratigraphic seismic interpretation along with conventional subsurface mapping.

Hydrocarbon reservoirs are inferred from anomalies in the seismic surveys. Characterization at this level begins from inter-well spacing and extends up to the field dimensions. Field-wide regional variation in the reservoir architecture is caused by either original depositional setting or subsequent structural deformation and modification due to the tectonic activity. Examples of types of information obtained from this megascopic heterogeneity are (1) division of reservoir into more than one producing zones or reservoirs; (2) position, size, shape, architecture, and connectivity of facies or reservoir units; (3) evaluation of the spatial distribution or lithologic heterogeneity that comprises barriers, baffles, widespread sealing bed unconformities, and high-permeability zones; (4) large-scale structural features of folds and faults; and (5) the relationship of lithofacies to depositional environment and hydraulic flow units.

DISTRIBUTION OF ROCK PROPERTIES

Reservoir rocks are seldom if ever found to be homogeneous in physical properties or uniform in thickness. Variation in the geologic processes of erosion, deposition, lithification, folding, faulting, etc. dictate that reservoir rocks be heterogeneous and nonuniform. Although engineers have been producing oil and gas from reservoirs for more than a century, they are still inadequately informed about the distribution of reservoir rock properties. For a few locations in a reservoir, the mineral composition of the rock is known. Beyond this point, the real knowledge becomes sparse. The overall problem, as stated partly by Hutchinson et al., can be best expressed by the following three questions [71]:

(1) How can heterogeneities be identified and classified as to extent and geometry?
(2) How can the extent and geometry of heterogeneities within a specific reservoir be predicted?
(3) How can the performance of heterogeneous reservoirs be predicted with confidence?

Considerable progress in the field of numerical methods and computer modeling during the past 50 years has provided very useful answers to these three questions. Unfortunately, a necessary condition for the practical use of these models is that the reservoir be adequately described. In spite of all the advances in core analysis, well logging, geostatistics, and in particular well testing, petroleum engineers are still unable to specify the nature and extent of heterogeneities at every point in the formation. Warren and Price stated that "In many cases, the predicted performance of a reservoir is so completely dominated by irregularities in the physical properties of the formation that the gratuitous assumption of a particular form for the variation can reduce the solution of the problem to a mere tautological exercise" [72]. Fortunately, however, although all porous media are microscopically heterogeneous, only macroscopic variations of the rock need to be considered because the fundamental concepts of fluid flow in porous media are based on macroscopic quantities. Inasmuch as rock samples are usually available only from a small portion of the total reservoir, it seems logical that if measurements from these samples were to be used to infer the properties of the actual reservoir, the data should be treated statistically.

PERMEABILITY AND POROSITY DISTRIBUTIONS

For many reservoir applications, e.g., predicting reservoir performance during waterflooding, a quantitative description of the vertical heterogeneity of the formation is necessary. Geostatistical methods are often used to describe the spatial distribution of many reservoir parameters, including permeability and porosity. Modern geostatistical procedures, using powerful computers, can be used to (a) interpolate and extrapolate the permeability of unsampled locations, and (b) provide the quantitative relationship describing the spatial distribution of permeability. Kelkar and Perez [73] describe fundamental concepts of geostatistics applicable in petroleum reservoirs. The two widely used statistical parameters used to describe the degree of heterogeneity are: the Lorenz coefficient L_K and the Dykstra–Parsons permeability variation V_K.

Lorenz Coefficient L_K

The first practical attempt to statistically analyze the fluctuations of rock properties was reported by Law [74]. He demonstrated that porosity has a normal frequency distribution and that permeability has a log-normal frequency distribution. Using Figure 3.57, Zahaf and Taib and Schmalz and Rahme proposed the Lorenz coefficient, L_K, for characterizing the permeability distribution [69,75]:

$$L_K = \frac{\text{Area ABCA}}{\text{Area ADCA}} \tag{3.112}$$

The value of L_K ranges from zero to one. The reservoir is considered to have a uniform permeability distribution if $L_K \approx 0$. The reservoir is considered to be completely heterogeneous if $L_K \approx 0$. This coefficient, however, is not unique to a particular reservoir because different permeability distributions can yield the same value of L_K.

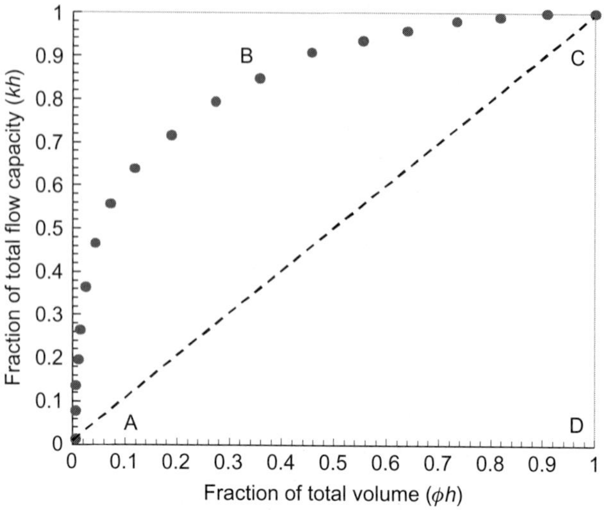

FIGURE 3.57 Flow capacity distribution [76].

The Lorenz coefficient is calculated as follows:

(1) Tabulate thickness h, permeability k, and porosity ϕ
(2) Arrange permeability data in a descending order
(3) Calculate the cumulative permeability capacity $\sum(kh)_i$ and cumulative capacity volume $\sum(\phi h)_i$
(4) Calculate the normalized cumulative capacities $C_k = \sum(kh)_i/\sum(kh)_t$ and $C_\phi = \sum(\phi h)_i/\sum(\phi h)_t$
(5) Plot C_k versus C_ϕ on a Cartesian graph as shown in Figure 3.57
(6) Use Eq. (3.112) to calculate the Lorenz coefficient

Dykstra–Parsons Coefficient V_K

The coefficient of variation, or Dykstra–Parsons coefficient (V_K), is another measure of heterogeneity. It is a dimensionless measure of sample variability or dispersion [77]. It is defined as the ratio of the sample standard deviation(s) to the mean. V_K is often applied in geological and engineering studies as an assessment of permeability heterogeneity [78,44]. For data from different populations, the mean and standard deviation often tend to change together such that V_K stays relatively constant. Any large changes in V_K between two samples indicate a dramatic difference in the populations associated with those samples [77]. The coefficient of variation is also used to provide a relative measure of data dispersion compared to the mean for the normal (bell-shaped) distribution. The coefficient of variation has no units. It may be reported as a simple decimal value, or it may be reported as a percentage and ($0 < V_K < 1$). When the V_K is small (near zero), the data scatter compared to the mean is small. When the V_K is large (near 1) compared to the mean, the amount of variation is large.

Dykstra and Parsons used the log-normal distribution of permeability to define the coefficient of permeability variation, V_K [76].

$$V_K = \frac{s}{\bar{k}} \tag{3.113}$$

where, s and $\bar{k}$ are the standard deviation and the mean value of k, respectively. The standard deviation of a group of n data points is:

$$s = \sqrt{\frac{\sum_{i=1}^{n}(k_i - \bar{k})^2}{n-1}} \tag{3.114a}$$

or

$$s = \sqrt{\frac{\sum_{i=1}^{n}k_i^2 - n\bar{k}^2}{n-1}} \tag{3.114b}$$

where $\bar{k}$ is the arithmetic average of permeability, n the total number of data points, and k_i the permeability of individual core samples. In a normal distribution, the value of k is such that 84.1% of the permeability values are less than $\bar{k} + s$ and 15.9% of the k values are less than $\bar{k} - s$.

The Dykstra–Parsons coefficient of permeability variation, V_K, can be obtained graphically by plotting permeability values on log-probability paper, as shown in Figure 3.58, and then using the following equation:

$$V_K = \frac{k_{50} - k_{84.1}}{k_{50}} \tag{3.115a}$$

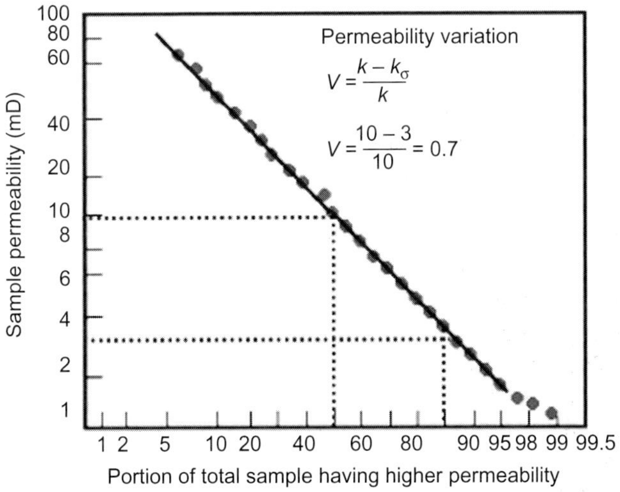

FIGURE 3.58 Log-normal permeability distribution [76].

where

k_{50} = permeability value with 50% probability
$k_{84.1}$ = permeability at 84.1% of the cumulative sample

For a log-normal permeability distribution, the Dykstra–Parsons coefficient can be estimated from (91):

$$V_K = 1 - \exp\left[-\sqrt{\ln\left(\frac{k_a}{k_h}\right)}\right] \quad (3.115a)$$

where k_a and k_h are, respectively, the arithmetic and harmonic average permeability.

The Dykstra–Parsons coefficient is an excellent tool for characterizing the degree of reservoirs heterogeneity. The term V_K is also called the *Reservoir Heterogeneity Index.*

The range of this index is $0 < V_K < 1$:

- $V_K = 0$, ideal homogeneous reservoir.
- $0 < V_K < 0.25$, slightly heterogeneous, can be approximated by a homogeneous model in reservoir simulation with minimal error.
- $0.25 < V_K < 0.50$, heterogeneous reservoir, geometric averaging technique is applicable. If the index is closer to 0.50, run the numerical simulator with the heterogeneous model.
- $0.50 < V_K < 0.75$, the reservoir is very heterogeneous, a combination of geometric and harmonic averaging technique is necessary.
- $0.75 < V_K < 1$, the reservoir is extremely heterogeneous, none of the conventional averaging techniques (arithmetic, geometric, and harmonic) are applicable in this range.
- $V_K = 1$, perfectly heterogeneous reservoir. It is unlikely that such reservoirs exist, as geologic processes of deposition and accumulation of sediments are not extreme.

Data on the probability axis are obtained by arranging permeability values taken from core analysis in a descending order and then computing the percent of the total number of k-values exceeding each tabulated permeability value. The best-fit straight line is drawn such that the central points, i–e, in the vicinity of the mean permeability, are weighted more heavily than the more distant points. The midpoint of the permeability distribution is the log mean permeability, or k_{50}. In Figure 3.46, $k_{50} = 10$, $k_{84.1} = 3$, and the coefficient is 0.70, which indicates that the reservoir is very heterogeneous. Unusually high values of V_k, i.e, >1, may indicate that the samples contain extreme values that may affect estimation of values at unsampled locations. V_k may sometimes be as high as 2 or even 5 [73]. This occurs when samples exhibit several order of magnitude variations.

Warren and Price presented an extensive study of fluid flow in heterogeneous porous media. They concentrated on understanding the effect of the deposition of heterogeneous permeability on single-phase flow for a known

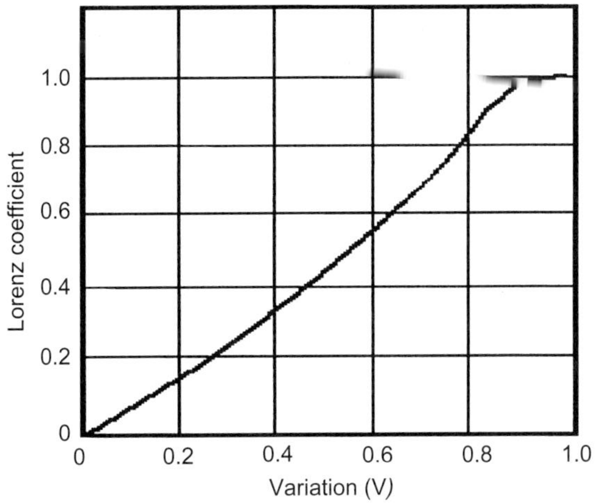

FIGURE 3.59 Correlation of Lorenz coefficient and permeability variation [79].

permeability distribution, determining if it is possible to infer the presence and probable configuration of heterogeneities from core analysis and conventional pressure transient tests. They showed that the Lorenz coefficient can be correlated with the permeability variation coefficient, as shown in Figure 3.59.

Using this figure, Ahmed [80] derived the following two correlations relating the Lorenz coefficient L_K and the Dykstra–Parsons coefficient V_K:

$$L_K = 0.0116356 + 0.339794\, V_K + 1.066405\, V_K^2 - 0.3852407\, V_K^3 \quad (3.115c)$$

$$V_K = -0.000505971 + 1.7475225\, L_K - 1.468855\, L_K^2 + 0.701023\, L_K^3 \quad (3.115d)$$

Example

The values of permeability, porosity, and formation thickness shown in Table 3.14 were obtained from several cored wells [80].

1. Calculate the Lorenz (L_K) and Dykstra–Parsons (V_K) coefficients.
2. Use the normalized cumulative to find the number of flow units.

Solution

(1) Table 3.15 shows the calculated values of the normalized cumulative capacities. Figure 3.60 is a Cartesian plot of C_K versus C_ϕ.

The areal summation method yields 0.5 for the area A and 0.21 for area B, then using Eq. (3.112), the Lorenz coefficient is $L_K = 0.21/0.5 = 0.42$.

The Dykstra–Parsons coefficient can be directly determined from Eq. (3.115):

$$\begin{aligned} V_K &= -0.000505971 + 1.7475225 L_K - 1.468855 L_K^2 + 0.701023 L_K^3 \\ &= -0.000505971 + 1.7475225 \times 0.42 - 1.468855 \times 0.42^2 + 0.701023 \times 0.42^3 = 0.52 \end{aligned}$$

(2) Figure 3.61 shows a plot of depth versus the normalized cumulative RQI. At least three distinct flow units are observed.

TABLE 3.14 Core Data from a Layered Reservoir [80]

Well No. 1			Well No. 2			Well No. 3		
Depth (ft)	k (mD)	ϕ (%)	Depth (ft)	k (mD)	ϕ (%)	Depth (ft)	k (mD)	ϕ (%)
5,389–5,391	166	17.4	5,397–5,398.5	72	15.7	5,401–5,403	28	14
5,389–5,393	435	18	5,397–5,399.5	100	15.6	5,401–5,405	40	13.7
5,389–5,395	147	16.7	5,397–5,402	49	15.2	5,401–5,407	20	12.2
5,389–5,397	196	17.4	5,397–5,404.5	90	15.4	5,401–5,409	32	13.6
5,389–5,399	254	19.2	5,397–5,407	91	16.1	5,401–5,411	35	14.2
5,389–5,401	105	16.8	5,397–5,409	44	14.1	5,401–5,413	27	12.6
5,389–5,403	158	16.8	5,397–5,411	62	15.6	5,401–5,415	27	12.3
5,389–5,405	153	15.9	5,397–5,413	49	14.9	5,401–5,417	9	10.6
5,389–5,406	128	17.6	5,397–5,415	49	14.8	5,401–5,419	30	14.1
5,389–5,409	172	17.2	5,397–5,417	83	15.2			

Averaging Techniques

There are three standard techniques used to estimate the average permeability of a reservoir: arithmetic, geometric, and harmonic.

(a) *Arithmetic average:* The unweighted arithmetic average permeability $\bar{k}$ is determined from:

$$\bar{k}_A = \frac{\sum k_i}{n} \tag{3.116}$$

If the analysis of pressure transient tests yields much lower permeability values than those obtained from core data, the lateral continuity of the producing formation may not be sufficient to justify the arithmetic averaging.

(b) *Geometric average:* In heterogeneous and anisotropic formations, a geometric average, which assumes random distribution of the matrix, is preferable:

$$\bar{k}_G = \sqrt[n]{k_1 k_2 k_3 \ldots k_n} \tag{3.117}$$

According to Warren and Price, the geometric mean permeability is more consistent with the distribution found in many porous rocks [72]. The main weakness of the geometric mean is if one individual value of

TABLE 3.15 Normalized Cumulative Capacities

h	kh	ϕh	$\sum kh$	$\sum \phi h$	$C_k = \sum kh/5,681$	$C_\phi = \sum \phi h/885.7$
2	18	21.2	18	21.2	0.003	0.024
2	40	24.4	58	45.6	0.010	0.052
2	54	25.2	112	70.8	0.020	0.080
2	54	24.6	166	95.4	0.029	0.108
2	56	28	222	123.4	0.039	0.139
2	60	28.2	282	151.6	0.050	0.171
2	64	27.2	346	178.8	0.061	0.202
2	70	28.4	416	207.2	0.073	0.234
2	80	27.4	496	234.6	0.087	0.265
2	88	28.2	584	262.8	0.103	0.297
2.5	122.5	38	706.5	300.8	0.124	0.340
2	98	29.8	804.5	330.6	0.142	0.374
2	98	29.6	902.5	360.2	0.159	0.407
2	124	31.2	1,026.5	391.4	0.181	0.442
1.5	108	23.55	1,134.5	414.95	0.200	0.469
2	166	30.4	1,300.5	445.35	0.229	0.503
2.5	225	38.5	1,525.5	483.85	0.269	0.547
2.5	227.5	40.25	1,753	524.1	0.309	0.592
1	100	15.6	1,853	539.7	0.326	0.610
2	210	33.6	2,063	573.3	0.363	0.648
2	256	35.2	2,319	608.5	0.408	0.688
2	294	33.4	2,613	641.9	0.460	0.725
2	306	31.8	2,919	673.7	0.514	0.761
2	316	33.6	3,235	707.3	0.569	0.799
2	332	34.8	3,567	742.1	0.628	0.839
2	344	34.4	3,911	776.5	0.688	0.877
2	392	34.8	4,303	811.3	0.757	0.917
2	508	38.4	4,811	849.7	0.847	0.960
2	870	36	**5,681**	**885.7**	1.000	1.000

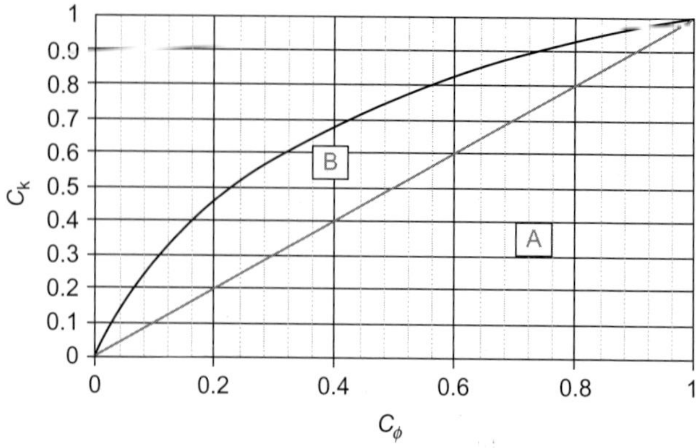

FIGURE 3.60 Cartesian plot of C_K versus C_ϕ

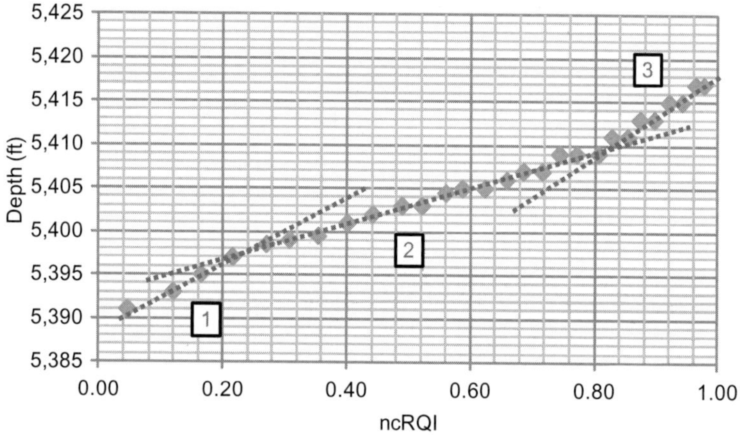

FIGURE 3.61 Normalized cumulative RQI versus depth showing the existence of three distinct zones.

k is zero, the entire average becomes zero. To avoid this zeroing effect in reservoir simulation, a relatively small value is assigned to the block that has zero permeability. It should be noted that even shale has permeability in the order of 10^{-7} mD.

(c) *Harmonic average:* The harmonic averaging technique is best suited for layers in series such as in composite systems. This technique is extensively used in reservoir simulation studies where different grid cells are in series.

$$\bar{k}_{\mathrm{H}} = \frac{n}{\sum_{i=1}^{n}(1/k_i)} \tag{3.118}$$

(d) *Weighted average:* Equations (3.116)–(3.118) assume the weight factors, w_i, are equal, and that the flow is one-dimensional. If the weight factors are not equal, then these equations become, respectively:

$$\bar{k}_A = \frac{1}{n}\left(\sum_{i=1}^{n} w_i k_i\right) \tag{3.119}$$

$$\bar{k}_G = \left(\prod_{i=1}^{n} k_i^{w_i}\right)^{1/n} \tag{3.120}$$

$$\bar{k}_H = \frac{n}{\left(\sum_{i=1}^{n} w_i/k_i\right)} \tag{3.121}$$

The thickness of the formation (or height of the core sample) corresponding to each permeability is a common weighting factor; thus, the thickness-weighted arithmetic mean is:

$$\bar{k}_A = \frac{\sum_{j=1}^{n} k_j h_j}{\sum_{j=1}^{n} h_j} \tag{3.122}$$

The width of each block arranged in series is used as a weight factor in harmonic averaging technique. The arithmetic average will yield the highest average permeability, while the harmonic averaging method will yield the lowest average permeability value. Thus:

$$\bar{k}_A < \bar{k}_G < \bar{k}_H$$

Tehrani et al. [81] investigated the practicality of these three averaging techniques in two and three-dimensional flow problems. They showed that although in some heterogeneous reservoirs the geometric averaging method yields representative values of the effective permeability, there are many field cases in which none of the three averaging techniques gives satisfactory results.

In two or three-dimensional heterogeneous flow systems, a combination of these three averaging techniques is necessary. For flow into a well in a two-dimensional layered system, the arithmetic average for horizontal permeability, k_h, and the harmonic average for the vertical permeability, k_V, are used to estimate the anisotropy index, $I_A = k_V/k_h$, from Eq. (3.96).

The average radial or horizontal permeability, k_r, is best determined from a pressure buildup or drawdown test. The average radial permeability of a heterogeneous and anisotropic system is estimated from:

$$k_r = \sqrt{k_x k_y - k_{xy}^2} \tag{3.123}$$

where k_x, k_y, and k_{xy} are components of the symmetrical permeability tensor aligned with the coordinate system. These components are best determined from a multiwell interference test. Three observation wells located on different rays extending from the active well, which is located at the origin of the coordinate system, are necessary to calculate k_x, k_y, and k_{xy}.

Effective Permeability from Core Data

The effective permeability, obtained from core data, may be estimated from [82,83]:

$$k_e = \left(1 + \frac{\sigma_k^2}{6}\right)\exp\left[\bar{k}_G\right] \tag{3.124}$$

where $\bar{k}_G$ is the geometric mean of the natural log of permeability, i.e.:

$$\bar{k}_G = \sqrt[n]{\ln k_1 \ln k_2 \ln k_3 \ldots \ln k_n} \tag{3.125}$$

and σ_k^2 is the variance of the natural log of the permeability estimates:

$$\sigma^2 = \frac{\sum_{i=1}^{n} (\ln k_i - \ln \bar{k})^2}{n-1} \tag{3.126}$$

where

$$\ln \bar{k} = \frac{\sum \ln k_i}{n} \tag{3.127}$$

The effective permeability obtained from Eq. (3.124) should be expected to be in the same range as the effective permeability obtained from the interpretation of a pressure transient test, if $V_K < 0.25$.

Core-derived permeability is an accurate representation of a particular core sample. Using this permeability value to represent reservoir formation permeability can, however, lead to erroneous predictions of well productivity, as core samples represent a small portion of the interval in a particular well and an even smaller portion of a reservoir [84]. The average effective permeability obtained from pressure transient test should be considered as an accurate representation of the reservoir, but only within the drainage area of the test. Beyond this drainage area, the average permeability could be different if the radial variation in permeability is significant. As long as the measurements are consistent, the core-derived permeability can be very useful in completion design, particularly in choosing the phasing and vertical spacing of perforation [83].

Example

Given the permeability data in Table 3.16 [83] for well HBK5, calculate:

1. The arithmetic, geometric and harmonic averages of the core-derived permeability values.
2. The effective permeability.
3. The Dykstra–Parsons coefficient.

TABLE 3.16 Permeability Data for Well HBK5

Interval	k (mD)
1	120
2	213
3	180
4	200
5	212
6	165
7	145
8	198
9	210
10	143
11	79
12	118
13	212
14	117

Solution

(1) Average values of permeability

The arithmetic, geometric, and harmonic averages of the core-derived permeability values are, respectively:

$$\bar{k}_A = \frac{\sum k_i}{n} = \frac{120 + 213 + \cdots + 117}{14} = 165 \text{ mD}$$

$$\bar{k}_G = \sqrt[n]{k_1 k_2 k_3 \ldots k_n} = \sqrt[14]{120 * 213 * \cdots * 117} = 158.7 \text{ mD}$$

$$\bar{k}_H = \frac{n}{\sum_{i=1}^{n}(1/k_i)} = \frac{14}{(1/120) + (1/213) + \cdots + (1/117)} = 151.4 \text{ mD}$$

The harmonic averaging technique yields, as expected, the lowest value of average permeability. But the difference between the three averages is not significant, implying that the formation is essentially homogeneous.

(2) The effective permeability of this 14-m thick formation is estimated from Eq. (3.124).

From Eq. (3.125), we calculate the geometric mean of the natural log of the core-derived permeability values:

$$\bar{k}_G = \sqrt[n]{\ln k_1 \ln k_2 \ln k_3 \ldots \ln k_n} = (7.173 \times 10^9)^{1/14} = 5.058 \text{ mD}$$

TABLE 3.17 Intermediate Results for Calculating Variance for Well HBK5

Interval	k (mD)	$\ln(k_i)$	$\prod(\ln k_1, \ln k_2, \ldots, \ln k_{14})$	$\sum(k_i - \bar{k})^2$
1	120	4.7875	4.7875E+00	0.0781
2	213	5.3613	2.5667E+01	0.1647
3	180	5.1930	1.3329E+02	0.1806
4	200	5.2983	7.0620E+02	0.2342
5	212	5.3566	3.7828E+03	0.3181
6	165	5.1059	1.9315E+04	0.3196
7	145	4.9767	9.6126E+04	0.3278
8	198	5.2883	5.0834E+05	0.3768
9	210	5.3471	2.7181E+06	0.4553
10	143	4.9628	1.3490E+07	0.4661
11	79	4.3694	5.8942E+07	0.9525
12	118	4.7707	2.8120E+08	1.0403
13	212	5.3566	1.5063E+09	1.1242
14	117	4.7622	7.1730E+09	1.2171

To calculate the variance σ_k^2, we need to use Eqs. (3.126) and (3.127) (Table 3.17):

$$\ln \bar{k} = \frac{\sum \ln k_i}{n} = \frac{70.938}{14} = 5.067 \text{ mD}$$

$$\sigma^2 = \frac{\sum_{i=1}^{n} (\ln k_i - \ln \bar{k})^2}{n-1} = \frac{1.2171}{14-1} = 0.093623$$

The arithmetic average of the natural log of the 14 permeability values is practically equal to the geometric mean of the same permeability values. This further indicates that this particular formation is practically homogeneous.

Using the geometric mean of the natural log of k values, the effective permeability is:

$$k_e = \left(1 + \frac{0.093623}{6}\right) \exp(5.067) = 161.17 \text{ mD}$$

The effective permeability is essentially equal to the geometric mean of core-derived permeability data. This should be expected, since the variance is very small.

(3) The Dykstra–Parsons coefficient is obtained from Eq. (3.115).

TABLE 3.18 Frequency Distribution for the Permeability Data (Well HBK5)

Intervals	k (mD)	Frequency	Number of Samples With Larger Permeability	Cumulative Frequency Distribution (% >k_i)
2	213	1	0	0.0
5 and 13	212	2	1	7.1
9	210	1	3	21.4
4	200	1	4	28.6
8	198	1	5	35.7
3	180	1	6	42.9
6	165	1	7	50.0
7	145	1	8	57.1
10	143	1	9	64.3
1	120	1	10	71.4
12	118	1	11	78.6
14	117	1	12	85.7
11	79	1	13	92.9

The procedure for graphically determining the Dykstra–Parsons coefficient is as follows:

(a) Arrange permeability data in descending order as shown in Column 2 of Table 3.18.

(b) Determine the frequency of each permeability value (Column 3).

(c) Find the number of samples with larger permeability (Column 4).

(d) Calculate the cumulative frequency distribution by dividing values in Column 4 with the total number of permeability points, *n*, which is 14 in this example (Column 5).

(e) Plot permeability data (Column 2) versus cumulative frequency data (Column 5) on a log-normal probability graph, as shown in Figure 3.48.

(f) Draw the best straight line through the data, with more weight placed on points in the central portion where the cumulative frequency is close to 50%. This straight line reflects a quantitative, as well as a qualitative, measure of the heterogeneity of the reservoir rock.

(g) From the graph (Figure 3.62), read the values: $k_{50} = 158.7$ mD and $k_{84.1} = 117.2$ mD. These values can also be interpolated from Table 3.18.

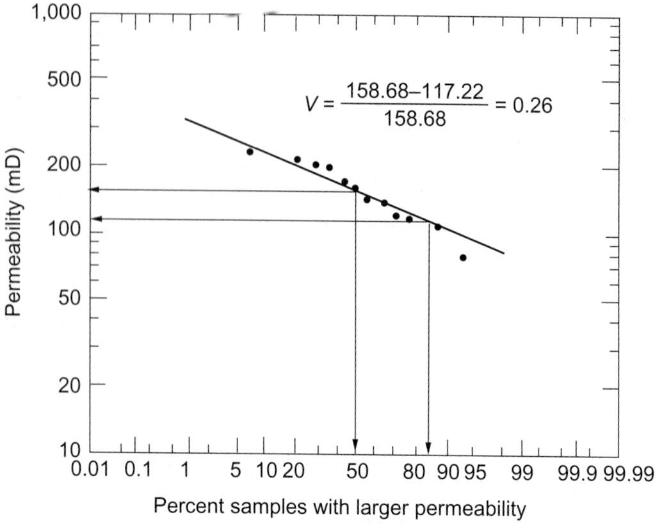

FIGURE 3.62 Dykstra–Parsons Coefficient for well HBK5 (Homogeneous Reservoir).

(h) Calculate the Dykstra–Parsons coefficient (Eq. (3.115)):

$$V_K = \frac{k_{50} - k_{84.1}}{k_{50}} = \frac{158.7 - 117.22}{158.7} = 0.26$$

This formation is slightly heterogeneous, but it can be treated as homogeneous for reservoir simulation purposes.

Average Porosity and Saturation

Amyx et al. [19] showed that the histogram distribution is also an excellent representation of porosity data obtained from core analysis. Most porosity histograms are symmetrical about the mean value, as shown in Figure 3.63. For classified data, i.e., arranged in increasing or decreasing order, the arithmetic mean porosity is given by:

$$\overline{\phi} = \sum_{i=1}^{n} \phi_i f_i \tag{3.128a}$$

where

ϕ_i = porosity at the midpoint of range, fraction
f_i = frequency for porosity range, fraction
n = number of porosity ranges

For unclassified data, the arithmetic mean porosity is:

$$\overline{\phi} = \frac{1}{n} \sum_{i=1}^{n} \phi_i \tag{3.128b}$$

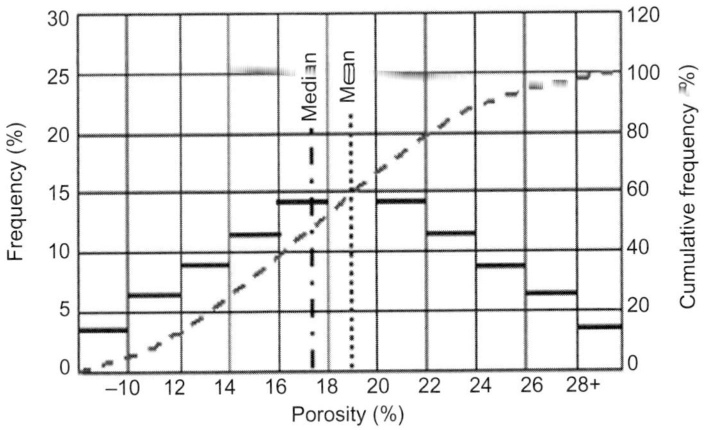

FIGURE 3.63 Typical porosity histogram [19].

One disadvantage of the arithmetic mean is that any gross error in a porosity value of one sample can have considerable effect on the value of the mean. To avoid this potential problem, the average porosity value can be obtained from another statistical measure called the "median," which is defined as the value of the middle variable of class data. It is also the value of the variable corresponding to the 50% point on the cumulative frequency curve. The mean and the median of a set of porosity values rarely coincide. Unlike the mean, the median is not sensitive to extreme values of a variable.

The thickness weighted average porosity is:

$$\bar{\phi} = \frac{\sum_{j=1}^{n} \phi_j h_j}{\sum_{j=1}^{n} h_j} \tag{3.128c}$$

The average water saturation from a stratified core section can be obtained by PV (pore volume) weighting [90,91]:

$$\bar{S}_w = \frac{\sum_{j=1}^{n} S_{wj} \phi_j h_j}{\sum_{j=1}^{n} \phi_j h_j} \tag{3.129}$$

A similar equation can be written for oil and gas.

Example

The petrophysical properties of the core samples including the porosity, permeability and formation resistivity factor actually measured in the laboratory are listed in Table 3.19. The tortuosity is calculated from Eq. (3.78). Calculate:

1. The arithmetic mean porosity and the median porosity.
2. The arithmetic, geometric and harmonic averages of the core-derived permeability values.

TABLE 3.19 Petrophysical Properties of a Heterogeneous Reservoir

Core No.	Cementing Material	Carbonates (%)	Clay (%)	Φ (%)	k (md)	F	τ
1	SiO_3	5.1	3	17	90	23.3	6
2	Clay and carbonate	6.1	9	14.7	7	51	7
3	SiO_2 and carbonate	21.9	6	6.7	4	67	15
4	SiO_2 and clay	7.2	4	17.6	220	16.6	6
5	SiO_2	2.3	1	26.3	1,920	8.6	4
6	SiO_2	0.7	0.2	25.6	4,400	9.4	4
10	SiO_2	1.9	2	13.9	145	33	7
11	SiO_2 and clay	4.9	4	18.6	25	22.9	5
12	SiO_2	2.8	2	18.8	410	18.6	5
13	Carbon, clay, and SiO_2	7	4	16.1	3	42	6
14	Clay and carbonate	8	7	15	9	41	7
15	Carbonate and clay	12.1	1	22.1	200	13.1	5
16	Carbonate and clay	14.1	2	20.6	36	16.6	5
17	SiO_2	5.2	4	30.7	70	8.4	3
20	SiO_2 and carbonate	9	1	16.4	330	21.1	6
22	SiO_2, carbon, and clay	6.8	0.4	18.8	98	19.3	5
23	SiO_2 and clay	1.1	0.2	24.8	1,560	10.8	4
25	Clay	7.7	6	19.1	36	17.2	5
28	Clay	3.8	5	29.8	1,180	8.4	3
31	Clay	0	2	27.1	3,200	11.7	4
32	SiO_2 and carbonate	1.4	1	28.2	2,100	10.9	4
33	Clay and SiO_2	–	–	19.4	8	24	5
34	SiO_2 and clay	1.2	7	19.7	18	20.8	5
35	Clay	2.2	3	31.5	2,200	6.9	3
36	Clay and SiO_2	3.9	7	19.3	19	24.4	5
37	Clay and SiO_3	0	4	27.3	88	12.4	4
38	Carbon, clay, SiO_2	21.2	5	25.1	370	11.6	4
39	Carbonate	–	–	15	115	37.3	7
40	SiO_2 and carbonate	6.1	1	18.4	130	19	5

After Ref. 85.

3. The effective permeability.
4. The Dykstra–Parsons coefficient.

Solution

(1) The arithmetic mean of porosity is obtained from Eq. (3.129):

$$\overline{\phi} = \frac{1}{n}\sum_{i=1}^{n} \phi_i = \frac{1}{29}(17 + 14.7 + 6.7 + \cdots + 15 + 19.4) = 20.81\%$$

The arithmetic mean can also be estimated graphically from a plot of the frequency and cumulative frequency (%) versus porosity, as long as the histogram is relatively symmetrical, as is the case in most porosity distributions.

Figure 3.64 is a porosity histogram and distribution (cumulative frequency) curve for the porosity data shown in Table 3.20. It is evident from this figure that the porosity histogram is not symmetrical. This lack of symmetry is further confirmed in Figure 3.65 that shows the plot of porosity data versus the cumulative frequency on an arithmetic probability graph. Theoretically, if the porosity data approximate a straight line, then a normal curve, which is completely defined by the arithmetic mean, is a reasonable fit of the data.

Figure 3.66 shows a significant deviation from the fitted straight line between 30% and 60% cumulative frequency. In this particular example the porosity distribution is not normal and, therefore, cannot be represented by the frequency function.

The "median" porosity corresponds to the 50% point on the cumulative frequency curve, if the porosity distribution is normal and the histogram is symmetrical. Theoretically, the median value divides the histogram into two equal areas, which is not the case in this example, as shown in Figure 3.65. The "median" value of porosity (19.3%), shown in this figure, is only an approximation. The cumulative volume capacity for the porosity data is

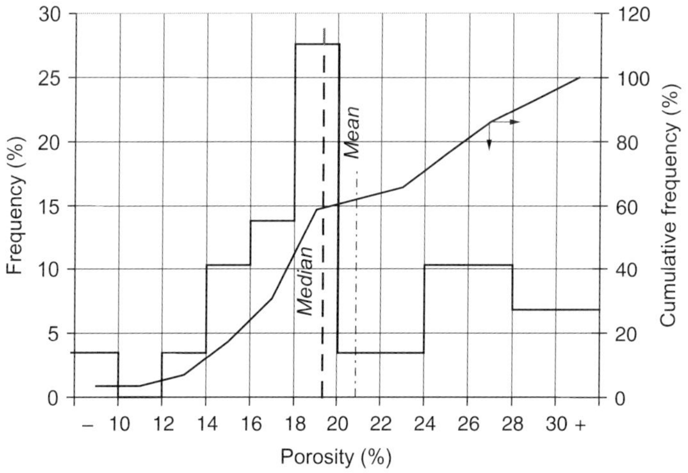

FIGURE 3.64 Porosity histogram and distribution for all samples.

TABLE 3.20 Classification of Porosity Data into Ranges of 2% Porosity for All Samples

Porosity Range (%)	Mid-Value of Range (%)	No. of Samples	Frequency, F (%)	Cumulative Frequency, F_c (%)
Less than 10	9	1	3.45	3.45
10–12	11	0	0.00	3.45
12–14	13	1	3.45	6.90
14–16	15	3	10.34	17.24
16–18	17	4	13.79	31.03
18–20	19	8	27.59	58.62
20–22	21	1	3.45	62.07
22–24	23	1	3.45	65.52
24–26	25	3	10.34	75.86
26–28	27	3	10.34	86.21
28–30	29	2	6.90	93.10
30+	31	2	6.90	100.00
Total		29	100	

calculated in Table 3.21 and plotted in Figure 3.66. This plot indicates that the distribution of porosity capacity is bi-modal.

(2) Average values of permeability

The arithmetic, geometric and harmonic averages of the 29 core-derived permeability values are, respectively:

$$\bar{k}_A = \frac{\sum k_i}{n} = \frac{90 + 7 + 4 + 220 + \cdots + 130}{29} = 655 \text{ mD}$$

$$\bar{k}_G = \sqrt[n]{k_1 k_2 k_3 \ldots k_n} = [90 \times 7 \times 4 \times 220 \times \ldots \times 130]^{1/29} = 123 \text{ mD}$$

$$\bar{k}_H = \frac{n}{\sum_{i=1}^{n}(1/k_i)} = \frac{29}{(1/90) + (1/7) + (1/4) + (1/220) + \cdots + (1/130)} = 23 \text{ mD}$$

The harmonic averaging technique yields, as expected, the lowest value of average permeability. In this case, the difference between the three averages is very significant, implying that the formation is extremely heterogeneous. Another reason for this large difference is that no values of permeability were cutoff. Generally, the amount of cementing material is high for low permeability values, and low for very high permeability values.

(3) The effective permeability of this formation is estimated from Eq. (3.124).

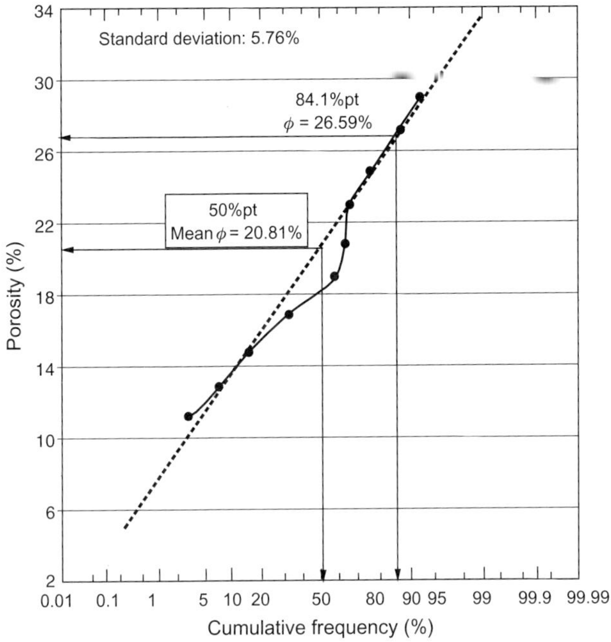

FIGURE 3.65 Porosity distribution on probability paper (heterogeneous reservoir).

From Eq. (3.125), we calculate the geometric mean of the natural log of the core-derived permeability values:

$$\overline{k}_G = \sqrt[n]{\ln k_1 \ln k_2 \ln k_3 \ldots \ln k_n} = (1.9855 \times 10^{18})^{1/29} = 4.275 \text{ mD}$$

To calculate the variance σ_k^2 we need to use Eqs. (3.126) and (3.127):

$$\ln \overline{k} = \frac{\sum \ln k_i}{n} = \frac{139.54}{29} = 4.812 \text{ mD}$$

$$\sigma^2 = \frac{\sum_{i=1}^{n} (\ln k_i - \ln \overline{k})^2}{n-1} = \frac{124.51}{29-1} = 4.44678$$

Using the geometric mean of the natural log of k values, the effective permeability is:

$$k_e = \left(1 + \frac{4.44678}{6}\right) \exp(4.812) = 214 \text{ mD}$$

(4) The Dykstra–Parsons coefficient is obtained from Eq. (3.115).

Using the same approach as in the previous example, we find:

$$V_K = \frac{k_{50} - k_{84.1}}{k_{50}} = \frac{122.95 - 8.38}{122.95} = 0.93$$

where $k_{84.1} = 8.38$ mD was obtained by interpolating in Table 3.22 or Figure 3.67.

The Dykstra–Parsons coefficient is very high, indicating an extremely heterogeneous reservoir.

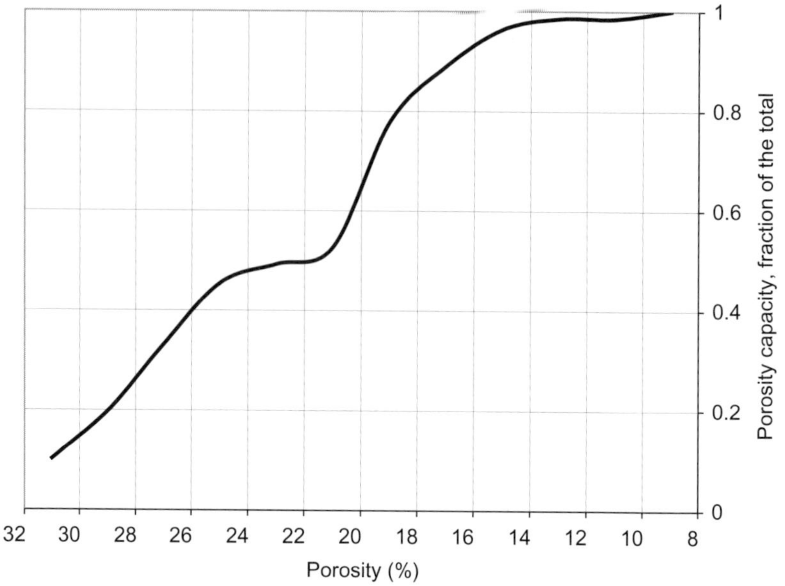

FIGURE 3.66 Distribution of porosity capacity.

TABLE 3.21 Calculation of Porosity Distribution from Classified Data for Determination of Net Pay Sand

Porosity Range	Mid-Value of Range (%), ϕ_i	No. of Samples	Frequency Fraction (F_i)	$\phi_i \cdot F_i$	$\frac{\phi_i F_i}{\phi_t}$	Cumulative Capacity, $\sum_{i=1}^{n} \frac{\phi_i F_i}{\phi_t}$
Less than 10	9	1	0.0345	0.31	0.01	1.00
10–12	11	0	0.0000	0.00	0.00	0.99
12–14	13	1	0.0345	0.45	0.02	0.99
14–16	15	3	0.1034	1.55	0.07	0.96
16–18	17	4	0.1379	2.34	0.11	0.89
18–20	19	8	0.2759	5.24	0.25	0.78
20–22	21	1	0.0345	0.72	0.03	0.53
22–24	23	1	0.0345	0.79	0.04	0.49
24–26	25	3	0.1034	2.59	0.12	0.45
26–28	27	3	0.1034	2.79	0.13	0.33
28–30	29	2	0.0690	2.00	0.10	0.20
30+	31	2	0.0690	2.14	0.10	0.10
Total		29	1.0000	20.9310		

TABLE 3.22 Frequency Distribution for the Permeability Data

Core No.	Permeability, *k* (mD)	Frequency	Number of Samples With Larger Permeability	Cumulative Frequency Distribution (% $>k_i$)
6	4,400	1	0	0.0
31	3,200	1	1	3.4
35	2,200	1	2	6.9
32	2,100	1	3	10.3
5	1,920	1	4	13.8
23	1,560	1	5	17.2
28	1,180	1	6	20.7
12	410	1	7	24.1
38	370	1	8	27.6
20	330	1	9	31.0
4	220	1	10	34.5
15	200	1	11	37.9
10	145	1	12	41.4
40	130	1	13	44.8
39	115	1	14	48.3
22	98	1	15	51.7
1	90	1	16	55.2
37	88	1	17	58.6
17	70	1	18	62.1
16, 25	36	2	19	65.5
11	25	1	21	72.4
36	19	1	22	75.9
34	18	1	23	79.3
14	9	1	24	82.8
33	8	1	25	86.2
2	7	1	26	89.7
3	4	1	27	93.1
13	3	1	28	96.6

Total samples, $n = 29$.

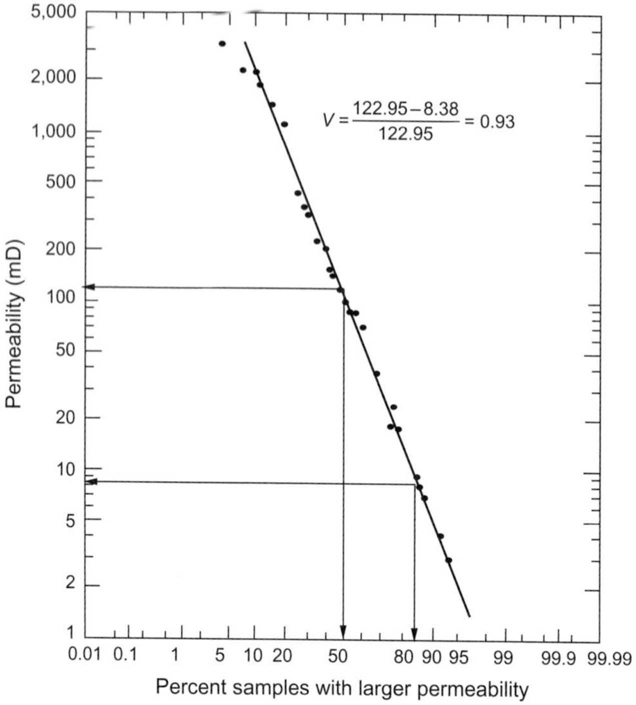

FIGURE 3.67 Dykstra–Parsons coefficient for a heterogeneous reservoir.

PERMEABILITY FROM WELL TEST DATA

Despite the considerable value of core analysis and well log interpretation, some doubt always remains concerning the potential productivity of a well, especially during the exploratory stage. This doubt is not dispelled until a sizable sample of formation fluids has been recovered during a production test, commonly known as drill-stem test. The recovery of fluids obtained is primarily dependent upon the permeability and porosity of the formation tested and the viscosity of the fluids contained in the zone. Various techniques have been developed for analyzing the fluid recovery and recorded pressure curves to determine whether or not a formation test has indicated that commercial production can be attained.

To utilize these pressure curves, some knowledge of the response of the curve to a given formation conditions is necessary. Basic to this knowledge is an understanding of various analytical equations describing the flow of fluids through porous media. These equations, which are solutions of diffusivity equations for different boundary conditions, express the relationship between characteristics of the porous rock, such as porosity and permeability, and properties of the fluids (oil, gas, and water) moving through the rock.

The basic well testing technique is to create a pressure drop in the bottom-hole pressure, which causes reservoir fluids to flow at a certain rate from the rock to the wellbore, followed by a shut-in period. The production period is generally referred to as the "pressure drawdown," whereas the shut-in period is called the "pressure buildup."

Practical information obtained from well testing includes permeability along with porosity, reservoir shape, average reservoir pressure, and the location of the reservoir boundaries, such as sealing faults, in the vicinity of the well. The most common method for obtaining permeability consists of plotting pressure data versus time on a semilog graph paper (e.g., Figure 3.68 for pressure drawdown test and Figure 3.69 for pressure buildup test). Upon determining the slope of the straight line, the following equation is used to calculate permeability:

$$k = 162.6\frac{q\mu B_o}{mh} \tag{3.130}$$

where

k = formation permeability, mD
q = flow rate, STB/D
μ = fluid viscosity, cP
B_o = formation volume factor, bbl/STB
h = formation thickness
m = slope of the straight line, psi/log cycle

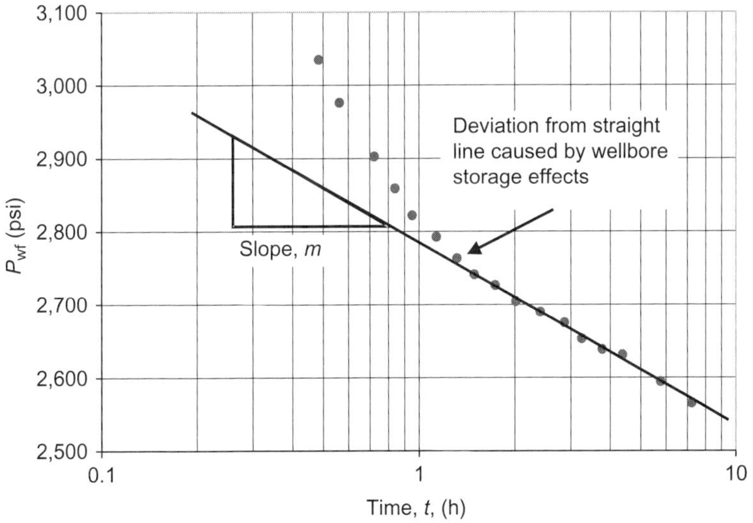

FIGURE 3.68 Semilog plot of pressure drawdown test.

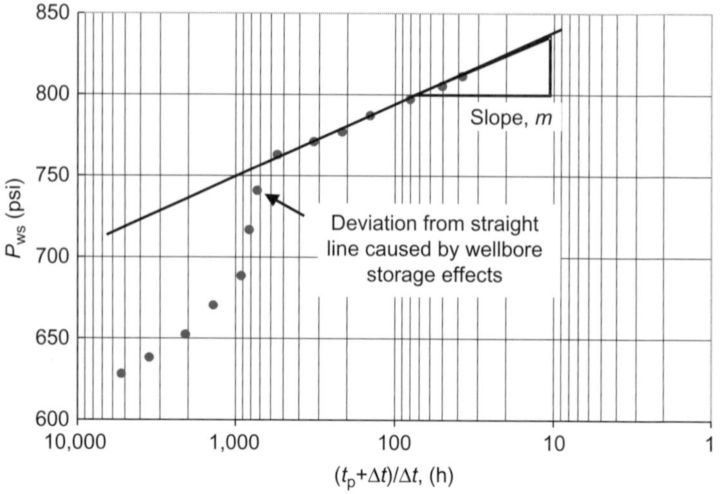

FIGURE 3.69 Semilog plot of pressure buildup test (Hornet plot).

Other parameters obtainable from these types of tests are degree of formation damage around the wellbore, number and type of reservoir boundaries, and degree of connectivity to other wells. A modern technique for analyzing pressure data is based on the log–log plot of pressure derivative $(t \times \Delta p')$ versus time, as shown in Figure 3.70. Using *the Tiab's Direct Synthesis* technique, the permeability is obtained from [86]:

$$k = \frac{70.6 q \mu B_o}{h(t \times \Delta p')_R} \tag{3.131}$$

where $(t \times \Delta p')_R$ is obtained from the horizontal straight line, which corresponds to the infinite acting radial flow regime. On the semilog plot, this flow regime corresponds to the straight line of slope m.

Example

A new well in a small bounded reservoir, North of Hobbs, New Mexico, was produced at a constant rate of 250 STB/D. The initial reservoir pressure is 4,620 psia. Other relevant data are as follows: $h = 16$ ft, $\mu_o = 1.2$ cP, and $B_o = 1.229$ bbl/STB.

Calculate the permeability from:

(a) The semilog plot of P_{wf} versus time.

(b) The pressure derivative, using *Tiab's Direct Synthesis* technique.

Solution

Figure 3.71 is a semilog plot of the flowing bottom-hole pressure versus time in Table 3.23. Figure 3.72 shows a log–log plot of $\Delta P = P_i - P_{wf}$ and the

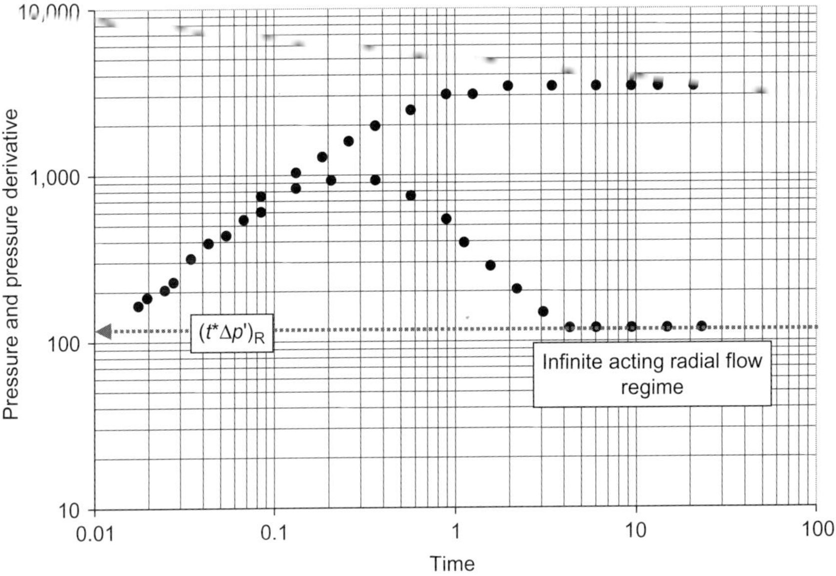

FIGURE 3.70 Pressure derivative plot showing infinite acting radial flow regime.

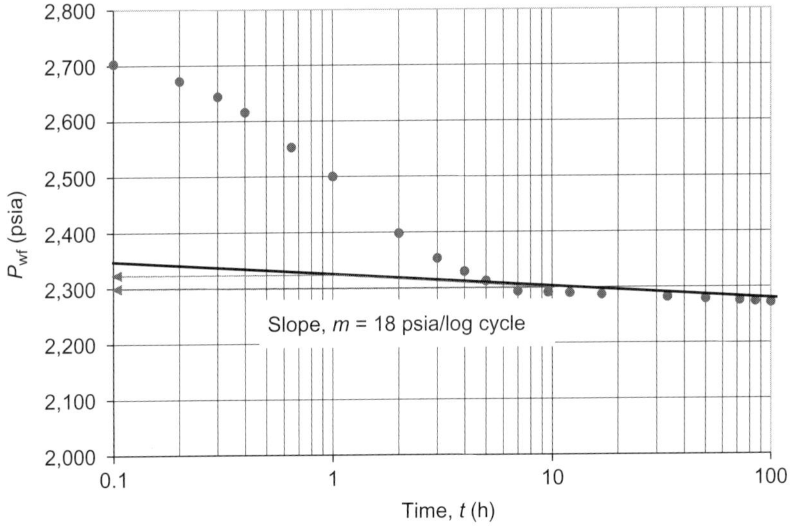

FIGURE 3.71 Semilog plot of pressure drawdown versus test time.

pressure derivative ($t \times \Delta P'$) versus test time. Eq. (3.97) was used to calculate the pressure derivative values.

(a) The absolute value of the slope of the straight line corresponding to the infinite acting line, i.e., radial flow regime, is 18.5 psi/log cycle.

TABLE 3.23 Pressure Drawdown Test

Time (h)	P_{wf} (psia)	ΔP (psi)	$t \times \Delta P'$ (psi)
0	2,733	0	0.00
0.1	2,703	30	31.05
0.2	2,672	61	58.95
0.3	2,644	89	84.14
0.4	2,616	117	106.30
0.65	2,553	180	129.70
1	2,500	233	125.69
2	2,398	335	144.39
3	2,353	380	102.10
4	2,329	404	81.44
5	2,312	421	65.42
7	2,293	440	34.47
9.6	2,291	442	5.62
12	2,290	443	6.32
16.8	2,287	446	7.63
33.6	2,282	451	7.99
50	2,279	454	7.94
72	2,276	457	10.50
85	2,274	459	12.18
100	2,272	461	13.36

Using Eq. (3.132):

$$k = 162.6 \frac{q \mu B_o}{mh} = 162.6 \frac{(250)(1.2)(1.229)}{(18.5)(16)} = 202 \text{ mD}$$

(b) From Figure 3.72, the value of $(t \times \Delta p')_R = 8$ psi is obtained by extrapolating the horizontal line portion of the pressure derivative curve to the vertical axis. This line corresponds to the radial flow regime. Using Eq. (3.133) we obtain:

$$k = \frac{70.6 q \mu B_o}{h(t * \Delta p')_R} = 70.6 \frac{(250)(1.2)(1.229)}{(16)(8)} = 203 \text{ mD}$$

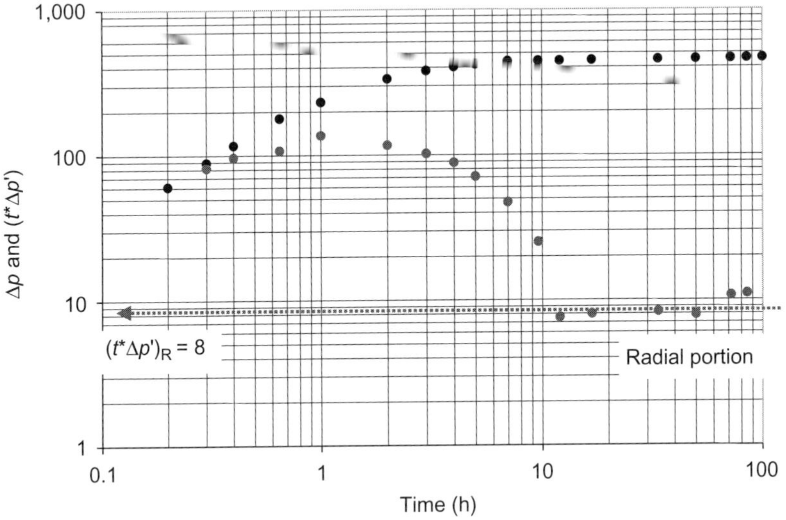

FIGURE 3.72 Log–log plot of ΔP and $t \times \Delta P'$ versus test time.

STATISTICAL ZONATION TECHNIQUE

Reservoir description may be approached in a number of ways. Hutchnison et al. developed a method for statistically predicting the probability of the presence of reservoir nonuniformities of a certain type in any reservoir by considering the cores, logs, and general geologic background of the reservoir [71]. Using a statistical analysis of laboratory measurements of air permeability in eight directions spaced at 45° intervals on 142 two-inch vertical plugs from 30 cores, Greenkorn et al. showed that there is a significant point anisotropy in about 60% of the core plugs [87]. Also, they found that the permeability of a heterogeneous anisotropic porous medium is a tensor consisting of a point-to-point variation that depends on grain size and a point variation that depends on bedding. Trudgen and Hoffmann proposed a procedure that employs Pearson's system of curve fitting for defining a frequency distribution of reservoir rock properties obtained from core data [88]. Many more statistical methods for describing various reservoir rock properties were proposed. However, only few of these methods were found to be practical.

Testerman described a statistical technique for identifying and describing porous and permeable zones in a reservoir, and for determining which ones are likely to be continuous between adjacent wells [45]. The technique is particularly useful in describing permeability distribution in a reservoir where crossflow between adjacent communicating reservoir strata, due to imbibition and gravity segregation, is important. Inasmuch as there are, however, no geological parameters concerning the depositional environment in the statistical evaluation, judgment is necessary to determine whether the zones so defined are, in fact, continuous and consistent with the geological model. Although it has been developed primarily for permeability zonation, the technique is general and can be applied to reservoir properties other than permeability. The reservoir zonation technique is a two-part operation.

Permeability data from the top to the bottom of the strata of a single well are divided into zones. These zones are selected such that the variation of permeability within the zones is minimized and maximized between the zones. The statistical equations used to zone the permeability data are:

$$s_{zz} = \frac{1}{N_z - 1}\left[\sum_{i=1}^{N_z} \frac{1}{N_{ki}}\left(\sum_{j=1}^{N_{kj}} N_{ij}\right)^2 - \frac{1}{N_k}\left(\sum_{i=1}^{N_z}\sum_{j=1}^{N_{kj}} N_{ij}\right)^2\right] \quad (3.132)$$

where

s_{zz} = variance between the zones
N_z = number of zones
n_k = total number of permeability data in the strata
N_{ki} = number of permeability data in the ith zone
k_{ij} = permeability data
i = summation index for number of zones
j = summation index for the number of data within the zone

The variance within any zone, s_z, is computed from:

$$s_z = \frac{1}{N_k - N_z}\left[\sum_{i=1}^{N_z}\left(\sum_{j=1}^{N_{ki}}\left(N_{ij}^2\right)\right) - \sum_{i=1}^{N_z}\left(\frac{1}{N_{ki}}\left(\sum_{j=1}^{N_{kj}} N_{ij}\right)^2\right)\right] \quad (3.133)$$

and the zonation index, I_z, is:

$$I_z = 1 - \frac{s_z}{S_{zz}} \quad (3.134)$$

This index is the criterion used to indicate the best division. I_z, which ranges between 0 and 1, indicates how closely the division corresponds to homogeneous zones. The closer I_z is to 1, the more homogeneous are the zones. Any negative value of I_z must be replaced by zero in order to conform to the definition of I_z.

The zonation of individual wells is a multi-step procedure:

(a) First, the permeability data, in their original order of depth, are divided into all possible combinations of two zones. Then, the zonation index is calculated from Eq. (3.134), and the larger value, which denotes the best division into two zones, is retained for comparison with other indices.

(b) The permeability data of the best two-zone combination are divided into all possible three-zone combinations. The index I_z is again calculated for determining the best three-zone division.

(c) The permeability data of the best three-zone combinations are divided into all possible four-zone combinations. Then the zonation index criterion is applied.

The division into additional zones continues until the difference between two successive indices, ΔI_z, is negligible. Testerman found that the difference is negligible if $\Delta I_z < 0.06$ [45].

After all the wells in the reservoir have been zoned, the zones between adjacent wells are correlated for determining which strata are likely to be continuous, i.e., connected. Zones are considered to be connected if the difference in mean permeability of two zones in adjoining wells is less than or equal to that expected from variations of measurements within zones.

Example

Figure 3.73 shows the location of four wells in a consolidated sandstone reservoir selected to illustrate the use of the statistical zonation technique. Table 3.24 lists the permeability data and the corresponding depth for each of the four wells. The number of zones and the corresponding average permeability for each well must be determined.

Solution

To illustrate the zonation of individual wells, the permeability data of well No. 11 were selected. Table 3.25 shows the division of the permeability data into two zones. Equations (3.132)–(3.134) were used to compute the variance factors s_{zz} and s_z and the zonation index I_z, respectively, for each division into

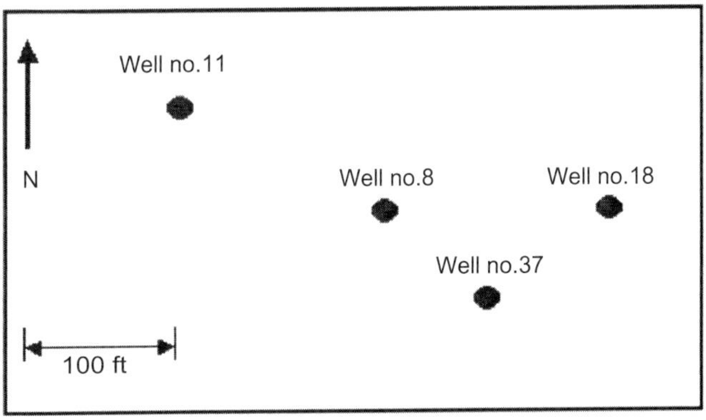

FIGURE 3.73 Location of wells [45].

TABLE 3.24 Reservoir Permeability Data [45]

Well No. 8		Well No. 11		Well No. 18		Well No. 37	
Depth (ft)	Permeability (mD)	Depth (ft)	Permeability (mD)	Depth (ft)	Permeability (mD)	Depth (ft)	Permeability (mD)
1,917.5*	11	1,906.5	10	1,973.5	20	1,922.5	34
1,918.5	27	1,907.5	52	1,974.5	40	1,923.5	67
1,919.5	157	1,908.5	276	1,975.5	190	1,924.5	20
1,920.5	234	1,909.5	140	1,976.5	146	1,925.5	197
1,921.5	390	1,910.5	139	1,977.5	53	1,926.5	186
1,922.5	90	1,911.5	165	1,976.5	4.8	1,927.5	33
1,923.5	192	1,912.5	342	1,979.5	0.0	1,928.5	30
1,924.5	218	1,913.5	87	1,980.5	45	1,929.5	21
1,925.5	42	1,914.5	0.0	1,981.5	14	1,930.5	117
1,926.5	120	1,915.5	0.0	1,982.5	0.0	1,931.5	27
1,927.5	158	–	–	1,983.5	84	1,932.5	27
1,928.5	315	–	–	1,984.5	28	1,933.5	26
1,929.5	20	–	–	1,985.5	0.0	1,934.5	61
1,930.5	99	–	–	1,986.5	0.0	–	–
1,931.5	121	–	–	1,987.5	0.0	–	–
1,932.5	43	–	–	1,988.5	0.0	–	–
1,933.5	88	–	–	–	–	–	–
1,934.5	7.4	–	–	–	–	–	–
1,935.5	149	–	–	–	–	–	–
1,936.5	0.0	–	–	–	–	–	–

*Top of the productive interval.

two zones. For example, the s_{zz}, s_z, and I_z values in the first column of Table 3.24 (i.e., depth = 1,917.5 ft and $k = 10$ mD) are computed as follows:

$$s_{ZZ} = \frac{1}{I}\left[\frac{(10)^2}{1} + \frac{(1{,}192)^2}{9} - \frac{(1{,}202)^2}{10}\right] = 13{,}493$$

TABLE 3.25 Division of Data of Well No. 11 into Two Zones [45]

Sample No. Per Group	Permeability (mD)	Cumulative Sum of Permeability	Grand Sum Minus Cum. Sum	D (mD^2)	W (mD^2)	R
1	0.0	10	1,192	13,493	13,600	0.0
2	52	63	1,140	19,892	12,800	0.35
3	276	338	864	243	15,256	0.0
4	140	478	724	3	15,286	0.0
5	139	617	585	102	15,273	0.0
6	156	773	429	1,118	15,146	0.0
7	342	1,115	87	35,646	10,830	0.69
8	87	1,202	0.0	36,120	10,771	0.70
9	0.0	1,202	0.0	16,053	13,280	0.17
10	0.0	1,202	0.0	–	–	–
Sum	1,202					

$$s_z = \frac{1}{8}\left[\begin{array}{l}(10)^2+(52)^2+(276)^2+(140)^2+(139)^2+(342)^2+\\(87)^2+(0)^2+(0)^2-\dfrac{(10)^2}{1}-\dfrac{(1,192)^2}{9}\end{array}\right]$$

$$= 13,600$$

$$I_z = 1-\frac{13,600}{13,493} = 0.008$$

Because $I_z < 0$, it is replaced by zero. The other lines are similarly calculated. Zone II, where $k = 0$ in samples No. 9 and No. 10, is easy to identify without any calculations.

Table 3.26 illustrates the next step in zonation of well No. 11, i.e., division of permeability data into three zones. Zone II of the two-zone division is now divided into two zones. Testerman attempted to divide the permeability into four zones [45]. It was found, however, that the largest four-zone index (0.79) is smaller than the three-zone index of 0.81, and concluded that well No. 11 is best described as three permeability zones.

TABLE 3.26 Division of Data of Well No. 11 into Three Zones [45]

Sample No. per Group	Permeability (mD)	Cumulative Sum of Permeability	Grand Sum Minus Cumulative Sum	*B* (mD^2)	*w* (mD^2)	*R*
1	10	10	1,192	29,300	9,098	0.68
2	52	63	1,140	37,021	6,893	0.81
3	276	338	864	21,450	11,341	0.47
4	140	478	724	218,242	11,229	0.48
5	139	617	585	22,866	10,937	0.52
6	156	773	429	23,564	10,737	0.54
7	342	1,115	87	20,346	11,657	0.42
8	87	1,202	0.0	–	–	–

The permeability data of wells 8, 18, and 37 were divided into zones using the same approach as for well No. 11. Table 3.27 indicates that each well is best described by three zones.

The results of the zonation between wells, which are not given here, define the existence of three continuous zones that have an average thickness, from top to bottom, of 2.1, 5.0, and 7.5 ft, and an average permeability of 33, 189, and 36 mD, respectively. Figure 3.74 is a cross-section showing the final zonation. The zonation technique is general and can therefore be applied to reservoir properties other than permeability, including porosity, formation resistivity factor, and fluid saturation. The statistical zonation technique assumes a priori that stratification exists within the reservoir. If the reservoir is not stratified, however, the statistical zonation technique will show it.

PROBLEMS

1. **(a)** Show that the porosity of a cubic packing of spherical grains of any diameter d_1 is 47.6%.

(b) Insert a spherical grain of diameter d_2 in the void space between the grains in (a) and recalculate the porosity.

(c) Insert a spherical grain of diameter d_3 in the void space between the grains in (a) and (b) and recalculate the porosity.

(d) Repeat the process above for n diameters and plot porosity versus the diameter of the grains.

TABLE 3.27 Final Zonation of Reservoir Permeability Data [45]

Well No. 8		Well No. 11		Well No. 18		Well No. 37	
Zone	Permeability (mD)	Zone	Permeability (mD)	Zone	Permeability (mD)	Zone	Permeablity (mD)
	11		10		20		34
(1, 8)		(1, 11)		(1, 18)		(1, 37)	
	27		52		40		67
	*		*		*		*
	157		276		190	*	
	234		140	(2, 18)			197
	390		139		146	(2, 37)	
	90	(2, 11)					186
	192		156		*		
(2, 8)			342				*
	218		87		53		
	42				4.8		33
	120		*		0.0		
	158				45		21
	316		0.0		14		117
		(3, 11)		(3, 18)	0.0	(3, 37)	
			0.0				20
	20				84		27
	99				28		26
	121				0.0		
	43				0.0		
(3, 8)					0.0		
	88				0.0		
	7.4						
	149						
	0.0						

Not enough data [37].

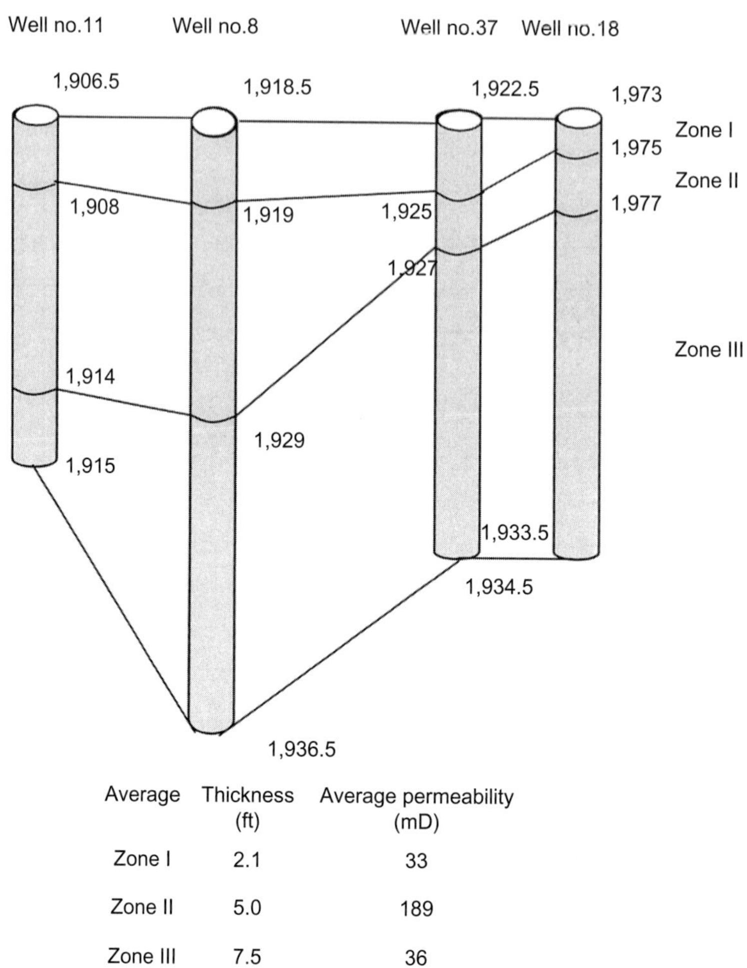

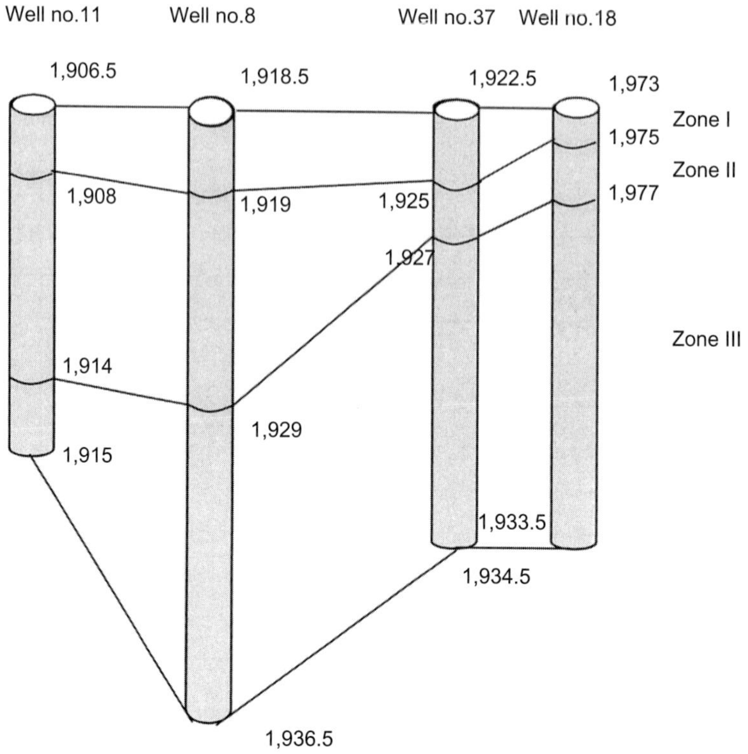

Average	Thickness (ft)	Average permeability (mD)
Zone I	2.1	33
Zone II	5.0	189
Zone III	7.5	36

FIGURE 3.74 Cross-section showing final zonation [45].

2. Calculate the porosity of the sample described below:

Dry sample mass	104.1 g
Mass of the water-saturated sample	120.2 g
Density of water	1.001 g/cm^3
Mass of saturated sample immersed in water	64.7 g

(a) Is this the effective or the total porosity of the sample?
(b) What is most probably the lithology of the sample? Explain why.

TABLE 3.28 Permeability and Porosity Data

Sample No.	k (mD)	ϕ
1	0.022	0.088
2	0.061	0.100
3	0.115	0.11
4	0.438	0.118
5	1.050	0.121
6	1.120	0.130
7	2.202	0.140
8	2.500	0.150
9	2.900	0.159

3. A core 2.54 cm in length and 2.54 cm in diameter has a porosity of 22%. It is saturated with oil and water. The oil content is 1.5 cm^3. What is the pore volume of the core? What are the oil and water saturations inside the core?

4. If a formation is 2.5 m thick, what is the volume of oil-in-place in 100 acres, if the core described in Problem 2 above is representative of the reservoir. Report the volume in cubic meters and barrels.

5. Assume that the permeability and porosity data in Table 3.28 are representative of several thousand data points taken for permeability–porosity measurements of a reservoir.

a. Plot the data and develop an equation that represents the relationship between permeability and porosity. Show the limits of the applicability of this equation.

b. Give an explanation for the deviation of the data that occurs for the high-permeability core.

c. Match your data to Figure 3.13 and use it to identify the most likely rock type of the reservoir.

d. Determine the correct mean value of the permeability.

6. If a core has a porosity of 18% and an irreducible water saturation of 24%, what is the permeability of the sample?

7. What is the matrix permeability of the core described in Problem 5 if the effective pore throat radius is 3.5 μm and the cementation factor is 2? Estimate the tortuosity of the sample.

8. Core analysis for permeability and porosity of 36 one-foot core samples obtained from a well located in a clean sandstone formation is provided in Table 3.29.

TABLE 3.29 Core Analysis Data for Permeability and Porosity of 36 One-Foot Core Samples Obtained from a Well Located in a Clean Sandstone Formation

Sample No.	k (mD)	ϕ	Cumulative h (ft)	Sample No.	K (mD)	ϕ	Cumulative h (ft)
1	100	0.268	1	19	1,720	0.266	19
2	822	0.354	2	20	500	0.275	20
3	436	0.264	3	21	495	0.269	21
4	220	0.26	4	22	612	0.206	22
5	348	0.258	5	23	897	0.264	23
6	256	0.272	6	24	974	0.272	24
7	150	0.256	7	25	790	0.351	25
8	127	0.255	8	26	955	0.358	26
9	36	0.272	9	27	1,030	0.273	27
10	779	0.257	10	28	784	0.266	28
11	945	0.263	11	29	491	0.262	29
12	815	0.295	12	30	623	0.313	30
13	1,190	0.277	13	31	557	0.255	31
14	928	0.355	14	32	937	0.358	32
15	238	0.286	15	33	854	0.279	33
16	78	0.274	16	34	818	0.272	34
17	1,780	0.262	17	35	363	0.285	35
18	1,510	0.269	18	36	306	0.315	36

a. Is the permeability distribution with depth in Table 3.29 linear, exponential, or logarithmic? Find the best curve-fit straight line.

b. Calculate the Dykstra–Parsons coefficient. Is the formation homogeneous or heterogeneous? Justify your answer.

c. Find arithmetic, geometric, and harmonic means of the permeability. Also, calculate the standard deviation, normalized mean, and dispersion of the three means.

d. Determine the arithmetic porosity mean and the median porosity.

9. Considering the porosity and permeability data in Table 3.29:

a. Determine the best permeability–porosity correlation.

b. What is the average grain diameter of each sample?

c. Calculate RQI and plot against porosity.
d. Determine the number of flow units and corresponding FZI.

10. The log surveys in a well indicated irreducible water saturation of 18% and an average porosity of 15%. Core analysis as well as well-test analysis indicated a reservoir rock permeability of 16 mD. Estimate the bound water irreducible and free fluid index for this formation.
11. Estimate the permeability (mD) of a reservoir rock which has a porosity of 13% and an irreducible water saturation of 20%. Assuming this permeability is representative of the flow unit:
    a. Calculate the reservoir quality index RQI (μm).
    b. Calculate the flow zone indicator FZI.
    c. Calculate the Tiab flow unit characterization factor H_T.
12. An NMR log was run in an oil well, indicating a porosity of 13% and log mean of relaxation time 250 μs at the depth of 7,500. Estimate the permeability at this depth for this well, using SDR correlations.
13. Core analysis in a new well in a formation with similar petrophysical characteristics to the TAGI formation revealed a horizontal permeability of 15 mD.
    A. Assuming that the formation has equal contents of Kaolinite and Illite, estimate the vertical permeability and anisotropy ratio for this well, using the models developed for the TAGI formation, for the following cases:
        a. Only horizontal permeability is known.
        b. Only horizontal permeability and porosity data are available (porosity = 13%).
        c. Horizontal permeability, porosity, and average grain diameter data are available (grain diameter = 0.025 cm).
    B. Rework the example, assuming that this formation contains a high percentage of Illite clay.
14. A large number of high quality back-scatter mode scanning electron images (BSE) for 14 core samples from 4 formations in Western Canada were used to investigate the relationship of sample permeability to core porosity. BSE image date were obtained from a section of a core plug, measuring 3.8 cm in diameter, while the core permeability and porosity were measured in the rest of the core sample. About 60 BSE images per core sample, each consisting of 765 × 573 picture elements (pixels) were taken for statistical image analysis. The conventional core analysis property values and the specific area per unit pore volume obtained by image analysis are summarized in Table 3.30. Calculate:
    (a) Specific area per unit-grain volume (S_{Vgr})
    (b) Pore-level effective zoning factor (K_T)
    (c) Lithology index (J_l)
    (d) Reservoir quality index
    (e) Number of low units

TABLE 3.30 Core analysis and BSE images results [83]

Core	Porosity	k,mD	Spv, μm^{-1}
58A	0.197	728	0.0241
45A	0.153	25.9	0.0522
45B	0.129	28	0.038
35B	0.101	3.51	0.0543
9B	0.152	5.29	0.0719
31B	0.129	1.78	0.0846
30B	0.122	2.09	0.0824
31A	0.102	0.47	0.0864
16	0.202	646	0.0284
15A	0.173	114	0.0538
7	0.134	412	0.0158
4B	0.069	1.65	0.0194
4A	0.198	6.5	0.0732
1	0.125	3	0.0288

(f) Formation resistivity factor, assuming the following form of the Kozeny-Carman model [92, 93] is applicable, where F_R = Formation resistivity factor and K_{SP} = Pore shape factor.

$$k = \frac{\phi^2}{K_{SP} F_R S_{Vp}^2}$$

NOMENCLATURE

A_i internal surface area
A_S surface area, acres
B formation volume factor (FVF), bbl/STB and SCF/ft^3
B^{-1} inverse of matrix coefficients b_{nj}
c_i volume vector
C conductivity
d_{gr} diameter of grain particles
d_i performance data
d_{ic} calculated values of performance data
d_{iOB} observed values of performance data
d depth

d_i deviation
f frequency
f_c cumulative frequency
F_R formation resistivity factor
G initial gas in place
h formation thickness
I_z zonation index
I_{s2} secondary porosity index
$\Delta I_{\phi 2}$ difference between two successive indices
k permeability
k_H horizontal permeability
k_n number of permeabilities
k_s permeability at 84.1% cumulative sample
k_v vertical permeability
K_S Kozeny shape factor
K_Z Kozeny constant
L_a core length, cm
L_k Lorenz permeability coefficient
m cementation factor or exponent
n number of capillary tubes
N initial oil in place
N_k number of permeability data
N_{ki} number of permeability data in zone i
N_z number of zones
P_D threshold pressure of capillary pressure curve
P pressure
ΔP pressure loss or difference
p_d displacement pressure
P_{WS} shut-in pressure
q flow rate
r radius
r_p pore radius
R random number
R_N uniformly distributed random number
R_t true formation resistivity, Ohm m
s standard deviation
s^2 sample variance
s_k^2 permeability variance
s_p^2 population variance
s_z^2 variance of zone
s_{zz}^2 variance between zones
s_1^2, s_2^2 sample variance in zones 1 and 2
S saturation
$S_{V_{gr}}$ specific surface area per unit grain volume
S_{iW} connate (irreducible)
S_{V_p} internal surface area per unit PV
t time
T temperature

u velocity
V volume
V_b bulk volume
V_k coefficient of permeability
V_P pore volume
V_{sh} shale content
V_{tot} volume of total shale, fraction
w fracture width
W water
x, y, z cartesian coordinates
x variable
x_M mode of variable

SUBSCRIPTS

ap apparent
e effective
f fracture
g gas
gr grain
h hydrocarbon
h highest value of variable x
i initial, index
irred irreducible
i, j, k, l, n, u indices
l lowest
ma matrix
o oil
r relative
sc standard conditions
t, tot total
w water

SUPERSCRIPTS

r run number

GREEK SYMBOLS

α angle between plane and pressure gradient
ε error
γ specific gravity
μ viscosity
θ contact angle
ϕ porosity
ϕ_o effective open porosity
τ tortuosity

REFERENCES

1. Fraser HJ, Graton LC. Systematic packing of spheres with particular relations to porosity and permeability. *J Geol* 1935;**November–December**:785–909.
2. Gatlin C. *Petroleum engineering–drilling and well completions*. Englewood Cliffs, NJ: Prentice-Hall Book; 1960. pp. 21–22.
3. Ellison Jr. SP. *Origin of porosity and permeability. Lecture notes*. Austin, TX: University of Texas; 1958.
4. Pittman EC. Microporosity in carbonate rocks. *Am Assoc Petrol Geol Bull* 1971;**55** (10):1873–81.
5. Hohlt RB. The nature and origin of limestone porosity. *Colorado School Mines Q* 1948;**43**:4.
6. Chilingarian GV, Mazzullo SV, Reike HH. *Carbonate reservoir characterization: a geologic engineering analysis, Part I. Developments in petroleum science 30*. Amsterdam, NY: Elsevier Science; 1992. 639 p.
7. Choquette PW, Pray LC. Geologic nomenclature and classification of porosity in sedimentary carbonates. *Am Assoc Petrol Geol Bull* 1970;**54**:207–50.
8. Archie GE. Classification of carbonate reservoir rocks and petrophysical considerations. *Am Assoc Petrol Geol Bull* 1952;**36**(6):278–98.
9. Lucia FJ. Petrophysical parameters estimated from visual descriptions of carbonate rocks: a field classification of carbonate pore space. *J Petrol Technol* 1983;**35**(March):626–37.
10. Kulander BR, Barton CC, Dean SL. The application of a fractography to core and outcrop fracture investigations. *US DOE METC/SP-79/3*. Springfield, VA: National Technical Information Service; 1979.
11. Anderson G. *Coring and core analysis handbook*. Tulsa, OK: Petroleum Publishing; 1975. 200 p.
12. Darcy HJ. *Les Fontaines Publiques de la Ville de Dijon*. Paris: Libraire de Corps Impériaux des Pontset Chaussées et des Mines; 1856. p. 590–594.
13. Klinkenberg LJ. The permeability of porous media to liquids and gases. *API DrillProd Pract* 1941;200.
14. Clark NJ. *Elements of petroleum reservoirs. Henry L. Doherty series*. revised ed. Dallas, TX: Society of Petroleum Engineers; 1969. p. 19–30.
15. Chilingarian GV. Relationship between porosity, permeability and grain size distribution of sands and sandstones. In: Van Straaten JU, editor. *Deltaic and shallow marine deposits*. Amsterdam, NY: Elsevier Science; 1963. p. 71–5.
16. Chilingarian GV, Wolf KH. *Compaction of coarse-grained sediments*. New York: Elsevier; 1975. Figs. 1–25, p. 33.
17. Kozeny J. Über kapillare Leitung des Wassers im Boden (Aufstieg Versikerung und Anwendung auf die Bemasserung). *Sitzungsber Akad., Wiss, Wein, Math- Naturwiss, KL* 1927;**136**(Ila):271–306.
18. Pirson SJ. *Oil Reservoir Engineering*. 2nd ed. New York: McGraw-Hill Book; 1958. 735 p.
19. Amyx JW, Bass Jr. DM, Whiting RL. *Petroleum reservoir engineering*. New York: McGraw-Hill Book; 1960. 610 p.
20. Wyllie MRJ, Spangler MB. Application of electrical eesistivity measurements to problems of fluid flow in porous media. *Am Assoc Petrol Geol Bull* 1952;**February**.
21. Carman PC. Permeability of saturated sands, soils and clays. *J Agric Sci* 1939;**29**. Also, *J. Soc. Chem. Ind.*, 1939, pp. 57–58.

22. Babadagli T, Al-Salmi SA. Review of permeability-prediction methods for carbonate reservoirs using well-log data. *SPE Reservoir Eval Eng* 2004;**7**(2):76.
23. Bear J. *Dynamics of fluids in porous media*. New York: Elsevier; 1972.
24. Ebanks WJ. The flow unit concept—an integrated approach to reservoir description for engineering projects. *Am Assoc Geol Annu Convention* 1987.
25. Hear CL, Ebanks WJ, Tye RS, Ranganatha V. Geological factors influencing reservoir performance of the Hartzog Draw Field, Wyoming. *J Petrol Technol* 1984; **August**:1335–44.
26. Gunter GW, Finneran JM, Hartman DJ, Miller JD. Early determination of reservoir flow units using an integrated petrophysical method. SPE 38679. In: *SSPE annual technical conference and exhibition*. San Antonio, TX; October 5–8, 1997.
27. Kolodzie Jr S. Analysis of pore throat size and use of the waxman-smits equation to determine OOIP in spindle field, Colorado. Paper SPE 9382. In: SPE AITC. Dallas, TX; September 1980.
28. Jennings JW, Lucia FJ. Predicting permeability from well logs in carbonates with a link to geology for interwell permeability mapping. *SPE Reservoir Eval Eng* 2003;**August**.
29. Aguilera R. Incorporating capillary pressure, pore aperture radii, height above free water table, and winland r_{35} values on pickett plots. *Am Assoc Petrol Geol Bull* 2002;**86**(4):605.
30. Kwon BS, Picken GR. A new pore structure model and pore structure interrelationships. *Trans, SPWLA* 1975.
31. Tiab D. *Advances in Petrophysics, Vol. 1—flow units. Lecture notes manual*. University of Oklahoma; 2000.
32. Tiab D. *Modern core cnalysis, Vol. I—theory*. Houston, TX: Core Laboratories; 1993. 200 p.
33. Tiab D, Marschall DM, Altunbay MH. Method for identifying and characterizing hydraulic unitsof saturated porous media: tri-kappa zoning process. U.S. Patent No. 5,193,059; March 9, 1993.
34. Amaefule JO, Altunbay MH, Tiab D, Kersey DG, Keelan DK. Enhanced reservoir description using core and log data to identify hydraulic (flow) units and predict permeability in uncored intervals/wells. *Soc Petrol Eng* 1993. Paper No. 26436.
35. Zemanek J. Low-resistivity hydrocarbon-bearing sand reservoirs. *SPEFE* 1989; **December**:515–21.
36. Sneider RM, King HR, Hawkes HE, Davis TB. Methods for detection and characterization of reservoir rocks, Deep Basin Gas Area, Western Canada. *SPE Trans* 1983;**275**:1725–34.
37. Coates G, Denoo S. The producibility answer product. *Tech Rev Schlumberger,* Houston, TX 1981;(2):55–66.
38. Leverett MC. Capillary behavior in porous solids. *Trans AIME* 1941.
39. Amyx JW, Bass DM, Whiting RL. *Petroleum reservoir engineering – physical properties*. New York: McGraw-Hill; 1960.
40. Rose W, Bruce WA. Evaluation of capillary characters in petroleum reservoir rocks. *Trans AIME* 1949.
41. Brown HW. Capillary pressure investigations. *Trans AIME* 1951.
42. Cuddy S. *The FOIL-function, a simple, convincing model for calculating water saturation in southern north sea gas fields*. Paper H1-17. Presented at the SPWLA Annual Logging Symposium. Calgary, Canada; 1993.
43. Amabeoku MO, Kersey DG, BinNasser RH, Belowi AR. Relative permeability coupled saturation-height models on the basis of hydraulic (flow) units in a gas field. *SPE Reservoir Eval Eng* 2008;**December**.

44. Siddiqui S, Okasha TM, Funk JJ, Al-Harbi AM. New Representative Sample Selection Criteria for Special Core Analysis. SCA 2003-40. *International Symposium of Society of Core Analysts*. Pau, France. September 21–24, 2003.
45. Testerman JDA. Statistical reservoir-zonation technique. *Soc Petrol Eng J* 1962; **August**:889–93.
46. Funk JJ, Balobaid YS, Al-Sardi AM, Okasha TM. *Enhancement of reservoir characteristics modeling, Saudi Aramco Engineering report no. 5684*. Dhahran: Saudi Aramco; 1999.
47. Slichter SC. Theoretical investigation of the motion of ground water. *US Geological Survey, 19th annual report, Part II*. Springfield, VA: NTIS; 1899. pp. 295–384.
48. Wyllie MRJ, Rose WD. Some theoretical considerations related to the quantitative evaluation of the physical characteristics of reservoir rock from electric log data. *Trans AIME* 1950;**189**:105–18.
49. Schlumberger Inc. *Log interpretations charts*. Houston, TX: Schlumberger Educational Services; 1977.
50. Timur A. An investigation of permeability, porosity and residual water saturation relation for sandstone reservoirs. *Log Analyst* 1968;**9**:4.
51. Langnes GL, Robertson Jr. OJ, Chilingar GV. *Secondary recovery and carbonate reservoirs*. Amsterdam, NY: Elsevier Science; 1972. 304 pp.
52. Al-Ajmi FA, Holditch SA. NMR permeability calibration using a non-parametric algorithm and data from formation in central arabia. SPE Paper 68112. SPE MEOS, Bahrain; March 17–20, 2001.
53. Hidajat I, Mohanty KK, Flaum M, Hitasaki G. Study of vuggy carbonates using NMR and X-ray CT scanning. *SPERE J* 2004;**October**:365 –377.
54. Shokir EMA. Novel model for permeability prediction in uncored wells. *SPE Reservoir EvalEng J* 2006;**June**:266–73.
55. Takagi T, Sugeno M. Fuzzy identification of systems and its applications to modeling and control. *IEEE Trans Syst, Man Cybern* 1985;**15**(1):116–32.
56. Mohaghegh S, Balan B, Ameri S. Permeability determination from well log data. *SPE Format Eval* 1997;**12**(3):170.
57. Craze RC. Performance of limestone reservoirs. *Trans AIME* 1950;**189**:287–94.
58. Bagrintseva KI. *Carbonate rocks, oil and gas reservoirs*. Moscow: *Izdatet' stvo Nedra*; 1977. 231.
59. Mortensen J, Engstrom F, Lind I. The relation among porosity, permeability, and specific surface of chalk from the Gorn Field, Danish North Sea. *SPE Reservoir Eval Eng* 1998;**I** (3):245.
60. Parsons RW. Permeability of idealized fractured rock. *Soc Petrol Eng I* 1966;**6**:126–36.
61. Murray GH. Quantitative Fracture Study—Sanish Pool, McKenzie County, North Dakota. *Am Assoc Petrol Geol Bull* 1968;**52**:57–65.
62. McGuire WJ, Sikora VJ. The effect of vertical fractures on wells productivity. *Trans AIME* 1960;**219**:401–3.
63. Steams DW, Friedman M. Reservoirs in fractured rock. *Am Assoc Petrol Geol (AAPG), Memoir 16* and *Soc Expl Geophys*, Special Publ. No. 10, 1972.
64. Watfa M, Youssef FZ. An improved technique for estimating permeability in carbonates. Societyof Petrololeum Engineering Paper No. 15732, *5th SPE Middle East Oil Show*. Bahrain; March 1987. pp. 7–10.
65. Ambastha AK, Almatar D, Ma E. Long-term field development opportunity assessment using horizontal wells in a thin, carbonate reservoir of the greater burgan field, Kuwait. *SPE Reservoir Eval Eng*, February, 2009.

66. Lucia FJ. Rock fabric/petrophysical classification of carbonate pore space for reservoir characterization. *Am Assoc Petrol Geol Bull* 1995;**79**(9):1275.
67. Moncada K. Application of TDS technique to calculate vertical and horizontal permeabilities for vertical wells with partial completion and partial penetration. M.Sc. thesis, University of Oklahoma; 2003.
68. Manseur S, Tiab D, Berkat A, Zhu T. *Horizontal and vertical permeability determination in clean and shaly reservoirs using in-situ measurement.* SPE Paper 75773. Presented at SPE WR/AAPG Pacific Section joint meeting. Anchorage, Alaska; May 20–22, 2002.
69. Zahaf K, Tiab D. Vertical permeability from in situ horizontal measurements in shaly-sand reservoirs. *JCPT* 2002;**August**:43–50.
70. Peffer MA, O'Callagan PJ. In-situ determination of permeability anisotropy and its vertical distribution—a case study. SPE 38942. *Proceedings, SPE ATCE.* San Antonio, TX; October 5–8, 1997.
71. Hutchinson CA, Dodge CF, Polasek TL. Identification and prediction of reservoir nonuniformities affecting production operations. *Soc Petrol Eng J* 1961;**March**:223–30.
72. Warren CA, Price HS. Flow in heterogeneous porous media. *Soc Petrol Eng J* 1961; **September**:153–69.
73. Kelkar M, Perez G. *Applied geostatistics for reservoir characterization.* Richardson, TX: SPE; 2002.
74. Law J. Statistical approach to the interstitial heterogeneity of sand reservoirs. Trans AIME 1944;**155**.
75. Schmalz JP, Rahme HD. The variation of waterflood performance with variation in permeability profile. *Producers Mon* 1950;**15**(9):9–12.
76. Dykstra, H, Parsons RL. The prediction of oil recovery in waterflood. *Secondary recovery of oil in the United States.* 2nd ed. American Petroleum Institute (API), 1950, p. 160–74.
77. Jensen JL, Lake LL, Corbett PWM, Goggin DJ. *Statistics for petroleum engineers and geoscientists.* Upper Saddle River, NJ: Prentice Hall; 1997. pp. 144–66.
78. Saner S, Sahin A. Lithological and Zonal porosity-permeability distributions in the Arab D Reservoir, Uthmaniyah Field, Saudi Arabia. *Am Assoc Petrol Geol Bull* 1999;**83**(2):230–43.
79. Craig FF. *The reservoir engineering aspects of waterflooding. Society of Petrololeum Engineering monograph*, 3. Dallas, TX; 1971.
80. Ahmed T. *Reservoir engineering handbook.* 2nd ed. Gulf Professional; 2001.
81. Tehrani DH, Chen GL, Peden JM. Effective permeability of a heterogeneous reservoir for calculating well-bore pressure. *Proceedings, regional symposium on IOR in the Gulf Region*; December 17–19, 1995. pp. 145–55.
82. Gelhar LW, Axness CL. Three-dimensional stochastic analysis of macrodispersion in aquifers. *Water Resour Res* 1983;**19**:161.
83. Chatzis I, Jewlal DM, Loannidis MA. Core sample permeability estimation using statistical image analysis. Paper SCA 9723. *Proceedings, international symposium of SCA.* Calgary, Canada; September 7–10, 1997.
84. Willhite G Paul. *Waterflooding.* 7th ed. Richardson, TX: SPE; 2001.
85. Winsauer WO, Shearin HM, Masson PH, Williams M. Resistivity of brine saturated sands in relation to pore geometry. *Am Assoc Petrol Geol Bull* 1952;**February**.
86. Tiab D. Analysis of pressure and pressure derivative without using type-curve matching—skin and wellbore storage. *J Petrol Sci Eng* 1994;**11**:323–33.
87. Greenkorn RA, Johnson CR, Shallenberger LK. Directional permeability of heterogeneous anisotropic porous media. *Soc Petrol Eng I* 1964;**June**:124–32.

88. Trudgen P, Hoffmann F. Statistically analyzing core data. *Soc Petrol Eng J* 1967; **April**:497–503.
89. Zahaf K. Vertical permeability from in-situ horizontal permeability measurement in shaly sand reservoirs. MS thesis, University of Oklahoma; 1999.
90. Larsen JK, Fabricius IL. Interpretation of water saturation above the transitional zone in chalk reservoirs. *SPE Reservoir Eval Eng* 2004;**7**(2):155.
91. Fanchi JR. *Principles of applied reservoir simulation.* 2nd ed. Houston, TX: Gulf Publishing; 2000.
92. Berryman JG, Blair SC. Kozeny-Carman relations and image processing methods for estimating Darcy's constant. *J Appl Phys* 1987;**62**(6):2221.
93. Berryman JG, Milton GW. Normalization constraint for variational bounds on fluid permeability. *J Chem Phys* 1985;**83**(2):754.

Chapter 4

Formation Resistivity and Water Saturation

Sedimentary formations are capable of transmitting an electric current only by means of their interstitial and adsorbed water content. They would be nonconductive if they were entirely dry. The interstitial or connate water containing dissolved salts constitutes an electrolyte capable of conducting current, as these salts dissociate into positively charged cations (such as Na^+ and Ca^{2+}) and negatively charged anions (such as Cl^- and SO_4^{2-}). These ions move under the influence of an electrical field and carry an electrical current through the solution. The greater the salt concentration, the greater the conductivity of connate water. Freshwater, for example, has only a small amount of dissolved salts and is, therefore, a poor conductor of an electric current. Oil and gas are nonconductors.

The electrical resistivity (reciprocal of conductivity) of a fluid-saturated rock is its ability to impede the flow of electric current through that rock. Dry rocks exhibit infinite resistivity. In electrical logging practice, resistivity is expressed in $\Omega\ m^2/m$ or simply Ω m. The resistivity of most sedimentary formation ranges from 0.2 to 2,000 Ω m. The resistivity of poorly consolidated sand ranges from 0.20 Ω m for sands containing primarily saltwater to several ohm-meter for oil-bearing sands. For well-consolidated sandstones, the resistivity ranges from 1 to 1,000 Ω m or more depending on the amount of shale interbedding. In nonporous carbonate rocks, resistivity may be as high as a few million ohm-meter. The resistivity of reservoir rocks is a function of salinity of formation water, effective porosity, and quantity of hydrocarbons trapped in the pore space [1]. Relationships among these quantities indicate that the resistivity decreases with increasing porosity and increases with increasing petroleum content. Resistivity measurements are also dependent upon pore geometry, formation stress, composition of rock, interstitial fluids, and temperature. Resistivity is, therefore, a valuable tool for evaluating the producibility of a formation.

FORMATION RESISTIVITY FACTOR

A rock that contains oil and/or gas will have a higher resistivity than the same rock completely saturated with formation water, and the greater the connate water saturation, the lower the formation resistivity. This relationship to saturation makes the formation resistivity factor an excellent parameter for the detection of hydrocarbon zones.

RESISTIVITY MEASUREMENT

The resistance of brine in a container of length L and cross-sectional area A to the flow of electricity is measured by applying a voltage E, in volts, across the liquid and recording the amount of current I, in amperes, that will flow, as shown in Figure 4.1. According to Ohm's law, the resistance, r_w, is given by:

$$r_w = \frac{E}{I_w} \tag{4.1}$$

The resistivity of the brine is:

$$R_w = r_w \frac{A}{L} = \frac{E}{I_w} \frac{A}{L} \tag{4.2}$$

Now consider a block of porous rock (clean sand) of the same dimensions A and L, and 100% saturated with the same brine (Figure 4.2). On applying the same voltage E across the block of sand, a current I_o will flow. The resistivity of this porous rock sample, R_o, is expressed as:

$$R_o = r_o \frac{A}{L} = \frac{E}{I_o} \frac{A}{L} \tag{4.3}$$

Dividing Eq. (4.3) by Eq. (4.2) and canceling similar terms gives:

$$\frac{R_o}{R_w} = \frac{I_w}{I_o} \tag{4.4}$$

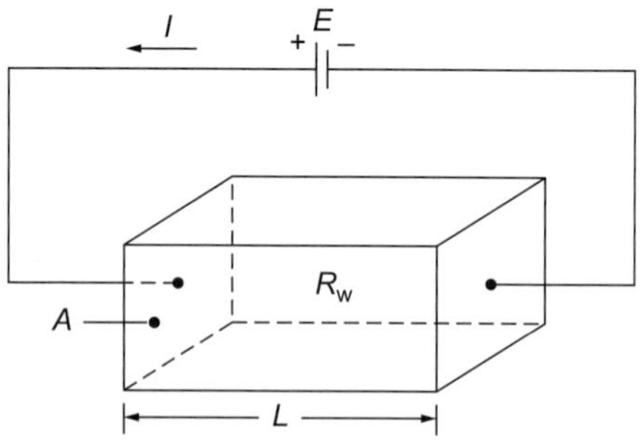

FIGURE 4.1 Resistivity measurement of salty water [1].

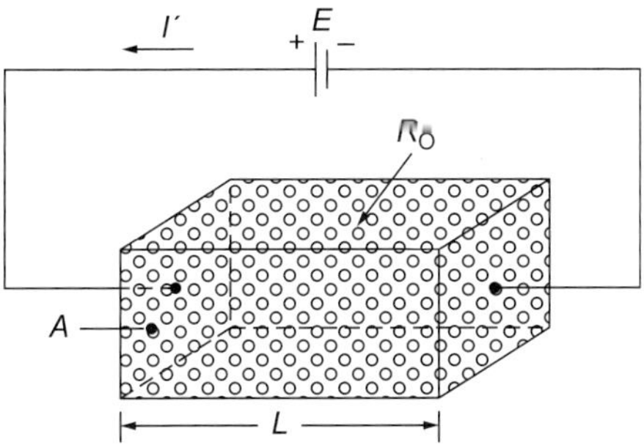

FIGURE 4.2 Resistivity measurement of a porous rock sample [1].

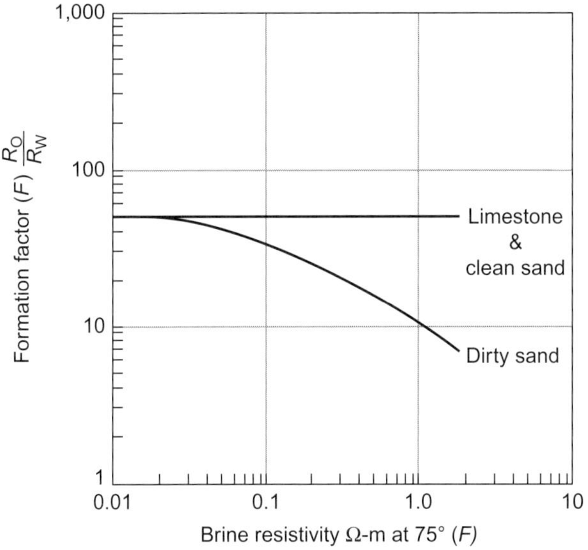

FIGURE 4.3 General relationship between formation factor F and brine resistivity factor R_w. Source: *Courtesy of Core Laboratories.*

Archie defined this ratio as the "formation resistivity factor F_R" or [2]:

$$F_R = \frac{R_o}{R_w} \tag{4.5}$$

For a given value of E, I_w will be greater than I_o. Hence, R_o will be greater than R_w and F_R will always be greater than unity. Figure 4.3 shows the qualitative effect of brine resistivity (assuming all other factors, such as porosity, cementation, and amount of shale, remain constant) on F_R for limestone and clean sand, and shaly ("dirty") sand. The formation factor is essentially constant for clean sand and limestone. For dirty or shaly sand, F_R decreases as brine resistivity, R_w, increases, and although R_o increases, it does not increase proportionately because the clay in the water acts as a conductor. This effect is dependent upon the type, amount, and manner of distribution of the clay in the rock. Equation (4.5) is an important relation in well log interpretation for locating potential zones of hydrocarbons.

Of all the rock parameters measured by modern well-logging tools, resistivity is essential because it is used to determine water saturation, which is then used to calculate the volume of oil- and/or gas-in-place. The producibility of the formation can also be estimated by comparing the resistivity close to the wellbore (i.e., flushed-zone resistivity), where mud filtrate has invaded the formation, and resistivity of the virgin portion of the formation (i.e., true resistivity, R_t.)

DETERMINATION OF FORMATION WATER RESISTIVITY

The value of R_w can vary widely from well to well in some reservoirs because parameters that affect it include salinity, temperature, freshwater invasion, and changing depositional environments. However, several methods for determining the reservoir water resistivity have been developed, including chemical analysis of produced water sample, direct measurement in a resistivity cell, water catalogs, spontaneous potential (SP) curve, resistivity–porosity logs, and various empirical methods.

Chemical Analysis

Although direct measurement of R_w in a resistivity cell is always preferred, chemical analysis of water samples is still performed. In many cases, R_w is estimated using a logging tool, for example, from the SP log. Chemical analysis of uncontaminated water samples yields representative values of formation water resistivity as a function of the salinity of the sample and reservoir temperature, using Figure 4.4. Salinity is a measure of the connection of dissolved salts, which is expressed in parts per million (ppm), grains per gallon, or grams per liter of sodium chloride. One grain per gallon is approximately equal to 17.2 ppm or 17.14×10^{-3} g/L. Inasmuch as sodium chloride is the most common salt present in the formation water, ionic concentration of other dissolved salts is generally converted to the NaCl concentration equivalent, using Figure 4.5. The abscissa of this chart is actually the sum of the concentrations of each ion. Once the weighting multipliers for the various ions, present in the water sample, are determined from Figure 4.5, the concentration of all ions, C_{si}, is multiplied by its "multiplier." The equivalent

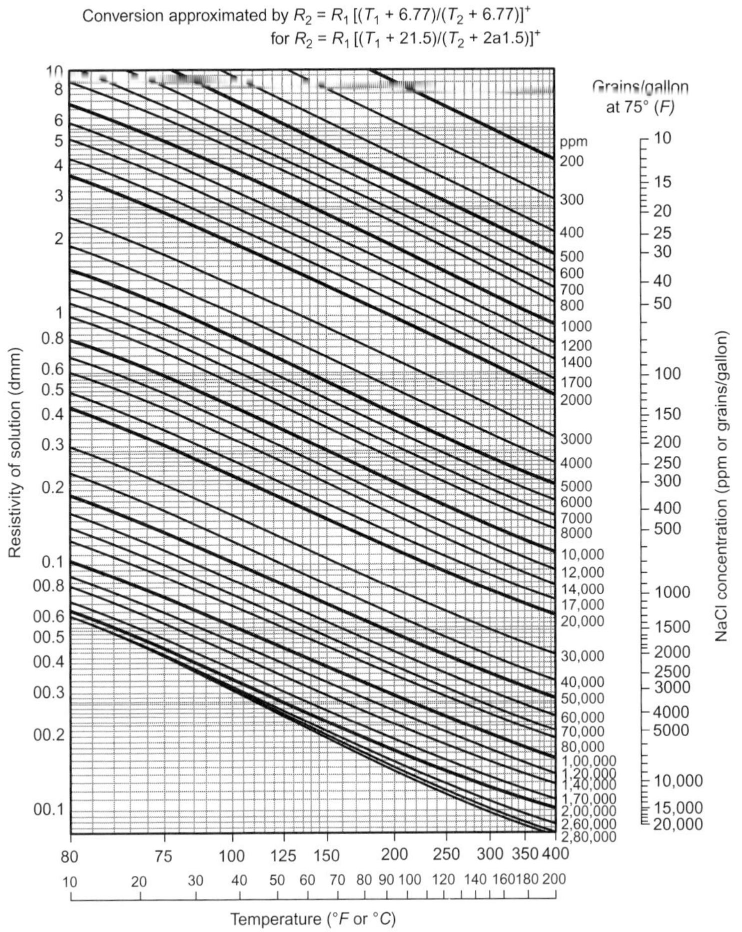

FIGURE 4.4 Resistivity of water as a function of salinity and temperature. Source: *Courtesy of Schlumberger Co.*

NaCl concentration in ppm, C_{sp}, is obtained by adding the products for all ions, or:

$$C_{sp} = \sum_{i=1}^{n} M_i C_{sii} \tag{4.6}$$

where

n = number of ions in the solution
M_i = weighting multiplier (see Figure 4.5)
C_{sii} = concentration of each ion, ppm

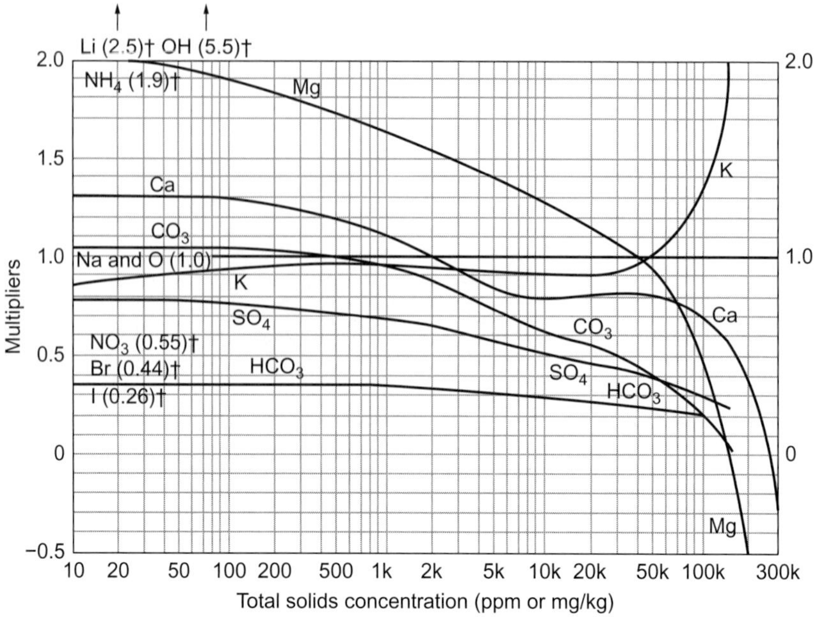

FIGURE 4.5 Multipliers for converting ionic concentrations of dissolved salts to NaCl concentration equivalents. Source: *Courtesy of Schlumberger Co.*

Figure 4.4 shows that the resistivity of formation water decreases as the temperature increases. To estimate the effect of temperature on water resistivity, Arp's formula can be used:

$$R_{wT_2} = R_{wT_1}\left(\frac{T_1 + 6.77}{T_2 + 6.77}\right) \tag{4.7}$$

where R_{wT_1} and R_{wT_2} are the water resistivities (Ω m) at formation temperature T_1 and T_2, respectively. In freshwater, the constant 6.77 in Eq. (4.7) is not necessary. The temperatures T_1 and T_2 are expressed in °F. If 75°F is used as a reference temperature, Eq. (4.7) can be used to find water resistivity at any reservoir temperature T:

$$R_{wT} = R_{wT}\left(\frac{81.77}{T + 6.77}\right) \tag{4.8}$$

The value of R_{w75} may be estimated from the following equation [3]:

$$R_{w75} = \frac{1}{2.74 \times 10^{-4} C_{sp}^{0.955}} + 0.0123 \tag{4.9}$$

Inasmuch as the slopes of the salinity lines in Figure 4.4 are not constant, and these lines are not perfectly straight, the following equation may be derived [4]:

$$R_{wT} = R_{wT_1}\left(\frac{T_1 + X_H}{T + X_H}\right) \tag{4.10}$$

where

$$X_H = 10^{X_R} \tag{4.11}$$

and

$$X_R = -0.3404\log(R_{wT_1}) + 0.6414$$

X_H is a function of salinity and compensates for the change in slope and accounts for the curving or deviation of the constant salinity lines below 75°F and above 300°F. Equation (4.10) yields more accurate values of R_w than does Eq. (4.7), for all ranges of salinity and temperature (see also Tables 4.1 and 4.2). Table 4.2 is very convenient for programming purposes.

TABLE 4.1 Solids Concentration and Multiplier

C_{st}	Multiplier*					
	HCO_3	SO_4	K	CO_3	Ca	Mg
10	0.35	0.78	0.84	1.05	1.3	2.2
20	0.35	0.78	0.88	1.05	1.32	2
50	0.35	0.78	0.92	1.06	1.32	1.95
100	0.35	0.77	0.94	1.05	1.29	1.89
200	0.35	0.76	0.95	1.02	1.28	1.82
500	0.35	0.72	0.96	0.98	1.2	1.7
1,000	0.34	0.69	0.95	0.95	1.1	1.62
2,000	0.33	0.65	0.94	0.89	0.989	1.53
10,000	0.29	0.5	0.9	0.635	0.789	1.28
20,000	0.25	0.462	0.91	0.55	0.818	1.14
50,000	0.2	0.38	1.01	0.39	0.798	0.92
100,000	0.18	0.3	1.3	0.2	0.71	0.52
150,000		0.23	2	0.03	0.572	
200,000		0.18			0.372	
300,000					−0.685	

*M = 1 for Na^+ and Cl^-

TABLE 4.2 Multiplier (M_i) as a Function of C_{st} (ppm), Used Instead of Figures 4.4 and 4.5: C_{st} = Total Salt Concentration, ppm; M = Multiplier

	Equation for HCO_3 ($C_{sp} \leq 100K$)
1	$M = 3.446 \times 10^{-1} + 9.121 \times 10^{-6} C_{st} - 7.8434 \times 10^{-9} C_{st}^{1.5} - 5.16 \times 10^{-4} \ln(C_{st}) C_{st}^{0.5} + 3.4326 C_{st}^{0.5}$
	Equation for SO_4 ($C_{st} \leq 140K$)
2	$M = \frac{1}{1.22235 + 6.6 \times 10^{-3} C_{st}^{0.5}}$
	Equation for K ($C_{st} \leq 100K$)
3	$M = 8.61 \times 10^{-1} + 7.26177 \times 10^{-3} [\ln(C_{st})]^2 + 2.6376 \times 10^{-4} \left(\frac{C_{st}}{\ln C_{st}}\right) - 8.558 \times 10^{-3} C_{st}^{0.5}$
	Equation for CO_3
4	$M = -9.306 \times 10^{-1} + 4.24 \times 10^{-1} e^{-2.486 \times 10^{-4} C_{st}} + 1.573 e^{-3.308 \times 10^{-6} C_{st}}$
	Equation for Ca ($C_{st} \leq 275K$)
5	$M = 1.026 - 9.7454 \times 10^{-5} C_{st} - 2.675 \times 10^{-2} [\ln(C_{st})]^2 + 1.255 \times 10^{-3} \left(\frac{C_{st}}{\ln(C_{st})}\right) + 1.8266 \times 10^{-1} \ln(C_{st})$
	Equation for Mg ($C_{st} \leq 200K$)
6	$M = 2.06566 - 1.6307 \times 10^{-8} C_{st}^{1.5} - 8.2966 \times 10^{-3} [\ln(C_{st})]^2$
7	For Na and Cl, $M = 1$ for any value of C_{st}
8	For Br, $M = 0.44$ for any value of C_{st}
9	For NO_3, $M = 0.55$ for any value of C_{st}

Example

The chemical analysis of an oil reservoir brine yielded the following ionic concentrations: 11,000 ppm Na^+, 15,000 ppm Cl^-, 8,000 ppm Mg^{2+}, 6,000 ppm Ca^{2+}, and 10,000 ppm SO_4^{2-}. Room temperature = 75°F. Calculate:

1. The equivalent NaCl concentration.
2. The resistivity of the formation brine at 150°F.

Solution

1. The total solids concentration, C_{st} is 11,000 + 15,000 + 8,000 + 6,000 + 10,000 = 50,000 ppm. Table 4.1 and Figure 4.5 show the following multipliers: 1 for Na^+ and Cl^-, 0.92 for Mg^{2+}, 0.798 for Ca^{2+}, and 0.38 for SO_4^{2-}. The equivalent NaCl concentration (Eq. (4.6)) is given by:

$$C_{sp} = 1 \times 11,000 + 1 \times 15,000 + 0.92 \times 8,000 + 0.798 \times 6,000 + 0.38 \times 10,000 = 41,948$$

2. The water resistivity at the reference temperature is calculated from Eq. (4.9):

$$R_{w75} = \frac{1}{2.74 \times 10^{-4} \times 41,948^{0.955}} + 0.0123 = 0.153\Omega\ \text{m}$$

The resistivity of the brine at 150°F is obtained from Eq. (4.8):

$$R_{w150} = 0.153\left(\frac{81.77}{150 + 6.77}\right) = 0.0797\Omega\ \text{m}$$

Equation (4.10) yields $R_{w150} = 0.08$ for $X_H = 8.3$.

Resistivity–Concentration Equations

Worthington et al. recommended that brine resistivity of reservoir waters that are not pure NaCl solution should be measured in a well-equipped standard laboratory, using a calibrated resistivity cell [3]. However, the resistivity of an accurately prepared NaCl solution can be calculated from accurately measured masses or volumes of components.

Using data in the literature, Worthington et al. obtained several correlations for conversion between resistivity and concentration of NaCl solutions. For a given concentration of NaCl solution in mol/L, C_{sm}, the water resistivity at 25°C is:

$$R_{w25} = 10^{-X_c} \tag{4.12}$$

$$X_c = a_1 + a_2(\log C_{sm}) + a_3(\log C_{sm})^2 + a_4(\log C_{sm})^3 + a_5(\log C_{sm})^4 + a_6(\log C_{sm})^5 \tag{4.13}$$

where

X_c = concentration correlation function

TABLE 4.3 Concentration Correlation Constants for Eq. (4.13)

Concentration Correlation Constant	Range of Concentration of NaCl Solution		
	10^{-4} to 10^{-1}	C_{sm} (mol/L) 0.09–1.1	1.0–5.35
a_1	9.42203×10^{-1}	9.33134×10^{-1}	9.33292×10^{-1}
a_2	8.889×10^{-1}	8.47181×10^{-1}	8.45971×10^{-1}
a_3	-2.72398×10^{-2}	-1.04563×10^{-1}	-910632×10^{-1}
a_4	-2.25682×10^{-3}	-7.05334×10^{-2}	-2.27399×10^{-1}
a_5	1.46605×10^{-5}	-2.35562×10^{-2}	2.27456×10^{-1}
a_6	0	0	-3.81035×10^{-1}

C_{sm} = concentration of NaCl solution, mol/L
$a_1, \ldots, a_6$ = correlation constants

Table 4.3 shows the value of the correlation constants for three ranges of concentration: $10^{-4} < C_{sm} < 0.1$, $0.09 < C_{sm} < 1.4$, and $1.0 < C_{sm} < 5.35$.

Arp's equation for calculating R_w at any reservoir temperature T in degrees Celsius is:

$$R_{wT} = R_{w25}\left(\frac{46.5}{T + 21.5}\right) \tag{4.14}$$

If the concentration of NaCl solution is given in ppm, then it is necessary to convert it to mol/L, in order to solve the concentration correlation function X_c and then R_{w25}. The conversion equation between ppm and mol/L is [3]:

$$C_{sm} = \left(\frac{\rho_{25}}{58.443 \times 10^3}\right) C_{sp} \tag{4.15}$$

where

C_{sp} = concentration of NaCl solution, ppm
ppm = mass of NaCl per 10^6 mass units of solution
ρ_{25} = density of NaCl solution at 25°C, g/cm^3
58.443 = molecular mass of NaCl, g

The value of the density of the NaCl solution as a function of the concentration C_{sm} is given by:

$$\rho_{25} = 0.99708 + 0.040785C_{sm} - 9.5818 \times 10^{-4}C_{sm}^2 + 5.1208 \times 10^{-5}C_{sm}^3 \tag{4.16}$$

To convert a given value of concentration in ppm to mol/L, and then to calculate the resistivity R_{w25}, one must employ Eqs. (4.15) and (4.16) in an iterative procedure:

(1) Substitute C_{sp} and $\rho_{25} = 1$ into Eq. (4.15), and calculate the value of C_{sm}.
(2) Substitute this value of C_{sm} into Eq. (4.16), and calculate a new density value.

(3) Substitute this value of density into Eq. (4.15), and calculate a new value of C_{sm}. Repeat these three steps until successive iterations yield $\Delta\rho_{25} < 10^{-4}$ g/cm^3.

(4) Select the correlation constants from Table 4.3 for the range in concentration that includes the value of C_{sm} calculated in Step 3, and then calculate the concentration correlation function X_c (Eq. (4.13)).

(5) Substitute this value of X_c into Eq. (4.12), and calculate the resistivity R_{w25}.

(6) Use Arp's equation (4.14) to convert this resistivity at 25°C to a resistivity at reservoir temperature R_{wT}.

An alternative method to Steps 1–3 is to use the following equation to convert C_{sp} to C_{sm}:

$$C_{sm} = \frac{17.061}{(10^6/C_{sp}) - 0.69787} \tag{4.17}$$

Equation (4.17) is obtained by combining Eqs. (4.15) and (4.16) and solving explicitly for C_{sm}. This equation assumes that the products 164×10^{-12} C_{sm} and 8.76×10^{-12} C_{sm}^2 are negligible. For the maximum value of $C_{sm} = 5.35$ mol/L these two terms are 8.77×10^{-12} and 4.69×10^{-11}, respectively. Conversion of ppm to mol/L, and vice versa, is straightforward when using Eq. (4.17). Once C_{sm} is calculated, X_c and then R_{w25} are calculated as discussed in Steps 4 and 5 of the previous procedure.

Another alternative to the iterative procedure for converting ppm to mol/L is to use the following correlation:

$$C_{sm} = 15.9604 \times 10^{-6}\, C_{sp}^{1.013575} \tag{4.18}$$

This correlation is obtained by substituting the assumed values of C_{sm} into Eq. (4.16) and then calculating the corresponding values of C_{sp} from Eq. (4.15). Values of C_{sm} and C_{sp} are plotted on a log–log graph and then curve-fitted, as shown in Figure 4.6. The relationship between C_{sp} and C_{sm} on a log–log graph is practically a straight line. In many cases water resistivity at reservoir temperature, R_{wT}, is estimated from wireline logs. To convert this resistivity value to concentration, first convert R_{wT} to R_{w25} using Arp's equation (4.14), and then use the following equation to calculate C_{sm}:

$$C_{sm} = 10^{X_R} \tag{4.19}$$

$$X_R = b_1 + b_2 \log\frac{1}{R_{w25}} + b_3\left(\log\frac{1}{R_{w25}}\right)^2 + b_4\left(\log\frac{1}{R_{w25}}\right)^3 + b_5\left(\log\frac{1}{R_{w25}}\right)^4 \tag{4.20a}$$

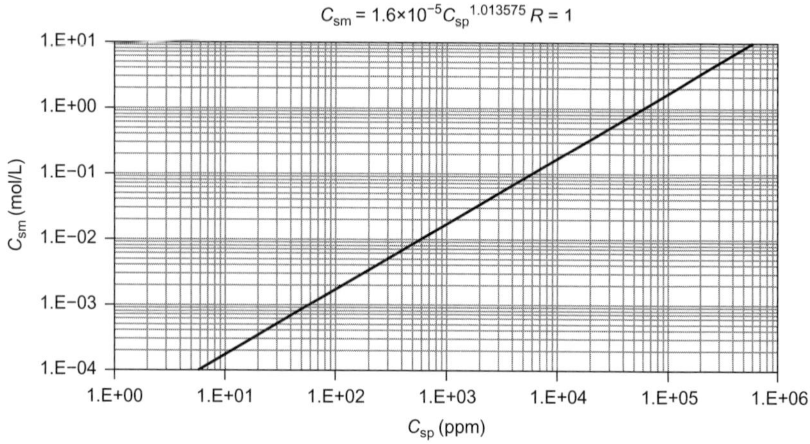

FIGURE 4.6 Conversion correlation between ppm and mol/L.

TABLE 4.4 Resistivity Correlation Constants for Eq. (4.20a)*

Resistivity Correlation Constants	Range of Resistivity, Ω m	
	$0.94 < R_{w25} < 796$	$0.09 < R_{w25} < 1.03$
b_1	−1.030224	−1.03015
b_2	1.06627	1.06090
b_3	2.41239×10^{-2}	5.66201×10^{-2}
b_4	3.68102×10^{-3}	-6.09085×10^{-2}
b_5	1.46369×10^{-4}	5.33566×10^{-2}

No direct conversion to concentration correlation, such as Eq. (4.19), is possible in the range 0.04–0.10.

where

X_R = resistivity correlation function
$b_1, \ldots, b_5$ = correlation constants

Table 4.4 shows values of the correlation constants b_1, b_2, b_3, b_4, and b_5 for two ranges of resistivity: $0.94 < R_{w25} < 796$ and $0.09 < R_{w25} < 1.03$. For these two ranges, the conversion of R_{w25} into concentration in mol/L is straightforward. First, substitute the resistivity value R_{w25} into Eq. (4.20a) for the appropriate R_{w25} range, and then calculate the resistivity correlation function X_R. Second, substitute this value of X_R and calculate the corresponding value of C_{sm} (mol/L).

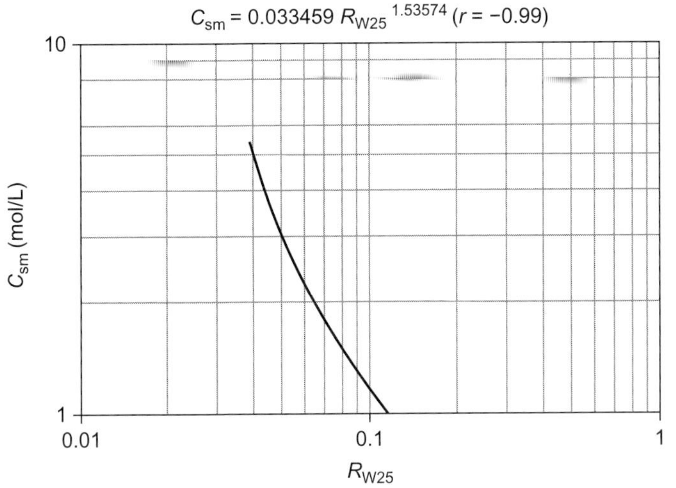

FIGURE 4.7 Conversion correlation between R_{w25} and concentration in ppm for the range $0.04 < R_{w25} < 0.1$ and $1 < C_{sm} < 5.35$.

For the case of low-resistivity range, that is, $0.04 < R_{w25} < 0.10$, Figure 4.7 or the curve-fit equation (4.20b) can be used to convert directly the resistivity value of R_{w25} to a concentration value:

$$C_{sm} = 0.03207(R_{w25})^{-1.54681} \tag{4.20b}$$

Equation (4.20b) is, for practical purposes, an acceptable representation of the data used to plot Figure 4.1. However, if a higher degree of accuracy is required for concentration, then the following iteration scheme will yield the desired value of C_{sm} for a given value of R_{w25}:

(a) Calculate an initial value of C_{sm} by substituting the value of R_{w25} into Eq. (4.20b).
(b) Substitute this value of C_{sm} into Eq. (4.13) and calculate the concentration correlation function X_c.
(c) Calculate a new value of R_{w25} by substituting the value of X_c into Eq. (4.12).
(d) Calculate the resistivity value and the calculated ΔR_{w25}, between the given resistivity value and the calculated R_{w25} value. If $\Delta R_{w25} < 10^{-6}$ Ω m, or some other required degree of accuracy, then the calculated value of C_{sm} in Step a is accurate. Otherwise, the value of C_{sm} is increased (for $\Delta R_{w25} > 0$) or decreased (for $\Delta R_{w25} < 0$) and Steps b–d are repeated until the criterion of accuracy is satisfied.

If concentration in ppm is required, substitute the calculated concentration value, C_{sm}, into Eq. (4.17) and calculate C_{sp}. A more accurate value of

C_{sp} can be obtained by substituting the value of C_{sm} into the density equation (4.16) and then calculating the value of C_{sp} from Eq. (4.15).

SPONTANEOUS POTENTIAL CURVE

The SP curve is a recording of naturally occurring physical phenomena in in situ rocks. The SP curve records the electrical potential or voltage (millivolts or mV) produced by the interaction of formation connate water, conductive drilling mud, and shale. Although relatively simple in concept, the SP curve is quite useful and informative.

Some of its uses are as follows:

- Differentiates porous and permeable rocks from clays and shales
- Gives a qualitative indication of bed shaliness
- Aids in lithology identification
- Determines R_w

The SP curve can be recorded simply by suspending a single electrode in the borehole and measuring the voltage difference between this electrode and a ground electrode, making electrical contact with the earth at the surface. The SP curve cannot be recorded in oil-base muds, which do not allow a conductive path. The SP curve is a good indicator of formation water salinity, and therefore an excellent tool for determining R_w.

When mud filtrate (R_{mf}) salinities are lower than connate water salinities, that is $R_{mf} > R_w$, the SP curve deflects to the left (i.e., the SP is negative). This is called a normal SP. When the salinities are reversed ($R_{mf} < R_w$), the SP deflects to the right. This is called a reverse SP. Other things being equal, there will be no SP curve at all when $R_{mf} = R_w$. In a clean formation, the SP curve response is given by:

$$\mathrm{SP} = -K \log\left(\frac{R_{mf}}{R_w}\right) \tag{4.21a}$$

Solving Eq. (3.20a) for R_w yields:

$$R_w = R_{mf} \times 10^{\mathrm{SP}/K} \tag{4.21b}$$

where SP is measured in mV, and K is a constant that depends on the formation temperature, which can be approximated by:

$$K = 0.133T + 61 \tag{4.21c}$$

where T is the formation temperature in °F. Mud resistivity, R_m, mud cake resistivity, R_{mc}, and mud filtrate resistivity, R_{mf}, are generally measured at the time of the survey on a mud sample from the flow line or mud pit. In the absence of measured values, R_{mf} and R_{mc} may be estimated from the following correlations, which are valid for $10 < W_m < 18$:

$$R_{mf} = (R_m^{1.065}) \times 10^{(9 - W_m)/13} \tag{4.21d}$$

where W_m = mud weight, lb/gal

$$R_{mc} = (R_m^{0.88}) \times 10^{(W_m - 10.4)/7.6} \tag{4.21e}$$

For predominantly NaCl muds, $R_{mf} = 0.75R_m$ and $R_{mc} = 1.5R_m$.

Shale-Properties Method

The shale-properties method estimates R_w from the shale beds associated with the formation of interest [5]. Assuming that the water of the shale completely occupies the total pore space of the shale, the shale formation factor F_{sh} is obtained from:

$$F_{sh} = \phi_{ash}^{-m_{sh}} \tag{4.22}$$

where ϕ_{ash} is the apparent porosity of the shale, and m_{sh} is the cementation exponent for shales. The water resistivity of the shale zone is calculated from:

$$R_{wsh} = \frac{R_{tsh}}{F_{sh}} \tag{4.23}$$

where R_{tsh} is the true resistivity of the shale zone, obtained from any resistivity log. If the zone is overpressured, R_{wsh} is approximately equal to the R_w of the formation. For normal-pressured zones, the formation water resistivity at reservoir temperature is obtained from Eq. (4.8), where the water resistivity of the zone at the reference temperature of 75°F is obtained from the following statistical correlation:

$$R_{w75} = 0.0123 + 3,647.5(a_p a_s)^{-0.955} \tag{4.24}$$

The empirical constant a_p is approximately equal to 7 for normal-pressured zones and 1 for overpressured zones. The total solids concentration, a_s, in ppm, is obtained from Figure 4.4 or from the following equation:

$$C_{sp} = a_s = 10^{1.047[3.562 - \log(R_{wsh}^{-0.0123})]} \tag{4.25}$$

The shale-properties method yields acceptable values of R_w, primarily in the U.S. Gulf Coast area.

The formation resistivity factor, F_R, of a reservoir rock is an extremely valuable tool in the area of formation evaluation. It depends on numerous parameters, including:

(1) Salinity of connate water
(2) Formation temperature
(3) Rock porosity
(4) Irreducible water saturation

(5) Amount, distribution, and type of clays
(6) Amount, distribution, and type of conductive minerals
(7) Number and type of fractures
(8) Layering of sand beds

Many researchers have investigated the effect of these factors on resistivity, and a large number of correlations were published. Only a few, however, survived the test of usefulness.

CORRELATION BETWEEN F_R AND POROSITY

Inasmuch as clean sedimentary rocks conduct electricity by virtue of the salinity of water contained in their pores, it is natural that the porosity is an important factor in controlling the flow of electric current. As a first approximation, one would expect that the current conductance would be no more than that represented by the fractional porosity; for example, a formation with 20% connate water saturation and 80% oil saturation would be expected to transmit no more than 20% of the current that would be transmitted if the entire bulk volume conducted to the same degree as the water [6].

Assuming that the saturated porous rock sample in Figure 4.2 can be represented by an equivalent system of n straight capillary tubes as in Figure 4.8, the relationship between A, the total cross-sectional area of the block sample of length L, and $A_n(=n\pi r_c^2)$, cross-sectional area of n capillary tubes of length L, is:

$$A_n = \phi A \tag{4.26}$$

where ϕ is the fractional porosity of the rock sample. The resistivity of saline solution in the capillaries is:

$$R_{wc} = \frac{E}{I_{wc}}\frac{A_n}{L} \tag{4.27}$$

Dividing Eq. (4.3) by Eq. (4.27) gives F_R or:

$$F_R = \frac{R_o}{R_{wc}} = \frac{A}{A_n}\frac{I_{wc}}{I_o} \tag{4.28}$$

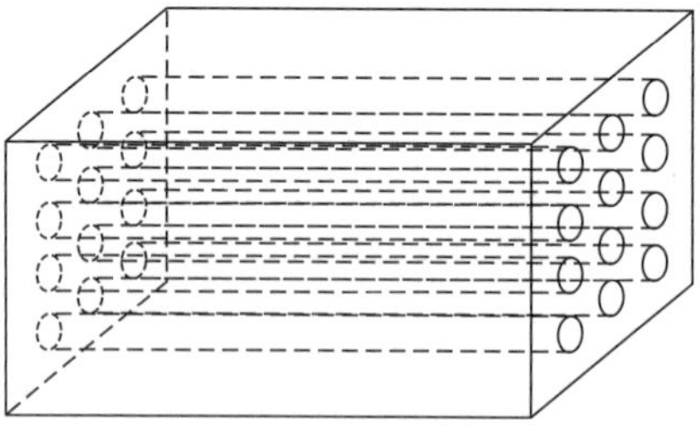

FIGURE 4.8 Ideal porous material of n straight cylindrical capillaries.

Assuming $I_{wc} = I_o$, because the system of n capillaries is supposed to be equivalent to the porous rock sample, and substituting for A_n (Eq. (4.26)) in Eq. (4.28), one obtains:

$$F_n = \frac{1}{\phi} \tag{4.29}$$

This is the simplest relationship between the formation resistivity factor, F_R, and porosity. Oil-bearing rock, however, is much more analogous to a container filled with sand than it is to parallel capillaries embedded in a rock matrix. Consequently, the actual formation resistivity factor value is considerably greater than that obtained using Eq. (4.29).

CORRELATIONS BETWEEN F_R AND TORTUOSITY

The departure of the porous system from being equivalent to a system made up of straight capillary tubes is measured by the tortuosity factor, τ, which is defined by Eq. (3.20), that is, $\tau = (L_a/L)^2$, where L is the length of the rock sample, and L_a is the actual length of the flow path as shown in Figure 4.9. Using Eq. (4.27), the resistivity of the brine in the capillaries of length L_a is:

$$R_{wc} = \frac{E}{I_{wc}} \frac{\phi A}{L_a} \tag{4.30}$$

Dividing Eq. (4.3) by Eq. (4.30) gives (for $I_{wc} = I_o$):

$$F_R = \frac{1}{\phi} \frac{L_a}{L} = \frac{\sqrt{\tau}}{\phi} \tag{4.31}$$

Cornell and Katz derived a slightly different expression. Their derivation was based on the inclined capillary tube model of porous media shown in Figure 4.10 [7]. For an inclined capillary tube, Eq. (3.13) becomes:

$$\phi = \frac{n\pi r^2 L_a}{AL} = \frac{A_n L_a}{AL} \tag{4.32}$$

or

$$A_n = \phi A \frac{L}{L_a} \tag{4.33}$$

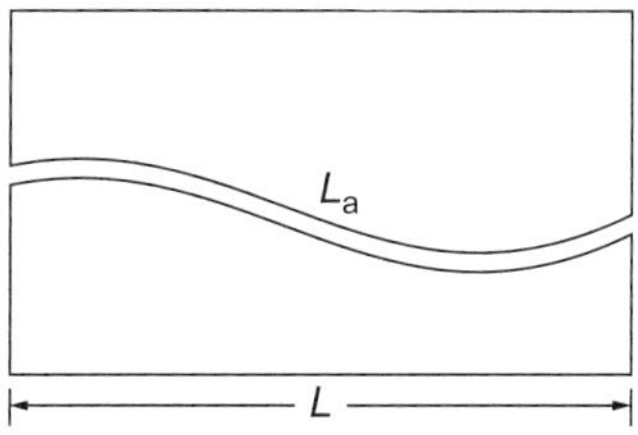

FIGURE 4.9 Actual flow path and tortuosity.

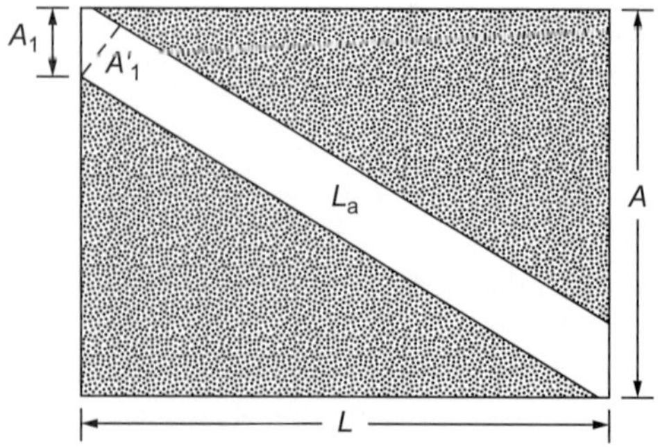

FIGURE 4.10 Inclined capillary tube model of porous media [7].

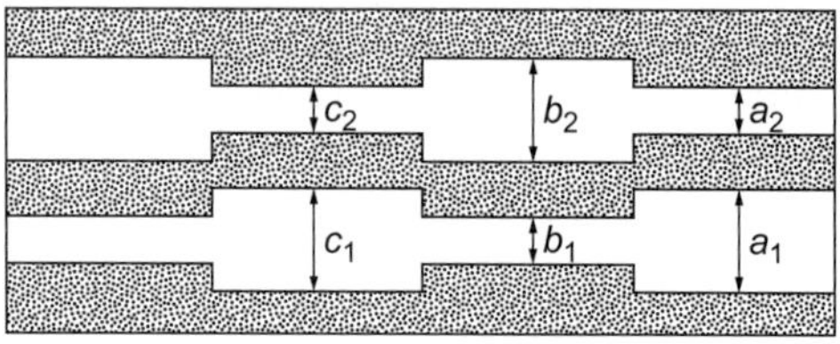

FIGURE 4.11 Two-size capillary tube model of porous media [8].

Substituting for A_n in Eq. (4.27), one obtains:

$$R_{wc} = \frac{E}{I_{wc}} \frac{\phi A(L/L_a)}{L} \tag{4.34}$$

Dividing Eq. (4.3) by Eq. (4.34) (assuming $I_{wc} = I_o$) gives:

$$F_R = \left(\frac{L_a}{L}\right)\frac{1}{\phi} = \frac{\sqrt{\tau}}{\phi} \tag{4.35}$$

Equation (4.35) gives the right order of magnitude for the formation resistivity factor in naturally fractured reservoirs. For the theoretical case of a horizontally fractured formation, the tortuosity factor τ is equal to 1, and consequently:

$$F = \frac{1}{\phi} \tag{4.36}$$

Wyllie and Gardner proposed the following relation for F_R based on the complex capillary tube model shown in Figure 4.11 [8]:

$$F_R = \frac{\sqrt{\tau}}{\phi^2} \tag{4.37}$$

Unfortunately, natural formations rarely have uniform pore geometry as shown in Figures 4.8–4.11. But it is evident from Eqs. (4.31)–(4.36) that the formation factor F_R is a function of porosity and pore structure. Equation (4.37) produces satisfactory results in carbonates and highly cemented sands.

CORRELATIONS BETWEEN F_R AND CEMENTATION

The degree of cementation of sand particles depends on the nature, amount, and distribution of numerous cementing materials including silica, calcium carbonate, and a variety of clays. The less-cemented sands normally have higher porosities and, from Eq. (4.36), lower resistivity factors. As the sand becomes more cemented, ϕ decreases and, therefore, F_R increases. Archie derived, from laboratory measurements of the formation resistivity factor F_R with porosity, a relation between these two variables, which seems to have survived the test of time and usefulness [2]. The general form of this expression is:

$$F_R = \frac{1}{\phi^m} \tag{4.38}$$

The exponent m, which is referred to as the "cementation" factor, is a function of the shape and distribution of pores. It is determined from a plot of the formation resistivity factor F_R versus porosity on a log–log graph. Such a plot generally can be approximated by a straight line having slope m. In chalky rocks and compacted formations, m is approximately equal to 2. For compact limestones, which are very highly cemented rocks, the value of m may be as high as 3. It is important to emphasize that these empirical values of m very often differ from well to well in the same or like formations. When the value of m cannot be determined, the following equation, commonly referred to as the Humble formula, can be used to estimate the formation factor:

$$F_R = \frac{0.62}{\phi^{2.15}} \tag{4.39a}$$

or

$$F_R = \frac{0.81}{\phi^2} \tag{4.39b}$$

Whereas the Humble formula is satisfactory for many types of rocks, better results can be obtained using the following generalized form of the Humble formula:

$$F_R = \frac{a}{\phi^m} \tag{4.40}$$

where values of the constant a and the cementation factor m depend on the types of rocks. The proper choices of a and m are best determined by laboratory measurements.

The Humble and Archie formulas for various values of the cementation factor m and constant a are compared graphically in Figure 4.12.

Example

A cylindrical core sample of a well-consolidated sand is 100% saturated with a NaCl brine of 34,000 ppm salinity. At 120°F, the resistance of the core is 85 Ω. The core is 3 in. in diameter and 12 in. in length. Calculate its porosity.

Solution

The resistivity of this core is (from Eq. (4.3)):

$$R_o = r_o \frac{A}{L} = 85 \frac{\pi(3/2)^2}{12} = 50 \Omega \text{ in.}$$

or

$$R_o = \frac{50}{39.4} = 1.27 \Omega \text{ m}$$

The resistivity of the brine having a salinity of 34,000 ppm and temperature of 120°F is obtained from Figure 4.4:

$$R_w = 0.12 \Omega \text{ m}$$

The formation resistivity factor F_R is (Eq. (4.5)):

$$F_R = \frac{R_o}{R_w} = \frac{1.27}{0.12} = 10.57$$

Using the Humble formula, Eq. (4.40), the porosity of the core sample is:

$$\phi = \left(\frac{0.81}{F_R}\right)^{0.5} = \left(\frac{0.81}{10.57}\right)^{0.5} = 0.28$$

The cementation exponent m is affected by a large number of factors, including shape, sorting and packing of the particulate system, pore configuration and size, constrictions existing in a porous system, tortuosity, type of pore system (intergranular, intercrystalline, vuggy, fractured), compaction due to overburden pressure, presence of clay minerals, and reservoir temperature. The main effect of these parameters is to modify the formation resistivity factor F_R. Consequently, their combination can produce a countless number of values of F_R and m for a given porosity. For instance, compaction due to overburden pressure generally causes a considerable increase in resistivity, especially in poorly cemented rocks and in low-porosity rocks. Figure 4.13 illustrates the effect of overburden pressure on the formation resistivity factor F_R on core samples from a reef-type limestone. The

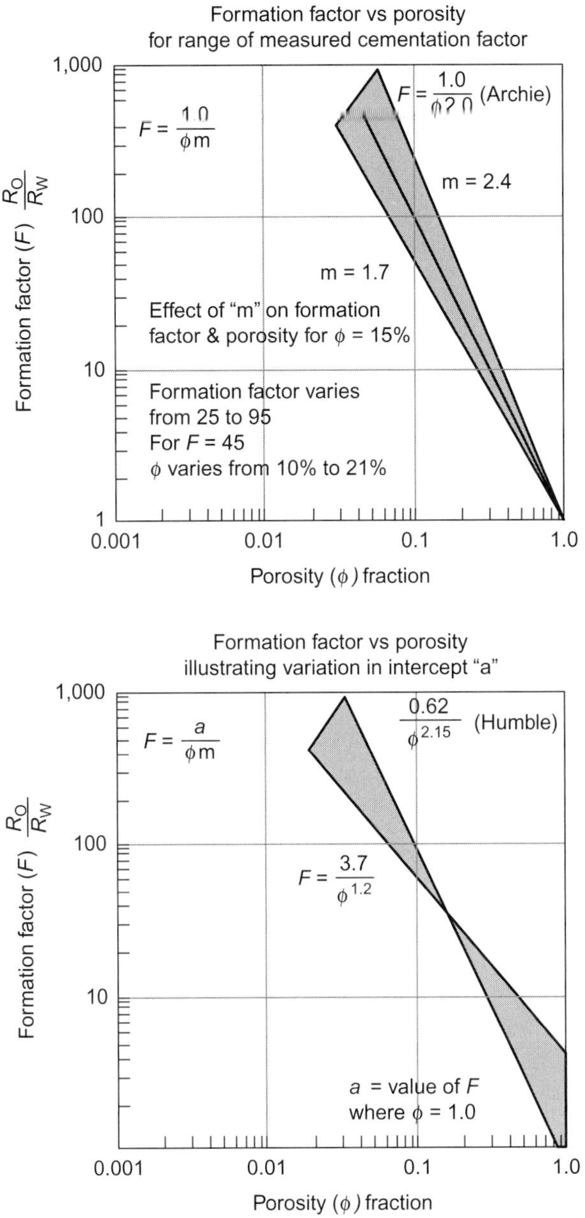

FIGURE 4.12 Comparison of Humble and Archie equations. Source: *Courtesy of Core Laboratories.*

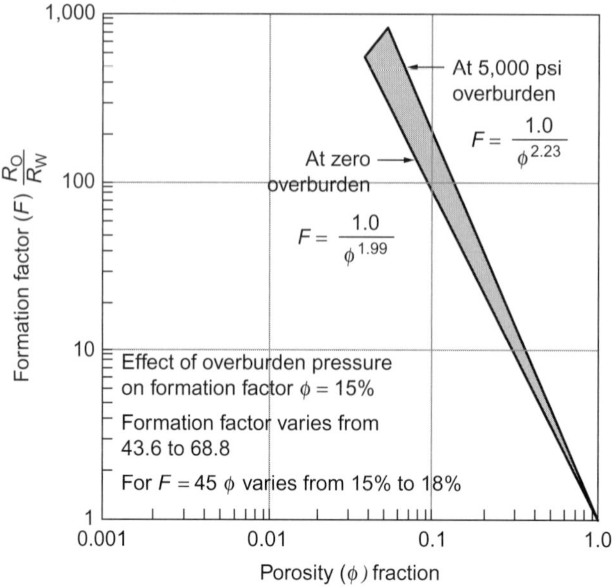

FIGURE 4.13 Effect of overburden pressure on formation factor. Source: *Courtesy of Core Laboratories.*

cementation factor m increases from 1.99 at zero overburden pressure to 2.23 at an overburden pressure of 5,000 psi. Nearly all of the values of m in widespread use today were determined on unconfined core samples. Of course, resistivity measurements determined under representative overburden pressures are strongly recommended for improved well log interpretation.

THEORETICAL FORMULA FOR F_R

Many attempts have been made to derive a general formula relating formation resistivity, porosity, and cementation factor. If an electric current is passed through a block of nonconducting porous rock saturated with a conducting fluid, only a portion of the pore space participates in the flow of electric current. Therefore, the total porosity ϕ can be divided into two parts such that [9]:

$$\phi = \phi_{ch} + \phi_{tr} \tag{4.41}$$

where ϕ_{ch} and ϕ_{tr} are, respectively, the flowing porosity associated with the channels and the porosity associated with the regions of stagnation (traps) in a porous rock. Apparently, ϕ_{ch} is equivalent to the "effective porosity" used by Chilingarian, and ϕ_{tr} is equivalent to the irreducible fluid saturation [10]. Figures 4.14 and 4.15 show that the electrical current can flow only through the channel indicated by C, whereas no current can flow through the traps

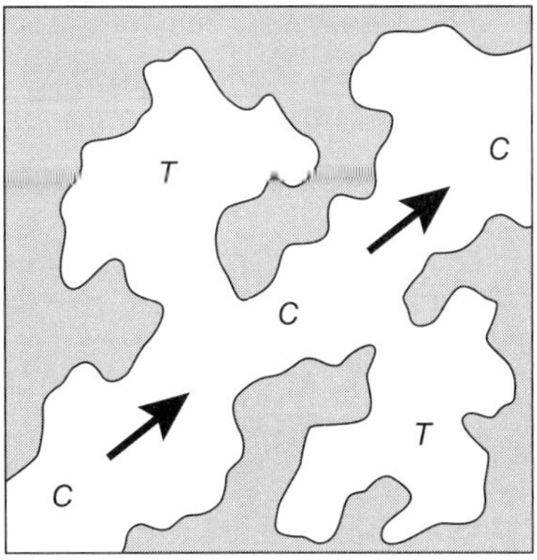

FIGURE 4.14 Portion of porous rock showing dead-end traps [9].

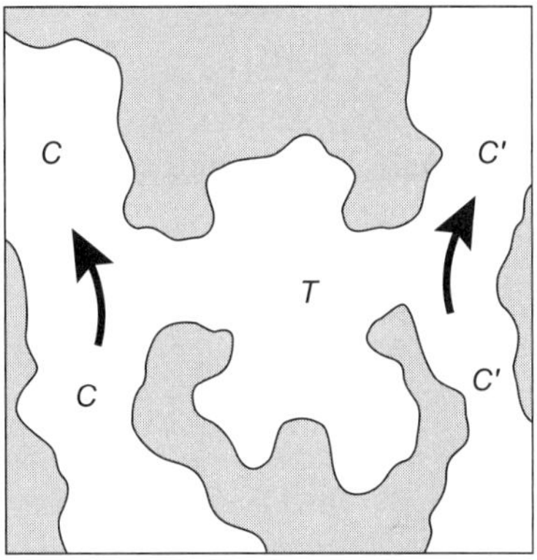

FIGURE 4.15 Portion of porous rock showing an open or symmetry trap [9].

indicated by T. In Figure 4.14 the traps are of the dead-end type. The trap in Figure 4.15 is called an open or symmetry trap. A general relationship between F and ϕ_f may be written as [11]:

$$F_R = 1 + f_G\left(\frac{1}{\phi_{ch}} - 1\right) \tag{4.42}$$

$$\phi_{ch} = \phi^m \tag{4.43}$$

where f_G is defined as the internal geometry parameter of the porous rock, and $m \geq 1$. Combining Eqs. (4.42) and (4.43) gives:

$$F_R = 1 + f_G\left(\frac{1}{\phi^m} - 1\right) \tag{4.44}$$

This is the Rosales relationship between formation resistivity, porosity, and cementation factor. If $f_G = 1$, Eq. (4.44) gives the Archie formula. Equation (4.44) can be expressed as:

$$F_R = \frac{f_G}{\phi^m} + (1 - f_G) \tag{4.45}$$

The value of f_G for most porous rocks is close to unity. Hence, $f_G/\phi^m \gg (1 - f_G)$ and Eq. (4.45) can be approximated by:

$$F_R = \frac{f_G}{\phi^m} \tag{4.46}$$

This expression is the Humble formula (Eq. (4.40)), where $f_G = a$. Thus, Archie and Humble formulas are special cases of Rosales' general formula. Rosales showed experimentally that, for sandstones, Eq. (4.46) can be written as follows [11]:

$$F_R = 1 + 1.03\left(\frac{1}{\phi^{1.73}} - 1\right) \tag{4.47}$$

This expression was compared graphically with the Humble formula, Eq. (4.34), and the Timur formula [12]:

$$F_R = \frac{1.13}{\phi^{1.73}} \tag{4.48}$$

Figure 4.16 is a log–log plot of Eqs. (4.39) (line A), (4.42) (line B), and (4.48) (line C). The three formulas give approximate results within the region practical interest, that is, $10 \le \phi \le 40$. As ϕ approaches unity, however, Eq. (4.42) gives a curved line that satisfies the condition $F_R = 1$ when $\phi = 1$, whereas the Humble formula (Eq. (4.39)) and the Timur formula (Eq. (4.48)) are straight lines for all values of ϕ, which does not satisfy that condition. From Eq. (4.35), the tortuosity is:

$$\tau = (\phi F_R)^2 \tag{4.49}$$

Substituting Eq. (4.44) into Eq. (4.49) yields a general expression for calculating tortuosity:

$$t = \phi\left[1 + f_G\left(\frac{1}{\phi^m} - 1\right)\right] \tag{4.50}$$

Inasmuch as the value of f_G is approximately equal to unity for most porous rocks, Eq. (4.50) can be written as follows:

$$\sqrt{\tau} = \frac{1}{\phi^{(m-1)}} \tag{4.51}$$

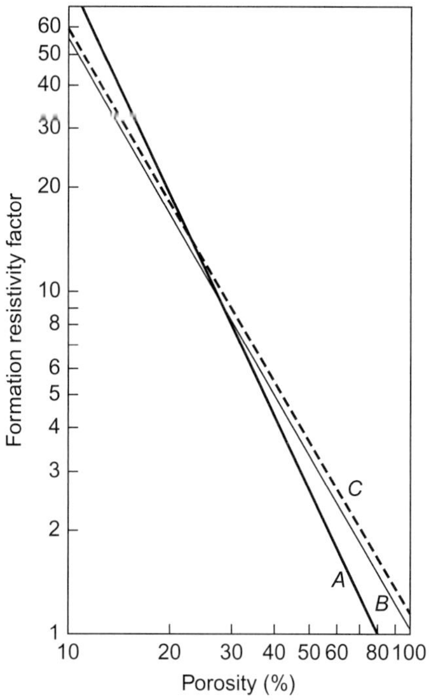

FIGURE 4.16 Graphical comparison of Humble equation (A), Rosales equation (B), and Timur equation (C) [9].

Combining Eqs. (4.41), (4.43), and (4.51) gives:

$$\sqrt{\tau} = 1 + \frac{\phi_{tr}}{\phi_{ch}} \tag{4.52}$$

This expression indicates the physical significance of tortuosity in terms of stagnant and flowing porosities. Equation (4.52) is an approximation valid only for consolidated porous rocks. For unconsolidated sands, the general expression (Eq. (4.50)) should be used, where $f_G = 1.49$ and $m = 1.09$.

Example

Laboratory measurements made on a Berea Sandstone core yielded $\phi = 0.216$ and $F_R = 13.7$. Compute:

1. the formation resistivity factor using Rosales, Humble, and Archie expressions and compare with the measured value;
2. the flowing and stagnant porosity;
3. tortuosity of the Berea core.

Solution

1. Formation resistivity:

Rosales' formula for consolidated sandstones, that is, Eq. (4.47), gives:

$$F_R = 1 + 1.03(0.216^{-1.73} + 1) = 14.57$$

Humbles formula (Eq. (4.39)) gives:

$$F_R = (0.62)0.216^{-2.15} = 16.72$$

Because Berea cores are generally moderately to strongly cemented, one can use $m = 1.7$ in the Archie formula:

$$F_R = 0.216^{-1.7} = 13.53$$

It is obvious that the value of F_R obtained from the Archie equation is the closest to the measured value. Thus, the Archie formula is a good model to use for calculating the formation resistivity factor of a Berea Sandstone core. The value of the cementation factor for this core is:

$$m = \frac{-\ln(13.7)}{\ln(0.216)} = 1.71$$

2. The flowing porosity is approximated from Eq. (4.43):

$$\phi_{ch} = 0.216^{1.71} = 0.073$$

and from Eq. (4.41), the stagnant porosity is equal to:

$$\phi_{tr} = 0.216 - 0.073 = 0.143$$

This means that only 34% [(0.073/0.216)(100)] of the total pore volume participates actively in the flow of electric current and 66% of the pore volume, corresponding to the dead-end and symmetrical traps, is neutral to the flow of electric current. If the flow of viscous fluids were considered, instead of electric current, different results would have been obtained, that is, higher ϕ_{ch} and lower ϕ_{tr} because viscous forces, such as capillary forces, promote the transfer of a fraction of the fluid in stagnant regions to the flowing regions [11].

3. The tortuosity of this Berea Sandstone core can be obtained from Eq. (4.52):

$$\tau = \left(1 + \frac{0.143}{0.073}\right)^2 = 8.76$$

If this Berea Sandstone core has a length $L = 6$ in., the actual mean length of the flow path of an electric current L_a can be estimated from the definition of tortuosity (Eq. (3.20)):

$$L_a = L\sqrt{\tau} = 6 \times \sqrt{8.76} = 17.76$$

For viscous fluid flow through this Berea Sandstone core, the value of L_a will be slightly closer to the value of L because the stagnant porosity will be <0.143 and the flowing porosity will be >0.073. In any case, the large value of τ

indicates that the internal geometry of porous systems is extremely complex, which is in accordance with microscopic observations indicating that regions of stagnation or traps should be the rule rather than the exception, especially from the standpoint of the flow of electric current.

CORRELATION BETWEEN F_R AND WATER SATURATION

In a formation containing oil and/or gas, both of which are nonconductors of electricity, with a certain amount of water, the resistivity is a function of water saturation S_w. For the same porosity, the true resistivity, R_t, of this formation is larger than R_o (the resistivity of a formation 100% saturated with brine), because there is less available volume for the flow of electric current. Archie determined experimentally that the resistivity factor F_R of a formation partially saturated with brine can be expressed by the following relation [2]:

$$S_w = \left(\frac{R_o}{R_t}\right)^{1/n} = \left(\frac{F_R R_w}{R_t}\right)^{1/n} \tag{4.53}$$

Substituting for F_R from Eq. (4.40) into Eq. (4.53) gives:

$$S_w = \left(\frac{aR_w}{\phi^m R_t}\right)^{1/n} \tag{4.54}$$

where

R_t = true resistivity of formation containing hydrocarbons and formation water
R_o = resistivity of formation when 100% saturated with water
n = saturation exponent

The water saturation in a hydrocarbon zone can be estimated from:

$$S_w \approx \sqrt{\frac{R_w}{R_{wa}}} \tag{4.54a}$$

Where R_{wa} is the apparent water resistivity, which can be calculated from:

$$R_{wa} = \frac{ILD}{F} = \frac{R_t}{F} \tag{4.54b}$$

The apparent true formation resistivity (R_t) of the uninvaded zone is obtained from the deep induction log (IDL). Water resistivity R_w is best obtained from the spontaneous potential curve or laboratory measurements on a water sample. In a zone where the water saturation is 100%, $R_{wa} \approx R_w$. A zone where $R_{wa} > 3R_w$ probably contains hydrocarbons.

The ratio R_t/R_o is commonly referred to as the resistivity index I_R. If the formation is totally saturated with brine (i.e., $R_o = R_t$), the resistivity index is equal to 1. I_R is greater than 1, when hydrocarbons are present in the formation. The resistivity index is then a function of the salinity and the amount of

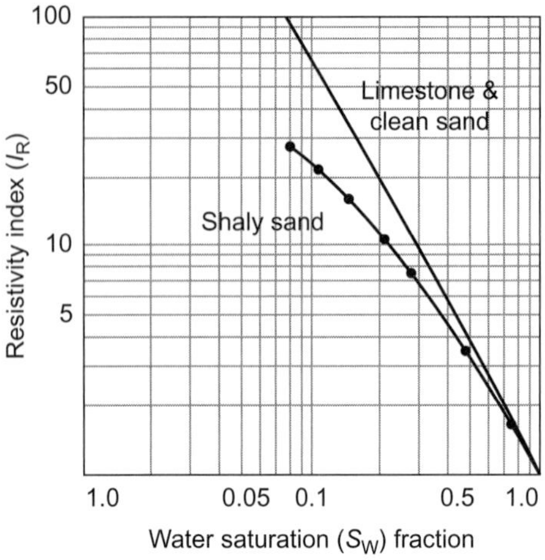

FIGURE 4.17 Effect of clay conductance on resistivity index. Source: *Courtesy of Core Laboratories.*

formation water. I_R is also a function of the amount, distribution, and type of clays present in the formation being evaluated. For instance, the presence of cation exchangeable clays, such as smectites, typically causes low I_R values to be observed, whether or not the rock contains hydrocarbons is illustrated in Figure 4.17. The saturation exponent n is determined experimentally by saturating a core sample with brine and measuring the rock resistivity R_o.

Then the brine is displaced with air, naphtha, or live crude oil, and the true resistivity, R_t, is measured after each increment of displacement. The water saturation S_w is determined by measuring the volume of water produced and applying the material balance equations.

A plot of the ratio R_t/R_o versus the water saturation on logarithmic scale gives a straight line of slope $-n$, as illustrated in Figure 4.18 and the following equation:

$$S_w = \left(\frac{R_o}{R_t}\right)^{1/n} = \left(\frac{R_t}{R_o}\right)^{-1/n} = (I_R)^{-1/n} \tag{4.55}$$

and

$$\log I_R = -n \log S_w \tag{4.56}$$

The slope of the straight line passing through $S_w = 1$, when $I_R = 1$, is $(-n)$, which can be calculated as follows:

$$n = \frac{\log I_{R1} - \log I_{R2}}{\log S_{w2} - \log S_{w1}} \tag{4.57}$$

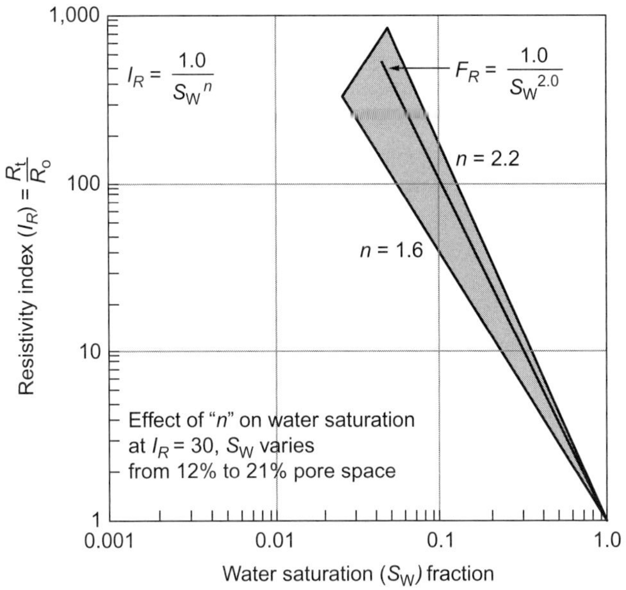

FIGURE 4.18 Resistivity index versus water saturation for a range of measured saturation exponents. Source: *Courtesy of Core Laboratories.*

The value of n is affected by wettability, overburden pressure, nature and microscopic distribution of the reservoir fluids, and types and amounts of conductive clays (or measuring the slope directly from the graph). Anderson examined the effects of wettability on the saturation exponent and found that [13]:

(1) n is essentially independent of wettability when the brine saturation S_w is sufficiently high to form a continuous film on the grain surfaces of the porous medium and, consequently, to provide a continuous path for a current flow. This continuity is common in clean and uniformly water-wet systems. The value of the saturation exponent n in these systems is ~ 2 and remains essentially constant as the water saturation is lowered to its irreducible value, S_{wi}.

(2) In uniformly oil-wet systems with low brine saturations, large values of the saturation exponent, 10 or higher, should be expected.

Table 4.5 shows what typically occurs in an oil-wet core as the water saturation drops [13]. An examination of this table shows that below a certain brine saturation, the exponent n begins to increase rapidly. For instance, in the case where the nonwetting brine saturation is reduced by oil injection, the value of n increases from 4 to 7.15 as S_w drops from 34.3 to 33.9. This rapid rise of n in oil-wet systems, as the brine saturation decreases, is due to an increase in the resistivity of the system. The resistivity increase is due to disconnection and eventual isolation and trapping of a portion of the brine by oil. This portion no

TABLE 4.5 Archie Saturation Exponents as a Function of Saturation for a Conducting Nonwetting Phase [13]

Air–Brine Brine Saturation (% PV)	*n*	Oil–Brine Brine Saturation (% PV)	*n*
66.2	1.97	64.1	2.35
65.1	1.98	63.1	2.31
63.2	1.92	60.2	2.46
59.3	2.01	55.3	2.37
51.4	1.93	50.7	2.51
43.6	1.99	44.2	2.46
39.5	2.11	40.5	2.61
33.9	4.06	36.8	2.81
30.1	7.50	34.3	4.00
28.4	8.90	33.9	7.15
		31.0	9

longer contributes to the current flow, because it is surrounded by nonconducting oil, thereby increasing electrical resistivity of the porous system.

The exponent n must be measured at reservoir wetting conditions, that is, on native or restored-state cores, otherwise the water saturation in the reservoir obtained from well logs would be underestimated. Figure 4.19 shows the effect of cleaning on the saturation exponent. Extraction lowered the value of n from 2.71 to 1.91. The effects of wettability on carbonate cores were investigated by Sweeney and Jennings [14]. They found that the saturation exponent ranged from 1.6 to 1.9 for water-wet cores, whereas the oil-wet cores exhibited two different types of behaviors as shown in Figure 4.20. In some cores, n was about 8 even when S_w was very high. In other cores, the behavior of n was similar to the water-wet and neutrally wet, that is, $1.5 < n < 2.5$, until a brine saturation of nearly 35% was reached, at which point n increased rapidly to 12. When a core is extracted with toluene, it is considered to have neutral wettability, but, in actuality, its wettability is somewhere between mildly water-wet and mildly oil-wet.

Longeron et al. and Lewis et al. showed that the saturation history of the formation has a considerable effect on the saturation exponent, particularly in water-wet porous systems [15,16]. Both studies found that there is a significant resistivity and saturation exponent hysteresis between drainage and imbibition saturation cycles. Drainage tests, that is, tests in which water

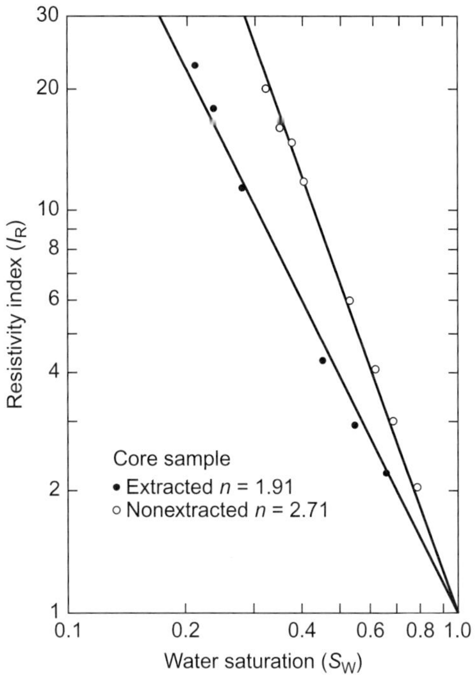

FIGURE 4.19 Effect of cleaning on the saturation exponent [13].

saturation decreases, describe the process that probably takes place when hydrocarbons migrate from the source rock to the reservoir trap. Imbibition tests, that is, tests in which water saturation increases, are useful for evaluating saturations in a reservoir that already has been subjected to water flooding. During drainage of water-wet sandstone cores, both studies found that n is essentially equal to 2. During imbibition, however, the value of n is a function of brine saturation: (1) for $0.28 < S_w < 0.40$, $n = 2.56$, and (2) for $0.40 < S_w < 0.58$, $n = 1.56$. Figure 4.21 shows the behavior of the resistivity index as a function of brine saturation during drainage and imbibition cycles, under low effective stress. The curve-fit equation for the imbibition cycle is of the general form $I_R = b_i S_w^{-n}$, where b_i is the imbibition correlation constant.

Figures 4.22 and 4.23 illustrate the influence of overburden pressure on the saturation exponent for water-wet sandstone core and Berea Sandstone core, respectively. The maximum change in the saturation exponent with overburden pressure was ~8% for water-wet cores and 4% for oil-wet cores [16].

The stress effect on the value of n was considerably higher in drainage tests where water was displaced with live crude oil [17]. In this case, the saturation exponent increased from 1.82 at ambient conditions to 2.09 at reservoir conditions, an increase of 15%.

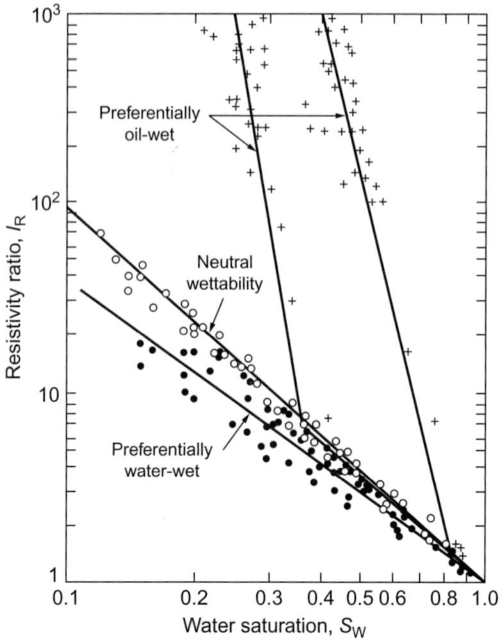

FIGURE 4.20 Resistivity index versus water saturation in carbonate cores [14].

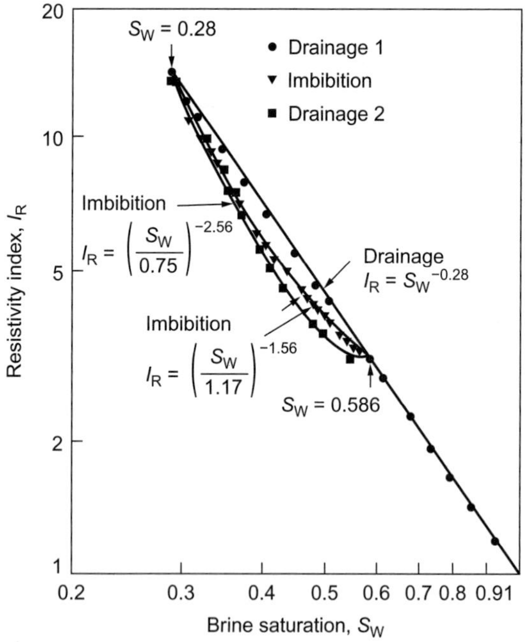

FIGURE 4.21 Resistivity index versus brine saturation—sandstone, low stress [15].

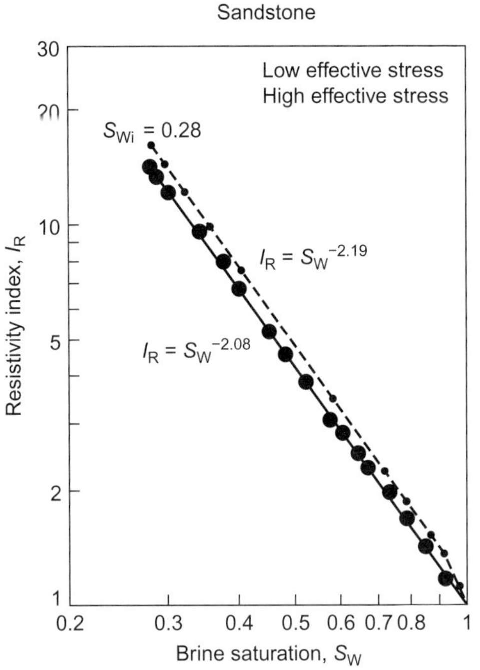

FIGURE 4.22 Effect of stress on resistivity index—drainage curves [15].

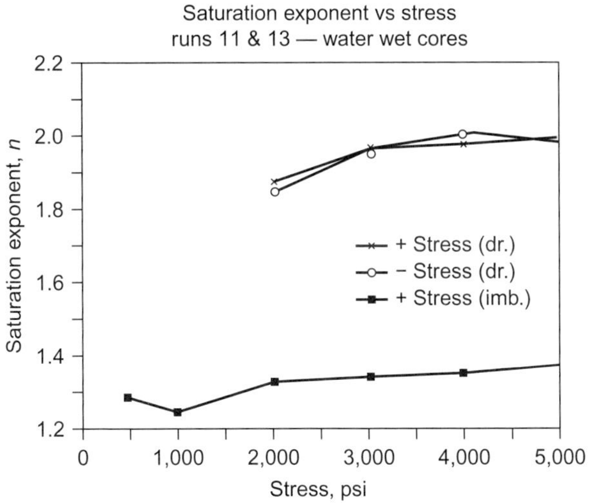

FIGURE 4.23 Influence of stress on the saturation exponent for water-wet cores during drainage and imbibition saturation cycles [16].

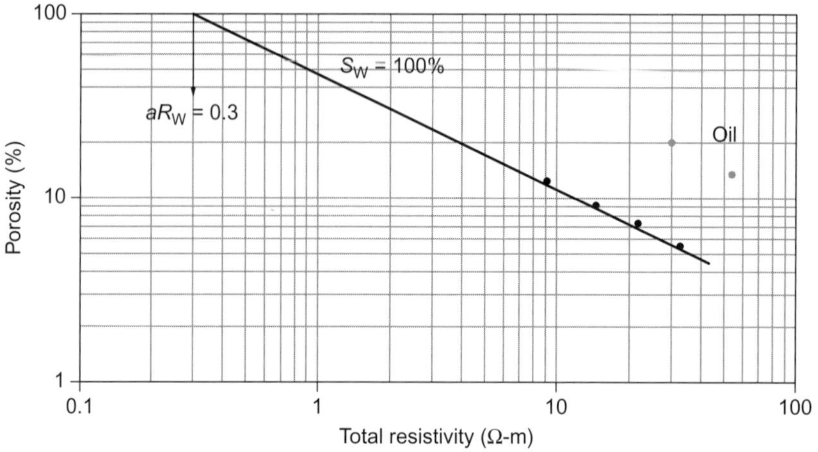

FIGURE 4.24 Porosity versus total resistivity for $S_w = 100\%$.

Manipulating Eqs. (4.40), (4.54), and (4.55) yields a general relationship, which can be used to estimate the cementation factor m and Humble constant a:

$$\log \phi = -\frac{1}{m}[\log R_t - \log (aR_w) - \log I_R] \tag{4.58}$$

Thus, a plot of ϕ versus R_t on a log–log graph should yield a straight line with a slope $(-1/m)$ for 100% water-saturated zones, having constant resistivity index, I_R and aR_w. Figure 4.24 shows such a plot for the case of a homogeneous sandstone oil reservoir.

The water zones, where $I_R = 1$, form a straight line with a slope equal to ~2. When $\phi = 100\%$ and $I_R = 1$, Eq. (4.58) reduces to $\log R_t = \log (aR_w)$. Therefore, the product aR_w can be determined directly from Figure 4.24 by simply extrapolating the straight line to $\phi = 100\%$ and reading the corresponding value of aR_w on the R_t axis. Knowing the water resistivity R_w, from Figure 4.24 for instance, one can calculate a. The log–log plot of ϕ versus R_t also can be used to estimate the water saturation of the reservoir from Eq. (4.55), that is, $S_w = I_R^{-1/n}$, by assuming the cementation factor m and the saturation exponent n are equal. The limitation of this equation is that a significant number of water-bearing zones of constant a, m, and R_w must be available [18].

Example

A sandstone core sample ($d = 1.90$ cm and $L = 3.20$ cm) is saturated with brine of 0.55 Ω m resistivity. The core was desaturated in steps, and the resistance (Table 4.6) was measured at each saturation.

1. Estimate the rock porosity.
2. Determine the saturation exponent, n, by the conventional technique.

Solution

1. The rock porosity may be estimated from the Humble equation (4.39b).

TABLE 4.6 Saturation and Resistance Values for Core Sample

S_w (Fraction)	r_o (Ω)
1.000	521
0.900	678
0.800	913
0.730	1,151
0.640	1,510
0.560	2,255
0.480	3,135
0.375	5,270
0.350	6,820
0.300	10,400

The resistivity of the rock sample, 100% water saturated, is obtained from Eq. (4.3):

$$R_o = r_o \frac{A}{L} = 521 \frac{2.835 \times 10^{-4}}{0.032} = 4.6162 \Omega \text{ m}$$

The formation resistivity factor is (Eq. (4.5)):

$$F = \frac{R_o}{R_w} = \frac{4.6162}{0.55} = 8.393$$

Assuming that the Humble equation is applicable, that is,

$$F = \frac{0.81}{\phi^2}$$

the porosity of the core sample is:

$$\phi = \sqrt{\frac{0.81}{F}} = \sqrt{\frac{0.81}{8.393}} = 0.31$$

2. The log–log plot of the resistivity index versus water saturation normally yields a straight line in a homogeneous core. The saturation exponent n is equal to the slope of this straight line, as indicated by Eq. (4.55).

   The true resistivity of the rock sample when $S_w = 90\%$ is:

$$R_t = r_o \frac{A}{L} = 678 \frac{2.835 \times 10^{-4}}{0.032} = 6.007 \Omega \text{ m}$$

   The resistivity index I_R for $S_w = 100\%$ is unity. For $S_w = 90\%$ the resistivity index is:

$$I_R = \frac{R_t}{R_o} = \frac{6.007}{4.616} = 1.301$$

TABLE 4.7 Example Results

S_w (Fraction)	r_o (Ω)	R_t (Ω m)	I_R
1.000	521	4.616	1.000
0.900	678	6.007	1.301
0.800	913	8.089	1.752
0.730	1,151	10.197	2.209
0.640	1,510	13.378	2.898
0.560	2,255	19.978	4.328
0.480	3,135	27.774	6.017
0.375	5,270	46.689	10.115
0.350	6,820	60.421	13.090
0.300	10,400	92.138	19.962

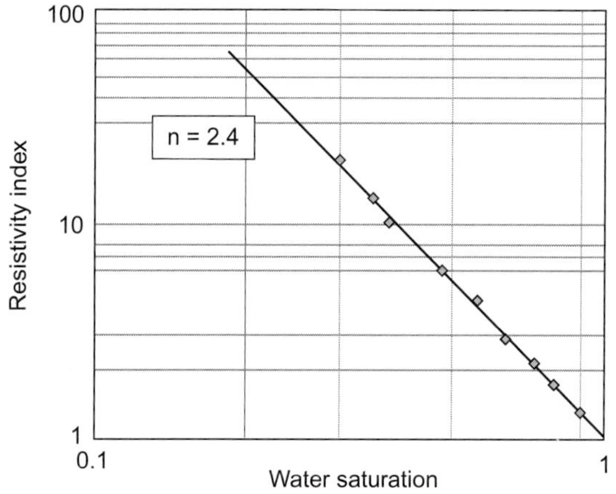

FIGURE 4.25 Resistivity index versus water saturation.

Table 4.7 shows the values of I_R for other values of S_w.

Figure 4.25 shows a plot of water saturation versus the resistivity index. The slope of the log–log straight line is −2.4 (found by regression). Thus, $n = 2.4$.

CORRELATION BETWEEN F_R AND PERMEABILITY

The following three factors are important in correlating F_R with the permeability of sedimentary rocks:

1. The range of grain sizes, which is characterized by a grain-size distribution factor and a geometric mean-grain diameter
2. The degree of packing of the sand particles, which is a function of several factors, such as the angularity of the particles, the current velocity during the deposition, and the grain-size distribution
3. The combination of cementation and compaction of the sediments

Of these three factors, only the range of particle sizes can be quantified with a reasonable accuracy, even though a reservoir core sample is too small to have a statistical meaning.

Once a sand deposit is accumulated and buried by subsequent depositions, the texture-controlled permeability of this sand is severely distorted by compaction and cementing materials, such as calcite, silica, and various types of clays. Clays generally affect the original distribution of permeability only slightly, because clays tend to deform by compaction to conform to the adjacent quartz grains and seem to be distributed in a relatively uniform manner throughout the sandy portion of the formation [19]. Silica cement, however, tends to distort qualitatively and quantitatively the texture-controlled permeability distribution of sand beds. In the loosely packed sandstones, packing appears to have little or no effect on this permeability distribution. In consolidated sandstones, the effect of packing on permeability distribution is considerable.

Hutchinson et al. proposed one of the earliest statistical relationships between the formation resistivity factor and permeability for unconsolidated sands (Figure 4.26) [19]:

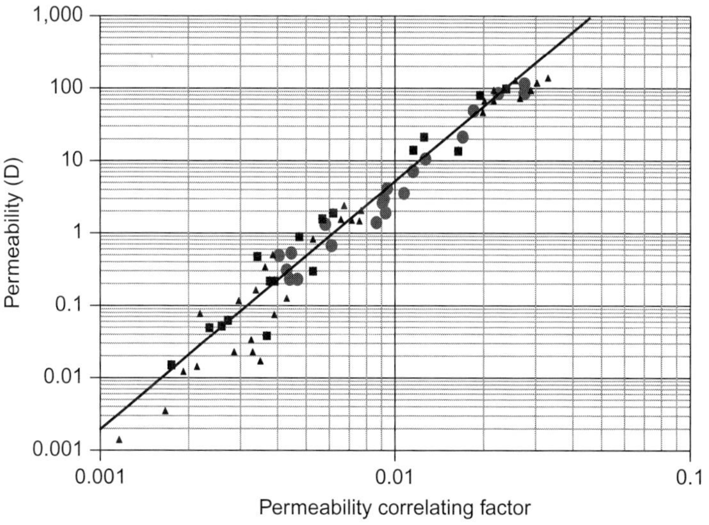

FIGURE 4.26 Correlation of permeability with the geometric mean-grain diameter and the formation factor [19].

$$k \propto \left(\frac{d_{gr}}{F_R \, e^{0.6\sigma_2}} \right) \tag{4.59}$$

where the permeability k is expressed in darcies (D) and the geometric mean-grain diameter d_{gr} is in millimeters. The term σ_2 is the standard deviation of $\log_2$ grain-size distribution; it is a measure of the grain sorting and an indicator of the primary texture. For a uniform grain-size distribution, $\sigma_2 = 0$, and for a nonuniform grain distribution, $\sigma_2 = 1$. This relationship was established after an extensive study of sandstone outcrops typical of producing horizons for the purpose of predicting the size, shape, and permeability contrast of reservoir nonuniformities. The study concentrated on several Cretaceous age outcrops in the Four Corners area (New Mexico, Arizona, Utah, and Colorado) and the Woodbine Formation outcrop near Dallas, Texas. The straight line in Figure 4.26 can be expressed as:

$$k = 2.53 \times 10^5 \left(\frac{d_{gr}}{F_R \, e^{0.6\sigma_2}} \right)^{2.75} \tag{4.60}$$

For a uniform grain-size distribution system, the standard deviation, $\sigma_2 = 0$ and Eq. (4.60) becomes:

$$k = 2.53 \times 10^5 \left(\frac{d_{gr}}{F_R} \right)^{2.75} \tag{4.61}$$

A correlation of porosity with respect to the formation resistivity and geometric mean-grain diameter, as shown in Figure 4.27, was developed in the same study:

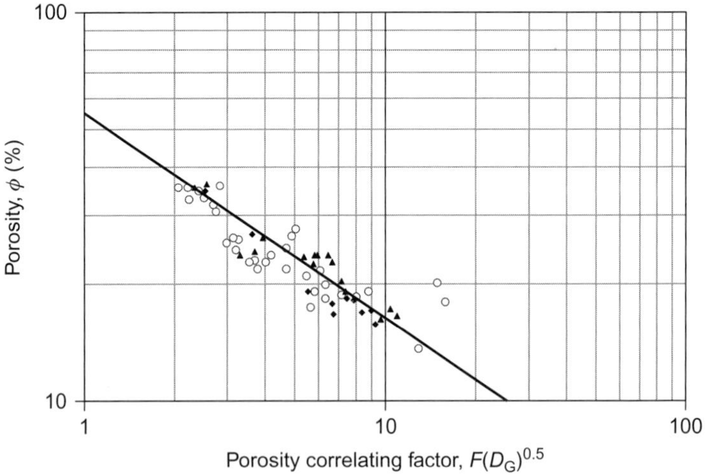

FIGURE 4.27 Correlation of porosity with the geometric mean-grain diameter and the formation factor [19].

$$\phi = \left(\frac{1}{F_R\sqrt{d_{gr}}}\right) \tag{4.62}$$

where the porosity, ϕ, is expressed as a percentage. The straight line in Figure 4.27 can be expressed as follows:

$$\phi = \left(\frac{51}{(F_R\sqrt{d_{gr}})^{0.50}}\right) \tag{4.63}$$

Combining this expression with Eq. (4.60) for a nonuniform grain-size distribution system, the following equation is obtained:

$$k = \frac{1.536 \times 10^{24}}{e^{1.65\sigma_2}}\left(\frac{1}{\phi^{11}F^{8.25}}\right) \tag{4.64}$$

This and similar empirical relationships are useful because they allow one (1) to understand, qualitatively and quantitatively, the interaction of various petrophysical parameters, and (2) to approximate the formation permeability in uncored wells knowing porosity and formation resistivity factor, which can both be derived from well logs. The value of σ_2 varies between zero, for a system with uniformly distributed grain sizes, and unity, for the theoretical case in which every single grain has a different size. For a laboratory sandpack, σ_2 is approximately equal to zero and Eq. (4.64) becomes:

$$k = 1.536 \times 10^{24}\left(\frac{1}{\phi^{11}F_R^{8.25}}\right) \tag{4.65}$$

For $\sigma_2 = 1$, the term $e^{1.65\sigma_2}$ in Eqs. (4.60) and (4.64) is equal to 5.2. Thus, the maximum effect of grain-size distribution is to reduce the permeability of the ideally uniform system (in which $\sigma_2 = 0$) by approximately fivefold. Typical values of the standard deviation term σ_s varied from 0.35 to 0.65 in the Dakota Sandstone outcrop near Cortez, Colorado, and from 0.40 to 0.75 in the Gallup Formation outcrop near Farmington, New Mexico. These outcrops are considered to be relatively clean consolidated sands and are not significantly modified by groundwater circulation.

RESISTIVITY OF SHALY (CLAYEY) RESERVOIR ROCKS

The presence of conductive clays and shales considerably complicates the interpretation of resistivity data of partially saturated formations. The shale type, the percentage present, and the mode of distribution in the formation have different effects on the resistivity. Generally, the presence of clay or shale in a sand bed lowers the true formation resistivity R_t and, if not corrected, will result in overestimating S_w, that is, interpreting as water-bearing zones that are actually oil-bearing. Shales contain, in various proportions, clay minerals such as illite, montmorillonite, and kaolinite, as well as silt, carbonates, and other nonclay minerals. Silt is a very fine-grained material that is predominantly quartz, but may includc feldspar, calcite, and other minerals. The silt fraction of the shales is at a maximum near the sand bodies and at a minimum in the shales far from the sands [20].

WATER SATURATION IN SHALY (CLAYEY) RESERVOIR ROCKS

Figure 4.28 shows three common modes of shale distribution within a reservoir rock (e.g., sand and carbonates).

(1) Laminar shales are thin beds of shale deposited between layers of clean sands. By definition, the sand and shale laminae do not exceed 0.5 in. thickness.

The effect of this type of shale on porosity and permeability of the formation is generally assumed to be negligible. Figure 4.29 shows an idealized series of laminar shales and sand beds. Inasmuch as,

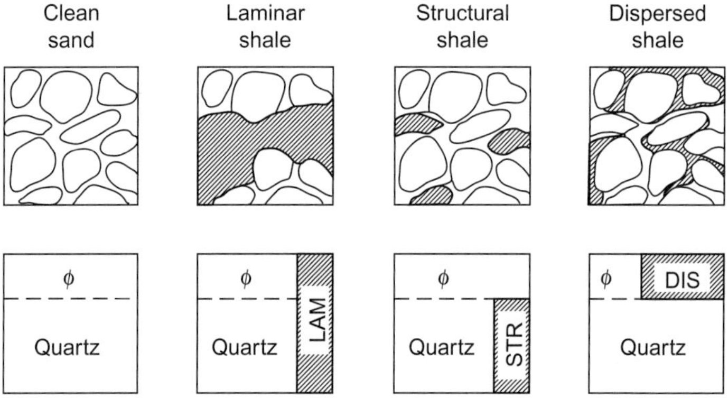

FIGURE 4.28 Modes of clay distribution [20].

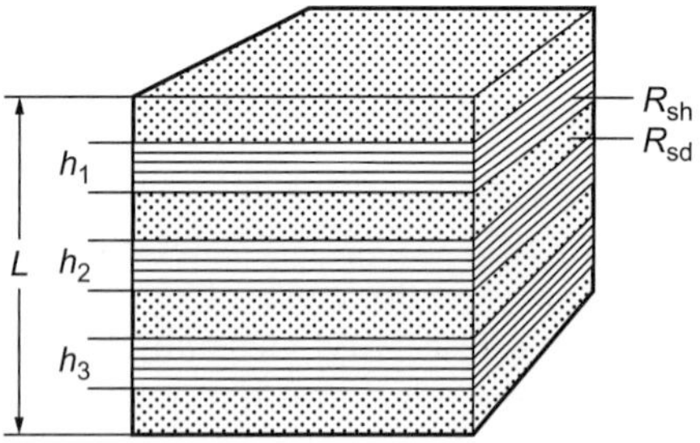

FIGURE 4.29 Distribution of resistivities in an idealized series of laminations [20].

electrically, laminar shales produce a system of conductive circuits in parallel with the porous beds, the total resistivity of the water-bearing formation can be expressed as [21]:

$$\frac{1}{R_t} = \frac{1 - V_{sh}}{R_{sd}} + \frac{V_{sh}}{R_{sh}} \tag{4.66}$$

where V_{sh} is the laminar shale volume estimated from:

$$V_{sh} = \frac{\sum h_{shi}}{h_t} \tag{4.67}$$

h_{shi} and h_t are defined in Figure 4.29. R_{sd} is the resistivity of the clean-sand layers:

$$R_{sd} = \frac{F_{sd} R_w}{S_w^n} \tag{4.68}$$

where F_{sd} is the formation resistivity factor of the sand beds. Using Eq. (4.35), F_{sd} is equal to:

$$F_{sd} = \frac{a}{\phi_{sd}^m} \tag{4.69}$$

where ϕ_{sd} is the sand-streak porosity:

$$\phi_{sd} = \frac{\phi}{1 - V_{sh}} \tag{4.70}$$

and ϕ is the bulk-formation porosity. Thus, the general expression for the formation resistivity factor for sands laminated with thin shale streaks is:

$$F_{sd} = \frac{a}{\phi^m}(1 - V_{sh1})^m \tag{4.71}$$

Combining Eqs. (4.66)–(4.71) and solving explicitly for the water saturation S_w, one obtains:

$$S_w^n = \left(\frac{1}{R_t} - \frac{V_{sh}}{R_{sh}}\right) \frac{a R_w (1 - V_{sh})^{m-1}}{\phi^m} \tag{4.72}$$

In practice, laminar shales are considered to have the same average properties (such as R_{sh}) as the closest thick shale body, because, in all probability, they have been subjected to the same geological process of deposition. The porous sand beds also are assumed to have the same effective porosities, permeability, and water saturation. Equation (4.72) can be simplified by assuming that the cementing factor $m = 2$, the saturation exponent $n = 2$, and the constant $a = 1$:

$$S_w^2 = \left(\frac{1}{R_t} - \frac{V_{sh}}{R_{sh}}\right) \frac{R_w (1 - V_{sh})}{\phi^2} \tag{4.73}$$

The shale volume V_{sh} is determined from various well logs, which are considered to be good clay indicators [22].

(2) Dispersed clays, which evolved from the in situ alteration and precipitation of various clay minerals, may adhere and coat sand grains or they may partially fill the pore spaces. Figure 4.30 shows three types of dispersed shales in a sandstone bed. This mode of clay distribution considerably reduces permeability and porosity of the formation, while increasing water saturation. This increase in S_w is due to the fact that clays adsorb more water than quartz (sand). Dispersed clays contain more bound water because they are subjected only to hydrostatic pore pressure rather than overburden pressure. In core analysis, much of this

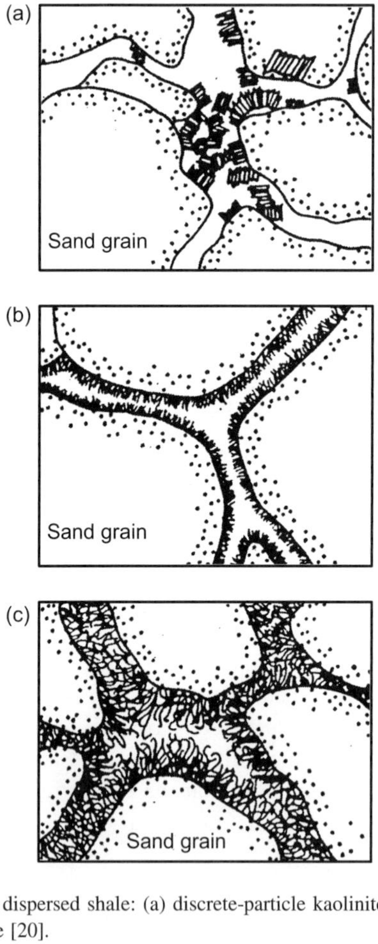

FIGURE 4.30 Types of dispersed shale: (a) discrete-particle kaolinite, (b) pore-lining chlorite, and (c) pore-bridging illite [20].

bound water is lost during the drying process, which results in an overestimation of the porosity of the core sample.

Electrically, the water-dispersed clay mixture or slurry may be approximated by a single electrolyte of resistivity R_t, expressed as [23].

$$\frac{1}{R_t} = \frac{f_{im}}{F_{im}}\left(\frac{f_{shd}}{R_{shd}} + \frac{f_{im} - f_{shd}}{R_w}\right) \tag{4.74}$$

where f_{im} is the fraction of total porosity ϕ_{im} occupied by the dispersed-clay and formation-water mixture, so that the fraction of the effective pore volume occupied by water $(1 - f_{shd})/f_{im}$ and the fraction of dispersed clays in the mixture f_{shd}/f_{im} are related by the following equation:

$$S_w\left(\frac{1 - f_{shd}}{f_{im}}\right) + \left(\frac{f_{shd}}{f_{im}}\right) = 1 \tag{4.75}$$

where f_{shd} is the fraction of the total porosity f_{im} occupied by the dispersed shale, that is,

$$f_{shd} = 1 - \frac{\phi_D}{\phi_s}$$

where

ϕ_D = density log porosity

ϕ_s = sonic log porosity

ϕ_{im} is the total or intermatrix porosity of the formation, which consists of all the space occupied by fluids and dispersed clays. R_{shd} is the resistivity of the dispersed clays, $\approx 0.40 R_{sh}$. F_{im} is the formation factor related to total porosity:

$$F_{im} = \frac{a}{\phi_{im}^m} = \frac{f_{im}^2 R_t}{R_{im}} \tag{4.76}$$

assuming the saturation exponent $n = 2$. Combining Eqs. (4.74)–(4.76) and solving for S_w gives:

$$S_w = \frac{1}{1 - f_{shd}}\left[\left\{\frac{aR_w}{\phi_{im}^m} + \left(\frac{f_{shd}(R_{shd} - R_w)}{2R_{shd}}\right)^2\right\}^{0.5} - \left(\frac{f_{shd}(R_{shd} + R_w)}{2R_{shd}}\right)\right] \tag{4.77}$$

The value of R_{shd} is difficult to evaluate. But because, in most shaly sands, it is much greater than R_w, its exact value is not too critical and Eq. (4.77) simplifies to (for $R_{shd} \gg R_w$):

$$S_w = \frac{1}{1 - f_{shd}}\left[\left(\frac{aR_w}{\phi_{im}^m R_t} + \frac{f_{shd}}{4}\right)^{0.5} - \left(\frac{f_{shd}}{2}\right)\right] \tag{4.78}$$

The total porosity ϕ_{im} is measured by the sonic log, while f_{shd} is determined from the sonic and density logs. If R_{shd} is not much larger than R_w, as is the case in the Rocky Mountain area where $R_w/R_{shd} \approx 0.25$, Eq. (4.78) overestimates S_w.

(3) Structural shale exists as grains of clay forming part of the solid matrix along with sand grains. This type of clay distribution is a rare occurrence. They are considered to have properties similar to those of laminar shale, as both are of depositional origin. They have been subjected to the same overburden pressure as the adjacent thick shale bodies and, thus, are considered to have the same water content.

Different clay distributions will affect the effective porosity and permeability in a drastically different manner. A porosity-dependent cutoff for reservoir permeability depends greatly on the distribution mode and type of clay minerals present [24]. Figure 4.31 illustrates the porosity–permeability relationship in fine-grained, well-sorted sandstones as a function of clay minerals present in the reservoir rock.

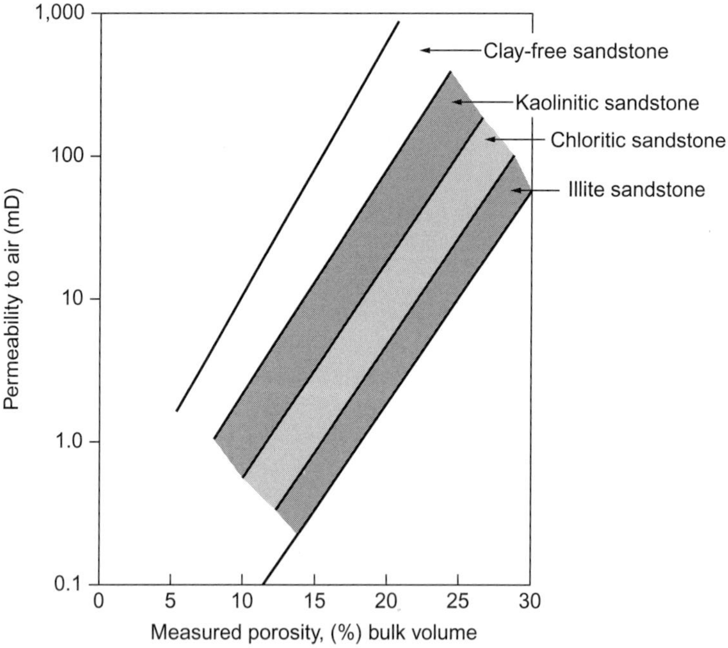

FIGURE 4.31 Permeability–porosity relationship in fine-grained and well-sorted sandstones as a function of various clay minerals [24].

For laminated shaly sands, in which sand and shale layers alternate, the effective porosity or the fractional volume occupied by the fluid in the sand–shale mixture, ϕ_m, is:

$$\phi_e = \phi_{cs}(1 - V_{sh}) \tag{4.79}$$

where ϕ_{cs} is the maximum or total clean-sand porosity and $(1 - V_{sh})$ is the fractional volume occupied by the porous clean-sand layer. The fraction of the clayey rock matrix occupied by sand grains is:

$$V_{gr} = (1 - \phi_{cs})(1 - V_{sh}) = \left(\frac{1}{\phi_{cs}} - 1\right)\phi_e \tag{4.80}$$

For the case of dispersed shales, the effective porosity of the rock matrix depends on the dispersed shale content. Inasmuch as shale fills the intergranular pores up to $V_{sh} = \phi_{cs}$, and for higher clay content the sand grains are in suspension, the effective porosity is [25]:

$$\phi_e = \begin{cases} \phi_{cs} - V_{sh}, & V_{sh} < \phi_{cs} \\ 0, & V_{sh} > \phi_{cs} \end{cases} \tag{4.81}$$

The fractional volume of sand grains is:

$$V_{gr} = \begin{cases} 1 - \phi_{cs}, & V_{sh} < \phi_{cs} \\ 1 - V_{sh}, & V_{sh} > \phi_{cs} \end{cases} \tag{4.82}$$

For structural shale, the effective porosity, ϕ_e, and the fractional volume of sand grains, V_{gr}, depend on the extent to which sand grains are replaced by clay particles. If the fractional volume of shale, V_{sh}, is less than $(1 - \phi_{cs})$, then effective porosity is equivalent to the clean-sand porosity and V_{gr} is:

$$V_{gr} = (1 - \phi_{cs}) - V_{sh} \tag{4.83}$$

If the fractional volume of shale, V_{sh}, is greater than $(1 - \phi_{cs})$, then the fractional volume of sand grains V_{gr} is practically zero, and the effective porosity is:

$$\phi_e = 1 - V_{sh} \tag{4.84}$$

The concept that shaly or clayey sands can be partitioned into two volume components gives satisfactory results only for the case of laminated shaly sands, because the clay content can vary without affecting the clean-sand porosity ϕ_{cs}. However, for the cases of dispersed and structural shales, two-volume model may not account fully for variations in clay contents [25]. If a multivolume model is used, it can be shown that the effective porosity of a shaly multicomponent lithology is:

$$\phi_e = \phi_{cs} + f_{mc} V_{sh} \tag{4.85}$$

where f_{mc} is the multicomponent sensitivity factor, which can be determined from a zone where $V_{sh} = 1$ and $\phi_e = 0$.

APPROXIMATE SHALE RELATIONSHIP

After comparing several models proposed in the literature, Hilchie [26] and Fertlan and Hammack [27] developed the following approximate shaly-sand equation, which generally is applicable to any shale (clay) distribution:

$$S_w = S_{wcs} - F_{sh} \tag{4.86}$$

where S_{wcs} is the clean sand–water saturation, which can be calculated from the Archie equation:

$$S_{wcs} = \left(\frac{F_R R_w}{R_t}\right)^{0.5} \tag{4.87}$$

where F_R reflects the effective porosity, ϕ_e, and F_{sh} is the shale correction factor [27]:

$$F_{sh} = \frac{V_{sh} R_w}{0.4\phi_e R_{sh}} \tag{4.88}$$

Equation (4.86) illustrates the practical aspect of the shale effect on the estimation of hydrocarbon in place. If, for instance, V_{sh} is neglected, the shale correction factor $F_{sh} = 0$. This will result in an overestimation of water saturation, and consequently, an underestimation of hydrocarbon in place. On the other hand, if V_{sh} is overestimated, it produces the opposite effect, that is, an overestimation of hydrocarbon [28].

In general, it is easier to interpret shaly (clayey) formations where ϕ_e and the salinity of water are both very high. Conversely, areas where ϕ_e is low and the water is fresher—such as the Cretaceous sands in Alberta and the Rocky Mountains, and parts of California—the shaly-sand interpretation can be very difficult. Sand interpretation problems, however, are not all caused by the presence of clays, and low-resistivity pay zones are not necessarily shaly-sand problems.

GENERALIZED SHALE RELATIONSHIP

Based on laboratory investigations and field experience, it has been found that, generally, all three forms of clay distribution exist in the same formation, and that the best formula for finding S_w in shaly sandstones is of the following general form:

$$AS_w^2 + BS_w + C = 0 \tag{4.89}$$

where A denotes the combined effect of the amount of sand, its porosity, cementation, and the resistivity of the saturating water. A always reduces to the Archie saturation equation, that is, Eq. (4.54), when the shale volume, $V_{sh}=0$. B denotes the combined effect of the amount of shale and its resistivity, and C is the reciprocal of the total resistivity of the shaly-sand system. For the range of S_w values encountered in the reservoirs, Eq. (4.89) can be expressed as follows:

$$\left(\frac{\phi^m}{aR_w(1-V_{sh})}\right)S_w^2+\left(\frac{V_{sh}}{R_{sh}}\right)S_w-\frac{1}{R_t}=0 \tag{4.90}$$

Based on what is implicitly being practiced in the field, a sandstone may be considered shaly only if the effective shale (clay) content, $V_{sh}>10\%$ [24]. For $V_{sh}=0$, Eq. (4.90) becomes:

$$\left(\frac{\phi^m}{aR_w}\right)S_w^2-\frac{1}{R_t}=0 \tag{4.91}$$

which is equivalent to Eq. (4.54), where the saturation exponent $n=2$. If $0<V_{sh}\leq 10$, the contribution of the content of clays to the term A of Eq. (4.89) is sometimes negligible, and the following form of Eq. (4.90) is recommended:

$$\left(\frac{\phi^m}{aR_w}\right)S_w^2+\left(\frac{V_{sh}}{R_{sh}}\right)S_w-\frac{1}{R_t}=0 \tag{4.92}$$

The positive root of this quadratic equation gives the water saturation of most shaly sandstones independent of the distribution of the shale:

$$S_w=\frac{aR_w}{2\phi^m}\left[-\frac{V_{sh}}{R_{sh}}+\left(\left(\frac{V_{sh}}{R_{sh}}\right)^2+\frac{4\phi^m}{aR_wR_t}\right)^{0.5}\right] \tag{4.93}$$

This expression, which is referred to as the total shale model or simply the Simandoux equation, gives good results in clean and uniformly water-wet systems. The saturation exponent in these systems is ~2. A common method of estimating the percentage of shale (clay) present in the reservoir rock, V_{sh}, and the shale-corrected porosity is to solve simultaneously the following pair of equations:

$$\phi_{DC}=\phi_D-V_{sh}\phi_{Dsh} \tag{4.94}$$

$$\phi_{NC}=\phi_N-V_{sh}\phi_{Nsh} \tag{4.95}$$

The formation porosity, ϕ, is obtained from the following root-square formula:

$$\phi=0.707\sqrt{\phi_{Nc}^2+\phi_{Dc}^2} \tag{4.96}$$

where ϕ_D and ϕ_N are, respectively, the uncorrected density and neutron porosities of the formation, and ϕ_{Dsh} and ϕ_{Nsh} are the density and neutron log readings in the shale portion or adjacent shale bed. If Eq. (4.94) yields a negative ϕ_{Dc}, then ϕ_{Dc} is assumed to be zero. The subscript c represents "corrected."

The two types of clay influences on logs defined by Hilchie are effective (montmorillonite and illite) and noneffective (kaolinite and chlorite). The influence of clays, effective or noneffective, on the density log is solely a function of the clay density. For instance, montmorillonite, which has a density (2.33 g/cm^3) lower than that of sandstone (2.65 g/cm^3), causes the density porosity to be higher than the true porosity. On the other hand, illite, which has a density (2.76 g/cm^3) greater than that of sandstone, causes the density porosity to be lower than the true porosity. Kaolinite, with a density (2.69 g/cm^3) approximately equal to that of a sandstone, cannot be detected by a density log when mixed with sand. The effect of chlorite (2.77 g/cm^3) on the density log becomes significant only when the reservoir porosity is lower than 9%. The influence of shale on the neutron log varies from one service company to another depending upon the instrumentation. In general, the apparent porosity derived from modern neutron logs is greater than the actual effective porosity of the reservoir rock.

Another common method for estimating the fraction of shale (clay), V_{sh}, present in reservoir rock is to use one of the following correlations between the shale volume and the gamma-ray index [29]. For tertiary sediments (<4,000 ft deep):

$$V_{sh} = 0.083(2^{3.7 I_{RA}} - 1) \tag{4.97}$$

For older rocks (4,000–8,000 ft deep):

$$V_{sh} = 0.33(2^{2 I_{RA}} - 1) \tag{4.98}$$

For very hard compacted formation (at a depth of 8,000 ft or more), $V_{sh} = I_{RA}$, where the gamma ray, or radioactive index I_{RA}, is defined as follows:

$$I_{RA} = \frac{GR - GR_{cs}}{GR_{sh} - GR_{cs}} \tag{4.99}$$

The gamma-ray deflection, GR, is obtained from the log at a zone of interest; GR_{cs} is the gamma ray, or radioactive log, reading in a clean (shale-free) sand zone and GR_{sh} is the radioactive log reading in a shale zone. A typical range of the GR deflection is 20 for clean sand and 140 for shaly zone.

Example

A well is drilled in an oil zone, where the pressure is above the bubble point. This shaly-sandstone formation has the following characteristics [28]:

$R_w = 0.05\ \Omega\ m$	$R_t = 5\ \Omega\ m$
$R_{sh} = 1\ \Omega\ m$	$\phi_e = 0.18$

Estimate the oil saturation, assuming (1) V_{sh} is negligible, (2) $V_{sh} = 0.20$, and (3) $V_{sh} = 0.40$. Use Eqs. (4.86) and (4.93), and compare.

Solution

1. For $V_{sh} = 0$, Eq. (4.86) reduces to the Archie equation, that is, Eq. (4.87). The formation factor is calculated from the Humble equation (4.39b):

$$F = \frac{0.81}{\phi^2} = \frac{0.81}{0.18^2} = 25 S_w = \sqrt{\frac{FR_w}{R_t}} = \sqrt{\frac{25 \times 0.05}{5}} = 0.50$$

$$S_o = 1 - S_w = 1 - 0.5 = 0.5$$

2. For $V_{sh} = 0.20$, Eq. (4.88) gives a shale correction factor of:

$$F_{sh} = \frac{V_{sh} R_w}{0.40 \times \phi_e R_{sh}} = \frac{0.20 \times 0.05}{0.40 \times 0.18 \times 1} = 0.14$$

The new estimate of water saturation is:

$$S_w = S_{wcs} - F_{sh} = 0.50 - 0.14 = 0.36 S_o = 1 - 0.36 = 0.64$$

Using Eq. (4.93) for $V_{sh} = 0.20$:

$$S_w = 0.5 FR_w \left[-\frac{V_{sh}}{R_{sh}} + \left(\left(\frac{V_{sh}}{R_{sh}} \right)^2 + \frac{4}{FR_w R_t} \right)^{0.5} \right] = 0.39$$

$$S_o = 1 - 0.39 = 0.61$$

3. Repeating the calculations in (b) for $V_{sh} = 0.40$, Eq. (4.88) gives $S_w = 22\%$ and $S_o = 78\%$. The Simandoux equation (4.93) gives: $S_w = 31\%$ and $S_o = 69\%$. Thus, the higher the shale fraction, the lower is the estimate of water saturation, and consequently the higher the estimate of hydrocarbon in place.

FLOW UNITS FOR SHALY SANDSTONES

The concept of flow units for clastic rocks with low shale content was discussed in Chapter 3. In formations with high shale content the flow unit equation (Eq. (3.47)) becomes:

$$\log(\text{RQI}) = H_{sh} \log \phi + \log(\text{FZI}_{sh}) \quad (4.100)$$

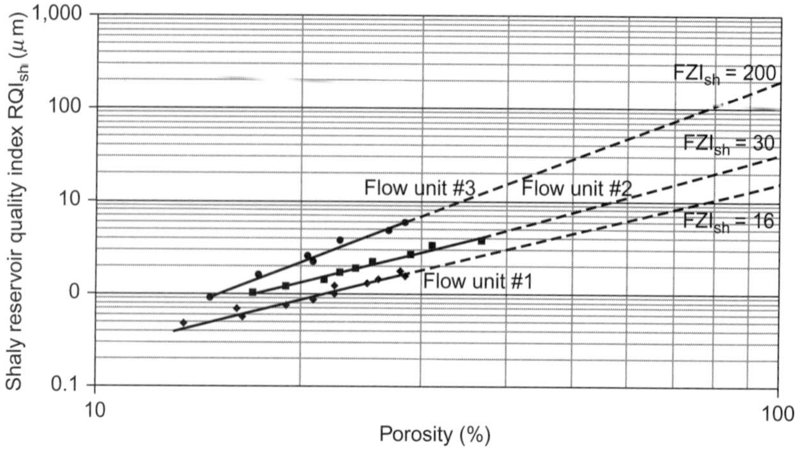

FIGURE 4.32 Shale reservoir quality index plotted against porosity. Three straight lines indicate the presence of three flow units.

where RQI (μm) is the reservoir quality index for the formation, and is defined as:

$$\mathrm{RQI} = 0.0314\sqrt{\frac{k}{\phi}} \tag{4.101}$$

The permeability, k, and porosity, ϕ, are expressed in mD and fraction, respectively. H_{sh} is the slope of a straight line obtained when plotting RQI versus porosity on a log–log graph and FZI_{sh} is the intercept of the straight line at $\phi = 1$, as shown in Figure 4.32. FZI_{sh} is the hydraulic unit factor that uniquely defines the flow unit. H_{sh} is influenced primarily by the degree of cementation of the pay zone, whereas the flow zone indicator FZI_{sh} is a function of the distribution and type of shale and resistivity.

The choice of the model depends on the type of distribution of shale and the types of data that are available. The total shale model or Simandoux equation can be written as:

$$\frac{R_t}{G_{sh}} = \left(\frac{aR_w}{S_w^n}\right)\phi^{-m} \tag{4.102}$$

where the total shale group G_{sh} is:

$$G_{sh} = 1 + \frac{\phi^m R_t}{aR_w[2C_{sh}^2 - 2C_{sh}^2\sqrt{aR_w/\phi^m R_t + C_{sh}^2}]} \tag{4.103}$$

$$C_{sh} = \frac{aR_w V_{sh}}{2\phi^m R_{sh}} \tag{4.104}$$

Taking the logarithm of both sides of Eq. (4.102) yields:

$$\log\left(\frac{R_t}{G_{sh}}\right) = -m\log(\phi) + \log\left(\frac{aR_w}{S_w^n}\right) \tag{4.105}$$

Thus a log–log plot or Pickett plot of the ratio R_t/G_{sh} versus porosity is a straight line of slope $-m$, assuming aR_w, water saturation S_w, and exponent n are constant within a given flow unit. In order to identify flow units, a permeability–porosity relationship is necessary. This relationship is then combined with a shale model that best describes the pay zone of interest, to derive the slope H_{sh} and intercept FZI_{sh}. If, for instance, the saturation exponent is equal to 2, the total shale model can simply be written as:

$$S_w = \sqrt{\frac{aR_w}{\phi^m R_t} + C_{sh}^2} - C_{sh} \tag{4.106}$$

The Timur permeability–porosity relationship is:

$$k = 8581\frac{\phi^{4.4}}{S_w^2} \tag{4.107}$$

The Wyllie and Rose relationship is:

$$k = 62{,}500\frac{\phi^6}{S_w^2} \tag{4.108}$$

The generalized form of the Timur or Wyllie and Rose permeability–porosity model is:

$$k = C_1\frac{\phi^{C_2}}{S_w^n} \tag{4.109}$$

where C_1 and C_2 are correlation constants. In the case of Timur (Eq. (4.107)), $n = 2$, $C_1 = 8{,}581$, and $C_2 = 4.4$, where k is in mD and porosity and saturation are expressed in fractions. For the Wyllie and Rose equation (Eq. (4.108)), $n = 2$, $C_1 = 62{,}500$, and $C_2 = 6$, where k is in mD and ϕ and S_w are fractions.

Assuming the pay zone has low shale content ($V_{sh} < 10\%$), that is, C_{sh} is approximately zero, then combining Eqs. (4.101), (4.106), and (4.109) and solving for RQI yields:

$$\log(\mathrm{RQI}) = \left(\frac{C_2 + m - 1}{2}\right)\log\phi + \log\left(0.314\sqrt{\frac{C_1 R_t}{aR_w}}\right) \tag{4.110}$$

Assuming the Wyllie and Rose equation is applicable, substituting for C_1 and C_2 gives:

$$\log(\mathrm{RQI}) = (0.5m + 2.5)\log\phi + \log\left(78.5\sqrt{\frac{R_t}{aR_w}}\right) \tag{4.111}$$

Thus, from Eq. (4.100), the values of H_{sh} and flow zone indicator FZI_{sh} are:

$$H_{sh} = 0.5m + 2.5 \tag{4.112}$$

$$\mathrm{FZI}_{sh} = 78.5\sqrt{\frac{R_t}{aR_w}} \tag{4.113}$$

This plot should result in the segregation of flow units represented by distinct clusters. If the Timur permeability model is used, H_{sh} and FZI_{sh} become:

$$H_{sh} = 0.5m + 1.7 \tag{4.114}$$

$$\mathrm{FZI}_{sh} = 29.09\sqrt{\frac{R_t}{aR_w}} \tag{4.115}$$

If a different permeability–porosity equation, but one of the same general form as Eq. (4.109), is more appropriate for a specific formation, then simply substitute the appropriate correlation constants C_1 and C_2 in Eq. (4.110).

In some field cases, it was found that the scatter in the log–log plot of RQI versus porosity was reduced by introducing the shale fraction term V_{sh} into RQI as follows:

$$\mathrm{RQI}_{sh} = 0.0314\sqrt{\frac{k}{\phi(1 - V_{sh})}} \tag{4.116}$$

In this case Eq. (4.110) becomes:

$$\log(\mathrm{RQI}_{sh}) = \left(\frac{C_2 + m - 1}{2}\right)\log\phi + \log\left(0.314\sqrt{\frac{C_1 R_t}{a(1 - V_{sh})R_w}}\right) \tag{4.117}$$

Equations (4.113) and (4.115) become:

$$\mathrm{FZI}_{sh} = 78.5\sqrt{\frac{R_t}{aR_w(1 - V_{sh})}} \tag{4.118}$$

$$\mathrm{FZI}_{sh} = 29.09\sqrt{\frac{R_t}{aR_w(1 - V_{sh})}} \tag{4.119}$$

If the pay zone has significant shale content, that is, $V_{sh} > 20\%$, then C_{sh} cannot be approximated to zero. Using the same approach as above, new equations must be derived for H_{sh} and FZI_{sh}. Assuming the total shale model and Timur or Walter and Rose permeability–porosity relationships are applicable, the procedure to identify and characterize flow units is as follows:

(1) First obtain all necessary data from core analysis and well logs.
(2) Obtain m from the Pickett plot or some other source.
(3) Plot RQI versus porosity on a log–log graph.
(4) Substitute the value of m into Eq. (4.112) and calculate H_{sh}.
(5) Draw a straight line of slope H_{sh} across a cluster or clusters. Clusters that best fit this straight line constitute distinct flow units. Data or clusters that do not fit this straight line may belong to a different flow unit. If the shale type distribution is homogeneous and the pay zone has several flow units, then the straight lines should be parallel with the same value of H_{sh} but different FZI_{sh} values (flow units 1 and 2, Figure 4.32). If the shale type distribution is heterogeneous or if the formation contains different types and volumes of shale, the straight lines will not be parallel, as shown in Figure 4.32 (flow unit 3).
(6) Obtain the flow zone indicator FZI_{sh} from the graph at $\phi = 1$. This value of FZI_{sh} should be approximately the same in all wells, where the flow unit is present.

This process of segregating flow units works best in formations with low shale content. The choice of the permeability–porosity relationship and the shale model heavily influence this process.

LAB-DERIVED EVALUATION OF SHALY (CLAYEY) RESERVOIR ROCKS

Not since Archie presented his classic empirical equation in 1942, relating F_R to porosity and cementation factor for clean sands, has there been another equation of equivalent impact in petrophysics as that of Waxman and Smits [30]. They used a simple physical model to derive an equation that relates the electrical conductivity of the water-saturated shaly (clayey) sand to the water conductivity and the cation exchange capacity (CEC) per unit pore volume of the rock. The model consists of two resistance elements in parallel: one element consisting of the free electrolyte contained in the pore volume of rock, C_{el}, and another resulting from the conductance contribution of the exchange cations associated with the clay C_c. The conductance of a rock is simply the sum of C_{el} and C_c. Thus, the specific conductance of a core, C_o, can be expressed as:

$$C_o = XC_s + YC_w \tag{4.120}$$

where C_s and C_w are the specific conductances of clay-exchange cations and equilibrating salt solution, respectively, and X and Y are geometric cell constants. C_o, C_s, and C_w are expressed in mho/cm. It is assumed that the brine solution in the porous rock has the same electrical conductivity as that for

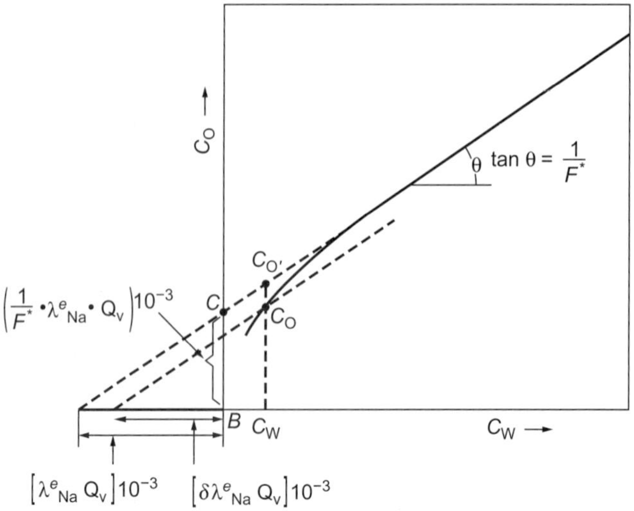

FIGURE 4.33 Core conductivity (C_o) as a function of equilibrating solution conductivity (C_w) [30].

the equilibrating solution. Figure 4.33 shows the behavior of the shaly-sand conductivity, C_o, as a function of C_w. In the range of dilute equilibrating electrolyte solutions, from 0 to 0.5 M NaCl, C_o increases sharply with increasing solution concentration at a greater rate than can be accounted for by the increase in C_w. This sharp increase is attributed to an increasing exchange cation mobility.

At some relatively high concentration of salt, the exchange cation mobility becomes constant, and further increase in concentration has no effect on this mobility. Beyond this dilute region, the sand conductivity, C_o, increases linearly with increases in solution conductivity C_w.

Assuming that the electric current transported by the counterions associated with the clay travels along the same tortuous path as the current attributed to the ions in the pore water, the geometric factors X and Y are equal, and Eq. (4.120) becomes:

$$C_o = X(C_s + C_w) \tag{4.121}$$

For clean sands, $C_s = 0$, and X becomes the reciprocal of the Archie formation resistivity factor defined as C_w/C_o. Thus, by analogy, for shaly sands, where $C_s = 0$,

$$C_o = \frac{1}{F^*}(C_s + C_w) \tag{4.122}$$

where F^* is the shaly (clayey) sand formation resistivity factor, related to porosity according to the Archie or Humble-type relationship, that is,

$$F^* = \frac{a^*}{\phi^{m^*}} \tag{4.123}$$

where a^* and m^* are the Humble coefficient and the cementation factor, respectively, for shaly sand. For the straight-line portion of the shaly-sand conductivity (Figure 4.33), the equation is:

$$C_{\mathrm{sc1}} = 10^{-3}\lambda_{\mathrm{Na}}Q_{\mathrm{v}} \tag{4.124}$$

where

C_{sc1} = specific conductance of the clay counterions for the straight line portion of the C_{o} versus C_{w} curve, mho/cm
Q_{v} = volume concentration of sodium exchange cations associated with the clay, eq/L
λ_{Na} = maximum equivalent ionic conductance of the sodium exchange ions, cm^2/eq/Ω

Equation (4.124) can be modified to include the curved portion of the conductivity curve in the low C_{w} region, by assuming an exponential rise of the counterion mobility up to its constant and maximum mobility.

$$C_{\mathrm{s}} = \delta C_{\mathrm{sc1}} \tag{4.125}$$

with

$$\delta = 1 - \alpha\, \mathrm{e}^{-C_{\mathrm{w}}/\gamma} \tag{4.126}$$

where

γ = constant, determined by the rate of increase of the counterion mobility from zero water conductivity up to its constant value at the higher water conductivities, mho/cm
α = dimensionless constant, $1 - \lambda'_{\mathrm{Na}}/\lambda_{\mathrm{Na}}$
λ'_{Na} = equivalent ionic conductance of the exchange cations at $C_{\mathrm{w}} = 0$

Combining Eqs. (4.122)–(4.126), one can obtain the Waxman–Smits general equation for water-saturated shaly (clayey) sands:

$$F^* = \frac{1}{C_{\mathrm{o}}}(C_{\mathrm{eq}}Q_{\mathrm{v}} + C_{\mathrm{w}}) \tag{4.127}$$

with C_{eq}, the equivalent conductance of clay-exchange cations, expressed in mho/cm^2/meq^{-1} and given by:

$$C_{\mathrm{eq}} = 10^{-3}\lambda_{\mathrm{Na}} \tag{4.128}$$

Equation (4.127) can be written as follows:

$$F^* = \frac{C_{eq}Q_v}{C_o} + F_R \tag{4.129}$$

or

$$\frac{F^*}{F_R} = \frac{C_{eq}Q_v}{C_w} + 1 \tag{4.130}$$

where F_R is the Archie formation resistivity factor for clean sand, C_w/C_o. Hoyer and Spann used Eq. (4.129) to quantify the effects of shaliness on electric log response [31]. They showed that if $C_{eq}Q_v/C_w < 0.10$, the shaliness effect is negligible, and conventional clean-sand relationships can be used to interpret well logs. Shaliness effect is significant when $C_{eq}Q_v/C_w > 0.10$, and shaly sand relationships must be used to interpret logs. Several methods are available for measuring the CEC of a rock, Q_v. Mortland and Mellow described a procedure for measuring CEC that gives results with an accuracy equivalent to that using the ammonium acetate method [32]. This procedure requires repeated equilibration of the crushed rock sample with concentrated barium chloride ($BaCl_2$) solution and washing to remove excess barium ions, followed by conductometric titration with standard magnesium sulfate ($MgSO_4$) solution. Table 4.8 gives porosity, ϕ; permeability, k; formation resistivity factor, F^*; CEC, Q_v; water conductivity, C_w; and rock conductivity at 100% water saturation, C_o, of some representative sandstone samples.

Waxman and Smits presented a lengthy but rigorous procedure to determine CEC [30]. Because the method used was destructive, conductivity measurements were made on rock samples adjacent to the locations where the cores originated. The rock samples were equilibrated by repeated flushing and storing in a desiccator filled with the appropriate salt solution. The CEC and conductivity measurements were made at 25°C on the shaly sands equilibrated with NaCl solutions at 10 different concentrations varying from saturated solutions of 6.144 M to 0.018 M. A plot of electrical conductivity, C_o, of three representative shaly cores versus water conductivity, C_w, is presented in Figure 4.34. Values of F^* and C_{sc1} (i.e., 0.001 $\lambda_{Na}Q_v$) were determined from the slopes and intercepts, respectively, of the straight-line portions of Figure 4.35. The values of C_{scl} were plotted against the independently determined values of Q_v, as shown in Figure 4.36. The data can be curve-fitted by a straight line passing through the origin. In this case the slope is ~0.0384. Table 4.9 shows values of C_{scl} and the independently measured Q_v values of 27 cores. Values of δ at each value of C_w of these cores were calculated from:

$$\delta = \left(\frac{C_o}{C'_o}\right)\frac{C_w}{C_{sc1}} + \frac{C_o}{C'_o} \tag{4.131}$$

TABLE 4.8 Conductivity Data for Various Sandstone Samples Using the Mortland and Mellow Method [32]

Rock Type	Sample Number	Porosity (%)	Air Permeability (mD)	Air Brine Q_v Brine permeability (mD)	Q_v Exp. (eq/L)	F^*	C_o (m mho cm^{-1})		
							37.3	**81.5**	**22.8**
Clean sandstone	1	21.7	–	48.1	0.093	17.6	2.26	4.91	13.1
	2	16.6	0.38	0.04	0.102	36.3	1.3	2.61	6.55
	3	24.4		166	0.083	13.7	2.96	6.31	16.9
	4	20.2		290	0.036	16.6	2.41	5.16	13.8
	5	24.2		119	0.051	12.9	2.99	6.66	17.8
	6	21.2	28	19.6	0.104	16.8	2.45	5.16	13.7
	7	20.5	34	21.5	0.097	17.7	2.36	4.97	13.1
	8	20.8	51	26.2	0.069	17.6	2.44	5.16	13.2
	9	20.4	59	40.1	0.076	17.2	2.33	5	13.4
	10	24.5		220	0.049	12.6	3.11	6.66	18.2
	11	14.5	0.19		0.454	49.7	1.11	2.05	4.92
Shaly sandstone	1	20.2	80	56.3	0.085	19.4	2.16	4.53	11.9
	2	21.1	141	70	0.102	18.2	2.32	4.97	12.7
	3	25.8		440	0.05	11.4	3.43	8.45	20.2
	4	19.3	67	40	0.112	25.7	1.65	3.52	9.34
	5	20.7	226	158	0.062	18.7	2.23	4.77	12.4

(Continued)

TABLE 4.8 (Continued)

Rock Type	Sample Number	Porosity (%)	Air Permeability (mD)	Air Brine Q_v Brine permeability (mD)	Q_v Exp. (eq/L)	F^*	C_o (m mho cm^{-1})		
							37.3	81.9	22.8
	6	20	146	83	0.067	20.8	2.03	4.31	11.2
	7	18.9	95	132	0.065	22.1	2.05	4.25	10.6
	8	18.2	48	33	0.123	28.2	1.66	3.35	8.34
	9	16.1	12	6.9	0.158	33	1.3	2.7	7.05
	10	14.9	3.3	1.1	0.298	36.3	1.29	2.57	6.51
	11	15.9	1.3	0.16	0.254	37.1	1.31	2.55	6.43
	12	11.6	0.3	0.05	0.281	74.3	0.67	1.27	3.2
	13	18.6	59	31	0.1	25.1	1.64	3.42	9.19
	14	17.4	5.9	2.1	0.206	31.2	1.46	2.93	7.52
	15	16.3	3.7	16.2	0.185	35.8	1.44	2.89	6.7

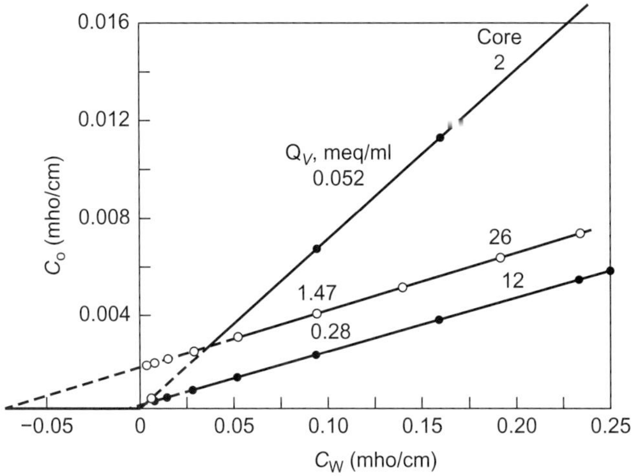

FIGURE 4.34 Electrical conductivity of three core samples versus brine conductivity [30].

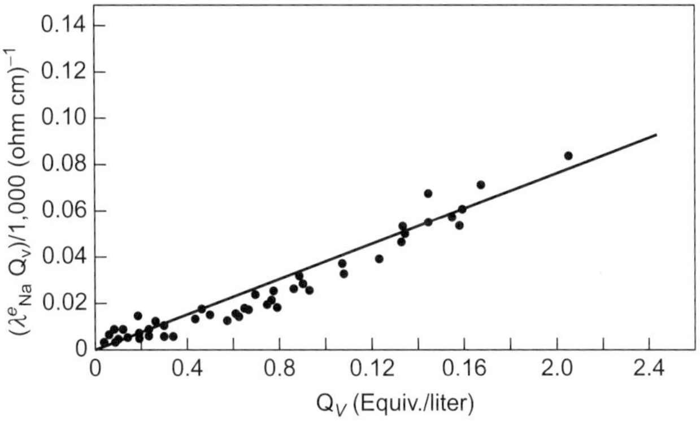

FIGURE 4.35 Plot of $(\lambda_{Na}Q_v)$ obtained from conductivity curves versus Q_v [30].

where C_o and C_o' are, respectively, the measured and hypothetical conductance of the core. C_o' is obtained from the straight-line extrapolation at the same C_w as is shown in Figure 4.33. Values of λ_{Na} were calculated from the adjacent Q_v determinations for each core, using Eq. (4.124), and values of C_{eq} in mho were calculated using Eq. (4.127) and average values of δ and λ_{Na}. Figure 4.36 shows a plot of these values of B as a function of C_w at 25°C, which can be represented by (see also Figure 4.35):

$$C_{eq} = \lambda_{Na}(1 - \alpha_1 e^{-C_w/\alpha_2}) \tag{4.132}$$

where $\alpha_1 = 0.6$, $\alpha_2 = 0.013$, and $\lambda_{Na} = 0.046$. The equivalent ionic conductance of the exchange cations, λ_{Na}', can be obtained from Eq. (4.126)

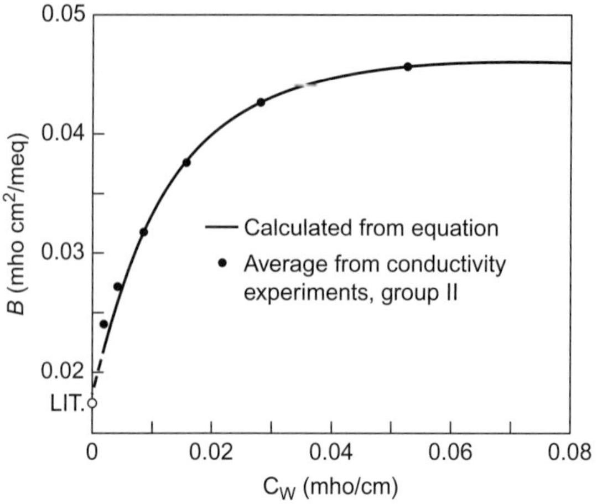

FIGURE 4.36 Equivalent conductivity of the counterions associated with clay as a function of equilibrating electrolyte conductivity [30].

at $C_w = 0$ or from the definition of the dimensionless constant a. Both cases give $\lambda'_{Na} \approx 18.5\,\text{cm}^2/\text{eq}\Omega$, which is in agreement with the actual value of equivalent conductance of sodium exchange ions of about $18\,\text{cm}^2/\text{eq}/\Omega$.

Waxman and Smits extended the conductivity equation for water-bearing shaly sands (Eq. (4.127)) to describe the conductivity of shaly sands containing both oil and brine. Assuming that the mobility of the exchange ions is unaffected by the partial replacement of water, Eq. (4.127) becomes:

$$C_t = \frac{1}{f_G^*}\frac{C_{eq}Q_v}{S_w} + C_w \tag{4.133}$$

where

C_t = specific conductance of a partially water-saturated sand
f_G^* = geometric factor
S_w = water saturation

The factor f_G^* is a function of porosity, water saturation, and pore geometry, but independent of clay content. f_G^* tends to increase with increasing oil saturation. For clean sands F^* and f_G^* become, respectively, $F_R = C_w/C_o = R_t/R_w$ and $f_G = C_w/C_t = R_t/R_w$, and the resistivity index is equal to:

$$I_R = \frac{R_t}{R_o} = \frac{C_o}{C_t} = \frac{f_G}{f_R} = S_w^{-1} \tag{4.134}$$

TABLE 4.9 Fraction of Maximum Equivalent Conductance as a Function of Water Conductivity [30]

Core No.	Q_v (meq/cm^3)	$(\lambda_{NA}Q_v)/1{,}000$ (mho/cm)	$(\lambda_{NA}Q_v)/1{,}000$ (mho/cm^2/meq)	0.05249	Values of δ for C_w in mho/cm				
					0.02822	**0.01492**	**0.007802**	**0.004049**	**0.002085**
1	0.017	0.00232	(0.136)	0.929	0.803	(0.383)	(0.254)	(0.182)	(0.127)
2	0.052	0.00264	0.0507	0.979	0.848	0.647	0.513	0.412	0.315
3	0.052	0.00268	0.0515	0.959	0.793	0.5873	0.417	0.327	0.239
4	0.26	0.00287	0.011	0.961	1.032	1	0.758	0.614	0.456
5	0.2	0.00412	0.0206	1.014	0.969	0.787	0.676	0.571	0.462
6	0.095	0.00415	0.0437	1.027	0.922	0.789	0.683	0.514	0.436
7	0.053	0.00589	(0.111)	0.921	0.78	0.685	0.574	0.472	0.381
8	0.053	0.00584	(0.110)	0.91	0.75	0.62	0.503	0.402	0.316
9	0.085	0.00443	0.0521	1.039	0.978	0.838	0.707	0.604	0.491
10	0.253	0.01376	0.0544	0.952	0.786	0.648	0.53	0.416	0.313
11	0.253	0.00857	0.0339	0.993	1	0.83	0.711	0.525	0.413
12	0.28	0.01243	0.0444	1.016	0.941	0.857	0.67	0.557	0.545
13	0.28	0.01617	0.0578	0.996	0.909	0.804	0.708	0.578	0.[illegible]15
14	0.28	0.01384	0.0494	0.957	0.888	0.813	0.733	0.61	0.548
15	0.41	0.02433	0.0593	0.968	0.868	0.726	0.604	0.452	0.353

(Continued)

TABLE 4.9 (Continued)

Core No.	Q_v (meq/cm^3)	$(\lambda_{NA}Q_v)/1,000$ (mho/cm)	$(\lambda_{NA}Q_v)/1,000$ (mho/cm^2/meq)	Values of δ for C_w in mho/cm					
				0.05249	**0.02822**	**0.01492**	**0.007802**	**0.004049**	**0.002085**
16	0.67	0.02898	0.0433	1.006	0.964	0.911	0.826	0.681	0.627
17	0.33	0.02947	0.0893	1.014	0.963	0.913	0.784	0.712	0.656
18	0.59	0.02354	0.0399	1.003	0.952	0.881	0.732	0.671	0.712
19	0.59	0.01853	0.0314	0.985	0.904	0.809	0.723	0.609	0.8
20	0.59	0.01463	0.0248	1.005	0.953	0.84	0.733	0.64	0.673
21	0.29	0.00872	0.0301	1.056	1.085	0.954	0.729	0.734	0.79
22	0.72	0.0374	0.0519	1.005	0.956	0.892	0.836	0.802	0.777
23	1.04	0.0454	0.0437	0.996	0.925	0.833	0.762	0.715	
24	0.81	0.0526	0.0649	1.002	0.975	0.896	0.793	0.774	0.778
25	1.27	0.0724	0.057	1.01	0.981	0.941			
26	1.47	0.0771	0.0524	1.008	0.969	0.907	0.844	0.806	0.781
27	1.48	0.0783	0.0529	1.012	1.004	0.964			
Average			0.0463	0.99	0.916	0.821	0.69	0.592	0.524
	S.D.		0.0158	0.034	0.034	0.111	0.11	0.13	0.159

Values between parentheses are considered not reliable because of extreme deviation from the rest of the group, probably due to Q_v determination. They were not used in calculating the average. For $C_w > 0.060$ mho/cm, $\delta = 1$.

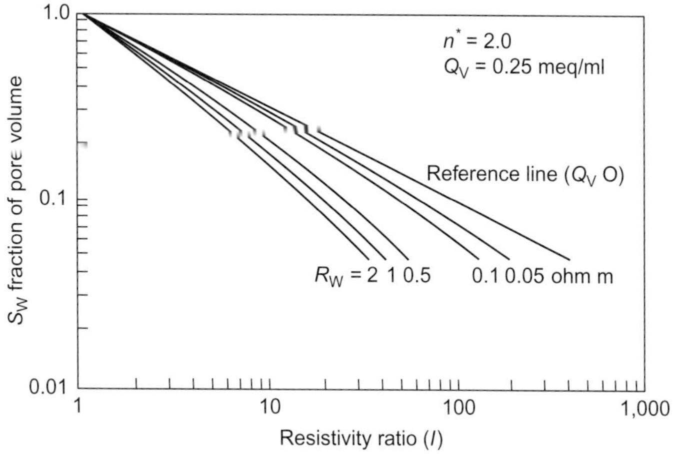

FIGURE 4.37 Water saturation as a function of resistivity index, with variable water resistivity [30].

By analogy, for shaly sands one can obtain:

$$\frac{f_G^*}{F^*} = S_w^{-n^*} \tag{4.135}$$

where n^* is the saturation exponent for shaly sand. Combining Eqs. (4.127) and (4.133)–(4.135) and solving for the resistivity index, one obtains:

$$I_R = S_w^{-n^*}\left(\frac{C_w + C_{eq}Q_v}{C_w + C_{eq}Q_v/S_w}\right) \tag{4.136}$$

or, in terms of water resistivity:

$$I_R = S_w^{-n^*}\left(\frac{1 + R_w C_{eq}Q_v}{1 + R_w C_{eq}Q_v/S_w}\right) \tag{4.137}$$

where R_w and $C_{eq}Q_v$ are expressed in Ω m and $(\Omega\ \text{m})^{-1}$, respectively. If Q_v is expressed in eq/L, C_{eq} can be correlated by:

$$C_{eq} = 4.6(1 - 0.6e^{0.77/R_w}) \tag{4.138}$$

Figures 4.37 and 4.38 show logarithmic plots of the resistivity index as a function of water saturation for different values of R_w and Q_v, respectively. Waxman and Smits observed that even small amounts of clay have a considerable effect on the resistivity index and that Eq. (4.137) predicts higher oil saturation estimates than are obtained from conventional clean-sand equations.

A laboratory study by Waxman and Thomas involving a large number of shaly rock samples from seven different fields demonstrated excellent agreement between experimental oil saturations and those calculated from the Waxman and Smits model [33]. In the same study, Waxman and Thomas

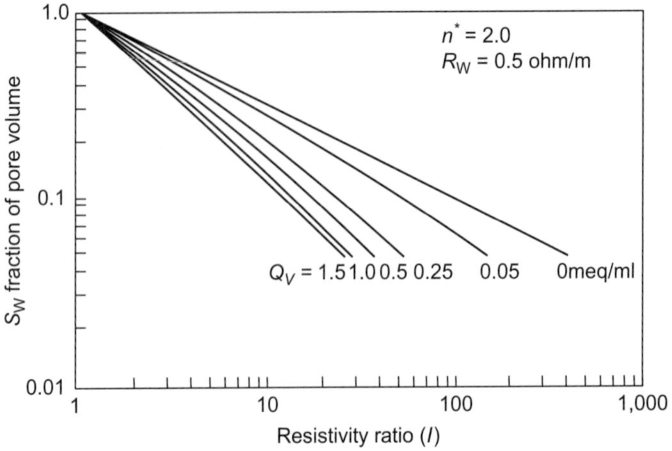

FIGURE 4.38 Water saturation as a function of resistivity index with variable Q_v [30].

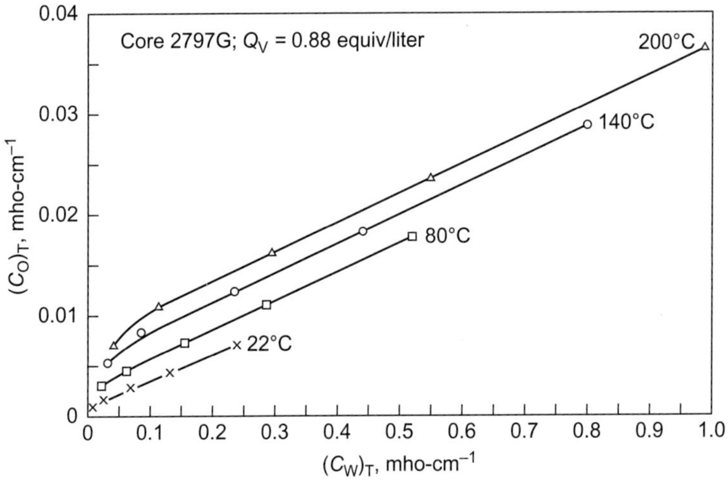

FIGURE 4.39 Electrical conductivity of a shaly sand versus equilibrating brine conductivity at various temperatures [33].

investigated the effect of temperature on the electrical conductivity of shaly cores, and showed that:

1. The shaly-sand formation resistivity factor, as defined by Waxman and Smits, is independent of temperature.
2. The observed increase in conductivity of shaly cores with increasing temperature, as illustrated by Figure 4.39, is due to two temperature-dependent parameters: the equivalent counterion conductance, C_{eq}, and the resistivity of the equilibrating brine, R_w, as shown in Figure 4.40.

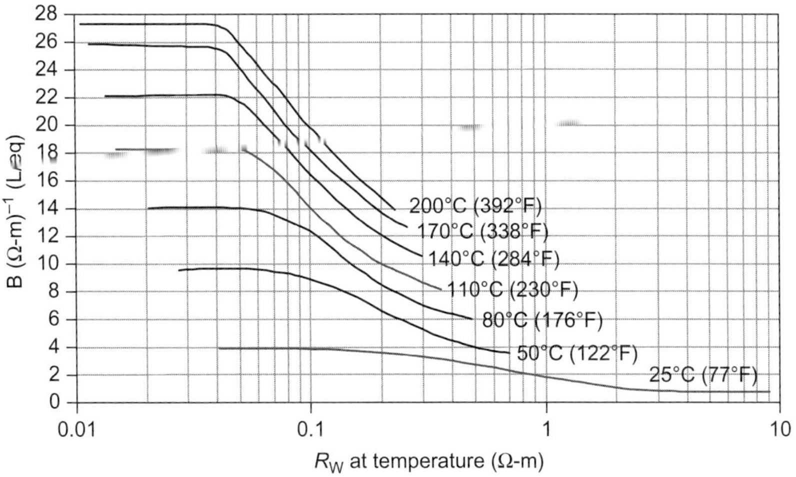

FIGURE 4.40 Equivalent counterion conductance versus resistivity of equilibrating brine at various temperatures [33].

C_{eq} versus R_w curve at $T = 25°C$ can be represented by:

$$C_{eq} = 3.83(1 - 0.83\,e^{0.5/R_w}) \tag{4.139}$$

This correlation is slightly different, but more accurate, than Eq. (4.138).

The effect of temperature on the relationship between the resistivity index I_R and water saturation S_w also was investigated assuming both Q_v and n^* are temperature-independent. Figure 4.41 shows a typical behavior of I_R versus S_w for various temperatures. This figure indicates that a decrease in the resistivity index at the constant water saturation is obtained with increasing temperature and, for temperatures >80°C, the I_R versus S_w relationship is virtually independent of temperature.

Ideally, the laboratory-determined electrical properties of cores would be made at reservoir conditions of temperature and water salinity. Using the Waxman and Thomas approach, however, laboratory properties can be adjusted to reflect reservoir conditions, according to the following equation:

$$\frac{(C_o)_T}{(C_o)_{T_L}} = \frac{C_{eqT}Q_v + (C_w)_T}{C_{eqT_L}Q_v + (C_w)_{T_L}} \tag{4.140}$$

where the subscript T_L denotes the laboratory temperature and T is the reservoir temperature. The effective clay-exchange cations, Q_v, for the shaly core samples can be determined from the C_o versus C_w plot (such as Figure 4.33) at T_L. According to Figure 4.33, the intercept of the extrapolated straight-line portion of the curve is equal to $10^{-3}\lambda'_{Na}Q_v/F^*$ and F^* is determined from the slope of this straight line; therefore, Q_v values can be calculated. For a laboratory temperature of 25°C, $\lambda'_{Na} = 38.3\ \text{cm}^2/\text{eq}\ \Omega$.

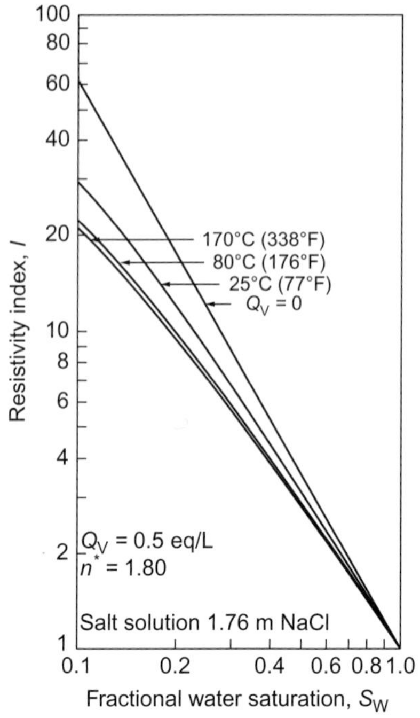

FIGURE 4.41 Temperature dependence of I versus S_w function.

Hoyer and Spann showed that measurements of electrical properties of a core sample are subject to considerable error if they are made before the rock and its saturating fluid reach equilibrium, which can lead to serious errors in calculating the reservoir fluid saturations [31]. The equilibrium problem was first detected when these authors observed the long-term electrical behavior of a sandstone core that was 100% saturated with 10,000 ppm NaCl brine and stored in this brine. The measured formation resistivity factor for this core increased continuously for 18 days, as shown in Figure 4.42. Even when the core was subjected to a continuous flow test, equilibrium was not reached for 5 more days.

The rock sample at reservoir conditions equilibrates the in situ brine with (1) the salts deposited in small cracks and fissures and (2) the clays. The difficulty in making the core reach equilibrium under laboratory conditions is due to two main causes. First, during coring, a fluid different from the native brine almost always comes in contact with and at least partially saturates the core. Second, during core handling, salts are deposited and clays are partially or totally dehydrated if the core is allowed to dry out. Hoyer and Spann

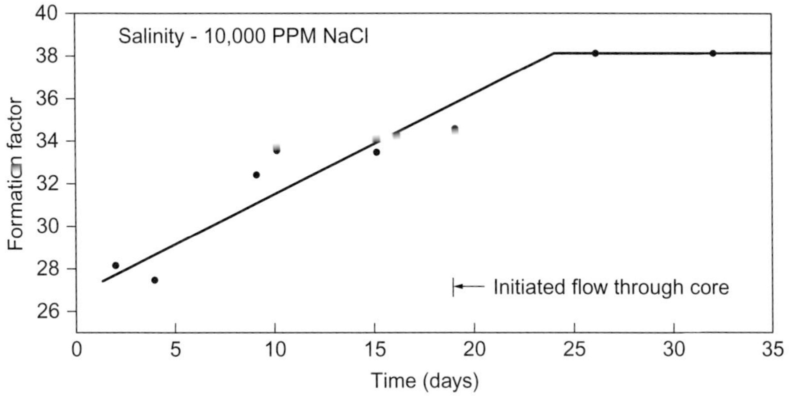

FIGURE 4.42 Formation factor equilibrium [30].

recommended the following procedure for obtaining electrical properties for cores [31]:

(1) Select cores that have not been dehydrated by excessive heat or cleaned excessively with solvents.
(2) Remove hydrocarbons from the core by flushing it with propane and then with water.
(3) Measure core electrical conductivity C_o versus water conductivity C_w for at least three different salinities.
(4) Plot C_o against C_w.
(5) Repeat Steps 3 and 4 until the points form a straight line as shown in Figure 4.34.
(6) If the term $C_{eq}Q_v/C_w$, evaluated at formation water salinity, is >0.1, the shaliness effect is significant.

Example

Estimate the water saturation of a sandstone formation which displays the following characteristics:

Porosity = 25%
Shaly-sand formation resistivity factor, $F^* = 12.43$
CEC = 0.08 meq/g
Formation temperature = 230°F
$R_w = 0.04\ \Omega\ m$
$R_t = 4\ \Omega\ m$
$\rho_{ma} = 2.65\ g/cm^3$
Shaly-sand saturation exponent, $n^* = 2$

Solution

Similarly to Eq. (4.92), the quadratic form of the Waxman and Smith model in terms of resistivity is:

$$(R_t)S_w^{n^*} + (R_t R_w B Q_v) S_w - F^* R_w = 0 \qquad (4.140a)$$

The effective volume concentration of clay-exchange cation, Q_v, can be estimated as follows (Eq. (4.141)):

$$Q_v = \frac{\mathrm{CEC}(1-\phi)\rho_{ma}}{\phi} = \frac{0.08(1-0.25)2.65}{(0.25)} = 0.636\ \mathrm{meq/cm^3}$$

The equivalent conductance of clay-exchange cations B can be estimated from Eq. (4.138), with $C_{eq} = B$:

$$B = 4.66\left(1 - 0.6\exp\left(\frac{-0.77}{R_w}\right)\right) = 4.66\left(1 - 0.6\exp\left(\frac{-0.77}{0.44}\right)\right) = 4.66$$

Substituting these values into Eq. (4.140a):

$$(4)S_w^2 + (4 \times 0.04 \times 4.66 \times 0.636)S_w - 12.43 \times 0.04 = 0$$
$$(4)S_w^2 + (0.4742)S_w - 0.4972 = 0$$

The positive solution of this quadratic equation is:

$$S_w = 29.8\%$$

Example

The Shannon sandstone of the Teapot Dome field is composed of fine- to medium-sized sand particles containing dispersed clay. The formation water is relatively fresh, and salinity varies from a low value of 3,700 ppm in the northern portion of the field to a high value of 13,000 ppm in the southern portion. Reservoir temperature also varies across the Shannon field, from a high value of 118°F in the northwest part to 70–95°F in the southern and eastern parts of the reservoir [34].

The adsorbed water technique was used to measure the CEC of a large number of cores obtained from several wells [35]. CEC values ranged from a low value of 1.58 meq/100 g in samples containing no visible clays to a maximum of 8.65 in cores described as highly shaly. Similar CEC values were found in all wells. A variety of correlations of the measured CEC values with the Shannon sandstone porosities were attempted (Figure 4.43), and though none was found acceptable, it was observed that CEC values increase with decreasing porosity. This is because the loss of porosity in the Shannon reservoir is caused by an infilling of clays with an associated increase in CEC. The salinity of the brine used to saturate core samples for the laboratory electrical property tests is 15,000 ppm NaCl and the resistivity of this brine at 25°C is 0.386 Ω m.

Table 4.10 shows the laboratory-derived values of ϕ, F_{RL}, and CEC, expressed in meq/g of rock sample, for 15,000 ppm NaCl at 25°C. The

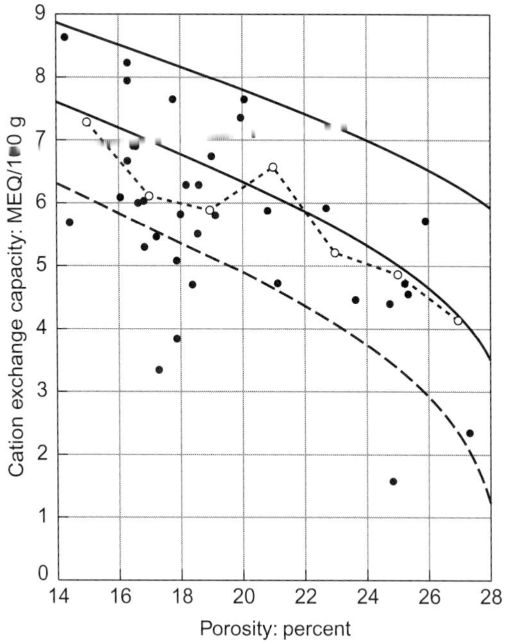

FIGURE 4.43 Cation exchange capacity versus porosity for Shannon sand, Wyoming [34].

TABLE 4.10 Laboratory Values of ϕ, F_{RL}, and CEC for the Shannon Sand Field

Sample No.	ϕ	F_{RL}	CEC
1	0.272	8.86	0.042
2	0.237	9.72	0.055
3	0.234	11.50	0.055
4	0.199	13.90	0.064
5	0.197	12.70	0.064
6	0.186	14.50	0.067
7	0.158	12.60	0.072

measured resistivity index values and the corresponding S_w values are presented in Table 4.11.

1. Calculate Q_v and F^* for each core. The average grain density (ρ_{ma}) for the seven samples is 2.65 g/cm^3.

TABLE 4.11 Laboratory Values of S_w and I_{RL}

S_w	I_{RL}
0.2	46.1
0.3	17.6
0.4	8.9
0.5	5.2
0.6	3.4
0.7	2.3
0.8	1.7
1.0	1.0

2. Determine a^* and m^*.
3. Calculate the saturation exponent n^*.

Solution

Keelan and McGinley presented a practical procedure for applying the Waxman–Smits model for calculating reservoir fluid saturations in the Shannon sandstone [34]. The following approach is based on this procedure.

1. The effective volume concentration of clay-exchange cations, Q_v, can be calculated from:

$$Q_v = \left(\frac{1}{\phi} - 1\right)\rho_{ma}(\text{CEC}) \tag{4.141}$$

and the formation resistivity factor for the shaly sands is obtained from the Waxman and Smits equation (4.130), or:

$$F^* = F_R(1 + R_w C_{eq} Q_v) \tag{4.142}$$

where the formation water resistivity is equal to 0.386 Ω m, and the specific counterion activity or equivalent conductance of clay-exchange cations is 2.9 L/eq Ω m from Figure 4.40. Using Eq. (4.139) one obtains:

$$C_{eq} = (1 - 0.83\,e^{-0.5/0.386})3.83 = 2.96$$

Inasmuch as $R_w C_{eq} = 0.386 \times 296 = 1.14$, the equation describing F^* for the Shannon sandstone field is:

$$F^* = F_R(1 + 1.14Q_v)$$

The values of Q_v and F^* for the first core sample in Table 4.12 are:

$$Q_v = \left(\frac{1}{0.272} - 1\right)(2.65)(0.042) = 0.30\ \text{L/eq}\ \Omega\ \text{m}$$

TABLE 4.12 Values of Q_v and F^* in the Shannon Sand Field

Sample No.	ϕ	F_{RL}	CEC	Q_v	F^*
0.2	46.10	2.99	103.00	0.2	46.10
0.3	17.56	2.94	32.00	0.3	17.56
0.4	8.85	2.91	14.00	0.4	8.85
0.5	5.21	2.88	7.36	0.5	5.21
0.6	3.37	2.86	4.35	0.6	3.37
0.7	2.34	2.84	2.80	0.7	2.34
0.8	1.70	2.82	1.90	0.8	1.70
0.9	1.28	2.77	1.35	0.9	1.28

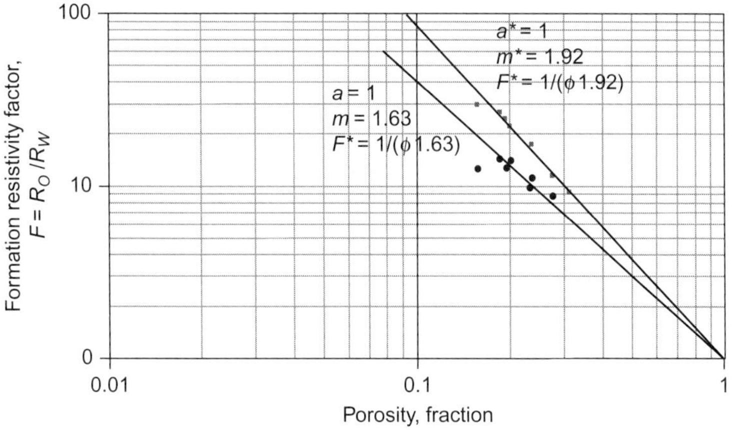

FIGURE 4.44 Formation resistivity factor for Shannon sand, Wyoming [34].

and

$$F^* = 8.86(1 + 1.14 \times 0.30) = 11.8$$

Values of Q_v and F^* for the seven core samples are included in Table 4.12. The laboratory brine salinity and temperature (and not salinity and temperature of the formation water at reservoir conditions) were used to calculate F^*. Subsequent calculations of reservoir water saturation, however, require use of the formation water salinity with its corresponding water resistivity and C_{eq} values at reservoir temperature [32].

2. From a logarithmic plot of F^* versus ϕ (Figure 4.44), one obtains a curve-fit straight-line portion having a slope $m^* = 1.92$ and intercept $a^* = 1$. Thus, the

formation resistivity factor for the Shannon shaly sands can be represented by the following expression:

$$F^* = \frac{1}{\phi^{1.92}}$$

Figure 4.44 also includes a plot of F versus ϕ. The slope m of the best-fit line is 1.62 and the coefficient a is unity. It is evident from this figure that the calculated F^* values fit an average line more closely than the measured F values, where conductivity of clays was not accounted for.

3. Two methods are available for calculating the saturation exponent n^*. Because each core sample has an individual laboratory-derived apparent saturation exponent n value, individual shaly n^* values can be calculated for each core sample, and these may then be averaged to yield a single n^* value [34]. The second approach requires obtaining the average n values from a log–log plot of the laboratory formation resistivity factor F versus brine saturation. This value is then corrected to n^* by using the measured ϕ, F, and CEC values shown in Table 4.12 to develop an average Q_v for all cores. This approach requires the following five steps:
   - **(a)** Plot laboratory values of I_R against S_w from Table 4.11, as shown in Figure 4.45.
   - **(b)** Draw a best-fit straight line, and calculate the average value of the saturation coefficient n. Using Eq. (4.134), n is approximately equal to 2.38. Thus, the equation of this straight line is:

$$I_R = \frac{1}{S_w^{2.88}}$$

   - **(c)** Using this equation, calculate the new I_R values for selected S_w values and tabulate (Table 4.13).

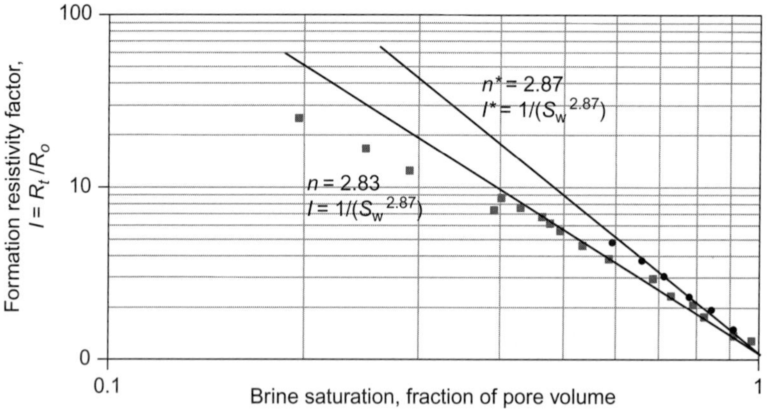

FIGURE 4.45 Formation resistivity index versus water saturation for Shannon sand [34].

TABLE 4.13 Summary Calculations of I_R, n^*, and I_R^*

(1) S_w	(2) $I_R = S_w^{-2.38}$	(3) n^* Eq. (4.143)	(4) $I_R^* = S_w^{-2.88}$
0.2	46.10	2.99	103.00
0.3	17.56	2.94	32.00
0.4	8.85	2.91	14.00
0.5	5.21	2.88	7.36
0.6	3.37	2.86	4.35
0.7	2.34	2.84	2.80
0.8	1.70	2.82	1.90
0.9	1.28	2.77	1.35

(d) Calculate the water saturation coefficient n^* from Eq. (4.137) for the same S_w values as in Step c:

$$n^* = \left(\frac{1}{\ln S_w}\right)\ln\left(\frac{1 + R_w C_{eq} Q_v}{I_R(1 + R_w C_{eq} Q_v / S_w)}\right) \tag{4.143}$$

From Table 4.12, the arithmetic average value of Q_v of the seven samples is 0.63 meq/mL. Hence, $R_w C_{eq} Q_v = 0.386 \times 2.96 \times 0.63 = 0.72$, and Eq. (4.143) for the Shannon sandstone becomes:

$$n^* = \left(\frac{1}{\ln S_w}\right)\ln\left(\frac{1.71}{I_R(1 + 0.72/S_w)}\right)$$

For $S_w = 0.4$, for instance, the corresponding resistivity index I_R from Table 4.11 is 8.85; thus:

$$n^* = \left(\frac{1}{\ln 0.40}\right)\ln\left(\frac{1.71}{8.85(1 + 0.72/0.40)}\right) = 2.91$$

Values of n^* at different water saturations were calculated and included in Table 4.13. The average saturation coefficient of the core samples is obtained by dividing the sum of the individual n^* by the number of S_w, points; thus, $n^* = (2.99 + 2.94 + \cdots + 2.77)/8 = 2.88$.

(e) Calculate I_R^* for several values of S_w from:

$$I_R^* = \frac{1}{S_w^{2.88}}$$

and plot these points as shown in Figure 4.44. Table 4.13 is a summary of these calculations.

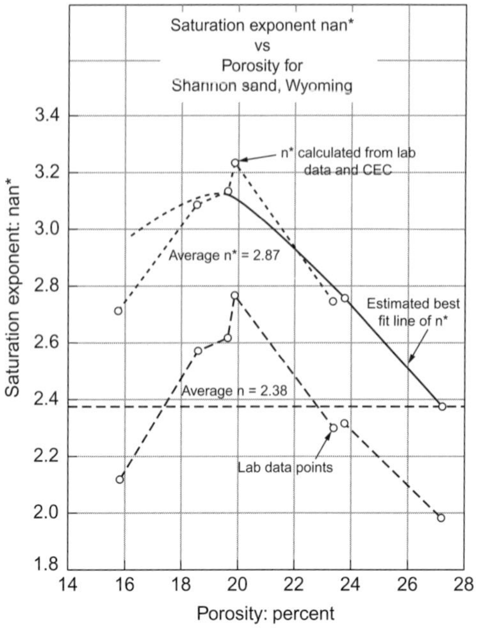

FIGURE 4.46 Saturation exponent versus porosity for Shannon sand [34].

Keelan and McGinley investigated the influence on calculated water saturations when an average coefficient n^* value for all cores is used instead of using individual core n^* values [34]. Figure 4.46 shows the correlation of n and n^* with porosity. Using the best-fit line for n^*, new n^* points were selected for individual sample porosities and new water saturations were calculated, as shown in Table 4.14. This approach, that is, using individual core values as a function of porosity, improves the calculations of S_w. In addition, being lengthy, however, this approach is not warranted in view of limited basic data necessary to develop a correlation between n^* and porosity.

Using the Shannon sand laboratory data and resistivity values read from the induction log, Keelan and McGinley also investigated some common errors encountered in utilizing the Waxman and Smits theory [29,34]. Table 4.15 shows a comparison of water saturation for three basic calculation approaches. These results indicate that the clean-sand approach, which ignores shallow effects, yields pessimistic results, that is, high S_w values. The use of the laboratory data as reported yields S_w values lower than those obtained from the clean-sand approach, but are much higher than those adjusted with CEC to "star" values using Waxman and Smits equations [30]. Table 4.16 illustrates two common misuses of the Waxman and Smits procedure. Column 2 shows S_w values calculated using CEC data with clean sand F_R, a, m, and n values, and column 3 gives S_w values calculated with the incorrect assumption that $F_R = F^*$, $m = m^*$, $a = a^*$, and $n = n^*$. These

TABLE 4.14 Comparative Calculations of Water Saturation Illustrating Influence of Common Errors Made in Utilizing Waxman and Smits [29] Equations and Use of Individual Sample n^* Values [34]

	Waxman et al. Correct Use		Incorrect Use of Reservoir Temp. ($T = 95°F$)		Individual Sample Value n^*	
	$a = 1.0$; $m = 1.63$; $n = 2.38$	$a^* = 1.0$; $m^* = 1.92$; $n^* = 2.87$	$a = 1.0$; $m = 1.63$; $n = 2.38$	$a^* = 1.0$; $m^* = 1.99$; $n^* = 2.98$	$a = 1.0$; $m = 1.63$; $n = f\phi$	$a^* = 1.0$; $m^* = 1.92$; $n^* = f\phi$
	(1)		(2)		(3)	
Sample No. 1	ϕ (%)	S_w (% PV)		S_w (PV)	n^*	S_w (% PV)
1	20.4	47		51	3.08	50
2	17.8	56		60	3.08	59
3	16.3	57		62	2.99	58
4	20.1	54		58	3.1	57
5	14.3	61		66	2.78	59
6	25.2	59		53	2.6	55
7	25.4	55		59	2.58	51
8	27.3	57		60	2.37	50
9	17.5	74		79	3.06	76
10	20.0	60		65	3.11	63
11	17.4	68		73	3.06	70
12	14.4	74		80	2.8	74

(1) Laboratory data correctly adjusted with CEC to * values.
(2) Incorrectly using formation R_w and temperature to compute * values.
(3) Individual sample n^* from Figure 4.46 as a function of porosity.

equalities are possible only when the laboratory brine is approaching 200,000 ppm salinity or greater, in which case the effects of clay conductivity are minimized. This is illustrated in Figures 4.47 and 4.48 where all three curves converge at high salinity. Because water of such high salinity cannot be prepared in the laboratory without precipitation of salts, measured values of F_R, m, a, and n will never be as high as "star" values.

Table 4.16 illustrates the incorrect practice of using CEC data and laboratory electrical properties, combined with formation water resistivity or

TABLE 4.15 Comparative Calculations of Water Saturation for Different Basic Calculation Approaches [33]

	Waxman–Smits [30] Waxman–Thomas [33]		Laboratory Data	Clean Sand
	$a = 1.0$; $m = 1.63$; $n = 2.38$	$a^* = 1.0$; $m^* = 1.92$; $n^* = 2.87$	$a = 1.0$; $m = 1.63$; $n = 2.38$	$a = 1.0$; $m = 2.0$; $n = 2.0$
		(1)	(2)	(3)
Sample No.	ϕ (%)	S_w (% PV)	S_w (PV)	S_W (% PV)
1	20.4	47	55	66
2	17.8	56	68	87
3	16.3	57	72	95
4	20.1	54	64	79
5	14.3	61	80	111
6	25.2	59	59	69
7	25.4	55	51	58
8	27.3	57	51	57
9	17.5	74	73	95
10	20.0	60	70	88
11	17.4	68	75	99
12	14.4	74	86	119

(1) Laboratory data correctly adjusted with CEC to * values.
(2) Laboratory values used as reported.
(3) Clean-sand values assumed correct and ignoring clay and shale effects.

salinity and reservoir temperature to calculate "star" values and subsequent water saturations. In this case, S_w values tend to be higher than those obtained using the Waxman and Smits approach [30]. If the cementation factor m is determined at downhole conditions by cross-plotting the formation resistivity R_o versus porosity, then the above practice is correct [34]. The effect of salinity and reservoir temperature is illustrated in Figure 4.48, where the laboratory data point F_R is increased to F^* and then adjusted to reflect reservoir conditions of temperature and salinity.

The Waxman and Smits equations are essentially valid for oven-dried cores where no formation water is left on the clay surfaces [30]. In many field

TABLE 4.16 Comparative Calculations of Water Saturation for Illustrating Influence of Common Errors Made in Utilizing the Waxman–Smits Equations [30,34]

Sample No.		Waxman–Smits Correct Usage	Clean Stand Values Corrected to * Values	Lab Values Assumed to Equal * Values
		$a = 1.0$; $m = 1.63$; $n = 2.38$ / $a^* = 1.0$; $m^* = 1.92$; $n^* = 2.87$	$a^* = 1.0$; $m^* = 2.0$; $n^* = 2.0$ / $a = 1.0$; $m = 2.29$; $n = 2.60$	$a = 1.0$; $m = 1.63$; $n = 2.38$ / $a^* = 1.0$; $m^* = 1.63$; $n^* = 2.38$
		(1)	(2)	(3)
	ϕ (%)	S_w (% PV)	S_w (PV)	S_w (% PV)
1	20.4	47	57	28
2	17.8	56	71	34
3	16.3	57	75	34
4	20.1	54	67	34
5	14.3	61	83	36
6	25.2	59	70	40
7	25.4	55	64	39
8	27.3	57	66	42
9	17.5	74	96	53
10	20.0	60	76	39
11	17.4	68	89	46
12	14.4	74	103	49

(1) Laboratory data correctly adjusted with CEC to * values.
(2) Incorrectly assuming clean-sand values should be corrected to * values.
(3) Incorrectly assuming laboratory *m* and *n* values equal * values.

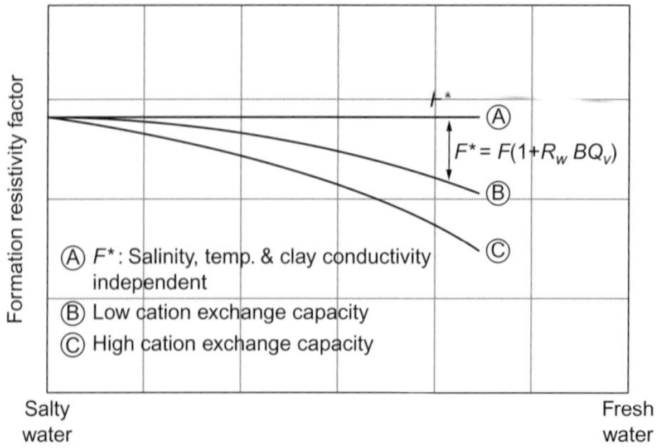

FIGURE 4.47 Salinity and cation exchange capacity effect on formation resistivity factor [33].

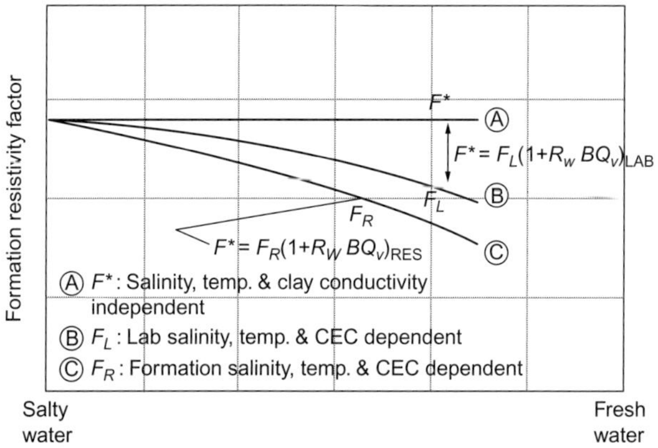

FIGURE 4.48 Salinity and temperature effect on formation resistivity factor where $B = C_{eq}$ [33].

laboratories, however, core samples are only partially dried at 145°F and 45% relative humidity, leaving molecular layers of water on the clay surfaces. This causes the measured porosity and grain density to be lower than those measured for totally dried cores. Table 4.17 illustrates the effects of using partially dried core porosity and grain density in the Waxman and Smits equations. Column 1 shows the correct results obtained by combining totally dried (180–240°F) core porosity and grain density with CEC to calculate Q_v and water saturation. Columns 2 and 3 give the results of combining partially or humidity-dried porosity and grain density with electrical properties calculated on the basis of oven-dried core porosity and humidity-dried core porosity, respectively. These results clearly show that partially dried core data yield erroneously low oil-in-place values and, therefore, it is recommended that this type of data should not be used in the Waxman and Smits equations [30].

TABLE 4.17 Comparison of Water Saturations and Oil Volumes Calculated for Normally Dried and Humidity-Dried Cores [34]

Variable	Oven-dried Core (180–240°F)	Humidity-dried Core (145°F and 45% Relative Humidity)	
		(1)	(2)
Porosity (%)	20	17.7	17.7
Grain density (g/cm³)	2.65	2.61	2.61
CEC (meq/100 g)	7	7	7
m	1.63	1.63	1.51
a	1	1	1
f	13.78	16.82	13.66
n	2.38	2.38	1.91
m^*	1.92	1.92	1.9
a^*	1	1	1
f^*	21.9	21.9	26.84
n^*	2.87	2.87	2.51
R_t	20	20	20
Q_v	0.74	0.85	0.85
R_v at 95°F	0.5	0.5	0.5
B	3	3	3
S_w	0.55	0.58	0.52
$(1 - S_w)$ (7,758)	698	577	659
ϕ		(17% low)	(5.6% low)

(1) Humidity-dried porosity and grain density used with electrical properties data calculated on basis of oven-dried core porosity.
(2) Humidity-dried porosity and grain density used with electrical properties data calculated on the basis of humidity-dried porosity.

Some clay-bearing cores crack when totally dried in an oven at high temperatures and, consequently, data measured in these cores are totally nonrepresentative of reservoir conditions. In this case, laboratory tests should be made on partially dried core samples for determining porosity, grain density, and electrical properties. Water saturation and oil-in-place are calculated as in Column 3 of Table 4.17. The calculated oil-in-place in the Shannon sand (659 bbl/acre ft, Column 3) is 5.6% lower than that obtained from the totally oven-dried cores (698 bbl/acre ft, Column 1). Thus, it is important that the testing laboratory indicates in the report whether saturation data are based on the partially (humidity) or totally (oven)-dried cores. Both partially and totally dried core data, including porosity, grain density, and electrical properties, should be reported where possible so that calculations of S_w and oil-in-place can be adjusted.

LOG-DERIVED EVALUATION OF SHALY (CLAYEY) RESERVOIR ROCKS

Several comprehensive reviews of the large number of studies on the evaluation of shaly sands using well logs can be found in References [33] through [36]. Important logging parameters were generated, including matrix density, hydrogen index (HI), CEC, and distribution of potassium, thorium, and uranium as shown by natural gamma-ray spectral-log information, for the three most common clay minerals, that is, illite, kaolinite and montmorillonite (smectites), and chlorite. However, because most shaly reservoir sands contain different clay minerals in various amounts, no single clay parameter can be used universally to characterize these sands [37]. Currently, the emphasis is on the application of the Waxman–Smits model based on the continuous computation of CEC per total volume, Q_v, and core data over the logged segments.

In many instances, however, core data over the logged zones of interest are not available. To overcome this limitation, digital shaly-sand analysis techniques based on the Waxman–Smits model and variations in the basic properties of various clay minerals were developed. Two of these digital techniques, CLASS and CLAY, developed by Ruhovets and Fertl, and Berilgen et al., respectively, provide information on total and effective porosity, total and effective fluid saturation distribution, silt volume, amounts, types and distribution modes of clay minerals present, and reservoir productivity [38,39]. The CEC and HI can be calculated knowing the three parameters: clay density, ρ_{cl}, neutron response to 100% clay, N_{cl}, and the clay volume, V_{cl}. ρ_{cl} and N_{cl} are best determined from density, neutron, and natural gamma ray–spectral data at every depth level over the interval of interest, so that the unrealistic assumption that clay properties in the adjacent shale beds and the reservoir rock are identical is not necessary [40]. The clay volume, V_{cl}, which is essentially independent of the clay types, is calculated from the potassium and thorium values.

FORMATION EVALUATION

The basic physical properties needed to evaluate a petroleum reservoir are its permeability, porosity, fluid saturation, and formation thickness. These parameters can be estimated from three common sources: core, well logging, and pressure test analyses. A less common source is geochemistry. The application of geochemical techniques to oil and gas exploration has only recently achieved widespread acceptance among exploration geologists; however, it is beyond the scope of this text. It is not the purpose of this section to make complete discussion of core analysis, well logging, and well testing but rather to highlight the significance of the measuring techniques.

CORE ANALYSIS

All phases of the petroleum industry rely directly or indirectly on the knowledge of reservoir rock properties. Analysis of rock samples yields valuable data basic to exploration, well completion, and evaluation of oil and gas reserves. Drill bit cuttings, because of their size and mode of recovery, essentially provide qualitative information. The necessity of recovering and examining large reservoir rock samples led to the development of coring techniques. The first coring tool appeared in 1908 in Holland. In 1921, H.E. Elliot of the United States introduced the first effective coring tool by successfully combining an inner core barrel with a toothed bit. Four years later, considerable improvements were made to Elliot's device to include a removable core head, a core catcher, and a stationary inner barrel, to which various refinements have been added [41]. Currently, several types of coring devices are available: diamond cores, rubber and plastic sleeve cores, percussion and continuous sidewall cores, and cores recovered in a pressure core barrel. Each one of these devices offers certain advantages. The selection is generally dictated by the type of reservoir rock and objectives of the core analysis.

Three coring methods are practiced: conventional, wireline, and sidewall. Conventional coring, which refers to core taken without regard to precise orientation, encompasses arrange of coring devices and core barrels. The main disadvantages of conventional coring is that coring equipment requires that the entire drill string be pulled to retrieve the core; however, the corresponding advantage is that large cores, 3–5 in. in diameter and 30–90 ft long, may be recovered. In the wireline coring method, the core may be retrieved without pulling the drill string by using an overshot run down the drill pipe on a wireline. The cores obtained by this method are small (i.e., approximately 1–2 in. in diameter and 10–20 ft in length). Other advantages include downhole durability and higher core recovery.

Sidewall coring is necessary when it is desirable to obtain core samples from a particular zone already drilled, especially in soft rock zones where hole conditions are not conducive to openhole drillstem testing. The sidewall coring device contains a hollow bullet which, when fired from an electric control panel at the surface, embeds itself in the formation wall. With the core sample caught in the bullet, a flexible steel cable retrieves the bullet and its contained core (~1 in. in diameter and 1 in. in length) to the surface. Sidewall diamond coring is necessary in hard rocks. A relatively new technique, known as directionally oriented coring, involves the scribing of grooves along the axis of the core in a gyroscopically controlled orientation [42]. This method requires periodic stops to take a measurement of orientation and is accomplished by replacing the conventional inner core barrel sub with the scribe shoe sub. The main purpose of oriented coring is to allow visualization of rock in its exact reservoir condition orientation, which may be useful in predicting reservoir continuity, especially in fluvial deposition systems [43].

The early methods of core analysis were more an art than a science, and the results were not taken seriously. The practice of breaking the core into small pieces to smell and taste for the presence of hydrocarbons was widespread, even though it was well known that the sweet gases (i.e., gases that do not contain hydrogen sulfide) have no apparent odor or taste [41].

Consequently, many gas formations were diagnosed as water productive because of the physical inability to detect gas. Today, core analysis is a highly specialized phase of petroleum reservoir engineering. Analysis of sidewall cores provides far more geological information than bit cuttings. Core data play an important role in exploration programs, well completion, and reservoir evaluation programs. Core analysis makes it possible to recognize the structure of the reservoir trap, determine its physical characteristics such as porosity and permeability, and estimate production possibilities of exploratory wells. Core data allow wells to be properly completed by selecting intervals for drillstem testing and evaluating the effectiveness of completion. In the field development stage, core measurements are employed to estimate hydrocarbon reserves, determine contacts between reservoir fluids such as water–oil contact line, and their variations across the field [44,45]. Table 4.18 shows typical data obtained from core analysis and their use. Routine core-analysis results are usually presented in tabular or in graphical form as shown in Figure 4.49. For the purpose of recognizing the stratification effect, the graphical form is preferred.

WELL LOG ANALYSIS

Well logging can be defined as a tabular or graphical portrayal of any drilling conditions or subsurface features encountered that relate to either the progress or evaluation of an individual well [46]. The ultimate aim of the well log interpretation, however, is the evaluation of potential productivity of porous and permeable formations encountered by the drill.

Electrical logging was introduced to the oil industry by Marcel and Conrad Schlumberger in 1927 in France. Since then, due to considerable technological and scientific advances, well logs have undergone constant and sweeping changes. The development of recording techniques compatible with the application of computers in well log interpretation has removed a large number of earlier assumptions and general estimates from well log computations. The result has been a change from a correlation tool for geologists to an indispensable data source for the oil industry. A successful logging program, along with core analysis, can supply data for subsurface structural mapping, define the lithology, identify the productive zones and accurately describe their depth and thickness, distinguish between oil and gas, and permit a valid quantitative and qualitative interpretation of reservoir characteristics, such as fluid saturation, porosity, and permeability. Unfortunately, these petrophysical properties cannot be measured directly and, therefore, they must be inferred from the measurement of other parameters of the reservoir rock, such as the resistivity of the rock, the bulk density, the interval transit time, the SP, the natural radioactivity, and the hydrogen content of the rock [47].

TABLE 4.18 Reservoir Characteristics Obtained from Core Analysis

Slabbed Core	**Thin Sections**
Photograph	Detail pore structure
Sedimentology	Diagenesis
Lithology	Porosity type
Samples	Environmental evidence
Small Samples	**Routine Core Plug Analysis**
Grain-size distribution	Porosity
Mineral analysis	Permeability
X-ray and SEM analysis	Grain density
Bio-dating and association	As-received saturations
Special Core Analysis	
Preserved/restored state	
Capillary pressure	
Relative permeability	
Electrical properties	
Acoustic properties	
Compressive properties	
Clay chemistry effects	
Specific tests	
Calibration of wireline log	

Water Saturation

Evaluation of the amount of hydrocarbons present in the reservoir is based on the ability of the log analyst to estimate the volume of water present in the pore space. This requires the solution of some form of Archie equation for the water saturation parameter S_w. Because of its simplicity and worldwide use, R_{wa} is the only method presented here for determining S_w. Water saturation in the uninvaded zone of a clean-sandstone formation having intergranular or intercrystalline porosity can be estimated from Eq. (4.53):

$$S_w^n = \frac{F_R R_w}{R_t} = \frac{R_o}{R_t} = \frac{1}{I_R} \tag{4.144}$$

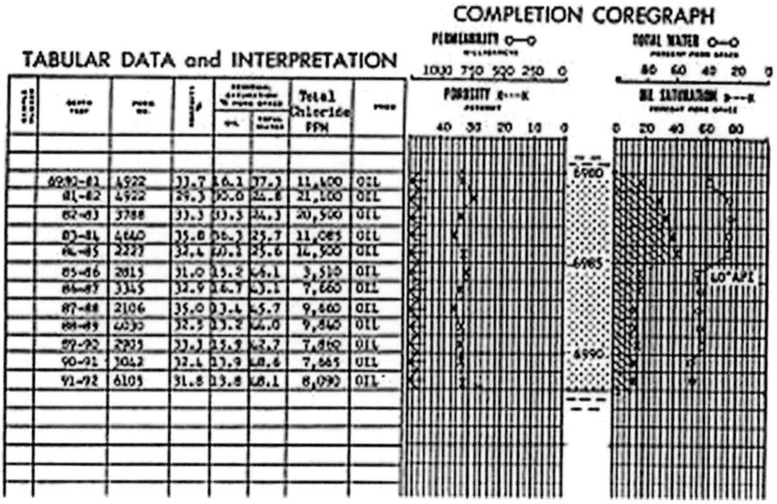

FIGURE 4.49 Typical presentation of core data. *Source: Courtesy of Core Laboratories.*

In a 100% water-saturated sand, that is, $I_R = 1$ and $R_t = R_o$, the water resistivity is equal to:

$$R_w = \frac{R_t}{E_R} \tag{4.145}$$

If the sand contains petroleum, the true resistivity factor R_t will increase, whereas the formation factor F_R will remain the same because it is a function of the formation porosity only. Therefore, Eq. (4.145) is of the general form

$$R_{wa} = \frac{R_t}{F_R} = R_w I_R \tag{4.146}$$

where R_{wa} is the apparent water resistivity. Thus, if R_{wa} is calculated from Eq. (4.146), one would actually be calculating the value of R_w/S_w^n. Then, if R_w is either experimentally measured from a sample or calculated from a chemical analysis using Figure 4.4, one can solve S_w. The value of R_{wa}, therefore, can be used to select hydrocarbon-bearing zones.

The following analytical procedure is recommended for selecting sand beds containing hydrocarbons [48].

Consider two adjacent porous and permeable zones: zone 1 containing hydrocarbons and zone 2 is 100% water saturated. Comparing the R_{wa} values of the two zones gives:

$$\frac{R_{wa1}}{R_{wa2}} = \frac{(R_w I_R)1}{(R_w I_R)2} \tag{4.147a}$$

Inasmuch as R_w is constant regardless of the value of S_w, the water saturation in zone 2 is 100%, that is, $I_{R2} = 1$, and R_{wa2} is the true R_w or a minimum observed value $(R_{wa})_{min}$, one obtains:

$$\frac{R_{wa1}}{R_w} = \frac{R_{wa1}}{(R_{wa})_{min}} = I_{R1} = \frac{1}{S_{w1}^n} \quad (4.147b)$$

Solving for S_w for any zone:

$$S_w = \left(\frac{R_w}{R_{wa}}\right)^{1/n} = \left(\frac{R_{w\,min}}{R_{wa}}\right)^{1/n} \quad (4.147c)$$

Assuming R_w remains fairly constant along the zones of interest, values of R_{wa} can be determined for each zone and compared. Zones where R_{wa} is greater than approximately $4R_w$ generally have $S_w < 50\%$ and should be considered potential hydrocarbon-bearing zones. Hilchie proposed an elaborate but complete and practical algorithm for selecting hydrocarbon-bearing zones using the R_{wa} method [4]. According to this algorithm, all zones where $R_{wa} > 3R_w$ have $S_w < 60\%$ and are potentially hydrocarbon bearing.

The R_{wa} technique gives excellent results when drilling fluid (mud) invasion is not deep, as in the high-porosity formation, or when low-water-loss mud is used. This technique, however, can yield pessimistic results when the formation is shaly and F_R is determined from the density log.

Another method for calculating the water saturation (S_w) uses a combination of the density log and the resistivity log [49]. The porosity is calculated from the density log:

$$\phi = \frac{\rho_{ma} - \rho_b}{\rho_{ma} - S_w\rho_w - (1 - S_w)\rho_b} \quad (4.148a)$$

where ρ_b, ρ_{ma}, ρ_w are, respectively, bulk density, matrix density, and water density. Archie's equation for calculating water saturation is:

$$S_w^n = F_R\frac{R_w}{R_t} = \frac{a}{\phi^m}\frac{R_w}{R_t} \quad (4.148b)$$

where R_w is the resistivity of formation water and R_t the true resistivity of the formation. For cementation factor $m = 2$, Archie's constant $a = 1$ and the saturation exponent $n = 2$:

$$S_w = \frac{1}{\phi}\sqrt{\frac{R_w}{R_t}} \quad (4.148c)$$

Substituting Eq. (4.148c) into (4.148a) and solving explicitly for porosity yields:

$$\phi = \left(\frac{\rho_{ma} - \rho_b}{\rho_{ma} - 1}\right) + \left(\frac{\rho_w - \rho_b}{\rho_{ma} - 1}\right)\sqrt{\frac{R_w}{R_t}} \quad (4.149a)$$

Substituting this equation into Archie's equation (4.148c) gives:

$$S_w = \frac{\rho_{ma} - 1}{((\rho_{ma} - \rho_b)/\sqrt{R_w/R_t}) + (\rho_w - \rho_b)} \quad (4.149b)$$

Porosity

Three logging porosity tools have been developed to determine porosity, namely, the sonic–acoustic log, the formation density log, and the neutron porosity log. In addition to porosity, these logs are affected by other parameters, such as lithology, nature of the pore fluids, and shaliness. Combinations of these logs are used to determine lithology and porosity as well as the fracture porosity.

The sonic log measures the interval transit time, Δt, that is, the shortest time required for a compressional sound wave to travel through 1 ft of formation parallel to the wellbore. The speed of sound in the formation depends on the nature of minerals making up the rock, porosity, pore space fluids, temperature, pressure, and rock texture. Inasmuch as for any given lithology, the zone of investigation of the sonic tool is essentially in the invaded zone containing mud filtrate, the speed of sound, that is, the interval transit time, is primarily a function of porosity.

The velocity of sound in the formation depends on the density and elastic properties of the medium, such as bulk and shear moduli of elasticity. It is faster in a hard substance than in a liquid. Hence, if one considers a rock composed of only solid and liquid, the following ratio of the transit times can be used to obtain porosity [50]:

$$\phi_s = \frac{t - t_{ma}}{t_{fl} - t_{ma}} \tag{4.150}$$

where

t = total transit time, μs/ft
t_{ma} = matrix travel time, μs/ft
t_{fl} = fluid travel time, μs/ft

Equation (4.150) is commonly used for determining the approximate value of porosity of clean consolidated sandstones as well as that of carbonate formations with intergranular porosity. The fluid travel time is ~190 μs/ft, whereas the matrix travel time can be obtained from the following equation:

$$t_{ma} = \frac{10^6}{V_{ma}} \tag{4.151}$$

where the velocity of sand (P wave) in the matrix, V_{ma}, is expressed as follows:

$$V_{ma} = \left[\frac{K + 0.75G}{\rho_{ma}}\right]^{0.5} \tag{4.152}$$

where K and G are the bulk and shear moduli, respectively, and ρ_{ma} is the density of matrix. Table 4.19 shows the velocity and matrix travel time for various rock types. The presence of shale, fractures, and gas complicates the sonic porosity measurements. In multiple-porosity rocks, such as vuggy or fractured carbonates, the travel time is often shorter than the time calculated for that

TABLE 4.19 Matrix Travel Time for various rock types

Formation	V_m (ft/s)	t_{ma} (μs/ft)
Sandstone		
Unconsolidated	17,000 or less	58.8 or more
Semiconsolidated	18,000	55.6
Consolidated	19,000	52.6
Limestone	21,000	47.6
Dolomite	23,000	43.5
Shale	6,000–16,000	167–62.5
Calcite	22,000	45.5
Anhydrite	20,000	50.0
Granite	20,000	50.0
Gypsum	19,000	52.6
Quartz	18,100	55.6
Salt	15,000	66.7
Water	5,300	189.0

given porosity. This is because vugs or fractures are irregularly located and the compressional sound wave goes through the formation with the least porosity, that is, shortest travel time. The secondary porosity is generally estimated by subtracting sonic porosity (Eq. (4.150)) from the neutron or density porosity (Eq. (4.157)). In some cases, this may lead to erroneous results.

Unconsolidated formations, almost always sandstones, tend to exhibit longer travel times than consolidated formations having the same porosity. Consequently, the Wyllie et al. correlation gives unacceptable high porosities [50]. In this case, Eq. (4.150) is modified to include a compaction correction factor, B_{cp}, as follows:

$$\phi_s = \left(\frac{t - t_{ma}}{t_{fl} - t_{ma}}\right)\frac{1}{B_{cp}} \tag{4.153}$$

The compaction correction factor is equal to:

$$B_{cp} = \left(\frac{t_{sh}}{100}\right)B_{sh} \tag{4.154}$$

where 100 is the travel time for compacted shale in μs/ft and t_{sh} is the sonic travel time of an adjacent shale. The normal range of B_{cp} for sandstone formations is from 1 to 2. When no compaction correction is used, $B_{cp} = 1$. The factor B_{sh}, which is empirically determined, is a function of shale (clay) type. The lack of compaction is indicated when adjacent shale beds exhibit a sonic travel time >100 μs/ft. In shaly (clayey) unconsolidated formations, the sonic porosity is calculated from the following equation:

$$\phi_s = \left(\frac{t - t_{ma}}{t_{fl} - t_{ma}}\right)\frac{1}{B_{cp}} - \left(\frac{t_{sh} - t_{ma}}{t_{fl} - t_{ma}}\right)V_{sh} \tag{4.155}$$

where V_{sh} is the shale (clay) volume. In formations, consolidated or unconsolidated, bearing oil or gas, the calculated sonic porosity tends to be high and the following empirical correction can be used:

$$\phi = B_{hc}\phi_s \tag{4.156}$$

where ϕ_s is obtained either from Eq. (4.153), for clean unconsolidated formations, or from Eq. (4.155), for shaly (clayey) unconsolidated formations. The factor B_{hc} may be empirically set at 0.90 for oil and 0.70 for gas. These constants seldom give good results, as B_{hc} depends on the type of mud, depth of mud invasion, pore pressure, etc.

From the formation evaluation standpoint, the main objective of the density log is the determination of formation porosity by measuring the bulk density of the reservoir rock. In the case of saturated porous rocks, bulk density includes the density of the fluid in the pore spaces as well as the grain density of the rock. For a clean formation of known matrix density, ρ_{ma}, having a bulk density ρ_b, and which contains a fluid (except gas and light hydrocarbons) of average density, ρ_{fl}, the formation porosity is equal to:

$$\phi_D = \left(\frac{\rho_{ma} - \rho_b}{\rho_{ma} - \rho_{fl}}\right) \tag{4.157}$$

The bulk density, ρ_b in g/cm^3, is read from the density log. The density of fluid in pores, generally mud filtrate, is 1.0 when fresh muds are used and 1.1 for salty drilling muds. If the formation is saturated with gas in the vicinity of the borehole, that is, little or no mud invasion, $\rho_{fl} = 0.7$ g/cm^3 [25]. In shaly (clayey) formations Eq. (4.157) becomes [43]:

$$\phi_D = \left(\frac{\rho_{ma} - \rho_b}{\rho_{ma} - \rho_{fl}}\right) - \left(\frac{\rho_{ma} - \rho_{sh}}{\rho_{ma} - \rho_{fl}}\right)V_{sh} \tag{4.158}$$

Determining porosity, especially in carbonate rocks, is one of the most important applications of neutron logs. These rocks generally contain smaller amounts of clay minerals than do the sandstones. Neutron logs can also be used to define bed boundaries and, when used in conjunction with other logs, as an indicator of lithology of gas-bearing zones. Modern neutron log data are recorded directly in apparent porosity units with only a minor correction, required to account for salinity, temperature, and tool positioning. Porosity can be determined from the combination of neutron and density logs using Eq. (4.96). The presence of shale (clay) in reservoir rocks (sandstone, limestone, or dolomite) will influence to some degree the measured response of all three porosity logs.

As previously stated, the standard practice at present for estimating the reservoir permeability distribution is to combine permeability values obtained from laboratory measurements on cores with log-derived parameters, such as porosity and water saturation. The following summary of a field case is a typical example of such practice [51].

Example

The Howard–Glasscock field is located south of Big Spring, Texas, and has produced oil from the lower Grayburg and San Andres carbonate formations since 1929. The 80-well field was unitized in 1972 and, during 1973, 40 additional wells were drilled to expand the ongoing waterflood, which was initiated in 1964. Ten wells were cored and 38 wells were logged. The objective is essentially to [51]:

1. calculate average values of permeability, porosity, and water saturation;
2. prepare contour maps of porosity and water saturations;
3. estimate oil reserves.

Core Interpretation of Data

The core recovery efficiency was 98.2% or an average of 404.5 ft of core recovered per 411.8 ft attempted per well. This recovery is exceptionally high for a carbonate formation. The core analysis involved five phases.

During the first phase, 15 lb of whole-core sections were analyzed. Because of the complex lithology of the San Andres Formation, the whole-core technique was selected to determine porosity, matrix permeability, and fluid saturations. Tables 4.20 and 4.21 give the values of permeability and porosity, respectively. A semilog plot of permeability versus porosity showed a considerable scattering of data points. Nevertheless, a best least-squares fit line is placed along the relative general trend of these points and the corresponding equation is derived:

$$\log k = 0.285\phi - 2.98 \qquad (4.159)$$

During the second phase, 162 unprocessed core samples, representative of the pay intervals in six of the ten cored wells, were examined, and lithology, texture, and type of porosity were determined. Also, 32 core samples representing the expected range of rock parameters, as determined from a preliminary computer correlation of whole-core and log data, were selected for special plug-core analysis. Samples with extremely large vugs were excluded from special core analysis in an attempt to determine more accurately the properties of the matrix. Following routine cleaning and drying procedures as described in the Appendix, air permeability and porosity (using Boyle's law porosimeter) were determined. Figures 4.50 and 4.51

TABLE 4.20 Permeability Data for the West Howard–Glasscock Unit, Zone D

Permeability Range (mD)	Sample		Percent of Samples		Arithmetic		Geometric		Percent Permeability Capacity
	Number	Cumulative	Range	Cumulative	Average *k*	*F* (Average *k*)	Log Average	Average *k*	
0.0–1.3	225	225	46.28	46.28	0.399	0.185	−0.39922	−0.18476	3.135
1.3–2.5	78	333	14.16	60.44	1.762	0.249	0.24589	0.03481	4.235
2.5–5.0	76	409	13.79	74.23	0.355	0.489	0.54975	0.07583	8.307
5.0–10.0	59	468	10.71	84.94	6.895	0.738	0.83853	0.08979	12.539
10.0–20.0	35	503	6.35	91.29	13.714	0.871	1.13717	0.07223	14.795
20.0–40.0	29	532	5.26	96.55	27.931	1.470	1.44609	0.07611	24.966
40.0–80.0	18	550	3.27	99.82	52.222	1.706	1.71786	0.05612	28.973
80.0–160.0	1	551	0.18	100	99.000	0.180	1.99564	0.00362	3.051

TABLE 4.21 Porosity Data for the West Howard–Galsscock Unit Zone D

Porosity Range (%)	Sample		Percent of Samples		Range Porosity = ϕ		Percent Porosity Capacity
	Number	Cumulative	Range	Cumulative	Average ϕ	F (Average ϕ)	
<2.0	0	0	0.00	0.00	0.000	0.00000	0.0000
2.0–4.0	15	15	2.72	2.72	3.300	0.08984	0.8632
4.0–6.0	62	77	11.25	13.97	5.152	0.57967	5.5695
6.0–8.0	70	147	12.70	26.68	6.933	0.88076	8.4621
8.0–10.0	97	244	17.60	4.28	9.012	1.58657	15.2433
10.0–12.0	114	358	20.59	64.97	10.959	2.26733	21.8945
12.0–14.0	101	459	18.33	83.30	12.818	2.34955	22.5745
14.0–16.0	56	515	10.16	93.47	14.850	1.50926	14.5009
16.0–18.0	25	540	4.54	98.00	16.728	0.76898	7.2923
18.0–20.0	99	549	1.63	99.64	18.489	0.30200	2.9016
20.0–22.0	0	549	0.00	99.64	0.000	0.00000	0.0000
22.0–24.0	1	550	0.18	99.82	20.000	0.03993	0.3836
24.0–26.0	1	551	1.18	100.00	24.300	0.04410	0.4237
26.0+	0	551	0.00	100.00	0.000	0.00000	0.0000

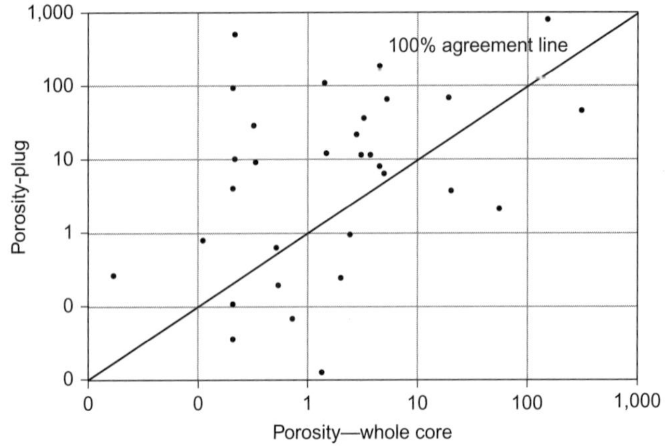

FIGURE 4.50 Relationship of whole-core permeability to plug permeability [51].

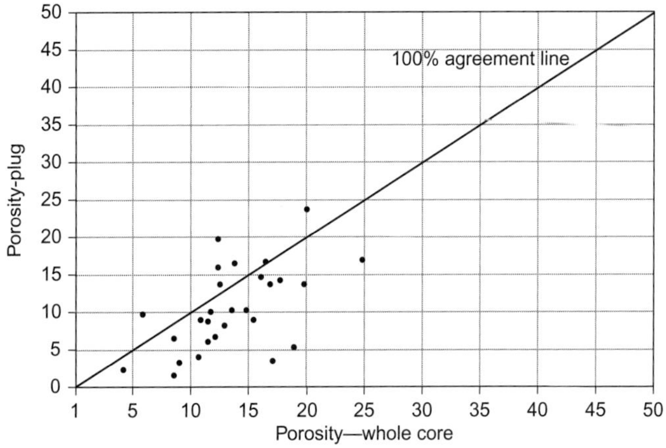

FIGURE 4.51 Relationship of whole-core porosity to plug porosity [51].

show a comparison of, respectively, whole-core permeability versus plug permeability and whole-core porosity versus plug porosity. It is evident from these figures that there are very low coefficients of correlation. This emphasizes the problem of obtaining representative core samples in carbonate reservoirs. Based on this limited (or lack of) agreement, it was concluded that the requirement of permeability and porosity agreement between the plug- and whole-core values on the sample-by-sample basis should not be the criterion for selecting samples for plug-core analysis.

During the third phase, resistivity measurements were made on 32 core plugs and formation resistivity factors were calculated using Eq. (4.5). Figure 4.52 is a

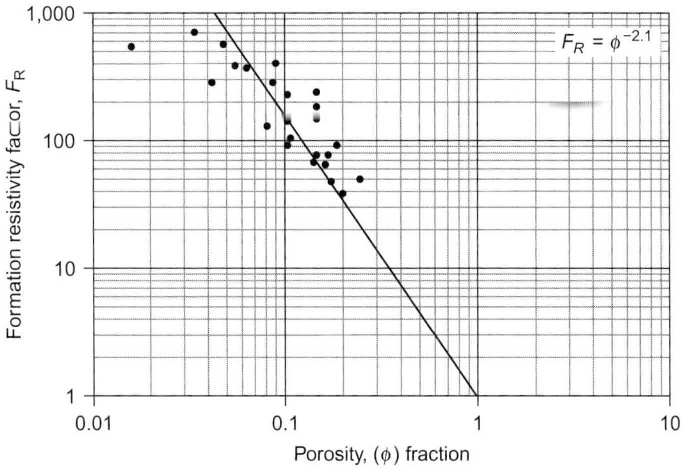

FIGURE 4.52 Relationship of formation resistivity factor to porosity, West Howard–Glasscock Unit [51].

log–log plot of F versus ϕ, and the equation of the best-fit line is:

$$F_R = \frac{1}{\phi^{2.1}} \tag{4.160}$$

Thus, the cementation factor m of this carbonate formation is 2.1. Similarly, resistivity index calculations were made and the results were plotted against water saturation on a log–log graph, as shown in Figure 4.53. The equation of the straight line is:

$$I_R = \frac{1}{S_w^{2.2}} \tag{4.161}$$

The resistivity index data appear to be more scattered than the formation resistivity factor data. This is attributed primarily to the heterogeneous nature of the carbonate formations.

The determination of the irreducible water saturation, S_{wi}, constitutes the fourth phase. Capillary pressure data published by Osborne and Hoga, and the correlative porosity data from 276 core samples in the San Andres reservoir were used with the following empirical equations for determining relative-permeability values and, subsequently, S_{wi} values [52]:

$$k_{rw} = \left(\frac{S_w - S_{wi}}{1 - S_{wi}}\right)^3 \tag{4.162}$$

$$k_{ro} = \left(\frac{0.9 - S_w}{0.9 - S_{wi}}\right)^3 \tag{4.163}$$

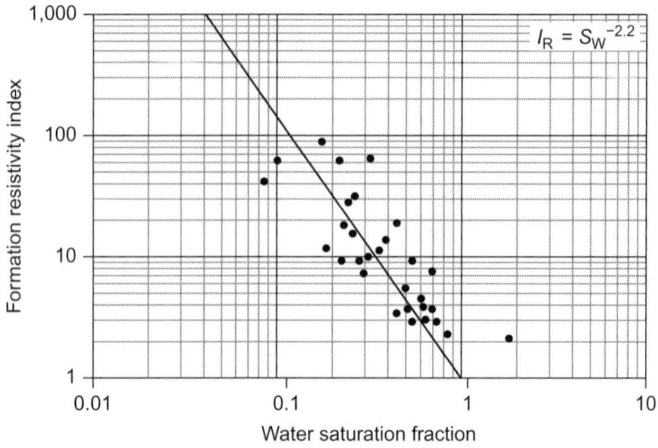

FIGURE 4.53 Relationship of resistivity index to water saturation, West Howard–Glasscock Unit [51].

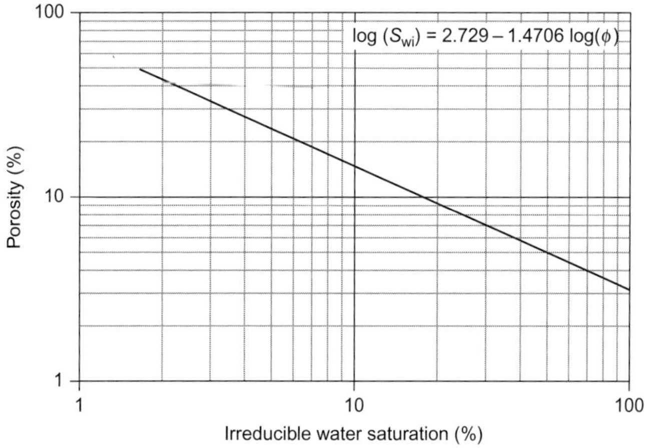

FIGURE 4.54 Relationship of porosity to irreducible water, West Howard–Glasscock Unit [51].

Figure 4.54 shows the relationship between the porosity and the irreducible water saturation, described by the following equation obtained by the least-squares method:

$$\log S_{wi} = -1.4706 \log \phi + 2.729 \tag{4.164}$$

where both S_{wi} and ϕ are expressed in percentage.

The fifth phase of the core analysis involved measurements of grain density for log analysis purposes. Rock fragments representative of the 32 samples, used for plug-core analysis, were crushed to grain size. Grain volume was determined on a helium porosimeter giving an average grain density of 2.85 g/cm^3.

Interpretation of Well Log Data

The analysis of logs obtained from the 38 new wells required several phases. First, the logs were digitized and corrections for hole size and invasion were determined. Then, cross-plots were made of the corrected data from the three porosity logs (acoustic, density, and neutron), and formation minerals were identified. The log data from the three porosity devices were corrected for shale (clay) volume by the following expressions:

$$t_c = \frac{t - (V_{sh})(t_{sh})}{1 - V_{sh}} \tag{4.165}$$

$$\rho_{bc} = \frac{\rho_b - (V_{sh})(\rho_{sh})}{1 - V_{sh}} \tag{4.166}$$

$$\phi_{nc} = \frac{\phi_n - (V_{sh})(\phi_{ah})}{1 - V_{sh}} \tag{4.167}$$

where

t, ρ_b, ϕ_n = acoustic, density, and neutron log readings, respectively
t_c, ρ_{bc}, ϕ_{nc} = three-device data values corrected for shale (clay) content
T_{sh}, ρ_{sh}, ϕ_{sh} = three-device readings for 100% shale

These corrected data become the input to a matrix solution for primary (matrix) and secondary (fractures and vugs) porosity indices, and three minerals, as shown in Table 4.22. The term "trilith" describes the porosity–lithology matrix obtained from the combination of the three porosity devices. In less complex formations (no secondary porosity), a simpler matrix called "bilith" is constructed from the combination of two porosity devices, namely, acoustic and density. Finally, these matrices were solved by a computer program called "Bitri." Figure 4.55 shows a good agreement between the profile of core-derived porosity and the Bitri-computed porosity. In addition to porosity, the program-computed permeability from Eq. (4.159), irreducible water saturation from Eq. (4.163), and water saturation from Eq. (4.54) for $m = 2.1$ and $n = 2.20$ are given by:

$$S_w = \left(\frac{R_w}{R_t} \frac{1}{\phi^{2.1}} \right)^{1/2.2} \tag{4.168}$$

where R_t is the true resistivity from a deep-reading resistivity log, corrected for borehole and invasion effects, and ϕ is the porosity obtained from Bitri. Figure 4.56 shows a comparison of the core-derived permeability and Biri-computed permeability. It was not evident from the report whether core data were shifted a few feet to obtain a better correlation with log data as is recommended by Sneider et al. [53]. Also, it was not reported whether the core data were corrected for the effects of overburden pressure [54].

The last phase of well log analysis is designed to produce the maps and grids required for visual presentation, for input to a field-wide simulator, and

TABLE 4.22 Bilith and Trilith Equations

Bilith Equations	
	$\Delta t = 189\phi + 43.5V_{dol} + 80V_{sh} + 50V_{anh}$ (acoustic equation)
	$r_b = 1.1\phi + 2.87V_{dol} + 2.65V_{sh} + 2.98V_{anh}$ (density equation)
Trilith Equations	
	$\Delta t_c = 189I_{f1} + 43.5I_{f2} + 43.5V_{dol} + 55.5V_{sd} + 50V_{anh}$ (acoustic equation)
	$r_{bc} = 1.1I_{f1} + 1.1I_{f2} + 2.87V_{dol} + 2.65V_{sd} + 2.98V_{anh}$ (density equation)
	$\phi NL_c = 1.0I_{f1} + 1.0I_{f2} + 0.06V_{dol} + 0.03V_{sd} + 0.01V_{anh}$ (neutron equation)
	Δt = acoustic log input data
	r_b = density log input data
	Δt_c = acoustic log data corrected for shale
	r_{bc} = density log data corrected for shale
	ϕNL_c = neutron log (limestone mode) data corrected for shale
	ϕ = porosity
	I_{f1} = primary porosity index
	I_{f2} = secondary porosity index
	V_{dol}, V_{sh}, V_{anh}, and V_{sd} are percentages of dolomite, shale, anhydrite, and sand, respectively

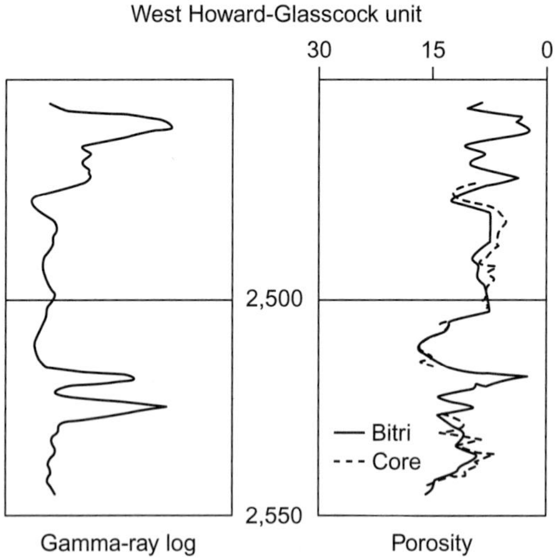

FIGURE 4.55 Agreement of Bitri-computed porosity with core-analysis porosity versus gamma-ray log [51].

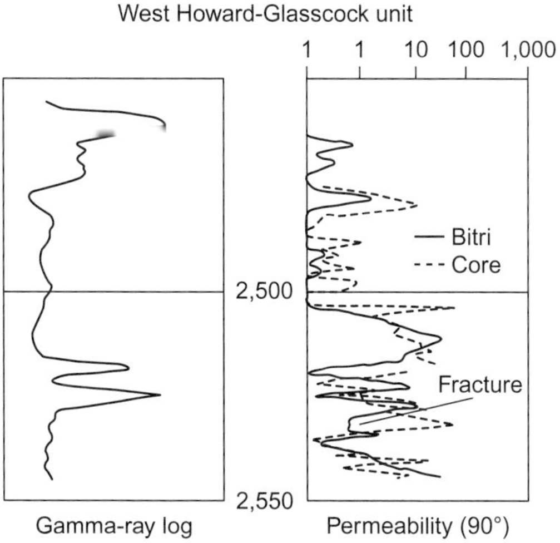

FIGURE 4.56 Permeabilities at 90° Bitri and core-analysis values [51].

TABLE 4.23 Summary of Results

Total footage	40 ft
Average porosity	12.2%
Average water saturation	35%
Average permeability	5.6 mD
Total hydrocarbons	3683 bbl/acre
Hydrocarbon feet	0.47
Porosity feet	0.73
Net average hydrocarbons	614 bbl/acre ft
Gross average hydrocarbons	92 bbl/acre ft
Porosity cutoff	8%
Water saturation cutoff	45%

for estimations of reserves. Table 4.23 shows a summary of calculated average values of porosity, permeability, water saturation, and reserve estimates.

PROBLEMS

1. The resistivity of a water sample is 0.35 Ω m at 25°C. What is its resistivity at 80°C?

TABLE 4.24

Sample No.	Porosity	R_o	R_t
1	0.204	0.665	30.0
2	0.178	0.830	24.0
3	0.163	0.960	22.0
4	0.201	0.680	21.0
5	0.143	1.190	20.0
6	0.252	0.470	16.5
7	0.254	0.460	20.0
8	0.273	0.410	23.0
9	0.175	0.850	20.0
10	0.200	0.680	16.0
11	0.174	0.860	17.0
12	0.144	1.170	17.0

2. Chemical analysis of an oil reservoir brine yielded the following results: 50,000 ppm Na^+, 60,000 ppm Cl^-, 15,000 ppm Mg^{2+}, 12,000 ppm SO_4^{2+}, 8,000 ppm HCO_3^-. Calculate:
  (a) the equivalent salinity in ppm sodium chloride;
  (b) the resistivity of the brine at 100, 175, and 250°F;
  (c) discuss the effect of temperature on the resistivity of water.

3. A normally pressured well located in offshore Louisiana is 5,900 ft deep. The producing interval is completely water-wet. The apparent porosity of the shale zone is ~0.39. The cementation exponent for shales in this field is 1.57. The true resistivity of the shale zone is 0.90 Ω m. Calculate:
  (a) Water resistivity of the formation at the reference temperature of 75°F;
  (b) water resistivity at the formation temperature of 140°F.

4. The results of laboratory measurements made on 12 water-wet clean-sandstone rock samples and well log analysis are shown in Table 4.24. The resistivity of a formation-water sample at 25°C is 0.056 Ω m. The formation temperature is 89°C.
  (a) Calculate the formation resistivity factor for each sample.
  (b) Estimate the cementation factor.
  (c) Determine the water saturation in each sample.
  (d) Find the best representative values of tortuosity.

TABLE 4.25

Sample No.	Permeability	Porosity	F	V_{sh} (%)	R_t (Ω m)
1	127.0000	0.151	29.58	4.46	31
2	237.0000	0.171	24.40	4.17	38
3	3.1600	0.098	74.68	7.32	19
4	17.6000	0.113	47.17	5.33	27
5	1.6200	0.093	87.53	8.14	12
6	0.2280	0.075	128.32	9.62	9
7	2.7600	0.098	78.41	7.68	18
8	0.0248	0.040	312.41	12.5	3

5. Table 4.25 shows values of permeability, porosity, formation resistivity factor, shale (clay) fraction, and resistivity R_o of eight zones of a shaly (clayey) formation. The resistivity of shale is 4 Ω m, and the formation water resistivity is 0.1 Ω m.

(a) Estimate the water saturation in each zone using (1) the generalized shale relationship, and (2) Hilchie's approximate shale relationship, and compare the results.

(b) Determine the tortuosity factor and correlate with the shale fraction. Explain.

6. The following data are obtained from a Texas Gulf Coast well:

	Zone A	Zone B
Porosity, %	28.40	25.2
Water resistivity, Ω m	0.06	0.06
True resistivity, Ω m	1.8	1.0

(a) Calculate water saturation in Zones A and B.

(b) Which zone is more likely to be producible?

7. Core analysis and well logging yielded the following data for a limestone formation: $\phi = 0.15$, $R_t = 25\ \Omega$ m, $R_w = 0.10\ \Omega$ m, and $n = 2.75$. Determine the water saturation.

8. A shaly (clayey) sandstone interval has the following characteristics:

$R_w = 0.02\ \Omega$ m	$m = 2.0$	GR = 40 API units
$R_w = 3.0\ \Omega$ m	$a = 1.0$	$GR_{sh} = 76$ API units
$R_w = 10.0\ \Omega$ m	$\phi = 17.9$	$GR_{cs} = 20$ API units

Calculate S_w, using the following three methods: (1) Archie, (2) Simandoux, and (3) approximate shale relationship. Compare results. Assume the saturation exponent $n = 2$.

9. (i) Show that the Waxman and Smits relationship for calculating S_w sands with dispersed clays may be written as [30]:

$$S_w^{-n^*} = \frac{R_t}{R_w \phi^{-m^*}} \left(1 + \frac{R_w C_{eq} Q_v}{S_w}\right)$$

(ii) The following parameters correspond to a shaly (clayey) sand interval:

$Q_v = 0.40$ meq/mL	$R_t = 22\ \Omega$ m
$\phi = 0.25$	$R_w = 6.1\ \Omega$ m
$m^* = 1.65$	$n^* = 2.0$

(a) Estimate the water saturation in this interval.
(b) Calculate the resistivity index and the formation factor.
(c) What is the value of R_o?

10. The total shale model, that is, the Simandoux equation, may be rewritten as:

$$\frac{R_t}{G_{sh}} = \frac{aR_w}{\phi^m S_w^n}$$

(a) Derive the equation of the total shale group G_{sh} and prove mathematically that, for intervals with constants aR_w and S_w, a log–log plot of R_t/G_{sh} versus ϕ should result in a straight line with a slope of $-m$.
(b) Develop a trial-and-error method for calculating a, m, and R_w.
(c) Table 4.26 shows data obtained from a well in a shaly (clayey) formation. Intervals 4 and 5 are known to be 100% water saturated.

TABLE 4.26

Zone	R_t	Porosity (%)	V_{sh} (%)
1	12.0	26.5	0.12
2	11.0	24.2	0.17
3	12.0	25.9	0.15
4	2.6	27.6	0.00
5	2.5	31.0	0.00

Analyses of rock samples indicate the existence of both laminar and dispersed clays in this formation. Calculate a, m, R_w, and S_w for Zones 1, 2, and 3, knowing $R_{sh} = 1.7\ \Omega$ m.

11. Determine the sonic porosity of a semiconsolidated sandstone given that.

$t = 84\ \mu s/ft$	$t_{ma} = \mu s/ft$
$t_f = 189\ \mu s/ft$	$B_{cp} = 1.3$

12. Given the following data and Table 4.27:

Irreducible water saturation	23.58%
Formation water resistivity	0.0531 Ω m
Cementation factor	1.89
Coefficients a, n	1.0, 2.0
GR_{max}	120
GR_{min}	8.0

TABLE 4.27

Interval No.	GR_{log} API	$I_{sh} = V_{sh}$ (%)	ϕ_{NC} (%)	ϕ_{DC} (%)	ϕ_{avg} (%)	Interval No.	GR_{log} API	$I_{sh} = V_{sh}$ (%)	ϕ_{NC} (%)	ϕ_{DC} (%)	ϕ_{avg} (%)
1	65	50.89	21	16.6	18.8	13	49	36.6	16.6	15.5	16.05
2	63.5	49.55	16.5	12.5	14.5	14	60	46.42	24	18.5	21.25
3	71.5	56.69	19.5	14.2	16.85	15	73	58.03	27.5	23.5	25.5
4	100	82.14	23	20.5	21.75	16	78	62.5	29	25	27
5	97	79.46	21.2	17	19.1	17	79	63.39	31	28.5	29.75
6	57	43.75	13.5	12	12.75	18	103	84.82	26	21	23.5
7	11	2.67	24	21.5	22.7	19	108	89.28	28.5	23	25.75
8	8	0	28.5	22.2	25.35	20	113	93.75	30.5	28.5	29.5
9	12	3.57	31	24.5	27.75	21	9	0.89	22.5	19	20.75
10	12	3.57	28.5	23.6	26.05	22	8	0	22.5	19.5	21
11	66	51.78	27.5	22.5	25	23	8	0	21	18	19.5
12	70	55.35	24	20.5	22.25	24	8	0	17.5	15	16.25

(a) Calculate the reservoir quality index and plot it against average porosity. Comment on the type of shale from the slope of the lines.

(b) Calculate vertical permeability for all the recognizable flow units by choosing proper permeability models. Use a porosity value of 15%, $V_{sh} = 40\%$, $R_{sh} = 50\ \Omega$ m, and $R_t = 100\ \Omega$ m for all flow units. Compare the results.

NOMENCLATURE

A cross-sectional area
a correlation constant
B_o formation volume factor
C_{sd} specific conductance of clay counterions
C_o specific conductance of a core
C_s specific conductance of clay-exchange cation
C_w specific conductance of water CEC cation exchange capacity
C_{eq} equivalent conductance of CEC
d diameter
d_{gr} grain diameter
Δt transit time
E voltage
f_g internal geometry factor
f_{im} fraction of total porosity occupied by a mixture of formation water and dispersed clay
f_{shd} fraction of total porosity occupied by dispersed shale
F_{sd} sand resistivity factor
F_R formation resistivity factor
G shear modulus
G_{sh} shale group in the total shale model
h thickness
I amount of current
I_o amount of current in oil
I_R resistivity index
I_{RA} radioactive or gamma-ray index
I_w amount of current in water
K bulk modulus
k permeability
L length
L_a actual length of the flow path
M weighing multiplier
m cementation factor
n saturation exponent
q flow rate
Q_v volume concentration
R resistivity
R_o resistivity of porous rock 100% saturated with brine
R_w resistivity of brine (water)
R_t true resistivity

R_{sh} resistivity of shale
R_{shd} resistivity of dispersed shale
R_{wT} water resistivity at temperature T
r_o resistance of oil
r_w resistance of water
S saturation of surface
S_{wsh} water saturation in shaly sand
T temperature
T_L laboratory temperature
t total transit time
t_{fl} fluid travel time
t_{ma} matrix travel time
t_{sh} acoustic reading for 100% shale
v velocity
v_{ma} velocity of sound in matrix
V_{sh} shale volume (fraction)
V volume
X, Y Cartesian coordinates, constants

SUBSCRIPTS

a actual
c cross section, corrected
D density log
e exchange, effective
F fracture
g gas
gr grain
im intermatrix
mf mud filtrate zone
m matrix
N neutron log
o oil, original
s stagnation
sd sand
sh shale
shd dispersed clays
t true
w water or solution
wo flushed zone

GREEK SYMBOLS

ϕ porosity
ϕ_{ch} porosity associated with channels
ϕ_D density log porosity
ϕ_e effective porosity

ϕ_N neutron log porosity
ϕ_s sonic log porosity
ϕ_{sd} sand-bed porosity
ϕ_{tr} porosity associated with traps
μ viscosity
ν partitioning coefficient
ρ density
ρ_b bulk density
ρ_{fl} fluid density
ρ_{ma} matrix density
σ_2 standard deviation of $\log_2$ grain-size distribution
τ tortuosity

REFERENCES

1. Pirson SJ. *Handbook of well log analysis.* Englewood Cliffs, NJ: Prentice-Hall; 1963. 326 pp.
2. Archie GE. The electrical resistivity log as an aid in determining some reservoir characteristics. *Trans AIME* 1942;**146**:54–61.
3. Worthington AE, Hedges JH, Pallat N. SCA guidelines for sample preparation and porosity measurement of electrical resistivity sample—Part I: Guidelines for preparation of brine and determination of brine resistivity for use in electrical resistivity measurements. *Log Analyst SPWLA* 1996;January–February:20–8.
4. Hilchie DW. A new water resistivity versus temperature equation. *Log Analyst* 1984; July–August:20–1.
5. Dusenbery RA, Osoba JS. Determination of formation water resist using shale properties. In: *Permian basin oil and recovery conference. Paper 15030.* Midland, TX: Society of Petroleum Engineers, March 13–14; 1986.
6. Calhoun Jr. JC. *Fundamentals of reservoir engineering,* 4th ed. Norman: University of Oklahoma Press; 1960. 426 pp.
7. Cornell D, Katz DL. Flow of gases through consolidated porous media. *Ind Eng Chem* 1953;**45**:2145.
8. Wyllie MRJ, Gardner GHF. The generalized Kozeny–Carman equation. *World Oil* 1958; March.
9. Rosales CP. On the relationship between formation resistivity factor and porosity. *Soc Petrol Eng J* 1982;August.
10. Chilingarian GV. Relationship between porosity, permeability and grain-size distribution of sands and sandstones. In: van Straaten LMJU, editor. *Deltaic and shallow marine deposits, I.* Amsterdam: Elsevier; 1964. pp. 71–5.
11. Rosales CP. Generalization of the Maxwell equation formation resistivity factors. *J Petrol Tech* 1976; July.
12. Timur A, Hemkins WB, Worthington AE. Porosity and pressure dependence of formation resistivity factor for sandstones. In: *Proceedings of the Canadian Well Logging Society*, 4th *Formation Evaluation Symposium.* Calgary, May 9–10; 1972.
13. Anderson WG. Effect of wettability on the electrical properties of porous media. *J Petrol Tech* 1986;December:1371–8.
14. Sweeney SA, Jennings HY. Effect of wettability on the electrical resistivity of carbonate rock from a petroleum reservoir. *J Phys Chem* 1960;**64**(May):551–3.

15. Longeron DG, Argaud MJ, Feraud JP. *Effect of overburden pressure, nature and microscopic distribution of the fluids on electrical properties of samples, Paper 15383*. Society of Petroleum Engineers; 1986.
16. Lewis MG, Sharma MM, Dunlap HF. Wettability and stress effect saturation and cementation exponents. In: *SPWLA 29th Annual Logging Symposium*, Paper K, June 5–8; 1988.
17. Suendenaa E, Brattel F, Kolltvelt K, Normann HP. A comparison between capillary pressure data and saturation exponent obtained at ambient conditions and at reservoir conditions. In: *64th annual conference*. San Antonio, TX: Society of Petroleum Engineers, Paper 19592, October 8–11; 1989, pp. 213–25.
18. Aguilera R. *Naturally fractured reservoirs*. Tulsa, OK: Petroleum Publication Company; 1980. 703 pp.
19. Hutchinson CA, Dodge CF, Polasek TL. Identification and prediction of reservoir nonuniformities affecting production operations. *Soc Petrol Eng J Petrol Tech* 1961;March.
20. Schlumberger, Inc.. *Log interpretation—principles*. Houston, TX: Schlumberger Educational Services; 1972.
21. Poupon A, Loy ME, Tixier MP. A contribution to electrical log interpretation in shaly sands. *Soc Petrol Eng J Petrol Tech* 1954;June.
22. Poupon A, Gaymard R. The evaluation of clay content from logs. In: *Society of Professional Well Log Analysts (SPWLA) Symposium*; 1970.
23. de Witte L. Relations between resistivities and fluid contents of porous rocks. *Oil Gas J* 1950;August.
24. Wilson MD. Origin of clays controlling permeability in tight gas sands. *J Petrol Tech* 1982; December:2871–6.
25. Katahara KW. Gamma ray log response in shaly sands. *Log Analyst SPWLA* 1995; July–August:50–6.
26. Hilchie DW. *Applied openhole log interpretation for geologists and engineers*. Golden, CO: D.W. Hilchie; 1982. 380 pp.
27. Fertlan WH, Hammack GW. A comparative look at water saturation computations in shaly pay sands, *Trans SPWLA*, Paper R, 1971.
28. Bassiouni Z. *Theory, measurement, and interpretation of well logs. SPE textbook Series*, vol. 4. Richardson, TX: SPE; 1994.
29. Larinov VV. *Borehole radiometry*. Moscow: Neclra; 1969.
30. Waxman MH, Smits LJH. Electrical conductivities in oil-bearing shale sands. *Soc Petrol Eng J* 1968;June:107–22 *Trans AIME*, 243.
31. Hoyer WA, Spann MM. Comments on obtaining accurate electrical properties of cores. In: *Society of Professional Well Log Analysts (SPWLA) Symposium*, June 4–7; 1975.
32. Mortland MM, Mellow JL. Conductometric titration of soils for cation exchange capacity. *Proc Soil Sci Soc Am* 1954;**18**:363.
33. Waxman MH, Thomas EC. Electrical conductivities in shaly sands—I. The relation between hydrocarbon saturation and resistivity index; II. The temperature coefficient of electrical conductivity. *J Petrol Tech* February 1974; February:213–25 *Trans. AIME*. 257, 1974; 257: 213–15.
34. Keelan DK, McGinley DC. Application of cation exchange capacity in study of the Shannon Sand of Wyoming. In: *Society of Professional Well Log Analysts (SPWLA) Symposium*; 1979:June 3–6.
35. Bush DC, Jenkins RD. Proper hydration of clays for rock property determinations. *J Petrol Tech* 1970;July:800–4.
36. Worthington PF. The evolution of shaly sand concepts in reservoir evaluation. *Log Analyst* 1985;January–February:23–40.

37. Fertl WH. Log-derived evaluation of shaly clastic reservoir. *J Petrol Tech* 1987; February:175–94.
38. Ruhovets N, Fertl WH. Digital shaly sand analysis based on Waxmar–Smiths model and log-derived clay typing. In: *Trans. SAID/SPWLA European Symposium*, Paris; 1981.
39. Berilgen BW, Sinha AK, Fertl WH. Estimation of productivity of Lobo 6 Sand (Lower Wilcox, TX) by identifying diagenetic clays using well log data. In: *SPE Annual Technical Conference*. Las Vegas, NV: Society of Petroleum Engineering, Paper No. 14278, September 22–25; 1985.
40. Howard JJ. Mixed-layer clays in Eocene interlaminated shales and sandstones. In: *Proceedings of the Annual AAPG/SEPM Conference*. Houston, TX, April 1–4; 1979.
41. Anderson G. *Coring and core analysis handbook*. Tulsa, OK: Petroleum Publication Company; 1975.
42. Bell HJ. Oriented cores guide Eliasville redevelopment. *Petrol Eng Int* 1979;**38**(December).
43. Archer JS, Wall CG. *Petroleum engineering—principles and practice*. London: Graham & Trotman; 1986. 362 pp.
44. Kersey DG. Coring. *World Oil* 1986;January.
45. Keelan DK. Core analysis for aid in reservoir description. *Soc Petrol Eng J Petrol Tech* 1982;November:2483–91.
46. Gatlin C. *Petroleum engineering—drilling and well completions*. Englewood Cliffs, NJ: Prentice-Hall; 1960. 241 pp.
47. Schlumberger, Inc.. *Log interpretation—principles/applications*. Houston, TX: Schlumberger Educational Services; 1987. 198 pp.
48. Helander DP. *Fundamentals of formation evaluation*. Tulsa, OK: Oil and Gas Consultants; 1983. 332 pp.
49. Larsen JK, Fabricius IL. Interpretation of water saturation above the transitional zone in chalk reservoirs. *SPE Reserv Eval Eng* 2004;**7**(2 (April)):155.
50. Wyllie MRJ, Gregory AR, Gardner GHF. An experimental investigation of factors affecting elastic wave velocities in porous media. *Geophys Soc Explor Geophys* 1958;**23**(3 (July)):459–93.
51. Wilson DA, Hensel Jr. WM. Computer log analysis plus core analysis equals improved formation evaluation in West Howard–Glasscock Unit. *Soc Petrol Eng J Petrol Tech* 1978; January:43–51.
52. Osborne CK, Hoga CA. *Corecomp—a practical application of core analysis*. Lubbock: Texas Technical University: Southwestern Petroleum Short Course; April 1972.
53. Sneider RM, Tinker CN, Richardson JG. Reservoir geology of sandstones. In: *Annual Conference*. New Orleans: Society of Petroleum Engineers, September 26–29; 1982, short course notes.
54. Elins LF. Evaluation. In: *Determination of residual oil saturation*. Oklahoma City, OK: Interstate Oil Compact Commission; 1978. pp. 177–254.

Chapter 5

Capillary Pressure

INTRODUCTION

Capillary pressure is the difference in pressure between two immiscible fluids across a curved interface at equilibrium. Curvature of the interface is the consequence of preferential wetting of the capillary walls by one of the phases. Figure 5.1 illustrates various wetting conditions. In Figure 5.1a, two immiscible fluids are shown in contact with a capillary. The water wets the walls of the capillary, but the oil is nonwetting and rests on a thin film of the wetting fluid. The pressure within the nonwetting fluid is greater than the pressure in the wetting fluid and, consequently, the interface between the fluids is curved convex with respect to the nonwetting fluid. The capillary pressure is defined as the pressure difference between the nonwetting and wetting phases:

$$P_c = p_{nw} - p_w \tag{5.1}$$

In Figure 5.1b, the two fluids wet the walls of the capillary to the same extent, and the pressure of each fluid is the same. Therefore, the interface between the immiscible fluids is straight across ($\sim 90°$) and the capillary pressure is equal to zero. If the pressure in the water is greater than that in the oil, the curvature of the interface is directed into the oil and the capillary pressure is positive (Figure 5.1c).

The radii of curvature between water and oil in the pores of the rock are functions of wettability, saturations of water and oil, pore geometry, mineralogy of the pore walls, and the saturation history of the system. Therefore, the radii of curvature and contact angle vary from one pore to another, and the average macroscopic properties of the rock sample apply.

DERIVATION OF THE CAPILLARY PRESSURE EQUATION

A fundamental property of liquids is the tendency to contract and yield the smallest possible surface area, producing a spherical form in small drops.

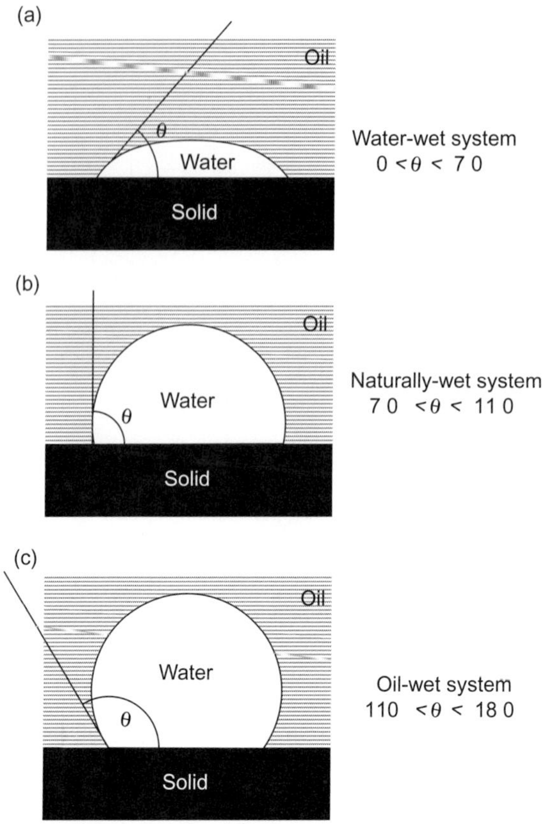

FIGURE 5.1 Various wetting conditions that may exist for water and oil in contact in a capillary, using the contact angle method.

The explanation for this behavior can be illustrated as an imbalance of molecular attractive forces at the surface of a liquid. Consider a liquid coexisting with a gas: molecules in the interior are surrounded by other molecules on all sides, subjecting them to uniform molecular attraction in all directions. On the surface, however, the molecules are attracted inward and on all sides, but there is no outward attraction to balance the inward tug on the surface molecules. This imbalance of forces causes the surface to contract to the smallest possible area and produces a surface tension (σ) expressed as newtons per meter. Work must be done to extend the surface in opposition to the surface tension by forcing molecules from the interior into the surface. This indicates that there is a free energy associated with the surface that has the same dimensions as the surface tension.

Capillary pressure is related to the curvature of the interface by the expression developed by Plateau and applied to porous media by Leverett [1, 2].

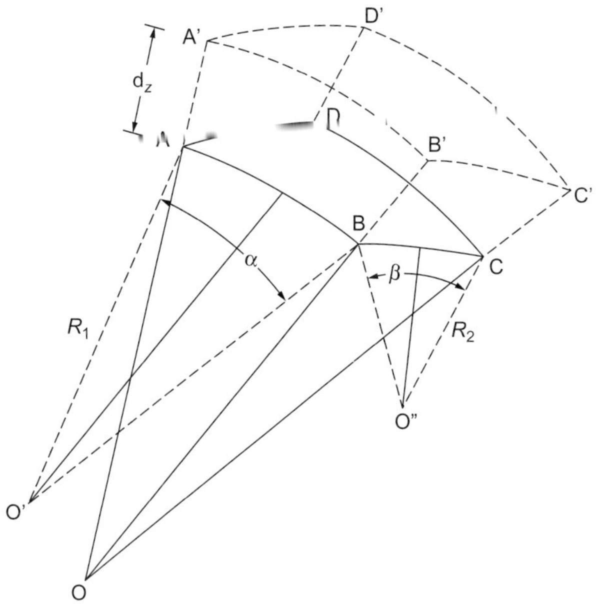

FIGURE 5.2 Radii of curvature of the interface between two fluids. Work is done on the interface to expand the surface against the interfacial tension.

Consider a segment of the interfacial surface separating two fluids with a pressure difference across the interface, producing a curvilinear rectangle as illustrated in Figure 5.2. Both centers of curvature are on the same side; therefore, R_1 and R_2 are both positive.

The work done in expanding the surface, by increasing the pressure on the convex side, is the work against the surface tension. The lengths of the arcs on the sides increase from L_1 to $L_1 + (L_1/R_1)(dz)$ and from L_2 to $L_2 + (L_2/R_2)(dz)$. The area of the original surface (ABCD) expands to the surface area A′B′C′D′ where:

$$\text{Area ABCD} = L_1 \times L_2 \tag{5.2}$$

$$\begin{aligned}\text{Area A}'\text{B}'\text{C}'\text{D}' &= \left[L_1 + \left(\frac{L_1}{R_1}\right)dz\right] \times \left[L_2 + \left(\frac{L_2}{R_2}\right)dz\right] \\ &= L_1L_2 \times \left(1 + \frac{dz}{R_1} + \frac{dz}{R_2} + \frac{dz^2}{R_1R_2}\right)\end{aligned} \tag{5.3}$$

Neglecting the small term dz^2/R_1R_2, the increase in the area is equal to

$$\text{A}'\text{B}'\text{C}'\text{D}' - \text{ABCD} = L_1L_2 \times dz \times \left(\frac{1}{R_1} + \frac{1}{R_2}\right) \tag{5.4}$$

The isothermal work $[(\mathrm{N/m}) \times \mathrm{m}^2 = \mathrm{Nm}]$ required to expand the area against the surface tension is:

$$\mathrm{Work}(1) = \sigma(L_1 L_2 \times \mathrm{d}z)\left(\frac{1}{R_1} + \frac{1}{R_2}\right) \tag{5.5}$$

The isothermal work done by the increase of pressure to advance the surface a distance dz is equal to:

$$\mathrm{Work}(2) = p(L_1 L_2 \times \mathrm{d}z) \tag{5.6}$$

Equating the two work quantities and cancelling common terms yield the capillary pressure as a function of interfacial tension and the radii of curvature [2]:

$$P_\mathrm{c} = \sigma\left(\frac{1}{R_1} + \frac{1}{R_2}\right) \tag{5.7}$$

When a porous medium is considered, R_2 in Eq. (5.7) may be negative; therefore, the more general equation for the capillary pressure is:

$$P_\mathrm{c} = \sigma\left(\frac{1}{R_1} \pm \frac{1}{R_2}\right) \tag{5.8}$$

If the radii of curvature are equal (in a capillary tube, for example), Eq. (5.8) reduces to:

$$P_\mathrm{c} = \frac{2\sigma}{R} \tag{5.9}$$

The special case of the Plateau equation (Eq. (5.9)) may be used to derive a relation from the interfacial geometry of a wetting fluid in a capillary. Figure 5.3 is an exaggerated view of a capillary tube containing water as the wetting phase in contact with a nonwetting fluid (gas or oil). The radius of the spherical interface is larger than the radius of the capillary, and the two radii are related by the cosine of the contact angle as follows:

$$\cos\phi = \frac{r_\mathrm{c}}{R_\mathrm{i}} \tag{5.10}$$

where

r_c = radius of the capillary tube
R_i = radius of the spherical interface

Substituting R_i for R in Eq. (5.9) yields the expression for capillary pressure in terms of the interfacial tension, contact angle, and radius of the capillary tube:

$$P_\mathrm{c} = \frac{2\sigma\cos\theta}{r_\mathrm{c}} \tag{5.11}$$

Microscopic observations of immiscible fluids using glass beads and sand grains have established the complex geometrical aspects of the liquid–liquid

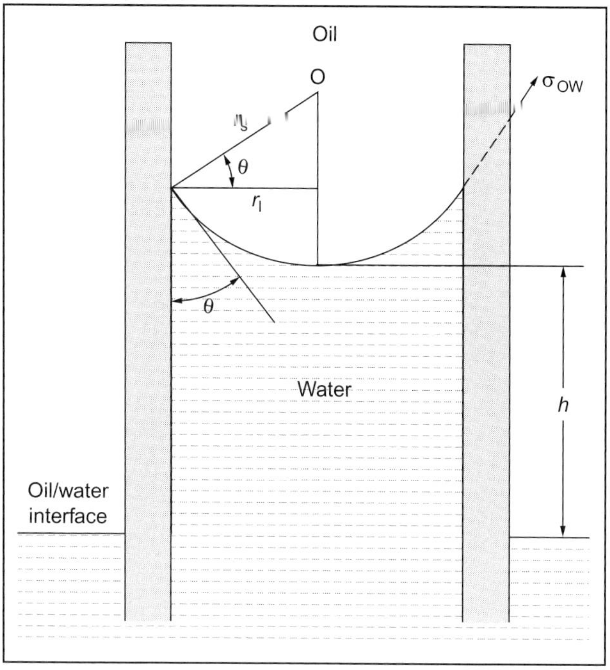

FIGURE 5.3 Capillary rise of water in a water-wet capillary tube.

and liquid–solid contacts. Figure 5.4 is a sketch of a three-phase system (water–oil–rock) at equilibrium. The curvature of the fluid interfacial boundary is a function of grain size, interstitial volume, fluid saturation, and surface tension. The angle of contact (θ) between the liquid interface and the solid, measured through the denser phase, is a function of the relative wetting characteristics of the two fluids with respect to the solid. Figure 5.4 illustrates the typical shapes of the interface and contact angle when the preferential wetting phase located between the solid grains is at a low saturation. When the radii of curvature have their centers of rotation on the same side of the interface, they are positive, but when the radii are on opposite sides, as shown in Figure 5.4, R_1 is positive and R_2 is negative [3, 4]. The radius of curvature on the side of the interface occupied by the preferential wetting fluid is given a negative sign [5–8].

CAPILLARY RISE

When a capillary tube is inserted below the interface of a two-phase system, the meniscus of the immiscible fluids in the capillary will be either:

1. concave with respect to the denser phase, which will rise above the interface between the two liquids outside the capillary;

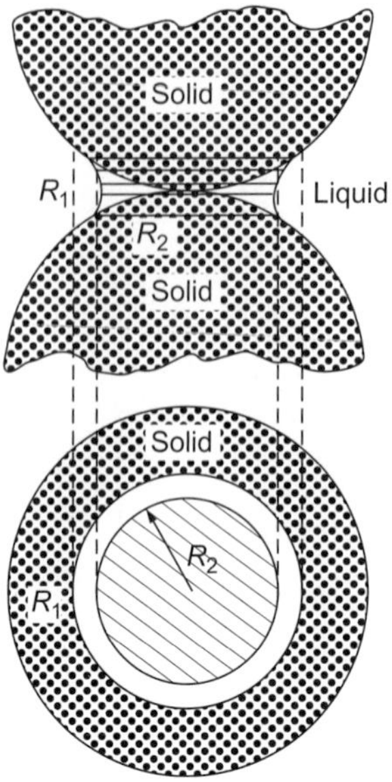

FIGURE 5.4 Three-phase water-wet system at equilibrium showing the radii of interfacial curvature.

2. straight across the capillary and level with the bulk fluids interface; or
3. convex with respect to the denser phase and below the bulk fluids interface as illustrated in Figure 5.5.

The shape and height of the meniscus depend on the relative magnitudes of the molecular cohesive forces and the molecular adhesive forces between the liquids and the walls of the capillary. The more dense liquid wets the solid preferentially when the contact angle is less than 90° (Figure 5.5a). When the contact angle is 0°, the molecular forces are balanced and the two fluids wet the walls equally (Figure 5.5b). When the contact angle is greater than 90°, the denser fluid wets the walls of the capillary to a lesser extent than does the lighter fluid (Figure 5.5c).

The denser fluid will rise in the capillary until the weight of the column of fluid balances the pressure difference across the meniscus. Consider the surface of the meniscus in a circular tube of radius r_c as a segment of a sphere with radius r_s (Figure 5.3). Then $\cos\theta = r_c/r_s$, and substitution into Eq. (5.9) yields Eq. (5.11).

The downward force (W), expressed in dynes (one dyne is 1.019716×10^{-3} g × cm/s^2) due to gravity, exerted by the cylindrical column (Figure 5.5a) is

$$\text{Force down } (W - B) = (\rho_w - \rho_o) g_c h \pi r_c^2 \tag{5.12}$$

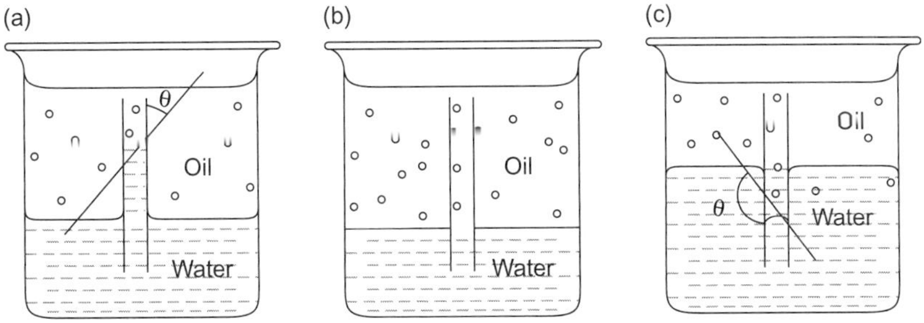

FIGURE 5.5 Menisci of capillaries having different wettabilities.

where

ρ_w = density of water in g/cm^3
ρ_o = density of oil in g/cm^3
g_c = gravitational acceleration = 981 cm/s^2

This downward force is opposed by the force due to the capillary pressure:

$$\text{Force up, } F_z = \left(\frac{2\sigma\cos\theta}{r_c}\right)\pi r_c^2 \tag{5.13}$$

Equating the two forces yields Eq. (5.14):

$$P_c = \Delta\rho g_c h = \frac{2\sigma\cos\theta}{r_c} \tag{5.14}$$

where P_c is expressed in dynes/cm^2 = mN/m^2 = Pa(10^{-1}).

Pressure (pascal) = N/m^2
Interfacial tension = dyne/cm = N*10^{-3}/m = mN/m
Dyne/cm^2 = Pa(10^{-1})

CAPILLARY PRESSURE J-FUNCTION

Leverett proposed the J-function of a specific reservoir, which describes the heterogeneous rock characteristics, more adequately by combining porosity and permeability in a parameter for correlation [2]. The J-function accounts for changes of permeability, porosity, and wettability of the reservoir as long as the general pore geometry remains constant. Therefore, different types of rocks exhibit different J-function correlations. All of the capillary pressure data from a specific formation usually can be reduced to a single J-function versus the saturation curve. This is illustrated in Figure 5.6, where Rose and Bruce prepared J-function correlations for six formations and compared them to data obtained from an alundum core and Leverett's correlation for an unconsolidated sand [9].

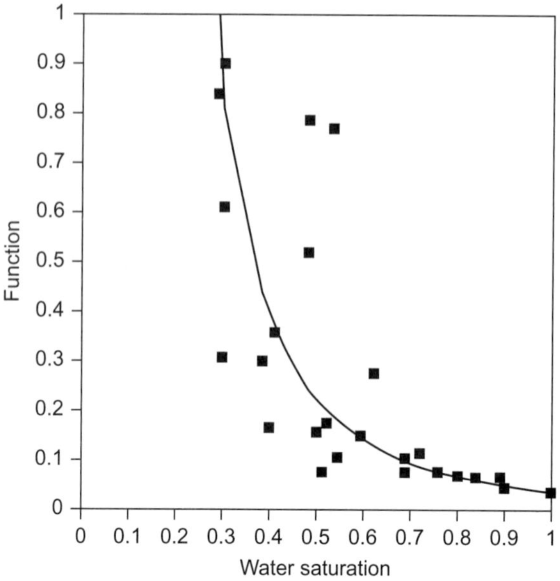

FIGURE 5.6 Typical behavior of dimensionless J-function versus saturation for cores from sandstone.

The J-function can be derived by dimensional analysis or by substitution of the capillary pressure equation into the Carman–Kozeny equation [10]. Permeability has the dimension L^2 and porosity is dimensionless; therefore, $(k/\phi)^{1/2}$ may be substituted for the radius in the capillary pressure equation (Eq. (5.11)) and rearranged as follows:

$$P_c = \frac{\sigma \cos\theta}{(k/\phi)^{1/2}}$$

or

$$J(S_w) = P_c \frac{(k/\phi)^{1/2}}{\sigma \cos\theta} \tag{5.15}$$

Alternatively, it may be derived from the Carman–Kozeny equation

$$u = \left(\frac{\bar{r}_H^2 \phi}{K_z \mu}\right)\left(\frac{\Delta p g_c}{L}\right)\left(\frac{L}{L_e}\right)^2 \tag{5.16}$$

where

K_z = Kozeny constant
L = length of the porous medium, cm
L_e = length of fluid path through the porous medium, cm
p = pressure, g/cm^2

$\bar{r}_H$ = mean hydraulic radius, cm
u = velocity, cm/s
μ = viscosity, $\frac{\text{dyne} \times \text{s}}{\text{cm}^2} = g/\text{cm} \times \text{s} = \text{Pa} \cdot \text{s}$.

Rearranging the above equation yields:

$$\frac{1}{\bar{r}_H} = \left(\frac{\phi \Delta p g_c}{u K_z \mu L_e}\right)^{1/2} = \frac{\phi}{A_s} = \frac{\bar{d}}{4} \quad (5.17)$$

where

A_s = area of the rock surface
$\bar{d}$ = mean pore diameter.

The mean hydraulic radius (r_H) is defined as the surface area divided by the porosity per cubic centimeter of the sample. Substituting ϕ/A_s for r_H, Darcy's equation for the fluid velocity, and rearranging yields:

$$\frac{A_s}{\text{ml}} = \left(\frac{981\phi^3}{K_z k}\right)^{1/2} = \frac{4\phi}{\bar{d}} \quad (5.18)$$

where k is the absolute permeability of the porous medium. Substituting the capillary pressure equation for the average pore diameter and rearranging result in:

$$31.3 = \sqrt{981}$$

$$\frac{1}{r_H} = \left(\frac{\phi * \Delta p * g_c}{u * K_z * \mu * L}\right)^{1/2} = \left(\frac{\phi}{A_c} = \frac{\bar{d}}{4}\right) = \frac{4\phi}{\bar{d}}$$

$$P_c = \frac{4\sigma Cos\ \theta}{\bar{d}}$$

$$\left[\frac{\phi * \Delta P * 981}{\left(\frac{k*\Delta P}{\mu * L}\right)(K_z * \mu * L)}\right]^{1/2} = \frac{4\phi * P_c}{4\sigma Cos\ \theta}$$

$$\left(\frac{981 * \phi}{K_z * k}\right)^{1/2} = \frac{\phi * P_C}{\sigma Cos\ \theta} \quad (5.19a)$$

$$J(S_w) = Const. = \frac{31.3}{\phi\sqrt{K_z}} = \frac{P_C}{\sigma}\left(\frac{k}{\phi}\right)^{1/2} \quad (5.19b)$$

Although Eq. (5.14) was derived from the physics of fluid equilibria in a straight vertical tube, it is applied for general analyses of capillary phenomena in porous media as indicated by applications using the J-function. Hence, the capillary pressure evaluations of porous media do not include the effects of tortuosity and alternating constrictions of the pores.

Example

The fluids in a straight tube have an interfacial tension equal to 32 mN/m and exhibit a contact angle of 80° and capillary pressure of 5.5 kPa. What is the radius of the tube?

Solution

$$r_c = \frac{2\sigma\cos\theta}{P_c} = \frac{2 \times 32 \times 10^{-3} \times 0.174\ \text{N/m}}{5.5 \times 10^3\ \text{N/m}^2} = 2.02 \times 10^{-6}\ \text{m} = 2.0\ \mu\text{m}$$

SEMIPERMEABLE DISK MEASUREMENT OF CAPILLARY PRESSURE

The derivation of capillary pressure equations thus far has been based on a single uniform capillary tube. Porous geologic materials, however, are composed of interconnected pores of various sizes. In addition, the wettability of the pore surfaces varies from point to point within the rock due to the variation in the mixture of minerals in contact with the fluids. This leads to variation of the capillary pressure as a function of fluid saturation and an overall mean description of the rock wettability.

Hydrocarbon reservoirs were initially saturated with water, which was displaced by migrating hydrocarbons. The water accumulated in the geologic structure and formed a trap for the oil, thus producing a petroleum reservoir. This process can be repeated in the laboratory by displacing water from a core with a gas or an oil. The pressure required for the equilibrium displacement of the wetting phase (water) with the nonwetting gas or oil is the water drainage capillary pressure, which is recorded as a function of the water saturation.

A core is saturated with water containing salts ($NaCl$, $CaCl_2$, or KCl) to stabilize the clay minerals, which tend to swell and dislodge when in contact with freshwater. The saturated core is then placed on a porous disk, which also is saturated with water (Figure 5.7). The porous disk has finer pores than does the rock sample. (The permeability of the disk should be at least 10 times lower than the permeability of the core.) The pore sizes of the porous disk should be small enough to prevent penetration of the displacing fluid until the water saturation in the core has reached its irreducible value.

The pressure of the displacing fluid is increased in small increments (Figure 5.8). After each increase of pressure, the amount of water displaced is monitored until it reaches static equilibrium. The capillary pressure is plotted as a function of water saturation as shown in Figures 5.9 and 5.10. If the pore surfaces are preferentially wet by water, a finite pressure (the threshold pressure, P_{ct}) will be required before any of the water is displaced from the core (Figure 5.9). If the core is preferentially oil wet, and oil is the displacing fluid, oil will imbibe into the core, displacing water at zero capillary pressure (Figure 5.10).

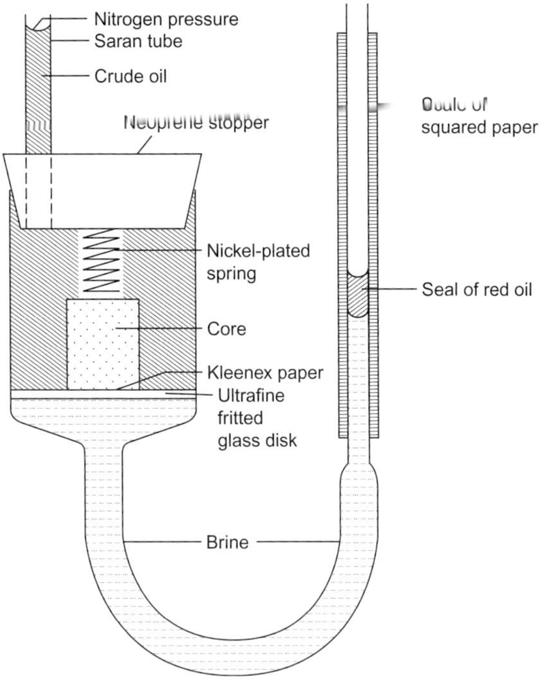

FIGURE 5.7 Porous disk method for measurement of capillary pressure using a manometer [4].

The displacement may be reversed by placing the core on another porous disk, which is saturated with oil, and the core is covered with water. If the core is preferentially wet by water, water will imbibe into the core and displace the oil toward residual oil saturation ($S_{or} = 1 - S_{wor}$), following a path such as curve 2 in Figure 5.9. If the core is preferentially wet by oil, a path similar to curve 2 in Figure 5.10 will be followed.

MEASUREMENT OF CAPILLARY PRESSURE BY MERCURY INJECTION

Capillary pressure curves for rocks have been determined by mercury injection and withdrawal because the method is simple to conduct and rapid. The data can be used to determine the pore size distribution, to study the behavior of capillary pressure curves, and to infer characteristics of pore geometry. In addition, O'Meara et al. showed that mercury injection capillary pressure data of water–oil systems (normalized using Leverett's J-function) are in good agreement with the strongly water-wet capillary pressure curves obtained by other methods [12]. However, water–oil–rock systems exhibit wide variations of wettability that play a decisive role in the behavior of the

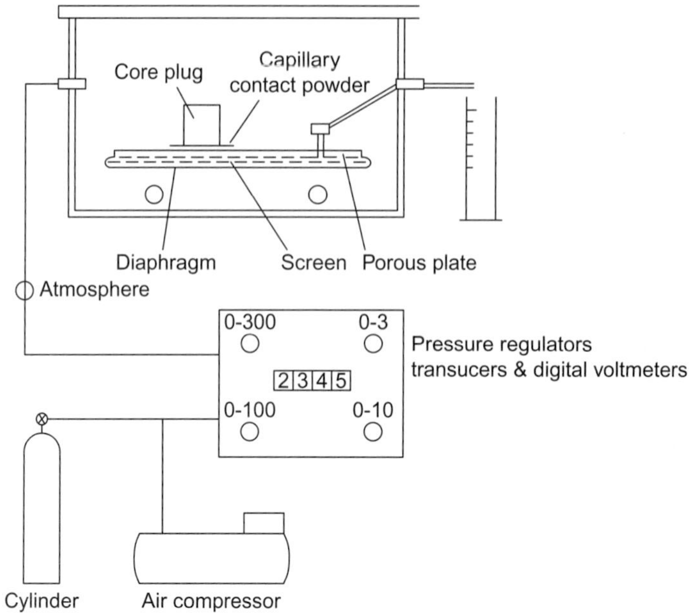

FIGURE 5.8 Porous disk method for measurement of capillary pressure using a pressure transducer [11].

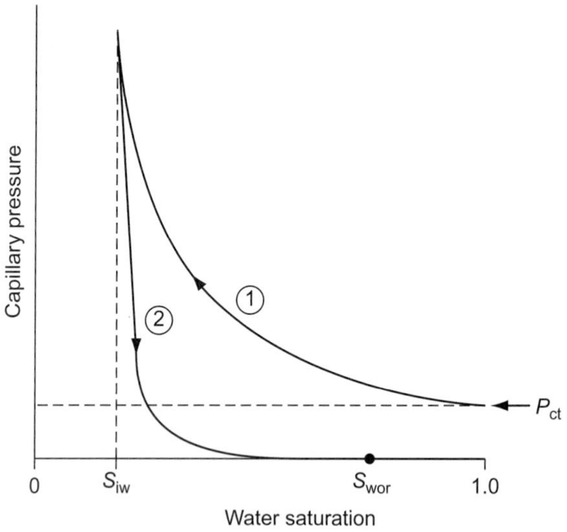

FIGURE 5.9 Typical method for plotting capillary pressure versus saturation for a water-wet system. Note the threshold pressure.

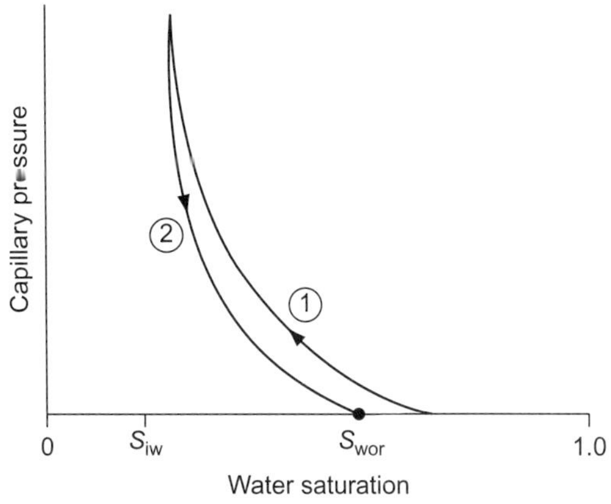

FIGURE 5.10 Typical method for plotting oil-wet capillary pressure curves. Note the imbibition of oil at zero capillary pressure. Capillary pressure is plotted versus the water saturation in most cases; however, it is frequently plotted against the wetting phase saturation, which is oil in this case.

capillary pressure curves. Therefore, when mercury injection capillary pressure data are normalized to represent water–oil systems, the state of wettability must be considered.

The mercury injection method has two disadvantages:

1. After mercury is injected into a core, it cannot be used for any other tests because mercury cannot be safely removed.
2. Mercury vapor is toxic, so strict safety precautions must be followed when using mercury.

To conduct a test, a core is cleaned, dried, and the pore volume and permeability are determined. If liquids are used in the core, it is dried once more before the capillary pressure is determined. The core is placed in the sample chamber of the mercury injection equipment (Figure 5.11). The sample chamber is evacuated, incremental quantities of mercury are injected, and the pressure required for injection of each increment is recorded. The incremental pore volumes of mercury injected are plotted as a function of the injection pressure to obtain the injection capillary pressure curve (Figure 5.12, curve 1). When the volume of mercury injected reaches a limit with respect to pressure increase (S_{imax}), a mercury withdrawal capillary pressure curve can be obtained by decreasing the pressure in increments and recording the volume of mercury withdrawn (Figure 5.12, curve 2). A limit will be approached where mercury ceases to be withdrawn as the pressure approaches zero (S_{wmin}). A third capillary pressure curve is obtained if mercury is reinjected by increasing the pressure incrementally from zero to the maximum pressure at S_{imax} (Figure 5.12, curve 3).

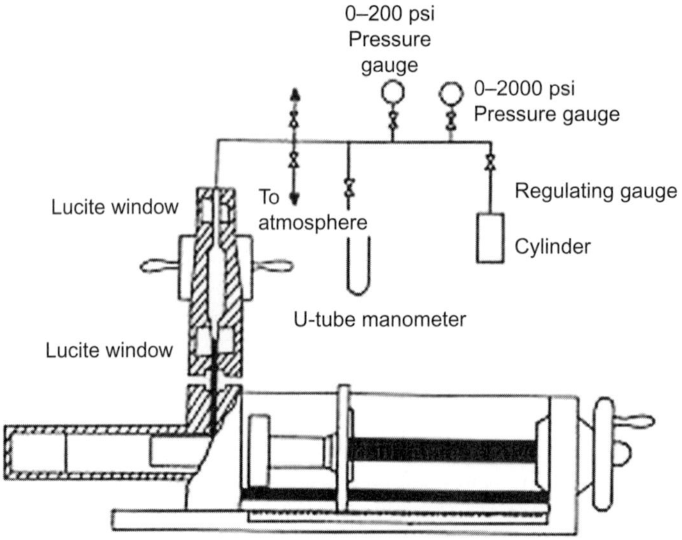

FIGURE 5.11 Equipment for mercury injection capillary pressure measurement.

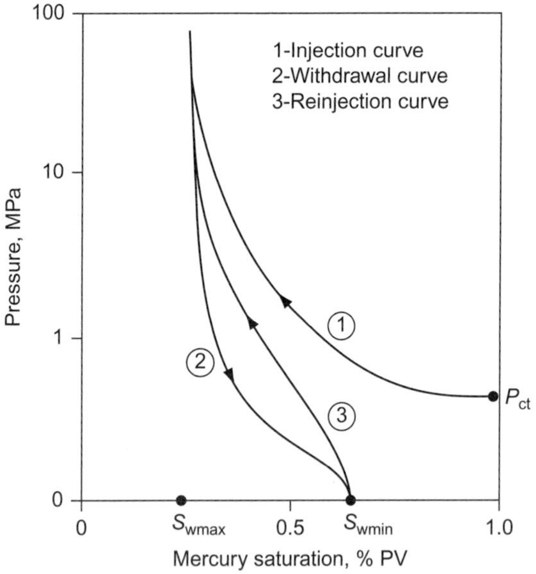

FIGURE 5.12 Mercury–gas capillary pressure curves showing the initial injection curve with its threshold pressure and the hysteresis loop. Note that very high pressures are required for mercury injection.

The closed loop of the withdrawal and reinjection curves (curves 2 and 3, Figure 5.12) is the characteristic capillary pressure hysteresis loop. Mercury is a nonwetting fluid; therefore, the hysteresis loop exhibits a positive pressure for all saturations—that is, the hysteresis loop is above the zero pressure line [13].

In order to transpose mercury injection data to represent water–oil or water–air capillary pressure curves, the mercury capillary pressure data are normalized using Leverett's J-function:

$$J = \frac{P_c(k/\phi)^{1/2}}{\sigma\cos\theta} \tag{5.20}$$

where

$\sigma_{Hg} = 480\ N(10^{-3})/m$
$\theta = 140°$
$k =$ darcies

$$\frac{P_{cw-o}}{\sigma_{w-o} \times \cos(0°)} = \frac{P_{cw-a}}{\sigma_{w-a} \times \cos(0°)} = \left(\frac{P_{cHg}}{\sigma_{Hg} \times \cos(140°)}\right)(k/\phi)^{1/2} \tag{5.21}$$

Capillary pressure transposed from mercury data to represent water-wet, water–oil, or water–air systems (p_{cw-o} or p_{cw-a}) can be obtained from Eq. (5.21). O'Meara et al. showed the close correspondence that can be obtained between J-function-normalized mercury capillary pressure curves and curves obtained for water–oil systems using a centrifuge (Figure 5.13) [12]. The core samples described in Figure 5.13 are cleaned sandstone cores taken from a one-foot interval.

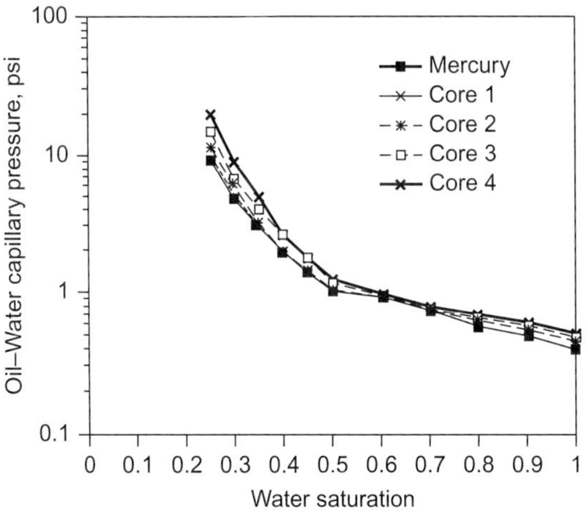

FIGURE 5.13 Oil–water primary drainage versus mercury, normalized with k/ϕ.

Inasmuch as it is an accepted practice to consider the contact angle for a water–air system to be equal to zero, one can use this to obtain a relationship between the contact angle and the saturation of water–oil systems as follows:

$$\cos\theta_{a-w} = 1.0 = \left(\frac{p_{c-aw}}{\sigma_{aw}}\right)\left(\frac{r}{2}\right) = f(S_w) \tag{5.22}$$

$$\cos\theta_{o-w} = \left(\frac{p_{c-ow}}{\sigma_{ow}}\right)\left(\frac{r}{2}\right) = f(S_w) \tag{5.23}$$

$$\cos\theta_{o-w} = \left(\frac{p_{c-ow}}{\sigma_{ow}}\right)\left(\frac{p_{c-aw}}{\sigma_{aw}}\right) = f(S_w) \tag{5.24}$$

An air-displacing water capillary pressure curve is obtained using Eq. (5.22), whereas an oil-displacing water capillary pressure curve is obtained using Eq. (5.23). The ratio of the capillary pressures at each saturation from $S_w = 1.0$ to $S_w = S_{iw}$ is obtained, and the contact angle for a water–oil system in a porous medium can then be plotted as a function of the wetting phase saturation. Implicit in Eq. (5.24) is the assumption that the pore size is the same for a given wetting phase saturation of the two fluids [14, 15].

Example

The following mercury injection capillary pressure data were obtained from a sandstone core with $k = 26$ mD and $\phi = 12\%$. Compute the corresponding water–oil capillary pressures for a strongly water-wet system if the water–oil interfacial tension is 36 mN/m.

Mercury–Air

S (Hg)	S (air)	P_c(Hg–air)
0.05	0.95	4.1
0.40	0.60	8.3
0.50	0.50	34.5
0.55	0.45	82.7
0.60	0.40	144.8
0.65	0.35	220.6

Solution

$$P_{c(w-o)} = (\sigma_{w-o} \times \cos 0°) \times \left(\frac{P_c(\text{Hg})}{\sigma_{\text{Hg}} \times \cos 140°}\right) \times \left(\frac{k}{\phi}\right)^{1/2}$$

$$= (36 \times 1.0) \times \left(\frac{P_c(\text{Hg})}{(480 \times 0.766)}\right) \times \left(\frac{26}{0.12}\right)^{1/2}$$

$$= 1.44 \times P_c(\text{Hg})$$

Water–Oil

S (oil)	S (water)	P_c(water–oil)
0.05	0.95	5.0
0.40	0.60	12.0
0.50	0.50	49.7
0.55	0.45	119.2
0.60	0.40	208.7
0.65	0.35	318.0

CENTRIFUGE MEASUREMENT OF CAPILLARY PRESSURE

Laboratory Procedure

The centrifuge procedure that is in general use now was introduced by Slobod et al. in 1951 [16]. The core is placed in a cup containing an extended calibrated small-diameter tube where fluids displaced from the core by centrifugal force are collected (Figures 5.14 and 5.15). A step-by-step procedure is presented by Donaldson et al. [17].

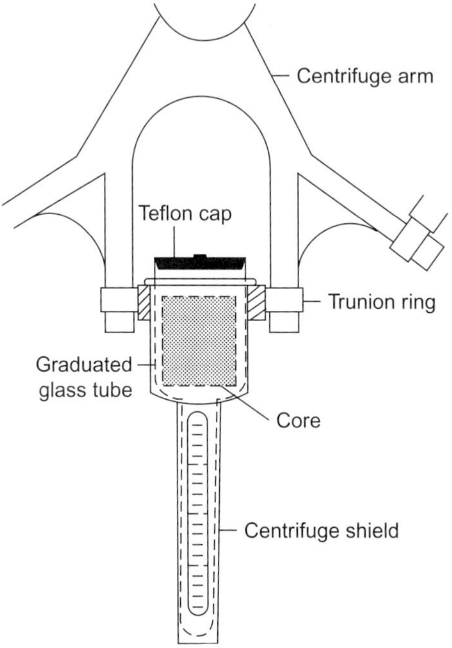

FIGURE 5.14 Positions of core and core holder in a centrifuge for measurement of oil-displacing water capillary pressure curves [17].

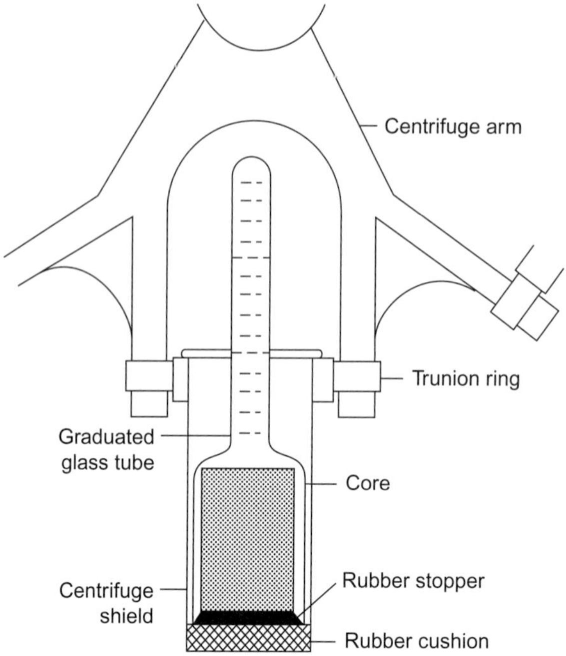

FIGURE 5.15 Positions of core and core holder in a centrifuge for measurement of water-displacing oil capillary pressure curves [17].

1. A weighed and measured core is saturated with brine under vacuum and then reweighed to determine the weight of the brine saturating the core (W_b). The volume of the brine in the saturated core (V_w) is then determined by dividing its weight by its density, and the porosity is determined by dividing the volume of water in the saturated core by the bulk volume of the core (V_b): $V_w = W_b/\rho$; $\phi = V_w/V_b$.
2. The core is placed in the core holder, which is filled with oil to cover the core. The core holder is placed in the centrifuge shield and then attached to the centrifuge arm.
3. When the rotor is filled with core holders containing cores placed on opposite sides, the centrifuge cover is locked, and the rotational speed (revolutions per minute) is increased in increments. At each incremental speed, the amount of fluid displaced is measured at successive intervals until fluid displacement stops. This process is continued until no more fluid is displaced when the rotational velocity is increased. This point is considered to represent a stabilized volume of displaced water and, thus, the irreducible fluid saturation of the core that is calculated from the amount of water displaced by the oil. The capillary pressure associated with the displacement of water by oil (curve 1, Figure 5.16) is calculated from the centrifugal force as described in the next step.

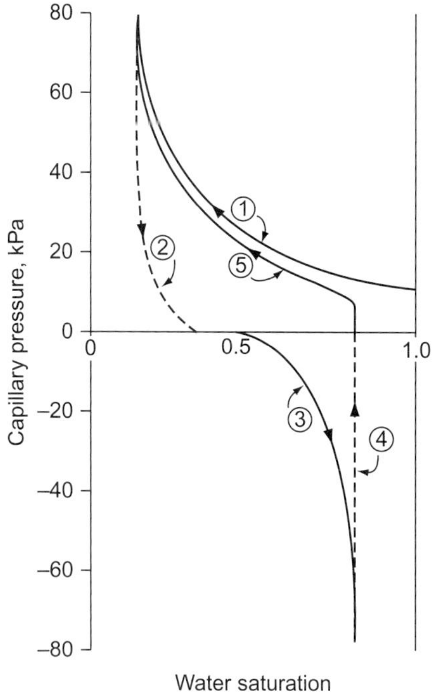

FIGURE 5.16 Capillary pressure curves from centrifugal data. Curves 2 and 4 show the estimated path because these cannot ordinarily be determined by centrifuge.

4. The core containing oil and water at the irreducible saturation is removed from the core holder and placed in another, similar core holder and is filled with water until the core is completely submerged in the brine. This procedure should be carried out as quickly as possible to avoid loss of fluid by evaporation during the period of transfer. The core holders are then assembled on the centrifuge rotor with the graduated end pointing to the center of the centrifuge for collection of oil, which will be displaced by water (Figure 5.15). The cores are once more centrifuged at incremental rotational velocities until oil can no longer be displaced from the core. This is the point of water saturation that corresponds to the residual oil saturation of the core ($S_{wor} = 1.0 - S_{or}$). The incremental rotational velocities and displaced oil are used to calculate the negative capillary pressure curve 3 (Figure 5.16).
5. The core, which is now at a saturation equal to S_{wor}, is placed in another core holder under oil and the displacement from S_{wor} to S_{iw} is conducted as described for the first displacement of water by oil. The curve obtained from this run is curve 5 (Figure 5.16). Curves 2 and 4 cannot be obtained with currently available equipment using the centrifuge method.

The centrifugal force affecting the core varies along the length of the core. Thus, the capillary pressure and the water saturation vary along the

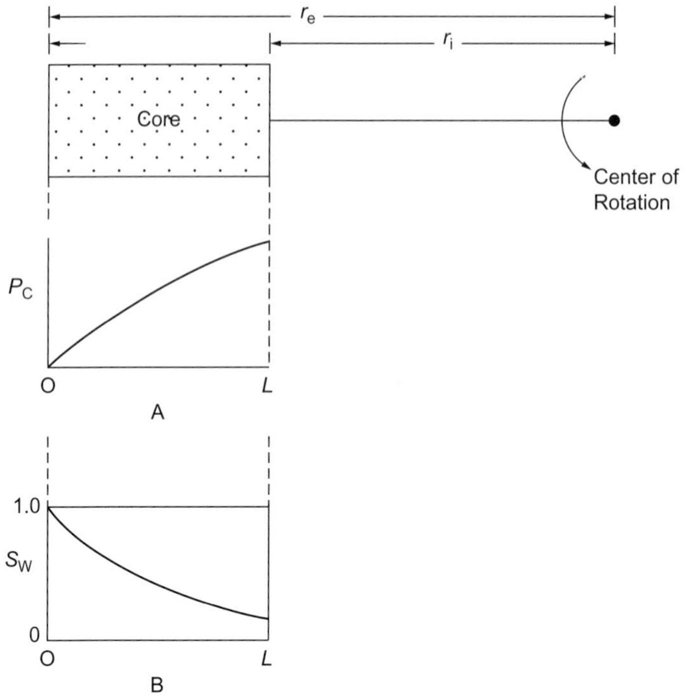

FIGURE 5.17 Centrifugal determination of capillary pressure showing the variation of pressure and water saturation as a function of length of the core.

entire length of the core (Figure 5.17). The capillary pressure at any position in the core is equal to the difference in hydrostatic pressure between the two phases (developed by the centrifugal force). The water saturation measured at each incremental rotational speed is the average saturation of the core at the time of measurement.

Calculation of Centrifuge Capillary Pressure Data

Slobod and Blum and Slobod et al. developed a method for computing the capillary pressure at the top of the core [15, 16]. The equation for capillary pressure in a centrifugal force field is derived beginning with Eq. (5.9), which was obtained for capillary rise in a straight tube. The centrifugal acceleration, a, is:

$$a = \frac{v_\theta^2}{r} \tag{5.25}$$

$$v_\theta = \frac{2\pi r N}{60} \tag{5.26}$$

where v_θ is the rotational velocity, cm/s. Dividing a by g_c to obtain the ratio of centrifugal acceleration to the gravitational acceleration and then substituting a/g_c for g_c in Eq. (5.14) yield:

$$\frac{a}{g_c} = \frac{4\pi^2 r N^2}{981 \times 3{,}600} \tag{5.27}$$

$$P_c = phg_c = ph\left(\frac{a}{g_c}\right) = 1.1179 \times 10^{-5} \Delta\rho N^2 hr \tag{5.28}$$

where P_c is expressed in g_f/cm^2. Equation (5.28) yields the capillary pressure in g_f/cm^2 at any height, h, in the core rotating at N revolutions per minute with a radius of rotation, r. Integrating across the total height of the core (from the inner radius, r_1, of the core to the outer radius, r_e) in order to take into account the variation of the centrifugal field within the core with respect to distance:

$$(P_c)_i = (P_c)_D + 1.1179 \times 10^{-5} \Delta\rho N^2[(r_e^2 - r_i^2)/2] \tag{5.29}$$

As expressed in Eq. (5.29), a capillary pressure gradient exists within the core; a saturation gradient also exists within the core, and the only measured quantities are the revolutions per minute, N, and the average saturation of the core, $\overline{S}_w$. Most centrifuge data reported in the literature adopt the boundary condition assumed by Hassler and Brunner, that is, the end face of the core remains 100% saturated with the wetting phase at all centrifugal speeds of the test [18]. Therefore, the capillary pressure at the end face, $(P_c)_D$, is equal to zero during the entire range of centrifuge speeds used. As long as there exists a continuous film on the surface of the rubber pad holding the core at the bottom, which is the most prevalent assumption, the condition of zero capillary pressure at the end face is correct.

Equation (5.29) is modified in practice to introduce the core length, L, because the lengths of cores used in the centrifuge vary slightly. In addition, the pressure is expressed in kPa rather than g_f/cm^2. These changes yield the final equation, which is used to obtain the capillary pressure (in kPa) at the inlet end, r_i, of the core:

$$(P_c)_i = (1.096 \times 10^{-6}) \Delta\rho N^2 (r_e - L/2) L \tag{5.30}$$

Limiting Centrifuge Speed

Melrose examined the Hassler–Brunner end-face boundary condition and concluded that the zero capillary pressure assumption is valid for the maximum centrifuge speeds used in practice [19,20]. This conclusion is reached by considering the mechanism of the wetting phase (water) displacement by the nonwetting phase (air or oil), commonly referred to as the drainage capillary pressure.

If the centrifuge speed reaches a sufficiently high value, the nonwetting phase will finger (or cavitate) through the largest pores to break through at the end face of the core. The capillary pressure at the end-face boundary will no longer be equal to zero if the nonwetting fluid breaks through. The nonwetting fluid will reach the end face when the capillary pressure at this point exceeds the displacement pressure required by the largest pore channel. This condition can be expressed in terms of the basic capillary pressure equations.

On considering a distance, $r = r_e - R_g$, where R_g is the radius of the largest grain determining the sizes of pore openings at the end face of the core, the capillary pressure in the core at this point must be equal to the displacement pressure, $(P_c)_D$, of the wetting fluid; thus from Eq. (5.29):

$$(P_c)_D = C\Delta\rho N^2[r_e^2 - (r_e - R_g)^2] \tag{5.31}$$

At the top face of the core:

$$(P_c)_i = C\Delta\rho N^2(r_e^2 - r_i^2) \tag{5.32}$$

Dividing Eq. (5.32) by Eq. (5.31) and neglecting the small term R_g^2 yield:

$$(P_c)_i = \left(\frac{r_e^2 - r_i^2}{2r_e R_g}\right) \times (P_c)_D \tag{5.33}$$

Breakthrough of the nonwetting phase will occur when $(P_c)_i > (P_c)_D$, which establishes the critical breakthrough capillary pressure, $(P_c)_{i\text{-crit}}$. To evaluate Eq. (5.33) quantitatively, R_g and $(P_c)_D$ must be expressed in terms that can be measured or estimated. $(P_c)_D$ can be expressed in terms of the capillary pressure equation (Eq. (5.11)) replacing $\cos\theta$ in Eq. (5.10), where $H = r_e/r_i$, and introducing the grain radius, R_g, in place of the mean pore radius:

$$(P_c)_D = \frac{2H\sigma}{R_g} \tag{5.34}$$

Melrose estimated that H assumes values between 4 and 6, which can occur when the fluid–fluid interface enters the constriction of a cone-shaped capillary between two grains of equal size [19].

The Leverett J-function can be expressed as follows [2]:

$$J(S_w) = \left(\frac{(P_c)_D}{\sigma}\right) \times \left(\frac{k}{\phi}\right)^{1/2} \tag{5.35}$$

Substituting Eq. (5.34) into Eq. (5.35) and rearranging yield:

$$R_g = \left(\frac{2H}{J(S_w)}\right) \times \left(\frac{k}{\phi}\right)^{1/2} \tag{5.36}$$

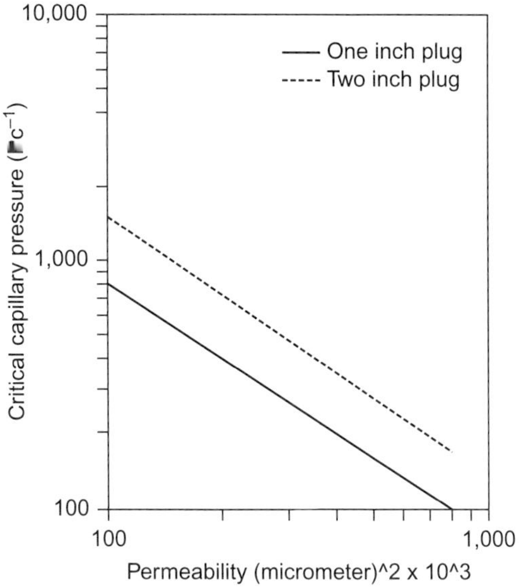

FIGURE 5.18 Estimated relationship between the critical breakthrough capillary pressure and permeability for the Beckman PIR-20 and RIR-16.5 rotors for an air–oil system with 25% porosity.

Then, substituting Eq. (5.36) into Eq. (5.33) yields the critical capillary pressure for breakthrough in terms that can be evaluated:

$$(P_c)_{i-crit} = \frac{(r_e^2 - r_i^2)J^2\sigma\phi}{4r_eHk} \tag{5.37}$$

Melrose used estimates of the terms in Eq. (5.37) to examine the range of critical capillary pressure with respect to the permeability of the rock. Given $J = 0.22$, $H = 5.55$, $\sigma = 25$, and $\phi = 0.25$, and using Eq. (5.37), he computed the values of $(P_c)_{i\text{-crit}}$ as a function of k as shown in Figure 5.18. For a 100-mD sample, the critical pressure exceeds the limitations of capillary pressure attainable with the Beckman core analysis centrifuge. Even at 1,000 mD, the critical pressure is 552 kPa (80 psi), which is still higher than capillary pressures expected for all but the most unusual reservoirs. Therefore, except for very unusual cases, the Hassler–Brunner boundary condition of zero capillary pressure at the outflow face of the core will be sustained.

APPROXIMATE CALCULATION OF THE INLET SATURATION

The capillary pressure calculated using Eq. (5.30) is the capillary pressure at the inlet end of the core; however, the saturation, measured from the amount of fluid displaced, is equal to the average saturation. In order to use the centrifuge-derived capillary pressure, it must be related to the saturation at the inlet.

The length of the core can be considered negligible with respect to the radius of rotation of the centrifuge; in other words, the distance to the top of the core is equal to the distance to the bottom of the core ($r_i = r_e$). Using this approximation, a method for calculating the inlet saturation can be derived directly from the mathematical definition of the average saturation, $\overline{S}$. Hassler and Brunner stated that if the ratio r_i/r_e is greater than 0.7, the error introduced by this assumption is negligible [18]. This ratio is 0.88 for the Beckman L5-50P Rock Core Ultracentrifuge and is even greater for the modified international centrifuge used by Donaldson et al. [17].

By definition, the average saturation in the core, $\overline{S}$, is:

$$\overline{S} = \frac{1}{L}\int S \times \mathrm{d}l \tag{5.38}$$

$$\overline{S} = \frac{1}{\rho g L}\int S \times \mathrm{d}(\rho g L) \tag{5.39}$$

The total pressure across the core is

$$P_c = \rho g_c L \tag{5.40}$$

Using the assumption that $r_i = r_e$, Eq. (5.39) becomes synonymous with the inlet pressure of the core, where $(P_c)_i = \rho g_c L$; therefore:

$$\overline{S} = \frac{1}{(P_c)_i}\int S \times \mathrm{d}(P_c)_i \tag{5.41}$$

By differentiating and rearranging, we get

$$\mathrm{d}[\overline{S} \times (P_c)_i] = S \times \mathrm{d}(P_c) \tag{5.42}$$

$$S_i = \mathrm{d}\left[\mathrm{d}(P_c) \times \overline{S} \times (P_c)_i\right] = \overline{S} \times (P_c)_i \times \frac{\mathrm{d}\overline{S}}{\mathrm{d}(P_c)_i} \tag{5.43}$$

To calculate the inlet saturation as a function of the inlet capillary pressure, the derivative of the average saturation with respect to the inlet capillary pressure, $\mathrm{d}S/\mathrm{d}(P_c)$, must be evaluated from the experimental data. This presents a difficult problem because the data have inherent errors that produce large errors in the derivatives. The various approaches for the analyses of centrifuge capillary pressure measurements differ in the way the derivative is evaluated.

Donaldson et al. found that a least-squares solution of a hyperbolic function represented all capillary pressure curves from the literature that were examined and curves obtained from samples that were treated to establish extremes of water-wet and oil-wet conditions [21]. Using the experimental data, the constants A, B, and C are obtained and then the derivative required by Eq. (5.43) are evaluated:

$$(P_c)_i = \frac{A + B\overline{S}}{1 + C\overline{S}} \tag{5.44}$$

Differentiating Eq. (5.44) yields:

$$d(P_c)_i = \left[\frac{B - AC}{(1 + C\overline{S})^2}\right] dS \tag{5.45}$$

Substituting into Eq. (5.43) yields:

$$S_i = S + (P_c)_i \times \left[\frac{(1 + C\overline{S})^2}{B - AC}\right] \tag{5.46}$$

Using this method and Eq. (5.44), the noise of experimental errors is removed by the least-squares fit of the experimental data. Thus, the saturation at the inlet face of the core, subject to the Hassler and Brunner assumption, may be readily calculated from Eq. (5.46). This saturation corresponds to the capillary pressure at the inlet face of the core, which is calculated using Eq. (5.31). The details of this procedure are presented in the example on page 353.

THEORETICALLY EXACT CALCULATION OF THE INLET SATURATION

Several attempts have been made to obtain an exact method for calculating the inlet-face saturation. Hassler and Brunner proposed a procedure that involves successive iterations to solve the basic equation without making the simplifying assumptions, but these iterations introduce approximations [18]. Van Domselaar showed the derivation of the basic equation beginning with Eq. (5.38) and replaced the length with the radial distances between the inlet and end of the core [22]:

$$\overline{S} = (r_e - r_i)^{-1} \int S \, dr \tag{5.47}$$

which is the average saturation, $\overline{S}$, between r and r_e.

The corresponding capillary pressure is

$$(P_c) = 0.5 \Delta\rho N^2 (r_e^2 - r^2) \tag{5.48}$$

Solving for r gives:

$$r = r_e[1 - P_c(0.5\Delta\rho N^2 r_e^2)^{-1}]^{1/2} \tag{5.49}$$

Differentiating Eq. (5.48) with respect to r and solving for dr gives:

$$dr = -(\Delta\rho N^2 r)^{-1} dP_c \tag{5.50}$$

Substituting Eqs. (5.49) and (5.50) into Eq. (5.47) and recognizing that the following conditions exist:

$$(P_c) = 0.5\Delta\rho N^2(r_e^2 - r_i^2)$$
$$P_c = 0 \text{ at } r = r_e$$
$$P_c = (P_c)_i \text{ at } r = r_i$$

algebraic reduction yields:

$$\overline{S} = \frac{1+R}{2(P_c)_i}\int (S\ dP_c)\left[1 - \left(\frac{P_c}{(P_c)_i}\right)(1-R^2)\right]^{-1/2} \tag{5.51}$$

This basic equation (Eq. (5.51)) provides the exact relationship between the average saturation, $\overline{S}$, the saturation at any point in the core, S, and the inlet capillary pressure, $(P_c)_i$. The inlet-face saturation, S_i, corresponding to $(P_c)_i$ is obtained by solving Eq. (5.51). Van Domselaar [22] attempted to derive a general solution for Eq. (5.51), but, as shown by Rajan [23], it involved an approximation.

Rajan developed a general solution for Eq. (5.51) without using the Hassler and Brunner simplifying assumptions [23]. Calculation of S_i using Rajan's expression, however, requires evaluation of the derivative, $dS/d(P_c)_i$, which can be obtained from the least-squares fit of the data using the hyperbolic expression (Eq. (5.44)). Refer to Rajan's paper for the details of the derivation [23]. The general solution is

$$S_i = \overline{S} + \left[\frac{2R}{1+R}\right](P_c)_i \times \left(\frac{dS}{d(P_c)_i}\right) + \frac{R}{1-R^2}\int_0^{(P_c)_i}\left(\frac{1-Z}{Z}\right)^2\left(\frac{dS}{d(P_c)_i}\right)dP_c \tag{5.52}$$

where

$$Z = \left[1 - \left(\frac{P_c}{(P_c)_i}\right) \times (1-R^2)\right]^{1/2} \tag{5.53}$$

Rajan used an analytic expression to obtain theoretical pseudocapillary pressure data that were used to compare the various solutions (Figure 5.19). Curve 2 describes the P_c versus S data from the analytic expression (shown on the illustration). The pseudo-$(P_c)_i$ versus S_i data obtained from Eq. (5.52) are illustrated by curve 2, which exactly match the pseudoexperimental data from the analytic expression. Curve 1 shows the capillary pressure data which were obtained using the Hassler and Brunner approximation (Eq. (5.43)), revealing an increasing negative error as the wetting-phase saturation decreases. Curve 3 shows the results of the Van Domselaar equation,

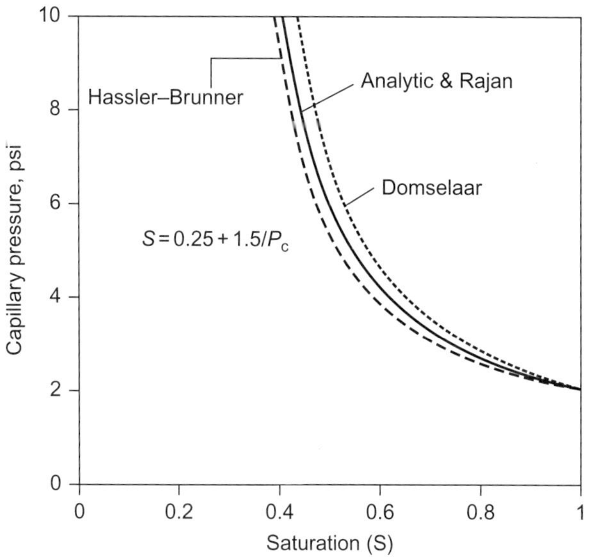

FIGURE 5.19 Agreement between various methods for calculation of the inlet saturation versus capillary pressure ($S = 1.0$ for $P_c < 2$; $S = 0.25 + 1.5/P_c$ for $P_c > 2$).

which is apparently more accurate than the Hassler and Brunner approximate solution at the higher wetting-phase saturations. But as the wetting-phase saturation decreases, it introduces an increasing positive error in the calculated capillary pressure. It is apparent that the accurate calculation of $(P_c)_i$ versus S_i data can be made by using Rajan's equation because it is theoretically exact, correctly models the physics of the problem, and does not contain simplifying assumptions.

The recommended procedure is to:

1. fit the experimental $(P_c)_i$ versus S data to the least-squares hyperbolic expression;
2. evaluate the derivatives at specific values of S and their corresponding $(P_c)_i$;
3. obtain S_i corresponding to each $(P_c)_i$, from a computer solution of Eq. (5.52); and
4. establish a table of values of $(P_c)_i$ versus S_i to plot the capillary pressure curves and to evaluate wettability and the thermodynamic energy required for immiscible fluid displacements.

Example

Prepare the capillary pressure versus inlet saturation curve based on the centrifuge displacement of water by air. Data: $L = 2.0$ cm; d = 2.53 cm; $V_p = 1.73$ cm^3; $k = 144$ mD; centrifuge arm = 8.6 cm; water–air density difference = 0.9988; porosity = 0.17. The experimental data for an air-displacing water capillary pressure experiment and the calculated capillary pressure (in psi) obtained from Eq. (5.31) are presented in Table 5.1.

Solution

Table 5.2 presents the least-squares regression procedure for calculation of P_c as a function of S, and Figure 5.20 shows a comparison between the raw data and the smoothed data obtained from the hyperbolic function.

Given the hyperbolic function $(A + B \times S)/(1.0 + C \times S)$ and using the Hassler–Brunner method, the inlet saturation was obtained. The data are presented in Table 5.3, and Figure 5.20 shows a comparison between the average saturation and the inlet saturation.

The unit psi was used to determine the three constants (A, B, and C) for Eq. (5.44) by the least-squares regression analysis of the experimental data in order to avoid very large numbers that would result from the use of kPa in these calculations.

The values of the constants for $P_{c(hy)}$ versus S are $A = -25.5296$, $B = 17.6118$, and $C = -4.5064$. The regression analysis was used once more to obtain the constants for $P_{c(hy)}$ versus S_i, which differ only with respect to the first constant (A); the second set of constants is $A_2 = -24.5296$, $B_2 = 17.6118$, and $C_2 = -4.5064$.

TABLE 5.1 Calculation of Capillary Pressure for Air Displacing Water from a Berea Sandstone Core

N (rpm)	V_d (H_2O)	S(ave)	p (psi)
1,300	0.30	0.827	4.135
1,410	0.40	0.769	4.865
1,550	0.50	0.711	5.879
1,700	0.60	0.653	7.071
1,840	0.70	0.595	8.284
2,010	0.75	0.566	9.885
2,200	0.80	0.538	11.843
2,500	0.90	0.480	15.293
2,740	1.00	0.422	18.370
3,120	1.05	0.393	23.818
3,810	1.10	0.364	35.518
4,510	1.20	0.306	49.769
5,690	1.25	0.277	79.219

TABLE 5.2 Least-Squares Calculation of P_c(Y) as a Function of S(ave) (X)[a]

$N=13$ X	Y	X×Y	X^2	Y^2	X×Y^2	X^2×Y	X^2×Y^2
0.827	4.135	3.418	0.683	17.099	14.134	2.825	11.683
0.769	4.865	3.740	0.591	23.664	18.192	2.875	13.986
0.711	5.879	4.180	0.505	34.557	24.569	2.972	17.468
0.653	7.071	4.619	0.427	50.004	32.661	3.017	21.334
0.595	8.284	4.932	0.354	68.624	40.857	2.936	24.325
0.566	9.885	5.600	0.321	97.722	55.357	3.172	31.358
0.538	11.843	6.366	0.289	140.248	75.394	3.422	40.530
0.480	15.293	7.337	0.230	233.866	112.202	3.520	53.831
0.422	18.370	7.751	0.178	337.450	142.392	3.271	60.085
0.393	23.818	9.362	0.154	567.316	222.991	3.680	87.650
0.364	35.518	12.934	0.133	1,261.557	459.411	4.710	167.300
0.306	49.769	15.247	0.094	2,476.926	758.827	4.671	232.473
0.277	79.219	21.980	0.007	6,275.611	1,741.210	6.098	483.110

NUM(1) = sum(X^2) × [sum(X × Y) × sum(X × Y^2)−sumY × sum(X^2 × Y^2)] + sum(X × Y) × [sumX × sum(X^2 × Y × 2)−sum(X × Y) × sum(X^2 × Y)] + sum(X^2 × Y) × [sumY × sum (X^2 × Y)−sumX × sum(X × Y^2)]
NUM(2) = N × [sum(X^2 × Y) × sum(X × Y^2) − sum(X × Y) × sum(X^2 × Y^2)] + sumX × [sumY × sum(X^2 × Y^2) − sum(X × Y) × sum(X × Y^2)] + sum(X × Y) × [sum(X × Y) × sum (X × Y)−sumY × sum(X^2 × Y)]
NUM(3) = N × [sum(X^2) × sum(X × Y^2) − sum(X × Y) × sum(X^2 × Y)] + sumX × [sumY × sum (X^2 × Y) − sumX × sum(X × Y^2)] + sum(X × Y) × [sumX × sum(X × Y)−sumY × sum(X^2)]
DENOM = N × [sum(X^2 × Y) × sum(X^2 × Y) − sum(X^2) × sum(X^2 × Y^2)] + sumX × [sumX × sum(X^2 × Y^2)] − sum[(X × Y) × sum(X^2 × Y)] + [sum(X × Y) × sum(X × 2)− sumX × sum(X × 2 × Y)]

A = NUM(1)/DENOM = −25.5296
B = NUM(2)/DENOM = 17.6118
C = NUM(3)/DENOM = −4.5064

[a] *The equation is* $Y=(A+B.X)/(1+C.X)$.

PORE SIZE DISTRIBUTION

An approximate pore size distribution of rocks can be obtained from capillary pressure curves if one of the fluid phases is nonwetting. If one phase is nonwetting, $\cos\theta$ in Eq. (5.11) is assumed to be equal to 1.0 at all saturations. The capillary pressure is then a function of only the interfacial tension and the radius of the pore. Equation (5.11) is based on uniform capillary tubes; however, a rock is composed of interconnected capillaries with varying pore throat sizes and pore volumes. The capillary pressure required to invade a given pore

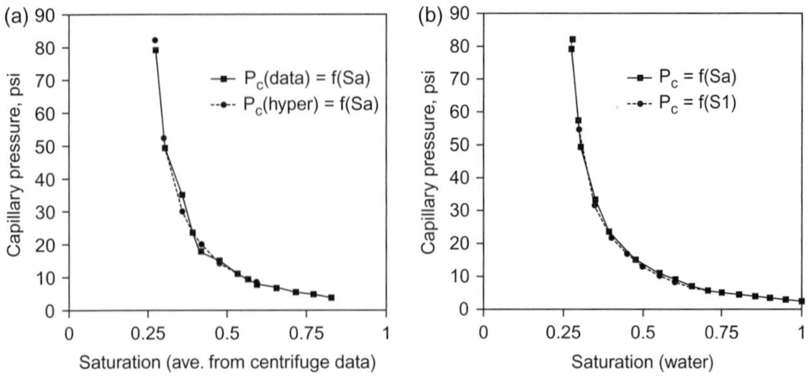

FIGURE 5.20 (a) Comparison of experimentally determined and smoothed curves of capillary pressure versus average saturation obtained by regression analysis. (b) Capillary pressure for a Berea sandstone as a function of the average and inlet saturations for centrifuge-derived data.

TABLE 5.3 Calculation of the Inlet Saturation, S_i, Using the Hasser–Brunner Method[a]

S_{av}	P(hy)	dS/dP	$P \times$ dS/dP	S_i
1.000	2.258	−0.036	−0.081	0.964
0.950	2.682	−0.034	−0.090	0.916
0.900	3.167	−0.031	−0.099	0.869
0.850	3.731	−0.029	−0.108	0.821
0.800	4.391	−0.027	−0.117	0.773
0.750	5.177	−0.024	−0.126	0.726
0.700	6.127	−0.022	−0.135	0.678
0.650	7.300	−0.020	−0.145	0.630
0.600	8.782	−0.017	−0.154	0.583
0.550	10.716	−0.015	−0.163	0.535
0.500	13.345	−0.013	−0.172	0.487
0.450	17.127	−0.011	−0.181	0.439
0.400	23.033	−0.008	−0.190	0.392
0.350	33.549	−0.006	−0.199	0.344
0.300	57.532	−0.004	−0.208	0.296
0.278	82.400	−0.003	−0.212	0.275

[a]$\{S_i = S_{(av)} + P_c(hy) \times (dS_{(av)}/dP_c)\}$ and $\{dS_{av}/dP_c = -(1 + C \times S_{av})^2/(B - A \times C)\}$

is a function of the size of pore throat. Although determination of the pore throat size distribution of rocks based on capillary pressure curves is only an approximation, the distribution is an important parameter for analysis of many fluid transport properties of porous media [24].

Ritter and Drake developed the theory for the penetration of a nonwetting phase into a porous medium [25]. Burdine et al. applied it to reservoir rocks using mercury injection capillary pressure curves [26]. The surface average area distribution of the pore, $D_{(\mathrm{ri})}$, by definition, is:

$$D_{(\mathrm{ri})}\mathrm{d}r = \mathrm{d}V_{\mathrm{p}} = V_{\mathrm{p}}\mathrm{d}S_{\mathrm{w}} \tag{5.54}$$

By differentiating Eq. (5.54) and rearranging, we obtain dr:

$$\mathrm{d}r = \left(\frac{r^2}{2\sigma}\right)\mathrm{d}P_{\mathrm{c}} \tag{5.55}$$

Substituting Eq. (5.55) into Eq. (5.54) yields

$$D_{(\mathrm{ri})} = \left(\frac{2\sigma V_{\mathrm{p}}}{r^2}\right)\left(\frac{\mathrm{d}S_{\mathrm{w}}}{\mathrm{d}P_{\mathrm{c}}}\right) \tag{5.56}$$

for one pore volume, that is, $V_{\mathrm{p}} = 1$,

$$D_{(\mathrm{ri})} = \left(\frac{2\sigma}{r^2}\right)\left(\frac{\mathrm{d}S_{\mathrm{w}}}{\mathrm{d}P_{\mathrm{c}}}\right) \tag{5.57}$$

Assuming that $\cos\theta$ in Eq. (5.11) is equal to 1.0, and substituting $P_{\mathrm{c}} \times r$ for 2σ into Eq. (5.56), the equation used for interpretation of the pore throat size distribution from the capillary pressure is obtained:

$$D_{(\mathrm{ri})} = \left(P_{\mathrm{c}} \times \frac{Vp}{r}\right)\left(\frac{\mathrm{d}S_{\mathrm{w}}}{\mathrm{d}P_{\mathrm{c}}}\right) \tag{5.58}$$

The maximum pore throat size for the sample occurs at $S_{\mathrm{w}} = 1.0$ where $r_{\mathrm{max}} = 2\sigma/P_{\mathrm{ct}}$, and the minimum pore size that will conduct fluid occurs at an irreducible water saturation (S_{iw}) where $r_{\mathrm{min}} = 2\sigma/P_{\mathrm{cmax}}$.

Pore size distributions are used to analyze reductions of permeability caused by clay swelling; precipitations of organic matter in pores—for example, asphaltenes and paraffins; particle migration; and growth of microbes in pores [27–30].

The procedure to determine the pore size distribution of cores is as follows:

1. Obtain air–brine inlet capillary pressure (Eq. (5.22)) versus average saturation data using the centrifuge method.
2. Obtain the three constants (A, B, and C) for the fit of the data to a hyperbola using least-squares method.
3. Obtain the inlet saturation (S_{i}) that corresponds to the inlet capillary pressure (P_{ci}), or obtain the exact solution (Eq. (5.46) or Eq. (5.52)); use the hyperbola to obtain the derivative ($\mathrm{d}S/\mathrm{d}P_{\mathrm{c}}$).

4. Use the least-squares fit of the P_{ci} versus S_i data to obtain the A, B, and C constants for the hyperbola $P_{ci} = (A + BS_i)/(1 + CS_i)$.
5. Extrapolate the P_{ci} versus S_i curve to $S_i = 1.0$ to obtain the correct threshold pressure (P_D) that corresponds to the largest theoretical pore entry size ($r_{max} = 2\sigma/P_D$).
6. Obtain a table of r_j versus P_{cj} for specific values of S_{ij} ($r_j = 2/P_{cj}$ from $S_i = 1$ to S_{iw}).
7. Differentiate the hyperbola to obtain dS_{ij}/dP_{cj} at each point (j).
8. Calculate the pore throat size distribution from Eq. (5.58).

Example

Use the capillary pressure data for air-displacing water from the Berea Sandstone core presented in the example on page 354 (Table 5.1) to calculate the pore throat size distribution.

The inlet saturation and capillary pressure from the earlier example are listed in Table 5.4 together with the pore throat size distribution, $D(r_i)$, as a function of the pore throat radius, r_i, and the results are shown in Figure 5.21.

Solution

The maximum pore entry size (10.59 μm) occurs at $S_w = 1.0$, and the minimum pore size that will conduct fluid occurs at the irreducible water saturation (0.27 μm).

VERTICAL SATURATION PROFILE IN A RESERVOIR

Welge and Bruce derived the capillary pressure equation from the equilibrium of vertical forces in a capillary tube [31, 32]. The weight of water (wetting phase) in the capillary tube, W, which is acting downward, is equal to

$$W = \pi r_c^2 h \gamma_w \tag{5.59}$$

The buoyant force (weight) of the displaced fluid (oil) is upward and equal to

$$B = \pi r_c^2 h \gamma_o \tag{5.60}$$

The vertical component of interfacial tension force (F_z), acting upward, is equal to

$$F_z = 2\pi r_c \sigma_{wo} \cos\theta_{wo} \tag{5.61}$$

Equating the forces and solving for h yield:

$$h = \frac{2\sigma_{wo}\cos\theta_{wo}}{r_c(\gamma_w - \gamma_o)} = \frac{P_c}{\Delta r} \tag{5.62}$$

Welge and Bruce showed that Eq. (5.62) can be used to calculate the water and oil saturations at any height above the free liquid surface if the capillary pressure versus saturation data are available [31, 32]. They applied

TABLE 5.4 Pore Throat Size Distribution from Data for a Berea Sandstone Core Used in Table 5.1

S_i	P_c(hy)	r_i	$D(r_i)$, m^2
1.000	1.973	10.586	0.043
0.950	2.377	8.787	0.054
0.900	2.840	7.353	0.067
0.850	3.377	6.184	0.081
0.800	4.008	5.211	0.097
0.750	4.757	4.390	0.114
0.700	5.663	3.688	0.133
0.650	6.781	3.080	0.153
0.600	8.195	2.549	0.174
0.550	10.039	2.080	0.196
0.500	12.547	1.665	0.220
0.450	16.154	1.293	0.246
0.400	21.787	0.959	0.273
0.350	31.817	0.656	0.301
0.300	54.691	0.382	0.330
0.278	78.408	0.266	0.344

$R_i = 144/P_c$
$D(r_i) = (P_c \times V_p/r) \times (dS_i/dP_c)$

this to the calculation of vertical water–oil–gas saturation distribution as a function of height for hydrocarbon reservoirs:

$$h(\text{m}) = \frac{0.102\ P_c}{\rho_w - \rho_o} \quad \text{and} \quad h(\text{ft}) = \frac{2.3 P_c}{\rho_w - \rho_o} \tag{5.63}$$

where

h = height of capillary rise in m (ft)ρ_w and ρ_o = densities of water and oil, respectively, in g/cm^3 (lb/ft^3) P_c = capillary pressure in kPa (psi)

Using the capillary pressure curve for oil displacing water from an initial water saturation of 100%, the oil saturation in the reservoir can be calculated at any height above the free water level, FWL, which occurs at zero capillary pressure, as shown in Figures 5.16 and 5.22.

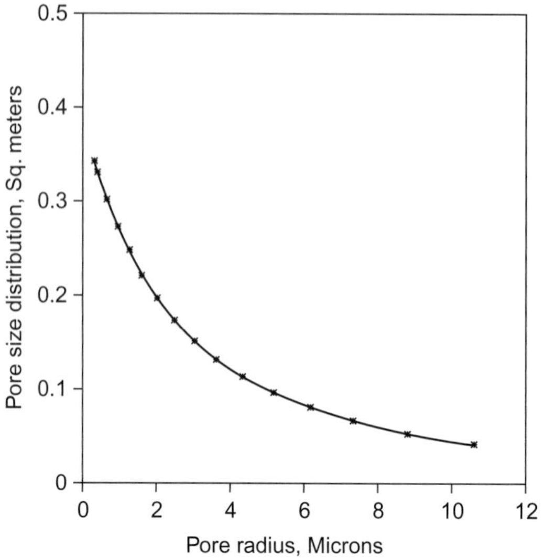

FIGURE 5.21 Pore entry size distribution for a Berea sandstone core.

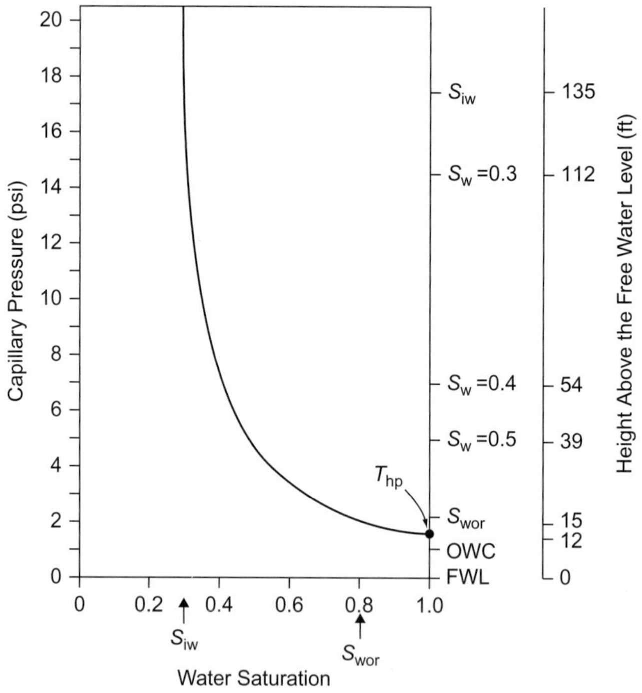

FIGURE 5.22 Vertical saturation profile of a reservoir calculated from a capillary pressure curve.

The FWL is difficult to locate in a reservoir, but the oil–water contact, OWC, is apparent in well logs. Knowing the threshold pressure, P_{ct}, from a capillary pressure curve obtained from reservoir cores and oil, the location of the FWL can be determined and the vertical saturations profile can then be calculated as a function of height above the FWL. The position of the maximum value of the residual oil saturation can be located; this is important because no mobile oil exists at depths below this point. A transition zone exists between S_{wor} and S_{iw}, where water is always produced along with the oil. If the reservoir thickness is large enough to exceed the value at S_{iw} (135 ft, Figure 5.22), only oil will be produced at heights above the point where S_{iw} occurs.

Mixed wettability and oil-wet reservoirs will not have a threshold pressure because the cores, when saturated 100% with water and contacted with oil, will imbibe oil at zero capillary pressure. Therefore, the OWC and the FWL are synonymous (occurring at the same point).

When the core data from a reservoir indicate variability of permeability and porosity throughout the sand, or an increasing or decreasing trend with respect to depth, the capillary pressure data for each core can be reduced to a single J-function curve, which is a function of saturation:

$$J_{ij}(S_{wij}) = \left(\frac{P_{cij}(S_{wij})}{\sigma}\right) \times \left(\frac{k_i}{\phi}\right)^{1/2} \tag{5.64}$$

where

1. subscript i indicates the individual cores tested
2. subscript j indicates values of J and P_c with respect to the specific core and data point (S_{wij}).

After the J versus S_w data have all been assembled, the single (average) curve may be obtained using the hyperbolic function:

$$J_j(S_w) = \frac{A + BS_{wi}}{1 + CS_{wj}} = \left(\frac{P_{cj}}{\sigma}\right) \times \left(\frac{\bar{k}}{\bar{\phi}}\right)^{1/2} \tag{5.65}$$

The geometric average of the permeability and the arithmetic average of the porosity are used in Eq. (5.65). Using these data and the average J-curve, the single correlated capillary pressure curve may be obtained from Eq. (5.64) and used to determine the reservoir saturation profile as discussed.

Example

The core data in Table 5.5 were obtained for four cores extracted from the same formation. Calculate the vertical saturation profile and determine the locations (heights above the FWL) of the oil–water contact, S_{or} and S_{iw}. The interfacial tension is $25\ N \times 10^{-3}/m$, the density difference between the brine and oil is $0.80\ g/cm^3$, the residual oil saturation is 30%, and S_{iw} occurs at $S_w = 0.20$.

TABLE 5.5 Core Analysis Data

Core 1			Core 2		
$k_1 = \mathbf{0.060}$			$k_2 = \mathbf{0.095}$		
$\phi_1 = \mathbf{0.18}$			$\phi_1 = \mathbf{0.20}$		
P_{c1j}	S_{w1j}	J_{1j}	P_{c2j}	S_{w2j}	J_{2j}
3	1.00	0.069	3	1.00	0.057
4	0.88	0.092	5	0.80	0.095
7	0.70	0.162	8	0.64	0.152
11	0.58	0.254	14	0.48	0.266
19	0.40	0.439	23	0.38	0.437
36	0.25	0.832	33	0.32	0.627
56	0.19	1.294	45	0.26	0.855
78	0.19	1.802	61	0.22	1.159
			74	0.21	1.406

Core 3			Core 4		
$k_3 = \mathbf{0.132}$			$k_4 = \mathbf{0.155}$		
$\phi_3 = \mathbf{0.21}$			$\phi_4 = \mathbf{0.23}$		
P_{c3j}	S_{w3j}	J_{3j}	P_{c4j}	S_{w4j}	J_{4j}
6	1.00	0.151	3	1.00	0.081
8	0.86	0.201	4	0.84	0.108
10	0.72	0.251	7	0.68	0.189
12	0.60	0.301	16	0.45	0.432
20	0.48	0.502	27	0.34	0.729
29	0.40	0.728	41	0.27	1.107
38	0.34	0.954	54	0.21	1.458
51	0.32	1.280	69	0.21	1.863
67	0.30	1.682	76	0.20	2.052
81	0.30	2.033			

Solution

The average permeability is

$$\log\overline{k} = \frac{1}{4} \times [\log(0.60) + \log(0.095) + \log(0.132) + \log(0.155)] \quad \overline{k} = 0.104\ D$$

The average porosity is:

$$\overline{\phi} = \frac{1}{4} \times [0.18 + 0.20 + 0.21 + 0.23] \quad \overline{\phi} = 0.21$$

J for each of the four sets of capillary pressure data were calculated and plotted to obtain the average J-curve. Then, the constants (A, B, and C) for the hyperbolic fit were obtained from a least-squares fit.

$$\overline{J} = \frac{(-0.0075 - 0.2856\ S_{wj})}{(1 + 0.0391\ S_{wj})}$$

The average capillary pressure versus water saturation was then calculated from J as follows (Table 5.6):

$$\overline{P}_{cj} = J_{j(sw)}\sigma \times \left(\frac{\overline{k}}{\overline{\phi}}\right)^{1/2} = 35.525\overline{J}_{j(sw)}$$

Height above the FWL was calculated from:

$$h(\text{m}) = \frac{0.102\ P_c}{0.8} = 0.128\ P_c$$

The OWC occurs at the threshold pressure where $S_w = 1.00$; therefore, OWC = 0.41 m above the FWL.

Height above the FWL at S_{or} occurs at $S_w = 0.70 = 0.77$ m.

Height above the FWL at $S_{iw} = 7.50$ m.

TABLE 5.6 Saturation Profile

S_{wj}	J_j	P_{cj}	h (m)
1.00	0.090	3.179	0.41
0.80	0.140	4.973	0.64
0.70	0.170	6.039	0.77
0.60	0.235	8.348	1.07
0.50	0.350	12.434	1.59
0.40	0.560	19.894	2.55
0.35	0.700	24.867	3.18
0.30	0.950	33.749	4.32
0.20	1.650	58.616	7.50

CAPILLARY NUMBER

The capillary number is a dimensionless group that represents the ratio of viscous forces to the interfacial forces affecting the flow of fluid in porous media. Applied to water displacement of oil, the capillary number is:

$$n_c = \frac{(\mu_w u_w)}{(\phi \sigma_{ow})} \tag{5.66}$$

where μ, u_w, and σ are expressed, respectively, in poises (or kg/m × s), m/s, and N/m (kg × m/s^2 × m).

In practice, the viscosity is expressed as cP, the velocity as m/D, and the interfacial tension as N × 10^{-3}/m; therefore

$$n_c = 1.16 \times 10^{-4} \left[\frac{\mu_w(\text{cP}) \times u_w(\text{m/D})}{\phi \sigma_{ow}(\text{mN/m})} \right] \tag{5.67}$$

The capillary number for field waterfloods ranges from 10^{-6} to 10^{-4}. Laboratory studies have shown that the value of the capillary number can have a marked influence on the ultimate recovery of oil. In order to study the effects of n_c on oil recovery, Melrose and Bradner defined the microscopic displacement efficiency as the ratio of the mobile oil saturation to the total oil saturation [33]:

$$E_D = \frac{(1 - S_{or} - S_{iw})}{(1 - S_{iw})} = 1 - \frac{S_{or}}{1 - S_{iw}} \tag{5.68}$$

As the value of n_c is increased to 10^{-4}, there is no noticeable effect on the displacement efficiency (the residual oil saturation for a given system remains constant). However, at values greater than 10^{-4}, a significant increase in displacement efficiency is observed. At values greater than 10^{-2}, E_D becomes 1.0 and complete displacement of oil occurs (Figure 5.23). Therefore, the critical value of n_c is established to be equal to 10^{-4}. At values of n_c less than 10^{-4}, the capillary forces are dominant and the oil displacement occurs by movement of oil ganglia several pore diameters in length. The residual oil saturation is distributed as isolated droplets and groups (depending on the wetting properties). When n_c exceeds 10^{-4}, the viscous displacement forces dominate and begin to increase the displacement efficiency.

As indicated in Figure 5.23, the pore throat size distributions can have considerable influence on the residual oil saturation, but the critical capillary number remains at approximately 10^{-4}. Small variations in the critical value are to be expected because of the wide variability of experimental procedures, rocks, and fluids. The direct influence of wettability on the critical value has not been explored because the experiments conducted thus far have been concentrated on strongly water-wet systems. Oil-wet systems will

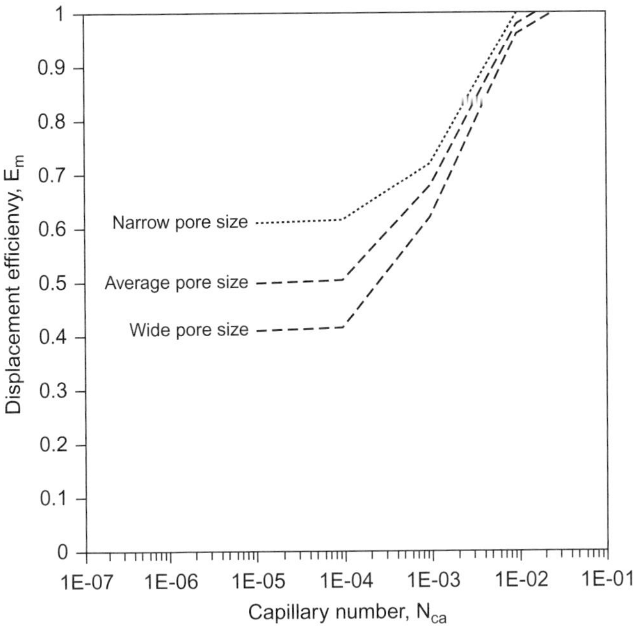

FIGURE 5.23 Effect of increasing the capillary number on the residual oil saturation at the end of a waterflood.

exhibit a lower displacement efficiency and probably will shift the critical capillary number to a higher value.

The capillary number is generally varied by increasing the flow rate and lowering the interfacial tension. The displacing-phase flow rate can be increased to a value near the inception of turbulent flow; turbulent flow produces eddy currents that create cross flows that are not accounted for in the capillary number. A large number of experiments have been conducted by lowering the interfacial tension, which can be precisely controlled for a given system, but for effective response, it must be lowered to a value less than 0.1 mN/m.

Using Darcy's law, the capillary number can be rearranged by substitution for the velocity to obtain:

$$n_c = \left(\frac{k_{rw} \times k}{\phi \times \sigma_{ow}}\right)\left(\frac{\Delta p}{L}\right) \tag{5.69}$$

This equation can be used to obtain the relationship between the capillary number and the pressure gradient for a given system, or it can be used to calculate the capillary number if all other conditions are known for the system being analyzed.

Example

Given the following data for a fluid-flow system, calculate the capillary number and the pressure gradient for the flow of the fluid at the velocity given: viscosity = 120 cP, interfacial tension = 36 mN/m, velocity = 0.68 m/D, porosity = 0.18, absolute permeability = 60 μm, and relative permeability = 0.21.

Solution

$$n_c = \frac{1.16 \times 10^{-4} \times 120 \times 0.68}{0.18 \times 36} = 1.46 \times 10^{-3}$$

$$\text{Pressure gradient } \left(\frac{\Delta p}{L}\right) = \frac{1.46 \times 10^{-3} \times 0.18 \times 36 \times 10^{-3}}{0.21 \times 60 \times 10^{-12}}$$

$$= 0.751 \times 10^{6} \ \text{Pa/m} = 109 \ \text{psi/m} = 33 \ \text{psi/ft}$$

PROBLEMS

1. Explain why the meniscus in a capillary inserted in mercury is depressed below the surface of the mercury. (See Figure 5.5c, where the meniscus is depressed below the water surface of a water–oil system.)
2. Explain how the equation for capillary pressure in straight capillaries is derived from the Plateau equation.
3. Assume that the following data were obtained by injection of mercury [$\sigma = 465 \times 10^{-3}$ N/m] into a core having a total pore volume of 2.8 cm^3 at 20°C:

P_c (psi)	S_w
3	1.00
4	0.88
7	0.70
11	0.58
19	0.40
36	0.25
56	0.19
78	0.19

   (a) What is the value of the threshold pressure, the maximum pore size, and the minimum pore size that can conduct a fluid?
   (b) Determine the pore size distribution and show all graphs used.
4. If the capillary pressure data given in problem 3 were obtained from oil displacing water in a centrifuge, where P_c is the core inlet pressure and S_{wa} is the average water saturation, carry out the following:
   (a) Plot the core inlet saturation versus the inlet capillary pressure, using Eq. (5.43).

(b) Plot P_{ci} versus S_i and determine the threshold pressure.
(c) Compute the vertical saturation profile if the water–oil contact occurred at 2,000 feet ($\rho_w - \rho_o = 0.180$ g/cm^3).
(d) What is the meaning of cavitation when used with respect to centrifuge determination of capillary pressure?

5. If porosity = 0.16 and permeability = 0.120 darcy for the core in problem 4, calculate the J-function versus saturation for the system and show the graph.

NOMENCLATURE

A area
A_s area of the surface of rocks or sands
cP viscosity (centipoise)
D day; darcy
d_{ave} average pore diameter
FWL free water level (in reference to capillary rise of fluids)
g gravitational constant; 980 cm/s^2; 32.2 ft/s^2
H_o length measured from the oil level to the bottom of the sample in the drainage measurement, cm
h height of capillary rise
J dimensionless function for correlation of capillary pressure to permeability and porosity
k absolute permeability (μm^2, darcy)
K_z Carman–Kozeny Constant
L length (m, cm, ft)
m meter
N centrifuge revolutions per minute; newton
n_c capillary number
o oil (used as subscript)
OWC oil–water contact
p pressure (Pa, atm, psi)
P_c capillary pressure (Pa, psi)
$(P_c)_D$ displacement capillary pressure at the effluent end of a core in a centrifuge
$(P_c)_i$ capillary pressure at the inlet of the core (centrifugal measurement of capillary pressure)
P_{ct} threshold capillary pressure
p_{nw} pressure of a nonwetting fluid
p_w pressure of a wetting fluid (generally taken as the water phase)
p_o pressure of the oil phase
r radius of curvature of an arc of an interface (cm)
R_g radius of a grain
r_c radius of a capillary tube
$\bar{r}_H$ mean hydraulic radius
r_i radius of a spherical interface

r_o radius of the centrifuge rotary axis measured to the oil level
$r_{(e)}$ exit (bottom) end of core (centrifugal measurement of capillary pressure)
$r_{(i)}$ inlet (top) end of core (centrifugal measurement of capillary pressure)
S_{imax} maximum saturation of mercury injection
S_{iw} irreducible wetting phase saturation
S_{or} residual oil saturation
S_{wa} average water saturation
S_{wmin} minimum saturation of mercury after withdrawal from a core
S_{wor} water saturation at residual oil saturation
$S_{w(i)}$ water saturation at the inlet end of the core (centrifugal measurement of capillary pressure)
u, v velocity (m/s, cm/s, ft/s)
v_θ angular velocity of centrifuge rotor
V volume
V_p pore volume

GREEK SYMBOLS

ϕ angle
Φ contact angle
σ interfacial tension (N/m)
μ viscosity (Pa · s, cP)
ρ_o density of oil
ρ_w density of water

REFERENCES

1. Plateau JAF. Statique Expérimentale et Théorique des Liquides Soumis aux Seules Forces Moléculaires. [Experimental and Theoretical Researches on the Figures of Equilibrium of a Liquid Mass Withdrawn from the Action of Gravity.] Smithsonian Institute Annual Reports: Series No. 1, 1863, 207–285; Series No. 2, 1864, 285–369; Series No. 50, 1865, 411–435; Series No. 6, 1866, 254–289.
2. Leverett MC. Capillary behavior in porous solids. *Trans AIME* 1941;**142**:152–69.
3. Leja J. *Surface chemistry of froth floatation*. New York: Plenum Press; 1982. 758 pp.
4. Purcell WR. Capillary pressures—their measurement using mercury and the calculation of permeability therefrom. *Trans AIME* 1949;**186**:39–110.
5. Adam NK. *The physics and chemistry of surfaces*, 3rd ed. London: Oxford University Press; 1941. pp. 7–12
6. Champion MA, Davy N. *Properties of matter*, 3rd ed. London: Blackie & Son; 1960. pp. 115–18
7. Moore WJ. *Physical chemistry*, 2nd ed. Englewood Cliffs, NJ: Prentice-Hall; 1955. pp. 500–2
8. Neumann HJ. Investigations on the wettability of formations and on oil migration. *Erdoel-und Kohle-Erdgas-Petrochemie* 1966;**19**(3):171–2.
9. Rose W, Bruce WA. Evaluation of capillary character in petroleum reservoir rocks. *Trans AIME* 1949;**186**:127–41.
10. Carman PC. Fluid flow through granular beds. *Trans Inst Chem Eng* 1937;**15**:150–66.

11. Cram PJ. Wettability studies with non-hydrocarbon constitutes of crude oil. Petroleum Recovery Research Institute, Report No. RR-17, Socorro, NM; December 1972.
12. O'Meara Jr DJ, Hirasaki GJ, Rohan JA. Centrifuge measurements of capillary pressure: Part 1—outflow boundary condition. In: *Proc. SPE 18290, 63rd Annual Conference*, Houston, TX; October 2–5, 1988, p. 14.
13. Wardlaw NC, Taylor RP. Mercury capillary pressure curves and the interpretation of pore structure and capillary behavior in reservoir rocks. *Bull Can Petrol Geol* 1976;**24**(2):225–62.
14. Lorenz PB. Notes on capillary pressure curves. Data obtained at U.S. Bureau of Mines, Bartlesville Energy Technology Center; 1972 (not previously published—used here with permission from P.B. Lorenz).
15. Slobod RL, Blum HA. Method for determining wettability of reservoir rocks. *Trans AIME* 1952;**195**:1–4.
16. Slobod RL, Chambers A, Prehn Jr. WL. Use of centrifuge for determining connate water, residual oil and capillary pressure curves of small core samples. *Trans AIME* 1951;**192**:127–34.
17. Donaldson EC, Kendall RF, Pavelka EA, Crocker ME. Equipment and procedures for fluid flow and wettability tests of geological materials. DOE/BETC/IC-79/5, National Technical Information Service, Springfield, VA; 1980:40.
18. Hassler GL, Brunner E. Measurement of capillary pressures in small core samples. *Trans AIME* 1945;**160**:114–23.
19. Melrose JC. Interpretation of mixed wettability states in reservoir rocks. In: *Proc. SPE 10971, Annual Conference*, New Orleans; September 1982: 26–9.
20. Melrose JC. Interpretation of centrifuge capillary pressure data. *The Log Analyst* 1988;**29**(1):40–7.
21. Donaldson D, Ewall N, Singh B. Characteristics of capillary pressure curves. *J Petrol Sci Eng* 1991;**6**:249–61.
22. Van Domselaar HR. An exact equation to calculate actual saturations from centrifuge capillary pressure measurement. *Rev Tech INTEVEP* 1984;**4**(1):55–62.
23. Rajan RR. Theoretically correct analytic solution for calculating capillary pressure-saturation from centrifuge experiments. *SPWLA Trans*, In: *Proc. 27th Annual Logging Symposium*, Houston, TX; June 9–12, 1986, Paper O, 18.
24. Obeida TA. Quantitative evaluation of rapid flow in porous media. General examination for Ph.D., Department of Petroleum and Geological Engineering, University of Oklahoma; 1988.
25. Ritter LC, Drake RL. Pore size distribution in porous material. *Ind Eng Chem Fundam* 1945;**17**:782–91.
26. Burdine NT, Gournay LS, Reichertz PP. Pore size distribution of petroleum reservoir rocks. *Trans AIME* 1950;**198**:195–204.
27. Chilingar GV, Yen TF. Some notes on wettability and relative permeability of carbonate reservoir rocks. *Energy Sources* 1983;**7**(1):67–75.
28. Donaldson EC, Baker BA, Carroll HB. Particle transport in sandstones. Society of Petroleum Engineers, Paper No. 6905, In: *Proc. 52nd Annual Fall Conference*, Denver, CO; October 9–12, 1977:20.
29. Donaldson EC, Chilingarian GV, Yen TF, editors. Amsterdam: Elsevier Science; 1985.
30. Zajic JE, Donaldson EC, editors. *Microbes and oil recovery*. El Paso, TX: Bioresources Publications; 1985.

31. Archer JS, Wall CG. *Petroleum engineering principles and practice*. Oxford: Graham & Trotman; 1986. 362 pp.
32. Welge HJ, Bruce WAA. Restored-state method for determination of oil in place and connate water. *Drilling and production practice. American Petroleum Institute* 1945;161–5.
33. Melrose JC, Bradner CF. Role of capillary forces in determining microscopic displacement efficiency for oil recovery by waterflooding. *J Can Petrol Technol* 1974;**13**(4):54–62.

Chapter 6

Wettability

Wettability is the term used to describe the relative adhesion of two fluids to a solid surface. In a porous medium containing two or more immiscible fluids, wettability is a measure of the preferential tendency of one of the fluids to wet (spread or adhere to) the surface. In a water-wet brine–oil–rock system, water will occupy the smaller pores and wets the major portion of the surfaces in the larger pores. In areas of high oil saturation, the oil rests on a film of water spread over the surface. If the rock surface is preferentially water-wet and the rock is saturated with oil, water will imbibe into the smaller pores, displacing oil from the core when the system is in contact with water.

If the rock surface is preferentially oil-wet, even though it may be saturated with water, the core will imbibe oil into the smaller pores, displacing water from the core when it is contacted with water. Thus, a core saturated with oil is water-wet if it will imbibe water and, conversely, a core saturated with water is oil-wet if it will imbibe oil. Actually, the wettability of a system can range from strongly water-wet to strongly oil-wet depending on the brine–oil interactions with the rock surface. If no preference is shown by the rock to either fluid, the system is said to exhibit neutral wettability or intermediate wettability, a condition that one might visualize as being equally wet by both fluids (50%/50% wettability).

Other descriptive terms have evolved from the realization that components from the oil may wet selected areas throughout the rock surface. Thus, fractional wettability implies spotted, heterogeneous wetting of the surface, labeled "dalmatian wetting" by Brown and Fatt [1]. Fractional wettability means that scattered areas throughout the rock are strongly oil-wet, whereas the rest of the area is strongly water-wet. Fractional wettability occurs when the surfaces of the rocks are composed of many minerals that have very different surface chemical properties, leading to variations in wettability throughout the internal surfaces of the pores. This concept is different from neutral wettability, which is used to imply that all portions of the rock have an equal preference for water or oil. Cores exhibiting fractional wettability will imbibe a small quantity of water when the oil saturation is high (e.g., at the irreducible water saturation, S_{wi}), and also will imbibe a small amount of oil when the water saturation is high (e.g., at the residual oil saturation, S_{or}).

The term "mixed wettability" commonly refers to the condition where the smaller pores are occupied by water and are water-wet, whereas the larger pores of the rock are oil-wet and a continuous filament of oil exists throughout the core in the larger pores [2–4]. Because the oil is located in the larger pores of the rock in a continuous path, oil displacement from the core occurs even at very low oil saturation; hence, the residual oil saturation of mixed-wettability rocks is unusually low. Mixed wettability can occur when oil containing interfacially active polar organic compounds invades a water-wet rock saturated with brine. After displacing brine from the larger pores, the interfacially active compounds react with the rock's surface, displacing the remaining aqueous film and, thus, producing an oil-wet lining in the large pores. The water film between the rock and the oil in the pore is stabilized by a double layer of electrostatic forces. As the thickness of the film is diminished by the invading oil, the electrostatic force balance is destroyed and the film ruptures, allowing the polar organic compounds to displace the remaining water and react directly with the rock surface [5].

The wettability of a rock–fluid system is an overall average characteristic of a heterogeneous system with microscopic relative wetting throughout the porous medium [6]. The rock pore surfaces have preferential wetting tendencies toward water or oil leading to establishment of the various states of overall wettability. This overall wettability has a dominant influence on the fluid flow and electrical properties of the water–hydrocarbon–rock system. It controls the capillary pressure and relative permeability behavior and thus the rate of hydrocarbon displacement and ultimate recovery [7–10].

INTERFACIAL TENSION

When two immiscible fluids (gas–liquid or liquid–liquid) are in contact, the fluids are separated by a well-defined interface, which is only a few molecular diameters in thickness. Within the fluid and away from the interface and the walls of the container, the molecules attract each other in all directions. At the surface between two immiscible fluids, there are no similar molecules beyond the interface and, therefore, there is an inward-directed force that attempts to minimize the surface by pulling it into the shape of a sphere. This surface activity creates a film-like layer of molecules that are in tension, which is a function of the specific free energy of the interface. The interfacial tension has the dimensions of force per unit length (newtons/meter), which is the modern standard expression of the units. In the earlier literature, however, it is expressed as dynes/centimeter, which is numerically equal to millinewtons per meter [$(N \times 10^{-3})$/m or mN/m].

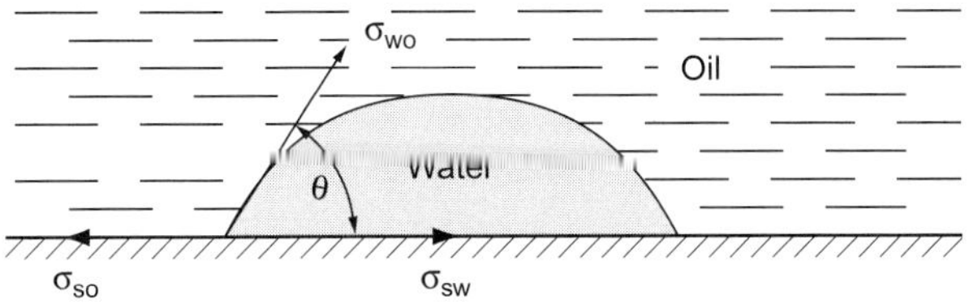

FIGURE 6.1 Relationships between the contact angle and interfacial tension expressed by the Young equation where σ_{so} = solid–oil, σ_{sw} = solid–water, and σ_{wo} = water–oil interfacial tensions.

CONTACT ANGLE

When the interface is in intimate contact with the walls of a container, for example a capillary tube, the interface intersects the solid surface at an angle, θ, which is a function of the relative adhesive tension of the liquids to the solid. This angle is described by Young's equation (Eq. (6.1)) [11]. The relationships are illustrated in Figure 6.1 where two liquids, water (w) and oil (o), are associated with a solid surface (s). The contact angle (measured through the denser phase) is:

$$\cos\theta = \frac{\sigma_{so} - \sigma_{sw}}{\sigma_{wo}} \tag{6.1}$$

where

σ_{so} = interfacial tension between the solid and oil
σ_{sw} = interfacial tension between the solid and water
σ_{wo} = interfacial tension between water and oil

Direct measurement of the solid–fluid surface tensions is not possible; however, by considering a three-phase system, one can eliminate the solid–fluid surface tensions to obtain a measurable relationship between the three contact angles. Writing the equation for the three conditions: (1) water, oil, solid; (2) water, gas, solid; and (3) gas, oil, solid, one obtains

$$\begin{aligned} \sigma_{so} &= \sigma_{sw} + \sigma_{wo}\cos\theta_{wo} \\ \sigma_{sg} &= \sigma_{sw} + \sigma_{wg}\cos\theta_{wg} \\ \sigma_{sg} &= \sigma_{so} + \sigma_{og}\cos\theta_{og} \end{aligned} \tag{6.2}$$

Algebraic elimination of the solid–fluid interfacial tensions yields:

$$\sigma_{wo}\cos\theta_{wo} = \sigma_{wg}\cos\theta_{wg} = \sigma_{og}\cos\theta_{og} \tag{6.3}$$

The conditions of wetting can be determined from the easily measured interfacial tensions and contact angles, using Eq. (6.3) [12, 13].

Adhesion tension, τ, is defined as the difference between the solid–oil and solid–water interfacial tensions:

$$\tau = \sigma_{so} - \sigma_{sw} = \sigma_{wo}\cos\theta \tag{6.4}$$

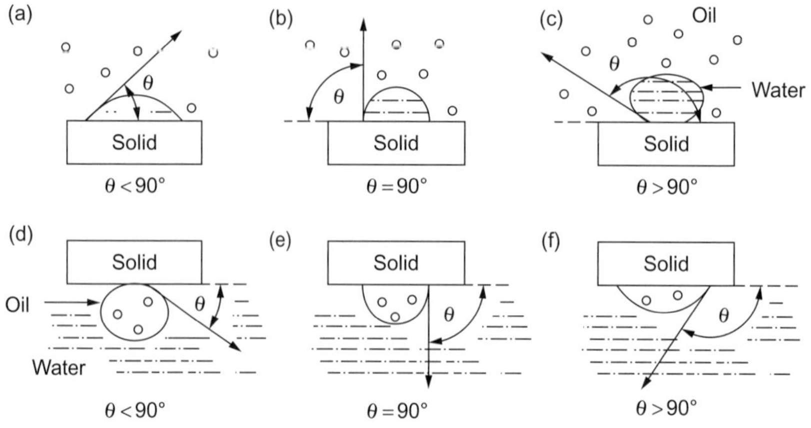

FIGURE 6.2 Measurement of contact angles for water–oil systems; (a–c) show measurements using a drop of water surrounded by oil, and (d–f) show drops of oil surrounded by water. The contact angle is measured through the denser phase.

Three conditions of wettability are apparent from Eq. (6.4) and are illustrated in Figure 6.2:

1. When the adhesion tension is positive, the system is water-wet ($\theta < 90°$, $\cos\theta = +$).
2. When τ is zero, the system is neutrally wet ($\theta = 90°$, $\cos\theta = 0$).
3. When τ is negative, the system is oil-wet ($\theta > 90°$, $\cos\theta = -$).

When a liquid spreads on a solid, or on the surface of another immiscible liquid, the imbalance of forces is defined as the spreading coefficient, C, as follows:

1. Water spread on a solid in the presence of oil (water-wet):

$$C_{\mathrm{swo}} = \sigma_{\mathrm{so}} - \sigma_{\mathrm{wo}} - \sigma_{\mathrm{sw}} \tag{6.5}$$

2. Oil spread on a solid in the presence of water (oil-wet):

$$C_{\mathrm{sow}} = \sigma_{\mathrm{sw}} - \sigma_{\mathrm{wo}} - \sigma_{\mathrm{so}} \tag{6.6}$$

3. Liquid spread on a solid surface in the presence of gas:

$$C_{\mathrm{slg}} = \sigma_{\mathrm{sg}} - \sigma_{\mathrm{lg}} - \sigma_{\mathrm{sl}} \tag{6.7}$$

Defining the relative wetting behavior of fluids in a rock is complex because there are variations of spreading behavior at points, or areas, within the rock and the measured wettability is an average of the physical and chemical interactions of the fluids. The relative amounts of rock surface wet by one fluid or the other define the overall wettability of the system [14–17, 55].

Assume that a preferentially water-wet rock core is saturated with 20% water and 80% oil. In this case, the adhesion tension is positive ($\sigma_{so} > \sigma_{sw}$) and the contact angle is less than 90°. If this water-wet core is contacted with water, some oil will be spontaneously expelled from the core as water is imbibed along the walls and into the smaller pores until a state of equilibrium is attained between the solid–fluid-specific surface energies (interfacial tensions). The wetting fluid entering the core will accumulate in the pores that create the greatest fluid–fluid interfacial curvature consistent with Eq. (6.1); thus, the wetting phase accumulates in the smallest pores.

SESSILE DROP MEASUREMENT OF CONTACT ANGLES

The sessile drop method is often used to make direct measurements of the contact angle to determine preferential wetting of a given solid by oil and water. A smooth, homogeneous surface is necessary for this test; a polished quartz surface is generally used to make contact angle measurements of water–oil systems [18–20]. Two procedures may be used, as shown in Figure 6.2. Figures 6.2 a–c illustrate the procedure where the solid plate is suspended horizontally below the surface of a clear (or refined) oil and a drop of water is placed on the solid. A photograph is taken of the system for accurate measurement of the contact angle. By convention, the contact angle is measured through the denser phase.

The second method is to suspend the plate horizontally in the water and place a drop of oil on the bottom of the plate (Figure 6.2 d–f). The contact angle is measured through the water phase and the same analysis is applied.

A modification of the sessile drop method was introduced by Leach et al. to measure the water-advancing contact angle [21]. Two polished mineral plates are mounted horizontally with a small gap between them; one plate is fixed and the other can be moved smoothly with a screw (Figure 6.3). A drop of oil is placed between the plates and allowed to age until the contact angle no longer changes; then the mobile plate is moved, creating the advancing contact angle. This angle changes gradually and eventually reaches a stable value after a few days.

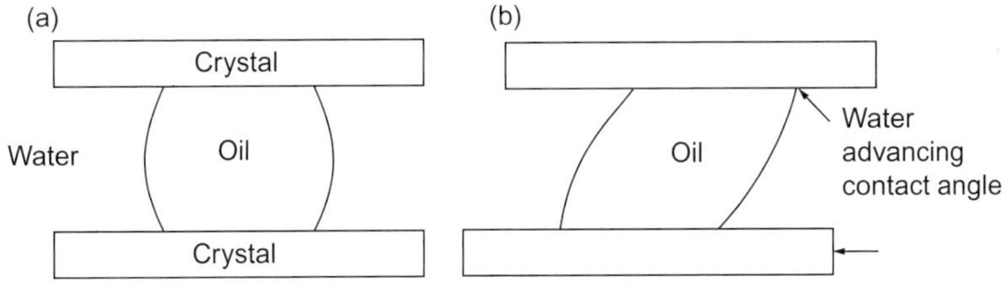

FIGURE 6.3 Method used to measure the advancing and receding contact angle.

The sessile drop method was used by Yan et al. to determine the mechanisms of contact angle hysteresis and advancing contact angles [22]. Mica surfaces equilibrated with crude oil, which was diluted with heptane to decrease the solvency of asphaltenes, exhibited larger contact angles ($\sim 140°$) than the surfaces equilibrated with undiluted crude oil ($\sim 75°$). Thus, the contact angle increases (the system becomes more oil-wet) when asphaltenes are deposited on the surfaces. This is important to oil production because it indicates that some deposition of asphaltenes on the rock surfaces around a producing well will cause the zone to become more oil-wet. This oil-wet zone will reduce the capillary end effects that cause a high water saturation and thus a high water/oil producing ratio.

WILHELMY PLATE MEASUREMENT OF CONTACT ANGLES

The Wilhelmy plate method yields direct measurements of the adhesion tension (Figure 6.4) acting on the perimeter of a plate, as well as the advancing and receding contact angles [23]. As the plate is moved into and out of a liquid, the change in force, F, due to the adhesion tension is:

$$F = l\sigma\cos\theta \tag{6.8}$$

and the contact angle is:

$$\theta = \text{arc } cos\left(\frac{F}{l\sigma}\right) \tag{6.9}$$

where

F = force, mN
$l = 2 \times (\text{width} + \text{thickness})$, m
σ = interfacial tension, mN/m

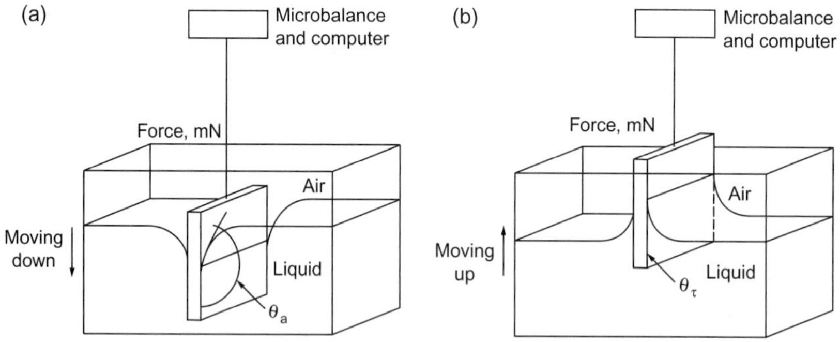

FIGURE 6.4 Wilhelmy plate method for measurement of advancing contact angle (a) and receding contact angle (b).

The plate is attached to a microbalance and its rate of movement is controlled by a computer that calculates the instantaneous value of the contact angle. The advancing contact angle is obtained when the plate is moved into the liquid, and the receding angle when the plate is pulled from the liquid [23–25].

The principal constituents of sandstones and limestones are quartz and calcite, respectively. Consequently, polished plates of quartz and calcite are used as representative surfaces for sessile drop and advancing contact angle measurements. These are not representative of reservoir rocks however, because the plates do not account for surface roughness, the large variety of minerals, or thin layers of organic materials. The wettability of these heterogeneous surfaces can be measured only by using one of the methods that measure the average core wettability, such as the Amott and USBM methods. The sessile drop and advancing contact angle measurements are therefore qualitative, rendering information on the behavior of the reservoir fluids and the gradual deposition of interfacially active compounds onto the solid surface. The methods are most useful for measurement of wettability effects of solutes in pure fluids. One can quickly observe wettability changes of the smooth plate toward different oils and aqueous solutions.

SURFACE CHEMICAL PROPERTIES

The chemical compositions of the fluids and the rock surfaces determine the values of the solid–fluid and fluid–fluid specific surface energies. Thus, the mineralogy of the rock surface has an influence on the relative adhesive tensions and contributes to the overall wettability of the fluid–rock system. Polar organic compounds in crude oil can react with the surface, forming a preferentially oil-wet surface. Interfacially active compounds—those that tend to accumulate at the interface—can lower the interfacial tension and affect the wetting characteristics of the fluid–rock system. Many of the surface properties of shales, sandstones, and carbonates that affect the relative wetting of the surfaces by water and crude oils are readily explained by examining the general chemical structures associated with the principal minerals.

Yaalon determined the composition of 10,000 shales and arrived at the following average composition: (1) 60% clay minerals—mostly illite, (2) 20% quartz, (3) 10% feldspar, (4) 6% carbonates, (5) 3% iron oxide, and (6) 1% organic matter [26]. The dominant characteristics of shales are their ion exchange properties, electrical conductivity, swelling, and dispersion when treated with freshwater (or low-salinity water). These properties are attributed to the dominant presence of the minerals. Many of these properties can be explained using an idealized, general structure of clays (Figure 6.5). This is not an exact structure; clay structure is three-dimensional and varies considerably from one type of clay to another. The silicon atom is small and has a very strong charge of $+4$; therefore, in the case of silicon dioxide, the

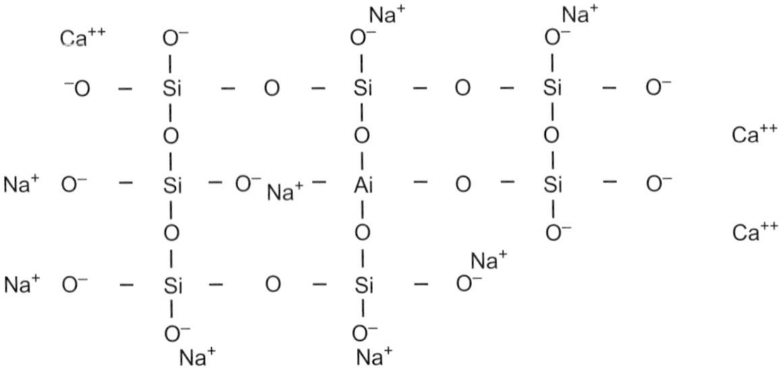

FIGURE 6.5 Idealized, planar illustration of clay structure and exchangeable cation association with the clay.

silicon atom's valence is always satisfied with strongly bonded oxygen. As a consequence of the strong silicon–oxygen bond, the clay mineral bonds are broken, leaving oxygen exposed with its negative charge. This negative charge is satisfied by association with positively charged cations, principally sodium and calcium as illustrated in Figure 6.5.

Cation exchange occurs when di- or trivalent ions enter in a stream of brine and displace the monovalent cations that are loosely associated with the clay mineral. The ion with the greatest charge attaches more strongly and cannot be displaced easily by monovalent ions. This can be demonstrated in the laboratory by treating the shale with strong hydrochloric acid to give rise to a "hydrogen"-based clay and then with a solution containing salts of the di- or trivalent cations.

Electrical properties develop because the associated cations are loosely held and therefore are mobile and can be displaced by a direct current electrical potential. Swelling occurs when freshwater is introduced and the H_3O^+ ion can enter the lattice structure of the clay mineral. The H_3O^+ ion is large and can enter into the lattice of some smectites (e.g., montmorillonite) causing them to swell into a gel-like mass, which may be 10 to 40 times the volume of the original clay. Dispersion of the clays occurs when the H_3O^+ ions loosen the clay particles, especially those lining the pore walls of the rock.

The sand SiO_2 molecule can react with hot water and water containing salts to form silanol groups, which are Brønsted acids (weak acids capable of freeing a proton):

$$
\begin{array}{l}
SiO_2 + 2H_2O \rightarrow Si(OH)_2 + 2OH^- \\
\text{Polymeric form is: } H - O - \underset{\underset{O^-}{|}}{\overset{\overset{O^-}{|}}{Si}} - O - \underset{\underset{O^-}{|}}{\overset{\overset{O^-}{|}}{Si}} - OH \\
= SiOH \rightarrow SiO^- + H^+
\end{array}
\tag{6.10}
$$

Because of their acid surfaces, sandstones react with and adsorb basic compounds readily, whereas acidic compounds are repelled. The major polar organic constituents of crude oils are weak acids. These do not adsorb readily on the SiO_2 surfaces and, therefore, sandstones generally exhibit neutral to water-wet characteristics, which have been observed by many investigators. Block and Simms furnished some direct experimental evidence of this [27]: they showed that octa-decyamine, an organic base, is strongly adsorbed on the surface of glass, whereas stearic acid is hardly adsorbed at all.

Silica and clay minerals mixed with the sand have negatively charged surfaces and, consequently, behave like weak acids in contact with water having a pH less than 7. Although these surfaces will form weak acid–base chemical bonds with the basic organic compounds present in crude oils, they are unaffected by the acidic compounds. The resins and asphaltene fractions of crude oils contain polar, polynuclear organic compounds that can be acidic or basic. The basic compounds can interact with the acidic silica and negatively charged clay surfaces, rendering the surface oil-wet to a degree depending on the amount and types of basic organic compounds available [8, 28–33].

The surfaces of carbonate rocks, on the other hand, are basic in character and, consequently, they react readily with the acid compounds in crude oils and exhibit neutral to oil-wet characteristics [30]. McCaffery and Mungan showed that stearic acid is strongly attached to calcite [34]. Lowe et al. also showed that acid compounds in crude oils become attached to the basic surfaces of carbonates, forming chemisorbed films [35].

The basic characteristics of carbonates may be due to Arrhenius–Oswalt calcium hydroxide-type bases or due to Lewis bases because of the electron pairs available in the exposed oxygens of the $-CO_3$ carbonate groups. If the characteristics are due to calcium hydroxide groups, the reaction is:

$$-\text{CaOH} + \text{HO}\overset{\overset{\text{O}}{|}}{\text{C}}\text{-R} \rightarrow -\text{Ca}-\text{O}\overset{\overset{\text{O}}{|}}{\text{C}}-\text{R} + \text{H}_2\text{O} \tag{6.11}$$

If the basic characteristics of the carbonates are due to Lewis-type bases, then the reaction is probably:

$$\text{CaCO}_2 - \text{O}^+ + \text{A}^- \rightarrow \text{CaCO}_2-\text{O:A} \tag{6.12}$$

Inasmuch as the carbonate surfaces are positively charged and consequently behave like weak bases, they are strongly affected by acidic components in crude oils, which are carboxylic acids, phenolic compounds, and ring structures containing sulfur and oxygen [35–41]. Apparently, acidic compounds are more prevalent in crude oils than basic compounds, which may account for the fact that carbonate rocks exhibit a range of wettability from neutral to strongly oil-wet.

Due to the acid–base interactions between rock surfaces and crude oils, the chemistry and pH of the brine associated with the crude oil is very important. If the pH is greater than 7, the dissociation of hydrogen ions is repressed and the surface will adsorb acidic organic compounds. If multivalent metallic cations such as Ca^{2+}, Ba^{2+}, Cu^{2+}, Fe^{3+}, and Al^{3+} are present in the brine, or added to it, these ions will be adsorbed on the negatively charged silica surface. The multivalent cations then provide positively charged sites that permit the adsorption of acidic compounds on the silica rock [42, 43]. Carbonate surfaces are positively charged at pH ranges less than 7 to 8, but become negatively charged at pH ranges greater than 8. Therefore, carbonate surfaces will adsorb positively charged (basic) organic compounds if the pH of the brine is greater than 8.

Example

A solution of calcium chloride is used to displace the monovalent cations from a 600 g sample of sandstone containing a small amount of clay. Using an atomic adsorption analytic unit to analyze the effluent from the core, 284 mg of sodium and 162 mg of potassium are found in the effluent liquid. Calculate the cation exchange capacity (CEC) of the rock.

Solution

$$\text{meq Na} = \frac{\text{mg} \times \text{valence}}{\text{atomic weight}} = \frac{284 \times 1.0}{23} = 12.3$$

$$\text{meq K} = \frac{162 \times 1.0}{39} = 4.2$$

$$\text{CEC} = \frac{12.3 + 4.2}{0.6} = 27.5 \frac{\text{meq}}{\text{kg}}$$

EVALUATION OF WETTABILITY

Evaluation of relative water/oil wetting of porous rocks is a very important aspect of petroleum reservoir characterization. Wettability has a decisive influence on oil production rates, the water/oil production ratio after water breakthrough, the oil production rates of enhanced oil production technologies, and the residual oil saturation of a reservoir at abandonment. A large amount of research has therefore been conducted on wettability, beginning in the 1930s. Several methods for evaluating wettability have been developed, based on the observable characteristic interactions of water, oil, and rocks. The direct measurement of wettability can be made by careful analysis of contact angles. In addition, several indirect methods provide indexes of the relative wetting properties: the Amott method, which is based on the amounts of fluids imbibed by a rock sample under various conditions; the USBM method, which is based on measurements of the areas under capillary pressure curves obtained using a centrifuge; the combined Amott–USBM method; and the spontaneous imbibition method, which is based on the rates of imbibition.

Amott Wettability Index

The Amott test for wettability is based on spontaneous imbibition and forced displacement of oil and water from cores [44]. This test measures the average wettability of the core, using a procedure that involves five stages:

1. The test begins at the residual oil saturation; therefore, the fluids are reduced to S_{or} by forced displacement of the oil.
2. The core is immersed in oil for 20 hours, and the amount of water displaced by spontaneous imbibition of oil, if any, is recorded as V_{wsp}.
3. The water is displaced to the residual water saturation (S_{iw}) with oil, and the total amount of water displaced (by imbibition of oil and by forced displacement) is recorded as V_{wt}.
4. The core is immersed in brine for 20 hours, and the volume of oil displaced, if any, by spontaneous imbibition of water is recorded as V_{osp}.
5. The oil remaining in the core is displaced by water to S_{or} and the total amount of oil displaced (by imbibition of water and by forced displacement) is recorded as V_{ot}.

The forced displacements of oil to S_{or}, and water to S_{iw} may be conducted using a centrifuge or by mounting the core in fluid-flow equipment and pumping the displacing fluids into the core.

The Amott wettability index is expressed as a relative wettability index defined as the displacement-by-oil ratio ($V_{osp}/V_{ot} = \delta_w$) minus the displacement-by-water ratio ($V_{wsp}/V_{wt} = \delta_o$):

$$I_w = \frac{V_{osp}}{V_{ot}} - \frac{V_{wsp}}{V_{wt}} = \delta_w - \delta_o \tag{6.13}$$

Preferentially water-wet cores are characterized by a positive displacement-by-water ratio, δ_o, and a value of zero for the displacement-by-oil ratio, δ_w. A value approaching 1.0 for the displacement-by-water ratio, δ_o, indicates a strongly water wet-sample, whereas a weakly water-wet sample is characterized by a value approaching zero. Neutral (or 50%/50%) wettability is characterized by a value of zero for both ratios. Cores that are oil-wet show a positive value for the displacement-by-oil ratio, δ_w, and zero for the displacement-by-water ratio, δ_o. A strongly oil-wet sample is characterized by a value approaching 1 for the displacement-by-oil ratio. Thus, the Amott wettability index varies from +1 for infinitely water-wet to −1 for infinitely oil-wet rocks, with zero representing neutral wettability [44, 45].

The 20-hour arbitrary time limit for the two periods of imbibition were probably chosen to allow completion of the test in a reasonable length of

TABLE 6.1 Results of the Amott Wettability Test on Three Cores: (1) Strongly Water-Wet, (2) Neutral Wet, and (3) Strongly Oil-Wet

No.	Displacement by Oil		Displacement by Water		I_w
	Spontaneous (mL)	Forced (mL)	Spontaneous (mL)	Forced (mL)	
1	0.00	1.24	0.79	0.85	+0.48
2	0.00	1.64	0.00	0.96	0.00
3	0.43	0.51	0.00	0.56	−0.46

time. Completion of imbibition, however, can sometimes take several weeks, and when the system is near neutral wettability, spontaneous imbibition may be very slow [46, 47]. If the imbibition is not allowed to go to completion, the values of δ_o and δ_w will be underestimated, leading to erroneous conclusions regarding the wettability of the rock sample. Rather than setting a 20-hour limit on the spontaneous imbibition periods, the amount of fluid displaced should be measured periodically and examined graphically until a stable equilibrium value is attained.

Results of tests on three cores presented by Amott are listed in Table 6.1 [44]:

1. A strongly water-wet fired Berea sandstone outcrop core
2. A sandpack in which the sand grains were bonded with epoxy resin and exhibited neutral wettability
3. A silane-treated Berea sandstone core that was strongly oil-wet

Amott showed that the method will yield a semi-quantitative measurement of wettability by treating unconsolidated sand samples with increasing percentages of a silane solution and measuring the resulting wettability. The results obtained by Amott are presented in Figure 6.6, showing a linear increase of preferential oil wettability with respect to the percentage of silicone solution used [48].

USBM Wettability Index

Donaldson et al. developed a method for determining a wettability index from the hysteresis loop of capillary pressure curves [49, 50]. The test is known as the USBM method. The capillary pressure curves are obtained by alternately displacing water and oil from small cores using a centrifuge. The areas under the capillary pressure curves represent the thermodynamic work required for the respective fluid displacements (Figure 6.7). Displacement of a non-wetting phase by a wetting phase requires less energy than

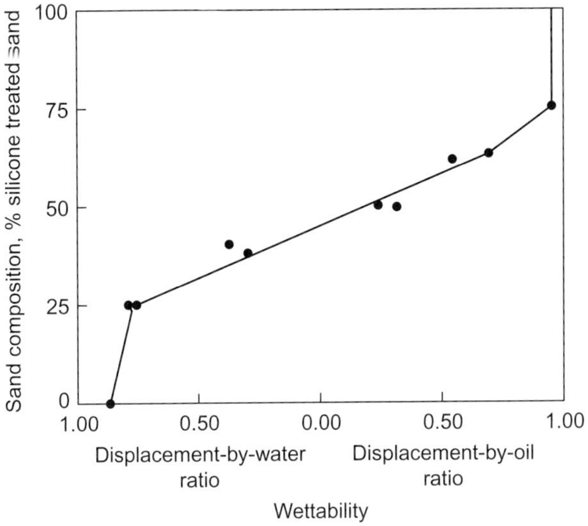

FIGURE 6.6 Wettability tests using samples of unconsolidated sand treated with silicone solutions.

displacement of a wetting phase by a non-wetting phase. Therefore, the ratios of the areas under the capillary pressure curves (between S_{iw} and S_{wor}) are a direct indicator of the degree of wettability. The logarithm of the area ratio of oil displacing water, A_1 (from S_{or} to S_{iw}), to water displacing oil, A_2 (from S_{iw} to S_{or}), was used as a convenient scale for the wettability index (I_u):

$$I_u = \log\left(\frac{A_1}{A_2}\right) \tag{6.14}$$

where

1. Increasing positive values to $+\infty$ indicate increasing preferential water wetting to infinite water wettability;
2. A value of zero represents equal wetting of rock by both fluids (neutral wettability); and
3. Increasing negative values to $-\infty$ indicate increasing preferential oil wetting to infinite oil wettability.

The USBM method does not depend on spontaneous imbibition and, therefore, is sensitive to wettability throughout the range from complete water-wetting ($+\infty$) to complete oil-wetting ($-\infty$). For example, if a water–oil–rock system being tested repeatedly becomes progressively more water-wet, A_1 will become larger while A_2 will decrease. Eventually, A_2 will vanish as the hysteresis loop rises above the line representing $P_c = 0$. In this case,

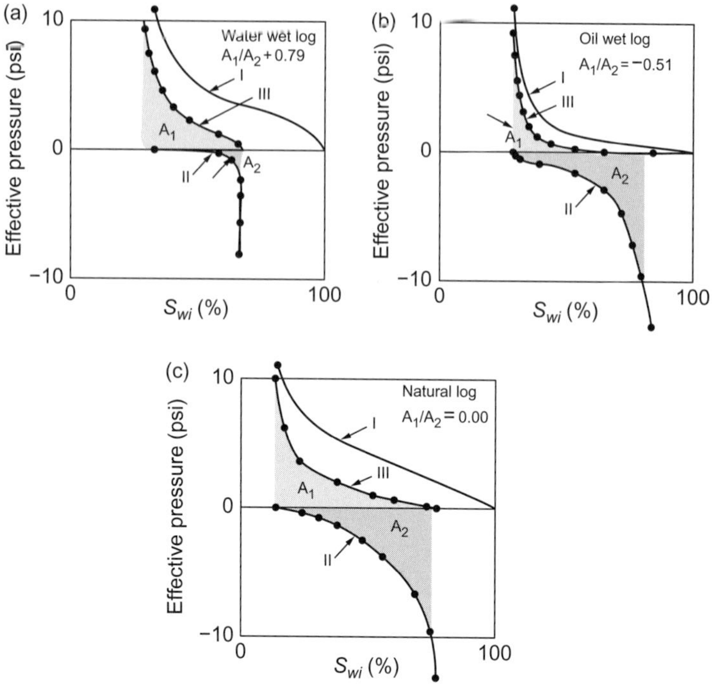

FIGURE 6.7 Method for determination of the USBM wettability index. The area under individual capillary pressure curves represents the thermodynamic work required for the fluid displacement.

A_2 is zero and the wettability index (defined in Eq. (6.14)) is infinite, meaning 100% wetting of the surface by water. Infinite oil wettability also is possible, in which case $A_1 = 0$ and the hysteresis loop is below the line where $P_c = 0$.

Kwan developed a centrifuge core holder for unconsolidated sands and used it to examine capillary pressure and wettability of viscous bitumen [51]. The tests were conducted with a heated (40°C) centrifuge to maintain mobility of the bitumen.

Combined Amott–USBM Wettability Test

A procedure has been developed for combining the Amott and USBM methods that yields both the USBM wettability index and the Amott ratio. According to several authors, the resolution of the USBM index is improved by being able to account for saturation changes that occur at zero capillary pressure [46, 47]. Figure 6.8 illustrates this combined method. At each point

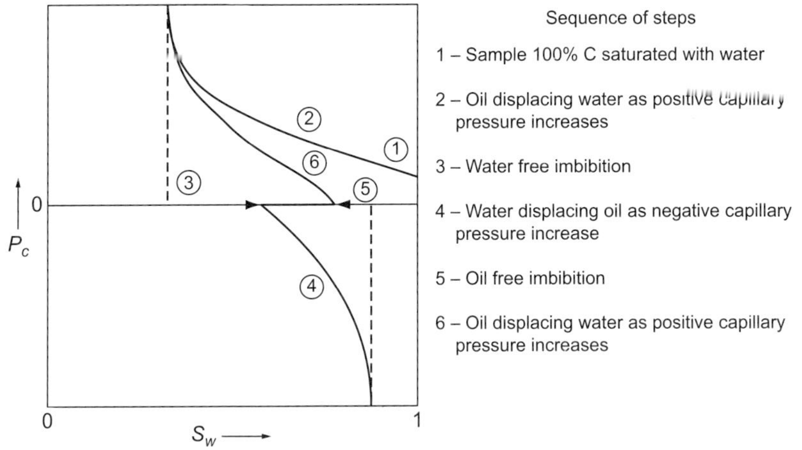

FIGURE 6.8 Illustration of the combined USBM–Amott method for determination of the wettability index.

where the capillary pressure is equal to zero, the sample is immersed in the displacing fluid for 20 hours and the amount of fluid imbibed is recorded and used to determine the Amott ratios. For the combined test, the capillary pressure data are plotted versus the average saturation (not the saturations at the inlet face of the core). Thus, the procedure has six steps:

1. Saturation of the core with water
2. Initial displacement of water to S_{iw} (oil drive)
3. Spontaneous imbibition of brine
4. Displacement of oil by brine (brine drive)
5. Spontaneous imbibition of oil
6. Final displacement of water by oil (oil drive)

The Amott index is calculated from the spontaneous imbibition and total water and oil displacements, whereas the USBM index is calculated from the areas under the curves.

Data presented by Sharma and Wunderlich are shown in Table 6.2 [48]. Spontaneous imbibition of oil does not occur in strongly water-wet samples that are 100% saturated by water (point 1, Figure 6.8). Therefore, the displacement of water-by-oil ratio for the Amott test is zero for the three water-wet cores. A USBM wettability index greater than 1.0 indicates very strong water wettability. The mixed wettability systems, however, imbibe oil at $S_w = 1.0$ and water at S_{iw}; the USBM index shows that these cores are near the point of "neutral" wettability (oil-wet cores have negative USBM wettability indices).

TABLE 6.2 Wettability Tests for Water-Wet and Mixed Wettability Cores (Close to "Neutral" Wettability) Using the Combined USBM/Amott Wettability Test and Contact Angle Measurements

No.	Displacement of Water (Amott)	Displacement of Oil (Amott)	USBM Index	Θ_A	Θ_R
Water-wet Samples					
1	0	0.92	1.6	—	—
2	0	0.80	1.2	26°	6°
3	0		0.89	—	1.4°
Mixed Wettability Samples					
1	0.24	0.015	0.3	—	—
2	0.24	0.073	0.3	167°	145°
3	0.24	0.065	0.3	—	—

Source: After Sharma and Wunderlich [48]—Tables III and IV.

SPONTANEOUS IMBIBITION WETTABILITY TEST

Observations of spontaneous imbibition led to the conclusion that the amount and rate of imbibition are related to the overall wettability of the porous system. The Amott wettability test is based in part on fluid displacements caused by imbibition. However, it was recognized that the rate of imbibition was influenced by several parameters: fluid viscosities, shape and boundary conditions of the core, porosity, permeability, and interfacial tension. Zhang et al. applied the scaling equation that was used by Mattox and Kyte for correlation of oil recovery by imbibition into fractured reservoirs to analyses of the rates of imbibition of cores at controlled states of wettability from strongly water wet to mixed wettability [25, 52]. The scaling equation was generalized by using the geometric average of the fluid viscosities and a characteristic core length, L_c, that compensates for size, shape, and boundary condition. Ma et al. correlated oil recovery by spontaneous imbibition to dimensionless time defined as [53]:

$$t_d = t\left(\frac{k}{\phi}\right)^{1/2} \times \frac{\sigma}{\sqrt{\mu_w \mu_o}} \times \frac{1}{L_c^2} \tag{6.15}$$

As the system becomes more oil-wet, the rate of imbibition decreases. Therefore, a graph of recovery versus dimensionless time indicates differences of wettability from strongly water-wet to mixed or neutral wettability.

Curves of oil recovery versus dimensionless time can be fit to the Aronofsky equation [54]:

$$\frac{R}{R_\infty} = 1 - e^{-0.05t_d} \qquad (6.16)$$

Equation (6.16) may be used to obtain an average curve of several repeated tests to examine the effect of the initial saturation on imbibition.

Ma et al. proposed a wettability index based on the pseudo work of imbibition, W_R [53]. A dimensionless curve of pseudo imbibition capillary pressure (vs. water saturation) that indicates the effect of wettability on the relative rate of imbibition was defined:

$$P_{c,p} = a\sqrt{t_d} \qquad (6.17a)$$

The pseudo work of imbibition, W, is the area under the $P_{c,p}$ versus S_w curve. The relative pseudo work of imbibition, W_R, was defined as the ratio of the pseudo work of a sample to the pseudo work of a very strongly water-wet system. The constant, a, in Eq. (6.17a) was set equal to 1.0; thus the pseudo work is:

$$W_R = \int_{S_{wi}}^{0.1 - S_{or,im}} P_{c,p} dS_w \qquad (6.17b)$$

Zhou and Blunt obtained a correlation between W_R and the Amott wettability index [55].

FLUID DISPLACEMENT ENERGY

If two immiscible phases (water and oil) are initially distributed equally throughout a column of porous material, they will adjust to capillary equilibrium and will coexist throughout the column. If an elemental volume, ΔV, of one of the phases is raised from height h to $h + dh$, the isothermal reversible work (the total free energy change), δF, is zero because capillary equilibrium is assumed to exist within the system. The total free energy change is a composite of two parts:

1. The free energy change accompanying the transfer of the fluid (say, water) from h to $h + dh$
2. The free energy change resulting from the change in pressure, which is experienced by the element of water when it is transferred from h to $h + dh$:

$$\frac{\delta F}{\delta h} = \rho_w g_c \Delta V \qquad (6.18)$$

$$\frac{\delta F}{\delta p_w} = \Delta V \qquad (6.19)$$

The expression for the total change in free energy, resulting from the change in the height of the element of water, is obtained by taking the total derivative with respect to the height and pressure and then substituting Eqs. (6.18) and (6.19) into it:

$$\delta F = \left(\frac{\delta F}{\delta h}\right)\mathrm{d}h + \left(\frac{\delta F}{\delta p_{\mathrm{w}}}\right)\mathrm{d}p_{\mathrm{w}} = 0 \tag{6.20}$$

$$\rho_{\mathrm{w}} g_{\mathrm{c}} \Delta V \mathrm{d}h + \Delta V \mathrm{d}p_{\mathrm{w}} = 0 \tag{6.21}$$

$$\rho_{\mathrm{w}} g_{\mathrm{c}} \mathrm{d}h + \mathrm{d}p_{\mathrm{w}} = 0 \tag{6.22}$$

Since the same derivation can be made for any phase:

$$p_{\mathrm{c}} = p_{\mathrm{o}} - p_{\mathrm{w}} = (\rho_{\mathrm{w}} - \rho_{\mathrm{o}}) g_{\mathrm{c}} h \tag{6.23}$$

$$\mathrm{d}P_{\mathrm{c}} = (\rho_{\mathrm{w}} - \rho_{\mathrm{o}}) g_{\mathrm{c}} \mathrm{d}hs \tag{6.24}$$

When the capillary pressure expressed by Eq. (6.24) is zero, one of two conditions exists: either the two phases are completely miscible or a single phase exists. Because two immiscible phases are considered in Eq. (6.24), the only condition in which the capillary pressure may be zero is when there exists only one phase at some point within the column. If the column is sufficiently high and enough time is allowed for equilibrium to be attained between the two fluids, no oil will be found below a certain level in the column due to gravity segregation. This is the free liquid surface ($h = 0$ at this point as a boundary condition for Eq. (6.19)).

Returning to Eq. (6.24) and noting that water is essentially incompressible leads to:

$$\frac{\delta F}{\delta V} = V, \text{ or } F_2 - F_1 = V(p_2 - p_1) = VP_{\mathrm{c}} \tag{6.25}$$

Therefore, for the transfer of a unit volume of water, the capillary pressure, P_{c}, represents the change in the isothermal, reversible work accompanying the process, or:

$$P_{\mathrm{c}} = \frac{\delta F}{\delta V} \tag{6.26}$$

If the element of volume of the porous system containing water and oil is such that it contains a unit pore volume of water, the fractional water saturation within the element multiplied by the pore volume is numerically equal to the volume of water. Thus, $\mathrm{d}V = -V_{\mathrm{p}} \times \mathrm{d}S_{\mathrm{w}}$ where V represents the water transferred out of the porous medium. Substituting this expression into Eq. (6.26) yields:

$$\mathrm{d}F = -P_{\mathrm{c}} \times V_{\mathrm{p}} \times \mathrm{d}S_{\mathrm{w}} \tag{6.27}$$

In Eq. (6.27), dF, expressed in Nm or J, represents the free energy change in water per unit of pore space accompanying a change in water saturation, dS_w. The integral of Eq. (6.27) is the area under the capillary pressure curve (Figure 6.7). The capillary pressure curves can be fit to hyperbolic equations by a least-squares fit of the capillary pressure versus saturation data:

$$P_c = \frac{(1 + A \times S_w)}{(B + C \times S_w)} \tag{6.28}$$

$$\text{Area} = \left(\frac{A + B \times S_w}{1 + C \times S_w}\right) \times dS_w = \frac{BS_w}{C} + \frac{AB - C}{C^2}\log(1 + CS_w) \tag{6.29}$$

where constants A, B, and C are obtained from the least-squares fit of the data.

The areas under the capillary pressure curves (Figure 6.7) can be readily calculated by integration of Eq. (6.30) below to yield the thermodynamic work required for fluid displacements:

$$\Delta w = \phi V_p \int_{S_{w1}}^{S_{w2}} P_c dS_w \tag{6.30}$$

As an example, displacements from a Cottage Grove sandstone core (Figure 5.16) were:

Curve 1 (oil displacing water from $S_w = 1.0$ to S_{iw}) = 0.165 J/mL = 24.82 BTU/bbl
Curve 3 (water displacing oil from S_{iw} to S_{wor}—waterflood) = 0.0089 J/mL = 1.34 BTU/bbl
Curve 5 (oil displacing water from S_{wor} to S_{iw}) = 0.771 J/mL = 116.25 BTU/bbl

Less energy is required for displacement of water from 100% saturation (Curve 1) than from S_{wor} (Curve 5), because a considerable amount of water is displaced at low pressure after the threshold pressure is exceeded. A very small amount of energy is required for displacement of oil because the Cottage Grove sandstone exhibits a strong water-wetting tendency; consequently, some oil is displaced by imbibition (at zero capillary pressure) upon initial contact of the oil-saturated core with water. Stating this in another way: if the water–oil–rock system is water-wet, A_1 is a large positive value and, therefore, considerable work must be done on the system to displace the water. On the other hand, the area under the water-displacing-oil curve is a very small positive value; hence, water will imbibe into the water-wet system spontaneously with simultaneous displacement of oil.

When a core is strongly water-wet (USBM $I_w > 0.7$), the core will imbibe water until the water saturation is essentially S_{wor} and the area under the curve is almost zero; hence, the work required for oil displacement is almost zero for a strongly water-wet system. The amount and rate of imbibition depend on a number of simultaneously acting properties of the water–oil–rock system: the rock and fluid chemical properties expressed as wettability, interfacial tension, saturation history of the system, initial saturation, fluid viscosities, pore geometry, and pore-size distribution.

As the system becomes less water-wet, the work required for displacement of oil increases and, consequently, the amount and rate of imbibition decreases. Thus, a smaller amount of water will imbibe at a lower rate as the system becomes less water-wet. At neutral wettability, water will not imbibe when the water saturation is at S_{iw} and oil will not imbibe when the water saturation is at S_{wor}. Thus, a positive initial displacement pressure is required for both fluids (for water-displacing-oil from S_{iw}, or oil-displacing-water from S_{wor}), which is the basis for determination of neutral wettability. If, however, a small amount of water will imbibe at S_{iw} and an almost equal amount of oil will imbibe at S_{wor}, the system is at a condition of fractional or mixed wettability. The distinction between these can be made only by microscopic observations of thin sections.

If the system is oil-wet, these conditions for the water-wet case are reversed: A_1 is small and A_2 is large. Oil will spontaneously imbibe into the system, displacing water. Water must be forced into the system and, therefore, A_2 is a large value.

The work required for displacement of oil by water is the theoretical work required for a waterflood and is one of the economic factors of oil production. For example, if the reserve estimates, from field and laboratory analyses of a small field, indicate that 1.6×10^5 m^3 (one million barrels of oil) will be recovered from a waterflood and the work required for displacement of the oil (from the current field saturation to S_{or}) is 10 kJ/m^3 (1.5 BTU/bbl), then 1.6 mJ (1.5 million BTU) of energy, in addition to friction losses in pumps and tubing, will be required for completion of the waterflood.

WATER–OIL–ROCK INTERFACIAL ACTIVITY

Surfactant-type compounds in crude oils, which are partially soluble in water, have been found to pass rapidly through the thin water film on water-wet surfaces and adsorb strongly on the rock [11, 35]. Asphaltenes (high-molecular-weight polynuclear aromatic compounds containing nitrogen, sulfur, and oxygen (NSO) in ring structures) penetrate the aqueous film to produce oil-wet surfaces in the rock. Thus, rocks containing asphaltic oils will exhibit oil-wetting tendencies.

The silicate–water interface is acidic. Acidic compounds in crude oils (those containing carboxylic and phenol groups) do not adsorb on silicate surfaces, but basic constituents (nitrogen-containing compounds such as amines and amides) adsorb readily, rendering the surface oil-wet. In contrast, the carbonate–water surface is basic and the acid compounds adsorb, whereas the basic compounds are repelled [28, 56–59]. Since crude oils generally contain polar compounds that are acidic, the wetting tendencies of brine–crude oil–rock systems is for silicate rocks to be neutral to water-wet and for carbonates to be neutral to oil-wet. Akhlaq treated quartz and kaolinite samples with crude oils and then characterized the adsorbed compounds with infrared spectroscopy [60]. Basic nitrogen compounds and organic esters were found adsorbed to quartz sand, whereas sulfonic acids together with carbonyl groups and phenols were adsorbed on kaolinite surfaces.

Crude oils contain surface-active compounds that can modify the wettability of the reservoir by changing the chemical species at the fluid and rock interfaces, depending on pH, salinity, and the nature of the surface-active compounds. Depending on the immediate environmental conditions, different types of surface-active compounds present in the crude oil will move to the fluid and rock interfaces and govern the wettability of the reservoir. Salinity and pH apparently control the aqueous-mineral interfacial cation binding and acid–base reactions of compounds. Binding of surface-active compounds present in the crude oils and precipitated asphaltene-type molecules occur at oil–rock interfaces [61].

All petroleum reservoirs were originally believed to be water-wet because clean rocks of all types exhibit preferential water-wetting tendencies. In addition, sedimentary rocks containing oil were originally saturated with water that was displaced when oil migrated into the geologic trap. Polar organic compounds in petroleum, however, are expelled from the bulk phase and react chemically with clay and other minerals in the rock to form neutral, mixed, or preferentially oil-wet systems. The Wilcox sandstone of the Oklahoma City field, the Tensleep sandstone in Wyoming, and the Bradford sands in Pennsylvania are well-known oil-wet reservoirs [3, 35, 62, 63]. Carbonate reservoirs have been found to range in wettability from neutral to strongly oil-wet [20, 64, 65].

Treiber et al. used contact angle measurements to examine the wettability of 30 silicate and 25 carbonate rocks (Table 6.3) [66]. Their contact angle criteria were as follows: water-wet = 0–75°, intermediate-wet = 75–105°,

TABLE 6.3 Relative Wetting Tendencies of Sandstones and Carbonates

Wettability	Treiber et al. [66]		Chilingarian and Yen [64]
	Silicates (%)	Carbonates (%)	Carbonates (%)
Water-wet	43	8	8
Intermediate-wet	7	4	12
Oil-wet	50	88	80

and oil-wet = 105–180°. A few of the silicate rocks were intermediate-wet, but the rest were almost equally divided between water-wet and oil-wet. On the other hand, the carbonate rocks were largely oil-wet. Chilingarian and Yen used contact angle measurements, with different criteria for the divisions of wettability, to measure the wettability of carbonate rocks from various parts of the world [64]: water-wet = 0–80°, intermediate-wet = 80–100°, and oil-wet = 100–180°. Using these criteria, they measured the wettability of 161 cores composed of limestone, dolomitic limestone, and calcitic dolomite, and found that 80% of the cores were oil-wet (Table 6.3).

Overall wettability and point-contact wettability are conditions imposed on the boundaries of the water–oil and fluid–rock interfaces by polar (NSO) compounds in the crude oil, depending on the chemical properties of the water and rock surface minerals. An equilibrium accumulation of surfactants at the interfaces can be destabilized by changes in pH, water-soluble surfactant, cationic concentration, and temperature. Once NSO compounds accumulate on mineral surfaces, strong adhesive properties immobilize them and the contact area now becomes oil-wet [14, 67]. If the condition is distributed in a fragmented (spotted) manner in the rock, a change in wettability from water-wet to fractional wetting occurs. If the condition (precipitation of NSO compounds) spreads through the rock, it establishes continuous oil-wet zones in the pores of the rock; the wettability change from water-wet will then tend toward an overall mixed wettability or, in an extreme case, the fluid–rock system will change from water-wet to oil-wet.

EFFECT OF WETTABILITY ON OIL RECOVERY

Primary oil recovery is affected by the wettability of the system because a water-wet system will exhibit greater primary oil recovery, but the relationship between primary recovery and wettability has not been developed. Studies of the effects of wettability on oil recovery are confined to waterflooding and analyses of the behavior of relative permeability curves. The changes in waterflood behavior as the system wettability is altered are clearly shown in Figure 6.9. Donaldson et al. treated long cores with various amounts of organochlorosilane to progressively change the wettability of outcrop cores from water-wet (USBM $I_u = 0.649$) to strongly oil-wet ($I_u = -1.333$) [49]. After determining the wettability, using a small piece of the core, they conducted waterfloods, using a crude oil. The results show that as the system becomes more oil-wet, less oil is recovered at any given amount of injected water. Similar results have also been reported by Emery et al. and Kyte et al. [8, 9].

Relative permeability curves are used for quantitative evaluation of waterflood performance, and the effects of wettability can be observed in changes that occur in the relative permeability curves (Figure 6.10). In mixed

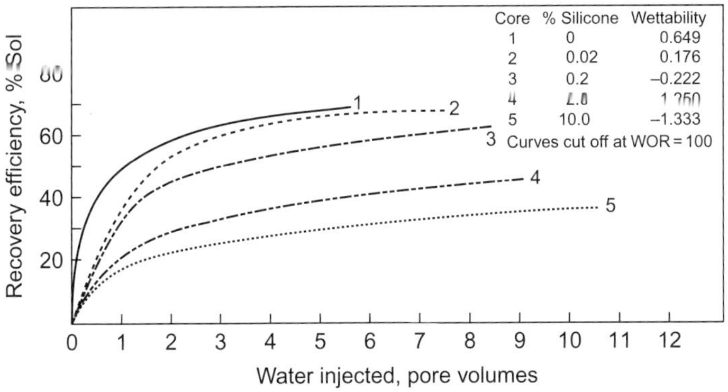

FIGURE 6.9 Recovery efficiency as a function of water injected and wettability.

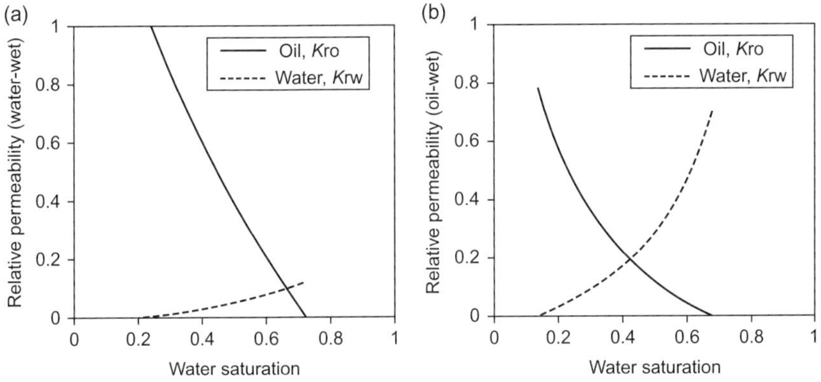

FIGURE 6.10 Typical oil–water relative permeability curves for (a) water-wet and (b) oil-wet systems.

wettability cases, however, the relative permeability of each phase is a function of the saturation distribution of the two phases in the rocks.

There are a number of other influences that disturb the normal trend of relative permeability curves, and they must not be confused with the effects due to wettability alone [72]. The relative flow of fluids is a function of pore-size distribution; therefore, any change in this distribution due to blocking will change the relative permeability curves. The overburden pressure applied to cores in the laboratory changes pore-size and pore-throat size distributions, reducing the size of the larger pores, which changes the porosity as well. Furthermore, smaller pore sizes may increase the irreducible water saturation in water-wet rocks and the residual oil saturation; thus, the mobile oil saturation is decreased. An increase of temperature causes the wettability to change to a more water-wet system. Thus, core floods for the determination of relative permeabilities should be conducted at simulated reservoir conditions of overburden pressure, pore pressure, and temperature for the resulting relative permeability curves to be representative of conditions in the reservoir [68–72].

The saturation history of the water–oil–rock system (or core) has a fundamental influence on the equilibrium wetting condition of the rock. Oil reservoirs are generally assumed to have been filled with water which was displaced into a trap by migrating oil. Thus oil reservoirs tend to be preferentially water-wet, although several major oil fields have been found to be oil-wet and heavy oil deposits (°API $<$ 20) are generally oil-wet. The wettability of the reservoirs probably changed gradually from strongly water-wet to some degree of intermediate wettability and finally to oil-wet as polar compounds in the oil diffused to the interface and adhered to the rock surface. Thus oil field rocks exhibit all degrees of wettability from strongly water-wet to strongly oil-wet [3, 46, 65, 73, 74]. Low-molecular-weight compounds and gas in the crude oil may accelerate the deposition of polar compounds by the deasphalting process (crude oils treated with light hydrocarbons precipitate asphalts and asphaltenes). Rocks also are wet by water and oil in a spotty (fractional and mixed wettability) fashion. Thus, certain regions of the surface may be wet by oil and the remainder by water; the overall wettability depends on the ratio of the surface area wet by water to that wet by oil [1, 57, 72, 75–77].

Anderson presented a thorough review of the literature on the effects of wettability on relative permeability curves [3, 46, 78–81]. In a water-wet system, water occupies the small pores and coats most of the large pores with a thin film. Inasmuch as most of the flow occurs through the larger pores where the oil is located and water is not present to impede the flow of oil, the oil-effective permeability, relative to water, is very high. On the other hand, the water-effective relative permeability is very low, even when the oil saturation has been reduced to S_{or}, because residual oil in the large pores remains to effectively block the flow of water (Figure 6.10). When a water-wet core is waterflooded from an initial saturation equal to the irreducible saturation (S_{iw}), only oil is produced until a critical average water saturation is attained where water breakthrough begins. Water breakthrough is indicated when water production first begins at the outlet. Prior to water breakthrough, piston-like displacement of oil occurs because for every volume of water injected an equal volume of oil is produced. Just after water breakthrough, the water-to-oil production ratio increases dramatically, reaching a point where oil production almost ceases and a practical residual oil saturation is reached. To attain the true (or ultimate) residual oil saturation, it is required that waterflooding continues until production of oil completely stops. This limit may require hundreds of pore volumes of injected water; therefore, the limiting S_{or} is only investigated for special research applications. For a strongly water-wet system with a moderate oil/water viscosity ratio, the three average saturations—breakthrough saturation, practical S_{or}, and ultimate S_{or}—are almost equal [82]. For intermediate or oil-wet systems, the three saturations can vary greatly.

In an oil-wet system, theoretically, the locations of the two fluids are reversed. Even at low water saturations, the effective permeability to oil is much lower than in water-wet systems (at any given saturation) because water in the larger pores is blocking the flow of oil. This becomes more pronounced as the water saturation increases during a waterflood, and it eventually results in a final residual oil saturation higher than it would be in a water-wet system (Figure 6.10). The effective permeability to water should be high in an oil-wet system because, theoretically, the oil is located in the small pores and is coating the larger pores with a thin film and is not interfering very much with the flow of water. The relative permeabilities are controlled by the distribution of the fluids in the pores of rock. The relative permeability of a fluid at any saturation is a function of its mobility, which in turn is a function of capillary size and wettability. The wetting phase has a lower mobility if it is located in the smaller pores and is adhering to the rock surface. In an oil-wet system, water breakthrough occurs very early in the flood; in fact, it may occur before oil is produced if the water/oil viscosity ratio is very low. After water breakthrough, production of oil continues with an ever increasing water-to-oil producing ratio until a decision is made with respect to the practical S_{or} for the waterflood (Figure 6.11).

The importance of the measurement of wettability is best illustrated by the performance of waterfloods for systems at various states of wettability [7]. Lorenz et al. [49] treated 30-cm-long sandstone cores with increasing concentrations of an organosilane compound to make progressively more oil-wet cores. The cores were then saturated with brine and reduced to the irreducible water saturation by displacement of the brine with oil. A small piece of each

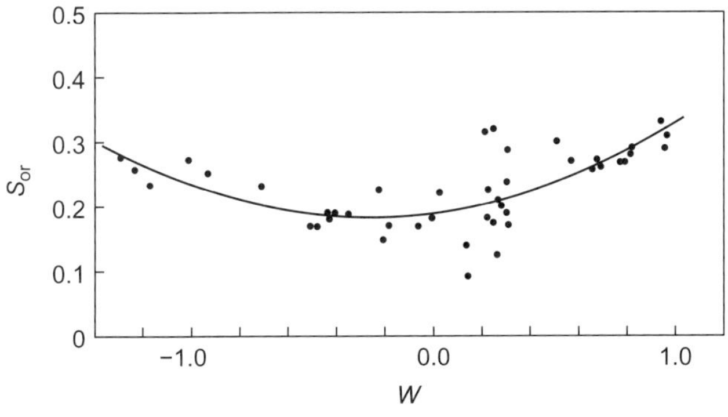

FIGURE 6.11 Ultimate recovery as a function of wettability. Maximum oil recovery apparently occurs in neutral or slightly oil-wet rocks [95].

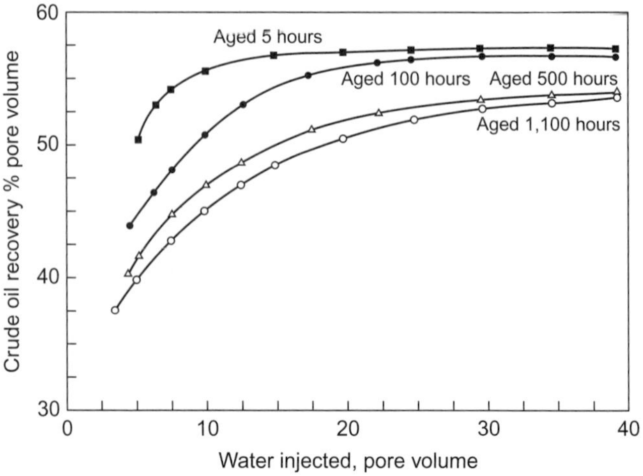

FIGURE 6.12 Effect of aging of water–oil–rock system on oil recovery efficiency.

core was removed and tested for wettability, and waterfloods were conducted using the remaining 25-cm cores (Figure 6.9). A wettability range from 0.649 (strongly water-wet) to −1.333 (strongly oil-wet) was achieved, and the waterfloods show that a strongly water-wet system will have breakthrough of water after most of the production of oil has taken place and that very little production of oil will take place after water breakthrough. As the system becomes more oil-wet, water breakthrough occurs earlier in the flood and production continues for a long period after water breakthrough at a fairly constant water/oil production ratio. Similar work also was presented by Emery et al., using packs of unconsolidated sand [8]. They obtained progressively more oil-wet sandpacks by varying the aging time of the cores, at 71°C and 7 MPa pressure, from 5 to 1,000 hours (Figure 6.12). Their results also show that, for a specific amount of water injected, less oil is recovered after water breakthrough as the system becomes more oil-wet.

Tweheyo et al. observed that water-wet systems exhibit greater oil recovery at water breakthrough with very little production thereafter [83]. Progressively more oil-wet systems, however, have a long period of significant production after water breakthrough. In addition, neutral-wet (or 50%/50% wettability) systems yield the largest amount of ultimate recovery.

Effect of Brine Salinity on Oil Recovery

Increased oil recovery has occurred in some cases when the injection brine salinity was substantially decreased. Tang and Morrow concluded from experimental data that several conditions are necessary [84]:

1. The reservoir should be a mixed wettability system where residual oil remains immobile in large oil-wet pores and the connate water principally occupies the smaller, water-wet pores.
2. The rock contains potentially mobile particles of clay and other minerals adhering to the walls of the pores.
3. The particle-size distribution is less than the pore-size distribution. Hence, when the particles are dislodged they can be transported through the rock by injection brine without damage to porosity and permeability.

In the mixed wettability system, surface-active compounds in the oil will tend to migrate to the oil–rock interface and coat the exposed area on the particles. When the injection brine has the same salinity as the connate water, the exposed oil-wet portions of the particles retain droplets of oil in some pores and filaments of oil fill a large number of pores. When the injection brine salinity is decreased, the equilibrium of the electrical double layer (in the water between the particles) is upset, causing expansion of the double layer and thus release of the particles from the pore walls [15]. When oil is resting on layers of particles, the injection brine displaces the particles with attached oil droplets (and the oil filaments filling the pores). The cumulative mobilization of particles and oil can produce a significant increase in oil recovery. A tenfold decrease of injection-brine salinity increased oil recovery at water breakthrough from 56.0% to 61.9% and ultimate waterflood recovery from 63.6% to 73.2% [84].

ALTERATION OF WETTABILITY

Wettability is perhaps the most important factor that affects the rate of oil recovery and the residual oil saturation, which is the target of enhanced oil recovery technology. Wettability controls the rate and amount of spontaneous imbibition of water and the efficiency of oil displacement by injection water, with or without additives.

The study of the effects of wettability on oil recovery is facilitated by using additives to treat the rock surface so as to produce direct changes through all degrees of wettability from water-wet, neutral, mixed, or fractional to strongly oil-wet. In addition, water- and oil-soluble additives are used to change, or establish, a particular state of wettability.

TREATMENT OF THE ROCK

Several methods have been used to alter wettability:

1. Treatment with organosilanes of general formula $(CH_4)_nSiCl_x$; the silanes chemisorb on the silica surface, producing HCl and exposing the CH_3^- groups that produce the oil wetting characteristics

2. Aging under pressure in crude oil
3. Treatment with naphthenic acids
4. Treatment with asphaltenes
5. Addition of surfactants to the fluids

Treatment of the cores or sand is conducted by first cleaning with solvents, acids, steam, or heating to 250°C to destroy organic materials; however, heating to such a high temperature dehydrates the clays and changes the surface chemistry of the rock. After cleaning, the core is treated with various concentrations of the additives mentioned above and dried once more at 110°C to fix the additive onto the surface of the rock.

Wettability alteration must be conducted under carefully controlled conditions because the final wettability depends on:

1. The mineral composition of the rock;
2. The cleaning procedure used;
3. The type of additive used (silane, asphaltene, etc.);
4. The concentration of the additive in the solvent used to permeate the core; and
5. The procedure used to evaporate the solvent and dry the core.

Completely uniform wettability throughout the core is not attained, but this method has been used successfully to obtain systems at various states of average wettability for examination of the effects of wettability on production [75, 85–87].

Addition of Fluid-Soluble Compounds to Water and Oil

Fractional wettability of unconsolidated sands and beads has been achieved by solute/solvent treatments. Generally, a portion of the cleaned, dry, sand is treated with a solvent containing the additive and then dried. The treated sand is then mixed in various proportions with untreated sand to produce different degrees of fractional wettability [76, 88–90]. Graue et al. found that chalk cores attain fractional wettability when they are aged for approximately 100 hours at 90°C by immersion in crude oil [91, 92]. The same method when used with sandstone cores (including Berea sandstone from Amherst, Ohio) did not produce consistent changes of wettability. They found that various degrees of change in wettability from water-wet toward neutral, or fractional, wettability are attained by immersion in different oils. Wettability was determined using the Amott wettability index and the imbibition-rate method.

Tweheyo et al. changed the wettability of sandstones from strongly water-wet to neutral and oil-wet by adding organic acids (*o*-toluic acid) and amines (dodecylamine and hexadecylamine) to the oil used to saturate the cores [83]. The amines produced the largest changes from water-wet to oil-wet. Waterfloods of the modified systems produced results that have been reported previously:

1. Water-wet samples exhibit rapid and almost complete production by the time water breakthrough occurs.
2. The greatest amount of recovery occurs with neutral-wet systems (8–10% higher).
3. The least amount of oil recovery was obtained from the oil-wet samples, which exhibited early water breakthrough followed by a low rate of oil recovery.

Kowalewski et al. found that they could control the changes from water-wet to neutral using Berea cores, *n*-decane, and oil-soluble hexadecylamine [93]. Waterfloods resulted in an almost linear correlation between the concentration of the amine and an increase of oil recovery (decrease of S_{or}) as the system changed from water-wet ($I_A = 0.7$) to neutral ($I_A = 0.05$). Langmuir isotherms were used to test the amounts of amine adsorbed on crushed rock from various concentrations used: the results ranged from 0.007 to 0.230 mg/g of rock. Thus the change in wettability was directly related to the amount of hexadecylamine that was adsorbed from the oil. Donaldson et al. developed a method for determining the Langmuir and Freunlich isotherms and calculating the thermodynamic heats of adsorption of organic compounds on sandstone cores [94]. The adsorption isotherms showed maximum amounts of adsorbed compounds that varied from 0.200 to 10 mg/g of sandstone. The rates of adsorption at various temperatures were also measured.

Many studies of the feasibility of using surfactants and caustics dissolved in water to enhance the rate and total recovery of oil from sandstone cores have been made [95, 96]. In addition, the U.S. Department of Energy conducted several field tests to evaluate the potential of surfactant/polymer water floods for mobilization of residual oil [97]. Surfactants and caustics lower the interfacial tension and, intuitively, this should result in economically enhanced oil recovery, but the results have generally been disappointing; enhanced recovery (recovery of more oil than the S_{or} of waterfloods) is usually less than 5% regardless of the applied technology (surfactant/water, surfactant/polymer, surfactant/CO_2, foam floods, and surfactant/thermal recovery). The poor results are attributed to adsorption and precipitation caused by divalent cations in the oil field brines. The early depletion of the surfactant from the injected water solution rapidly diminishes the effectiveness of the surfactants.

Standnes and Austad made a careful study of changes of wettability from oil-wet to water-wet in chalk cores, using spontaneous imbibition with anionic and cationic surfactants [98]. The anionic surfactants were ineffective; however, the cationic surfactants changed the wettability from oil-wet to water-wet and produced as much as 70% of the original oil in place compared to a maximum of 10% production using brine alone. The enhanced production and change in wettability caused by cationic surfactants were attributed to ionic reaction of the cations with organic carboxylates adsorbed from the oil.

Water-wet silica cores are produced by successively cleaning the rock with toluene to remove organic compounds, and steam to remove the residual toluene and heavy crude oil components or humic acids (in the case of outcrop sandstone), followed by treatment with hydrochloric acid. This sequence will produce a water-wet core whose wettability depends on the mineralogy of the rock and the composition of the fluids. Excellent reproducibility is therefore possible under carefully controlled conditions.

The surfaces of carbonate rocks may be made more oil-wet by treatment with naphthenic acids, which react with the calcium carbonate to produce a stable oil-wet surface [56]. Another way to control the wettability of clean cores is to add surface-active compounds to the fluids. Owens and Archer used barium dinonyl sulfonate dissolved in oil to achieve an extreme oil-wet condition with a contact angle of 180° [99]. Mungan used hexylamine and *n*-octylamine dissolved in water to change the advancing contact angle on a silica surface from 60° (slightly water-wet) to 120° (slightly oil-wet) [69]. Kowaleswki et al. changed the wettability of sandstone cores from water-wet to neutral by adding hexadecylamine to *n*-decane [93]. The degree of wettability change was controlled by the concentration of the amine dissolved in the oil. Grattoni et al. altered wettability with oil-soluble tetramethyl orthosilicate, which reacts with water in the pores to form a silicate gel [100]. The gel initially produces a water-wet system that changes with respect to time to a neutral or oil-wet system.

The wettability of reservoir cores may be altered by penetration of drilling fluids containing surface-active compounds or possessing a pH which is either acidic or basic. Other aspects that must be controlled for proper evaluation of oil field cores are the packaging at the wellhead, the length of storage prior to use, and laboratory core-cutting and handling procedures. As the core is extracted to the surface, the decrease of pressure results in expansion and loss of low-molecular-weight components. This loss of lighter components can result in precipitation of paraffins and asphaltenes that can alter the wettability toward a more oil-wet condition. Several investigators have succeeded in preserving long-term (years) wettability of oil field cores by placing the core in a glass jar containing oil from the formation as soon as the core is available from the driller; the cores are then transported as soon as possible and kept in storage at about 5°C [101, 102]. Wrapping the cores in foil and coating with plastic (polyethylene, polyvinylidene) or paraffin is adequate for about 6 months. The problem with this method is that the light components slowly diffuse through the coating, leaving the high-molecular-weight compounds to gradually precipitate, which causes a gradual change toward a more oil-wet condition. Any method of storage that allows even a small amount of evaporation will result in alteration of wettability.

AGING THE OIL–BRINE–ROCK SYSTEM

Reservoir rocks that have been cleaned, and outcrop rocks that have not been in contact with oil, generally exhibit a water-wet condition, especially if refined oils are selected for experiments. Cleaned carbonate rocks, however, have neutral to slightly oil-wet tendencies. A small change toward more oil-wet is observed if the core is first saturated with water, and then the water is displaced to the point of irreducible water saturation with the oil. Immersion in the crude oil at an elevated temperature (60–90°C) will change the wettability to an oil-wet system. Stable core wettability is usually obtained after the core is aged in crude oil at an elevated temperature for at least 100 hours (Figure 6.12). Surface-active heterogeneous NSO compounds in the crude oil slowly migrate to the oil–rock interface and are adsorbed strongly on rock mineral surfaces. The oil should be centrifuged to remove suspended sedimentary particles and high-molecular-weight compounds that have been precipitated by changes of temperature, pressure, and storage time after production from the reservoirs. Some oils have compounds that react with atmospheric oxygen to form cross-linked compounds that precipitate from the oil. Micelles holding asphaltenes in suspension can be broken by air oxidation, causing precipitation of the asphaltenes. When an oil is encountered that is sensitive to air oxidation, even after repeated filtration, it must be collected in the field under a blanket of nitrogen and maintained under nitrogen for all transfers between containers when used in the laboratory [103].

Emery et al. investigated the effects of aging cores with water and Singleton crude oil for varying lengths of time [8]. Being saturated with oil, the system behaved like a water-wet system. Most of the production occurred shortly after breakthrough, and the practical residual oil saturation was attained just after one pore volume of water was injected. As the aging time was increased, water breakthrough occurred sooner and there was a considerable amount of subsequent production, with the S_{or} occurring after two or three pore volumes had been injected. These results are similar to those obtained by Donaldson et al. [49], as shown in Figure 6.9, and they show that one must equilibrate a core and its fluids before running waterflood tests to determine the amount of production, relative permeability curves, or wettability.

Cores with fractional, or mixed, wettability have other effects that have been discussed previously. Experiments conducted to evaluate the effect of wettability on residual oil saturation show that the residual oil saturation is less for systems that are at neutral wettability (probably fractional or mixed wettability). Residual oil saturation (at its minimum at neutral wettability) increases as the system becomes more water-wet or oil-wet (Figure 6.11) [44, 104–108].

A strongly water-wet core will produce most of the oil before water breakthrough, which will occur soon after one pore volume of water has been injected. The water/oil ratio will increase rapidly after water breakthrough to an infinite value; thus production will diminish to an insignificant amount.

An oil-wet core will produce water early at a low water/oil ratio which will continue to increase gradually. After about two pore volumes of water have been injected, production will continue for a long time with gradually increasing water/oil ratio.

EFFECTS OF TEMPERATURE AND PRESSURE

The wettability of a water–oil–rock system becomes progressively more water-wet as the temperature of the system is increased. Lorenz et al., working with outcrop cores saturated with brine and crude oil, observed an average USBM wettability index increase of 0.3 for water–oil–outcrop sandstone systems when the temperature was changed from 25°C to 65°C (Table 6.4) [95].

Work conducted by Donaldson and Siddiqui to examine the effect of wettability on the Archie saturation exponent at two temperatures also showed the change to a more water-wet system that occurs when the temperature is increased [109, 110]. Figures 6.13 and 6.14 show an increase of the USBM I_u for water–oil–rock systems when the test temperature is increased from 25°C to 78°C. The observed wettability index change with respect to temperature is strongly influenced by the chemical and physical properties of the rock surface; the Berea sandstone ($k = 325$ mD), registered an increase of I_u of about 0.4 over that of the Elgin sandstone ($k = 1,900$ mD), which exhibited a change

TABLE 6.4 Effect of Temperature on the USBM Wettability Index*

Core Samples	Crude Oil Samples	Average I_u at 25°C		Average I_u at 65°C	
		I_u	No. of Samples	I_u	No. of Samples
Cottage Grove sandstone	Bartlesville, OK	−0.16	3	0.13	3
Cottage Grove sandstone	Muddy J, CO	−0.39	3	−0.03	3
Cottage Grove sandstone	Singleton, NB	−0.32	3	0.21	3
Torpedo sandstone	Squirrel, OK	0.00	6	0.35	18
Torpedo sandstone	Squirrel, OK	0.09	2	0.22	6

Source: After Lorenz et al. [95].
**The tests were conducted using separate cores of the designated outcrop.*

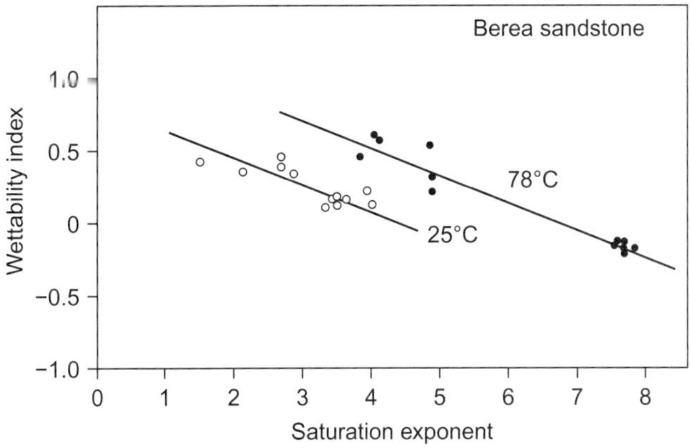

FIGURE 6.13 Change in the Archie saturation exponent as a function of wettability and temperature for Berea outcrop sandstone, Ohio.

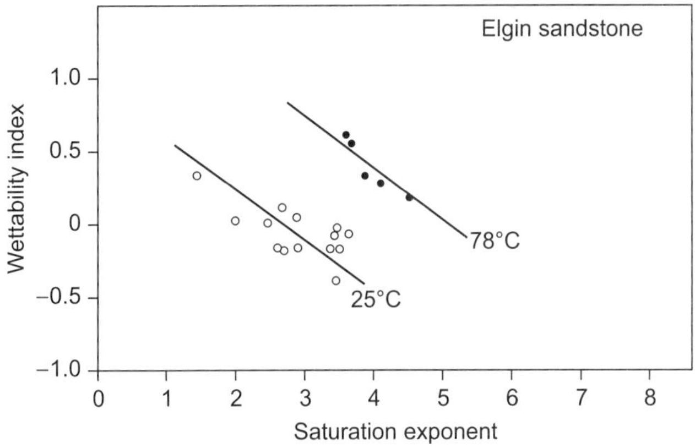

FIGURE 6.14 Change in the Archie saturation exponent as a function of wettability and temperature for Elgin outcrop sandstone, Oklahoma.

of 0.9. The change in wettability to a more water-wet system as the temperature is increased has also been observed, directly and indirectly, by other investigators [102, 111–114].

Donaldson and co-workers have shown conclusively that an increase of temperature produces a more water-wet system [7, 109]. The test for wettability should therefore be conducted at reservoir temperature using reservoir fluids. The presence of trace metals in the actual formation water, if used, renders the cores more water-wet.

Jadhunandan and Morrow examined the effects of variations of aging with temperature and time and established approximately 240 hours of aging to obtain stable wettability conditions [115]. Waterfloods following the aging periods resulted in greater oil recovery at neutral wettability systems.

Zhou and Blunt used variable aging times to obtain cores with different degrees of wettability to develop a correlation between spontaneous imbibition and wettability [55]. The rate and amount of water imbibition decreases as the cores change from strongly water-wet toward neutral or oil-wet conditions. Thus a correlation to the wettability index could be made from the area under the imbibition capillary pressure curve (displacement energy or pseudo work) and the advancing contact angle at the point of 50% oil recovery by imbibition.

Restoration of Original Wettability

After the wettability of an oil field core has been determined and perhaps waterflood tests conducted with reservoir fluids, the core is generally cleaned to determine permeability, porosity, and other parameters. Many methods have been suggested for cleaning cores, but by far the most used method is to place the core in a Soxhlet extractor and extract with toluene; this is frequently followed by extraction with ethanol to remove the toluene. The cores are then dried and used for various tests. If the tests require restoration of the original wettability, the cores are generally saturated with the oil field brine and oil and aged by various procedures prior to use. The principal problem with cores treated in this manner is that not all of the adsorbed high-molecular-weight resins and asphaltenes are removed. Therefore, the restored cores have variable wettabilities that are more oil-wet than the original. Extraction with chloroform and methanol for 3 weeks, followed by aging in crude oil for 30–40 days, produces better results, but the long time required may be a constraint.

McGhee et al. and Donaldson et al. found that reproducible restoration of wettability can be attained repeatedly, using the same cores, with the following treatment sequence [77, 80]:

1. Cleaning with toluene
2. Cleaning with steam (they found that steam does not disturb the clay minerals)
3. Saturating with brine and crude oil to S_{wi}
4. Aging in the crude oil for at least 100 hours at 65°C

EFFECT OF WETTABILITY ON ELECTRICAL PROPERTIES

Keller showed that different values of resistivity can be obtained at the same water saturation in rocks if the wettability is changed [116]. His values of the saturation exponent n ranged from 1.5 to 11.7 for the same rock. Oil-wet rocks have a high resistivity because oil is an insulator. Even at very low water saturations, a water-wet sand will have a continuous water film along the surfaces of the sand grains from the entrance to the exit, which furnishes a conductive path for the electric current. In an oil-wet sand, however, oil is the continuous phase and is in contact with the pore walls. Since water is the discontinuous phase in this case, the electrical path is interrupted by the insulating oil. Consequently, the resistivity of an oil-wet sand is very high, and the Archie saturation exponent n is considerably greater than 2.0.

Sweeney and Jennings obtained variations of n from 1.6 to 5.7 for carbonate rocks treated with acids to make them preferentially water-wet [117]. Even after cleaning the carbonate surfaces with acid, polar organic compounds from the crude oil apparently adsorbed on the surface of many of the samples, resulting in high values of n. Morgan and Pirson reported a very wide range of values for n, from 2.5 for strongly water-wet samples to 25.2 for strongly oil-wet packs of glass beads treated with progressively higher concentrations of silicone solutions [47]. Donaldson and Siddiqui confirmed previously reported results showing that Archie's saturation exponent increases from values near 2.0 for strongly water-wet to values higher than 8.0 for strongly oil-wet systems [109]. A linear relationship was observed between the USBM wettability index and the saturation exponent. They showed that the water–oil–rock systems become more water-wet when the temperature is increased. Figures 6.13 and 6.14 show a wettability index increase of 0.4, corresponding to a temperature increase from 25°C to 78°C. The difference in the slopes of the wettability index saturation exponent lines for Berea and Elgin sandstones is attributed to a wide difference in the physical properties of these two sandstones (average permeability and porosities are: Berea sandstone, $k = 258\ \mu m^2$, $\phi = 0.210$; Elgin sandstone, $k = 1727\ \mu m^2$, $\phi = 0.239$).

The Archie saturation exponent was obtained from the logarithm of the resistivity index I_R versus log (S_w) line; n is the slope of this line (Figure 6.15). The exponent n was determined by linear regression of the resistivities measured at water saturations of 1.00, S_{iw}, and S_{wor} as the USBM wettability test was conducted.

The significance of errors in the value of saturation exponent is very clear from examination of Figure 6.16. When the value of n is less than 8.0, small errors of this parameter result in large errors of the calculated water saturation from resistivity data. For example, where $F_{R_w}/R_t = 0.36$, if the correct value of n is 3.0, but if 2.0 is used to calculate the water saturation, an optimistic (higher oil saturation) error of 10% will result. This could lead to loss of considerable investment if a decision to conduct an enhanced oil recovery process, or some other production stimulation procedure, is based on such a large error of oil saturation.

Wettability plays an important role in all aspects of fluids associated with rocks. Therefore, for laboratory tests of water–oil–rock systems to be representative of the in situ subsurface conditions, equipment and procedures that simulate the subsurface temperature and pressure conditions for specific

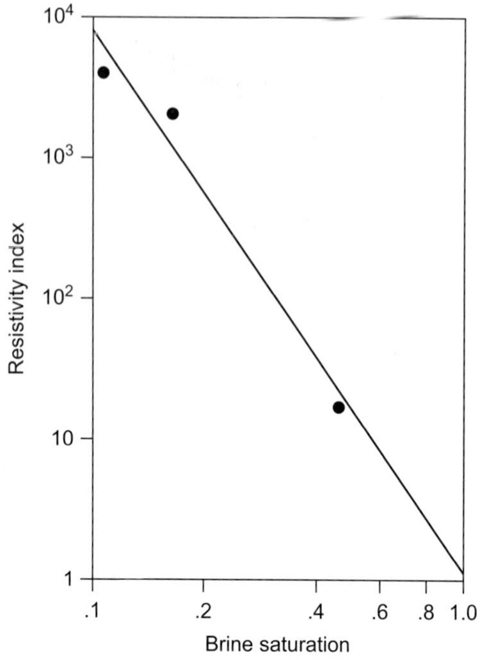

FIGURE 6.15 Resistivity index versus brine saturation.

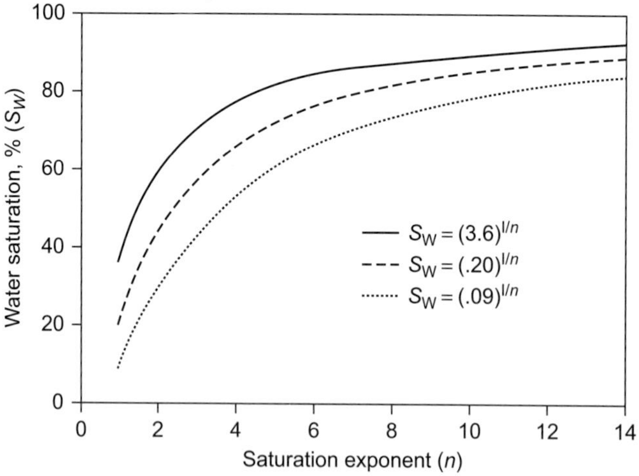

FIGURE 6.16 Variation of water saturation calculated from Archie's equation for various values of the saturation exponent.

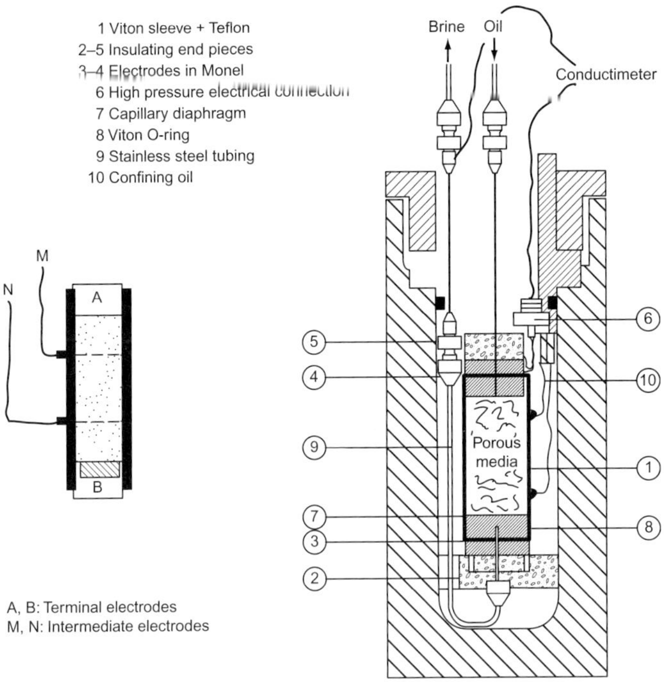

FIGURE 6.17 Design of coreholder for measurement of porosity, formation resistivity factor, and capillary pressure at simulated reservoir conditions.

depths must be used. In addition, the saturation history of the test core and aging of the sample with the oil must be carefully considered because they influence the relative wetting of the surface by water and oil [118–120].

Longeron et al. designed experimental equipment for measuring the electrical properties of cores at reservoir conditions (Figure 6.17) [71]. The coreholder is equipped with four electrical contacts that enable resistivity measurements to be made continuously along the length of the core (electrodes A and B at the ends) and for each section by making resistivity measurements between electrodes A–M, M–N, and N–B. The combination of measurements allowed continuous monitoring of the saturation distribution in the core after any changes were made by injection of oil or water. After uniform saturation was attained (by capillary force equilibration), resistivity and capillary pressure measurements were recorded.

Measurement of drainage (oil-displacing-water) and imbibition (water-displacing-oil) capillary pressures were made possible by placing a low-permeability porous plate, or diaphragm, at the bottom of the core (item 7 in Figure 6.17). The semipermeable disk enabled measurement of capillary pressure at reservoir conditions of overburden pressure and temperature. Injection and withdrawal of fluids from this type of system must be done

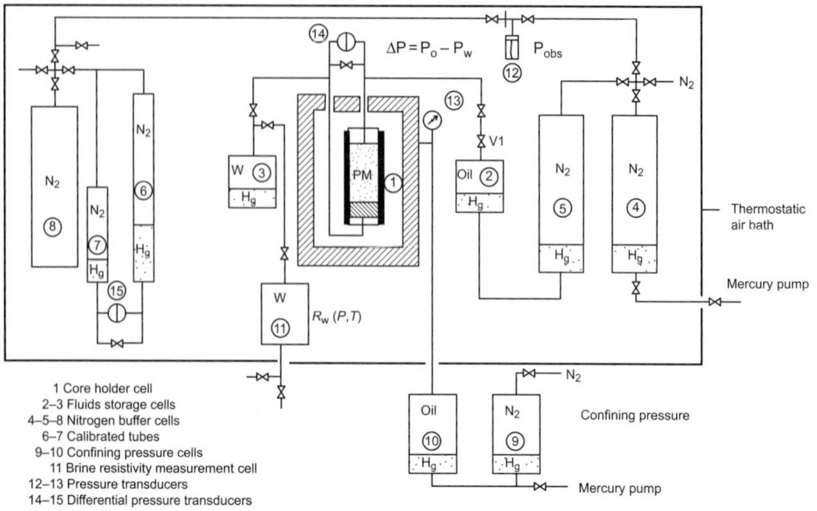

FIGURE 6.18 Fluid flow equipment for measurement of petrophysical of cores at simulated reservoir conditions.

very carefully using the type of fluid flow equipment described by Longeron et al. in Figure 6.18 [71]. Free-piston displacement (with mercury as the displacing piston, or an actual free piston), where the cylinder containing the displacing fluid is driven by a separate metering pump, is required to maintain close control of the injection and withdrawal of fluids.

The changes of porosity and formation resistivity factor obtained by Longeron et al. for sandstone and limestone cores, when stressed to a moderate pressure of 20 MPa, are listed in Table 6.5. At this total overburden pressure, the deformation was found to be completely elastic—that is, the cores returned to their original porosities when the overburden stress was removed. If greater stress is applied, however, inelastic deformation will take place and a reduction in porosity will result.

Although the relative changes of porosity and formation resistivity factors were approximately the same for the stress change, their responses to the step increases of pressure were much different. The sandstone cores deformed immediately in response to the applied stress, whereas the limestone cores exhibited gradual deformation at each step increase of stress.

A 15% underestimation of F_R will lead to underestimation of S_w by 7.5% for a clean sand with a saturation exponent of 2.0. At greater depths, the error will be more serious because expanded reservoir cores develop microfractures that contribute to the measurement of porosity at ambient conditions (in addition to the expanded matrix porosity). Thus, measurements of the petrophysical properties of reservoir rocks must be conducted at simulated reservoir conditions.

TABLE 6.5 Change in Porosity (ϕ) and Formation Resistivity Factors (F_R) for Sandstone and Limestone Cores Accompanying a Change in Pressure from 3 MPa [71]

	Sandstone Cores			
	1	2	3	4
ϕ at 3 MPa	0.144	0.150	0.195	0.184
ϕ at 20 MPa	0.137	0.143	0.189	0.18
% Change	4.9	4.7	3.1	1.6
F_R at 3 MPa	33.7	33.1	21.3	22.1
F_R at 20 MPa	40.4	41.4	23.2	23.6
% Change	19.9	25.1	8.9	6.8

	Limestone Cores					
	1	2	3	4	5	6
ϕ at 3 MPa	0.204	0.231	0.246	0.241	0.260	0.261
ϕ at 20 MPa	0.197	0.225	0.236	0.231	0.249	0.249
% Change	3.4	2.6	4.1	4.1	4.2	4.6
F_R at 3 MPa	17.4	13.3	11.4	11.3	13.3	10.4
F_R at 20 MPa	19.1	14.6	12.4	13.6	15.8	11.7
% Change	9.8	9.8	8.1	10.6	18.8	12.5

Example

The areas under the water-displacing-oil (from S_{iwl} to S_{wor}), and the oil-displacing-water (from S_{wor} to S_{iw2}) capillary pressure curves are 56 and 30, respectively. Compute the USBM wettability index and the energy required for the displacement of oil by water (in J/m^3). Core properties: 2.54 cm in diameter by 2.54 cm long, porosity = 22%, $S_{iw} = 0.28$, and $S_{wor} = 0.77$.

Solution

From Eq. (6.14), the USBM wettability index is:

$$I_u = \log\left(\frac{56}{30}\right) = 0.271 \text{ (slightly water-wet)}$$

$$V_p = 3.14 \times \left(\frac{2.54}{2}\right)^2 \times 2.54 \times 0.22 = 2.83 \text{ cm}^3$$

$$\text{Volume of oil displaced} = 2.83 \times (0.77 - 0.28) = 1.39 \text{ cm}^3$$

$$\text{Displacement energy} = 30{,}000\ \text{N/m}^2 \times 1.39(10^{-6})\ \text{m}^3$$
$$= 0.0416\ \text{N-m/m}^3 = 0.0416\ \text{J/m}^3$$

PROBLEMS

1. If the contact angle of a drop of water placed on a smooth plate submerged in a clear oil is 120°, what will be the value of the contact angle for a drop of the oil on the plate if it is submerged in water (see Figure 6.2)?
2. A 0.655 kg sample of rock is flooded with a tetravalent cation solution that displaces the following quantities of cations: 30 mg Mg^{2+}, 50 mg K^{+}, 70 mg Ca^{2+}, and 120 mg Na^{+}. Calculate the cation exchange capacity of the rock. Report the value as milliequivalents of cations per kilogram of rock.
3. **(a)** If a carbonate rock surface is in equilibrium with an acetic brine, will the rock adsorb acetic polar organic compounds from an oil? Explain your answer using chemical formulas.
   (b) If a sandstone is saturated with a brine containing multivalent cations, will the surface adsorb acetic polar organic compounds from an oil? Explain your answer using chemical formulas.
4. **(a)** Explain three methods that can be used to artificially render a water-wet sandstone surface oil-wet.
   (b) What is the difference between the terms mixed wettability and fractional wettability?
   (c) If the Amott wettability index is negative, does this mean that the rock is water-wet or oil-wet?
   (d) If the USBM wettability index is positive, does this mean that the rock is water-wet or oil-wet?
   (e) Draw two hypothetical (negative) water-displacing-oil capillary pressure curves, one for a water-wet system and the other for an oil-wet system. Why is the displacement energy less for the displacement of oil from the water-wet sand?
5. What is the principal reason why fluid/flow experiments should be conducted at reservoir conditions of temperature, pressure, and salinity? What is the influence of salinity?
6. If two water–oil–rock systems are tested and one is found to be water-wet while the other is oil-wet, will the Archie saturation index (n) be greater for one of the systems? Which one? Justify your answer.
7. If the formation resistivity factor (F_R) for a core is measured at ambient conditions (25°C, 100 kPa) and again at reservoir conditions (62°C, 28 MPa):

(a) Would you expect F_R to be greater at ambient conditions or lower? Give a sample calculation to prove your answer.
(b) What took place to produce the change in F_R, if any?

8. Draw two hypothetical oil recovery versus water injected curves, one for a water-wet system and the other for an oil-wet system.
   (a) Which one has the greater rate of oil recovery?
   (b) Which one exhibits water breakthrough first?
   (c) Which one shows greater recovery after 1.5 pore volumes of water have been injected?
   (d) Explain, theoretically, why the systems are different.

NOMENCLATURE

A area; constant (as defined in text)
B constant
C constant
C_{swo} spreading coefficient
F force (mN)
F_E free energy
F_R formation resistivity factor
g gravitational constant, g
h height
I_A Amott wettability index
I_u USBM wettability index
I_w resistivity index
k absolute permeability (μm^2, darcy)
L_C characteristic length for fluid imbibition
l length
P_c capillary pressure
P_o pressure of the oil phase
P_w pressure of the water phase
$P_{c,p}$ pseudo capillary pressure
R oil recovery (percent of original oil in place)
R_∞ ultimate oil recovery by waterflood or imbibition
S_o oil saturation
S_{or} residual oil saturation
S_w water saturation
S_{wi} irreducible water saturation
t time (minutes)
t_d dimensionless time
v_φ angular velocity of centrifuge rotor
V volume
V_b bulk volume
V_p pore volume
V_{ot} total oil displaced by imbibition and centrifugal displacements (Amott)

V_{wt} total water displaced by imbibition and centrifugal displacements (Amott)
V_{osp} oil displaced by spontaneous water imbibition (Amott)
V_{wsp} water displaced by spontaneous oil imbibition (Amott)
W pseudo work of imbibition
W_R wettability index from pseudo work of imbibition
w pseudo work of imbibition; thermodynamic work

GREEK SYMBOLS

δ partial derivative, or fluid displacement ratio
θ contact angle
μ viscosity
ρ density
σ interfacial tension
τ adhesion tension
ϕ porosity

SUBSCRIPTS

s solid
l liquid
o oil
w water
Φ contact angle

REFERENCES

1. Brown RJS, Fatt I. Measurement of fractional wettability of oil field rocks by the nuclear magnetic relaxation method. *Trans AIME* 1956;**207**:262–4.
2. Alba P. Discussion on effect of fractional wettability on multiphase flow through porous media. *Trans AIME* 1958;**216**:426–32.
3. Anderson WG. Wettability literature survey—Part 1: Rock/oil/brine interactions and the effects of core handling on wettability. *Soc Petrol Eng JPT* 1986;**38**:1124–44.
4. Salathiel RA. Oil recovery by surface film drainage in mixed-wettability rocks. *Soc Petrol Eng JPT* 1973;**25**:1216–24.
5. Melrose JC. Interpretation of mixed wettability states in reservoir rocks. SPE 10971. In: *Annual Conference*, New Orleans, LA; September 1982. pp. 26–9.
6. Iwankow EN. A correlation of interstitial water saturation and heterogeneous wettability. *Producers Monthly* 1960;**24**(12):18–26.
7. Donaldson EC, Thomas RD. Microscopic observations of oil displacement in water wet and oil wet systems. SPE 3555. In: *Annual Conference*, New Orleans, LA; 1971.
8. Emery LW, Mungan N, Nicholson RW. Caustic slug injection in the singleton field. *Soc Petrol Eng JPT* 1970;**22**:1569–76.
9. Kyte JR, Naumann JO, Mattox CC. Effects of reservoir environment on water-oil displacements. *Soc Petrol Eng JPT* 1961;**13**:579–82.

10. Masalmeh SK. Studying the effect of wettability heterogeneity on the capillary pressure curves using the centrifuge technique. *JPSE* (special issue on wettability—III) 2002;**33**:29–38
11. Collins SH, Melrose JC. Adsorption of asphaltenes and water on reservoir rock minerals Society of Petroleum Engineers Paper 11800, 1983, In: *International Symposium on Oilfield and Geothermal Chemistry*, Denver, CO; June 1–3, 1983.
12. Bartell FE, Osterhof JJ. Determination of the wettability of a solid by a liquid. *Ind Eng Chem* 1927;**19**(11):1277–80.
13. Van Dijke MIJ, Sorbie KS. The relation between interfacial tensions and wettability in three-phase systems: consequences for pore occupancy and relative permeability. *JPSE* 2002;**33**:39–48.
14. Chesters AK, Elyousfi A, Cazabat AM, Vilette S. The influence of surfactants on the hydrodynamics of surface wetting. *SPSE* (special issue on wettability—I) 1988;**20**:217–22.
15. Drummond C, Israelachevili J. Surface forces and wettability. *JPSE* (special issue on wettability—III) 2002;**33**:123–33.
16. Pam DT-K, Hirasaki GJ. Wettability/spreading of alkanes at the water-gas interface at elevated temperatures and pressures. *JPSE* (special issue on wettability—I) 1998;**20**:239–49.
17. Vizika O, Rosenberg E, Kalaydjian F. Study of wettability and spreading impact in three-phase gas injection by cryo-scanning electron microscopy. *JPSE* (special issue on wettability—I) 1998;**20**:189–202.
18. Raza SH, Treiber LE, Archer DL. Wettability of reservoir rocks and its evaluation. *Producers Monthly* 1968;**32**(4):2–7.
19. Richardson JG, Perkins FM, Osoba JS. Differences in the behavior of fresh and aged East Texas woodbine cores. *Soc Petrol Eng JPT* 1955;**7**:866–91.
20. Treiber LE, Archer DL, Owens WW. A laboratory evaluation of the wettability of fifty oil-producing reservoirs. *Soc Petrol Eng J* 1972;**12**(6):531–40.
21. Leach RO, Wagner OR, Wood HW, Harpke CF. A laboratory study of wettability adjustment in waterflooding. *Soc Petrol Eng JPT* 1962;**14**:206–12.
22. Yan S-Y, Hirasaki GJ, Basu S, Vaidya R. Statistical analysis on parameters that affect wetting for the crude oil/brine/mica system. *JPSE* (special issue on wettability—III) 2002;**33**:203–15.
23. Wilhelmy L. Ueber die abhangigkeit der capillaritäts-constanten des alkohols von substanz und gestalt des benetztenfesten korpers. *Ann Phys* 1863;**119**(6):177–217.
24. Mennella A, Morrow NR, Xie X. Application of the dynamic wilhelmy plate to identification of slippage at a liquid-liquid-solid three phase line of contact. *JPSE* 1995;**13**:179–92.
25. Zhang X, Tian J, Wang L, Zhou Z. Wettability effect of coatings on drag reduction and paraffin deposition prevention in oil. *JPSE* 2002;**36**:87–95.
26. Yaalon DH. Mineral composition of average shale. *Clay Miner Bull* 1962;**5**(27):31–6.
27. Block A, Simms BB. Desorption and exchange of adsorbed octadecylamine and stearic acid on steel and glass. *J Colloid Interface Sci* 1967;**25**:514.
28. Cram PJ. Wettability studies with non-hydrocarbon constituents of crude oil. Report RR-17, Petroleum Recovery Research Institute, Socorro, NM; 1972.
29. Craig FF. The reservoir engineering aspects of waterflooding. *Soc Petrol Eng* Monograph 3, Richardson, TX; 1971. p. 141.
30. Cuiec LE. Restoration of the natural state of core samples. Society Of Petroleum Engineers Paper 5634, In: *Annual Technical Conference*, Dallas, TX; September 28–October 1, 1975.

31. Denekas MO, Mattax CC, Davis GT. Effect of crude oil components on rock wettability. *Trans AIME* 1959;**216**:330–3.
32. Dunning NH, Moore JW, Denekas MO. Metalliferous substances adsorbed at crude petroleum-water interfaces. *Ind Eng Chem* 1952;**44**(11):1759–65.
33. Dunning NH, Moore JW, Myers AT. Properties of porphyrins in petroleum. *Ind Eng Chem* 1954;**46**(9):2000–7.
34. McCaffery FG, Mungan N. Contact angle and interfacial tension studies of some hydrocarbon water-solid systems. *J Can Petrol Tech* 1970;**8**(3):185.
35. Lowe AC, Phillips MC, Riddiford AC. On the wettability of carbonate surfaces by oil and water. *J Can Petrol Tech* 1973;**12**(44):33–40.
36. Collins RE, Cook Jr. CE. Fundamental basis for the contact angle and capillary pressure. *Trans Faraday Soc* 1959;**55**:1602.
37. Cuiec LE. Study of problems related to the restoration of the natural state of core samples. *J Can Petrol Tech* 1977;**16**(4):68–80.
38. Rall HT, Thompson CJ, Coleman HJ, Hopkins RL. Sulfur compounds in crude oil. U.S. Bureau of Mines Bulletin 659, U.S. Government Printing Office, Washington, DC; 1972, 193 pp.
39. Seifert WK, Howells WG. Interfacially active acids in a California crude oil. *Anal Chem* 1969;**41**(4):554–62.
40. Seifert WK, Teeter RM. Identification of polycyclic aromatic and heterocyclic crude oil carboxylic acids. *Anal Chem* 1970;**42**(7):750–8.
41. Somasundaran P. Interfacial chemistry of articulate floatation. In: Somasundaran P, Grieves RB, editors, *Advances in interfacial phenomena of particulate/solution/gas systems: applications to floatation research. AICHE symposium series*, vol. 71, no. 150. New York: American Institute of Chemical Engineers; 1975. pp. 1–15.
42. Leach RO, Wagner OR, Wood HW, Harpke CF. A laboratory study of wettability adjustment in waterflooding. *Soc. Petrol. Eng. JPT* 1962;206–12.
43. Somasundaran P, Agar GE. The zero point of charge of calcite. *J Colloid Interface Sci* 1967;**24**(4):433–40.
44. Amott E. Observations relating to the wettability of porous rock. *Trans AIME* 1959;**216**:156–62.
45. Trantham JC, Clampitt RL. Determination of oil saturation after water-flooding in an oil-wet reservoir—The North Burbank Unit, Track 97 Project. *Soc Petrol Eng JPT* 1977;**29**:491–500.
46. Anderson WG. Wettability literature survey—Part 2: wettability measurement. *Soc Petrol Eng JPT* 1986;**38**:1246–62.
47. Morgan WB, Pirson SJ. The effect of fractional wettability on the Archie saturation exponent. In: *Transactions of SPWLA, 5th Annual Symposium*, Sec. B, May 13–15, 1964, Midland, TX, pp. 1–13.
48. Sharma MM, Wunderlich RW. The alteration of rock properties due to interactions with drilling fluid components. *JPSE* 1989;**1**(2):127–43.
49. Donaldson EC, Lorenz PB, Thomas RD. Wettability determination and its effect on recovery efficiency. *Soc Petrol Eng J* 1969;**9**(1):13–20.
50. Donaldson EC, Kendall RF, Pavelka EA, Crocker ME. *Equipment and procedures for fluid flow and wettability tests of geological materials. U.S. Department of Energy report DOE/BETC/IC-79/5*. Springfield, VA: National Technical Information Service; 1980. 40 pp.
51. Kwan MY. Measuring wettability of unconsolidated oil sands using the USBM method. *JPSE* 1998;**21**:61–78.

52. Mattox CD, Kyte JR. Imbibition oil recovery from fractured water drive reservoir. *Soc Petrol Eng J* 1962; 177–84. June.
53. Ma S, Morrow NR, Zhang X. Characterization of wettability from spontaneous imbibition measurements. *J Can Petrol Tech* (special edition) 1999;**30**(10):56.
54. Aranofsky JS, Masse L, Natanson SG. A model for the mechanism of oil recovery from the porous matrix due to water invasion in fractured reservoirs. *Trans AIME* 1958;**213**:17–9.
55. Zhou D, Blunt M. Wettability effects on three-plane gravity drainage. *JPSE* [special issue on wettability—I], 1998;**20**:203–11.
56. Benner FC, Bartell FE. The effect of polar impurities upon capillary and surface phenomena in petroleum production. In: *Drilling and production practices*. New York: API; 1941. p. 34.
57. Holbrook OC, Bernard CC. Determination of wettability by dye adsorption. *Trans AIME* 1958;**213**:261–4.
58. Morrow NR, Cram PJ, McCaffery FG. Displacement studies in dolomite with wettability control by octanoic acid. *Soc Petrol Eng J* 1973;**13**(4):221–32.
59. Mungan N. Interfacial effects in immiscible liquid-liquid displacements in porous media. *Soc Petrol Eng J* 1966;**6**(3):247–53.
60. Akhlaq MS. Characterization of the isolated wetting crude oil components with infrared spectroscopy. *JPSE* 1999;**22**:229–35.
61. Buckley JS, Liu Y. Some mechanisms of crude oil/brine/solid interactions. *JPSE* (special issue on wettability—I) 1998;**20**:155–60.
62. Katzkeller DL. Possibility of secondary recovery for the Oklahoma City Wilcox sand. *Trans AIME* 1942;**146**:28–43.
63. Marsden SS, Nikias PA. The wettability of the Bradford sand (Parts I and II). *Producers Monthly*1962;**26**(5):2–5, 1965;**29**(6):10–14.
64. Chilingarian GV, Yen TF. Some notes on wettability and relative permeabilities of carbonate rocks, II. *Energy Sources* 1983;**7**(1):67–75.
65. Cuiec LE, Longeron D, Paesirszky J. On the necessity of respecting reservoir conditions in laboratory displacement studies. In: *SPE 7785, SPE Middle East Oil Conference*, Bahrain; March 25–29, 1979.
66. Treiber LE, Archer DL, Owens WW. A laboratory evaluation of the wettability of fifty oil-producing reservoirs. *Soc Petrol Eng J* 1972;**12**(6):531–40.
67. Standal S, Haavik J, Blokhus AM, Skauge A. Effect of polar organic components on wettability as studied by adsorption and contact angles. *JPSE* (special issue on wettability—II) 1999;**24**:131–44.
68. Odeh AS. Effect of viscosity ratio on relative permeability. *Trans AIME* 1959;**216**:346.
69. Mungan N. Role of wettability and interfacial tension in waterflooding. *Soc Petrol Eng J* 1964;**4**(2):115–23.
70. Mungan N. Relative permeability measurements using reservoir fluids. *Soc Petrol Eng J* 1972;**12**(5):398–402.
71. Longeron DG, Argaud MH, Feraud JP. Effect of overburden pressure and the nature of microscopic distribution of fluids on electrical properties of rock samples. *SPEFE* 1989;**4**(2):194–202.
72. Geffen TM, Owens WW, Parrish DR, Morse RA. Experimental investigation of factors affecting laboratory relative permeability measurements. *Trans AIME* 1951;**192**:99–110.
73. Schmid C. The wettability of petroleum rocks and results of experiments to study the effects of variations in wettability of core samples. *Erdoel Kohle-Erdgas–Petrochem* 1964;**17**(8):605–9.

74. Donaldson EC, Crocker ME. *Characterization of the crude oil polar compound extract. U.S. Department of Energy report DOE/BETC/RI-80/5.*, Springfield, VA: National Technical Information Service; 1980.
75. Gatenby WA, Marsden SS. Some wettability characteristics of synthetic porous media. *Producers Monthly* 1957;**22**(1):5–12.
76. Fatt I, Klinkoff WA. Effect of fractional wettability on multiphase flow through porous media. *Trans AIME* 1959;**216**:426–32.
77. McGhee JW, Crocker ME, Donaldson EC. *Relative wetting properties of crude oil in Berea sandstone. U.S. Department of Energy report BETC/RI-79/9.* Springfield, VA: National Technical Information Service; 1979.
78. Anderson WG. Wettability literature survey—Part 3: The effects of wettability on the electrical properties of porous media. *Soc Petrol Eng JPT* 1986;**38**:1371–8.
79. Anderson WG. Wettability literature survey—Part 4: Effects of wettability on capillary pressure. *Soc Petrol Eng JPT* 1987;**39**:1283–300.
80. Anderson WG. Wettability literature survey—Part 5: The effects of wettability on relative permeability. *Soc Petrol Eng JPT* 1987;**39**:1453–68.
81. Anderson WG. Wettability literature survey—Part 6: The effects of wettability on waterflooding. *Soc Petrol Eng JPT* 1987;**39**:1605–22.
82. Rathmell JJ, Braun PH, Perkins RK. Reservoir waterflood residual oil saturation from laboratory tests. *Soc Petrol Eng JPT* 1973;**25**:175–85.
83. Tweheyo MT, Holt T, Torsaeter O. An experimental study of the relationship between wettability and oil production characteristics. *JPSE* (special issue on wettability—II) 1999;**24**:179–88.
84. Tang G-Q, Morrow NR. Influence of brine composition and fines migration on crude oil/brine/rock interactions and oil recovery. *JPSE* (special issue on wettability—II) 1999;**24**:99–111.
85. Coley FH, Marsden SS, Calhoun JC. A study of the effect of wettability on the behavior of fluids in synthetic porous media. *Producers Monthly* 1956;**20**(8):29–45.
86. Singhal AK, Dranchuk PM. Wettability control of glass beads. *Can J Chem Eng* 1975;**53**:3–8.
87. Warren JE, Calhoun JC. A study of waterflood efficiency in oil-wet systems. *Soc Petrol Eng JPT* 1955;**7**:22–9.
88. Talash AW, Crawford PB. Experimental flooding characteristics of unconsolidated sands. In: *SPE Permian Basin Oil Recovery Conference*, Midland, TX; May 4–5, 1961.
89. Talash AW, Crawford PB. Experimental flooding characteristics of 75 percent water-wet sands. *Producers Monthly* 1961;**25**(2):24–6.
90. Talash AW, Crawford PB. Experimental flooding characteristics of 50 percent water-wet sands. *Producers Monthly* 1962;**26**(4):2–5.
91. Graue A, Viksund BG, Eilertsen T, Moe R. Systematic wettability alteration by aging sandstone and carbonate rock in crude oil. *JPSE* (special issue on wettability—II) 1999;**24**:85–97.
92. Graue A, Aspenes E, Bogno T, Moe RW, Ramsdal J. Alteration of wettability and wettability heterogeneity. *JPSE* (special issue on wettability—III) 2002;**33**:3–17.
93. Kowalewski E, Holt T, Torsaeter O. Wettability alterations due to an oil soluble additive. *JPSE* (special issue on wettability—III) 2002;**33**:19–28.
94. Donaldson EC, Crocker ME, Manning FS. *Adsorption of organic compounds on cottage grove sandstone. US DOEBERC/RI-75/4.* Springfield, VA: National Technical Information Service; 1975. 16 pp.

95. Lorenz PB, Donaldson EC, Thomas RD. *Use of centrifugal measurement of wettability to predict oil recovery. U.S. Bureau of Mines RI 7873*. Springfield, VA: National Technical Information Service; 1974. 26 pp.
96. Sharma MK, Shaw DO. Use of surfactants in oil recovery. In: Donaldson EC, Chilingarian GV Yen TF, editors. *Enhanced oil recovery—II*. Amsterdam: Elsevier Science; 1989. pp. 255–315.
97. McCormick C, Hester R. *Responsive copolymers for enhanced oil recovery. DOE/BC/14882-5*. Springfield, VA: U.S. Department of Commerce and National Technical Information; 1994. 184 pp.
98. Standnes DC, Austad T. Wettability alteration in chalk 2. Mechanism for wettability alteration from oil wet to water wet using surfactants. *JPSE* 2000;**28**:123–43.
99. Owens WW, Archer DL. The effect of rock wettability on oil-water relative permeability relationships. *Soc Petrol Eng JPT* 1971;**71**:873–8.
100. Grattoni CA, Jing XD, Zimmerman RW. Wettability alteration by aging of a gel placed within a porous medium. *JPSE* (special issue on wettability—III) 2002;**33**:135–45.
101. Melrose JC, Bradner CF. Role of capillary forces in determining microscopic displacement efficiency for oil recovery by waterflooding. *J Can Petrol Tech* 1974;**13**(4):54–62.
102. Mungan N. Enhanced oil recovery using water as a driving fluid: Part 2—Interfacial phenomena and oil recovery: capillarity. *World Oil* 1981;**192**(5):149–58.
103. Xie X, Morrow NR, Buckley JS. Contact angle hysteresis and the stability of wetting changes induced by adsorption from crude oil. *JPSE* 2002;**33**:147–59.
104. Kennedy HT, Bruja EO, Boykin RS. An investigation of the effects of wettability on the recovery of oil by waterflooding. *J Phys Chem* 1955;**59**:867–9.
105. Loomis AG, Crowell DC. *Relative permeability studies: gas-oil and water-oil systems. U.S. Bureau of Mines Bull. 599*. Washington, DC: U.S. Government Printing Office; 1962. 39 pp.
106. Moore TF, Slobod RL. The effect of viscosity and capillarity on the displacement of oil and water. *Producers Monthly*, Vol. 20, August 1956. 20 pp.
107. Morrow NR, McCaffery FG. Displacement studies in uniformly wetting porous media. In: GF Paddy, editor. *Wetting, spreading and adhesion*. New York: Academic Press; 1978. pp. 289–319.
108. von Engelhardt W, Lubben H. Study of the influence of interfacial stress and contact angle on the displacement of oil by water in porous material, II—Test results. *Erdol Kohle* 1957;**10**(12):826–30.
109. Donaldson EC, Siddiqui TK. Relationship between the Archie saturation exponent and wettability. Society of Petroleum Engineers Paper 16790. In: *62nd Annual Technical Conference*, Dallas, TX; September 27–30, 1987. 8 pp.
110. Archie GE. The electrical resistivity as an aid in determining some reservoir characteristics. *Trans AIME* 1942;**146**:54–62.
111. Donaldson EC, Civan F, Alam MWU. Relative permeabilities at simulated reservoir conditions. Society of Petroleum Engineers Paper 16970. In: *62nd Annual Technical Conferernce*, Dallas, TX; September 27–30, 1987. 10 pp.
112. Edmondson TA. Effect of temperature on waterflooding. *J Can Petrol Tech* 1965; **4**(4):236–42.
113. Lo HY, Mungan N. Effect of temperature on water-oil relative permeability in oil-wet and water-wet systems. Society of Petroleum Engineers Paper 4505. In: *48th Annual Conference*, Las Vegas, NV; September 30–October 3, 1973. 12 pp.
114. Poston SW, Ysrael SC, Hossain AKMS, Montgomery EF, Ramey Jr. HJ. The effect of temperature on irreducible water saturation and relative permeability of unconsolidated sands. *Soc Petrol Eng J* 1970;**10**(2):171–80.

115. Jadhunandan PP, Morrow NR. Effect of wettability on waterflood recovery for crude oil/brine/rock systems. *SPEFE* 1995;**10**(1):40–6.
116. Keller GV. Effect of wettability on the electrical resistivity of sands. *Oil Gas J* 1953; **51**(1):65.
117. Sweeney SA, Jennings Jr. HY. The electrical resistivity of preferentially water-wet and preferentially oil-wet carbonate rocks. *Producers Monthly* 1960;**24**:29–32.
118. Moss AK, Jing XD, Archer JS. Laboratory investigation of wettability and hysteresis effects on resistivity index and capillary pressure characteristics. *JPSE* (special issue on wettability—II) 1999;**24**:231–42.
119. Moss AK, Jing XD, Archer JS. Wettability of reservoir rock and fluid systems from complex resistive measurements. *JPSE* (special issue on wettability—III) 2002;**33**:75–85.
120. Stalheim SO, Eidesmo T, Rueslatten H. Influence of wettability on water saturation modelling. *JPSE* (special issue on wettability—II) 1999;**24**:243–53.

Chapter 7

Applications of Darcy's Law

This chapter describes the characteristics of the flow of fluids through porous geological materials. The pores, or flow conduits, are complex, interconnected capillaries and channels of various sizes as described in previous chapters. The flow of compressible and incompressible fluids through porous rocks is described by Darcy's law and its derivatives. The simplest case of fluid flow through porous media is the linear flow of a single-phase fluid under a constant pressure gradient, which is known as linear steady-state flow. When two fluids are present in a porous medium, steady-state flow occurs under a constant pressure gradient only when the fluid saturations remain constant. If the saturations change with respect to time (e.g., if the water saturation is increasing while the oil saturation is decreasing), the flow of fluids is characterized as unsteady-state flow.

Steady-state and pseudosteady-state flow rate equations, based on Darcy's law for linear and radial flow of compressible and incompressible fluids, can be used to predict the production performance of porous and permeable flow systems of simple geometry. In steady-state flow systems, the pressure and fluid velocity at every point throughout the porous system adjust instantaneously to changes in pressure or flow rate in any part of the system [1]. This flow condition occurs only when the rock is 100% saturated with a fluid and the pressure of the porous media is effectively maintained constant by either an active aquifer or the injection of a displacing fluid; that is, fluid withdrawal from the porous rock is exactly balanced by fluid entry across the open boundary and $\delta p/\delta t = 0$. If there is no flow across the reservoir boundary and the well is produced at a constant flow rate for a long time, the pressure decline throughout the reservoir becomes a linear function of time and $\delta p/\delta t = \text{constant}$. When this flow regime occurs, it is referred to as pseudosteady state or semi-steady state.

Natural reservoir systems do not ordinarily conform to any simple geometrical shape. The two most practical geometries are the linear flow and the radial flow systems. In the linear system, the flow occurs through a constant cross-sectional area and the flow lines are parallel. In the radial system, the flow occurs between two concentric cylindrical surfaces, the well being the inner cylinder and the reservoir boundary the outer cylinder. Another flow system of interest is the spherical geometry. Finally, reservoir fluids are classed either as incompressible or slightly compressible liquid, or gas. A compressible liquid is defined as one whose change of volume is small with respect to the change of pressure.

DARCY'S LAW

To express the quantity of fluid that will flow through a porous rock system of specified geometry and dimensions, such as the one shown in Figure 7.1, it is necessary to integrate Darcy's law over the boundaries of the porous system. This law, in its simple differential form, is:

$$v = -\frac{k}{m}\frac{\mathrm{d}p}{\mathrm{d}x} \tag{7.1}$$

where

v = apparent fluid flowing velocity, cm/s
k = permeability of the porous rock, darcy
μ = viscosity of the flowing fluid, centipoise
$\mathrm{d}p/\mathrm{d}x$ = pressure gradient in the direction of flow, atm/cm
x = distance in the direction of flow, always positive, cm

This one-dimensional empirical relationship was developed by Henry Darcy, a French engineer, in 1856 while he was investigating the flow of water through sand filters for water purification [2]. The experimental variation in this investigation is the type of sandpack, which had the effect of changing the value of the permeability. All of the experiments were carried out with water; therefore, the effects of fluid density and viscosity in Eq. (7.1) were not investigated [3, 4]. In addition, Darcy's law holds only for conditions of viscous flow; that is, the rate of the flowing fluid is sufficiently low to be directly proportional to the potential gradient. Another requirement of this law is that the flowing fluid must not react chemically with the porous

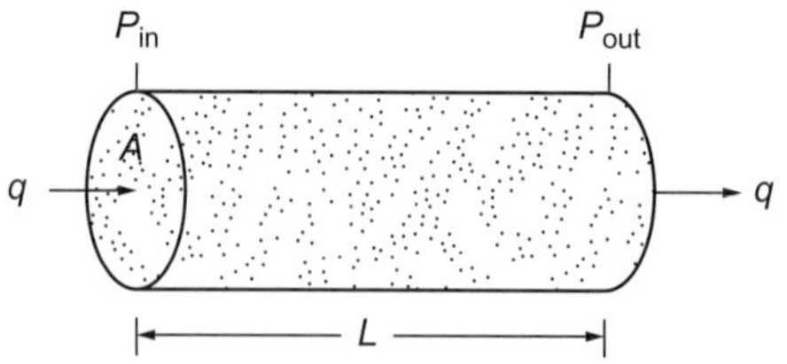

FIGURE 7.1 Typical linear flow system in a core sample.

medium. Such a reaction can alter the permeability of the sand body as flow continues. The sandpack in Darcy's original experiment was always maintained in the vertical position. Subsequent researchers repeated this experiment under less-restrictive conditions and found the following.

1. Darcy's law could be extended to fluids other than water.
2. The constant of proportionality is actually the mobility ratio k/μ.
3. Darcy's law is independent of the direction of flow in the Earth's gravitational field.

The gradient dp/dx is the driving force, and may be due to fluid pressure gradients and/or hydraulic (gravitational) gradients [5]. Generally, the hydraulic gradients are small compared with the fluid pressure gradients, and are, therefore, negligible. In oil reservoirs with a large expanding gas cap and considerable gravity drainage characteristics, however, the gravitational gradients are important and must be taken into account when analyzing reservoir performance.

LINEAR FLOW OF INCOMPRESSIBLE FLUIDS

The following assumptions are necessary for the development of the basic equations describing linear flow of incompressible or slightly compressible fluids through porous media:

1. Steady-state flow conditions exist.
2. The porous rock is 100% saturated with the flowing fluid; however, a fixed and immobile phase may be present and often is, as in the case of oil flow in a porous rock containing irreducible water saturation, or in the case of oil flow with an immobile gas phase of saturation less than critical gas saturation [5].
3. The viscosity of the flowing fluid is constant.
4. Isothermal conditions prevail.
5. The porous rock is homogeneous and isotropic.
6. Porosity and permeability are constant, that is, independent of pressure.
7. The flow is laminar, that is, there are negligible turbulence effects.
8. Gravity forces are negligible.

With these restrictions in mind, the apparent velocity is

$$v = \frac{q}{A} \tag{7.2}$$

where q is the volumetric flow rate (cm^3/s) and A is the cross-sectional area perpendicular to flow direction (cm^2). The actual velocity (v_a) is determined by dividing the apparent velocity (v) by the porosity of the rock (ϕ). If an irreducible water saturation, S_{iw}, is present, the actual velocity in a water-wet reservoir is:

$$v_a = \frac{v}{\phi(1 - S_{iw})} \tag{7.3}$$

Combining Eqs. (7.1) and (7.2) yields:

$$q = \frac{kA}{m}\frac{dp}{dx} \tag{7.4}$$

Separating variables and integrating between limits 0 and L, and p_1 and p_2, one obtains the following expression for the volumetric flow rate:

$$q = \frac{kA(p_1 - p_2)}{\mu L} \tag{7.5}$$

Equation (7.5) is the conventional linear flow equation used in fluid-flow calculations. This expression is written in the fundamental units that define the Darcy unit. Transforming it into the commonly used oil field units, such that q = bbl/day, A = ft^2, p = psi, L = ft, and k = mD, gives:

$$q\left(\frac{5.615 \times 30.48^3}{24 \times 60 \times 60}\right) = A(30.48^2)\frac{k(10^{-3})}{\mu}\frac{\Delta p}{L}\frac{(1/14.7)}{30.48}$$

or

$$q = 1.127 \times 10^{-3}\frac{kA}{\mu}\frac{\Delta p}{L} \tag{7.6}$$

In SI units, a flow rate of 1 m^3/s will result for a fluid flowing through a porous medium with a permeability of 1 μm^2, a cross-sectional area of 1 m^2, and fluid viscosity of 1 Pa × s under a pressure gradient of 10^{12} Pa/m.

Example

A 10-cm long cylindrical core sample was subjected to a laboratory linear flow test under a pressure differential of 3.4 atm using a fluid of viscosity of 2.5 cP. The diameter of the core is 4 cm. A flow rate of 0.35 cc/s was obtained. Calculate the permeability of this core sample.

Solution

Figure 7.1 is a schematic representation of the core sample. Using Eq. (7.5) the permeability of the core sample is:

$$k = \frac{0.35 \times 2.5 \times 10}{12.57 \times 3.4} = 0.204 \text{ darcy} = 204 \text{ mD}$$

To use Eq. (7.6), one must first convert the data to oil field units.

$$\Delta p = 3.4(\text{atm}) \times 14.7\ \text{psi/atm} = 50\ \text{psi}$$

$$q = (0.35\ \text{cm}^3/\text{s})\left(\frac{1}{30.48^3}\text{ft}^3/\text{cm}^3\right)\left(\frac{1}{5.615}\text{bbl/ft}^3\right)(24 \times 60 \times 60\ \text{s/D})$$

$$= 0.19\ \text{bbl/day}$$

$$L = (10\ \text{cm})\left(\frac{1}{30.48}\text{ft/cm}\right) = 0.328\ \text{ft}$$

$$A = \pi\frac{d^2}{4} = \pi\frac{4^2}{4} = 12.57\ \text{cm}^2 = 0.0135\ \text{ft}^2$$

The permeability of this core sample using Eq. (7.6) is:

$$k = \frac{q\mu L}{1.127 \times 10^{-3} A\Delta P} = \frac{0.19 \times 2.5 \times 0.328}{1.127 \times 10^{-3} \times 0.0135 \times 50} = 204\ \text{mD}$$

Because of the many unit systems employed by the industry, it is very important that petroleum engineers be able to convert units from one system to another.

To estimate the pressure at any point in a linear flow system, Eq. (7.4) is integrated between the limits of 0 and x, and p_1 and p, respectively, yielding:

$$p_1 - p = \left(\frac{q\mu}{kA}\right)x \tag{7.7}$$

From Eq. (7.5), one obtains:

$$p_1 - p_2 = \left(\frac{q\mu}{kA}\right)L \tag{7.8}$$

Dividing Eq. (7.7) by (7.8) and solving for the variable pressure, p, yields:

$$p = (p_2 - p_1)\frac{x}{L} + p_1 \tag{7.9}$$

This equation indicates that the pressure behavior of a linear flow system during steady-state flow is a straight line as a function of distance.

LINEAR FLOW OF GAS

Consider the same linear flow system of Figure 7.1, except that the flowing fluid is now natural gas. Because the gas expands as the pressure declines, however, the pressure gradient increases toward the downstream end and, consequently, the flow rate q is not constant, but is a function of p. Assuming that Boyle's law is valid (gas deviation factor $z = 1$) and a constant mass flow rate, that is, pq is constant, one can write:

$$p_1 q_1 = p_2 q_2 = pq = \overline{pq} \tag{7.10}$$

where subscripts denote point of measurement, $\overline{q}$ is the mean flow rate, and $\overline{p}$ is the mean pressure. Combining this relationship with Darcy's law, that is, Eq. (7.5), gives:

$$q = \frac{q_2 p_2}{p} = \frac{kA}{\mu_g}\frac{dp}{dx} \tag{7.11}$$

where μ_g is the viscosity of gas in cP. Separating variables and integrating between p_1 and p_2, and 0 and L, gives:

$$q_2 p_2 \int_0^L dx = -\frac{kA}{\mu_g}\int_{p_1}^{p_2} p\, dp \tag{7.12}$$

or

$$p_2 q_2 = \frac{kA}{\mu_{gL}}\left(\frac{p_1^2 - p_2^2}{2\overline{p}}\right) \tag{7.13}$$

The mean flow rate expression that follows can be derived by combining Eqs. (7.10) and (7.13):

$$\overline{q} = \left(\frac{kA}{\mu_{gL}}\right)\frac{(p_1 - p_2)(p_1 + p_2)}{2} \tag{7.14a}$$

If one assumes the mean pressure $\overline{p}$ is equal to $(p_1 + p_2)/2$, Eq. (7.14a) reduces to:

$$\overline{q} = \left(\frac{kA}{\mu_g}\right)\frac{(p_1 - p_2)}{L} \tag{7.14b}$$

Equation (7.14b) is the same as Eq. (7.5), which gives the volumetric flow rate of incompressible fluids. Therefore, the law for the linear flow of ideal gas is the same as for a liquid, as long as the gas-flow rate is expressed as a function of the arithmetic pressure.

To include the effect of changes in the gas deviation factor, z, from standard conditions of pressure, p_{sc}, and temperature, $\overline{p}$, to average pressure, and temperature, $\overline{T}$, let:

$$\frac{p_{sc} q_{sc}}{z_{sc} T_{sc}} = \frac{\overline{pq}}{\overline{z}\overline{T}} \tag{7.15}$$

Combining Eqs. (7.14) and (7.15), and solving for q_{sc}:

$$q_{sc} = \left(\frac{z_{sc} T_{sc}}{p_{sc}}\right)\left(\frac{\overline{p}}{\overline{z}\overline{T}}\right)\frac{kA}{\mu_g}\left(\frac{p_1 - p_2}{L}\right) \tag{7.16}$$

where q_{sc} is in cm^3/s. Inasmuch as $\bar{p} = (p_1 + p_2)/2$, Eq. (7.16) becomes:

$$q_{sc} = \left(\frac{z_{sc} T_{sc}}{p_{sc}}\right)\left(\frac{1}{\bar{z}\bar{T}}\right)\frac{kA}{\mu_g}\left(\frac{p_1^2 - p_2^2}{2L}\right) \tag{7.17}$$

Converting from Darcy's units to practical field units and assuming $z_{sc} = 1$ at $p_{sc} = 14.7$ psia and $T_{sc} = 60°F$ or $520°R$ gives:

$$q_{sc} = \frac{0.112\, kA}{\bar{z}\bar{T}\mu_g L}(\Delta p^2) \tag{7.18}$$

where

q_{sc} = volumetric flow rate at standard conditions, SCF/day
k = permeability of the reservoir rock, mD
μ_g = gas viscosity, cP
A = cross-sectional area, ft^2
$\bar{T}$ = mean temperature of the gas reservoir, °R
$\bar{z}$ = mean gas deviation factor at $\bar{T}$ and $\bar{p}$ dimensionless
L = length of the sand body, ft

$$\Delta p^2 = p_1^2 - p_2^2 \text{ psia}^2$$

If the mean flow rate is expressed in terms of ft^3/day at the mean pressure $\bar{p}$ and mean temperature $\bar{T}$ and other variables are expressed in oil field units, Eq. (7.14) becomes:

$$\bar{q} = \frac{6.33 \times 10^{-3} kA(p_1 - p_2)}{\mu_g L} \tag{7.19}$$

The following equation is useful in determining the outlet volumetric flow rate q_2 at the pressure p_2, which is generally the atmospheric pressure in a laboratory experiment:

$$q_2 = \frac{kA}{\mu_g L}\bar{p}\frac{(p_1 - p_2)}{p_2} \tag{7.20}$$

where q_2 is in cm^3/s. If practical oil field units are used in Eq. (7.20), where q_2 is expressed in ft^3/day:

$$q_2 = \frac{6.33 \times 10^3 kA}{\mu_g L}\bar{p}\frac{(p_1 - p_2)}{p_2} \tag{7.21}$$

In general, most equations used to study steady-state flow of incompressible fluids may be extended to gas-flow systems by simply squaring the pressure terms, and expressing the gas-flow rates as SCF/day and the gas formation volume factor in bbl/SCF.

Example

A horizontal pipe having 2 in. inside diameter and 12 in. long is filled with a sand of 24% porosity. This sandpack has an irreducible water saturation of 28% and a permeability to gas of 245 mD. The viscosity of the gas is 0.015 cP.

1. What is the actual velocity of the gas (in cm/s) under 100 psi pressure differential?
2. What is the average flow rate of the gas in ft^3/day and cm^3/s?

Solution

1. The actual velocity, v_a, of the flowing gas can be calculated from Eq. (7.3) where $\phi = 0.24$, $S_{wi} = 0.28$, and the apparent velocity, v, can be obtained from Darcy's law. Inasmuch as $k = 0.245$ darcy, $\mu_g = 0.015$ cP, $L =$ 12 in. $\times$ 2.54 cm/in. = 30.48 cm, and $\Delta p =$ (100 psi)/(14.7 psi/atm) = 6.80 atm, the apparent velocity is:

$$v = \frac{0.245}{0.015} \times \frac{6.8}{30.48} = 3.64$$

and the actual velocity is:

$$v_a = \frac{3.64}{0.24(1-0.28)} = 21.1 \text{ cm/s}$$

2. The mean volumetric flow rate of gas through this sandpack in ft^3/day is obtained from Eq. (7.19), where $k = 254$ mD, $L = 1$ ft, and $A = \pi(1/12)^2 =$ 0.0128 ft^2:

$$\overline{q} = \frac{6.33 \times 10^3 \times 245 \times 0.0218 \times 100}{0.015 \times 1} = 225.5 \text{ ft}^3/\text{day}$$

or

$$\overline{q} = \left(225.5 \text{ ft}^3/\text{day}\right)\left(\frac{\text{Day}}{24 \times 60 \times 60s}\right)\left(30.48^3 \text{ cm}^3/\text{ft}^3\right) = 73.9 \text{ cc/s}$$

Assuming constant steady flow rate, q, a pressure distribution equation along a linear sand body also can be derived by combining Boyle's and Darcy's laws, and integrating between p_1 and p, and 0 and x. Replacing L with x in Eq. (7.12) and integrating yields:

$$p_1^2 - p_2^2 = \frac{2\mu_g q_2 p_2}{kA}(x) \tag{7.22}$$

From Eq. (7.13), one can obtain:

$$p_1^2 - p_2^2 = \frac{2\mu_g q_2 p_2}{kA}(L) \tag{7.23}$$

Dividing Eq. (7.22) by (7.23) and solving for the variable pressure p gives:

$$p^2 = (p_2^2 - p_1^2)\frac{X}{L} + p_1^2 \tag{7.24}$$

This expression indicates that the pressure decline versus distance during steady-state flow of gas through a linear system follows the parabolic curve.

It also indicates that pressure is maintained near the inlet because of the release of energy stored in the gas, but it is still independent of fluid and rock properties. Generally, the use of linear steady-state flow of compressible and incompressible fluids is limited to laboratory testing.

DARCY'S AND POISEUILLE'S LAWS

Darcy's law for the linear flow of incompressible fluids in porous and permeable rocks and Poiseuille's law for liquid capillary flow are quite similar. The general form of Poiseuille's law for the viscous flow of liquid through capillary tubes is:

$$q = \frac{\pi r^4}{8\mu}\frac{\Delta p}{L} \tag{7.25}$$

where

r = radius of capillary tube, cm
Δp = pressure drop, dynes/cm^2 ($= 1.0133 \times 10^6$ atm)
L = length of capillary tube, cm.
μ = viscosity of flowing fluid, poise

If the fluid-conducting channels in a porous medium could be represented by a bundle of parallel capillary tubes of various diameters, then the flow rate through this system is:

$$q = \left(\frac{\pi}{8}\sum_{j=1}^{N} n_j r_j^4\right)\frac{\Delta p}{\mu L} \tag{7.26}$$

where

n_j = number of tubes of radius r_j
N = number of groups of tubes of different radii

This expression can be rewritten as:

$$q = \frac{C}{\mu}\frac{\Delta p}{L} \tag{7.27}$$

where C is the flow coefficient $\left(\pi/8\sum_{j=1}^{N} n_j r_j^4\right)$. It is evident that Eq. (7.26) is similar to Eq. (7.7) (Darcy's law) where the coefficient C is equivalent to the permeability. Thus:

$$k = \frac{\pi}{8A}\sum_{j=1}^{N} n_j r_j^4 \tag{7.28}$$

where A is the total cross-sectional area, as illustrated in Figure 7.2.

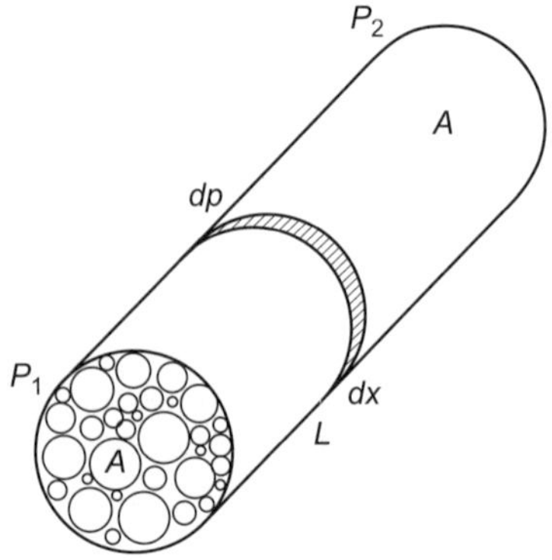

FIGURE 7.2 Poiseuille's flow system for straight capillaries.

Substituting $A = \pi R^2$ (R is the radius of total cross-sectional area) in Eq. (7.28) yields:

$$k = \frac{1}{8R^2} \sum_{j=1}^{N} n_j r_j^4 \tag{7.29}$$

If the radii r_j are the same for all tubes, this equation becomes:

$$k = \frac{nr^4}{8R^2} \tag{7.30}$$

The dimension of permeability is, therefore, L^2 (length squared). Thus, if L is in cm, $k = \text{cm}^2$. This measure is, however, too large to use with porous media, and the units of darcy, or mD, are preferred by the oil and gas industry.

This approach is, of course, an oversimplification of fluid flow in porous media, as the pore spaces within rocks seldom resemble straight, smooth-walled capillary tubes of constant diameter.

LINEAR FLOW THROUGH FRACTURES AND CHANNELS

Oil reservoirs with fracture–matrix porosity also contain solution channels. The matrix (intergranular porosity) is usually of low permeability and contains most of the oil (96–99%). Although these fractures and solution channels may not contain a significant volume of oil, generally less than 4% of the total oil in a reservoir, they are very important to the attainment of economic production rates [6]. Fracture porosity is common in many sedimentary rocks and is formed by structural failure of the rock under loads caused by various forms of diastrophism, such as folding and faulting [8]. Solution or vuggy porosity results from leaching of carbonate rocks by circulating acidic waters. Figure 7.3a,b shows porosity derived from fracturing and

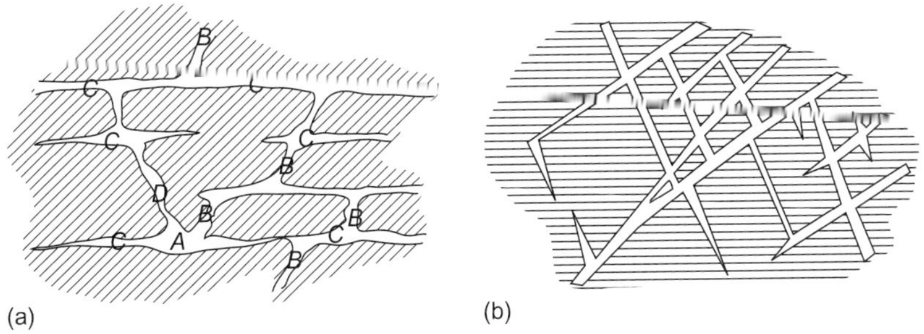

FIGURE 7.3 (a) Carbonate rock showing porosity: A, vugs; B, joint channels; C, bedding plane channels; D, solution channel [5]. (b) Carbonate rock showing porosity derived from fracturing and fissuring [5].

fissuring, and from solution along joints and bedding planes, respectively. Reservoir performance of most carbonates is considerably different than that of sandstone reservoirs due to the presence of strong directional permeability. In sandstone reservoirs, vertical permeability, k_v, is generally much less than horizontal permeability, k_n. In contrast, k_v, in carbonate reservoirs commonly exceeds k_h due to the dissolving effects of hot and acidic compaction-derived fluids moving upward, creating channels and vugs and enlarging existing fractures [7]. In sucrosic dolomite reservoirs with intergranular porosity, k_v is often approximately equal to k_h. Performance of sucrosic dolomites with intergranular interrhombohedral porosity is similar to that of sandstones [5].

FLOW THROUGH FRACTURES

The significance of the fractures as fluid carriers can be evaluated by considering a single fracture extending for some distance into the body of the rock and opening into the wellbore, as shown in Figure 7.4 [9]. Recalling the classical hydrodynamics equation for flow through slots of fine clearances and unit width as reported respectively by Croft and Kotyakhov [10, 11]:

$$q = \frac{h^3 w_f \Delta p}{12\,\mu L} \tag{7.31}$$

where

h = height (or thickness) of fracture, cm
w_f = width of fracture, cm
L = length of fracture, cm
μ = fluid viscosity, poise
Δp = pressure drop $(p_1 - p_2)$, dynes/cm^2.

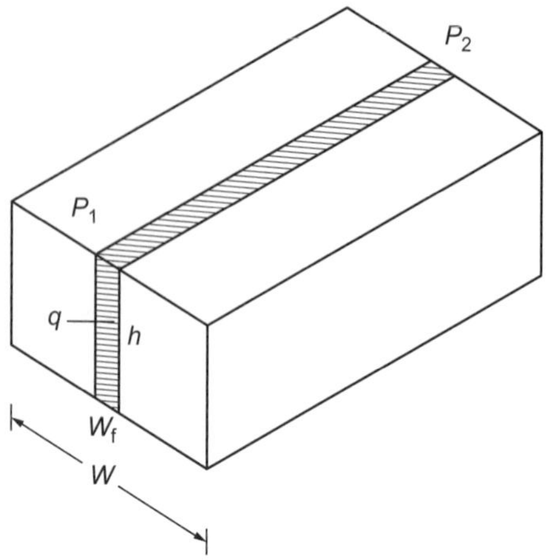

FIGURE 7.4 Linear model for fracture flow.

The actual velocity of the fluid flowing through the fracture is thus:

$$v = \frac{q}{w_f h} = \frac{h^2}{12} \frac{\Delta p}{\mu L} \tag{7.32}$$

Assuming that the porosity of the fracture is unity and the connate water saturation within the fracture is zero, the actual velocity (according to Darcy's law where Δp is expressed in dynes/cm^2, k in darcy, μ in poise, and L in cm) is:

$$v = (9.869 \times 10^{-9} k_f) \frac{\Delta p}{\mu L} \tag{7.33}$$

Combining Eqs. (7.32) and (7.33) and solving for the permeability of the fracture k_f (where w_f is expressed in cm and k_f in darcy):

$$k_f = 8.444 \times 10^6 w_f^2 \tag{7.34}$$

Fractures are classified as open (visible open space), closed (no visible open space in thin section), partially filled, or completely filled [8]. Many carbonate reservoirs exhibit fractures with some degree of filling, which may consist of crystals of calcite, dolomite, pyrite, gypsum, etc. Precipitates from the leaching solution, which circulates through the carbonate rock and deposits fine particles inside the fractures and vugs, contribute to the filling process. This leads to fractures with porosities ranging from a few percent to 100%. In addition, the connate water saturation in these fractures can be zero or 100% depending on the preferential wettability of the reservoir rock.

Equation (7.34) is, therefore, valid only for the case where the fracture is totally open and clean of any filling particles, that is, $\phi_f = 1$. It also assumes that the connate water saturation in the secondary pores is zero, or

$S_o = 100\%$ such as in reservoirs where the oil entered into a tight, oil-wet formation by upward migration along fractures from deeper zones. The Ain Zalah oil field, Iraq, appears to be such a reservoir [12]. In cases where $\phi_f < 1$ and $S_{wcf} > 0$, Eq. (7.34) must be modified. However, determining the values of the fracture porosity and the connate water saturation within the fracture is difficult even with whole core analysis, because cores tend to break along the natural fracture plane as they are brought to the surface. In addition, many fractures form during the process of core recovery. The most common laboratory technique for estimating directly the matrix and fracture porosity was presented in 1950 by Locke and Bliss [13]. The actual permeability of the fracture can be determined from the equation of actual velocity of the fluid flowing through the fracture:

$$v_a = \frac{v}{\phi_f(1 - S_{wcf})} \tag{7.35}$$

where ϕ_f is the fractional porosity of the fracture and S_{wcf} is the connate water saturation in the fracture. By definition:

$$\phi_f = \frac{w_f h}{A} = \frac{w_f h}{Wh} = \frac{w_f}{W} \tag{7.36}$$

The apparent velocity from Eq. (7.35) is:

$$v = v_a(1 - S_{wcf})\phi_f \tag{7.37}$$

where the actual velocity is expressed as the actual rate of fluid flow through the fracture divided by the fracture area, or:

$$v_a = \frac{q}{A_f} = \frac{q}{w_f h} \tag{7.38}$$

and the flow rate q is expressed by Eq. (7.31). Substituting for v_a and q in Eq. (7.37) gives:

$$v = w_f^2 \phi_f (1 - S_{wcf}) \frac{\Delta p}{12\mu L} \tag{7.39}$$

Equating this expression with Darcy's law (Eq. (7.33)) and solving for the actual permeability of the fracture (in darcy) yields:

$$k_f = 8.444 \times 10^6 (1 - S_{wcf}) \phi_f w_f^2 \tag{7.40}$$

Example

A cubic block of a carbonate rock with an intercrystalline–intergranular porosity system has a matrix porosity of 19%. The permeability of the matrix is 1 mD. Calculate:

1. The permeability of the fracture if each square foot contains one fracture in the direction of fluid flow

2. The flow rate in field units through the fracture and the fracture–matrix system

The width of the fracture is 2.5×10^{-3} in., the viscosity of the flowing fluid is 1.5 cP, and Δp across this block is 10 psia.

Solution

1. The permeability of a fracture is estimated from Eq. (7.34), where

$$w_f = 2.5 \times 10^{-3} \times 2.54 = 6.35 \times 10^{-3} \text{ cm}$$

$$k_f = 8.444 \times 10^6 \times (6.35 \times 10^{-3})^2 = 340.5 \text{ darcy}$$

It is obvious from this extremely high value of permeability that fractures contribute substantially to the recovery of oil from tight formations that otherwise would be noncommercial. This contribution is actually even higher as 1 ft^2 of carbonate rock is generally likely to contain more than one fracture.

The flow rate through the fracture only can be estimated from Darcy's law (Eq. (7.6)), where $L = 1$ ft, $\Delta p = 10$ psia, $k = 340.5$ darcy, $\mu = 1.5$ cP, and $A_f = 0.0025 \times 1 = 2.08 \times 10^{-4}$ ft^2. Thus:

$$q = 1.127 \frac{340.5 \times 2.08 \times 10^{-4} \times 10}{1.5 \times 1} = 0.533 \text{ bbl/day}$$

2. The flow rate through the matrix only is also obtained from Eq. (7.6), where the permeability of the matrix is 1 mD and $A_m = A_f = 1 - 2.08 \times 10^{-4} \approx 1$ ft^2. Thus:

$$q = 1.127 \times 10^{-3} \frac{1 \times 1 \times 10}{1.5 \times 1} = 0.0075 \text{ bbl/day}$$

The total flow rate through the block is:

$$q = 0.533 + 0.0075 = 0.54 \text{ bbl/day}$$

The importance of the fracture to the productivity of reservoirs can be better appreciated in terms of percentage contribution to the total flow rate, which for this case is 0.533/0.54 = 98.6%.

The volume of oil contained in the fractures and matrix is [6]:

$$V_o = V_{om} + V_{of} \tag{7.41}$$

where V_{om} is the volume of oil contained in the matrix, and V_{of} is the volume of oil contained in the fractures, which can be estimated from the following equations:

$$V_{om} = \frac{A_s h \phi_m (1 - \phi_f)(1 - S_{wm})}{B_o} \tag{7.42}$$

$$V_{of} = \frac{A_s h \phi_f (1 - S_{wf})}{B_o} \tag{7.43}$$

where

V_o = oil-in-place, m^3

A_s = surface area of producing formation, m^2

h = average thickness of formation, m
ϕ_m = fractional porosity of matrix only
ϕ_f = fractional porosity of fractures only
S_{wm} = water saturation in matrix
S_{wf} = water saturation in fractures

The recoverable volume of oil is:

$$V_{oR} = V_{om} E_m + V_{of} E_f \tag{7.44}$$

where
E_m = recovery factor for the matrix, fraction
E_f = recovery factor for the fractures, fraction

If the permeability of the matrix is negligible, that is, less than 0.1 mD, $V_{oR} = V_{of} E_f$.

The average permeability of the fracture–matrix flow system can be obtained from:

$$k_{mf} = \left(\frac{n_f w_f h}{A}\right) k_f + \left(1 - \frac{n_f w_f h}{A}\right) k_m \tag{7.45}$$

where
k_f = fracture permeability
k_m = matrix permeability
A = total cross-sectional area
n_f = number of fractures per unit area
w_f = fracture width
h = fracture height

The average permeability of the carbonate reservoir of the above example can be estimated from Eq. (7.45): where

$$w_f = 0.0025 = 2.08 \times 10^{-5} \text{ ft}, A = 1 \text{ ft}^2, n_f = 1, \text{ and } h = 1 \text{ ft}$$

Inasmuch as:

$$\frac{n_f w_f h}{A} = 2.08 \times 10^{-5}$$

Therefore:

$$k_{mf} = (2.08 \times 10^{-5})(340.5) + (1 - 2.08 \times 10^{-5})(10^{-3})$$
$$= 0.072 \text{ darcy}$$

FLOW THROUGH SOLUTION CHANNELS

Craft and Hawkins, and Aguilera, combined Poiseuille's law for viscous flow of liquids through capillary tubes with Darcy's law for steady-state linear flow of incompressible fluids to estimate the permeability of solution

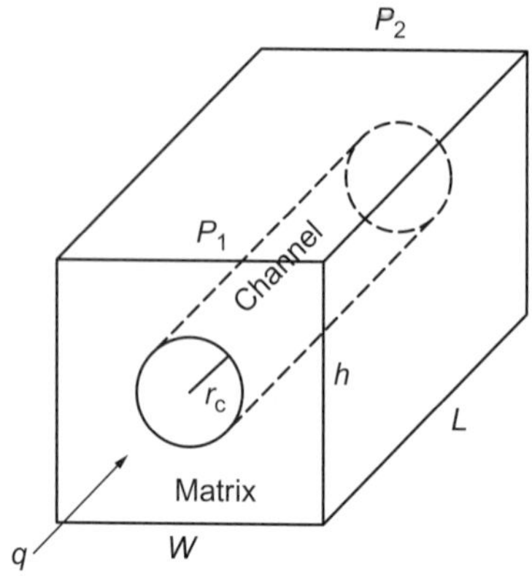

FIGURE 7.5 Channel–matrix system.

channels (Figure 7.5) [1, 15]. The actual volumetric rate of the fluid flowing through a capillary tube of radius r_c and length L is:

$$q = \frac{\pi r_c^4}{8} \frac{\Delta p}{\mu L} \tag{7.46}$$

From Darcy's law, assuming the channel porosity, ϕ_c, and the irreducible water saturation, S_{iwc}, are equal to unity and zero, respectively, the flow rate is:

$$q = 9.369 \times 10^{-9} \pi r_c^2 k \frac{\Delta p}{\mu L} \tag{7.47}$$

Equating Eqs. (7.46) and (7.47), and solving for the permeability of the solution channel yields:

$$k_c = 12.6 \times 10^6 {r_c}^2 \tag{7.48}$$

where

k_c = solution channel permeability, darcy
r_c = radius of tubular channel, cm

Porosity development in some carbonate reservoirs is due to leaching of carbonate rocks by mineralizing waters. Precipitates from this circulating water may be responsible for filling previously existing pores and channels with a variety of fine particles (salt, chert, anhydrite, and gypsum), making the porosity of the solution channel less than unity [8, 15]. Furthermore, the water saturation in these channels, which formed due to circulating water, is unlikely to be zero. Thus, the actual area open to flow is:

$$A_a = \phi_c (1 - S_{iwc}) \pi r_c^2 \tag{7.49}$$

and Eq. (7.48) becomes:

$$k_c = 12.6 \times 10^6 (1 - S_{iwc}) \phi_c r_c^2 \tag{7.50}$$

where

ϕ_c = solution channel porosity
S_{iwc} = irreducible water saturation in the channel

The average permeability of a channel–matrix flow system can be calculated from the following equation [13]:

$$k_{mc} = \left(\frac{n_c \pi r_c^2}{A}\right) k_c + \left(1 - \frac{n_c \pi r_c^2}{A}\right) k_m \tag{7.51}$$

where

k_c = permeability of channels, darcy
k_m = permeability of matrix, darcy
A = cross-sectional area, cm^2
n_c = number of channels per unit area
r_c = solution channel radius, cm

Carbonate reservoirs dominated by a vugular-solution porosity system exhibit a wide range of permeability. The permeability distribution may be relatively uniform, or quite irregular.

Example

A cubic sample of a limestone formation has a matrix permeability of 1 mD and contains five solution channels per square foot. The radius of each channel is 0.05 cm. Calculate:

1. The solution-channel permeability assuming a vug-porosity of 3% and an irreducible water saturation in these channels equal to 18%
2. The average permeability of this rock

Solution

1. The permeability of the solution-channel can be obtained from Eq. (7.50):

$$k_c = 12.6 \times 10^6 (1 - 0.18)(0.03)(0.05^2) = 775 \text{ darcy}$$

Using Eq. (7.48), that is, assuming $\phi_c = 1$ and $S_{iwc} = 0$, the permeability of a channel is 31,500 darcy, which is more than 40 times the value of k_c obtained from using Eq. (7.50) and, therefore, unrealistic.

2. The average permeability of this block containing five channels is estimated from Eq. (7.51), where $A = 1 \text{ ft}^2 = 929 \text{ cm}^2$, and $n_c \pi r^2/A = 5\pi(0.05^2)/929 = 42 \times 10^{-6}$:

$$k_{mc} = 42 \times 10^{-6} \times 0.775 \times 10^6 + (1 - 42 \times 10^{-6})(1) = 36.5 \text{ mD}$$

This example illustrates the importance of estimating the actual irreducible water saturation and porosity of the solution channels and fractures. These parameters are important in determining oil-in-place within vugular pores and fractures, and ignoring them can lead to overestimating the production capacity of wells in carbonate reservoirs [14–17].

RADIAL FLOW SYSTEMS

Figure 7.6 illustrates a single producing well located in a radial reservoir system. Flow in this system converges from the external boundary of radius r_e and pressure p_e to the well of radius r_w and pressure p_w. The flow rate at any radius r and pressure p, according to Darcy's law for radial incompressible fluid flow, is:

$$q = \frac{2\pi rhk}{\mu}\frac{\partial p}{\partial r} \tag{7.52}$$

In radial flow, the minus sign in Darcy's law is no longer required as the radius r increases (from r_w to r_e) in the same directions as pressure.

Combining Darcy's law, the law of conservation of mass, and the equation of state, the following general mathematical expression describing the flow of fluids in porous media, known as the diffusivity equation, can be derived:

$$\frac{\partial^2 p}{\partial r^2} + \frac{1}{r}\frac{\partial p}{\partial r} = \frac{\phi\mu c_t}{k}\frac{\partial p}{\partial t} \tag{7.53}$$

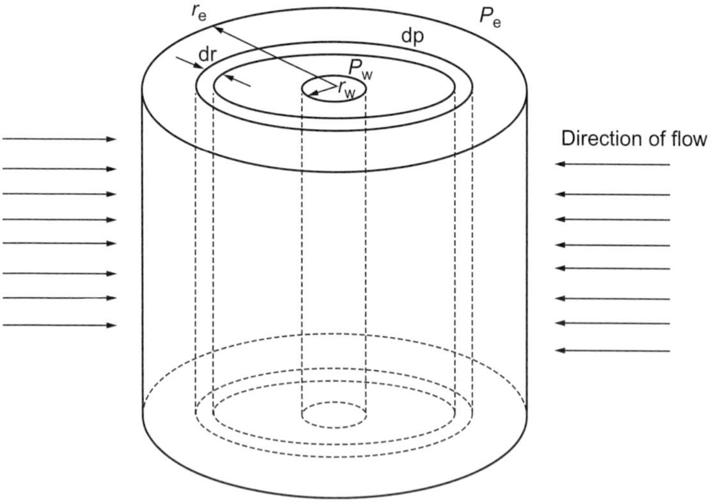

FIGURE 7.6 Ideal radial flow system [47].

The ratio $k/\phi\mu c_t$ is called the hydraulic diffusivity constant. Equation (7.53) is for the case of unsteady-state flow because it is time dependent. This flow regime is beyond the scope of this book and, therefore, will not be discussed. The solution of differential Eq. (7.53), of interest to the development of steady-state and pseudosteady-state flow equations, is for the case of a centrally located well producing at a constant volumetric rate. The exact form of these flow equations depends on the nature of external reservoir boundaries. Three basic outer boundary conditions exist: infinite pressure, constant pressure, and no-flow.

STEADY-STATE FLOW

Strictly speaking, steady-state flow can occur only if the flow across the drainage boundary, r_e, is equal to the flow across the wellbore wall at well radius r_w, and the fluid properties remain constant throughout the reservoir. These conditions may never be met in a reservoir; however, in petroleum reservoirs produced by a strong water drive, whereby the water influx rate at r_e equals the well producing rate, the pressure change with time is so slight that it is practically undetectable. In such cases, the assumption of steady state is acceptable [18]. Steady-state flow equations are also useful in analyzing the reservoir conditions in the vicinity of the wellbore for short periods of time, even in an unsteady-state system [19].

Mathematically, true steady-state flow occurs when $\partial p/\partial t = 0$, which reduces the diffusivity equation to:

$$\frac{\partial^2 p}{\partial r^2} + \frac{1}{r}\frac{\partial p}{\partial r} = \frac{\partial}{\partial r}\left(r\frac{\partial p}{\partial r}\right) = 0 \tag{7.54}$$

Integrating this differential equation gives:

$$r\frac{\partial p}{\partial r} = C_i \tag{7.55a}$$

where C_i is a constant of integration. For constant flow rate at the wellbore, one can impose the following condition on the pressure gradient at the well (from Darcy's law):

$$\frac{\partial p}{\partial r} = \left(\frac{q\mu}{2\pi kh}\right)\frac{1}{r_w} \tag{7.55b}$$

Combining these two expressions and solving for C_i at the well:

$$C_i = \frac{q\mu}{2\pi kh} \tag{7.55c}$$

Substituting this term in Eq. (7.55a), separating variables, and integrating between r_w and r_e, where the pressures are p_w and p_e, respectively:

$$\frac{q\mu}{2\pi kh}\int_{r_w}^{r_e}\frac{dr}{r}=\int_{p_w}^{p_e}dp \tag{7.55d}$$

Integrating Eq. (7.55d) and solving explicitly for the volumetric flow rate, q, results in the following equation:

$$q=\frac{2\pi kh(p_e-p_w)}{\mu\ln(r_e/r_w)} \tag{7.56}$$

Expressing all the terms in practical oil field units, this relationship becomes:

$$q_{sc}=\frac{0.00708kh(p_e-p_w)}{\mu B_o\ln(r_e/r_w)} \tag{7.57}$$

where

q_{sc} = surface production rate at T_{sc} = 60°F and p_{sc} = 14.7 psia, STB/day
k = formation permeability, mD
h = formation thickness, ft
μ = oil viscosity, cP
p_e = external pressure, psia
p_w = well pressure, psia
r_w = wellbore radius, ft
r_e = external radius $(43{,}560\ A/\pi)^{1/2}$, ft
A = drainage area, acres
B_o = oil formation volume factor, bbl/STB

This equation is only valid for the case where the well is located at the center of a circular drainage area.

The external pressure p_e is generally approximated by the static pressure of the reservoir, especially in the case of an infinite reservoir. In strong water-drive reservoirs, p_e is equivalent to the initial reservoir pressure p_i. If the pressure p_e cannot be determined with some reasonable accuracy, Eq. (7.57) should be expressed in terms of the average reservoir pressure, $\overline{p}$, which can be easily obtained from a pressure buildup or drawdown test [20].

Inasmuch as $r_e \gg r_w$, the volumetric average reservoir pressure may be expressed as [1]:

$$\overline{p}=\frac{2}{r_e^2}\int_{r_w}^{r_e}pr\,dr \tag{7.58}$$

where p is the reservoir pressure at any radius r. From Eq. (7.56):

$$p=p_w+\frac{q\mu}{2\pi kh}\ln\left(\frac{r}{r_w}\right) \tag{7.59}$$

Substituting the above expression into Eq. (7.58), integrating, and solving explicitly for the volumetric flow rate, one obtains (in oil field units):

$$q_{sc} = \frac{0.00708kh(\bar{p} - p_w)}{\mu B_o[\ln(r_e/r_w) - 0.5]} \tag{7.60}$$

assuming that the term r_w^2/r_e^2 is negligible. It is important to emphasize that Eqs. (7.56) and (7.60) are strictly valid for the case of a single well in an infinite reservoir and strong water-drive reservoir producing at steady-state flow conditions. These equations also apply equally well in an oil reservoir experiencing pressure maintenance by water injection or gas injection.

PSEUDOSTEADY-STATE FLOW

In bounded cylindrical reservoirs, the pseudosteady-state flow regime is common at long producing times. In these reservoirs, also called volumetric reservoirs, there can be no flow across the impermeable outer boundary, such as a sealing fault, and fluid production must come from the expansion and pressure decline of the reservoir. This condition of no-flow boundary is also encountered in a well that is offset on four sides.

If there is no flow across the external boundary, then after sufficiently long producing time elapses the pressure decline throughout the drainage volume becomes a linear function of time. Therefore, for a well producing at a constant production rate, the rate of pressure decline is constant:

$$\frac{\delta p}{\delta t} = \frac{-q}{cV_p} \tag{7.61}$$

where V_P is the drainage pore volume, which is equal to $\pi r_e^2 h\phi$, and c is the compressibility of the fluid at the average reservoir pressure. Substituting Eq. (7.61) into the diffusivity equation (Eq. (7.53)), integrating twice and solving for the flow rate (in oil field units) gives [4]:

$$q_{sc} = \frac{0.00708\, kh(p_e - p_w)}{\mu B_o[\ln(r_e/r_w) - 0.5]} = \frac{0.00708\, kh(p_e - p_w)}{\mu B_o \ln(0.606 r_e/r_w)} \tag{7.62}$$

If the external pressure, p_e, is unknown, Eq. (7.62) should be derived in terms of the average reservoir pressure, $\bar{p}$. The pressure p at any radius r of the bounded reservoir is obtained from Eq. (7.62):

$$p = p_w + \frac{2\pi q\mu}{kh}\left[\ln(r/r_w) - 0.5(r/r_e)^2 + 0.5(r_w/r_e)^2\right] \tag{7.63}$$

If Eq. (7.63) is used in Eq. (7.58) and the integration carried out, the following expression is obtained for q_{sc} (in oil field units), assuming $r_e \gg r_w$:

$$q_{sc} = \frac{0.00708kh(\bar{p} - p_w)}{\mu B_o[\ln(r_e/r_w) - 0.75]} = \frac{0.00708kh(\bar{p} - p_w)}{\mu B_o \ln(0.472 r_e/r_w)} \tag{7.64a}$$

TABLE 7.1 Values of the Shape Factor for Various Well Locations and Drainage Area Shapes

System	C_A	System	C_A
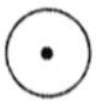	31.62	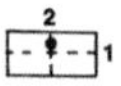	21.83
	30.88	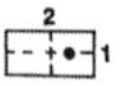	4.51
	31.6	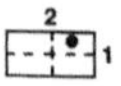	2.08
	27.6	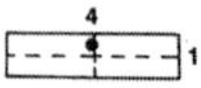	2.69
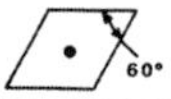	27.1	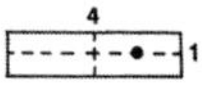	0.232
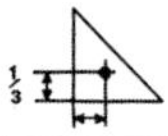	21.9	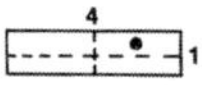	0.115
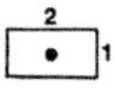	21.83	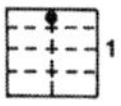	3.335
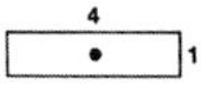	5.38	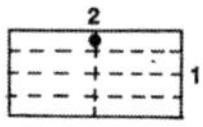	3.157
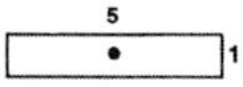	2.36	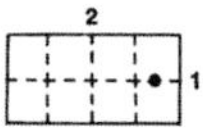	0.581
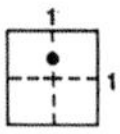	12.98	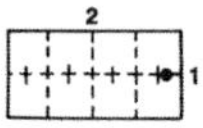	0.111
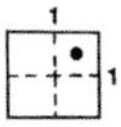	4.51		

For other well locations, drainage area shapes, and external boundary conditions, the general form of Eq. (7.64a) is:

$$q_{sc} = \frac{0.00708kh(\overline{p} - p_w)}{\mu B_o \ln(r'_e / r_w)} \tag{7.64b}$$

where r'_e is an effective drainage radius that includes the effect that a well placement in a given drainage area will have on the performance of the well. The effective radius can be written as:

$$r'_e = 1.498\sqrt{\frac{A}{C_A}} \tag{7.64c}$$

where A is the drainage area (ft^2), and C_A is the shape factor, as shown in Table 7.1 [20, 25].

When external reservoir boundaries are mixed, the methods of obtaining flow equations become more complex, especially during unsteady state. During steady-state flow, however, this system can be approximated by a radial cylindrical reservoir where only a fraction f of the reservoir periphery is open to water encroachment. The fraction f is referred to here as the drainage boundary index. This partial water-drive reservoir is produced by two processes:

1. Expansion of the reservoir fluid
2. Displacement of the reservoir fluid by water

Kumar presented an equation giving the flow rate at any radius r between r_w and r_e [21]:

$$q_r = q\left[1-(1-f)\frac{r^2}{r_e^2}\right] \tag{7.65}$$

where q is the wellbore flow rate. If Eq. (7.65) is substituted in Darcy's law (Eq. (7.52)), and integration is carried out between r_w and r_e, where the pressures are p_w and p_e, respectively, and assuming $r_w^2/r_e^2 = 0$, one obtains:

$$q_{sc} = \frac{0.00708kh(p_e - p_w)}{\mu B_o[\ln(r_e/r_w) - 0.50(1-f)]} \tag{7.66}$$

It is evident from Eq. (7.66) that:

1. $f=0$ represents no-flow condition at r_e, because for this value of f, Eq. (7.66) becomes similar to Eq. (7.62), which is specifically derived for the case of bounded reservoirs under pseudosteady state.
2. $f=1$ represents a full active water-drive reservoir, and Eq. (7.66) becomes similar to Eq. (7.57). $f=1$ also represents a balanced five-spot water injection pattern with unit mobility ratio.
3. $f>1$ indicates that the fluid volume entering a reservoir at r_e is greater than the fluid volume entering the wellbore at r_w, such as under excess fluid injection.

Equation (7.66) provides a way to determine the strength of water drive f, if the producing rate and pressure drop are known. The parameter f can be determined more accurately from transient well test analysis [22].

If the average reservoir pressure, $\bar{p}$, is used instead of the external pressure, p_e, which is practically impossible to establish in such a mixed boundary system, Eq. (7.66) becomes:

$$q_{sc} = \frac{0.00708kh(\bar{p} - p_w)}{\mu B_o[\ln(r_e/r_w) - 0.75 + 0.25f]} \tag{7.67}$$

This expression is similar to Eqs. (7.60 and (7.64) for $f=1$ and $f=0$, respectively, and can easily be derived by substituting p into Eq. (7.58) and integrating. p is obtained from Eq. (7.66) by assuming $r_e > r_w$ and replacing p_e with p

and r_e with r. Combining Eqs. (7.66) and (7.67) yields a very useful relationship for determining the external pressure of a mixed boundary system:

$$p_e = \bar{p} + \frac{q_{sc}\mu B_o}{141.2kh}\left(\frac{1+f}{4}\right) \quad (7.68)$$

For a well in a closed outer boundary reservoir $f = 0$, and because at $t = 0$ the external pressure, p_e, is equivalent to the initial reservoir pressure, p_i, Eq. (7.68) can be written in oil field units as follows:

$$p_i - \bar{p} = \frac{q\mu B_o}{141.2kh}\left(\frac{1}{4}\right) \quad (7.69)$$

The right-hand side of this equation corresponds to the amount of fluid produced, causing the reservoir pressure to drop from p_i to $\bar{p}$. It can be demonstrated that this pressure drop is also expressed as:

$$p_i - \bar{p} = \left(\frac{0.2339 q_{sc} B_o}{A\phi c_t h r_e^2}\right)t = \left(\frac{0.0744 q_{sc} B_o}{\phi c_t h r_e^2}\right)t \quad (7.70)$$

If Eqs. (7.69) and (7.70) are combined, one obtains for time t:

$$t = 474.5 \frac{\phi c_t r_e^2 \mu}{k} \quad (7.71)$$

where t is in hours, c_t is the total compressibility in psi^{-1}, and the permeability is in mD. Craft and Hawkins defined this time as the readjustment time, t_r, or the time required to establish a logarithmic pressure distribution between r_w and r_e [1]. For a well in a fully active water drive reservoir, that is, $f = 1$, the constant 0.25 in Eq. (7.69) is replaced by 0.50, and the constant 474.5 in Eq. (7.71) is replaced by 949. Generally, steady-state flow equations should be used only when t_r is small compared to the total producing life of the reservoir. If t_r is too large, as it is often the case in fully active water-drive reservoirs, unsteady-state flow equations must be used.

Skin Zone

In many cases, it has been found that the permeability in the vicinity of the wellbore differs from that in the major portion of the reservoir as shown in Figure 7.7. This zone of altered permeability, k_s, and radial extent, r_s, is called the "skin," and the degree of alteration is expressed in terms of the skin factor s [23, 24]. The permeability of the skin zone can be reduced ($s > 0$) as a result of drilling and well completion practices as discussed in the next chapter.

The average permeability of the formation in the vicinity of the wellbore also can be higher ($s < 0$) than that in the major portion of the reservoir after fracturing or acidizing the well at completion. Therefore, all the radial flow rate equations in this section, which were derived on the basis that the

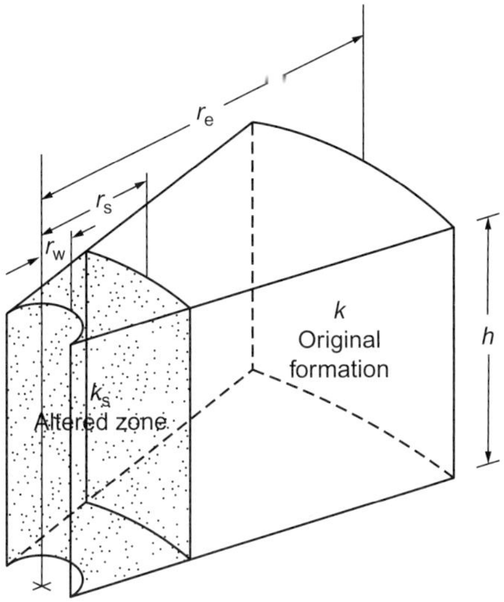

FIGURE 7.7 Skin zone.

permeability of the formation is the same between r_w and r_e, must be modified to include the effect of skin. This can be done either by subtracting the additional pressure drop (due to wellbore damage) or the pressure increase (due to stimulation), Δp_s, from the $(p_e - p_w)$ or $(\bar{p} - p_w)$ terms, where

$$\Delta p_s = \left(\frac{q_{sc}\mu B_o}{0.00708kh}\right)s \tag{7.72}$$

or by replacing the wellbore radius term, r_w, by an effective or apparent wellbore radius, r_{wa}, which is estimated from [24]:

$$r_{wa} = r_w e^{-s} \tag{7.73}$$

For example, if the well is damaged or stimulated, Eq. (7.67) becomes:

$$q_{sc} = \frac{0.00708kh(\bar{p} - p_w - \Delta p_s)}{\mu B_o(\ln(r_e/r_w) - 0.75 + 0.25f)} \tag{7.74}$$

or

$$q_{sc} = \frac{0.00708kh(\bar{p} - p_w)}{\mu B_o(\ln(r_e/r_{wa}) - 0.75 + 0.25f)} \tag{7.75}$$

or, because $\ln(r_e/r_{wa}) = \ln(r_e/r_w) + s$:

$$q_{sc} = \frac{0.00708kh(\bar{p} - p_w)}{\mu B_o(\ln(r_e/r_{wa}) - 0.75 + 0.25f)} \tag{7.76}$$

Equation (7.76) is very general and more accurate than Eqs. (7.60) and (7.64) because it includes the effect of wellbore condition, as well as the effect of the outer boundary of the reservoir. Equation (7.76) is valid for both steady and pseudosteady states, depending on the value of the parameter f. The skin factor is best obtained from pressure transient tests [18, 20].

Example

1. Calculate the production rate for an oil well in a 160-acre drainage area where the average pressure is partially maintained at 1,850 psia by water injection at the boundary. The following parameters are available:

$r_w = 0.5$ ft	$\mu = 2.2$ cP
$h = 16$ ft	$B_o = 1.1$ bbl/STB
$k = 180$ mD	$p_w = 1{,}230$ psia
$s = 2$	$f = 0.25$

2. What is the ideal production, that is, no skin damage?

Solution

In oil field units, Eq. (7.76) can be expressed as

$$q_{sc} = \frac{0.00708kh(\bar{p} - p_w)}{\mu B_o[\ln(r_e/r_w) - 0.75 + 0.25f + s]} \tag{7.77}$$

1. The radius of the drainage area is:

$$r_e = (43{,}560A/\pi)^{0.5} = (43{,}560 \times 160/\pi)^{0.5} = 1{,}489 \text{ ft}$$

The production rate of this well is equal to:

$$q_{sc} = \frac{(0.00708)(180)(16)(1850 - 1{,}230)}{(2.2)(1.1)[\ln(1{,}489/0.5) - 0.75 + (0.25)(0.25) + 2]} = 561 \text{ STB/day}$$

2. The ideal production rate of this well is obtained by letting $s = 0$ in Eq. (7.77), which gives $q_{sc} = 715$ STB/day. Thus, if this well is treated to remove the skin damage, an additional 154 STB/day will be produced, an increase of approximately 27%.

Dimensionless Pressure

Steady-state and pseudosteady-state radial flow equations previously presented are strictly applicable to the case where the well is located at the center of a circular drainage area. For other well locations, drainage area shapes, and external boundary conditions, the dimensionless form of these flow equations is given by:

$$q_{sc} = \left(\frac{kh}{141.2\mu B_o}\right)\left(\frac{\Delta p}{p_D + s}\right) \tag{7.78}$$

In reservoir systems where the pressure change with time is negligible and the assumption of steady-state radial flow is applicable, the dimensionless pressure drop, p_D, is:

$$p_D = \begin{Bmatrix} \ln(r_e/r_w) \text{ if } \Delta p = p_e - p_w \\ \ln(0.606 r_e/r_w) \text{ if } \Delta p = \overline{p} - p_w \end{Bmatrix} \quad (7.79)$$

As producing time increases, the pressure decline throughout the reservoir becomes a linear function of time, and the assumption of pseudosteady-state flow becomes applicable. When this flow regime occurs, and Δp is equal to $(p_e - p_w)$ or $(p_i - p_w)$, the dimensionless pressure, P_D, is:

$$p_D = 2\pi t_{DA} + \frac{1}{2}\ln\left(\frac{2.2458A}{r_w^2 C_A}\right) \quad (7.80)$$

where C_A is a dimensionless shape factor whose value depends on reservoir shape and well location as shown in Table 7.1. The dimensionless time, t_{DA}, is defined by:

$$t_{DA} = \left(\frac{0.0002637k}{\phi\mu C_t A}\right)t \quad (7.81)$$

where the drainage area, A, is expressed in ft^2 and producing time, t, is in hours.

For pseudosteady state and $\Delta p = \overline{p} - p_w$, the dimensionless pressure P_D in Eq. (7.78) is given by:

$$p_D = \frac{1}{2}\ln\left(\frac{2.2458A}{r_w^2 C_A}\right) \quad (7.82)$$

For drainage areas with mixed outer boundaries and $\Delta p = p_e - p_w$, the expression for p_D during pseudosteady state is [22]:

$$p_D = 2\pi(1-f)t_{DA} + \frac{1}{2}\ln\left[\frac{2.2458A(4^f)}{r_w^2 C_A}\right] \quad (7.83)$$

As noted earlier, after an extended time of production at a constant rate, the bottom-hole flowing pressure, p_w, becomes a linear function of time. A Cartesian plot of p_w versus time should yield a straight line with slope m^*. For bounded drainage areas:

$$m^* = -\frac{0.234 q_{sc} B_o}{c_t V_p} \quad (7.84)$$

where m^* is expressed in psi/h. The pore volume of the drainage area, V_P (ft_3), is:

$$V_p = \phi h A = \pi r_e^2 h \phi \quad (7.85)$$

Equation (7.84) is commonly used to calculate the pore volume, V_P, of a bounded reservoir. For mixed-boundary systems, such as in reservoirs under the influence of a partial water drive or in unbalanced injection patterns, the pseudosteady-state flow regime occurs only for small values of the drainage boundary index *f*. In this case, the slope of the straight line portion that corresponds to pseudosteady state, m^*, can be obtained from the derivative of Eq. (7.83) with respect to dimensionless time t_{DA}:

$$\frac{\partial p_D}{\partial t_{DA}} = p'_D = 2\pi(1-f) \tag{7.86}$$

Substituting for p_D and t_{DA}, and solving explicitly for $m^* = dp/dt$ gives:

$$m^* = -\left(\frac{0.234 q_{sc} B_o}{c_t V_p}\right)(1-f) \tag{7.87}$$

Equation (7.87) can be used to calculate V_p if *f* is known from pressure transient testing [22]. If the pore volume is known from other sources, then Eq. (7.87) provides a way to calculate the index *f*. Note that for $f = 0$, that is, the drainage boundary is closed, Eqs. (7.84) and (7.87) are identical. For the rare case where $f = 1$, the rate of change of pressure with time, dp/dt, is zero, and steady-state flow becomes the dominant regime.

The dimensionless pressure of a well producing at a constant rate in the infinite-acting portion of the pressure versus time curve, that is, during unsteady state, is approximated by:

$$p_D = \frac{1}{2}\ln\left(\frac{5.92 \times 10^{-4} kt}{\phi \mu c_t r_w^2}\right) \tag{7.88}$$

In this case, the pressure drop, Δp, in Eq. (7.78) is actually $(p_i - p_w)$, where p_i is the initial reservoir pressure. Equation (7.88) is applicable only if $(0.0002637 kt/\phi \mu c_t r_w^2) > 100$.

Example

A Cartesian plot of pressure data recorded during a constant rate well test in a water-driven oil reservoir yielded a pseudosteady-state straight line with slope −0.26 psi/h. Other pertinent reservoir and well data include the following:

$A = 7.72$ acres	$q_{sc} = 350$ STB/day
$h = 49$ ft	$B_o = 1.136$ bbl/STB
$\phi = 0.23$	$c_t = 17 \times 10^{-6}$ psi^{-1}

Determine the drainage boundary index *f*. Is the pressure at the drainage boundary constant?

Solution

The drainage boundary index f is calculated from a rearranged form of Eq. (7.87):

$$f = 1 + \left[\frac{c_t V_p}{0.234 q_{sc} B_o}\right] m^* \tag{7.89}$$

where the drainage pore volume is:

$$V_P = Ah\phi = 7.72 \times 43560 \times 49 \times 0.23 = 3.79 \times 10^6 \text{ft}^3$$

The boundary index is:

$$f = 1 + \left[\frac{17 \times 10^6 \times 3.79 \times 10^{-6}}{0.234 \times 350 \times 1.136}\right](-0.26) = 0.82$$

Since $f < 1$, the pressure at the drainage boundary is not constant because either the water drive is not very strong or only a fraction of the reservoir boundary is open to water drive.

RADIAL LAMINAR FLOW OF GAS

Three approaches are available for describing gas flow through porous rock.

1. If the reservoir pressure is high ($\overline{p} > 3,000$ psia), the radial flow equations of the previous section, even though they were developed strictly for the case of liquid flow, can be used to analyze gas flow by converting gas-flow rates from SCF/day to STB/day and calculating the formation volume factor in bbl/SCF from:

$$B_g = 0.00504 \frac{zT}{p} \tag{7.90}$$

   where the gas deviation factor z is estimated at the average reservoir pressure $\overline{p}$, and the reservoir temperature T is expressed in °R. Using this procedure can lead to large errors under certain conditions, as the diffusivity equation describing liquid flow in porous rock (Eq. (7.53)) was derived on the assumption that small pressure gradients are negligible. In low-permeability gas reservoirs, however, these gradients can be considerably high.
2. If the average reservoir pressure is low ($\overline{p} < 2,000$ psia), the radial gas-flow equations can be derived in terms of the pressure-squared function, p^2. This classical approach is discussed in the next section.
3. If the reservoir pressure is intermediate ($2,000 < \overline{p} < 3,000$ psia), the real gas pseudopressure function, $m(p)$, is more accurate than the pressure or the pressure-squared approach. Actually, in tight gas formations the $m(p)$

approach must be used, especially if the reservoir is produced at high rates. This function is defined as [26, 27]:

$$m(p) = 2\int_{p_b}^{p} \frac{p}{\mu(p)z(p)}\,dp \tag{7.91}$$

where p_b is an arbitrary base pressure, and $m(p)$ is expressed in psi^2/cP. Equation (7.91) only accounts for changes in μ and z, and fails to correct for changes in gas compressibility, c, and kinetic energy. When the real gas pseudopressure is used, the diffusivity equation (Eq. (7.54)) becomes:

$$\frac{\partial^2 m(p)}{\partial r^2} + \frac{1}{r}\frac{\partial m(p)}{\partial r} = \frac{\phi m_g c_g}{k}\frac{\partial m(p)}{\partial t} \tag{7.92}$$

The steady- or pseudosteady-state solutions of this diffusivity equation can be obtained using essentially the same mathematical procedure as that used to solve Eq. (7.54) for the flow of incompressible fluids. Thus, Eq. (7.77) is equivalent to:

$$q_{sc} = \frac{kh[m(\overline{p}) - m(p_w)]}{1{,}422T[\ln(0.472r_e/r_w) + 0.25f + s)]} \tag{7.93}$$

where

q_{sc} = gas-flow rate at T_{sc} = 60°F and p_{sc} = 14.7 psia, MSCF/day
k = permeability, mD
h = thickness of formation, ft
T = absolute reservoir temperature, °R
r_e = drainage radius, ft
r_w = wellbore radius, ft
f = drainage boundary index, dimensionless
s = total skin factor, dimensionless
$m(p_w)$, $m(\overline{p})$ = real gas pseudopressure at the well pressure and the average reservoir pressure, respectively, psi^2/cP

The real gas pseudopressure terms at any pressure, $m(p)$, can be obtained from published tables or by numerical integration (trapezoidal rule) [26, 28]. p can be converted to $m(p)$ by plotting the group $2p/\mu_g z$ versus p on a Cartesian graph. This group is calculated for several values of p using experimental values of μ_g and z. The area under the curve from any convenient reference pressure, generally zero, to p is the value of $m(p)$ corresponding to p.

Example

Given the following properties of a dry sweet gas (gravity = 0.61) at a temperature of 120°F, calculate the corresponding values of $m(p)$ by the numerical integration.

1	2	3
p, psia	z	μ, cP
400	0.955	0.0118
800	0.914	0.0125
1,200	0.879	0.0134
1,600	0.853	0.0145
2,000	0.838	0.0156

Solution

1	2	3	4	5	6	7
p (psia)	z	μ (cP)	$2p/\mu z$ (M psi/cP)	$(2p/\mu z)$ avg (M psia/cP)	$\Delta p x(2p/\mu z)$ avg (MM psi^2/cP)	$m(p)$ = Cum. Col. 6 (MM psi^2/cP)
400	0.955	0.0118	71.0	35.50	14.20	14.20
800	0.914	0.0125	140.0	105.52	42.21	56.41
1,200	0.879	0.0134	203.8	171.90	68.76	125.17
1,600	0.853	0.0145	258.7	231.24	92.50	217.66
2,000	0.838	0.0156	306.0	282.35	112.94	330.60

The $m(p)$ approach is theoretically a better method than p and p^2 approaches because it is valid for all pressure ranges, especially during the unsteady-state flow regime when μ_g and z may vary considerably. Inasmuch as only the radial flows of gas during the steady and pseudosteady states are considered, all the gas-flow equations in the remainder of this chapter will be expressed in terms of the pressure-squared approach, and μ_g and z will be assumed to remain constant at the average reservoir pressure. Several approaches are available in the literature for deriving radial gas-flow equations during the steady-state flow regime [27–29]. By treating natural gas as a highly compressible fluid, radial flow equations may be developed by combining Darcy's law (assuming laminar flow):

$$v = \frac{k}{\mu_g}\frac{\partial p}{\partial r} \tag{7.94}$$

the continuity equation:

$$\frac{1}{r}\frac{\partial}{\partial r}(\rho r v) = \phi\frac{\partial p}{\partial t} \tag{7.95}$$

and the equation of state for real gas:

$$\rho = \frac{M}{RT}\frac{p}{z} \tag{7.96}$$

Assuming c_g is approximately equal to $1/p$, the diffusivity equation describing the real gas flow in cylindrical porous and permeable rock is:

$$\frac{\partial^2 p^2}{\partial r^2} + \frac{1}{r}\frac{\partial p^2}{\partial r} = \frac{\phi \mu_g c_g}{k}\frac{\partial p^2}{\partial t} \tag{7.97}$$

where μ_g is estimated at the average reservoir pressure $\overline{p}$. This differential equation has essentially the same form as the diffusivity equation (Eq. (7.54)), which was derived for incompressible fluids, except that the dependent variable, p, has been replaced by p^2. This similarity suggests that the solutions to Eq. (7.97) also will be of the same form as those for Eq. (7.54).

Real gas-flow equations differ from incompressible fluid-flow equations because the gas-flow rate, q, varies with pressure due to the compressibility of gas. To make the wellbore flow rate a constant, let:

$$q = q_{sc} B_g \tag{7.98}$$

where q and q_{sc} are expressed, respectively, in bbl/day and MSCF/day, and the gas formation volume factor is defined by Eq. (7.90) in bbl/SCF. Thus:

$$q = (5.04 T q_{sc})\frac{z}{p} \tag{7.99}$$

and the volumetric flow rate, in bbl/day, at any radius in the reservoir, according to Darcy's law, where the permeability is expressed in mD, is:

$$q_r = 0.00708\frac{kh}{\mu_g}\left(r\frac{dp}{dr}\right) \tag{7.100}$$

If Eqs. (7.99) and (7.100) are substituted in Eq. (7.65) and the variables separated, one obtains:

$$\left[1-(1-f)\frac{r^2}{r_e^2}\right]\frac{dr}{r} = \left(\frac{1.404\times 10^{-3}\, kh}{T q_{sc}}\right)\frac{p\, dp}{\mu_g z} \tag{7.101}$$

Assuming μ_g and z remain constant at the average reservoir pressure and integrating between r_w and r_e, where the pressures are p_w and p_e, one finds:

$$\int_{r_w}^{r_e}\left[1-(1-f)\frac{r^2}{r_e^2}\right]\frac{dr}{r} = \left(\frac{1.404\times 10^{-3}\, kh}{q_{sc}\mu_g z T}\right)\int_{p_w}^{p_e} p\, dp \tag{7.102}$$

It can be shown from Eq. (7.102) that for $r_e \gg r_w$, the volumetric flow rate at standard conditions, including the skin factor s, is as follows:

$$q_{sc} = \frac{kh(p_e^2 - p_w^2)}{1422\mu_g z T(\ln(r_e/r_w) - 0.50(1-f) + s)} \tag{7.103}$$

The analysis of this equation with respect to f is similar to that of Eq. (7.66) for incompressible fluids; that is, if $f = 1$, the gas reservoir is under the influence of a full active water drive and the dominant flow regime is the steady state; if $f = 1$, then the reservoir is bounded and, therefore, the dominant flow regime is the pseudosteady state. Equation (7.103) can, of course, be used to determine the strength of the water drive, f, if the producing rate and pressure are known.

If the external pressure, p_e, is not known, an equation similar to Eq. (7.65) can be derived by expressing this equation in terms of the reservoir pressure p at any radius r:

$$p = \left\{ p_w^2 + \frac{1422\mu_g zTq_{sc}}{kh}\left[\ln\left(\frac{r}{r_w}\right) - 0.50(1-f) + s\right]\right\}^{0.5} \tag{7.104}$$

If the above expression is substituted in Eq. (7.58) and integration is carried out between two radii r_w and r_e, one obtains—after some algebraic manipulations—the following equation for q_{sc}:

$$q_{sc} = \frac{kh(\bar{p}^2 - p_w^2)}{1422\mu_g zT(\ln(r_e/r_w) - 0.75 + 0.25f + s)} \tag{7.105}$$

A useful relationship between the external boundary pressure and the average reservoir pressure can be obtained by equating Eqs. (7.103) and (7.105), and solving for p_e:

$$p_e = \left[\bar{p}^2 + \frac{\mu_g zTq_{sc}}{0.702 \times 10^{-3} kh}\left(\frac{1+f}{4}\right)\right]^{0.5} \tag{7.106}$$

The value of the water-drive (or drainage boundary) index is significant only if the petroleum reservoir is small, especially when $f = 0.50$.

For other well locations inside closed (or bounded) drainage area shapes ($f = 0$), the general form of Eq. (7.105) is:

$$q_{sc} = \frac{kh(\bar{p}^2 - p_w^2)}{1422\mu_g zT[\ln(r'_e/r_w) + s]} \tag{7.107}$$

where r'_e is given by Eq. (7.64c).

Example

A well is producing 275 MSCF/day from a gas reservoir under the influence of a partial water drive with an index of 0.5. Calculate the wellbore pressure and the pressure at the drainage boundary. The following reservoir and fluid properties are known:

$\mu_g = 0.035$ cP	$k = 5$ mD	$S_{wi} = 17\%$
$z = 0.95$	$h = 35$ ft	$\phi = 12\%$
$T = 130°$F	$r_e = 2{,}640$ ft	
$\bar{p} = 2{,}720$ psia	$r_w = 0.5$ ft	
$s = 0$		

Solution

The wellbore pressure can be obtained from Eq. (7.105):

$$p_w = \left[\bar{p}^2 - m\left(\ln\left(\frac{r_e}{r_w}\right) - 0.75 + 0.25f\right)\right]^{0.5}$$

where

$$m = \frac{q_{sc}\mu_g zT}{0.702 \times 10^{-3} kh}$$

Substituting the values of the fluid and reservoir properties gives:

$$m = \frac{(275)(0.035)(0.95)(460 + 130)}{(0.702 \times 10^{-3})(5)(35)} = 43{,}895$$

and

$$p_w = \left\{2{,}720^2 - (43{,}895)\left[\ln\left(\frac{2{,}640}{0.5}\right) - \frac{3}{4} + \frac{0.25}{2}\right]\right\}^{0.5} = 2{,}655 \text{ psia}$$

The pressure at the drainage boundary is obtained from Eq. (7.106):

$$p_e = \left[2{,}720^2 + (43{,}895)\left(\frac{1 + 0.5}{4}\right)\right]^{0.5} = 2{,}723 \text{ psia}$$

which is only 3 psia higher than the average reservoir pressure.

TURBULENT FLOW OF GAS

As the velocity of the gas flowing through the porous rock is increased, that is, the well is produced at higher flow rate, deviation from Darcy's law is observed. Various explanations for this deviation are presented in the literature [30–45]. The generally accepted explanation of this phenomenon is attributed to Wright, who demonstrated that, at very high velocities, the deviation from Darcy's law is due to inertial effects followed by turbulent effects [28, 32]. Actually, this phenomenon was observed by Reynolds in 1901 for flow in pipes [35]. Hubbert demonstrated that the transition from laminar flow to turbulent flow in porous media covers a wide range of flow rates [36].

LINEAR TURBULENT FLOW

The quadrangle relationship suggested by Forchheimer is generally found to be acceptable for expressing fluid flow under both laminar and turbulent conditions [41]. For horizontal, steady-state flow, this equation is:

$$-\frac{dp}{dL} = \frac{\mu_g v}{k} + \beta \rho v^2 \tag{7.108}$$

where

p = pressure, atm
L = length, cm
μ_g = viscosity of fluid, cP
k = permeability, darcy

v = velocity, cm/s
ρ = density of fluid, g/cm^3
β = turbulence or non Darcy factor, atm-s^2/g

If β is given in atm-s^2/g use the following expression to convert it to ft^{-1}:

$$\beta(\text{ft}^{-1}) = \beta\left(\frac{\text{atm}-s^2}{\text{g}}\right) \times 1013420 \times 12 \times 2.54 = \beta\left(\frac{\text{atm}-s^2}{\text{g}}\right) \times 3.0889 \times 10^7$$

where $g = 1{,}013{,}420$ gm-cm/(atm/cm^2)(s^2)

For gases, Katz expressed Eq. (7.108) in terms of the mass flow rate, q_m, because the mass flow rate is a constant when the cross-sectional area, A, is constant, permitting integration of Forchheimer equation [35]. Let

$$q_m = \rho q = \rho v A \tag{7.109}$$

where ρ is the density of the fluid and q is the volumetric flow rate.

If the equation of state for real gases (Eq. (7.96)) is substituted in Eq. (7.109), the mass flow rate is:

$$q_m = \left(\frac{pM}{zRT}\right) vA \tag{7.110}$$

Solving for v and substituting in Eq. (7.108) gives:

$$-\frac{dp}{dL} = \frac{\bar{z}RT}{pM}\left[\frac{\mu_g q_m}{kA} + \beta\left(\frac{q_m}{A}\right)^2\right] \tag{7.111}$$

If the variables are separated and integration is carried out over the length of the porous body, such as a core of length L and where the inlet and outlet pressures are p_1 and p_2, respectively, one obtains:

$$-\frac{M}{\bar{z}RT} = \int_{p_1}^{p_2} p\,dp = \left[\frac{\bar{\mu}_g q_m}{kA} + \beta\left(\frac{q_m}{A}\right)^2\right]\int_0^L dL \tag{7.112}$$

The gas deviation factor $\bar{z}$ is kept outside the integrand because it is assumed to remain constant at the average pressure $\bar{p}$, which is equal to $(p_1 + p_2)/2$. The integration gives:

$$\frac{M(p_1^2 - p_2^2)A}{2\bar{z}RT\bar{\mu}_g L} = \left(\frac{\beta}{A\bar{\mu}_g}\right) q_m^2 + \left(\frac{1}{k}\right) q_m \tag{7.113}$$

In practical oil field units, Eq. (7.113) can be written as follows:

$$\left(\frac{1.254 \times 10^{-10} T\bar{z}\gamma_g}{A^2}\right)\beta q_m^2 + \left(\frac{T\bar{z}\gamma_g}{11.9 \times 10^{-5} A}\right)\frac{1}{k} q_m - \frac{p_1^2 - p_2^2}{L} = 0 \tag{7.114}$$

where γ_g is the specific gravity of gas.

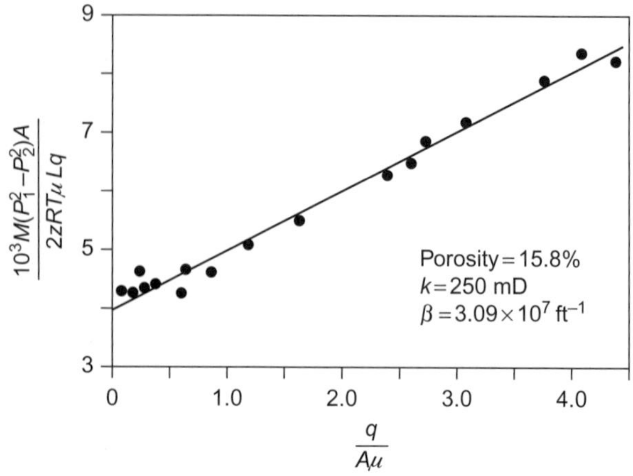

FIGURE 7.8 Evaluation of β and permeability.

Cornell used Eq. (7.113) to evaluate the permeability, k, and non-Darcy factor, β, for a large number of core samples from a variety of rocks by dividing the left-hand side of Eq. (7.113) by q_m and plotting it against $q_m/A\mu_g$, as shown in Figure 7.8 [37]. Equation (7.113) may be written in the general form:

$$Y_{ck} = \beta X_{ck} + \frac{1}{k} \tag{7.115a}$$

where Y_{ck} (atm-s/cm^2-cP) is given by:

$$Y_{ck} = \frac{M(p_1^2 - p_2^2)A}{2zRT\mu_g L q_m} \tag{7.115b}$$

and the variable X_{ck} (g/cm^2-s-cP) is given by:

$$X_{ck} = \frac{q_m}{A\mu_g} \tag{7.115c}$$

The subscript ck stands for Cornell–Katz. Equation (7.115a) plots as a straight line, with slope b and intercept $1/k$. Such a plot can be used to estimate the non-Darcy flow coefficient, as well as the permeability of the sample. This apparent permeability must be corrected for the Klinkenberg effect to obtain the absolute permeability as explained in the Appendix, especially at very low pressures. At high pressure, the Klinkenberg effect is negligible. Figure 7.9 shows a log–log plot of the turbulence factor versus permeability, which can be used to estimate β knowing k for any reservoir:

$$\beta = \frac{4.11 \times 10^{10}}{k^{4/3}} \tag{7.116a}$$

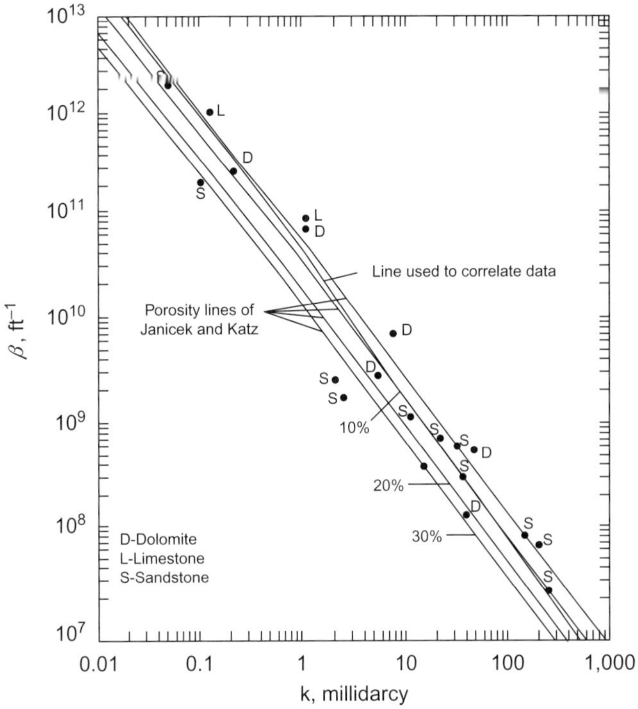

FIGURE 7.9 Correlation of β with permeability k and porosity ϕ.

Figure 7.9 also shows a correlation of β with porosity and permeability for carbonate and sandstone reservoirs. A reasonable estimation of the non-Darcy flow coefficient, β (ft^{-1}), can be obtained from the following equation [34]:

$$\beta = \frac{4.85 \times 10^4}{\phi^{5.5}\sqrt{k}} \tag{7.116b}$$

where

$\underline{k}$ = average permeability, mD
ϕ = porosity, fraction

Using experimental flow data on consolidated and unconsolidated sandstone, carbonate, and dolomite core samples, Liu et al. obtained a correlation of β (ft^{-1}) with respect to permeability (mD), porosity (fraction), and tortuosity τ [38]:

$$\beta = \frac{8.91 \times 10^8 \tau}{\phi k} \tag{7.116c}$$

Experimental and theoretical work of Firoozabadi et al. clearly shows that:

- changes in fluid properties over the length of the porous system do not account for the inadequacy of Darcy's law at high-velocity flow.
- the non-Darcy or high-velocity flow coefficient β is a function of rock properties and does not depend on the length of the porous system [45].

It is important to emphasize that the values of β obtained from Figure 7.9 and Eqs. (7.116a–7.116c) are only approximations. The non-Darcy flow coefficient β is best determined from laboratory measurements on cores obtained from the gas reservoir of interest. Values of β measured on several core samples obtained from different layers need to be averaged as follows before being applied in radial flow equations:

$$\overline{\beta} = \frac{\sum(\beta_i k_i^3 f_i)}{\sum(k_i f_i)^2 \times \sum(k_i f_i)} \tag{7.116d}$$

where f_i is the fraction of the total cross-sectional area or height of the core plug associated with the ith layer of height h_i, or, expressed mathematically, $f_i = h_i/h_t$, where h_t is the total thickness. Jones' original equation for calculating β did not include the term $\sum k_i f_i$, which is actually the average permeability [33].

Example

1. Determine the average or effective non-Darcy flow coefficient of a two-layer gas reservoir, using the following data.

Layer	k (mD)	ϕ	β (ft^{-1})	h (ft)
1	35.6	0.166	5.22×10^8	10
2	155	0.138	3.75×10^7	15

2. In the absence of lab-derived values of β, which of the correlations would have been applicable in this reservoir?

Solution

1. The fraction f is $10/25 = 0.40$ in layer 1, and $15/25 = 0.60$ in layer 2. The summation terms in Equation (7.116d) are:

$$\sum \beta_i k_i^3 f_i = 5.22 \times 10^8 \times 35.6^3 \times 0.40 + 3.75 \times 10^1 \times 155^3 \times 0.60$$
$$= 93.2 \times 10^{12}$$

$$\sum (k_i f_i)^2 = (35.6 \times 0.40)^2 + (155 \times 0.60)^2$$
$$= 8851.77$$

$$\sum k_i f_i = k = 35.6 \times 0.40 + 155 \times 0.60$$
$$= 107.24 \text{ mD}$$

From Eq. (7.116d), the average non-Darcy flow coefficient is:

$$\bar{\beta} = \frac{93.2 \times 10^{12}}{8851.77 \times 107.24} = 9.82 \times 10^{7}\ \text{ft}^{-1}$$

2. The arithmetic average porosity of the two layers is (0.166 + 0.138)/2 = 0.152, and the average permeability is 95.3 mD. Substituting these values into Eqs. (7.116a and 7.116b), we find, respectively:

$$\beta = \frac{4.11 \times 10^{10}}{95.3^{4/3}} = 9.44 \times 10^{7}\ \text{ft}^{-1}$$

$$\beta = \frac{4.11 \times 10^{10}}{0.152^{5.5} \times \sqrt{95.3}} = 15.7 \times 10^{7}\ \text{ft}^{-1}$$

The value of β obtained from Eq. (7.116a) compares well with the lab-derived value.

Example

A consolidated sand core 2 cm in diameter and 5 cm long has a permeability of 225 mD and a porosity of 20%. Air at 75°F is injected into this core. The inlet pressure is 100 psia and the outlet pressure 14.7 psia. The viscosity of air is 0.02 cP, and the compressibility is assumed to be equal to 1.0. Calculate the mass flow rate.

Solution

The mass flow rate for air can be calculated from Eq. (7.113). Let

$$a = \frac{\beta}{\mu_g A}$$

$$b = \frac{1}{k}$$

$$c = \frac{-MA(p_1^2 - p_2^2)}{2zRT\mu_g L}$$

Thus, Eq. (7.113) becomes:

$$aq_m^2 + bq_m + c = 0$$

This is, of course, a quadratic equation with two solutions: one negative and one positive. Inasmuch as the negative value has no physical meaning, the solution is:

$$q_m = \frac{1}{2a}\left[-b + (b^2 - 4ac)^{0.5}\right]$$

For $k = 225$ mD and $\phi = 20\%$ the value of β from Eq. (7.116a) is $3 \times 10^7\ \text{ft}^{-1}$. In laboratory units [$g_c = 1{,}013{,}420$ (g-cm)/(atm/cm^2)(s^2)], β is:

$$\beta = \frac{3 \times 10^{7}}{(2.54)(12)(1{,}013{,}420)} = 0.97\,\text{atm}{-}s^2/g$$

The cross-sectional area is:

$$A = \frac{(\pi)(2^2)}{4} = 3.14 \text{ cm}^2$$

The other variables in laboratory units for q_m in g/s are:

$M = 29$ g/g-mol
$\mu_g = 0.02$ cP
$R = 82.06$ cm^3-atm/(g-mol)(K)
$T = 297.2$ K
$L = 5$ cm

$$p_1^2 - p_2^2 = 6.80^2 - 1 = 45.28 \text{ atm}^2$$

$k = 0.225$ darcy.

The values of the constants a, b, and c are:

$$a = \frac{0.97}{(0.02)(3.14)} = 15.44$$

$$b = \frac{1}{0.225} = 4.44$$

$$c = \frac{(29)(3.14)(45.28)}{(2)(1)(82.06)(297.2)(0.02)(5)} = -0.85$$

Thus, the equation describing the mass flow rate through the core is:

$$15.44q_m^2 + 4.44q_m - 0.85 = 0$$

and q_m is:

$$q_m = \frac{1}{(2)(15.44)}\{-4.44 + [4.44^2 + (4)(15.44)(0.85)]^{0.5}\} = 0.13 \ g/s$$

To change this mass flow rate to volumetric flow rate, which is more commonly used, the density of the fluid at some pressure must be calculated and the mass flow divided by the fluid density. At an average pressure $\overline{p} = (100 + 14.7)12 = 57.35$ psia or 3.9 atm, the density of the fluid $[\rho = Mp/zRT]$ is:

$$\rho = \frac{(29)(3.9)}{(1)(82.06)(297.2)} = 4.64 \times 10^{-3} g/\text{cm}^3$$

and the volumetric flow rate, q, at the average pressure is:

$$q = \frac{0.13}{4.64 \times 10^{-3}} = 28 \text{ cm}^3/s$$

FRICTION FACTOR OF POROUS ROCKS

In flow of fluids in pipes, it is important to know if the flow is laminar or turbulent. The laminar flow regime is dominant if the fluids move along smooth streamlines parallel to the wall of the pipe. The velocity of the flowing fluid is virtually constant in time during laminar flow. The turbulent flow regime is dominant if the fluid velocity at any point in the pipe varies randomly with time. The differences between these two flow regimes were first investigated by Reynolds. His experimental and theoretical work showed that the nature of the flow regime in pipes depends on the Reynolds number ($Re = Dv\rho/\mu$), where D is the pipe inside diameter. In engineering practices if:

1. $Re < 2{,}100$, flow is in the laminar region;
2. $2{,}100 < Re < 4{,}000$, the nature of the flow regime is unpredictable, that is, flow passed through a transition region in which both laminar and turbulent flow regimes can be present;
3. $Re > 4{,}000$, the flow is fully turbulent

The flow of gas in very rough pipes can be considered fully turbulent because gas flows at high velocities and therefore high Re. Dimensionless analysis of energy loss in pipe flow of gas led to the concept of the friction factor. Moody showed that the friction factor, $2D\Delta p/\rho L v^2$, where L is the pipe length, is a function of Re and the relative roughness of the pipe [43]. Using a similar approach, Cornell and Katz investigated the flow of gas through porous media in terms of the Reynolds number and the friction factor [40]. They found that in order to analyze the gas-flow rate in porous media at very high velocities, that is, under the turbulent flow regime, the friction factor must be plotted versus Reynolds number, Re, as shown in Figure 7.10. The friction factor of a porous rock is given by:

$$f_{\mathrm{pr}} = \frac{64 g_{\mathrm{c}} \rho (p_1 - p_2)}{\beta L q_{\mathrm{m}}^2} A^2 \tag{7.117}$$

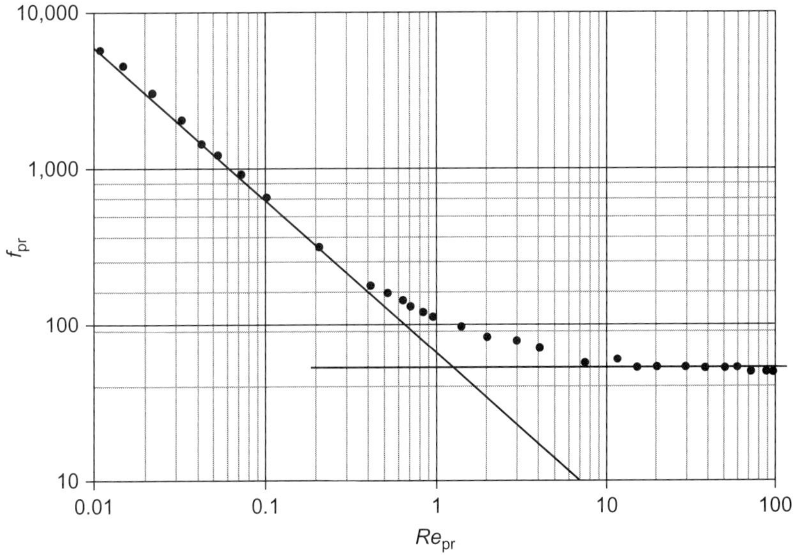

FIGURE 7.10 Friction factor for porous rock [40].

where

The subscript pr stands for porous rocks
g_c = conversion factor, 32.17 lbm-ft/(lb$_f$)(s^2)
ρ = fluid density, lbm/ft^3
p = pressure, lb$_f$/ft^2.
A = cross-sectional area, ft^2
β = turbulence factor, ft^{-1}
L = length of flow, ft
q_m = mass flow rate, lbm/s
f_{pr} = friction factor, dimensionless

Using the same units, the Reynolds number of a porous rock is:

$$Re_{pr} = \frac{\beta k q_m}{6.33 \times 10^{10} A \mu_g} \tag{7.118}$$

where the permeability k is expressed in mD and the viscosity in cP. The porous media Reynolds number, Re_{Pr}, is unitless.

Figure 7.10 shows three regions [40]:

1. For $Re_{pr} < 0.08$, the curve is a straight line of slope equal to -1 (laminar flow)
2. A transition region for $0.08 < Re_{pr} < 8$
3. A horizontal line for $Re_{pr} > 8$ (turbulent flow)

It is important to note that the unit-slope line and the horizontal line intercept at $Re_{pr} = 1$. The existence of a straight line for small Reynolds numbers indicates that the pressure drop $(p_1 - p_2)$ for a given porous medium is directly proportional to the flow rate (q_m), and that the laminar flow regime is dominant. Darcy's law is applicable during this portion of the curve only because the magnitude of the group of terms $(\beta \rho v^2)$ in Eq. (7.108) is too small to be detected in experimental data [42]. As the flow rate increases and Re_{pr} becomes larger, the turbulent flow regime becomes increasingly dominant. The horizontal portion of the f_{pr} versus Re_{pr} curve corresponds to the so-called non-Darcy flow, or fully turbulent flow regime. Katz and Lee and Firoozabadi and Katz suggested abandoning the concept of Darcy and non-Darcy flow [42, 44]. They recommended the use of "viscous Darcy flow" to describe the flow regime observed at low flow rates, and for high velocity flow to use "quadratic Darcy flow." Viscous Darcy flow theoretically occurs only when the flow rate is infinitely small [44].

Example

Solve the previous example using the friction factor plot for porous and permeable rock (Figure 7.10).

Solution

The variables have to be converted to the units used to derive Eqs. (7.117) and (7.118), which were used to generate the log–log plot of f_{pr} versus Re_r. Thus, from the previous example,

$\beta = 3 \times 10^7 \text{ ft}^{-1}$
$A = 3.5/30.48^2 = 3.767 \times 10^{-3} \text{ ft}^2$
$\rho = (4.64 \times 10^{-3})(30.48^3/453.6) = 0.29 \text{ lb/ft}^3$
$L = 5/30.48 = 0.164 \text{ ft}$
$\Delta p = 85.3 \text{ psi} = 85.8 \times 144 = 12{,}283 \text{ psf (or lb/ft}^2)$

The Reynolds number and the friction factor are:

$$Re_{pr} = \frac{(3 \times 10^7)(225)q_m}{(6.33 \times 10^{10})(3.767 \times 10^{-3})(0.02)} = 1{,}415.38\, q_m \tag{7.119}$$

$$f_{pr} = \frac{(64)(32.17)(0.29)(12{,}283)(3.767 \times 10^{-3})^2}{(3 \times 10^7)(0.164)q_m^2} = \frac{2.115 \times 10^{-5}}{q_m^2} \tag{7.120}$$

A trial-and-error method is necessary to solve for the flow rate. The correct value of q_m is such that the calculated values of Re_{pr} and f_{pr} behave according to the f_{pr} versus Re_{pr} curve in Figure 7.10. A practical first guess is $Re_{pr} = 1$. From Eqs. (7.119) and (7.120), $f_{pr} = 42.3$ for $Re_{pr} = 1$. However, for $Re_{pr} = 1$, Figure 7.10 gives $f_{pr} = 120$. After several trials, it was determined that $Re_{pr} = 0.48$ and $f_{pr} = 182$ are correct; thus, $q_m = 3.40 \times 10^{-4}$ lb/s or 0.158 g/s, which is approximately the same value obtained from Eq. (7.116).

This trial-and-error method can be simplified by using Tiab's correlations relating the friction factor directly to the Reynolds number [45]. These correlations are based on a large number of data of f_{pr} and Re_{pr} presented by Cornell and Katz for various sandstones, dolomites, and limestones (Figure 7.10) [40]:

1. For $Re_{pr} < 0.08$

$$f_{pr} = \frac{63.5}{Re_{pr}} \tag{7.121}$$

2. For $0.08 < Re_{pr} < 8$

$$f_{pr} = e^{3.1423 + 1.7534 Re_{pr}^{-0.2805}} \tag{7.122}$$

3. For $Re_{pr} > 8$

$$f_{pr} = 63.5 \tag{7.123}$$

A general formula that covers the laminar and turbulent flow regimes, and the transition range in between is:

$$f_{pr} = e^{3.5528 + 1.4253 Re_{pr}^{-0.2956}} \tag{7.124}$$

Equation (7.122) has a maximum error of 1%. Equation (7.124) has a maximum error of 5%.

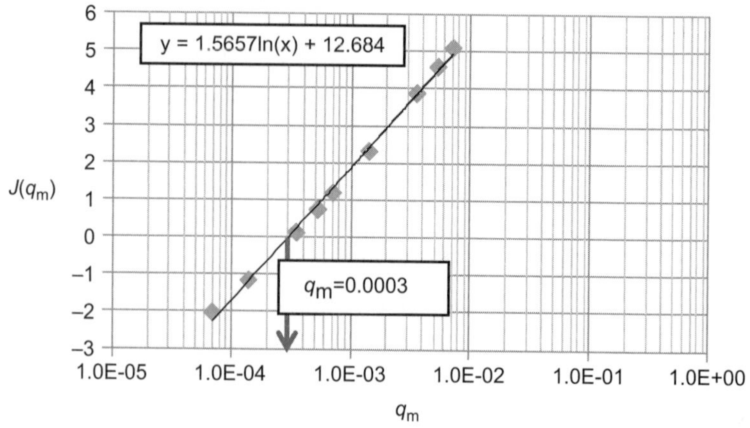

FIGURE 7.11 Semilog plot of $J(q_m)$ versus mass flow rate.

Inasmuch as the range of Reynolds number is not known until the flow test is completed, the generalized correlation (Eq. (7.124)) must be used first to estimate the mass flow rate, q_m. Substituting Eqs. (7.117) and (7.118) into Eq. (7.124), one can derive the following relations:

$$\ln(q_m^2) + 1.4253(J_2 q_m)^{-0.2956} + \ln\left(\frac{33.115}{J_1}\right) = 0 = J(q_m) \tag{7.125}$$

where

$$J_1 = \frac{64 g_c \rho A^2 \Delta p}{\beta L} \tag{7.125a}$$

and

$$J_2 = \frac{\beta k}{6.33 \times 10^{10} A \mu_g} \tag{7.125b}$$

The mass flow rate is determined from the following procedure:

1. Assume several values of Reynolds number Re_{pr} in the range of 0.01−100, and calculate the corresponding values of q_m from Eq. (7.118).
2. Calculate values of the function $J(q_m)$ using Eq. (7.125), which has a maximum error of 5%.
3. Plot $J(q_m)$ versus q_m on a semilog graph (q_m is plotted on the log-axis).
4. The correct value of the mass flow rate q_m corresponds to $J(q_m) = 0$ on the graph, as shown in Figure 7.11.
5. For more accuracy—that is, less than 1% error in the value of q_m—calculate Re_{pr}, which corresponds to q_m obtained in Step 4 using Eq. (7.118).
6. If this value of Re_{pr} is less than 0.08, then use the following equation to calculate a new value of q_m:

$$q_m = (5.122 \times 10^{10} \rho)\left(\frac{kA}{\mu_g}\frac{\Delta p}{L}\right) \tag{7.126}$$

Equation (7.126) is Darcy's law, which can be derived by substituting Eqs. (7.117) and (7.118) into Eq. (7.121).

7. If the value of Re_{pr} obtained in Step 5 is greater than 8, the following equation should be used to calculate a more accurate value of the mass flow rate:

$$q_m = 5.7\sqrt{\frac{\rho A^2 \Delta p}{\beta L}} \tag{7.127}$$

Equation (7.127) can be derived by substituting Eq. (7.117) into (7.123), which corresponds to the so-called fully turbulent flow regime.

8. If the Reynolds number obtained in Step 5 is in the range of 0.08, repeat Steps 1 through 4 using the following expressions for the function $J(q_m)$:

$$J(q_m) = \ln(q_m^2) + 1.7534(J_2 q_m)^{-0.2805} + \ln\left(\frac{23.157}{J_1}\right) = 0 \tag{7.128}$$

Equation (7.128) can be derived by substituting Eqs. (7.117) and (7.118) into Eq. (7.125). The terms J_1 and J_2 are given by Eqs. (7.125a) and (7.125b), respectively.

Example

Calculate the mass flow rate for the core flow test in the previous example using the above procedure.

Solution

The values of J_1 and J_2 are first calculated from Eqs. (7.125a) and (7.125b), respectively. By substituting values of β, A, ρ, L, and Δp, one finds:

$$J_1 = 2.115 \times 10^{-5}$$

and

$$J_2 = 1415.38$$

Substituting these values of J_1 and J_2 into Eq. (7.125), one obtains the following expression for the function $J(q_m)$:

$$J(q_m) = \ln(q_m^2) + 1.4253(1415.30q_m)^{-0.2956} + 14.2638 \tag{7.129}$$

Table 7.2 shows the assumed values of Re_{pr} and the corresponding values of q_m, f_{pr}, and $J(q_m)$, which are obtained from Eqs. (7.130), (7.120), and (7.129), respectively:

$$q_m = \frac{Re_{pr}}{J_2} \tag{7.130}$$

Table 7.2 also shows the values of the function $J(q_m)$ obtained from Eq. (7.129). A semilog plot of $J(q_m)$ versus q_m, as shown in Figure 7.11, yields a

TABLE 7.2 Calculation of [$J(q_m)$] Values for Example

Re_{pr}	q_m	f_{pr}	$J(q_m)$
0.1	0.0000706	4,236.98	−2.03
0.2	0.0001413	1,059.24	−1.17
0.5	0.0003532	169.47	0.116
0.75	0.0005298	75.032	0.73
1	0.0007065	42.036	1.178
2	0.001413	10.59	2.301
5	0.003532	1.69	3.858
7.5	0.005298	0.75	4.569
10	0.007065	0.42	5.08

mass flow rate value of 0.0003 (lbm/s) at $J(q_m) = 0$. To change this mass flow rate to volumetric flow rate, the following expression is used:

$$q = \frac{q_m(\text{lbm}/s)}{2.205 \times 10^3(\text{lbm}/g)\rho(g/\text{cm}^3)} \tag{7.131}$$

At an average pressure of 3.9 atm, the density is 4.64×10^{-3} g/s, and the volumetric flow rate is calculated from Eq. (7.131):

$$q = \frac{0.0003}{0.002205 \times 0.00464} = 29.32\ \text{cm}^3/s$$

Substituting the values of q_m at $J(q_m) = 0$, that is, 0.0003 lbm/s, and $J_2 = 1{,}415.38$ into Eq. (7.130) yields a Reynolds number of 0.467. Using the recommendation in Step 8 of the procedure yielded no significant change of the mass flow rate.

TURBULENT RADIAL FLOW

All of the radial steady-state flow equations presented so far are based on the assumption that Darcy's law is applicable at all times and throughout the reservoir. As shown in the previous sections, at high gas-flow rates, inertial and/or turbulent flow effects are of significance and should be accounted for. Non-Darcy flow effect is most significant near the wellbore because, in radial flow systems, the velocity of the flowing fluid increases as the fluid approaches the well. Under these circumstances, Eq. (7.111) becomes:

$$\frac{dp}{dr} = \frac{zRT}{pM}\left[\frac{\mu_g q_m}{2\pi rhk} + \beta\left(\frac{q_m}{2\pi rh}\right)^2\right] \tag{7.132}$$

Let:

$$Q_1 = \left(\frac{zRT\mu_g}{2\pi khM}\right)q_m$$

and

$$Q_2 = \left(\frac{zRT\beta}{(2\pi h)^2 M}\right)q_m^2$$

Substituting Q_1 and Q_2 into Eq. (7.125) gives:

$$\frac{dp}{dr} = \frac{Q_1}{rp} + \frac{Q_2}{r^2 p} \tag{7.133}$$

Then, separating the variables and integrating between r_w and r_e, where the pressures are p_w and p_e, respectively:

$$\int_{p_w}^{p_e} p\,dp = Q_1 \int_{r_w}^{r_e} \frac{dr}{r} + Q_2 \int_{r_2}^{r_e} \frac{dr}{r^2} \tag{7.134}$$

or

$$\frac{p_e^2 - p_w^2}{2} = Q_1 \ln\frac{r_e}{r_2} + Q_2\left(\frac{1}{r_w} - \frac{1}{r_e}\right)$$

Substituting Q_1 and Q_2, and assuming $r_w \ll r_e$ yields:

$$p_e^2 - p_w^2 = \frac{\mu_g zRTq_m}{M\pi kh}\ln\left(\frac{r_e}{r_w}\right) + \frac{zRT\beta q_m^2}{2\pi^2 Mh^2 r_w} \tag{7.135}$$

If Eq. (7.135) is expressed in field units and q_m is set equal to $q_{sc}\, P_{sc}M/T_{sc}R$, one finds:

$$p_e^2 - p_w^2 = \frac{1422\mu_g zT\ln(r_e/r_2)}{kh}q_{sc} + \frac{3.161\times10^{-12} zT\gamma_g\beta}{r_w h^2}q_{sc}^2 \tag{7.136}$$

where q_{sc} is expressed in MSCF/day. After some algebraic manipulation, Eq. (7.136) can be written as:

$$p_e^2 - p_w^2 = \frac{1422\mu_g zTq_{sc}}{kh}\left[\ln\left(\frac{r_e}{r_w}\right) + \left(\frac{2.22\times10^{15}\gamma_g\beta k}{\mu_g r_w h}\right)q_{sc}\right] \tag{7.137}$$

If D is the non-Darcy flow coefficient:

$$D = \left(\frac{2.22\times10^{-15}\gamma_g}{\mu_g r_w h}\right)\beta k \tag{7.138}$$

Eq. (7.137) becomes:

$$p_e^2 - p_w^2 = \frac{1424\mu_g zTq_{sc}}{kh}\left[\ln\left(\frac{r_e}{r_w}\right) + Dq_{sc}\right] \tag{7.139}$$

or, in a more familiar form:

$$q_{sc} = \frac{kh(p_e^2 - p_w^2)}{1422\mu_g ZT[\ln(r_e/r_w) + Dq_{sc}]} \tag{7.140}$$

Equation (7.140) is similar to Eq. (7.103) for the bounded reservoir case, that is, $f=1$ and no skin damage. Again, if the pressure at the outer boundary of the drainage area is not known, Eq. (7.140) is generally expressed in terms of the average reservoir pressure $\overline{p}$. If the wellbore and external boundary conditions are taken into account by introducing the skin factor s and water-drive index (WDI) f, in addition to non-Darcy flow effect, Eq. (7.140) becomes:

$$q_{sc} = \frac{kh(\overline{p}^2 - p_w^2)}{1.422\mu_g zT\left[\ln(r_e/r_w) - \frac{3}{4} + \frac{f}{4} + s + Dq_{sc}\right]} \tag{7.141a}$$

For volumetric or closed systems, that is, $f=0$, Eq. (7.141a) becomes:

$$q_{sc} = \frac{kh\left(\overline{p}^2 - p_w^2\right)}{1422\mu_g zT\left[\ln\left(0.472 r_e/r_w\right) + s + Dq_{sc}\right]} \tag{7.141b}$$

This equation forms the basis of most techniques for predicting the performance of gas wells. Inasmuch as the velocity of the flowing fluid in a cylindrical or radial flow system increases as the well is approached, because of the decrease in the area crossed by the fluid moving from r_e to r_w, turbulent flow is most pronounced in the vicinity of the wellbore. The additional pressure drop due to turbulence is equivalent to a skin effect. The non-Darcy flow coefficient D is best estimated from pressure-transient tests such as buildup and drawdown tests. Both the skin factor, s, and non-Darcy flow coefficient, D, are concentrated in the vicinity of the well, so they are generally detected during pressure testing as a single factor, that is, the total skin $s_t = s + Dq_{sc}$.

Because pressure in the skin zone changes, turbulence also changes with time. Consequently, the product Dq is a variable, making s_t also a variable. But, under steady-state conditions (which are rarely encountered), one can assume s_t to be approximately constant.

Whereas the factor s can be either positive (damaged well) or negative (stimulated well), the non-Darcy flow coefficient D is theoretically always positive and, therefore, always results in a pressure loss. An excellent approximation of D may be made from Eq. (7.138). Substituting Eq. (7.116) into (7.138) gives:

$$D = \frac{9.12 \times 10^{-5}\gamma_g}{\mu_g r_w h}\frac{1}{k^{1/3}} \tag{7.142}$$

where

D = non-Darcy flow coefficient $(\text{MSCF/day})^{-1}$
γ_g = gas gravity (air = 1)
h = formation thickness or, preferably, the perforated interval of the well, ft
r_w = wellbore radius, ft
μ_g = gas viscosity, cP

MULTIPLE-PERMEABILITY ROCKS

The foregoing fluid-flow equations were developed on the assumption that the reservoir is homogeneous. In reality, homogeneous reservoirs are seldom, if ever, encountered. Practically, every producing clastic formation is stratified to some extent, that is, it contains layered beds of differing petrophysical rock properties. This stratification resulted from variations in texture, dimensions of sand particles and composition, or temporary cessation of deposition, which allowed already-deposited sediments to undergo some changes before renewal of deposition.

Stratification has been classified as direct and indirect. The former occurs when the sediments are first deposited over extremely long periods of time. Indirect layering develops when sediments already deposited are thrown into suspension and redeposited. Sediments deposited in deep or very shallow, quiet water tend to yield regular stratification, whereas sediments deposited in agitated water tend to produce highly irregular layering. Sedimentary clastic units deposited in channels and deltas of rivers are likely to show considerable variation in thickness and areal extent over very short distances. In most clastic oil and gas reservoirs, therefore, permeability varies both laterally and vertically. Inasmuch as the foregoing derived steady-state flow equations require only a single permeability value, it is important to know how to recombine the permeability of various portions of the reservoir into an average value. Layered reservoirs are divided into two general types: layered reservoirs with crossflow and layered reservoirs without crossflow.

LAYERED RESERVOIRS WITH CROSSFLOW

Frequently, overlying reservoir beds, which have different thicknesses and petrophysical properties (such as permeability and porosity), are hydrodynamically communicating at the contact plane (Figure 7.12).

Russel and Prats investigated the practical aspects of layered reservoirs with crossflow and concluded that the flow equations in these systems are similar to those developed for a homogeneous reservoir with the permeability term k_t representing the sum of permeabilities of all layers, k_1, k_2, k_3, ..., k_4, that is, [46]:

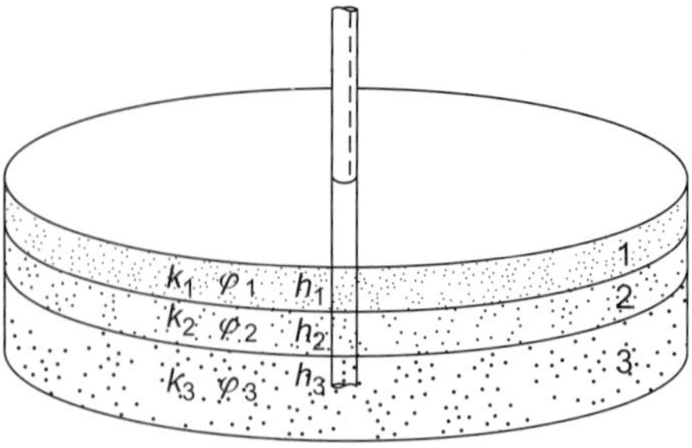

FIGURE 7.12 Three-layer reservoir with crossflow.

$$k_t = \sum_{i=1}^{n} k_i \tag{7.143}$$

If the thickness also varies, Eq. (7.143) becomes:

$$(kh)_t = \sum_{i=1}^{n} (kh)_i \tag{7.144}$$

where n is the total number of layers. If the porosity, thickness, and compressibility vary from layer to layer, then:

$$(\phi c_t h)_t = \sum_{i=1}^{n} (\phi c_t h)_i \tag{7.145}$$

If the total permeability–thickness product $(kh)_t$ is known from a pressure test, individual layer permeabilities may be estimated from the following equation:

$$k_i = \frac{q_i}{q_t}\frac{(kh)_t}{h_t} \tag{7.146}$$

where q_i is the volumetric rate of fluid flow through each layer. Equation (7.146) is valid only if the individual skin factors are all equal or negligible.

LAYERED RESERVOIRS WITHOUT CROSSFLOW

In many oil and gas pools, the reservoir rocks are interbedded with impermeable shales beds and silt laminations, such that there is no crossflow between the oil- and gas-saturated sand beds.

1. Consider a simple linear flow model shown in Figure 7.13. The total volumetric flow rate through the entire system is equal to the sum of flow rates through the individual beds, separated from one another by thin impermeable barriers:

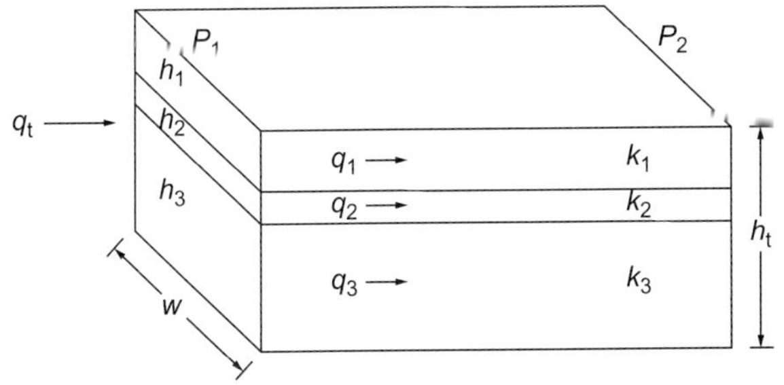

FIGURE 7.13 Linear flow in parallel beds.

$$q_t = q_1 + q_2 + \cdots + q_n = \sum_{j=1}^{n} q_t \tag{7.147}$$

where, according to Darcy's law, the rate of fluid flow through each bed of equal width w and length L is:

$$q_j = \frac{w\Delta p}{\mu L}(k_j h_j) \tag{7.148}$$

The total flow rate through the system, with an average permeability $\bar{k}$, is:

$$q_j = \frac{w\Delta p}{\mu L}(\bar{k} h_t) \tag{7.149}$$

If the last three expressions are combined and the identical terms, w, Δp, μ, and L, are cancelled, one finds:

$$\bar{k} h_t = k_1 h_1 + k_2 h_2 + \cdots + k_n h_n \tag{7.150}$$

Thus, the average permeability of a system with n parallel beds is:

$$\bar{k} = \frac{\sum_{j=1}^{n} k_j h_i}{\sum_{j=1}^{n} h_j} \tag{7.151}$$

The product kh is commonly referred to as the "flow capacity" of the producing zone. Equation (7.151) is used to determine the average permeability of a reservoir from core analysis data.

Example

Consider the graph of permeability versus depth shown in Figure 7.14 for a 27-ft-long core from an oil well in the Rodessa, Texas, fields [47]. Determine the average permeability in the vicinity of this well for the 6,204- to 6,208-ft core segment. What is the average permeability of the entire sand core?

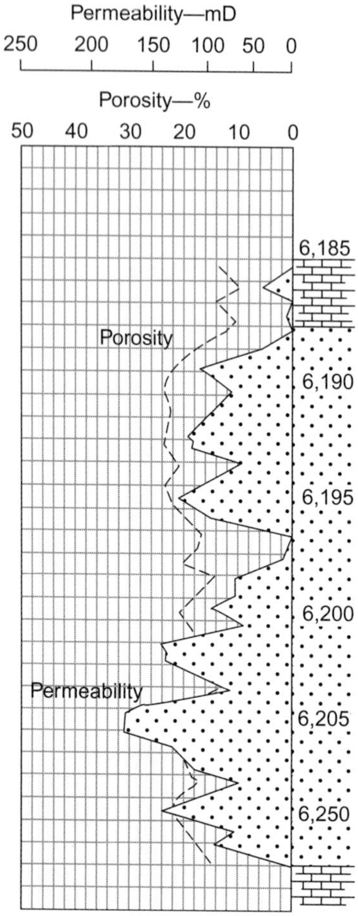

FIGURE 7.14 Core graph of permeability and porosity (Rodessa, Texas field).

Solution

The average permeability of the formation in this oil well can be approximated by that of the core, even though it is commonly known that the properties of a core will change. This is from slightly to very considerably once the core is brought to the surface, due essentially to the drop in pressure from thousands of psia in the reservoir to 14.7 psia at the surface. The range of this change is dependent on the lithology of the reservoir rock (sandstone, limestone, etc.), degree of consolidation and fracturing, etc. Table 7.3 is obtained by recording permeability values versus depth from Figure 7.15.

TABLE 7.3 Core Data From an Oil Well

Depth (ft)	Permeability (mD)	Flow Capacity (mD-ft)
6,204–6,205	105	105
6,205–6,206	150	150
6,206–6,207	132.5	132.5
6,207–6,208	95	95
		482.5

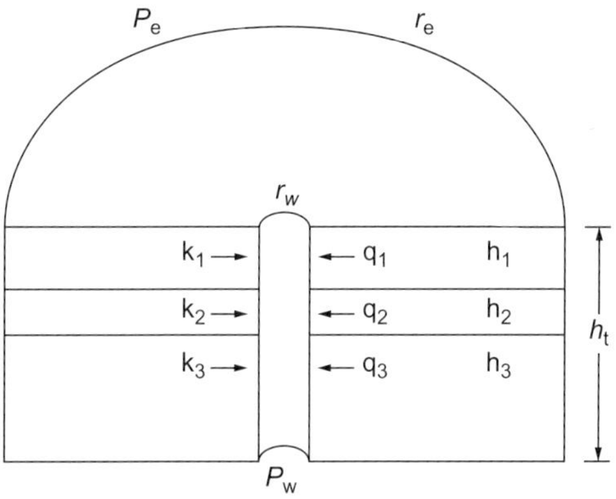

FIGURE 7.15 Radial flow in parallel beds.

The average permeability in the 6,204- to 6,208-ft segment is 482.5/4 = 120.5 mD. Using Eq. (7.151) for the entire core, one obtains:

$$\sum_{j=1}^{27} k_j h_j = 1886 \text{ mD-ft}$$

and the average permeability is given by:

$$\bar{k} = 1886/27 = 70 \text{ mD}$$

The in situ average permeability obtained from pressure tests gives a better representation of the flow capacity of the reservoir.

2. Figure 7.15 illustrates a horizontal radial system made up of several homogeneous layers each having its own thickness h_j and permeability k_j. The total fluid-flow rate through this stratified drainage area is the sum of flow rates through the individual layers separated by infinitely thin impermeable barriers, as expressed by Eq. (7.147). Assuming steady-state conditions, the individual flow rates (Eq. (7.149)) can be written as:

$$q_j = \frac{2\pi\Delta p}{\mu \ln r_e / r_w}(k_j h_j) \tag{7.152}$$

and the total flow rate through the drainage area with an average permeability $\bar{k}$ and the total thickness h_t is given by:

$$q_t = \frac{2\pi\Delta p}{\mu \ln r_e / r_w}(\bar{k} h_t) \tag{7.153}$$

If Eqs. (7.152) and (7.153) are substituted into Eq. (7.147), the identical terms, Δp, μ, $\ln(r_e/r_w)$, and 2π, are canceled, and solving for the average permeability $\bar{k}$, one obtains the same solution as obtained in the linear case (Eq. (7.151)):

$$\bar{k} = \frac{1}{h_t}\sum_{j=1}^{n} k_j h_j \tag{7.154}$$

This equation is applicable to both oil and gas reservoirs.

COMPOSITE RESERVOIRS

Earlougher defined composite reservoirs as systems where fluid-of-rock properties vary in a step-like fashion radially away from the well [20]. In most cases, variations of the rock and fluid properties are artificially induced as a result of drilling, well completion, and fluid injection practices. Figure 7.16 gives a schematic diagram of fluid distribution around an injection well.

Linear discontinuities in porosity and permeability in the horizontal direction frequently occur within reservoirs. The effect of a change in rock properties, especially permeability, in the horizontal direction is an important consideration in predicting reservoir performance and field development techniques. The magnitude of such a change is determined by arranging the zones of different permeability in series.

Linear Flow Systems

Figure 7.17 depicts a linear flow system, which consists of a number of homogeneous segments of different permeability arranged in series.

The total pressure drop across this system is equal to the sum of the pressure drops across each segment. Thus:

$$\Delta p_t = p_1 - p_n = \Delta p_1 + \Delta p_2 + \cdots + \Delta p_n \tag{7.155}$$

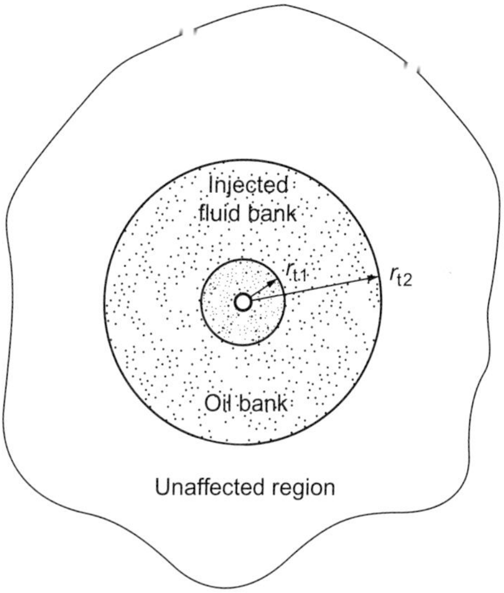

FIGURE 7.16 Typical composite reservoir.

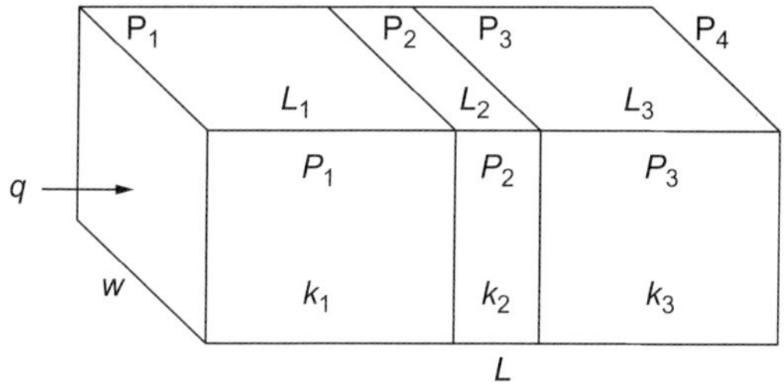

FIGURE 7.17 Linear flow through segments in series.

The individual pressure drops can be expressed, assuming flow of incompressible fluid under steady state, as:

$$\Delta p_i = \left(\frac{q\mu}{A}\right)\frac{L_i}{k_i} \tag{7.156}$$

Similarly, the total pressure drop is:

$$\Delta p_t = \left(\frac{q\mu}{A}\right)\frac{L_t}{\overline{k}} \tag{7.157}$$

Substituting Eqs. (7.156) and (7.157) into Eq. (7.155), and canceling the identical terms q, m, and A yields:

$$\frac{L_t}{\overline{k}} = \frac{L_1}{\overline{k}_1} + \frac{L_2}{\overline{k}_2} + \cdots + \frac{L_n}{\overline{k}_n} \tag{7.158}$$

and the average reservoir permeability of n segments arranged in series is:

$$\bar{k} = \frac{\sum_{i=1}^{n} L_i}{\sum_{i=1}^{n} (L_i / k_i)} \tag{7.159}$$

It is important to emphasize that the fluid properties are assumed to be constant in the entire system.

Radial Flow Systems

A similar analysis can be made to determine the average permeability when radial flow of fluids is through a series of homogeneous concentric segments with different permeabilities, as shown in Figure 7.18. Using the radial incompressible fluid-flow equation (Eq. (7.57)), the total pressure drop between radii r_w and r_e, where the pressures are p_w and p_e, respectively, is:

$$\Delta p_t = p_e - p_w = \left(\frac{q\mu}{2\pi h}\right) \frac{\ln(r_e/r_w)}{\bar{k}} \tag{7.160}$$

and the pressure drops in the individual segments are:

$$\Delta p_i = \left(\frac{q\mu}{2\pi h}\right) \frac{\ln\left(r_i/r_{i-1}\right))}{k_i} \tag{7.161}$$

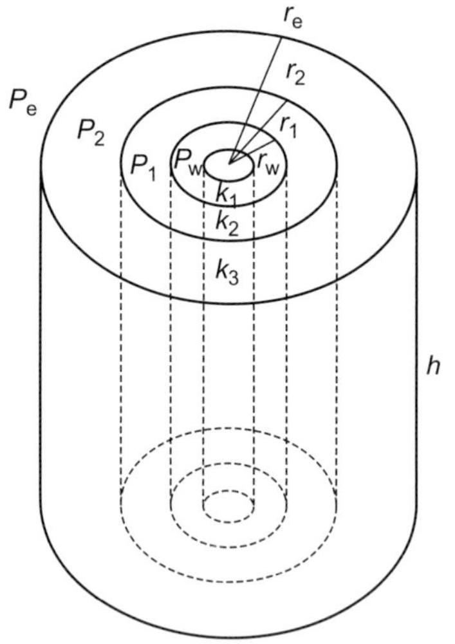

FIGURE 7.18 Radial flow through segments in series.

If Eqs. (7.155), (7.160), and (7.161) are combined, and the identical term $(q\mu/2\pi h)$ is canceled, one finds:

$$\frac{\ln(r_e/r_w)}{\bar{k}} = \frac{\ln(r_2/r_1)}{\bar{k}_1} + \frac{\ln(r_3/r_2)}{\bar{k}_2} + \dots + \frac{\ln(r_n/r_{n-1})}{\bar{k}_n} \tag{7.162}$$

If Eq. (7.162) is solved for the average permeability, one can derive the following general equation (where $r_n = r_e$ and $r_o = r_w$):

$$\bar{k} = \frac{\ln(r_e/r_w)}{\sum_{i=1}^{n} \frac{1}{k_j} \ln(r_i/r_{i-1})} \tag{7.163}$$

Equations (7.159) and (7.163) apply to gas as well as oil reservoirs. This can be demonstrated by using the linear gas-flow equation (7.18) with Eqs. (7.147) and (7.155) to derive Eqs. (7.151) and (7.159) respectively. Equations (7.154) and (7.163) can be derived by using the radial gas-flow equation (7.103) for $f=1$ and $s=0$ with Eqs. (7.147) and (7.155) respectively.

Example

The permeability of a 160-acre light-gas formation drained by a single well is 15 mD. The well was heavily acidized to a permeability of 25 mD and a radius of 30 ft, and then completed. During well completion, a 2-ft thick damaged zone developed in the vicinity of the wellbore. The permeability of this damaged segment is 4 mD. The wellbore radius is 0.50 ft. Calculate the average permeability of this drainage area.

Solution

Figure 7.19 is a schematic diagram of the drainage system, where $r_w = 0.5$ ft and $r_e = (43{,}560 \times 160/\pi)^{0.5} = 1{,}490$ ft. The average permeability of this system is calculated from Eq. (7.163):

$$\bar{k} = \frac{\ln(1{,}490/0.50)}{\frac{1}{4}\ln(2.5/0.5) + \frac{1}{25}\ln(30/2.5) + 1/15\ln(1{,}490/30)} = 10.5 \text{ mD}$$

This example illustrates how the beneficial effect of a successful acidizing job is counteracted by the negative effect of improper well completion, and how important it is to calculate accurately the average permeability of reservoirs. If the well was completed properly without zonal damage, the average permeability would have been:

$$\bar{k} = \frac{\ln(1{,}490/0.50)}{(1/25)\ln(23/0.5) + (1/15)\ln(1{,}490/30)} = 18.9 \text{ mD}$$

which is nearly double the permeability of the formation prior to well completion.

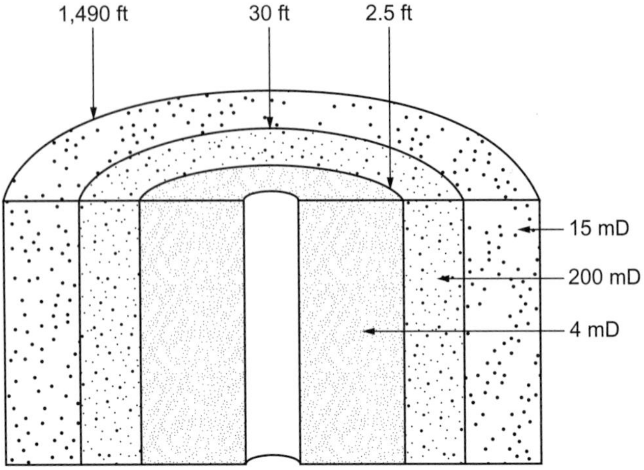

FIGURE 7.19 A system of three radial segments in series.

PROBLEMS

1. Oil is flowing through a 10-cm core sample prepared for reservoir rock properties evaluation at a rate of 20 cm^3/s. The pump pressure is 18 atm and the pressure at the end of the core chamber is 1 atm. If the diameter of the core is 5 cm and oil viscosity is 0.45 cP, calculate:
   (a) Oil permeability of the core.
   (b) Pressure drop through the entire core, if the pump rate is changed to a new value of 13 cm^3/s using permeability from (a).
   (c) Pressure drop in (b), if the core length is reduced to 75% of the original length.
   (d) Diameter of the core to cause a pressure drop of 2 atm, if the original length is reduced by half.
   (e) Apparent and actual velocity of oil, if core has a porosity of 20% and the pressure drop through it is 7 atm.
2. A core sample from a gas reservoir is to be tested for various properties using nitrogen gas. A 12-cm-long core has a diameter of 4 cm. Calculate:
   (a) Linear pressure drop across the core if gas-flow rate is 5 cm^3/s, nitrogen viscosity is 0.012 cP, and core permeability is 50 mD.
   (b) Actual velocity of the gas if core porosity is 15%, irreducible water saturation is 20%, and inlet and outlet pressures are 10 and 2 atm, respectively.
   (c) Average flow rate through the core, if linear pressure gradient through core is 0.8 atm/s and outlet pressure is 2 atm.

3. A newly completed well in the center of an oil reservoir drains 60 acres at initial reservoir pressure of 3,000 psia, estimated from DST. Following are the additional rock and fluid data in oil field units:

$P_{wf} = 1{,}500$ psia
$k = 30$ mD
$h = 50$ ft
$r_w = 7$ in. $= 0.58$ ft
$B_o = 1.15$ RB/STB
$\mu_o = 2.1$ cP
$\phi = 0.25$

(a) Calculate the flow rate from this well assuming that the well is flowing at steady state.
(b) If the well is allowed to flow at 1,200 STB/day, calculate the bottomhole pressure.
(c) If the bottomhole pressure is 1,000 psia, estimate the reservoir pressure 50 ft away from the well keeping a flow rate of 1,200 STB/day assuming pseudosteady state.
(d) Estimate the average reservoir pressure.
(e) Calculate the distances from the well where the value of average reservoir pressure would exist for both pseudosteady and steady states.

4. A 1 ft^3 sample of limestone has a matrix permeability of 2.5 mD and a matrix porosity of 13%, and contains seven solution channels. These channels have average radius of 0.045 cm, vug porosity of 15%, and irreducible water saturation of 15%. Calculate:

(a) Average permeability of solution channels.
(b) Average permeability of the block.
(c) The amount of fluid stored in channels and matrix, respectively.
(d) Fluid-flow rate if fluid viscosity is 1.3 cP, and inlet and outlet pressures are 90 and 28 psia, respectively.

5. A 2-ft^3 carbonate reservoir rock sample from the Permian basin, Texas, was analyzed for different rock properties. It was found that the rock is naturally fractured and contains four parallel fractures of 0.003-in. width. The matrix has a permeability of 2 mD. If flow through the sample is maintained in the direction of the fractures, calculate:

(a) Permeability of the fractures.
(b) Average permeability of the rock.
(c) Flow rate through fractures and matrix if pressure gradient through sample is 7.5 psia and the viscosity of the fluid is 0.85 cP.

6. In the second phase of field development, ZBN #5 was completed at the depth of 5,578 ft in the center of an oil reservoir with the following rock, oil, and well characteristics:

$k = 225$ mD
$B = 1.2$ RB/STB
$f = 0.16$

$d = 7$ in.
$r_w = 0.29$ ft
$P_{wf} = 1{,}450$ psia
$S_{wc} = 0.24$
$r_e = 745$ ft
$\phi = 16\%$
$\mu = 4.2$ cP
If the well drains 40 acres:

(a) Calculate ideal flow rate from this well if pressure gradient in the well location area is 0.45 psi/ft.

(b) From production optimization studies, it was realized that the well could have been completed with larger wellbore diameter to produce at a rate 134% of the rate from step (a). Determine the diameter.

(c) How long will oil take to reach the wellbore from the external boundary of 745 ft?

(d) After some time, a pressure buildup test indicated a skin of +2.5, skin zone of 8 ft, and permeability of the damaged zone of 125 mD. A fraction of reservoir periphery is open to water encroachment. If the bottomhole pressure is 2,000 psia, reservoir pressure is 2,510 psia, and well flow rate is 250 STB/day, calculate drainage boundary index and drawdown to produce 500 STB/day.

7. A newly discovered TDM #4 is completed in a sandstone oil reservoir at a depth of 11,324 ft with 7-in. production casing and drains 80 acres. After thorough economic feasibility evaluation of the project, it is decided to develop the field with 14 wells. The net production from the field is expected to be 24,500 STB/day (1,750 STB/day-well). Following are the rock and fluid data from the well:

$P_i = 8{,}000$ psia
$B_o = 1.13$ RB/STB
$c_t = 2.5 \times 10^{-5}$ psi^{-1}
$h = 55$ ft
$k = 33$ mD
$r_w = 0.58$ ft
$\mu = 3.5$ cP
$\phi = 17\%$

(a) Estimate the time to reach the bottomhole pressure equal to bubble point pressure of 2,000 psia.

(b) What will be the average reservoir pressure at the time estimated in (a)?

(c) After 6 months of smooth production, a slight decline in production was observed. A drawdown pressure test indicated a skin factor of +3. Calculate P_{wf}, pressure drop due to skin, apparent wellbore radius, and flow efficiency of the well.

(d) Determine annual revenue loss due to skin if the current black oil price is $10/bbl

(e) The well was acidized and pressure tested. The new skin factor is −4. Calculate the increase in production if the well is currently flowing at 5,000 psia.

8. A gas reservoir drains 120 acres and is partially pressured by a water aquifer with a WDI of 0.35. The reservoir rock permeability is 7 mD, porosity is 13%, average pressure is 3,500 psia, temperature is 140°F, and formation thickness is 27 ft. Overbalanced drilling had damaged the formation, and a well test indicated a skin factor of +2. Gas properties include viscosity of 0.012 cP and gas deviation factor of 0.9. If the well is flowing at 3,150 psia, and $r_w = 0.5$ ft, calculate:
   (a) Flow rate.
   (b) Bottomhole pressure if the rate is increased to 7.5 MMSCF/day.
   (c) Pressure at the boundary.

9. A consolidated core 3 cm in diameter and 8 cm long has a permeability of 174 mD and porosity of 15%. This core sample was subjected to a linear flow test using air ($\mu_g = 0.023$ cP).
   (a) What is the mass flow rate of air in g/s if the inlet and outlet pressures are 125 and 14.7 psia, respectively? Assume ideal gas behavior and a temperature of 75°F.
   (b) Find the volumetric flow rate at the mean pressure.
   (c) Repeat (a) using the friction factor plot for porous and permeable rock, Figure 7.10, and solve by trial and error method.
   (d) Calculate the mass flow rate (q_m) using appropriate correlations.

10. Given the following information on a well located in a three-layer gas reservoir:

$r_w = 0.5$ ft	$h_2 = 13$ ft
$r_e = 1,320$ ft	$h_3 = 6$ ft
$q_{sc} = 5,850$ MSCF/day	$\phi_1 = 0.175$
$P_{wf} = 1,950$ psia	$\phi_2 = 0.142$
$k_1 = 16$ mD	$\phi_3 = 0.110$
$k_2 = 8$ mD	$\mu_g = 0.15$ cP
$k_3 = 5$ mD	$Z = 0.92$
$h_1 = 5$ ft	$\gamma_g = 0.65$

Calculate:
   (a) The average permeability of the reservoir.
   (b) The turbulence factor.
   (c) The non-Darcy flow coefficient.
   (d) The pressure drop due to turbulent flow.
   (e) The pressure drop due to Darcy flow.
   (f) The pressure at the radius r_e.

NOMENCLATURE

A area
B_g gas formation volume factor
B_o oil formation volume factor
c_t total compressibility
D non-Darcy flow coefficient
E_f recovery factor in fracture
E_m recovery factor in matrix
f drainage boundary index
f_s specific density of fractures
g acceleration of gravity
h formation thickness
k permeability
k_c permeability of channels
k_f fracture permeability
k_m matrix permeability
k_{mf} average permeability of fracture-matrix system
L length
M molecular weight
$m(p)$ real gas pseudopressure function
n constant
n_c number of channels
n_f number of fracture
p pressure
p_e external boundary pressure
$\overline{p}$ mean pressure
p_w well pressure
q flow rate
$\overline{q}$ flow rate at mean pressure
q_m mass flow rate
q_r flow rate at any radius
q_m average mass flow rate
r radius
r_c radius of channels
Re Reynolds number
R_f fraction of total pore volume in secondary porosity
r_w wellbore radius
r_{wa} effective wellbore radius
s skin factor

GREEK SYMBOLS

β turbulence factor
ϕ porosity
γ_g gas specific gravity
μ viscosity

ρ density
ω Warren–Root fluid capacitance factor

REFERENCES

1. Craft BC, Hawkins MF. *Applied petroleum reservoir engineering*. Englewood Cliff, NJ: Prentice-Hall; 1959.
2. Darcy H. *Les fontaines publiques de la ville de Dijon*. Paris: Victor Dalmont; 1856.
3. Hubbert MK. Entrapment of petroleum under hydrodynamic conditions. *Bull Am Assoc Petrol Geol (AAPG)* 1953;**37**:1954–73.
4. Dake LP. *Fundamentals of reservoir engineering*, 2nd ed. New York: McGraw-Hill; 1978.
5. Pirson SJ. *Oil reservoir engineering*, 2nd ed. New York: McGraw-Hill; 1958.
6. Langnes GL, Robertson JO, Chilingar GW. *Secondary recovery and carbonate reservoirs*. Amsterdam, New York: Elsevier Science; 1972.
7. Chilingarian GV, Yen TF. Note on carbonate reservoir rocks, No. 5–interrelationships among various properties of carbonates. *Energy sources* 1987;**9**:51–65.
8. Pirson SJ. *Handbook of well log analysis*. Englewood Cliffs, NJ: Prentice Hall; 1963.
9. Muskat M. *Physical principles of oil production*. New York: McGraw-Hill; 1949.
10. Croft HO. *Thermodynamics, fluid flow and heat transmission*. New York: McGraw-Hill; 1938.
11. Kotyakhov FI. Approximate method of determining petroleum reserves in fractured rocks. *Neftyanoe Khozyaystvo* 1956;**4**:40–6.
12. Daniel EJ. Fractured reservoirs of Middle East. *Bull Am Assoc Petrol Geol (AAPG)* 1954;**38**:774–815.
13. Locke LC, Bliss JE. Core analysis technique for limestone and dolomite. *World Oil* 1950; Sept.
14. Chilingarian GV. Approximate method of determining reserves and average height of fractures in fractured rocks. An interim report. *Compass and Sigma Gamma Epsilon* 1959;**36**:202–5.
15. Aguilera R. *Naturally fractured reservoirs*. Tulsa, OK: Petroleum Publishing Company; 1980.
16. Black JL, Lacik HA. History of a Scurry County, Texas reef unit. *Proc. Southwest Petrol. Short Course* 1964;35–9.
17. Cargile LL. A case history of the Pegasus Ellenburger Reservoir. *J Petrol Technol* 1969;1330–6.
18. Matthews CS, Russel DG. *Pressure buildup and flow tests in wells*. SPE of AIME, Monograph, Vol. 1, Dallas, TX, 1967.
19. Slider HC. *Practical petroleum reservoir engineering methods*. Tulsa, OK: Petroleum Publishing Company; 1976.
20. Earlougher RC. *Advances in well test analysis*. SPE of AIME, Monograph, Vol. 5, Dallas, TX, 1977.
21. Kumar A. Steady flow equations for wells in partial water-drive reservoirs. *J Petrol Technol* 1977;**11**:1654–6.
22. Kumar A. Strength of water drive or fluid injection from transient well test data. *J Petrol Technol* 1977;**11**:1497–508.
23. van Everdingen AF. The skin effect and its influence on the productive capacity of a well. *Trans AIME* 1953;**198**:171–6.

24. Hurst W. Establishment of the skin effect and its impediment to fluid flow into a wellbore. *Petrol Eng* 1953;**25**:B6–16.
25. Brons F, Miller WC. A simple method for correcting spot pressure readings. *J Petrol Technol* 1961;**222**:803–5. Trans AIME
26. Al-Hussain R, Ramey JH Jr, Crawford PB. The flow of real gas through porous media. *J Petrol Technol* 1966;620–36. May.
27. Russel DG, Goodrich JH, Perry GE, Bruskotter JF. Methods for predicting gas well performance. *J Petrol Technol* 1966;99–108. January.
28. Energy Resources Conservation Board. *Theory and practice of the testing of gas wells*, 2nd ed. 1975. 495 pp.
29. Ikoku CU. *Natural gas reservoir engineering*. New York: Wiley; 1984. 503 pp.
30. Bakhmeteff BA, Feodoroff NV. Flow through granular media. *Trans ASME* 1937;**59**:A97.
31. Rowan G, Clegg MW. An approximate method for non-Darcy radial gas flow. *Soc Petrol Eng J* 1964;**4**:96–114. June.
32. Wright DE. Nonlinear flow through granular media. *J Hydraul Div Am Soc Civil Eng Proc* 1968;**94**:851–72.
33. Jones SC. Using the inertial coefficient, β to characterize heterogeneity in reservoir rock. SPE 16949, In: *62nd Annual Technical Conference*, Dallas, TX, Sept. 27–30, 1987, pp. 165–176.
34. Geertsma J. Estimating the coefficient of inertial resistance in fluid flow through porous media. *Soc Petrol Eng J* 1974;**14**:445–50.
35. Katz DL. *Handbook of natural gas engineering*. McGraw-Hill; 1959. 802 pp.
36. Hubbert M. Darcy's law and the field equations of flow of underground fluids. *Trans AIME* 1956;**207**:222–39.
37. Cornell D. *Flow of gases through consolidated porous media*. Ann Arbor, MI: University of Michigan; 1952. Ph.D. Dissertation
38. Liu X, Civan F, Evans RD. Correlation of the non-Darcy flow coefficient. *J Can Petrol Technol* 1995;**34**(10):50–4.
39. Janicek J, Katz DL. *Application of unsteady state gas flow calculations*. Ann Arbor, MI: Preprint, University of Michigan Publishing Services; 1955.
40. Cornell D, Katz DL. Flow of gases through consolidated porous media. *Ind Eng Chem* 1953;**45**:2145–52.
41. Forchheimer PH. Wasserbewegung durch boden. *Z Ver Dsch Ing* 1901;**49**:1781–93.
42. Katz DL, Lee RL. *Natural gas engineering—production and storage*. Houston, TX: McGraw-Hill; 1990. 389 pp.
43. Moody LF. Friction factors for pipe flow. *Trans ASME* 1944;**66**:671.
44. Firoozabadi A, Katz DL. An analysis of high-velocity flow through porous media. *J Petrol Technol* 1979;**31**(2):211–16.
45. Firoozabadi A, Thomas LK, Todd B. High-velocity flow in porous media. *SPE Reservoir Eng* 1995;149–52. May.
46. Russel DG, Prats M. The practical aspects of interlayer crossflow. *J Petrol Technol* 1962;589–94. June.
47. Calhoun Jr. John C. *Fundamentals of reservoir engineering,* 4th ed. Norman, OK: University of Oklahoma Press; 1960.
48. Amaefule JO, Kersey DG, Marschall DM, et al. Reservoir description—a practical synergistic engineering and geological approach based on analysis of core data. SPE 18167, In: *63rd Annual Technical Conference*, Houston, TX, Oct. 2–5, 1988. 30 pp.

49. Warren JE, Root PJ. The behavior of naturally fractured reservoirs. *Soc Petrol Eng J* 1963;**3**:245–55.
50. Pollard T. Evaluation of acid treatments from pressure buildup analysis. *Trans AIME*, 1959; 216: 38–43.
51. Pirson RS, Pirson SJ. An extension of the Pollard analysis method of well pressure buildup and drawdown tests. SPE 101, In: *36th Annual Fall Meeting*, Dallas, TX, Oct. 1961.
52. Kazemi H. Pressure transient analysis of naturally fractured reservoirs with uniform fracture distribution. *Soc Petrol Eng J* 1969;**12**:451–62.
53. De Swaan AO. Analytic solutions for determining naturally fractured reservoirs properties by well testing. *Soc Petrol Eng J* 1976;**16**:117–22.

Chapter 8

Naturally Fractured Reservoirs

INTRODUCTION

Fractures are displacement discontinuities in rocks, which appear as local breaks in the natural sequence of the rock's properties. Most geological formations in the upper part of the earth's crust are fractured to some extent. The fractures represent mechanical failures of the rock strength to natural geological stresses such as tectonic movement, lithostatic pressure changes, thermal stresses, high fluid pressure, drilling activity, and even fluid withdrawal, since fluid also partially supports the weight of the overburden rock. Although petroleum reservoir rocks can be found at any depth, at the deeper depths, the overburden pressure is sufficient to cause plastic deformation of most of the sedimentary rocks. Such rocks are unable to sustain shear stresses over a long period and flow toward an equilibrium condition.

Fractures may appear as either microfissures with an extension of several micrometers or continental fractures with an extension of several thousand kilometers. They may be limited to a single rock formation or layer, or may propagate through many rock formations or layers. In geological terms, a fracture is any planar or curvi-planar discontinuity that results from the process of brittle deformation in the earth's crust. Planes of weakness in rock respond to changing stresses in the earth's crust by fracturing in one or more different ways, depending on the direction of the maximum stress and the rock type. A fracture may consist of two rock surfaces of irregular shape, being more or less in contact with each other. The volume between the surfaces is the fracture void.

On the basis of their porosity systems, naturally fractured rocks can be geologically categorized into three main types:

1. intercrystalline–intergranular, such as the Snyder field in Texas, the Elk Basin in Wyoming, and the Umm Farud field in Libya;

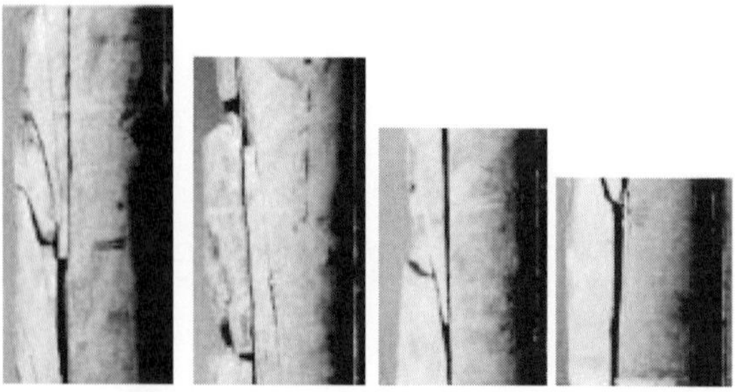

FIGURE 8.1 Naturally fractured rock cores taken from wells.

2. fracture–matrix, such as the Spraberry field in Texas, the Kirkuk field in Iraq, the Dukhan field in Qatar, and Masjid-i-Sulaiman and Haft-Gel fields in Iran; and
3. vugular–solution, such as the Pegasus Ellenburger field and the Canyon Reef field in Texas [1].

The accumulation and migration of reservoir fluids within a naturally fractured formation having the first type of porosity system are similar to those found in sandstone formations. Consequently, the techniques developed to determine the physical properties of sandstone porous media in Chapter 3 could be directly applied to formations having intercrystalline–intergranular porosity. Unfortunately, this is not the case for reservoirs having the other two types of porosity system. The pores in the matrix of a fracture–matrix formation are poorly interconnected, yielding a pattern of fluid movement that is very different from that of sandstone formations. Rocks with vugular–solution porosity systems exhibit a wide range of permeability distributions varying from relatively uniform to extremely irregular as shown in Figure 8.1.

ORIGIN OF PERMEABILITY IN CARBONATE ROCKS

Carbonate reservoirs, which are typically dual or triple porosity systems, produce a major portion of the world's oil and gas and hold more than half of the world's largest crude oil and natural gas reserves. The best known Jurassic carbonate reservoirs are the Arab-D and Arab-C systems, Hanifa and Hadriya. These reservoirs account for large quantities of crude oil production in Saudi Arabia. The Permo-Triassic Khuff formation holds the world's largest known natural gas reserves in five Arabian Gulf countries. Some of the world's best quality crude oil is located in two Cretaceous carbonate reservoirs: the Shuaiba and Tharmama reservoirs. Many of these carbonate reservoirs are naturally fractured and/or have an extensive and extremely complex secondary porosity system. Accurately predicting amounts and types of porosity, locally for field development and regionally for stratigraphic traps, is a major challenge. Three methods are used to estimate fracture porosity: well-test analysis, well logging, and core analysis.

A natural fracture is a planar discontinuity in reservoir rock due to deformation or physical diagenesis. Diagenesis—chemical and physical changes after deposition—strongly modifies the reservoir properties possessed at the time of deposition. The dominant diagenetic process consists of early cementation, selective dissolution of aragonite and reprecipitation as calcite, burial cementation, dolomitization, and compaction-driven microfracturing [2]. Cementation and compaction forces completely eradicate any porosity available at the time of deposition. However, chemical changes, usually dissolution, especially in carbonate rocks, modify the initial porosity and recover it partially. Depositional facies, their architecture, systems, and tracts are predominant driving factors in the distribution and quality of current reservoir properties, which are completely different from the properties at the time of deposition.

High-permeability vugs, molds, natural fractures, and caverns in carbonate rocks are the result of intense dissolution, which took place before burial as a result of nonreservoir or seal units. Dissolution is also caused by meteoric diagenesis, which is related to subaerial exposure of carbonate rocks and is explained by the general aggressiveness of meteoric water toward sedimentary carbonate minerals. Aragonite is metastable, which dissolves and precipitates into cement, whereas calcite is stable and less affected by dissolution. Such a type of dissolution causes significant variation in the distribution of porosity and permeability in the reservoirs, thereby defining reservoir quality.

GEOLOGICAL CLASSIFICATIONS OF NATURAL FRACTURES

Natural fracture patterns are frequently interpreted on the basis of laboratory-derived fracture patterns corresponding to models of paleostress fields and strain distribution in the reservoir at the time of fracture [3].

Classification Based on Stress/Strain Conditions: Stearns and Friedman proposed the classification of natural fractures on the basis of stress/strain conditions in laboratory samples and fractures observed in outcrops and subsurface settings. On the basis of their work, fractures are classified as follows [4]:

(a) Shear fractures that exhibit a sense of displacement parallel to the fracture plane. They are formed when the stresses in the three principal directions are all compressive. They form at an acute angle to the maximum principal stress and at an obtuse angle to the direction of minimum compressive stress.

(b) Extension fractures that exhibit a sense of displacement perpendicular to and away from the fracture plane. They are formed perpendicular to the minimum stress direction. They result when the stresses in the three principal directions are compressive, and can occur in conjunction with shear fracture.

(c) Tension fractures that also exhibit a sense of displacement perpendicular to and away from the fracture plane. However, in order to form this type of fracture, at least one of the principal stresses has to be tensile. Since rocks exhibit significantly reduced strength in tension tests, this results in increased fracture frequency.

Classification Based on Paleostress Conditions: The geological classification of fracture systems is based on the assumption that natural fractures depict the paleostress conditions at the time of the fracturing. On the basis of geological conditions, fractures can be classified into the following three categories.

Tectonic Fractures: The orientation, distribution, and morphology of these fracture systems are associated with local tectonic events. Tectonic fractures form in networks with specific spatial relationships to faults and folds. Fault-related fracture systems could be shear fractures formed either parallel to the fault or at an acute angle to it. In the case of the fault-wedge, they can be extension fractures bisecting the acute angle between the two fault shear directions [2, 5]. The intensity of fractures associated with faulting is a function of lithology, distance from the fault plane, magnitude of the fault displacement, total strain in the rock mass, and depth of burial.

Fold-related fracture systems exhibit complex patterns consistent with the complex strain and stress history associated with the initiation and growth of a fold [6]. Fracture types in fold-related systems are defined in terms of the dip and strike of the beds.

Regional Fractures: These fracture systems are characterized by long fractures exhibiting little change in orientation over their length. These fractures also show no evidence of offset across the fracture plane and are always perpendicular to the bedding surfaces. Regional fracture systems can be distinguished from tectonic fractures in that they generally exhibit simpler and more consistent geometry and have relatively larger spacing.

Regional fractures are commonly developed as orthogonal sets with two orthogonal orientations parallel to the long- and short axes of the basin in which the fractures are formed. Many theories have been proposed on the origin of the regional fractures, such as plate tectonics and cyclic loading/unloading of rocks associated with earth tides. As in the case of tectonic fractures, small-scale variation in regional fracture orientation of up to $\pm 20^{\circ}$ can result from strength anisotropies in reservoir rocks due to sedimentary features such as cross-bedding.

Contractional Fractures: These types of fractures result from bulk volume reduction of the rock. Desiccation fractures may result from shrinkage upon loss of fluid in subaerial drying. Mud cracks are the most common fractures of this type. Syneresis fractures result from bulk volume reduction within the sediments by subaqueous or surface dewatering. Dewatering and volume reduction of clays or of a gel or a colloidal suspension can result in syneresis fractures. Desiccation and syneresis fractures can be either tensile or extension fractures and are initiated by internal body forces. The fractures tend to be closely spaced and regular and isotropically distributed in three dimensions. Syneresis fractures have been observed in limestone, dolomites, shales, and sandstones [7].

Thermal contractional fractures may result from contraction of hot rock as it cools. Depending on the depth of burial, they may be either tensile or extension fractures. The generation of thermal fractures is predicted on the existence of a thermal gradient within the reservoir rock material. A classic example of thermally induced fracture is the columnar jointing observed in igneous rocks.

Fractures may also result from mineral changes in the rock, especially in carbonates and clay constituents in sedimentary rocks. Phase changes such as the chemical change from calcite to dolomite result in changes in bulk volume, and this leads to complex fracture patterns (Figure 8.2).

(a) Regional fracture patterns found in Jurassic Navajo sandstone. Lake Powell, southern Utah.

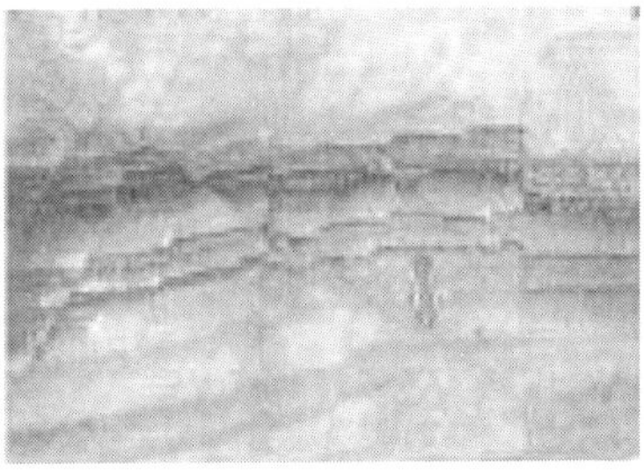

(b) Conjugate shear fractures corresponding to a tectonic feature system in an outcrop from Wyoming.

(c) Crack network observed in mud.

FIGURE 8.2 Different fracture systems in mud, and rocks. (a) Regional fracture patterns found in Jurassic Navajo sandstone, Lake Powell, southern Utah. (b) Conjugate shear fractures corresponding to a tectonic fracture system in an outcrop from Wyoming. (c) Crack network observed in mud. After Lui et al. [8].

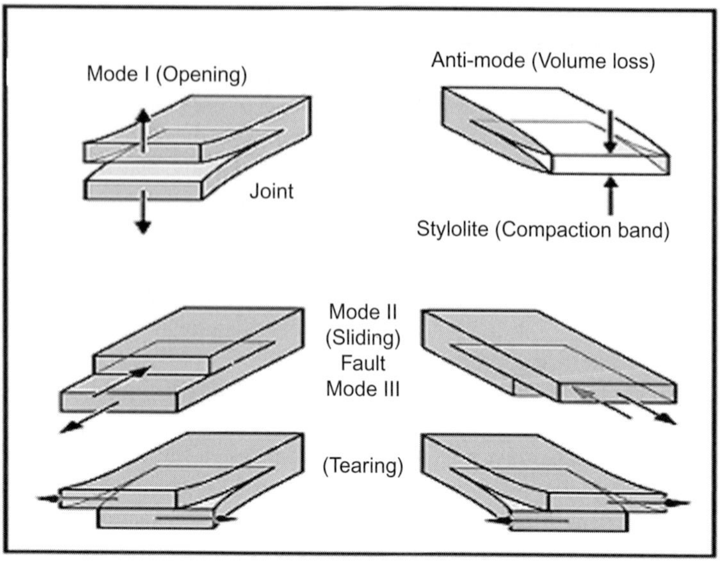

FIGURE 8.3 Modes of fracture formation.

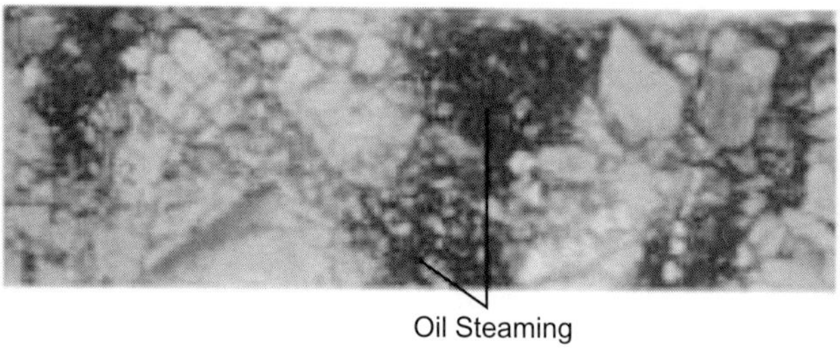

FIGURE 8.4 Reservoir fluids in shear fractures, Monterey Formation, California [9].

It is clear from the above discussion that the complex stress/strain distribution in reservoir rocks results in complex fracture patterns. Fracture patterns corresponding to different geological systems have key characteristics that can be used to classify and index natural fracture networks observed in outcrops and subsurface samples (Figures 8.3 and 8.4).

ENGINEERING CLASSIFICATION OF NATURALLY FRACTURED RESERVOIRS

Fractures may have either a positive or a negative impact on fluid flow, depending on whether they are open or sealed as a result of mineralization.

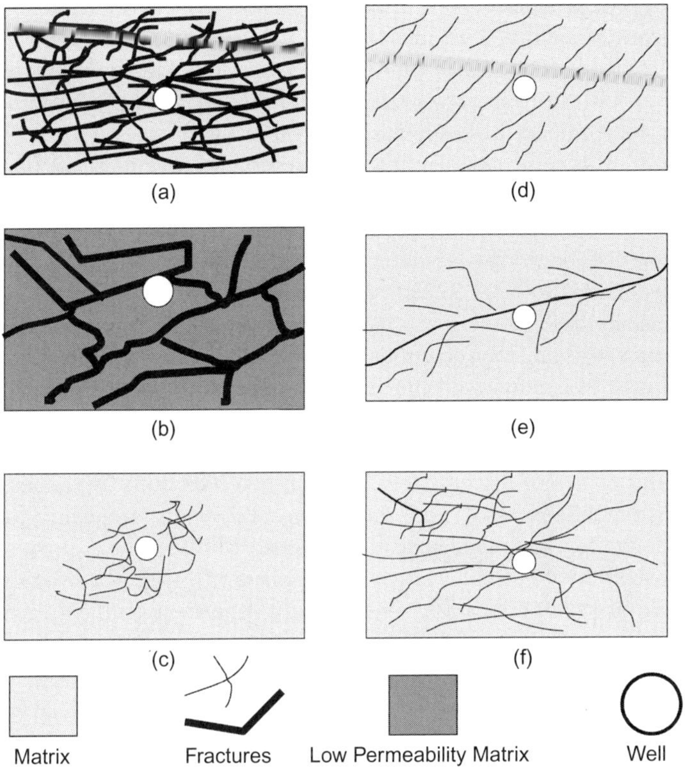

FIGURE 8.5 Types of naturally fractured reservoirs [10].

However, in most fracture modeling studies, fractures are considered open and have a positive impact on the fluid flow. A sealed, small natural fracture may even be undetectable (Figure 8.5).

Nelson identified four types of naturally fractured reservoirs, based on the extent to which fractures have altered the porosity and permeability of the reservoir matrix [1]:

Type 1: In type 1 naturally fractured reservoirs, fractures provide the essential reservoir storage capacity and permeability. Typical type 1 naturally fractured reservoirs are the Amal field in Libya, La Paz and Mara fields in Venezuela, and Precambrian basement reservoirs in eastern China. All these fields contain high fracture density.

Type 2: In type 2 naturally fractured reservoirs, fractures provide the essential permeability, and the matrix provides the essential porosity, such as in the Monterey fields of California, the Spraberry reservoirs of West Texas, and Agha Jari and Haft Kel oil fields of Iran.

Type 3: In type 3 naturally fractured reservoirs, the matrix has an already good primary permeability. The fractures add to the reservoir permeability and can result in considerable high flow rates, such as in Kirkuk field of Iraq, Gachsaran field of Iran, and Dukhan field of Qatar. Nelson includes the Hassi Messaoud (HMD) field of Algeria in this list. Although, indeed, there are several low-permeability zones in HMD that are fissured; in most zones, however, the evidence of fissures is not clear or unproved.

Type 4: In type 4 naturally fractured reservoirs, fractures are filled with minerals and provide no additional porosity or permeability. These types of fractures create significant reservoir anisotropy and tend to form barriers to fluid flow and partition formations into relatively small blocks. They are often uneconomic to develop and produce.

Nelson discusses three main factors that can create reservoir anisotropy with respect to fluid flow: fractures, cross-bedding, and stylolite. The anisotropy in Hassi Messaoud field, for instance, appears to be the result of a non-uniform combination of all three factors with varying magnitude from zone to zone. Stylolites, like fractures, are a secondary feature. They are defined as irregular planes of discontinuity between two rock units. Stylolites, which often have fractures associated with them, occur most frequently in limestone, dolomite, and sandstone formations. Mineral-filled fractures and stylolites can create strong permeability anisotropy within a reservoir. The magnitude of such permeability is extremely dependent on the measurement direction, thereby requiring multiple well testing. Interference testing is ideal for quantifying reservoir anisotropy and heterogeneity, because they are more sensitive to directional variations of reservoir properties, such as permeability, which is the case for type 4 naturally fractured reservoirs.

It is important to take this classification into consideration when, for instance, interpreting a pressure transient test for the purpose of identifying the type of fractured reservoir and its characteristics. A reservoir management program of naturally fractured formations also must take this classification into account. Each type of naturally fractured reservoir may require a different development strategy. Ershaghi [44] reports that (a) type 1 fractured reservoirs, for instance, may exhibit sharp production decline and can develop early water and gas coning; (b) recognizing that the reservoir is a type 2 will impact any infill drilling or the selection of improved recovery process; and (c) in type 3 reservoirs, unusual behavior during pressure maintenance by water or gas injection can be observed because of unique permeability trends.

INDICATORS OF NATURAL FRACTURES

Stearns and Friedman reviewed the multiple roles fractures played in exploration and exploitation of naturally fractured reservoirs [4]. They showed that fractures could alter the matrix porosity or the permeability, or both. If the fractures or connected vugs are filled with secondary minerals, they may restrict the flow. However, even in rocks of low matrix porosity, fractures and solution channels increase the pore volume by both increasing porosity

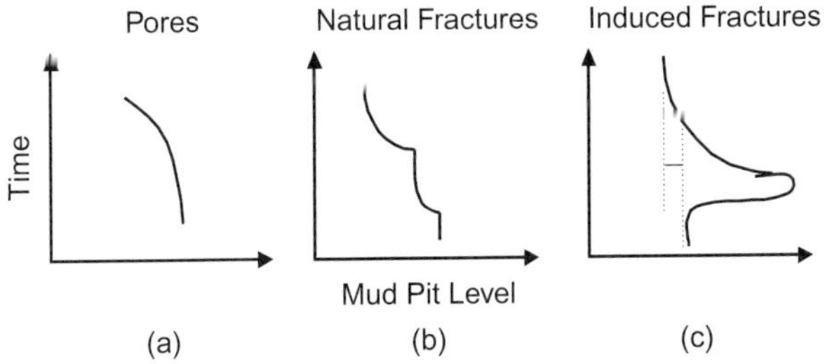

FIGURE 8.6 Mud loss indication and pit level behavior in pores, natural fractures, and induced fractures: (a) gradual buildup in loss ratio with pressure; (b) sudden start and exponential decline; and (c) loss that can occur on increase in ECD as pumps are turned off and on [15].

and connecting isolated matrix porosity and, therefore, help the recovery of petroleum fluids economically. Hence, the ability to estimate a fracture's density and its distribution of porosity is essential for reservoir evaluation. One should keep in mind, however, that fractures alone constitute less than 1% of the porosity [11,12]. Early recognition of a fractured reservoir and an estimate of its rock characteristics, such as porosity and permeability, will influence the location and number of subsequent development wells and, therefore, is of major economic significance. Stearns and Friedman [4], Aguilera [13], Saidi [14], and Nelson [1] reviewed many of the approaches used to detect and analyze naturally fractured reservoirs. Some of these methods are as follows:

1. Loss of circulating fluids (Figure 8.6a) and an increase in penetration rate during drilling are positive indications that a fractured and/or cavernous formation has been penetrated (Figure 8.6b, c).
2. Fractures and solution channels in cores provide direct information on the nature of a reservoir. A detailed systematic study of the cores must be made by the geologist in order to distinguish natural fractures from those induced by the core handling process. Careful examination of fracture faces and determination of density, length, width, and orientation of fractures may lead to the ability to distinguish fractures induced during coring from natural fractures. Preferably, a naturally fractured formation should be analyzed with full-diameter cores. Plug data, which do not reflect the permeability of fractures, often indicate a nonproductive formation, whereas full-diameter core data indicate hydrocarbon production. If actual production rates are several fold higher than those calculated from permeability determined by core analysis, natural fractures not observed in the core are suspected [16]. Low core recovery efficiency—less than 50%—suggests a highly fractured carbonate formation.
3. Logging tools are designed to respond differently to various wellbore characteristics, such as lithology, porosity, and fluid saturations, but not to natural fractures [17,18]. The presence of a large number of open

fractures, however, will affect the response of some logging tools. Well-logging measurements based on sonic wave propagation, which are negligibly affected by the borehole conditions, are used as fracture indicators. Measurements by the caliper log, density log, or resistivity log, under proper conditions, can be very effective in locating fractured zones. Identification of fractured intervals is straightforward using resistivity images. The fractures fill with conductive drilling fluids and represent a strong resistivity contrast to the surrounding rock matrix. Dipmeter data on FIL (fracture identification log) and FMI (fracture micro imager) provide the most effective methods for fracture detection. FMI images have allowed quantitative fracture evaluation and well comparison, providing data for completion design and reserve calculation. Combination of FMI and dual packer MDT (modular dynamics tester) adds a new dimension to fracture evaluation with the ability to straddle fracture intervals.

4. The subject of pressure buildup and flow tests in naturally fractured reservoirs has received considerable attention in the petroleum literature. Warren and Root assumed that the formation fluid flows from the matrix to fractures under pseudosteady state and showed that a semilog pressure buildup curve similar to that shown in Figure 8.7 is typical of a fractured formation [19]. If the existing fractures dominantly trend in a single

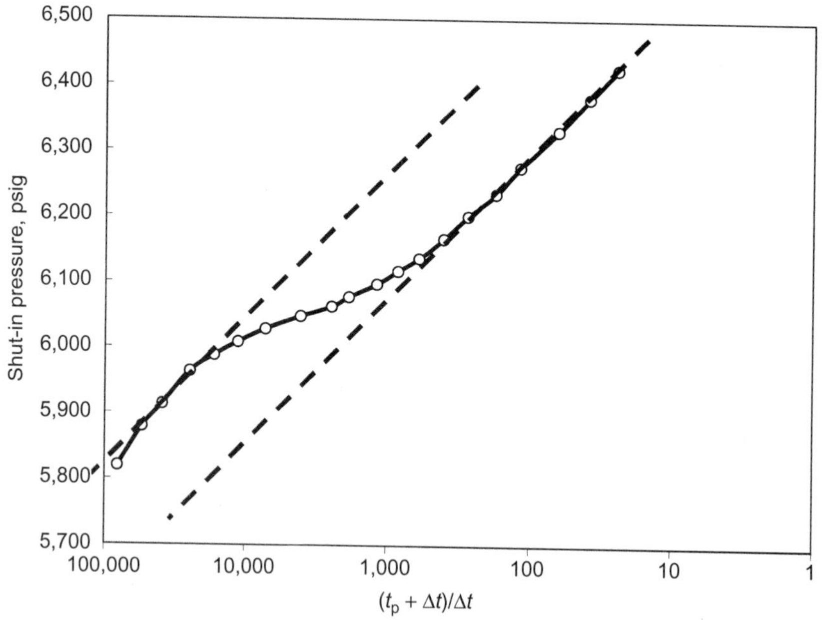

FIGURE 8.7 Pressure buildup curve from a naturally fractured reservoir [19].

direction, the reservoir may appear to have anisotropic permeability. If enough observation wells are used, pressure interference and pulse tests provide the best results.

5. Natural vertical fractures in a nondeviated borehole can be identified as a high-amplitude feature that crosses other bedding planes.
6. Downhole direct and indirect viewing systems, including downhole photographic and television cameras, are also used to detect fractures and solution channels on the borehole face. The borehole televiewer is an excellent tool that provides useful pictures of the reservoir rock, especially with the recent developments in signal processing. Vertical fractures appear as straight lines whereas dipping fractures tend to appear as sinusoidal traces because the televiewer shows the wellbore sandface as if it were split vertically and laid flat. Another useful televiewer tool for detecting natural fractures is the formation microscanner (FMS) device. This tool can detect fractures that range from few millimeters to several centimeters long, distinguishes two fractures as close as 1 cm apart (Figure 8.8), and distinguishes between open and closed fractures. Only fractures that are at least partially open contribute to production.

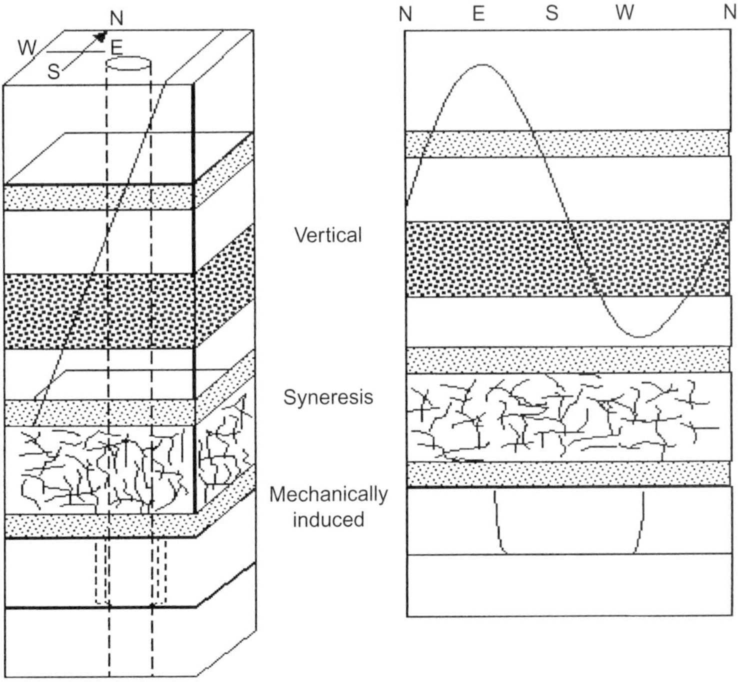

FIGURE 8.8 Visually detectable fractures.

7. Very high productivity index. A productivity index of 500 STB/D/psi or higher is typical of naturally fractured wells produced under laminar flow. Some wells in Iranian oil fields reported a productivity index of 10,000 STB/D/psi. In these wells, 95% of flow is through fractures [14].
8. A considerable increase in productivity of the well flowing after an artificial stimulation by acidizing is a strong indication of a naturally fractured formation. Acidizing is done essentially to increase the width of fractures and channels.
9. Because of the high permeability of the fractures, the horizontal pressure gradient is typically small near the wellbore as well throughout the reservoir [14]. This is primarily true in type 1 and to a lesser degree in type 2 fractured reservoirs.

Other indicators of the existence of the natural fractures in the reservoir are:

(a) local history of naturally occurring fractures;
(b) lack of precision in seismic recordings;
(c) extrapolation from observations on outcrops; and
(d) pressure test results that are incompatible with porosity and permeability values obtained from core analysis and/or well logging.

As can be deduced from the preceding discussion, no method used alone provides a definite proof of the presence of fractures. FMS logs and borehole televiewers often give a reliable indication of the presence of major features; however, they do not resolve the full complexity of many of the smaller scale fracture systems. Fracture detection is most certain when several independent methods confirm their presence. Different naturally fractured reservoirs require different combinations of methods of analysis. A combination of core analysis, pressure transient test analysis, and various fracture-finding logs is strongly recommended for detecting and locating fractures. Table 8.1 summarizes the many techniques available for detecting natural fractures.

VISUAL IDENTIFICATION OF FRACTURES

Nelson [1] defined, for consistency, four useful terminologies to describe cracks in a rock:

(a) fracture: any break in the rock;
(b) fissure: an open fracture;
(c) joint: one or a group of parallel fractures that has no detectable displacement along the fracture surface; and
(d) fault: a fracture with detectable displacement.

TABLE 8.1 Summary of Detection Techniques for Natural Fractures [15]

	Core	FMS/FMI	BHTV	Litho Density	Stoneley Wave	Mud Loss
What is detected?	Localized fracture porosity	Mud invasion into fracture	Contrast in acoustic properties	Invasion density of drill solids into fracture	Stoneley energy reflected by the fracture	Flow of mud into fractures
How narrow a fracture aperture can be detected?	On the order of microns	On the order of microns given sufficient electrical conductivity contrast	1 mm	5 mm	1 mm	0.20 mm
Fracture mistaken for permeable fractures	Fracture porosity. Induced fractures	Fracture porosity. Induced fractures. Drilling damage	Fracture porosity. Induced fractures. High impedance bedding and certain healed fractures	Fracture porosity. Drilling damage. Mineralization	Washout bed boundaries	None
Depth of investigation	Diameter of core	10 mm	3 mm	100 mm	Less than 1.8 m	Depth of mud invasion >1 m
Can it provide strike and dip?	Yes	Yes	Yes	No	No	No
Mud limitations	None	Water-based muds only	Mud weight must be less than 14 lb/gal	Mud weight must be greater than 10 lb/gal	None	None
Additional comments	Highly fractured "rubble zones" not recovered	Has difficulty in distinguishing high-permeability fractures from low-permeability ones	Has difficulty in distinguishing high-permeability fractures from low-permeability ones		Fractures plugged with mud solids are often not detected	Yields information on degree of formation damage and stimulation requirements

All these features can be visually identified on core or borehole electrical images. Figure 8.8 shows three types of fracture that may be visually detected:

(a) Natural vertical fractures in a nondeviated borehole can be identified as a high-amplitude feature that crosses other bedding planes. They occur in all lithologies. Fractures may be open, mineral filled, or vuggy. Visual inspection of cores and borehole electric images may be used only as a guide for interpretation. Core flow tests and actual production tests are recommended for interpreting the morphology of natural fractures. Production and recovery efficiency in reservoirs is influenced by the angle. The angle most often used by oil companies as a criterion is 75°. Fractures with dip angles of more than 75° are treated as vertical fractures, while those less than 75° are treated as high-angle fractures. Vertical fractures are more common in sandstone rocks.

(b) Syneresis fractures have a braided appearance and are often referred to as "chicken wire" fractures. They normally occur only in carbonate formations.

(c) Mechanically induced fractures are sometimes unintentionally created during the drilling operations, or by hydraulic fracturing to stimulate the formation.

Fracture morphology can also be visually detected on cores and/or borehole images. Figure 8.9 shows four detectable fracture morphologies: vuggy, mineral filled, partially mineral filled, and open.

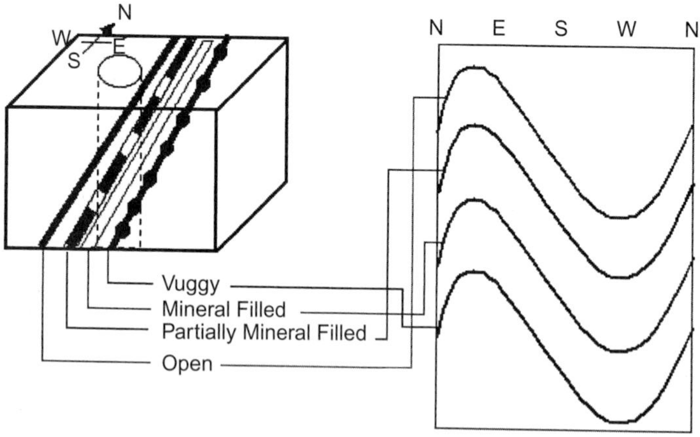

FIGURE 8.9 Fracture morphology showing vuggy, mineral filled, partially mineral filled, and open fractures and their log response sequence.

PETROPHYSICAL PROPERTIES OF NATURALLY FRACTURED ROCKS

Although advanced well-logging tools such as nuclear magnetic resonance (NMR) are currently being used to estimate rock permeability downhole, the technology is not yet fully developed. The only method used to estimate the permeability reliably is a combination of core-derived parameters with computer-processed log data that establishes, statistically, a relationship between the permeability of the fracture–matrix system and other parameters such as porosity and irreducible water saturation. Efforts have also been made to incorporate grain diameter and shale fraction in such models to reduce the scatter in the data. With such a relationship established, the formation petrophysical parameters, including permeability distribution, can be deduced from log data alone in wells or zones without core data. However, in carbonate formations, where structural heterogeneities and textural changes are common and, unfortunately, only a small number of wells are cored, the application of statistically derived correlations is extremely limited. Such correlations cannot be used to identify hydraulic flow units or bodies in naturally fractured reservoirs.

FRACTURE POROSITY DETERMINATION

The range of fracture porosity, ϕ_f, is 0.1–5%, depending on the degree of solution channeling, as shown in Figure 8.10, and on fracture width and spacing, as shown in Tables 8.2 and 8.3. In some fields, like the La Paz and Mara fields in Venezuela, fracture porosity may be as high as 7%. Accurate measurement of fracture porosity is essential for the efficient development

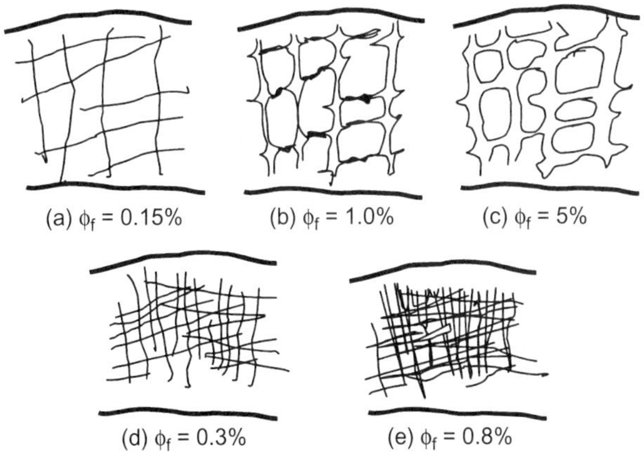

FIGURE 8.10 Development of fracture porosity in carbonate rocks that have low insoluble residue, (a), (b), (c), and high insoluble residue, (d) and (e) [20].

TABLE 8.2 Porosity of Various Naturally Fractured Reservoirs [21]

Field	Porosity Range (%)
Beaver Gas Field	0.05–5
Austin chalk	0.2
General statement	1
South African Karst Zone	1–2
CT scan examples	1.53–2.57
Epoxy injection examples	1.81–9.64
Monterey	0.01–1.1

TABLE 8.3 Fracture Width and Spacings of Various Naturally Fractured Formations [21]

Field	Width Range (mm)	Average	Spacing/Frequency
Spraberry	0.33 max.	0.051	Few inches to a few feet
Selected dam sites	0.051–0.10		4–14 ft
La Paz-Mara field	6.53 max.		
Small joints	0.01–0.10		
Extension fractures	0.1–1.0	0.2	
Major extension fractures	0.2–2		
Monterey		0.01	3–36 ft

and economical exploitation of naturally fractured reservoirs. If oil is trapped in both the matrix and fissures, then the total oil in place in the reservoir is given by the following equation [10]:

$$N_{ot}(\mathrm{STB}) = N_{om} + N_{of} \tag{8.1}$$

where N_{om} and N_{of} are, respectively, the oil volumes trapped in the matrix and fractures. Assuming a volumetric system, these two volumes, expressed in STB, are calculated as follows:

$$N_{om} = \frac{7{,}758 A h \phi_m (1 - S_{wm})}{B_o} \tag{8.2}$$

$$N_{\text{of}} = \frac{7,758Ah\phi_{\text{f}}(1 - S_{\text{wf}})}{B_{\text{o}}} \tag{8.3}$$

where

A = surface area of the reservoir, acres
h = average reservoir thickness, ft
ϕ_{f} = fracture porosity, fraction
ϕ_{m} = matrix porosity, fraction
S_{wf} = water saturation in fractures, fraction. In fractures, capillary pressure is practically negligible, which means S_{wf} will be very small ($\approx 0.25\ S_{\text{wm}}$)
S_{wm} = water saturation in matrix, fraction
B_{o} = oil formation volume factor, bb1/STB

Fracture porosity can be expressed as the ratio of the fracture pore volume (V_{Pf}) over the total bulk volume (V_{bt}):

$$\phi_{\text{f}} = \frac{V_{\text{Pf}}}{V_{\text{bt}}} \tag{8.4a}$$

The total porosity is:

$$\phi_{\text{t}} = \phi_{\text{f}} + \phi_{\text{m}} = \frac{V_{\text{pf}}}{V_{\text{bt}}} + \frac{V_{\text{pm}}}{V_{\text{bm}}} = \frac{V_{\text{pf}}}{V_{\text{bt}}} + \frac{V_{\text{pm}}}{(1 - \phi_{\text{f}})V_{\text{bt}}} \tag{8.4b}$$

The sonic log only measures the matrix porosity. However, neutron porosity is the combination of both the matrix and fracture porosity. Thus, fracture porosity can be estimated from well logs as [14]:

$$\phi_{\text{f}} = \phi_{\text{Neu}} - \phi_{\text{Son}} \tag{8.5}$$

Fracture porosity can also be estimated with the help of well-test analysis in such reservoirs, using Eq. (8.76).

On the basis of Pirson model of fully water-saturated rocks, Aguilera developed the following equation that relates the total formation resistivity factor, F_{t}, for dual porosity systems, to the total porosity [22]:

$$F = \frac{R_{\text{o}}}{\nu\phi_{\text{t}}R_{\text{o}} + (1 - \nu)R_{\text{w}}} \tag{8.6}$$

where R_{o} is the resistivity of porous rock 100% saturated with brine and R_{w} is the formation water resistivity, both expressed in ohm-m.

If only the matrix porosity is present in the system, the porosity partitioning coefficient, ν, is equal to zero. Thus, Eq. (8.6) simplifies Eq. (8.7), which is the same as for a consolidated matrix:

$$F = \frac{R_{\text{o}}}{R_{\text{w}}} \tag{8.7}$$

If only fracture porosity is present in the system, such as in type 1 naturally fractured reservoirs, the porosity partitioning coefficient is equal to unity. In this case, the formation resistivity factor can be expressed as:

$$F = \frac{1}{\phi_f^{m_f}} \tag{8.8a}$$

Laboratory tests indicate that the tortuosity factor, τ, and the fracture porosity exponent, m_f, are approximately unity in systems with open and well-connected fractures. In type 2 and type 3 naturally fractured reservoirs, the formation resistivity factor can be more generally expressed as:

$$F = \frac{\tau}{(1-\phi_f^{m_f})\phi_m^{m_m} + \phi_f^{m_f}} \tag{8.8b}$$

where m_m is the matrix porosity. If only matrix porosity is present, that is, $\phi_f = 0$, Eq. (8.8b) simplifies Eq. (4.40), where $m = m_m$ and $a = \tau$. On the other hand, if only fracture porosity is present such as in type 1, Eq. (8.8b) simplifies Eq. (8.8a). If only the total porosity is known, then F can be estimated from:

$$F = \frac{a}{\phi_t^m} \tag{8.8c}$$

The fractures should be considered as being well connected if the interporosity coefficient, λ, which is determined from a pressure transient test, is high, that is, 10^{-4} or 10^{-5}. If the interporosity factor is low, that is, λ is approximately 10^{-8} or 10^{-9}, the fractures are poorly interconnected and/or partially mineral filled. In this case, m_f and τ may be as high as 1.75 and 1.5, respectively. For $10^{-6} \geq \lambda \geq 10^{-7}$, $1.75 \geq m_f \geq 1$, and $1.5 \geq \tau \geq 1$.

Example

The following characteristics of a type 2 naturally fractured formation were obtained from core analysis:

$\phi_f = 0.034$	$\phi_m = 0.106$	$m_f = 1$	$m_m = 2$	$\nu = 0.22$
$R_o = 4.77$ ohm-m	$R_w = 0.045$ ohm-m			

Estimate the tortuosity factor for this formation.

Solution

Using Eq. (4.5), the formation resistivity factor is:

$$F = \frac{0.045 \times 4.77}{0.22 \times (0.034 + 0.106) \times 4.77 + (1 - 0.22) \times 0.045} = 26.2$$

The tortuosity is calculated from Eq. (8.8b):

$$\tau = F((1-\phi_f^{m_f})\phi_m^{m_m} + \phi_f^{m_f}) = (26.2)((1-0.034)(0.0.106^2) + 0.034) = 1.17$$

This value of tortuosity indicates the fractures are well connected.

CEMENTATION FACTOR FOR CARBONATES

Doveton [23] investigated the sonic, neutron, and density logs responses in carbonate formations. These logs normally respond to pores of all sizes. However, field observation has shown that the sonic log is a measure of interparticle (intergranular and intercrystalline) porosity but is largely insensitive to either fractures or vugs. According to Doveton, this discrimination can be explained largely by the way the sonic tool measures transit time by recording the first arrival waveform that often corresponds to a route in the borehole wall free of fractures or vugs. When sonic porosities are compared with neutron and density porosities, the total porosity can be subdivided between "primary porosity" (interparticle porosity) recorded by the sonic log and "secondary porosity" (vugs and/or fractures) computed as the difference between the sonic porosity and the neutron and/or density porosity [23]:

$$\phi_{f+v} = \phi_t - \phi_m = \phi_N - \phi_S \quad (8.8d)$$

$$\phi_{f+v} = \phi_t - \phi_m = \phi_N - \phi_D \quad (8.8e)$$

where

ϕ_{f+v} = secondary porosity (fractures + vugs)
ϕ_m = primary (matrix) porosity
ϕ_t = total porosity
ϕ_N = neutron log porosity
ϕ_S = sonic porosity
ϕ_D = density log porosity

Typically, moderate values in secondary porosity are caused by vugs, because fracture porosity does not usually exceed 1–2% by volume, and, therefore, the total porosity can be estimated from the average of the neutron and density porosities and is distinctly higher than the sonic porosity in oomoldic zones because of significant "secondary" porosity [23]:

$$\phi_t = 0.5(\phi_N + \phi_D) \quad (8.8f)$$

The Nugent equation [24] is often used to estimate water saturation in oomoldic zones, by computing an apparent cementation factor for each zone and then applying the Archie equation for water saturation:

$$m = \frac{2\log(\phi_S)}{\log(\phi_t)} \quad (8.8g)$$

The estimates of the cementation exponent by Eq. (8.8g) are good at low porosities (<10%). At high porosities, however, the estimates may be lower than their true values, which are obtained from a Picket plot.

In the absence of log data, typical values of m quoted in the log analysis literature for various carbonate pore systems are:

1. intergranular/intercrystalline, $m = 2.0$;
2. fractures, $m = 1.4$;
3. vugs, $m \geq 2.3$; and
4. oomoldic, $m \geq 3$.

If the fractures are plane fractures in a naturally fractured carbonate reservoir, the cementation factor m is between 2.0 and 1.0. Obviously, these numbers will be controlled by the degree of fracturing, vugginess, and proportion of moldic porosity, and can lead to a wide variety of m values observed in the field. Rasmus [25] developed the following relationship for calculating a value for m in fractured carbonate reservoirs:

$$m = \frac{\log[\phi_S^3 + \phi_S^2(1-\phi_t) + (\phi_t - \phi_S)]}{\log(\phi_t)} \tag{8.8h}$$

In order to use this formula, it is important to know that the reservoir is indeed fractured. The presence of fractures can be verified from a fracture identification log or cores. The value of m for vuggy porosity without fractures should be greater than 2.0, not less than, as it would be if it were calculated by Eq. (8.8h). The following is an example [26] of what would happen if the type of reservoir porosity were not taken into account.

Example

Given $R_w = 0.04$ ohm-m, $R_t = 20$ ohm-m, $\phi_s = 0.05$, $\phi_t = 0.15$ calculate the following:

1. cementation factor,
2. formation resistivity factor, and
3. water saturation.

Solution

Case 1—only vuggy porosity is assumed.

1. Using the above data and the original Nugent formula for m ((Eq. 8.8g)), we find:

$$m = \frac{2\log[\phi_S]}{\log\phi_t} = \frac{2\log[0.05]}{\log(0.15)} = 3.16$$

2. The formation resistivity factor is:

$$F = \frac{1}{\phi_t^m} = \frac{1}{\phi_t^{3.16}} = 401.4$$

3. The water saturation is obtained from the Archie equation:

$$S_w = \sqrt{\frac{FR_w}{R_t}} = \sqrt{\frac{401.4 \times 0.04}{20}} = 89.6\%$$

Case 2—only fracture porosity is assumed.

Using the above data and the Rasmus formula for m (Eq. (8.8h)), we find:

$$m = \frac{\log[\phi_S^3 + \phi_S^2(1-\phi_t) + (\phi_t - \phi_S)]}{\log \varphi_t}$$

$$= \frac{\log[0.05^3 + 0.05^2(1-0.15) + (0.15-0.05)]}{\log(0.15)} = 1.2$$

The formation resistivity factor and water saturation are:

$$F = \frac{1}{\phi_t^m} = \frac{1}{\phi_t^{1.2}} = 9.7$$

$$S_w = \sqrt{\frac{FR_w}{R_t}} = \sqrt{\frac{9.7 \times 0.04}{20}} = 14\%$$

From the above calculations, it is clear that if the difference in sonic porosity and total porosity was assumed to be due to fractures, that is:

$$\phi_f = \phi_t - \phi_S = 0.15 - 0.05 = 0.10$$

and there were no fractures, but rather vuggy porosity in the reservoir, then one would calculate a water saturation of only 14% when the true water saturation was actually 89.6%. Conversely, if the difference in sonic porosity and total porosity was assumed to be due to vuggy porosity only, but fracturing was present, then a water saturation of 89.6% would have been calculated when actually the true water saturation was only 14%. This example illustrates how important it is to look at the rocks first to determine porosity type before undertaking field development of carbonate reservoirs.

Ohen et al. [41] introduced the cementation factor into the definition of the reservoir quality index to account for the presence of natural fractures and vugs as follows:

$$\mathrm{RQI_f} = 0.0314\sqrt{\frac{k}{\phi_t^{2m-1}}} \tag{8.8i}$$

Therefore, as was discussed in Chapter 3 for sandstone, a plot of $\mathrm{RQI_f}$ versus ϕ_z on a log–log paper will delineate the flow units. Ohen et al. used the following relationships to estimate the cementation factor from logs and to account for the variable pore types (fracture, bi-modal, vuggy and intercrystalline):

$$m = \frac{\log(R_w/R_t \times S_{WR}^2)}{\log(\phi_t)} \tag{8.8j}$$

$$S_{WR} = \left(\frac{R_{xo}/R_t}{R_{inf}/R_w}\right)^{5/8} \tag{8.8k}$$

R_{mf} = mud filtrate resistivity
R_{xo} = resistivity of the mud-invaded zone
ϕ_{xo} = porosity of the mud-invaded zone

According to Ohen et al., using the above equations, the pore type of a formation can be recognized as follows:

$m = 2$, pore type is intercrystalline
$m > 2$, pore type is vuggy
$m < 2$, pore type is fractured (if $\phi_{xo} < \phi_t$) or bi-modal (if $\phi_{xo} > \phi_t$)

Several other correlations of m with total porosity have been published. Shell proposed the following formula for m that is applicable for low-porosity (<10%), no-fractured carbonates:

$$m = 1.87 - 0.019\phi \qquad (8.8l)$$

Borai [42] proposed another formula for low porosity carbonates that is based on core and log studies from the Persian Gulf:

$$m = 2.2 - 0.035(\phi + 0.042) \qquad (8.8m)$$

The Asmari formation is one of the most significant oil-producing reservoirs in the Persian Gulf. Hassani-giv and Rahimi [43], working with data from the Asmari formation, identified five classes of rock type and corresponding correlations of m with total porosity. The characteristics of the classes are as follows:

Class 1 rock type: Coarse crystalline sucrosic dolostone (100–150 μm) mostly with intercrystalline and vuggy pore types. The pore network is well connected and the *connectivity intensifies with the occurrence of occasional fractures.*
Class 2 rock type: Fine to medium crystalline dolostone. The most common pore type is large irregular vugs and enlarged molds. Low intercrystalline porosity.
Class 3 rock type: Fine to medium size crystal dolostone (15–20 μm). Irregular, medium vugs and rare molds, which are rarely interconnected through microfractures, are the most common pore types. The tightly interlocking dolomite crystals reduce the visible intercrystalline pore types.
Class 4 rock type: Consists mostly of packstone. The most common porosity types are interparticle, mold, enlarged mold, and vugs. Interparticle pores are generally well connected and contribute effectively to fluid flow. The narrower pore throats connecting enlarged mold pores provide a less effective pore network for fluid flow.
Class 5 rock type: Consists of dolomitized packstone–wackestone. The most frequent pore types are: mold, enlarged mold, and rare intercrystalline and interparticle pores.

The best correlation of m with total porosity (fraction) for Classes 1 and 3 is:

$$m = e^{2.25\phi} + 0.6 \tag{8.8n}$$

The best correlation of m with total porosity (fraction) for Classes 2 and 5 is:

$$m = 2.48 - \frac{0.048}{\phi + 0.01} \tag{8.8o}$$

The best correlation of m with total porosity (fraction) for Class 4 is:

$$m = 2.52 - \frac{0.045}{\phi + 0.001}$$

According to Hassani-giv and Rahimi, applying the m values determined from these three correlations in the Archie equation yields representative values of water saturations.

POROSITY PARTITIONING COEFFICIENT

Reservoirs with a fracture–matrix porosity system—such as found in many carbonate rocks due to the existence of vugs, fractures, fissures, and joints—differ considerably from reservoirs having only one porosity type. The secondary porosity strongly influences the movement of fluids, whereas the primary pores of the matrix, where most of the reservoir fluid is commonly stored (more than 96% in type 3 naturally fractured reservoirs), are poorly interconnected. The Spraberry field of West Texas is an example of a naturally fractured sandstone oil reservoir, which is composed of alternate layers of sands, shales, and limestones [38]. The Altamont trend oil field in Utah is another naturally fractured sandstone reservoir with a porosity of 3–7% and an average matrix permeability less than 0.01 mD [13].

Laboratory-measured values of permeability for naturally fractured cores can be significantly different from the in situ values determined by well-pressure analysis. The difference is attributed to the presence of fractures, fissures, joints, and vugs, which are not adequately sampled in the core analysis. One of the earliest methods used to analyze full-sized naturally fractured cores was developed by Locke and Bliss [27]. The method consists of injecting water into a core sample and measuring the pressure values as a function of the cumulative injected volume of water (Figure 8.11). The secondary pore space, V_f, because of its high permeability, will be the first to fill up with water. A sharp increase in pressure is recorded later, indicating that the matrix porous space, V_m, has to fill up. The total pore volume, $V_t = V_f + \phi_m V_m$, is considered to be filled up when a pressure of 1,000 psi is

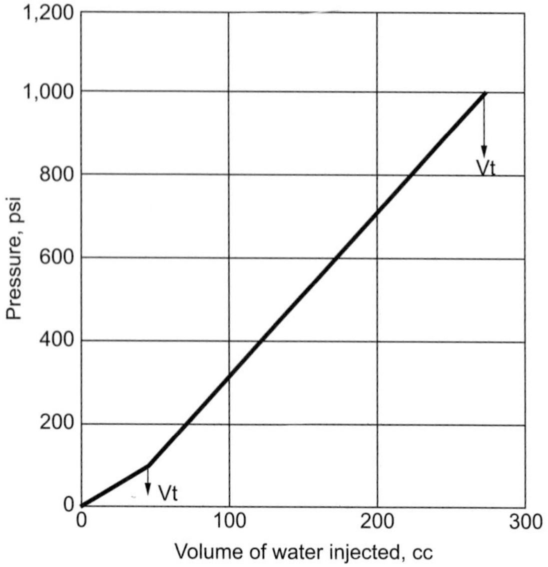

FIGURE 8.11 Locke and Bliss method for estimating the pore space of fractures.

reached in the test. If the fraction of total pore volume in the secondary porosity is ν, then:

$$\nu = \frac{V_f}{V_t} = \frac{V_f}{V_f + \phi_m V_m} \tag{8.9}$$

The term ν is commonly referred to as the *porosity partitioning coefficient*. This coefficient represents the apportioning of total porosity (ϕ_t) between the matrix (intergranular) porosity, ϕ_m, and secondary pores (vugs, fractures, joints, and fissures), ϕ_f.

The value of ν ranges between zero and unity for dual porosity systems. For total porosity equal to matrix porosity (absence of fracture porosity), $\nu = 0$. For total porosity equal to fracture porosity, $\nu = 1$. This coefficient can be estimated from core analysis using the Locke and Bliss method, pressure analysis, and well-logging data. By assuming that the fractures and matrix are connected in parallel and the drilling fluid used is nonconductive, as shown in Figure 8.20, Pirson suggested the following equations for short and long normal or induction tools [28]:

(a) Short normal:

$$\frac{1}{R_{xo}} = \frac{v\phi_t S_{xo}}{R_w} + \frac{(1-v)\phi_t S_w^2}{R_{mf}} \tag{8.10a}$$

$$R_{xo} = \frac{FR_{mf}}{S_{xo}^2} \tag{8.10b}$$

(b) Long normal:

$$\frac{1}{R_t} = \frac{\nu\phi_t S_w}{R_w} + \frac{(1-\nu)\phi_t S_w^2}{R_{mf}} \tag{8.11}$$

where

R_{xo} = borehole-corrected invaded zone, short normal, resistivity, ohm-m
R_{mf} = mud filtrate resistivity, ohm-m
R_t = borehole-corrected true, long normal, resistivity, ohm-m
R_w = water resistivity, ohm-m
ϕ_t = total porosity of the formation, fraction
S_w = water saturation, fraction
S_{xo} = saturation of mud filtrate in the flushed zone, fraction

Most of these parameters can be measured on either cores or well logs. The water saturation of the flushed zone can be estimated from correlations. Each formation seems to require a slightly different correlation. One of these correlations is:

$$S_{xo} = S_w^{C_X} \tag{8.12a}$$

The exponent C_X (typically 0.20 to 0.25) is assumed arbitrarily, depending on the experience of the mud-log analyst and results obtained in nearby wells. In a high-porosity and high-permeability formation, $C_X \approx S_W$. Because mud filtrate and formation water are miscible, in a water-bearing zone, $S_{xo} = 1$. In an oil-bearing zone, $S_{xo} = 1 - S_{oxo}$, where S_{oxo} is the residual oil saturation in the flushed zone, typically in the range of 0.20–0.30. It is generally assumed that the amount of residual oil or gas is the same in both the flushed zone and the invaded zone. The flushed zone is that which immediately surrounds the wellbore (3 to 6 in. radius). The invaded zone is that which is beyond the flushed zone (several feet thick). The presence of fissures near the borehole may increase the radius of both zones.

In low-porosity ($\phi_t < 10\%$) and low-permeability formations ($k < 5$ mD), any mud invasion would be very limited; but if k is high, then mud filtration could be high and deep into the formation. In this case, the range of the residual oil saturation is 10–20%. In high-porosity ($\phi_t > 15\%$) and high-permeability ($k > 100$ mD) formations, a low mud invasion results, with residual oil saturation of approximately 30%. In the case of high porosity and low permeability, S_{oxo} is in the order of 20%. These ranges of S_{oxo} are applicable primarily in water-wet sandstone formations. The presence of fractures near the wellbore and their density are factors that must be taken into account when estimating S_{oxo}.

If R_{xo} is obtained from a shallow resistivity log, microlaterolog, and the mud filtrate resistivity is corrected for temperature, Eq. (8.10b) can also be used to estimate S_{xo}:

$$S_{xo} = \sqrt{\frac{FR_{mf}}{R_{xo}}} \tag{8.12b}$$

If deep resistivity tools are used, the flushed zone water saturation is calculated from:

$$S_{xo} = \frac{S_w}{\sqrt{R_{xo}R_w/R_{mf}R_t}} \tag{8.12c}$$

The significance of knowing the value of S_{xo} through a porous interval as well as S_w is that it permits the determination of the degree of hydrocarbon flushing by the invading filtrate, that is, whether the oil in place is likely to flow or not. For instance, if $S_{xo} > S_w$, it can be inferred that there are movable hydrocarbons present; if $S_{xo} = S_w$, it can be inferred that there are no movable hydrocarbons present. The invasion process acts like a localized waterflood. Invading filtrate displaces not only connate water but any movable hydrocarbons. Let:

$$\begin{aligned} N_{obi} &= \phi(1 - S_w) \\ N_{oai} &= \phi(1 - S_{xo}) \end{aligned} \tag{8.12d}$$

where

N_{obi} = fractional pore volume occupied by oil before invasion
N_{oai} = fractional pore volume occupied by oil after invasion

The difference between the two values is the fractional pore volume that contained movable oil (N_{om}); thus:

$$N_{om} = N_{obi} - N_{oai} = \phi(S_{xo} - S_w) \tag{8.12e}$$

The fraction of the original oil in place that has moved is:

$$N_{fm} = \frac{N_{om}}{N_{obi}} = \frac{S_{xo} - S_w}{1 - S_w} \tag{8.12f}$$

N_{fm} can be used as a measure of the quality of the pay zone. In formations where the relative permeability to oil is low, $S_{xo} \approx S_w$ and, therefore, $N_{fm} \approx 0$, which may indicate that no movable hydrocarbons are present.

Combining Eqs. (8.10a) and (8.11) and solving explicitly for the porosity partitioning coefficient, ν, yields:

$$\nu = \frac{R_w}{\phi_t(S_w - S_{xo})}\left(\frac{1}{R_t} - \frac{1}{R_{xo}}\right) \tag{8.13}$$

If the total porosity, ϕ, is known from logs or cores, the matrix porosity and fracture porosity may be estimated from:

$$\phi_m = \phi_t(1 - \nu) \tag{8.14a}$$

$$\phi_f = \phi_t - \phi_m \tag{8.14b}$$

Aguilera and Aguilera [29] published rigorous and practical equations for dual porosity systems that were shown to be valid for all combinations of matrix and fractures or matrix and nonconnected vugs. For a dual porosity system made out of matrix and fractures and saturated 100% with water:

$$\frac{1}{R_{fo}} = \frac{\nu\phi}{R_w} + \frac{(1 - \nu\phi)}{R_o} \tag{8.15a}$$

where

R_{fo} = resistivity of the system made out of matrix and fractures at reservoir temperature when it is 100% saturated with water of resistivity R (ohm-m)
R_w = water resistivity at reservoir temperature (ohm-m)
R_o = resistivity of the matrix system at reservoir temperature when it is 100% saturated with water of resistivity R_w (ohm-m)
ϕ = total porosity, fraction

Solving explicitly Eq. (8.15a) for the porosity partitioning coefficient yields:

$$\nu = \left(\frac{1}{\phi}\right)\frac{R_w(R_o - R_{fo})}{R_{fo}(R_o - R_w)} \tag{8.15b}$$

Assuming a triple porosity reservoir, that is, a reservoir composed at the same time of matrix, fractures, and nonconnected vugs, can be modeled as a parallel resistance network for matrix and fractures and a series resistance network for the nonconnected vugs, Aguilera and Aguilera [29] derived the following equation:

$$\phi^{-m} = \nu_{nc}\phi + \frac{(1 - \nu_{nc}\phi)}{\nu\phi + ((1 - \nu\phi)/\phi_b^{-m_b})} \tag{8.15c}$$

where

ν_{nc} = nonconnected vug porosity ratio (ϕ_{nc}/ϕ), fraction
ϕ_b = matrix block porosity attached to the bulk volume of the matrix system. It is equivalent to porosity from unfractured plugs, fraction
ϕ_{nc} = porosity of nonconnected vugs attached to the bulk volume of the composite system, fraction
m_b = porosity exponent (cementation factor) of the matrix block from unfractured plugs (or based on lithology)
m = cementation factor of the composite system
ϕ = total porosity or

$$\phi = \phi_m + \phi_f + \phi_{nc} \tag{8.15d}$$

ϕ_m = matrix block porosity attached to the bulk volume of the composite system, fraction
ϕ_f = porosity of natural fractures attached to the bulk volume of the composite system, fraction

If the reservoir has only matrix and fracture porosities and no vugs, then nonconnected vug porosity ϕ_{nc} will be equal to zero, which means that the vug porosity ratio v_{nc} in Eq. (8.15c) will also be equal to zero. The morphology of the fracture can be open, partially mineralized, or composed of diagenetic channels by connected vugs (vuggy fractures). In this case, Eq. (8.15c) becomes equal to [29]:

$$\phi^{-m} = \frac{1}{\nu\phi + (1 - \nu\phi/\phi_b^{-m_b})} \tag{8.15e}$$

Solving for the porosity partitioning coefficient, Eq. (8.15e) yields:

$$\nu = \frac{\phi^m \phi_b^{-m_b} - 1}{\phi(\phi_b^{-m_b} - 1)} \tag{8.15f}$$

The porosity partitioning coefficient, ν, commonly used by the petrophysicist, is physically equivalent to the storage capacity ratio, ω, which is more commonly used in well-test analysis. But, because of the difference in scale, it is unlikely that the two values would ever be equal for the same formation. Note that even Eqs. (8.13) and (8.9) will yield slightly different values of ν because one is obtained from well logs (Eq. (8.9)) and the other is measured in cores.

Logs seem to yield slightly lower values of ν because the measurements are done under in situ conditions.

In the absence of experimental and well-log data, the following correlation is recommended for estimating the porosity partitioning coefficient ν in a naturally fractured reservoir. The permeability and porosity data used to generate this correlation are included in Table 8.4:

$$\nu = C_1 + C_2 k + \frac{C_3}{\phi_t} + C_4 k^2 + \frac{C_5}{\phi_t^2} + C_6 \frac{k}{\phi_t} \tag{8.16}$$

where

ϕ_t = total porosity, %
k = formation permeability, mD
$C_1 = -0.1254$
$C_2 = 6.95 \times 10^{-5}$
$C_3 = 4.204$
$C_4 = -3.016 \times 10^{-9}$
$C_5 = -9.675$
$C_6 = -6.88 \times 10^{-5}$

TABLE 8.4 Results of Core Analysis of Fractured Reservoir

k_f (mD)	ϕ_t (%)	ν	ϕ_f (%)	ϕ_m (%)
1,800	8.33	0.328	2.73	5.60
19,250	8.63	0.302	2.61	6.02
15,220	4.25	0.435	1.85	2.40
1,704	4.67	0.4	1.87	2.80
8,520	3.03	0.396	1.20	1.83
386	4.34	0.26	1.13	3.21
824	9.93	0.266	2.64	7.29
514	5.54	0.462	2.56	2.98
226	4.06	0.37	1.50	2.56
302	7.69	0.223	1.71	5.98

Example

A newly drilled well in a type 3 naturally fractured reservoir was logged. The average permeability from a pressure test is 1637 mD. The average total porosity of the system was estimated from cores as 14%. Other known characteristics are:

$A = 3{,}000$ acres,	$h = 52$ ft,	$S_w = 0.22$,
$B_o = 1.25$ bbl/STB,	$R_w = 0.19$ ohm-m,	$R_t = 95$ ohm-m,
$R_{mf} = 0.17$ ohm-m,	$m = 1.75$,	$k = 1673$ mD

1. Estimate the porosity partitioning coefficient.
2. Estimate the matrix porosity and fracture porosity.
3. Calculate the total oil in place, STB.

Solution

1. In order to calculate the porosity partitioning coefficient ν from Eq. (8.13), we need to determine first the resistivity in the flushed zone. Using Eq. (8.10b):

$$R_{xo} = \frac{FR_{mf}}{S_{xo}^2} = \frac{(31.2)(0.17)}{0.738^2} = 9.7 \text{ ohm-m}$$

The formation resistivity factor F and the water saturation in the invaded zone S_{xo} are estimated from Eqs. (8.8c) and (8.12), respectively, assuming that $\tau \cong 1$ and $C_X = 0.2$:

$$F = \frac{1}{0.14^{1.75}} = 31.2$$

$$S_{xo} = 0.22^{0.20} = 0.738$$

Using Eq. (8.13), the porosity partitioning coefficient is:

$$\nu = \frac{0.19}{0.14(0.22 - 0.738)}\left(\frac{1}{95} - \frac{1}{9.72}\right) = 0.24$$

This value indicates that fractures contribute 24% of the total pore space. Using Eq. (8.16), the porosity partitioning coefficient is:

$$\nu = -0.1254 + 6.95 \times 10^{-5} \times 1673 + \frac{4.204}{14} - 3.016 \times 10^{-9} \times 1673^2 - \frac{9.675}{14^2} - 6.88 \times 10^{-5}\frac{1673}{14} = 0.22$$

The correlation approximately yields a similar value of ν as Eq. (8.13).

2. Now we can estimate the matrix porosity and fracture porosity from Eqs. (8.14a) and (8.14b):

$$\phi_m = \phi_t(1 - \nu) = 0.14(1 - 0.24) = 0.106$$
$$\phi_f = \phi_t - \phi_m = 0.14 - 0.106 = 0.034$$

3. Assuming the water saturation in the fractures is equal to the water saturation in the matrix, the initial oil in place in the matrix and fractures is calculated from Eqs. (8.2) and (8.3), respectively:

$$N_{om} = \frac{(7,758)(3,000)(52)(0.106)(1 - 0.22)}{1.25} = 80.05 \times 10^6 \text{ STB}$$

$$N_{of} = \frac{(7,758)(3,000)(52)(0.034)(1 - 0.22)}{1.25} = 25.54 \times 10^6 \text{ STB}$$

The total oil in place in this naturally fractured reservoir is:

$$N_{ot} = 80.05 \times 10^6 + 25.54 \times 10^6 = 105.6 \times 10^6 \text{ STB}$$

This total oil volume is correct, assuming the porosity partitioning coefficient is the same in the entire reservoir. This is highly unlikely in naturally fractured formations, where porosity varies over short distances.

FRACTURE INTENSITY INDEX

Tension stress causes rock failure along major faults, giving rise to fracture porosity (ϕ_f), and fractures of decreasing width (w_f) and length (h_f) and frequency of occurrence (FII) away from the fault plane, as shown in Figures 8.12 and 8.13 [28]. Thus, permeability is much more affected by fracture dimensions than the matrix or total porosity. The curve-fit equations of the lateral distance to the surface of the fault with respect to FII for both upthrown and downthrown blocks are as follows:

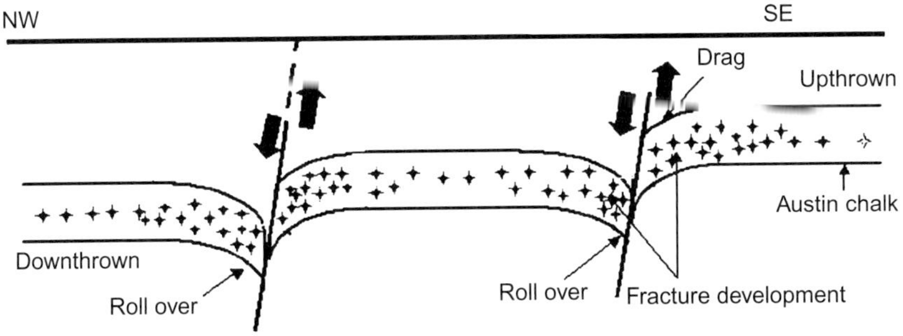

FIGURE 8.12 Frequency of occurrence of natural fractures near major faults [28].

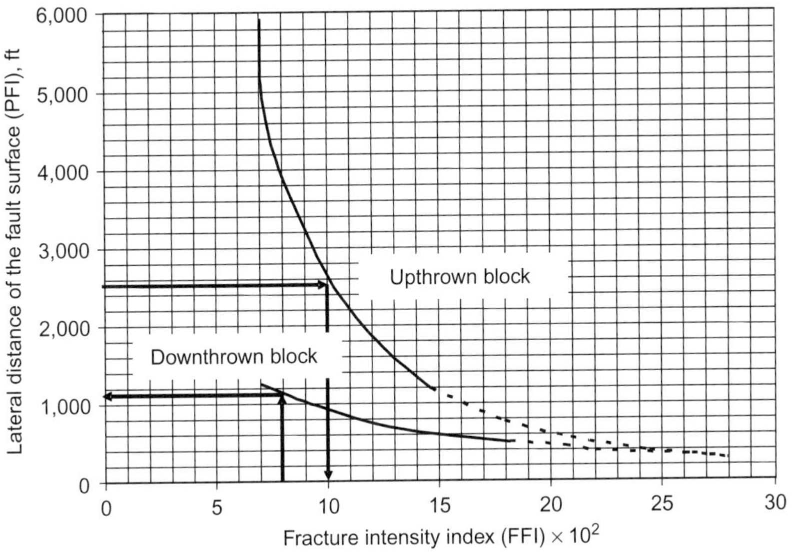

FIGURE 8.13 Fault proximity index (FPI) as a function of fracture intensity index (FII) in Austin chalk [28].

Upthrown block:

$$d_{LU} = \frac{1{,}000}{42\text{FII}^2 - 0.02.43} \tag{8.16}$$

Downthrown block:

$$d_{LD} = \frac{1{,}000}{9.44e^{\text{FII}} - 9.3} \tag{8.17}$$

where

d_{LU} = lateral distance to the fault for upthrown block, ft
d_{LD} = lateral distance to the fault for downthrown block, ft
FII = fracture intensity index, fraction

Assuming the drilling fluid used is nonconductive, the in situ value of the fracture intensity index is estimated from:

$$\text{FII} = \frac{(1/R_{xo}) - (1/R_t)}{(1/R_{mf}) - (1/R_w)} \tag{8.18}$$

Equation (8.16) ($R^2 = 0.989$) is applicable for a distance range of 250 to 5,000 ft and an FII range of 7–25%, and Eq. (8.17) ($R^2 = 0.998$) is applicable for a distance range of 250 to 1,250 ft and an FII range of 7–25%. These correlations were developed from well-log data obtained by Pirson near the Luling–Mexia fault in the Austin chalk. The primary application of these two correlations is in the exploration stage and when the presence of a nearby fault is known a priori from seismic data, as they provide only an order of magnitude of the distance to the fault. It is important to emphasize that (a) FFI is influenced by several factors, including the number of fractures and fracture geometry, and (b) not all natural fractures are the result of faulting.

The following equations can be used to estimate fracture width and fracture permeability in a type 1 naturally fractured reservoir:

$$w_f = \frac{0.064}{\phi_t}[(1 - S_{iw})\text{FII}]^{1.315} \tag{8.19a}$$

$$k_f = 1.5 \times 10^7 \phi_t[(1 - S_{wi})\text{FII}]^{2.63} \tag{8.19b}$$

where porosity, FII, and irreducible water saturation are expressed as fractions, and fracture width and fracture permeability in cm and mD, respectively. The fracture porosity can be directly estimated using the following empirical correlation [14]:

$$\phi_f = \left[R_{mf}\left(\frac{1}{R_{LLS}} - \frac{1}{R_{LLD}}\right)\right]^{C_T} \tag{8.20}$$

where the range of the coefficient C_T is between 2/3 (typical for type 1 fractured reservoir) and 3/4. R_{mf}, R_{LLS}, and R_{LLD} are, respectively, the mud filtrate, laterolog shallow, and laterolog deep resistivities in ohm-m. R_{LLS} and R_{LLD} are equivalent to R_{xo} and R_t, respectively.

Example

Seismic surveys and geological studies have indicated that the well is located in an upthrown layer of a faulted reservoir. Visual inspection of cores revealed the presence of stress fractures. Knowing the resistivity of the invaded zone R_{xo} is

9.7 ohm-m, the mud filtrate resistivity R_{mf} is 0.17 ohm-m, the true resistivity R_t is 95 ohm-m, and the water resistivity R_w is 0.19 ohm-m, estimate the following:

1. the fracture intensity index (FII) and
2. the distance to the nearest fault.

Solution

1. Using Eq. (8.18), the fracture intensity index is:

$$\text{FII} = \frac{(1/R_{xo}) - (1/R_t)}{(1/R_{mf}) - (1/R_w)} = \frac{(1/9.7) - (1/95)}{(1/0.17) - (1/0.19)} = 0.15$$

2. The distance to the nearest fault is estimated from the correlation corresponding to the upthrown block, that is, Eq. (8.16):

$$d_{LU} = \frac{1{,}000}{42\text{FII}^2 - 0.0243} = \frac{1{,}000}{42(0.15)^2 - 0.0243} \approx 1{,}100 \text{ ft}$$

The distance to the fault can be directly estimated using Figure 8.13. For the FII value of 15%, the distance is approximately 1,100 ft.

Example

Resistivity survey in a well yielded the following data: wellbore-corrected mud filtrate resistivity = 0.165 ohm-m, water resistivity = 0.18 ohm-m, invaded zone resistivity = 12 ohm-m, and deep formation resistivity = 85 ohm-m. The reservoir average porosity (17%) was determined from a neutron log. Substantial mud loss was observed during drilling of this well as well as in neighboring wells. Pressure test analysis as well as cores confirmed the presence of extensive natural fractures in the well.

1. Estimate the fracture intensity index and the porosity partitioning coefficient. Note that the coefficient C_X in Eq. (8.12a) is typically 0.25 in this field.
2. If the average irreducible water saturation estimated from log analysis is 24%, determine the fracture width.
3. Estimate the fracture width, fracture permeability, and fracture porosity of the formation.
4. Estimate the distance to the closest fault knowing the well is located in a downthrown block.

Solution

1. Knowing that $R_{LLS} = R_{xo} = 12$ and $R_{LLD} = R_t = 85$, the fracture intensity index is estimated from Eq. (8.18):

$$\text{FII} = \frac{(1/R_{xo}) - (1/R_t)}{(1/R_{mf}) - (1/R_w)} = \frac{(1/12) - (1/85)}{(1/0.165) - (1/0.18)} = 0.1417$$

The saturation of mud filtrate in the flushed zone is:

$$S_{xo} = S_w^{C_X} = 0.24^{0.25} = 0.70$$

The porosity partitioning coefficient is calculated using Eq. (8.13):

$$\nu = \frac{R_w}{\phi_t(S_w - S_{xo})}\left(\frac{1}{R_t} - \frac{1}{R_{xo}}\right)$$
$$= \frac{0.18}{0.17(0.24 - 0.70)}\left(\frac{1}{85} - \frac{1}{12}\right) = 0.165$$

2. Using Eq. (8.19a), the fracture width is:

$$w_f = \frac{0.064}{\phi_t}[(1 - S_{iw})FII]^{1.315}$$
$$= \frac{0.064}{0.17}[(1 - 0.24)(0.1417)]^{1.315} = 0.02 \text{ cm}$$

3. Using Eq. (8.19b), the fracture permeability is:

$$k_f = 1.5 \times 10^7 \phi_t[(1 - S_{wi})FII]^{2.63}$$
$$= 1.5 \times 10^7(0.17)[(1 - 0.24)0.1417]^{2.63} = 7,265 \text{ mD}$$

Using Eq. (8.20), where $C_T = 3/4$, the fracture porosity is:

$$\phi_f = \left[R_{mf}\left(\frac{1}{R_{LLS}} - \frac{1}{R_{LLD}}\right)\right]^{C_T} = \left[0.165\left(\frac{1}{12} - \frac{1}{85}\right)\right]^{3/4} = 0.0358$$

For $C_T = 2/3$, the fracture porosity is 0.052; thus, the value of ϕ_f is between 0.036 and 0.052.

The matrix porosity is:

$$\phi_m = \phi_t(1 - v) = 0.17(1 - 0.165) = 0.142$$

Note that the sum of ϕ_f (for $C_T = 3/4$) and ϕ_m is 0.177, which is approximately equal to the total porosity obtained from well logs. Therefore, the fracture porosity of the reservoir is 3.6%.

4. The distance to the nearest fault can be estimated from Eq. (8.17):

$$d_{LU} = \frac{1,000}{9.44e^{0.1417} - 9.3} \approx 634 \text{ ft}$$

PERMEABILITY–POROSITY RELATIONSHIPS IN DOUBLE POROSITY SYSTEMS

Petroleum reservoirs can be divided into three broad classes based on their porosity systems:

1. intergranular;
2. Intercrystalline–intergranular; and
3. solution channels and/or natural fractures.

Reservoirs with vugular-solution channels and/or fractures differ from those having intercrystalline–intergranular porosity in that the double porosity system strongly influences the movement of fluids. The double porosity can be the result of fractures, joints, and/or solution channels within the reservoirs. Carbonate reservoirs with a vugular–solution porosity system, such as the Pegasus Ellenburger field and Canyon Reef field in Texas, exhibit a wide range of permeability. The permeability distribution may be relatively uniform or quite irregular. The double porosity reservoir with a uniform permeability distribution is analyzed as follows.

Consider a rock sample with two dominant pore radii, as shown in Figure 8.14. The total flow through such systems is the sum of individual flow rates through each system, the systems having different petrophysical properties such as porosity and permeability.

$$q_t = q_1 + q_2 \tag{8.21}$$

Using Darcy's law (for q_t) and Poiseuille's law (for q_1 and q_2), we have:

$$kA_t \frac{\Delta P}{\mu L} = \left[\frac{n_1 \pi r_{c1}^4}{8} + \frac{n_2 \pi r_{c2}^4}{8}\right] \frac{\Delta P}{\mu L} \tag{8.22}$$

The total area for the system is:

$$A_t = \frac{n_1 \pi r_{c1}^2}{\phi_1} + \frac{n_2 \pi r_{c2}^2}{\phi_2} \tag{8.23}$$

Also, we know from Chapter 3 that:

$$r_c = \frac{2}{S_{vp}} \tag{8.24}$$

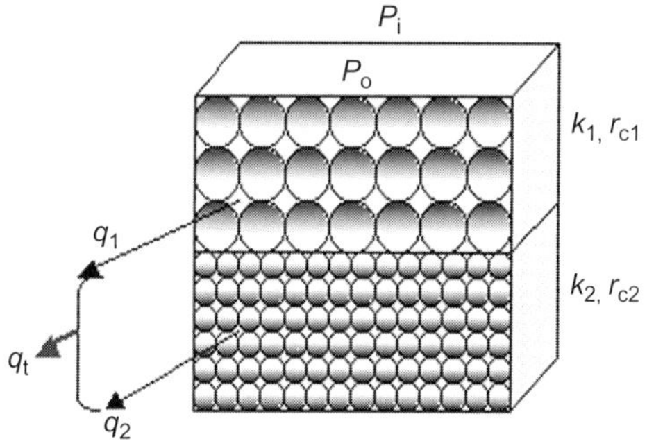

FIGURE 8.14 Unit model with two dominant pore radii. The systems possess different petrophysical properties such as porosity and permeability.

Substituting Eq. (8.23) into Eqs. (8.22) and Eq. (8.24) results in:

$$k = \frac{1}{2}\frac{\left[(1/S_{\text{vp1}}^4) + (1/S_{\text{vp2}}^4)\right]}{\left[(1/\phi_1 S_{\text{vp1}}^2) + (1/\phi_2 S_{\text{2P2}}^4)\right]} \tag{8.25a}$$

The general form of this equation is:

$$k = \frac{1}{2}\frac{\sum_{i=1}^{n}(1/S_{\text{vpi}}^4)}{\sum_{i=1}^{n}(1/\phi_{\text{i}} S_{\text{vpi}}^2)} \tag{8.25b}$$

For a single porosity system, this equation reduces to:

$$k = \frac{1}{2}\frac{(1/S_{\text{vp}}^4)}{(1/(\phi S_{\text{vp}}^2))} = \frac{\phi}{2S_{\text{vp}}^2} \tag{8.26}$$

The constant 2 in Eq. (8.26) is related to the shape of the capillaries and their tortuosity and can be replaced by K_{T}:

$$k = \frac{\sum_{i=1}^{n}(1/S_{\text{vpi}}^4)}{\sum_{i=1}^{n}(K_{\text{Ti}}/(\phi_{\text{i}} S_{\text{vpi}}^2))} \tag{8.27}$$

where

$$K_{\text{T}} = 6 f_{\text{sp}} \tau \tag{8.28}$$

Methods for estimating the pore shape factor f_{sp} and the tortuosity of the capillaries τ are discussed in Chapter 3. In the case of formations containing a very small number of channels per unit pore volume, such as in reservoirs with high storage capacity in a rock matrix, and very low storage capacity in channels, $n_1 \gg n_2$, Eq. (8.25a) can be written as:

$$k = \frac{\phi_1}{2S_{\text{pv1}}^2} = \frac{\phi_1 r_{\text{c1}}^2}{8} \tag{8.29a}$$

where the subscript 1 stands for primary pore space, which stores most of the fluid. In the case of $n_2 \gg n_1$, that is, rocks in which the fluid is stored mainly in secondary pore spaces such as fissures and vugs, Eq. (8.29a) becomes:

$$k = \frac{\phi_2}{2S_{\text{pv2}}^2} = \frac{\phi_2 r_{\text{c2}}^2}{8} \tag{8.29b}$$

where the subscript 2 stands for secondary pore space. Thus, in cases where $n_1 \gg n_2$ and $n_2 \gg n_1$, double porosity systems may be approximated by a single pore space system, and consequently the methods developed in Chapter 3 for clastic rocks can be used in carbonate formations. In the case where n_1 is approximately equal to n_2, and since it is impossible to determine n_1 and n_2, an alternative to the above approach is to take the geometric mean of the two capillary systems, that is:

$$k = \sqrt{\left(\frac{\phi_1 r_{c1}^2}{8}\right)\left(\frac{\phi_2 r_{c2}^2}{8}\right)} = \frac{r_{c1} r_{c2}}{8}\sqrt{\phi_1 \phi_2} \tag{8.30}$$

Using an average value of r_{c1} and r_{c2}, and an average value of ϕ_1 and ϕ_2, Eq. (8.30) becomes similar to the Kozeny equation.

POROSITY AND PERMEABILITY RELATIONSHIPS IN TYPE 1 NATURALLY FRACTURED RESERVOIRS

As mentioned previously in regard to type 1 reservoirs, fractures provide all the storage capacity and permeability and the fluid flow behavior is controlled by the fracture properties. The equation for volumetric flow rate, combined with Darcy's law, provides the basic approach for estimating fracture permeability.

Consider a block of naturally fractured rock with n fractures, as shown in Figure 8.15. Assuming the fractures are rectangular and smooth, and do not contain any mineral, the Hagen–Poiseiulle equation gives:

$$q = \left(\frac{n h_f w_f^3}{12}\right)\frac{\Delta P}{\mu L} \tag{8.31a}$$

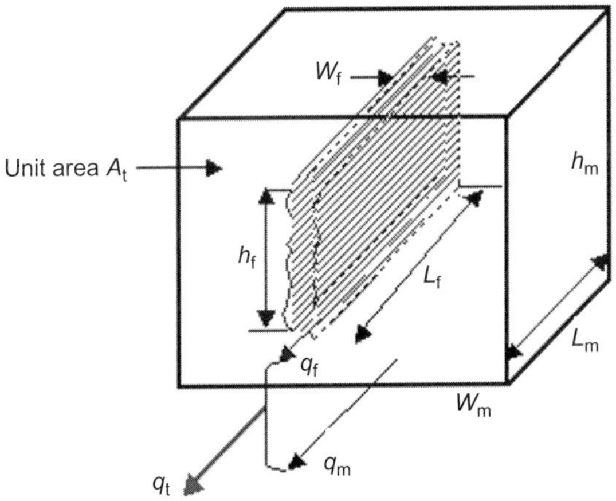

FIGURE 8.15 Unit model used in calculation of fracture permeability in type 1 naturally fractured reservoirs.

and Darcy's law is:

$$q = kA\frac{\Delta P}{\mu L} \tag{8.31b}$$

Equating these two equations and solving for permeability results in:

$$k = \frac{nh_f w_f^3}{12A} \tag{8.32}$$

The physical difficulty in using Eq. (8.32) is that the number of fractures, fracture height, and fracture width has to be known. Since, by definition:

$$\phi = \frac{V_p}{V_b} = \frac{nh_f w_f L}{AL} \tag{8.33}$$

and

$$A = \frac{nh_f w_f}{\phi} \tag{8.34}$$

substituting Eq. (8.34) into Eq. (8.32) yields:

$$k = \frac{\phi w_f^2}{12} \tag{8.35}$$

Equation (8.35) is similar to Eq. (3.14), $k = \phi r^2/8$, where the capillary radius r and constant 8 have been replaced by the fracture width w_f and 12, respectively. Equation (8.35) is commonly used to calculate the fracture permeability.

Equation (8.35) can be used to calculate w_f if the porosity and permeability are known from well logs or well testing:

$$w_f = \sqrt{12\frac{k}{\phi}} \tag{8.36}$$

Expressing fracture porosity in percent and fracture width in micrometers (μm), Equation (8.35) becomes:

$$k_f = 8.33 \times 10^{-4} w_f \phi_f \tag{8.37}$$

Where k_f is expressed in darcies.

FRACTURES POROSITY AND APERTURE FROM CORES

Oil-bearing fractured granite is a major productive formation in some parts of the world, such as in the Bach Ho field, offshore Vietnam. Fractured granite consists of three main elements: macrofractures, low-permeability matrix with microfractures, and tight nonpermeable matrix [30]. Tuan et al. selected 10 of the most representative whole cores ($D = 6.7$ cm) from this oil field, with total porosity from 3.03% to 9.93% and permeability from 226 to

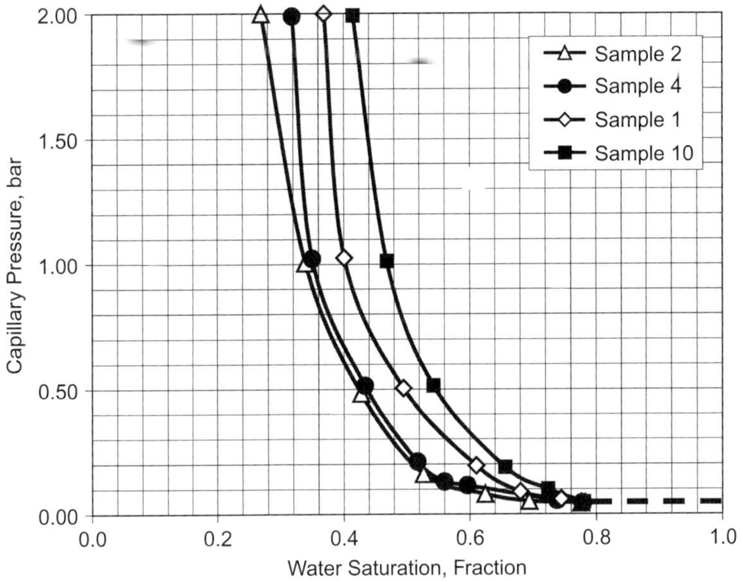

FIGURE 8.16 Air/water capillary pressure curves [30].

19,250 mD, as shown in Table 8.4 [30]. After trimming and cleaning, the cores were saturated with a brine and the total porosity (ϕ_f) was determined. The saturated samples were then loaded into a capillary cell (porous plate technique) and P_c was increased in steps from 0.05 to 5 bars. The water saturation S_w was recorded at each step. Tuan et al., observed that the larger the fracture width, the lower the capillary forces. From this observation, they demonstrated that the sudden change in the slope of P_c versus S_w, as shown in Figure 8.16, corresponds to the volume of fractures. The fracture porosity and porosity partitioning coefficients were then calculated from:

$$\phi_f = \phi_t(1 - S_{WS}) \tag{8.38a}$$

$$\nu = \frac{\phi_f}{\phi_t} \tag{8.38b}$$

where S_{WS} is the water saturation corresponding to the sudden change in slope of the P_c curve. Values of ϕ_f and ν are shown in Table 8.4. The table shows very high values of ν, which indicates a very high density of microfractures in the matrix.

Tuan et al. also performed simultaneous measurement of permeability (to water) and resistivity on naturally fractured core samples, using a Hassler core holder equipped with two silver-coated electrodes. For each core sample, resistivity and permeability were determined at various overburden

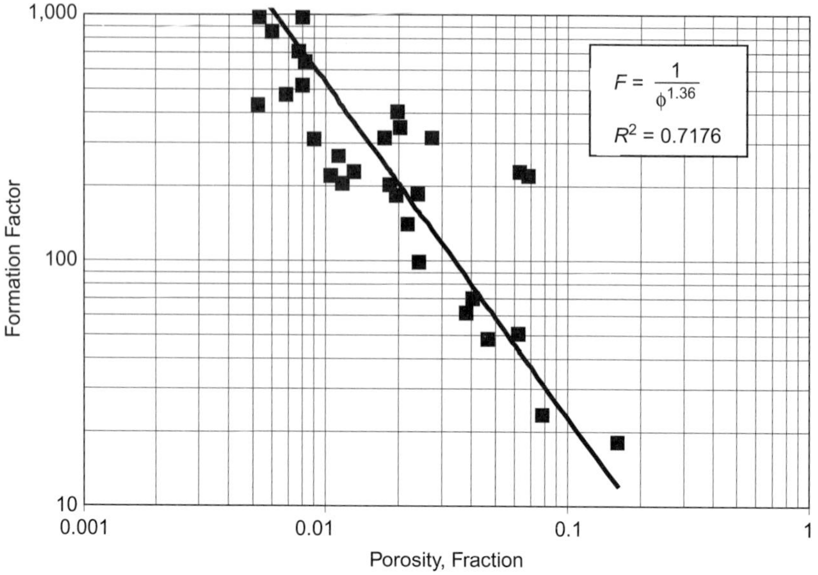

FIGURE 8.17 Formation resistivity factor versus porosity [30].

pressure from 15 to 400 bars. Samples of two saturation states were investigated: full brine saturation and partial brine saturation. Resistivity measurements were performed on 31 fully brine saturated cores with $k < 3$ mD and then the calculated formation resistivity factor ($F = R_O/R_W$) was plotted against fracture porosity (Figure 8.17). A curve fit of the data points shows that the cementation factor is significantly low, which is typical of systems with high porosity partitioning coefficient. The log–log plot of resistivity versus permeability (Figure 8.18) allowed Tuan et al., to investigate the relationship between fracture permeability and fracture width. They concluded that the fracture porosity and the fracture width (aperture) can be accurately calculated from core analysis using the following equations:

$$\phi_f = \frac{0.04 n_f w_f}{\pi D} \tag{8.39}$$

$$w_f = \left(\frac{\pi D k_f}{33.32 \times 10^{-6} n_f}\right) \tag{8.40a}$$

Or, assuming the fracture length is equal to the length of the core sample:

$$w_f = \frac{10^6 R_w L}{n_f r_{of} D} \tag{8.40b}$$

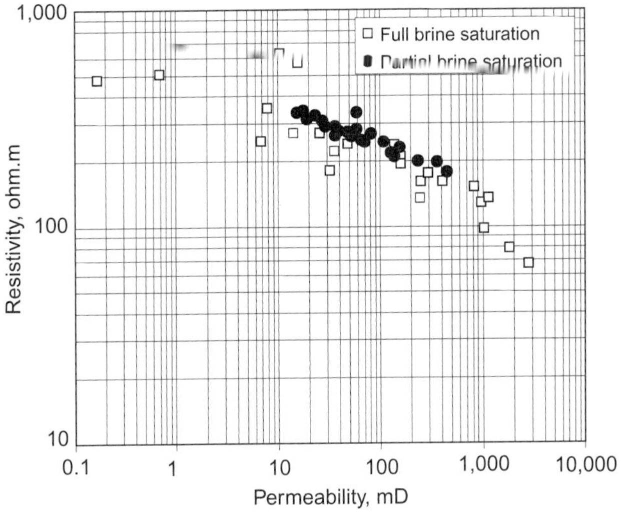

FIGURE 8.18 Cross-plot of resistivity versus permeability [30].

where

n_f = number of fractures in fractured core sample
R_w = brine or water resistivity, ohm-m
r_{of} = electrical resistance of the fractured core sample, ohm
D = diameter of fractured core sample as shown in Figure 8.28a, cm
L = length of fractured core sample as shown in Figure 8.28a, cm
k_f = fracture permeability, darcy
ϕ_f = fracture porosity, %
w_f = fracture width or aperture, micrometer (μm)

Example

Resistivity measurements were performed on a 100% water-saturated core sample containing five fractures. Given that the following results were obtained:

$r_{of} = 675$ ohm,	$R_w = 0.15$ ohm-m,	$D = 7.62$ cm,
$L = 30.48$ cm,	$\phi_t = 10.3\%$,	$S_{WS} = 0.52$

calculate

1. fracture width,
2. fracture porosity,
3. fracture permeability, and
4. porosity partitioning coefficient and matrix porosity.

Solution

1. The fracture width or aperture is obtained from Eq. (8.40b):

$$w_f = \frac{10^6 R_w L}{n_f r_{of} D} = \frac{10^6 \times 0.15 \times 30.48}{5 \times 675 \times 7.62} = 178 \ \mu m$$

2. The fracture porosity is estimated from Eq. (8.39):

$$\phi_f = \frac{0.04 n_f w_f}{\pi D} = \frac{0.04 \times 5 \times 178}{7.62\pi} = 1.5\%$$

3. The fracture permeability is estimated from Eq. (8.38):

$$k_f = 8.44 \times 10^{-4} \times 178^2 \times 1.5 = 39.6 \ \text{darcy}$$

4. The porosity partitioning coefficient and matrix porosity are determined from Eqs. (8.37b) and (8.15):

$$\nu = \frac{\phi_f}{(1 - S_{WS})\varphi_t} = \frac{1.5}{(1 - 0.52)10.3} = 0.30$$

$$\phi_m = \phi_t - \phi_f = 8.8\%$$

Several correlations have been published, which relate fracture porosity to permeability and other petrophysical parameters. Nelson [31] obtained the following correlation for fracture porosity as a function of fracture spacing and flow test permeability:

$$\phi_f = 0.493\left(\frac{k}{d_f}\right)^{1/3} \tag{8.41a}$$

where

ϕ_f = fracture porosity, %
d_f = average spacing between parallel fractures, cm
k = average permeability from flow tests, darcy
For instance, if $d_f = 5$ cm and $k = 8$ darcy, the fracture porosity is:

$$\phi_f = 0.493\left(\frac{8}{5}\right)^{1/3} = 0.57\%$$

This correlation assumes all flow is due to fractures, such as in type 1 naturally fractured reservoir. Fracture porosity may also be estimated from:

$$\phi_f = \frac{k - k_m}{k_f} \tag{8.41b}$$

where

ϕ_f = fracture porosity, fraction
k = average permeability from flow tests, mD
k_m = matrix permeability from core analysis, mD
k_f = fracture permeability from flow tests or correlation, mD

SPECIFIC AREA OF FRACTURES

Let S_{pv} be the internal surface area per unit of pore volume, where the surface area for n fractures is $n(2w_f L + 2h_f L) = 2n(w_f + h_f)L$, and the pore volume is $n(w_f h_f L)$, assuming that the fracture provides all of the storage and permeability. The specific surface area per unit pore volume is:

$$S_{vp} = \frac{2n(w_f + h_f)L}{n2w_f h_f L} = 2\left(\frac{1}{h_f} + \frac{1}{w_f}\right) \tag{8.42a}$$

Using the same assumptions, the specific surface area per unit grain volume is:

$$S_{gv} = \frac{2n(w_f + h_f)L}{\mathrm{AL}(1-\phi)} \tag{8.42b}$$

Multiplying and dividing by $w_f h_f$, and simplifying, yields:

$$S_{gv} = \frac{2n w_f h_f}{A(1-\phi)}\left(\frac{1}{h_f} + \frac{1}{w_f}\right) \tag{8.43}$$

Substituting for A from Eq. (8.37) and simplifying results in:

$$S_{gv} = 2\left(\frac{\phi}{1-\phi}\right)\left(\frac{1}{h_f} + \frac{1}{w_f}\right) \tag{8.44}$$

The term $1/h_f$ is very small in comparison to $1/w_f$ because $h_f \gg w_f$. Thus, Eq. (8.44) reduces to:

$$S_{gv} = \frac{2}{w_f}\left(\frac{\phi}{1-\phi}\right) \tag{8.45}$$

Combining Eqs. (8.42a) and (8.44) yields:

$$S_{gv} = \left(\frac{\phi}{1-\phi}\right)S_{pv} \tag{8.46}$$

Since $1/w_f \gg 1/h_f$, Eq. (8.42a) reduces to:

$$S_{vp} = \frac{2}{w_f} \quad \text{or} \quad w_f = \frac{2}{S_{vp}} \tag{8.47}$$

Substituting for w_f, Eq. (8.35) becomes:

$$k = \frac{\phi}{3S_{pv}^2} \tag{8.48}$$

Combining Eqs. (8.46) and (8.36) yields:

$$k = \frac{1}{3S_{gv}^2}\left(\frac{\phi^3}{(1-\phi)^2}\right) \tag{8.49}$$

The derivation of Eqs. (8.33) through (8.49) assumes that the fractures are rectangular, smooth, uniform, and that fracture length is equal to the length of the rock sample. The constant 3 is specific to the shape of the fracture. Equations (8.48) and (8.49) can be generalized for all fracture shapes as follows:

$$k = \frac{\phi}{K_{\mathrm{Tf}} S_{\mathrm{pv}}^2} \tag{8.50}$$

$$k = \frac{1}{K_{\mathrm{Tf}} S_{\mathrm{gv}}^2} \left(\frac{\phi^3}{(1-\phi)^2} \right) \tag{8.51}$$

where $K_{\mathrm{Tf}} = K_{\mathrm{sf}}\tau$, K_{sf} being the fracture shape factor and τ the tortuosity. This equation is similar to the generalized Kozeny equation. Unlike sandstone formations, identification and characterization of flow units in carbonate formations are not possible because of extreme variations of fissures, in terms of both geometry and intensity. However, in reservoirs where the geometry and distribution of fissures are uniform throughout, one could use the same concepts of reservoir quality index (RQI), flow zone index (FZI), and Tiab's hydraulic unit characterization factor (H_{T}) as presented in Chapter 3.

EFFECT OF FRACTURE SHAPE

Consider a fracture with an elliptical cross section as shown in Figure 8.19. Assuming a type 1 naturally fractured reservoir, the specific surface area per unit pore volume, S_{pv}, is:

$$S_{\mathrm{pv}} = \frac{A_{\mathrm{se}}}{V_{\mathrm{p}}} \tag{8.52}$$

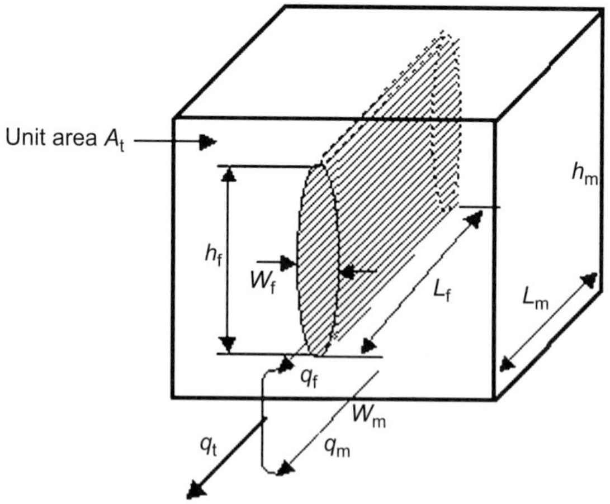

FIGURE 8.19 Effect of fracture shape on the permeability–porosity relationship.

where A_{se} is the surface area of the elliptical fracture and is given by:

$$A_{se} = \pi \left[0.75(w_f + h_f) - 0.5\sqrt{w_f h_f} \right] L \tag{8.53}$$

and

$$V_p = \frac{\pi}{4} w_f h_f L \tag{8.54}$$

Combining the above three equations, and simplifying, yields:

$$S_{vp} = 3\left(\frac{1}{w_f} + \frac{1}{h_f}\right) - \frac{2}{\sqrt{w_f h_f}} \tag{8.55}$$

Since $1/h_f \ll 1/w_f$, the above equation reduces to:

$$S_{vp} = \left(\frac{3}{w_f}\right) - \frac{2}{\sqrt{w_f h_f}} \tag{8.56}$$

Assuming $w_f h_f \gg w_f$, Eq. (8.56) further reduces to (with less than 5% error):

$$S_{vp} = \frac{3}{w_f} \tag{8.57}$$

It is clear from the above equations that the value of the fracture shape factor K_{sf} changes with fracture shape.

HYDRAULIC RADIUS OF FRACTURES

The effective or hydraulic radius of a fracture (r_{hf}) can be obtained by representing the fracture as a capillary tube. Equating Eq. (3.10), which is valid for a capillary tube system, and Eq. (8.31a), which accounts for fracture geometry, yields:

$$\frac{\pi r_{hf}^4}{8} = \frac{h_f w_f^3}{12} \tag{8.58a}$$

Solving for the radius results [32] in:

$$r_{hf} = \left(\frac{2}{3\pi} h_f w_f^3\right)^{1/4} \tag{8.58b}$$

Equation (8.58b) is very important in the sense that it interprets the fracture geometry in terms of equivalent hydraulic radius and, thus, can be incorporated into any tube model (Figure 8.20).

Substituting Eq. (8.58b) into Eq. (3.14) (where $r = r_{hf}$) yields:

$$k_f = \frac{\phi_f}{8}\sqrt{\frac{2}{3\pi} h_f w_f^3} = 0.05758\ \phi_f \sqrt{h_f w_f^3} \tag{8.59}$$

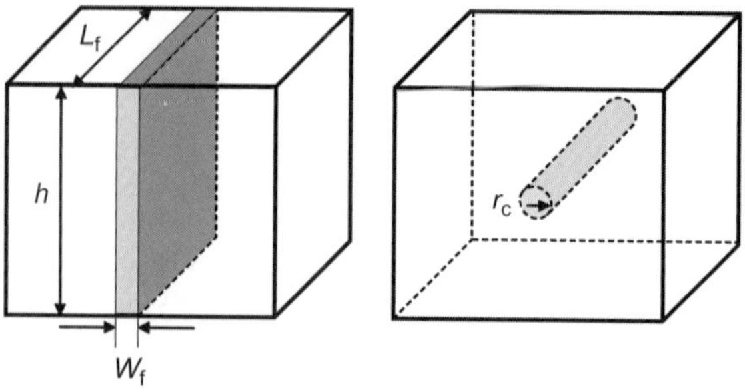

FIGURE 8.20 Fracture hydraulic radius in fractured and unfractured rocks. A rock sample with one fracture of hydraulic radius equal to 1 cm is equivalent to a rock sample with one solution channel of radius 1 cm.

The fracture permeability in this equation is in square centimeters, porosity is a fraction, and fracture height and width are in centimeters. If fracture width, w_f, and fracture porosity, ϕ_f, are determined from core analysis and permeability is determined from well testing, fracture height can be determined from Eq. (8.59). Another application of Eq. (8.59) is to decide what radius a horizontal well has to have in order to get the same benefit of a hydraulic fracture of width w_f and height h_f.

Example

Given that core analysis, well logs, and pressure data yielded the following values:

$\phi_f = 0.027$,	$w_f = 0.015$ cm,	$k_f = 51.3$ darcies

Estimate the following:

1. fracture height, and
2. hydraulic radius of the fractures.

Solution

1. Equation (8.59) can be rearranged for fracture height as follows:

$$h_f = \left(\frac{k_f}{0.05758\phi_f}\right)^2 \frac{1}{w_f^3} \tag{8.60}$$

Since 1 darcy $= 9.87 \times 10^{-7}$ cm^2, therefore 51.3 darcies (51.3)(9.87×10^{-7}) $= 5.06 \times 10^{-5}$ cm^2:

$$h_f = \left(\frac{5.06 \times 10^{-5}}{0.05758(0.027)}\right)^2 \frac{1}{(0.015)^3} = 314 \text{ cm} = 10.3 \text{ ft}$$

2. Hydraulic radius can be calculated using Eq. (8.58b):

$$r_{hf} = \left(\frac{2}{3\pi} h_f w_f^3\right)^{1/4}$$

$$r_{hf} = \left(\frac{2}{3\pi}(314)(0.015)^3\right)^{1/4} = 0.12 \text{ cm}$$

This value of r_{hf} implies that a fracture with height 314 cm and width 0.015 cm is equivalent to a cylindrical channel with a hydraulic radius of 0.12 cm.

TYPE 2 NATURALLY FRACTURED RESERVOIRS

In this type of reservoir, the matrix has a good porosity and permeability. Oil is trapped in both the matrix and fractures. Consider a representative block containing two parallel layers, as shown in Figure 8.21.

The average permeability in the matrix can be modeled using the capillary tube model and equations developed in Chapter 3. The average permeability in the fracture system can be expressed by the equations developed in previous sections in this chapter.

For n_c capillaries and n_f fractures, the following approach can help estimate permeability in type 2 naturally fractured reservoirs. Total flow rate from both matrix and fractures can be expressed as:

$$q_t = q_f + q_m \tag{8.61}$$

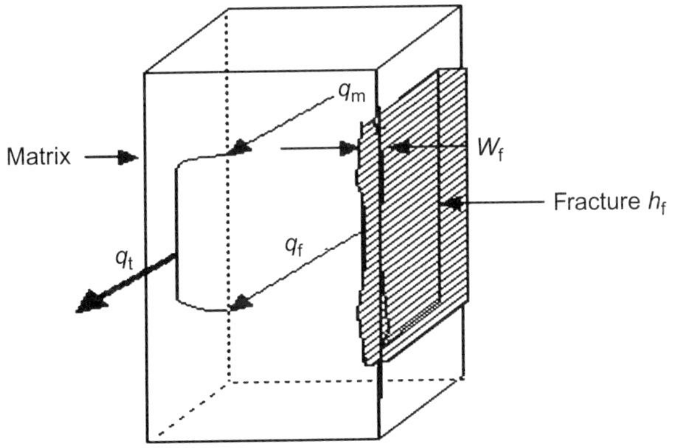

FIGURE 8.21 Representative elementary rock volume containing two parallel systems of matrix and fracture. The fluid is stored in both matrix and fractures (type 2 reservoirs).

Using Darcy's law (for q_t) and Poiseuille's law (for q_m and q_f) gives:

$$kA_t \frac{\Delta P}{\mu L} = \left[\frac{n_c \pi r_c^4}{8} + \frac{n_f h_f w_f^3}{12} \right] \frac{\Delta P}{\mu L} \tag{8.62}$$

The total area of matrix and fracture can be expressed as:

$$A_t = \frac{n_c \pi r_c^2}{\phi_c} + \frac{n_f h_f w_f}{\phi_f} \tag{8.63}$$

Assuming equal storage capacity of both systems (matrix and fracture), that is, the porosity partitioning coefficient ν is approximately 0.50 and therefore $n_c = n_f$ and $\phi_f = \phi_c$, Eq. (8.63) is simplified as:

$$A_t = \frac{n}{\phi}(\pi r_c^2 + h_f w_f) \tag{8.64}$$

Thus, the average permeability can be extracted first by substituting Eq. (8.64) into Eq. (8.62) and then solving for k:

$$k = \frac{\phi}{\pi r_c^2 + h_f w_f} \left[\frac{\pi r_c^4}{8} + \frac{h_f w_f^3}{12} \right] \tag{8.65}$$

For a unit block area, $h_f = 1$. Although h_f and w_f can be relatively easily measured, this is not always the case with r_c. A rather simplistic approach to determine average permeability in type 2 reservoirs is to calculate the geometric mean of the two systems:

$$k = \sqrt{\left(\frac{\phi_c r_c^2}{8} \right) \left(\frac{\phi_f w_f^2}{12} \right)} \tag{8.66}$$

Assuming the average porosity $\phi = \sqrt{\phi_f \phi_c}$, Eq. (8.66) becomes:

$$k = \left(\frac{r_c w_f}{9.8} \right) \phi \tag{8.67}$$

It is obvious from this discussion that in naturally fractured carbonate formations, where structural heterogeneities and textural changes are common and a small number of wells are cored, the practice of using statistical core permeability–porosity relations to characterize flow units is not recommended. The main parameters that influence the flow units in naturally fractured reservoirs include secondary porosity (fractures, fissures, and vugs), matrix porosity, fracture intensity index, fracture dimensions (shape, width, and height), tortuosity, porosity, partitioning coefficient, specific surface area, and irreducible water saturation. These parameters must be incorporated into the definition of flow units in order to effectively characterize them.

FLUID FLOW MODELING IN FRACTURES

Fractures are modeled as flow channels or cracks. From the fluid flow point of view, fractures have two main properties: storage capacity and fluid transmission or transfer capacity, also known as fracture conductivity. These two properties are dependent on the dimensions of length, width, and height.

FRACTURE AREA

Fracture area is determined by the shape and relative dimensions of the fracture and influences the mechanical behavior of the rock mass. Fractures are usually assumed to be circularly shaped, with constant radius, or parallelogram shaped, using a rectangle or square shape. Fracture area is influenced by the extent of the fracture. There are three cases: (1) fractures are infinitely laterally extensive, (2) fractures terminate on other fractures, and (3) fractures terminate in intact rock. However, from the fluid transfer point of view, they are modeled as rectangular planes of a certain width w, height h, and length L or x, as shown in Figure 8.22.

Three-dimensional fracture geometry systems can be represented in:

1. three principal planes: defining matrix blocks, Figure 8.23a;
2. two principal planes: defining matches, Figure 8.23b, b′; and
3. one series of parallel planes: defining sheet, Figure 8.23c.

FRACTURE STORAGE CAPACITY

In contrast to the matrix porosity, fracture porosity contributes only a few percent to the total porosity. Fracture aperture is typically up to a few millimeters in width, and typical fracture spacing is in the range of centimeter to

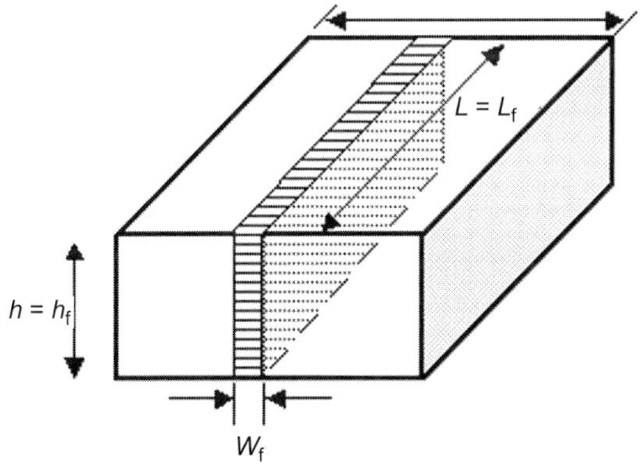

FIGURE 8.22 Fracture dimensions from the flow modeling point of view.

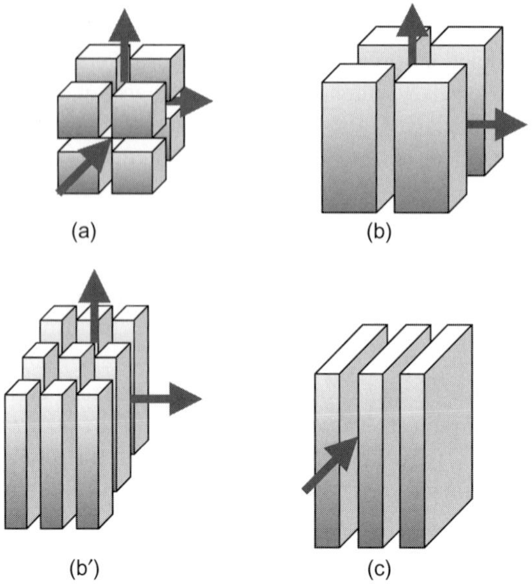

FIGURE 8.23 An idealized schematic of elementary blocks [33]. (a) Cubic blocks with permeable horizontal planes. (b) Cubic blocks with impermeable horizontal planes. (b′) Matches. (c) Sheet.

meter. Because fracture apertures are significantly greater than typical matrix pore-throat sizes, they contribute the major portion of the total transmissivity of the petroleum rocks and, consequently, are an important factor in the movement of fluids. Fracture porosity initially is very high, but, over time, fractures may become partially filled with fines. This filling process considerably reduces the fracture porosity to less than 5%. Since only fracture conductivity is necessary in flow calculations, not much attention has been given to fracture porosity or storage capacity. The overall fracture storage capacity, which indicates how much fluid is held within the fracture network of a particular reservoir, is best estimated from pressure buildup tests.

FRACTURE CONDUCTIVITY

In reservoir engineering, fractures have been categorized on the basis of their fluid transmission capacity or conductivity as follows:

1. Finite conductivity: Finite conductivity fractures allow a limited amount of the fluid to flow. If the fracture has dimensionless conductivity $F_{CD} = (k_f w_f)/(k_r x_f) < 300$, it is termed a finite conductivity fracture.
2. Infinite conductivity: Infinite conductivity fractures are highly conductive and their fluid transferring capacity is greater than that of the finite

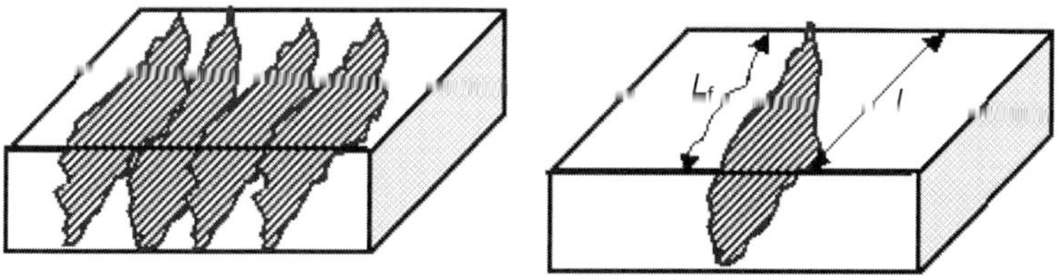

FIGURE 8.24 Examples of (a) fracture frequency and (b) fracture tortuosity.

conductivity fractures. If $F_{CD} = (k_f w_f)/(k_r x_f) > 500$, the fracture is infinitely conductive. This number is accepted by many researchers; however, some works assume $F_{CD} > 300$ for infinite conductivity.

3. Uniform flux: Uniform flux fractures allow the fluid to flow through them such that there occurs a certain pressure drop and the amount of the fluid entering and leaving the fracture remains constant.

These three categories of fractures were developed for hydraulic fractures since physical dimensions of hydraulic fractures can be controlled by increasing the injection pressure, and the amount of fluid and propant control the fracture opening. Natural fractures, on the other hand, rarely show infinite conductivity behavior. This is because no propant is present in natural fractures and the fracture surface with time develops a skin due to the chemical and physical changes that take place with time and due to the presence of reservoir fluids.

Total reservoir conductivity is controlled by the fracture frequency, width or aperture, and length. Fracture frequency is the number of fractures per unit length (depth). Fracture frequency determines the fracture volume in a rock and is needed to determine the porosity caused by the fractures.

Fracture aperture or width is the fracture opening and is a critical parameter in controlling fracture porosity and permeability. Fracture length determines the distance the fracture is penetrating the reservoir rock from the wellbore. Fractures are rarely straight, as shown in Figure 8.24. They are curvilinear and create a tortuous path as compared with straight tubes. The term "fracture tortuosity" is frequently used to define the irregular shape of the fractures and flow paths in reservoir rocks. Tortuosity is the ratio of the actual fracture length connecting two points and the minimum fracture length; therefore, the more fractures are interconnected, the less the value of τ.

CHARACTERIZING NATURAL FRACTURES FROM WELL-TEST DATA

Warren and Root first modeled the transient flow of fluids in naturally fractured rocks, assuming that the rock consists of a fracture network as shown

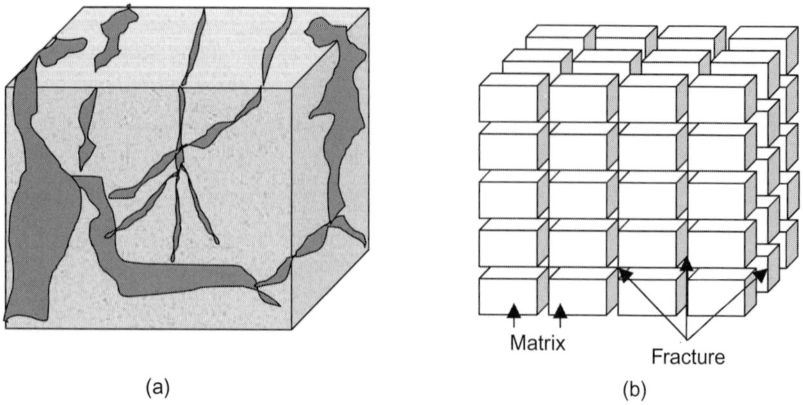

FIGURE 8.25 Realistic and idealized naturally fractured rocks, regenerated after Warren and Root [19].

in Figure 8.25 [19]. This widely popular model for flow analysis in naturally fractured reservoirs is referred to as the "sugar cube model." Warren and Root assumed that the entire fluid flows from the matrix to fractures and that only fractures feed the wellbore.

Since not all naturally fractured reservoirs behave similarly, the degree of fluid flow is controlled by the matrix and the fracture properties. Thus, Warren and Root introduced two key parameters to characterize naturally fractured reservoirs:

1. the storage capacity ratio, ω, which is a measure of the fluid stored in fractures as compared with the total fluid present in the reservoir; and
2. the interporosity flow parameter, λ, which is a measure of the heterogeneity scale of the system and quantifies the fluid transfer capacity from matrix to the fracture and vice versa.

A value of unity for λ indicates the absence of fractures or, ideally, that fractures behave like the matrix such that there is physically no difference in petrophysical properties; in other words, the formation is homogeneous. Low values of λ, on the other hand, indicate slow fluid transfer between the matrix and the fractures. However, the actual range of λ is 10^{-3}, which indicates a very high fluid transfer, to 10^{-9}, which indicates poor fluid transfer between the fractures and the matrix. The storage factor ω has a value between zero and unity. A value of 1 indicates that all the fluid is stored in the fractures, whereas a value of 0 indicates that no fluid is stored in the fractures. A value of 0.5 indicates that the fluid is stored equally in the matrix and fractures.

Mathematically, the storage capacity ratio and the interporosity flow parameter are defined as follows:

$$\omega = \frac{(\phi c_t)_f}{(\phi c_t)_t} = \frac{(\phi c_t)_f}{(\phi c_t)_f + (\phi c_t)_m} \tag{8.68}$$

and

$$\lambda = \alpha \frac{k_m r_w^2}{k_f} \tag{8.69}$$

where α is the geometry parameter, given by:

$$\alpha = \frac{4n}{(n+2)X_m^2} \tag{8.70}$$

where n is 1, 2, and 3 for sheet, matches, and cube models, respectively, as shown in Figure 8.23. For cubical and spherical geometries [34]:

$$\alpha = \frac{60}{X_m^2} \tag{8.71a}$$

where X_m represents the side length of the cube or the diameter of the sphere block (Figure 8.26). For long cylinders:

$$\alpha = \frac{32}{X_m^2} \tag{8.71b}$$

where X_m is the diameter of the cylinder. For layered or slab formations:

$$\alpha = \frac{12}{h_f^2} \tag{8.71c}$$

where h_f is the fracture height, usually taken as the formation thickness of the fractured zone stacked in between the other layers. Knowing the interporosity parameter λ from well-test analysis, the fracture height h_f can be calculated from:

$$h_f = r_w \sqrt{\frac{12 k_m}{\lambda k_f}} \tag{8.72}$$

For the sugar cube model, the side length of each matrix block is obtained from:

$$X_m = r_w \sqrt{\frac{60 k_m}{\lambda k_f}} \tag{8.73}$$

Example

A well is completed in a naturally fractured reservoir. The following data were obtained from core analysis and a single pressure drawdown test:

$\lambda = 2.5 \times 10^{-6}$,	$r_w = 0.3$ ft,	$k_f = 39{,}000$ mD,	$k_m = 0.185$ mD

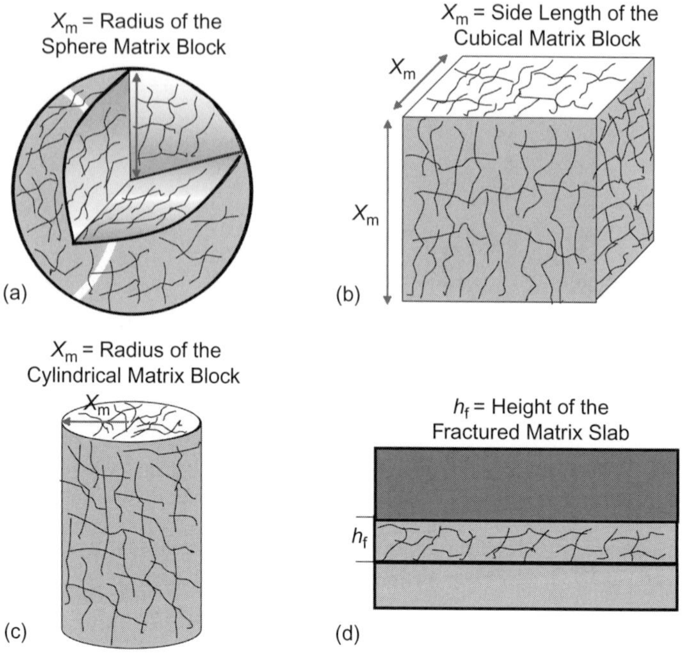

FIGURE 8.26 (a) Spherical, (b) cubical, (c) cylindrical, and (d) layered or stacked matrix blocks with natural fractures.

1. Calculate the side length of the matrix blocks.
2. What would be the height of the fracture zone if the system were layered?

Solution

1. The side length X_m is calculated from Eq. (8.73):

$$X_m = r_w\sqrt{\frac{60k_m}{\lambda k_f}} = 0.3\sqrt{\frac{60 \times 0.185}{2.5 \times 10^{-6} \times 39{,}000}} = 3.2 \text{ ft}$$

2. The fracture height is calculated from Eq. (8.72):

$$h_f = r_w\sqrt{\frac{12k_m}{\lambda k_f}} = 0.3\sqrt{\frac{12 \times 0.185}{2.5 \times 10^{-6} \times 39{,}000}} = 1.4 \text{ ft}$$

FRACTURE POROSITY FROM PRESSURE TESTS

Both of the Warren and Root parameters, λ and ω, are preferably obtained from well-test data by using either the conventional semilog analysis or the type-curve matching techniques. Using the Tiab direct synthesis (TDS) technique, both parameters can be determined from the log–log plot of the pressure derivative versus time without using the type-curve matching technique [35].

Figure 8.27 shows the semilog pressure test, with a typical two parallel lines indicating presence of natural fractures. The storage capacity ratio can be estimated from the figure, using the following equation:

$$\omega = \exp\left(-2.303\frac{\delta P}{m}\right) \tag{8.74}$$

or

$$\omega = 10^{-\delta P/m} \tag{8.74a}$$

where δP is the pressure difference between the two parallel lines in Figure 8.27 and m is the slope of either line. The degree of fracturing in each segment of the reservoir can influence the estimated value of ω; consequently, testing different wells can yield different values of ω.

The slope is used to estimate the formation permeability, k, from:

$$k = \frac{162.6 q\mu B_o}{mh} \tag{8.75}$$

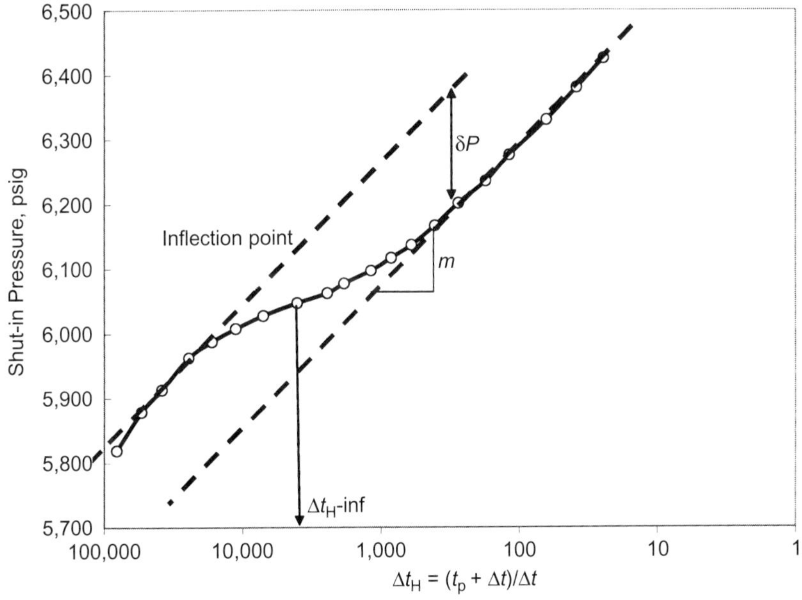

FIGURE 8.27 A typical pressure test curve showing two parallel lines, a strong indication of the presence of natural fractures in petroleum reservoir rock.

The units of pressure and slope are psi and psi/log cycle, respectively. Once ω is estimated, the fracture porosity can be estimated, if matrix porosity ϕ_m, total matrix compressibility c_m, and total fracture compressibility c_f are known, as follows:

$$\phi_f = \frac{\omega \phi_m c_m}{c_f(1-\omega)} \quad (8.76)$$

Fracture compressibility may be different from matrix compressibility by an order of magnitude. Naturally fractured reservoirs in the Kirkuk field (Iraq) and Asmari field (Iran) have fracture compressibility that ranges from 4×10^{-4} to 4×10^{-5} psi^{-1}. In the Grozni field (Russia), c_f ranges from 7×10^{-4} to 7×10^{-5}. In all these reservoirs, c_f is 10 to 100 fold higher than c_m. Therefore, the practice of assuming $c_f = c_m$ is not acceptable.

The fracture compressibility can be estimated from the following expression [14]:

$$c_f = \frac{1-(k_f/k_{fi})^{1/3}}{\Delta P} \quad (8.77)$$

where

k_{fi} = fracture permeability at the initial reservoir pressure, p_i
k_f = fracture permeability at the current average reservoir pressure, $\overline{p}$

In deep naturally fractured reservoirs, fractures and the stress axis on the formation generally are vertically oriented. Thus, when the pressure drops due to reservoir depletion, the fracture permeability reduces at a lower rate than one would expect, as indicated by Eq. (8.77). In type 2 naturally fractured reservoirs, where matrix porosity is much greater than fracture porosity, as the reservoir pressure drops, the matrix porosity decreases in favor of fracture porosity [14]. This is not the case in type 1 naturally fractured reservoirs, particularly if the matrix porosity is very low.

A representative average value of the effective permeability of a naturally fractured reservoir may be obtained from:

$$k = \sqrt{k_{max} k_{min}} \quad (8.78)$$

where

k_{max} = maximum permeability measured in the direction parallel to the fracture plane (Figure 8.28a); thus $k_{max} \approx k_f$
k_{min} = minimum permeability measured in the direction perpendicular to the fracture plane (Figure 8.28b); thus $k_{min} \approx k_m$

Equation (8.78) becomes:

$$k = \sqrt{k_f k_m} \quad (8.79)$$

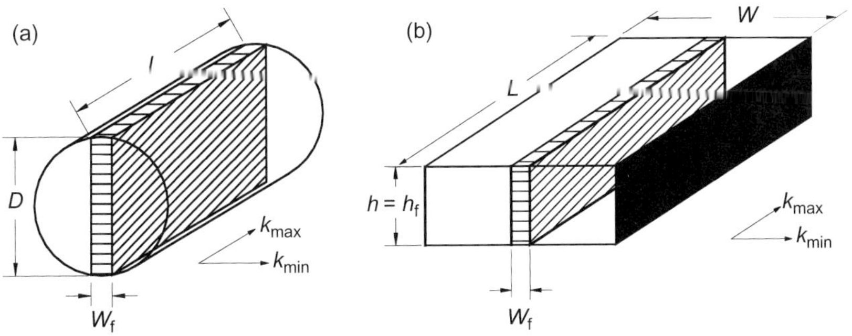

FIGURE 8.28 Maximum (a) and (b) minimum permeability.

The fracture permeability (plus connected vugs) can therefore be estimated from:

$$k_f = \frac{k^2}{k_m} \tag{8.80}$$

where k_m is the matrix permeability, which is measured from representative cores, and k is the mean permeability obtained from pressure transient tests. Combining Eqs. (8.77) and (8.80) yields:

$$c_f = \frac{1 - (k/k_i)^{2/3}}{\Delta P} \tag{8.81}$$

where

k_i = average permeability obtained from a transient test run when the reservoir pressure was at or near initial conditions p_i
k = average permeability obtained from a transient test at the current average reservoir pressure, $\bar{p}$

Thus, $\Delta p = p_i - \bar{p}$.

Matrix permeability is assumed to remain constant between the two tests. Note that Eqs. (8.77) and (8.81) are valid for any two consecutive pressure transient tests, and therefore $\Delta p = \bar{p}_1 - \bar{p}_2$. Knowing the matrix compressibility (c_m) and the compressibility of the system (c_{mf}), and assuming the porosity partitioning (ν) is equivalent to the storage capacity ratio (ω), the total compressibility of the fractures (c_f) can be determined from:

$$c_f = \frac{c_{mf} - (1 - \omega)c_m}{\omega} \tag{8.81a}$$

The time between the two tests must be long enough for the fractures to deform significantly in order to determine an accurate value of c_f.

The fracture permeability can also be estimated from the following correlation [36] if the fracture width (or aperture) w_f is known from logs or core measurements:

$$k_f = 33\omega\phi_t w_f^2 \tag{8.82}$$

where w_f is in microns (1 micron = 10^{-6} m), and the storativity ratio ω and total porosity ϕ_t are expressed as a fraction and k_f in mD.

If the fracture width cannot be measured from logs or core analysis and k_f can be calculated from Eq. (8.80), then Eq. (8.82) may be used to estimate w_f:

$$w_f = \sqrt{\frac{k_f}{33\omega\phi_t}} \tag{8.84}$$

The interporosity fluid transfer coefficient is then estimated as:

$$\lambda = \frac{3,792(\phi c_t)_{f+m}\mu r_w^2}{k_f \Delta t_{inf}}\left(\omega \ln\left(\frac{1}{\omega}\right)\right) \tag{8.85}$$

The reservoir permeability, k, is expressed in mD, fluid viscosity, μ, in cP, wellbore radius, r_w, in ft, inflection time, Δt_{inf}, in hrs, porosity in fraction, and total compressibility, ct, in psi^{-1}.

The test time corresponding to the inflection point, Δt_{inf}, is obtained from the semilog plot of the pressure drop ΔP versus shut-in time Δt. Sometimes, however, the inflection point is not obvious on a semilog plot due to the presence of a nearby boundary or near-wellbore effects such as wellbore storage and skin. It is thus recommended that a pressure derivative plot be used as a guide for locating this inflection point.

If a Horner plot is used, that is, a plot of the shut-in pressure versus Horner time, $\Delta t_H = (t_p + \Delta t)/\Delta t$, then the point of inflection is obtained from:

$$\Delta t_{inf} = \frac{t_p}{(\Delta t_H)_{inf} - 1} \tag{8.86}$$

where $(\Delta t_H)_{inf}$ is simply $((t_p + \Delta t)/\Delta t)_{inf}$ as shown in Figure 8.27, t_p is the production time, and Δt is the test time during a pressure buildup test.

On the log–log plot of the pressure derivative $(t^* \times \Delta P')$ versus test time Δt, the inflection point Δt_{inf} is easily recognized on the pressure derivative plot. It corresponds to the time at which the minimum value of the trough is reached. Applying the TDS technique, k, ω, and λ can be obtained from the log–log plot without using type-curve matching [35]. The permeability is obtained from:

$$k = \frac{70.6 q\mu B_o}{h(t \times \Delta p')_R} \tag{8.87}$$

where $(t \times \Delta P')_R$ is obtained from the horizontal line of the pressure derivative, which corresponds to the infinite acting radial flow regime.

The interporosity fluid transfer coefficient is given by [39]:

$$\lambda = \left(\frac{42.5h(\phi c_t)_{m+f} r_w^2}{qB_o}\right)\frac{(t \times \Delta p')_{min}}{\Delta t_{min}} \tag{8.88}$$

where $(t \times \Delta P')_{min}$ and Δt_{min} are the coordinates of the minimum point, as shown in Figure 8.32. The fracture storage ratio is given by [39]:

$$\omega^{\omega} = e^{-\lambda t_{D\,min}} \tag{8.89}$$

where the dimensionless time corresponding to the minimum time, t_{Dmin}, is calculated from:

$$t_{D\,min} = \left(\frac{0.0002637k}{\mu r_w^2 (\phi c_t)_{m+f}}\right)\Delta t_{min} \tag{8.90}$$

where Δt_{min} is the time coordinate of the minimum point of the trough on the pressure derivative curve. Δt_{min} on the log–log plot of the derivative curve is equivalent to Δt_{inf} on the semilog plot of pressure versus time.

Equation (8.89) is plotted in Figure 8.29. Curve fitting the points and solving explicitly for ω yield:

$$\omega = \left(2.9114 - \frac{3.5688}{\ln(N_S)} - \frac{6.5452}{N_S}\right)^{-1} \tag{8.91}$$

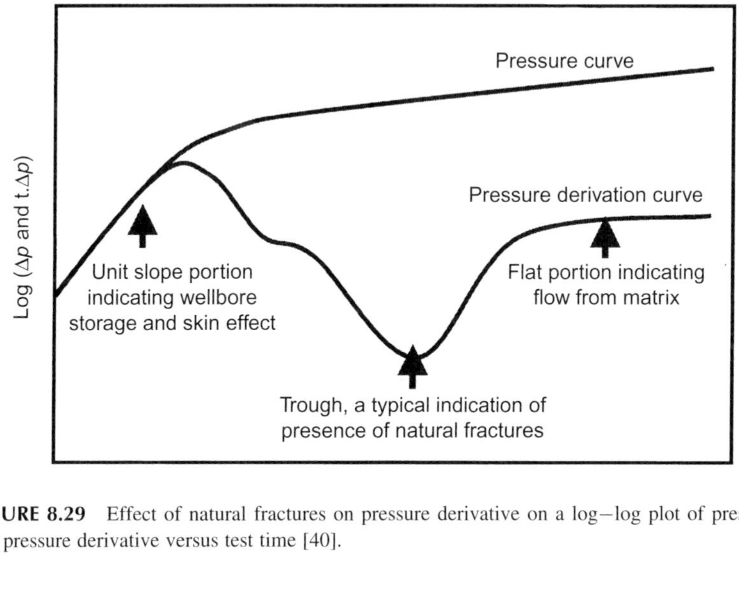

FIGURE 8.29 Effect of natural fractures on pressure derivative on a log–log plot of pressure and pressure derivative versus test time [40].

where

$$N_S = e^{-\lambda t_{Dmin}} \tag{8.92}$$

where the total porosity–compressibility product is:

$$(\phi c_t)_{f+m} = (\phi c_t)_m \left(1 + \frac{\omega}{1-\omega}\right) \tag{8.93}$$

The fracture storage ratio ω can be directly determined from Figure 8.30 or Eq. (8.91).

Equation (8.91) is obtained by assuming values of ω, from 0 to 0.5; then values of $\omega^{\omega} = N_S$ are plotted against ω. The resulting curve is curve-fitted. Combining Eqs. (8.89) through (8.92), and solving explicitly for ω yields (Tiab et al., 2007):

$$\omega = \left(2.9114 + 4.5104\frac{(t \times \Delta P')_R}{(t \times \Delta P')_{min}} - 6.5452e^{0.7912\frac{(t\times\Delta P')_{min}}{(t\times\Delta P')_R}}\right)^{-1} \tag{8.94}$$

where $(t \times \Delta P')_r =$ value of the pressure derivative during the infinite acting radial flow line. It is of emphasis that Eq. (8.93) assumes wellbore storage and boundary effects do not influence the trough, and the infinite acting radial flow line is well defined [37].

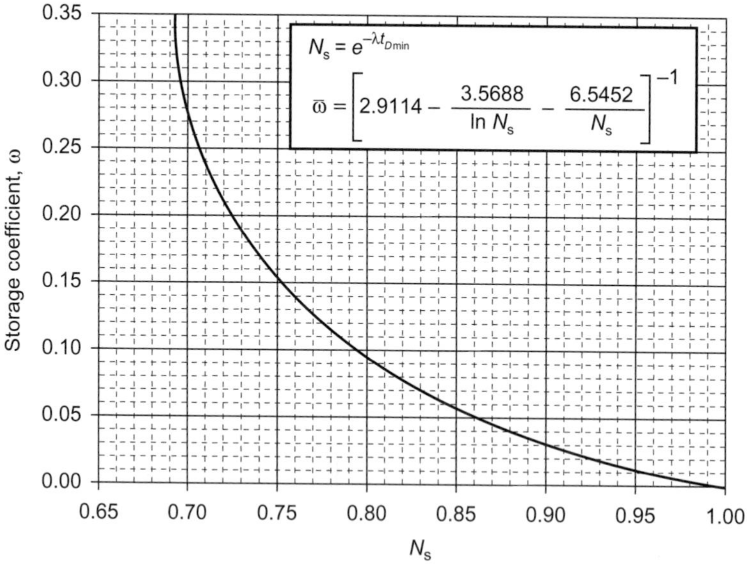

FIGURE 8.30 Storage coefficient from the time coordinate of the minimum point of the pressure derivative curve.

Example

Pressure tests in the first few wells located in a naturally fractured reservoir yielded the same average permeability of the system of 85 mD. An interference test yielded the same average reservoir permeability, which implies that fractures are uniformly distributed. Only the porosity, permeability, and compressibility of the matrix could be determined from the recovered cores.

Figures 8.31 and 8.32 show the behavior of pressure and pressure derivative of a recent pressure buildup test conducted in a well. The pressure drop from the initial reservoir pressure to the current average reservoir pressure is 974.5 psia. The characteristics of the rock, fluid, and well are given below:

$h = 1{,}150$ ft,	$r_w = 0.292$ ft,	$\mu = 0.47$ cP,
$B_o = 1.74$ RB/STB,	$q = 17{,}000$ STB/D,	$k_m = 0.15$ mD,
$c_{tm} = 4.15 \times 10^{-5}$ psi^{-1},	$\phi_m = 14\%$	

1. Using conventional semilog analysis and the TDS technique, calculate the current:
   a. formation permeability,
   b. storage capacity ratio, and
   c. fluid transfer coefficient.
2. Estimate the four fracture properties: (a) permeability, (b) porosity, (c) width, and (d) matrix block dimensions.

Solution

1. Conventional method:

   From Figure 8.31, $\delta P = 15$ psi, $m = 25$ psi/log cycle, and $\Delta t_{inf} = 0.22$ hr

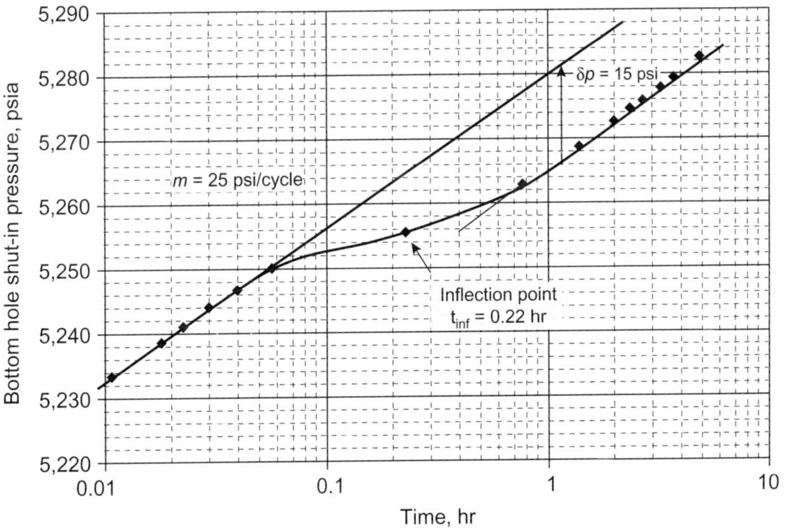

FIGURE 8.31 Pressure buildup test data plotted against shut-in time.

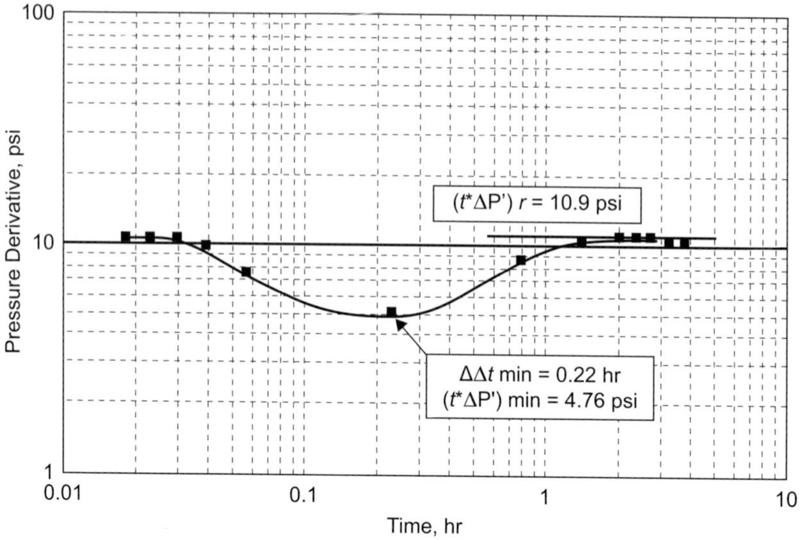

FIGURE 8.32 Pressure derivative group plotted against shut-in time.

a. The average permeability of the formation is estimated from the slope of the semilog straight line. Using Eq. (8.75) yields:

$$k=\frac{162.6q\mu B_o}{mh}=\frac{162.6\times 17{,}000\times 0.47\times 1.74}{25\times 1{,}150}=78.6\text{ mD}$$

b. Fluid storage coefficient is estimated using Eq. (8.74):

$$\omega=\exp\left(-2.303\frac{\partial P}{m}\right)=\exp\left(-2.303\frac{15}{25}\right)=0.25$$

The storage coefficient of 0.23 indicates that the fractures occupy 23% of the total reservoir pore volume. The total porosity–compressibility product is:

$$(\phi c_t)_{f+m}=(\phi c_t)_m\left(1+\frac{\omega}{1-\omega}\right)$$
$$=(0.14\times 0.0000415)\times\left(1+\frac{0.23}{1-0.23}\right)=7.545\times 10^{-6}\ \text{psi}^{-1}$$

c. The interporosity fluid transfer coefficient is given by Eq. (8.85):

$$\lambda=\frac{3{,}792\times 7.545\times 10^{-6}\times 0.47\times 0.292^2}{78.6\times 0.22}\left(0.23\ln\left(\frac{1}{0.23}\right)\right)=2.24\times 10^{-5}$$

The high value of fluid transfer coefficient indicates that the fluid transfer from matrix to fractures is very efficient.

TDS technique:

From Figure 8.32, the following characteristic points are read:

$\Delta t_{min}=0.22$ hr $\quad (t\times\Delta P')_R=10.9$ psia $\quad (t\times\Delta P')_{min}=4.76$ psia

Using the TDS technique, the value of k is obtained from Eq. (8.87):

$$k = \frac{70.6 q \mu B_o}{h(t \times \Delta p')_R} = \frac{70.6 \times 17{,}000 \times 0.47 \times 1.74}{1{,}150 \times 10.9} = 78.3 \text{ mD}$$

b. The interporosity fluid transfer coefficient is given by Eq. (8.88):

$$\lambda = \left(\frac{42.5 h(\phi c_t)_{m+f} r_w^2}{q B_o}\right) \frac{(t \times \Delta p')_{min}}{\Delta t_{min}}$$

$$= \left(\frac{42.5 \times 1{,}150 \times 1.4 \times 10^{-6} \times 0.292^2}{17{,}000 \times 1.74}\right) \frac{4.76}{0.22} = 2.30 \times 10^{-5}$$

c. The storage coefficient ω is calculated from Eqs. (8.91) and (8.94):

$$t_{D\,min} = \left(\frac{0.0002637\, k}{\mu r_w^2 (\phi c_t)_{m+f}}\right) \Delta t_{min} = \left(\frac{0.0002673 \times 78.3}{0.47 \times 0.292^2 \times 7.545 \times 10^{-6}}\right) \times 0.22 = 15{,}227.7$$

$$N_S = e^{-\lambda t_{D\,min}} = e^{-2.30 \times 10^{-5} \times 15222.7} = 0.7045$$

$$\omega = \frac{1}{2.9114 - 3.5668/\ln(0.7045) - 6.5452/0.7045} = 0.26$$

$$\omega = \left(2.9114 + 4.5104 \frac{10.9}{4.76} - 6.5452 \times e^{0.7912(4.76/10.9)}\right)^{-1} = 0.25$$

This may be interpreted that the fracture pore volume is 25% of the total pore volume, provided $(\phi c_t)_m = (\phi c_t)_f$, which is often not the case. Equation (8.93) is not applicable in this case because the trough is influenced by wellbore storage effect. The conventional semilog analysis yields the same values of k, ω, and λ as the TDS technique. The main reason for this match is that both parallel straight lines are well defined.

2. Current properties of the fracture

a. The fracture permeability is calculated from Eq. (8.80):

$$k_f = \frac{k^2}{k_m} = \frac{78.6^2}{0.15} = 41{,}186 \text{ mD}$$

The fracture permeability at initial reservoir pressure is:

$$k_{fi} = \frac{k^2}{k_m} = \frac{85^2}{0.15} = 48{,}166 \text{ mD}$$

b. The fracture porosity is obtained from Eq. (8.76).

In fractured reservoirs with deformable fractures, the fracture compressibility changes with declining pressure. The fracture compressibility can be estimated from the following expression [14]:

$$c_{tf} = \frac{1 - (k_f/k_{fi})^{2/3}}{\Delta P} = \frac{1 - (41{,}216/48{,}166)^{2/3}}{974.5} = 1.01 \times 10^{-4} \text{ psi}^{-1}$$

The total compressibility of the fracture–matrix system is estimated from [26]:

$$c_{tmf} = c_{tf} + (1 - \omega) c_{tm} = 1.01 \times 10^{-4} \times 0.23 + (1 - 0.23) \times 4.15 \times 10^{-5}$$

$$= 5.52 \times 10^{-5} \text{ psi}^{-1}$$

The compressibility ratio is:

$$\frac{c_{tf}}{c_{tm}} = \frac{1.01 \times 10^{-4}}{4.15 \times 10^{-5}} = 2.45$$

Thus, the fracture compressibility is more than 24 fold higher than the matrix compressibility, or $c_{tf} = 2.43 c_{tm}$. The fracture porosity is:

$$\phi_f = \left(\frac{\omega}{1-\omega}\right)\frac{c_{tm}}{c_{tf}}\phi_m = \left(\frac{0.25}{1-0.25}\right)\frac{0.14}{24.4} = 0.017 = 1.7\%$$

The total porosity of this naturally fractured reservoir is:

$$\phi_t = \phi_m + \phi_f = 0.14 + 0.017 = 0.157$$

c. The fracture width or aperture may be estimated from Eq. (8.84):

$$w_f = \sqrt{\frac{k_f}{33\omega\phi_t}} = \sqrt{\frac{41{,}186}{33 \times 0.25 \times 0.157}} = 178 \text{ microns} = 0.178 \text{ mm}$$

The side length X_m of the cubic block is calculated from Eq. (8.73b):

$$X_m = r_w\sqrt{\frac{60k_m}{\lambda k_f}} = 0.298\sqrt{\frac{60 \times 0.15}{2.3 \times 10^{-5} \times 41{,}186}} = 2.9 \text{ ft}$$

This interpretation assumes that (a) the change in matrix compressibility and porosity of this naturally fractured reservoir is negligible and (b) the Warren and Root sugar cube model is applicable.

PROBLEMS

1. What are the major factors in the creation of natural fractures in the reservoir rock?
2. Discuss (a) the geological classification and (b) the engineering classification of natural fractures. What are these classifications based on?
3. What are the major indicators of natural fractures?
4. Name the most prolific naturally fractured oil fields of the world.
5. What are the major petrophysical characteristics of the natural fractures. How do these characteristics affect the flow of fluids through the fractures?
6. Name the most common techniques used to characterize natural fractures in petroleum-bearing rocks.
7. What are the two main parameters involved in the Warren and Root sugar cube model? Discuss their significance and physical meaning.
8. Differentiate among fault, joint, and fracture. How do they affect the fluid flow in petroleum reservoirs?
9. A 5-in-long, 2-in-thick rock sample has only one fracture. The fracture width is measured as 0.03 cm and fully penetrates the rock sample over its entire thickness.

(a) Calculate the surface area of the space created by the fracture assuming rectangular and elliptical fracture shapes.
(b) Calculate the hydraulic radius of the fracture.

10. A resistivity survey in a well showed a wellbore-corrected mud filtrate resistivity of 0.1 ohm-m, water resistivity 0.19 ohm-m, invaded zone resistivity 2 ohm-m, and deep formation resistivity 115 ohm-m. The average porosity of 22%, estimated from log data, well matches the porosity estimated from cores. Pressure test analysis as well as cores indicated the presence of natural fractures in the well. Substantial mud loss was also observed during drilling of this well, and in neighboring wells. Several outcrops also indicate the presence of natural fractures in the area. Using log data, estimate the fracture intensity index and the porosity partitioning coefficient.
11. A newly drilled well in an oil reservoir was logged. Seismic surveys and geological studies indicated that the well is located in a faulted naturally fractured zone, which is a downthrown layer. The average total porosity (15%) of the system was estimated from cores. Other known characteristics are:

$A = 4{,}500$ acres	$h = 70$ ft
$S_w = 0.25$	$B_o = 1.1$ bbl/STB
$R_w = 0.11$ ohm-m	$R_t = 80$ ohm-m
$R_{mf} = 0.15$ ohm-m	$m = 1.30$

(a) Estimate the porosity partitioning coefficient.
(b) Estimate the matrix porosity and fracture porosity.
(c) Calculate the total oil in place, STB.
(d) Calculate the FII.
(e) Estimate the distance to the nearest fault, if the resistivity of the invaded zone is 6.5 ohm-m.
(f) Does the presence of a nearby fault change the estimate of total oil in place?

NOMENCLATURE

A area, cm^2
B_o formation volume factor, RB/STB
c compressibility, psi^{-1}
d distance, ft
FII fracture intensity index, unitless
F_t total fracture intensity index, unitless
h_f fracture height, cm or ft
h formation thickness, ft
H_T Tiab's hydraulic unit characterization factor, unitless

k permeability, mD, darcy, or cm^2
L length, cm
m cementation factor or slope of semilog straight line
m_f fracture or double porosity cementation factor, unitless
m_m matrix cementation or porosity factor, unitless
N_o oil in place, STB
P pressure, psi or $dynes/cm^2$
q flow rate, STB/D or cm^3/S
R resistivity, ohm-m
r radius, cm
r_w wellbore radius, ft
S specific surface area, cm^2
S saturation, fraction
V volume, cm^3
w width, cm or μm

SUBSCRIPTS

b bulk
c capillary
c characterization
e ellipsoidal
f fracture
gr grain
h hydraulic
m matrix
o oil
mf mud filtrate
f+m fracture and matrix
p pore; producing
pv pore volume
s surface
sh shape
t true
t total
w wellbore or water
Wf wellbore flowing
Xo flushed zone

GREEK SYMBOLS

ϕ porosity, fraction
μ viscosity, cP
τ tortuosity, unitless
ν porosity partitioning coefficient, unitless
δP vertical separation on the two pressure curves

REFERENCES

1. Nelson RA. Fractured reservoirs: turning knowledge into practice. *Soc Pet Eng J* April 1987;**39**(4):407−14.
2. Massonnat G, Pernarcic E. Assessment and modeling of high permeability areas in carbonate reservoirs, Paper SPE 77591. In: *Proc. SPE/DOE improved oil recovery symposium (IOR)*, Tulsa, OK; April 13−17, 2002.
3. Handin J, Hager RV. Experimental determination of sedimentary rocks under confining pressure: tests at room temperature in dry samples. *AAPG Bull* 1957;**41**:1−50.
4. Stearns DW, Friedman M. Reservoirs in fractured rock. *Am Assoc Pet Geol (AAPG) Memoir 16 and Soc Expl Geophys, Special Publ.* 1972;(10):82−100.
5. Yamaguchi T. Tectonic study of rock fractures. *J Geol Soc Jpn* 1965;**71**(837):257−75.
6. Charlesworth KAK. Some observations on the age of jointing in macroscopically folded rocks. In: Baer AJ, Norris DK, editors. *Kink bands and brittle deformation.* Geological Survey of Canada; 1968, paper 68−52 p. 125−35.
7. Picard MD. Oriented linear shrinkage cracks in green river formation (eocene), raven ridge area, Uinta Basin, Utah. *J Sediment Pet*, **36**(4):1050−1057.
8. Lui X, Srinivasan S, Wong D. Geological characterization of naturally fractured reservoirs using multiple point geostatistics. Paper SPE 75246, In: *Proc. SPE/DOE improved oil recovery symposium*, Tulsa, OK; April 13−17, 2002.
9. Dholakia SK, Aydin A, Pollard D, Zoback MD. Development of fault controlled hydrocarbon migration pathways in monterey formation, California. *AAPG Bull* 1998;**82**:1551−74.
10. Chilingarian GV, Mazzullo SJ, Rieke HH. *Carbonate reservoir characterization: a geologic-engineering analysis, part i.* New York: Elsevier Science; 1992.
11. Choquette PW, Pray LC. Geologic nomenclature and classification of porosity in sedimentary carbonates. *Am Assoc Pet Geol (AAPG) Bull* 1970;**54**:207−50.
12. Chilingarian GV, Chang J, Bagrintseva KI. Empirical expression of permeability in terms of porosity, specific surface area, and residual water saturation. *J Pet Sci Eng* 1990;**4**:317−22.
13. Aguilera R. *Naturally fractured reservoirs.* Tulsa: Petroleum Publishing Company; 1980. p. 703
14. Saidi AM. *Reservoir engineering of fractured reservoirs.* Paris: Total Edition Press; 1987.
15. Dyke CG, Wu B, Tayler MD. Advances in characterizing natural fracture permeability from mud log data. Paper SPE 25022, In: *Proc. European petroleum conference*, Cannes, France; November 16−18, 1992.
16. Keelan DK. Core analysis for aid in reservoir description. *Soc Pet Eng J* November 1982;**34**(11):2483−91.
17. Schlumberger Inc. *Log interpretation—principles/applications.* Houston: Schlumberger Educational Services; 1987.
18. Pirson SJ. *Oil reservoir engineering.* 2nd ed. New York: McGraw-Hill; 1978.
19. Warren JE, Root PJ. The behavior of naturally fractured reservoirs. *Soc Pet Eng J* Sept. 1963;245−55.
20. Tkhostov BA, Vezirova AD, Vendel'shtyen BY, Dobrynin VM. *Oil in fractured reservoirs.* Leningrad: Izd. Nedra; 1979, p. 219.
21. Hensel Jr. WM. A perspective look at fracture porosity. *SPE Form Eval* Dec. 1989.
22. Aguilera R. Analysis of naturally fractured reservoirs from conventional well logs. *J Pet Technol* July 1976;764−72.
23. Doveton JH. Development of an Archie equation model for long analysis of Pennsylvanian oomoldic zones in Kansas, KGS. Open File Report 67; 2001.
24. Nugent WH. Letter to the editor. *The Log Analyst* 1984;**24**(4):2−3.

25. Rasmus JC. A variable cementation exponent m for fractured carbonates. *The Log Analyst* November–ecember, 1983;**24**(6):13 –23.
26. Pulido H, Samaniego F, Munoz GG, Rivera JR, Velez C. Petrophysical characterization of carbonate naturally fractured reservoirs for use in dual porosity simulators. In: *Proc. 32nd workshop on geothermal reservoir engineering (GRE)*, Stanford, California; 2007.
27. Locke LC, Bliss JE. Core analysis technique for limestone and dolomite. *World Oil*, September. 1950.
28. Pirson SJ. *Geologic well log analysis*. Houston, TX: Gulf Publishing Company; 1970.
29. Aguilera MS, Aguilera R. Improved models for petrophysical analysis of dual porosity reservoirs. *Petrophysics* 2003;**44**(1):21–35.
30. Tuan PA, Martyntsiv OF, Dong TL., Evaluation of fracture aperture and wettability, capillary properties of oil-bearing fractured granite. Paper SCA-9410, In: *Proc. 1994 international symposium of SCA*, Stavanger, Norway; September 12–14, 1994.
31. Nelson RA. *Geologic analysis of naturally fractured reservoirs*. second edition Boston, MA: Gulf Professional Publishing; 2001.
32. Tiab D. *Modern core analysis, vol. i – theory*. Houston, Texas: Core Laboratories; May 1993. p. 200
33. Elkewidy T. Characterization of hydraulic flow units in heterogeneous clastic and carbonate reservoirs. Ph.D. dissertation, School of Petroleum and Geological Engineering, University of Oklahoma, Norman, OK; 1996.
34. Horne RN. *Modern well test analysis*. 2nd ed. Petroway; 1995.
35. Engler T, Tiab D. Analysis of pressure and pressure derivative without type curve matching, 4. Naturally fractured reservoirs. *J Pet Sci Eng* 1996;**15**(2–4):127–38.
36. Bona N, Radaelli F, Ortenzi A, De Poli A, Peduzzi C, Giorgioni M. Integrated core analysis for fractured reservoirs: quantification of the storage and flow capacity of matrix, vugs, and fractures. *SPE Res Eval Eng* August 2003;**6**:226–33.
37. Tiab D, Igbokoyi A, Restrepo D. Fracture porosity from pressure transient data. IPTC 11164, In: *Proc. SPE international petroleum technology conference*, Dubai, UAE; December 4–6, 2007.
38. Lake W, Carroll Jr. HB. *Reservoir characterization*. New York: Academic Press; 1968. p. xi.
39. Tiab D. *Advances in pressure transient analysis*. Lecture notes manual, Norman, OK, 2003.
40. Stewart G, Ascharsobbi F. Well test analysis for naturally fractured reservoirs. Paper SPE 18173, In: *Proc. 63rd SPE annual technical conference and exhibition (ATCE)*, Houston, Texas; October 2–5, 1988.
41. Ohen HA, Enwere P, Daltaban S. The role of core analysis data in the systematic and detailed modeling of fractured carbonate reservoir petrophysical properties to reduce uncertainty in reservoir simulation. *SCA* 2002-49.
42. Borai AM. A new correlation for cementation factor in low-porosity carbonates. *SPE Formation Evaluation* 1987;4:495–9.
43. Hassani-giv M, Rahimi M. New correlations for porosity exponent in carbonate reservoirs of Iranian oil fields in Zagros Basin. *JSUT* 2008;34(4):1–7.
44. Ershaghi I. Evaluation of naturally fractured reservoirs. IHRDC, PE 509, 1995.

Chapter 9

Effect of Stress on Reservoir Rock Properties

Fairhurst defines rock mechanics as "the field of study devoted to understanding the basic processes of rock deformation and their technological significance" [1]. The significance of these processes to petroleum engineers is considerable. For instance, being able to predict the mechanical behavior of underground formations is a key to avoiding borehole instabilities during drilling. If rock deformation results in a noticeable contraction of the wellbore due to the state of induced stress in the rock formation immediately adjacent to the wellbore, the motion of the drill bit may be restricted or the emplacement of the casing, after drilling ceases, may be hampered. If the deformation results in a large expansion of the wellbore, the rock formation may fracture and result in lost circulation [2]. Predicting the mechanical behavior of reservoir rock is essential for well completion or stimulation programs. Reservoir compaction, which may lead to surface subsidence, is a critical factor with respect to the design of the casing platforms and to the overall reservoir performance [3]. Figure 9.1 illustrates casing failure resulting from compaction of reservoir rock. The production of oil, natural gas, and/or water from underground rock formations results in a local change in the stress and strain field in the formation due to the decline in pore pressure [4]. In order to predict the compaction or compressibility behavior of petroleum reservoirs due to this decline in pore pressure, it is necessary to know the compressibility characteristics of the reservoir rock.

Rock mechanical properties such as Poisson's ratio, shear modulus, Young's modulus, bulk modulus, and compressibility can be obtained from two different sources:

1. *Laboratory measurements*, which allow for direct measurements of strength parameters and static elastic behavior with recovered core material from discrete depths.
2. *Downhole measurements* through wireline logging, which allow the determination of dynamic elastic constants from the continuous measurement of compressional and shear velocities.

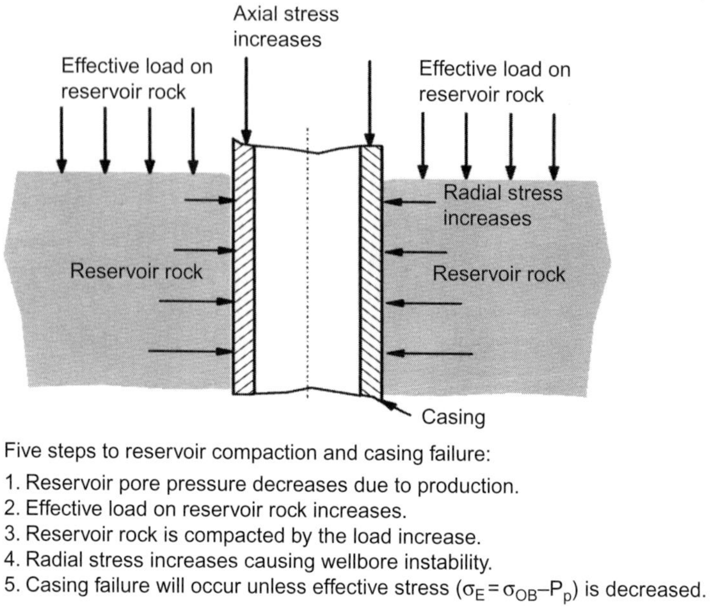
Axial stress increases
Effective load on reservoir rock
Effective load on reservoir rock
Radial stress increases
Reservoir rock
Reservoir rock
Casing
Five steps to reservoir compaction and casing failure:
1. Reservoir pore pressure decreases due to production.
2. Effective load on reservoir rock increases.
3. Reservoir rock is compacted by the load increase.
4. Radial stress increases causing wellbore instability.
5. Casing failure will occur unless effective stress ($\sigma_E = \sigma_{OB} - P_p$) is decreased.

FIGURE 9.1 Illustration of casing failure resulting from compaction of reservoir rock.

However, it is important to remember that, because reservoir rocks are often layered, fractured, faulted, and jointed, rock masses sometimes may be controlled more in their reactions to applied loads by the heterogeneous nature of the overall rock mass than by the microscopic properties of the rock matrix. Consequently, the mechanical properties obtained from laboratory core tests may be considerably different from those existing in situ. Core alteration during and after drilling may also influence the results. Nevertheless, mechanical properties determined under laboratory test conditions are a source of valuable information for most projects in rock mechanics because knowledge of deformational characteristics of rock is essential in locating and extracting mineral resources, and in the design and construction of any structure in the rock [5].

STATIC STRESS–STRAIN RELATION

Poulos and Davis developed the following analytical model of the evaluation of the static stress–strain relation [6]. Consider a cubic rock sample in a three-dimensional stress field, as shown in Figure 9.2 [5]. To understand the significance of the diagram, a number of important physical concepts associated with the mechanical behavior of rocks must be defined.

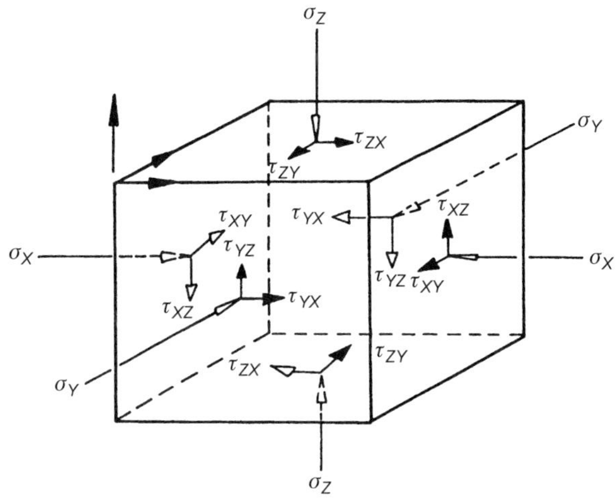

FIGURE 9.2 Three-dimensional stress field of a cubic element.

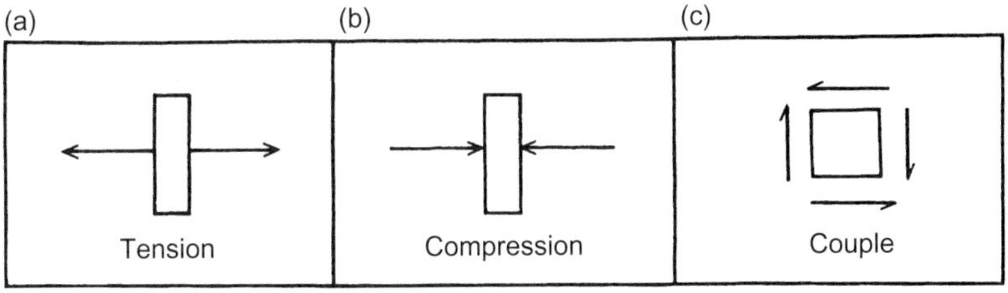

FIGURE 9.3 Representation of tension, compression, and shear or couple.

STRESS ANALYSIS

If a rock body is subjected to an external load or force, internal stresses are developed. If these stresses are strong enough, the rock deforms. Deformation refers to changes in shape (distortion) accompanied by change in volume (dilation). Three basic internal stress conditions are recognized: compressive, shear, and tensile, as illustrated in Figure 9.3 [7]. Compressive stresses occur when external forces are directed toward each other along the same plane. If the external forces are parallel and directed in opposite directions along the same plane, tensile stress develops. Shear stress occurs when the external forces are parallel and directed in opposite directions, but in different planes.

If any plane is taken within a solid body, as shown in Figure 9.4 along the *yz*-plane, then the internal components of stress may be resolved into normal stress (σ_{xx}), which acts at a right angle to the plane, and shear stress

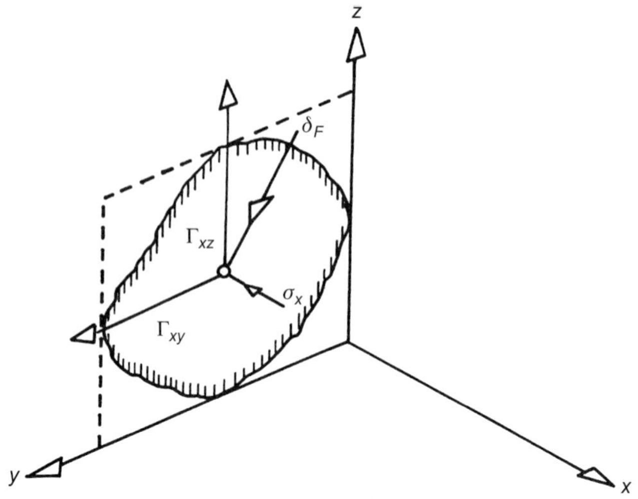

FIGURE 9.4 Stress at a point zero in a plane [5].

components, which act parallel to the plane (τ_{xy} and τ_{xz}). If the solid plane is taken along the xz-plane, the normal and shear stress components at point 0 are τ_{yx}, σ_{yy}/τ_{yz}, and in the xy-plane the three components are τ_{zx}, τ_{xy}, and σ_{xx}. Therefore, nine components of stress are required to fully define the forces acting on the cubic element shown in Figure 9.2. The stress matrix is:

$$\sigma_{xyz} = \begin{bmatrix} \sigma_{xx} & \tau_{xy} & \tau_{xz} \\ \tau_{yx} & \sigma_{yy} & \tau_{yz} \\ \tau_{zx} & \tau_{zy} & \sigma_{zz} \end{bmatrix} \tag{9.1}$$

The notation τ_{ij} should be read as the "shear stress acting in the j direction on a plane normal to the i axis." By convention, the normal stresses σ_{xx}, σ_{yy}, and σ_{zz}—or for convenience σ_x, σ_y, and σ_z—are positive when directed into the plane. If the body is at equilibrium, then $\tau_{xy} = \tau_{yx}$, $\tau_{yz} = \tau_{zy}$, and $\tau_{zx} = \tau_{yz}$. In matrix operations, it is convenient to express the stress tensor as:

$$\sigma_{123} = \begin{bmatrix} \sigma_{11} & \sigma_{12} & \sigma_{13} \\ \sigma_{21} & \sigma_{22} & \sigma_{23} \\ \sigma_{31} & \sigma_{32} & \sigma_{33} \end{bmatrix} \tag{9.2}$$

It is possible to show that there is one set of axes with respect to which all shear stresses are zero and the normal stresses have their extreme values. The three mutually perpendicular planes where these conditions exist are called the principal planes, and the three normal stresses on these planes are the principal stresses (Figure 9.5): σ_1 or σ_{11} (maximum or major), σ_2 or σ_{22} (intermediate), and σ_3 or σ_{33} (minimum or minor). The principal stresses

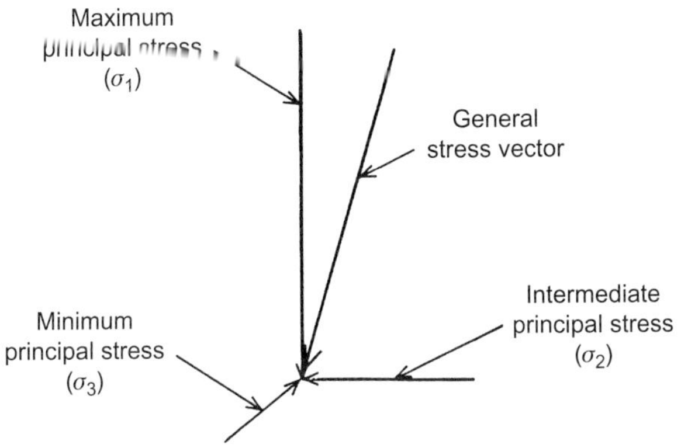

FIGURE 9.5 Principal stresses acting on a point [7].

may be determined from the roots of the equation developed by Poulos and Davis [6]:

$$\sigma_i^3 + J_1\sigma_i^2 + J_2\sigma_i - J_3 = 0 \tag{9.3}$$

where

$$J_1 = \sigma_{xx} + \sigma_{yy} + \sigma_{zz}$$
$$J_2 = \sigma_{xx}\sigma_{yy} + \sigma_{yy}\sigma_{zz} + \sigma_{zz}\sigma_{xx} - \tau_{xy}^2 - \tau_{yz}^2 - \tau_{xz}^2$$
$$J_3 = \sigma_{xx}\sigma_{yy}\sigma_{zz} - \sigma_{xx}\tau_{yx}^2 - \sigma_{yy}\tau_{zx}^2 - \sigma_{yy}\tau_{yx}^2 + 2\tau_{xy}\tau_{yz}\tau_{zx}$$

J_1 (or bulk stress), J_2, and J_3 are defined as the first, second, and third stress invariants (because they remain constant) and are independent of the coordinate system. In terms of the principal stresses:

$$\begin{aligned} J_1 &= \sigma_1 + \sigma_2 + \sigma_3 \\ J_2 &= \sigma_1\sigma_2 + \sigma_2\sigma_3 + \sigma_3\sigma_1 \\ J_3 &= \sigma_1\sigma_2\sigma_3 \end{aligned} \tag{9.4}$$

The principal stress tensor is represented as:

$$\sigma_p = \begin{bmatrix} \sigma_{11} & 0 & 0 \\ 0 & \sigma_{22} & 0 \\ 0 & 0 & \sigma_{33} \end{bmatrix} \tag{9.5}$$

The maximum shear stress at a point, τ_{max}, occurs on a plane at an angle of 45° with the σ_1 and σ_3 directions, and is given by:

$$\tau_{max} = \frac{1}{2}(\sigma_{11} - \sigma_{33}) \tag{9.6}$$

If the magnitudes and directions of the principal stresses can be easily obtained, it is convenient to use them as reference axes.

STRAIN ANALYSIS

Strain is defined as the compression (positive) or extension (negative) resulting from the application of external forces, divided by the original dimension. Two types of strain can be recognized: homogeneous and heterogeneous. When every part of a body is subjected to a strain of the same type and magnitude in any direction of the displacement, the strain is considered homogeneous [6]. The strain is heterogeneous if it is not the same throughout the body. Strain resulting from extended application of large stresses at high temperatures is described as finite. If, however, the strain results from the application of an increment of stress and can be treated mathematically, then it is defined as infinitesimal strain. The strain is responsible for inducing body displacement, rotation, and strain. Shear strain, γ, is defined as the angular change in a right angle at a point in a body and is related to the displacements in the x, y, and z directions. Assuming that a negative shear strain represents a decrease in the right angle and a positive shear strain represents an increase in the right angle:

$$\varepsilon_{xyz} = \begin{bmatrix} \varepsilon_{xx} & \frac{\gamma_{xy}}{2} & \frac{\gamma_{xz}}{2} \\ \frac{\gamma_{yx}}{2} & \varepsilon_{yy} & \frac{\gamma_{yz}}{2} \\ \frac{\gamma_{zx}}{2} & \frac{\gamma_{zy}}{2} & \varepsilon_{zz} \end{bmatrix} \tag{9.7}$$

where ε_{xx}, ε_{yy}, and ε_{zz} are the normal strains.

In matrix operations, it is convenient to use the double suffix notation and to define $\gamma_{ij}/2$ as ε_{ij}. The strain matrix is then [6]:

$$\varepsilon_{xyz} = \begin{bmatrix} \varepsilon_{xx} & \varepsilon_{xy} & \varepsilon_{xz} \\ \varepsilon_{yx} & \varepsilon_{yy} & \varepsilon_{yz} \\ \varepsilon_{zx} & \varepsilon_{zy} & \varepsilon_{zz} \end{bmatrix} \tag{9.8}$$

The shear strains in the three principal planes of strain are zero, and the normal strains are the principal strains. The greatest and least normal strains at a point are preferably referred to as the major and minor principal strains. The principal strains are determined in a similar manner to principal stresses, that is, as the roots of Eq. (9.3), in which σ and τ are replaced by ε and $\gamma/2$, respectively.

The maximum shear strain, γ_{max}, occurs on a plane whose normal makes an angle of 45° with the ε_1 and ε_3 directions, and is calculated from:

$$\gamma_{max} = \varepsilon_1 - \varepsilon_3 \tag{9.9}$$

where

ε_1 = maximum principal normal strain
ε_3 = minimum principal normal strain

The principal strain tensor is represented as:

$$\varepsilon_p = \begin{bmatrix} \varepsilon_1 & 0 & 0 \\ 0 & \varepsilon_2 & 0 \\ 0 & 0 & \varepsilon_3 \end{bmatrix} \quad (9.10)$$

The sum of the principal strains is the volumetric strain or dilatation, ΔV

$$\Delta V = (\varepsilon_1 + \varepsilon_2 + \varepsilon_3)V \quad (9.11)$$

where V is the initial volume of the rock. Inasmuch as strain is a ratio of volumes or lengths, it is dimensionless.

TWO-DIMENSIONAL STRESS–STRAIN SYSTEMS

Many situations in rock mechanics can be treated as two-dimensional problems in which only the stresses or strains in a single plane need be considered. Poulos and Davis showed that the normal and shear stresses on a plane making an angle θ with the z-direction, as shown in Figure 9.6, are [6]:

$$\sigma_\theta = \frac{1}{2}(\sigma_x + \sigma_z) + \frac{1}{2}(\sigma_x - \sigma_z)\cos 2\theta + \tau_{xz}\sin 2\theta \quad (9.12)$$

$$\tau_\theta = \tau_{xz}\cos 2\theta + \frac{1}{2}(\sigma_x - \sigma_z)\sin 2\theta \quad (9.13)$$

and the principal stresses are:

$$\sigma_1 = \frac{1}{2}(\sigma_x + \sigma_z) + \frac{1}{2}[(\sigma_x - \sigma_z)^2 - 4\tau_{xz}^2]^{0.5} \quad (9.14a)$$

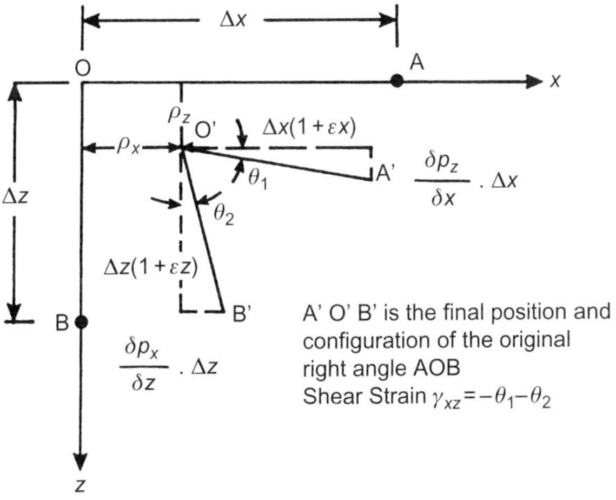

FIGURE 9.6 Shear strain [6].

$$\sigma_2 = \frac{1}{2}(\sigma_x + \sigma_z) - \frac{1}{2}[(\sigma_x - \sigma_z)^2 + 4\tau_{xz}^2]^{0.5} \tag{9.14b}$$

The principal planes are inclined at an angle ($\theta_1 + 90$) to the z-axis, where

$$\theta_1 = \frac{1}{2}\tan^{-1}\left(\frac{2\tau_{xz}}{\sigma_x - \sigma_z}\right) \tag{9.15}$$

The maximum shear stress occurs on a plane inclined at 45° to the principal plane, and is given by:

$$\tau_{\max} = \frac{1}{2}[(\sigma_x - \sigma_z)^2 + 4\tau_{xz}^2]^{0.5} \tag{9.16}$$

The normal strain, ε_q, and shear strain, γ_θ, in a plane inclined at θ_1 to the x-axis, as shown in Figure 9.6 are:

$$\varepsilon_\theta = \frac{\varepsilon_x + \varepsilon_z}{2} + \frac{\varepsilon_x - \varepsilon_z}{2}\cos 2\theta + \frac{\gamma_{xz}}{2}\sin 2\theta \tag{9.17}$$

$$\gamma_\theta = \gamma_{xz}\cos 2\theta - (\varepsilon_x - \varepsilon_z)\sin 2\theta \tag{9.18}$$

It is clear from these equations that in order to completely describe strain, it is necessary to specify not only its magnitude, direction, and sense but also the plane upon which it acts.

ROCK DEFORMATION

The relationship between stress and strain for reservoir rocks is influenced by a large number of factors. Some of these factors are the composition and lithology of rocks, their degrees of cementation and alteration, type of cementing material, amount and type of fluids in the porous space, compressibility of the rock matrix and fluids, porosity and permeability, and reservoir pressure and temperature. Many of these factors are interdependent, and their separate and combined effect on the stress–strain relationship can be measured only in the laboratory, using an actual rock sample from the reservoir and controlling the testing parameters to accurately simulate the in-situ condition. Three measuring and loading techniques are commonly used: hydrostatic, uniaxial, and triaxial. These techniques, which are discussed later, essentially involve applying a specified load and measuring the corresponding strain according to the theory of linear elasticity.

HOOKE'S LAW

If a rock body is subjected to directed forces lasting for a few minutes, hours, or days, it usually passes through four stages of deformation: elastic,

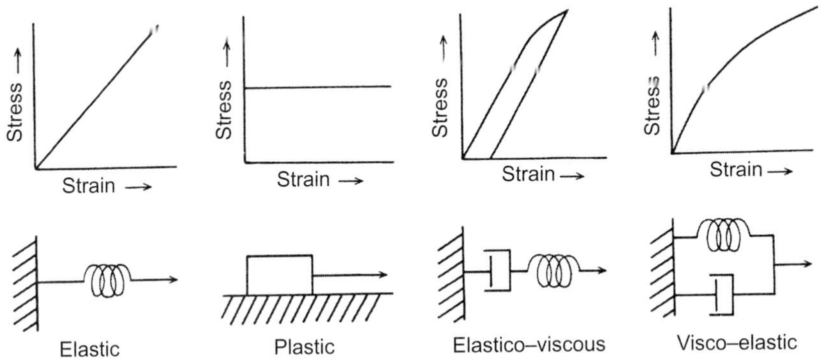

FIGURE 9.7 Stress–strain relationships with related mechanical models [7].

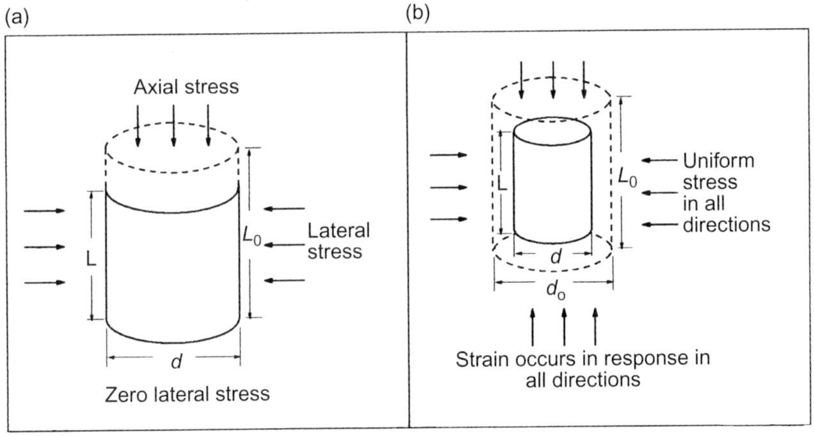

FIGURE 9.8 Compaction tests. (a) Uniaxial-strain compaction test; (b) hydrostatic compaction test.

elastico-viscous, plastic, and rupture. The stages are dependent upon the elasticity, viscosity, and rigidity of the rock, as well as on its stress history, temperature, time, pore pressure, and anisotropy.

At first, the deformation is elastic—that is, if the stress is withdrawn, the body returns to its original shape and size. With purely elastic deformation, the strain is a linear function of stress; that is, the material obeys Hooke's law, as shown in Figure 9.7.

$$\sigma = E\varepsilon \tag{9.19}$$

where E is the modulus of elasticity. E, which is also known as Young's modulus, is a measure of the property of the rock to resist deformation. If a cylindrical rock sample is subjected to stress parallel to its long axis, it will lengthen and the diameter of the cylinder becomes smaller under tension as shown in Figure 9.8. Under compression parallel to the axis, the rock sample will shorten, while its diameter becomes greater. The ratio of transverse or lateral strain to axial strain is known as Poisson's ratio, *n*, or:

$$V = -\frac{\varepsilon_{\text{lat}}}{\varepsilon_{\text{ax}}} = \frac{\Delta d/d_o}{\Delta L/L_o} \tag{9.20}$$

where

d_o = original diameter of cylindrical core sample
Δd = change in diameter
L_o = original length of core
ΔL = change in length
ε_{lat} = strain in the lateral direction
ε_{ax} = strain in the axial direction

Using these terms, Young's modulus can be expressed as:

$$E = \frac{\sigma}{\varepsilon_{\text{ax}}} = \frac{F/A}{\Delta L/L_o} \tag{9.21}$$

where *F/A* is the load per unit area. Another important elastic constant is the modulus of rigidity, *G*, which is a measure of the resistance of a body to change in shape, and is expressed as:

$$G = \frac{\text{shear stress}}{\text{shear strain}} = \frac{\tau}{\gamma} \tag{9.22}$$

Another elastic constant of rocks is the bulk modulus *K*, which is the ratio of change in hydrostatic pressure (stress) to the corresponding volumetric strain:

$$K = \frac{\Delta p}{\Delta V/V_o} \tag{9.23}$$

where Δp is the change in hydrostatic pressure, ΔV is the change in volume, and V_o is the original volume. The bulk modulus is the reciprocal of matrix compressibility, c_r:

$$K = \frac{1}{c_r} \tag{9.24}$$

Table 9.1 shows typical rock elastic constants for several formations in the United States under in situ conditions [8]. Table 9.2 summarizes the physical properties of a formation from the Piedras Negras field in Mexico at a depth of 6,550 ft (1,996 m) under various confining pressures and temperatures. The values of *E* and *n* presented in Table 9.2 are obtained by testing the rock samples both vertically and horizontally to the borehole axis. Table 9.3 shows typical values of *E* and *n* for some typical sedimentary rocks [9].

TABLE 9.1 Typical Rock Elastic Constant [8]

Formation*	Depth (ft)	P_c (psi)	T (°F)	C_o (psi)	E (10^6 psi)	ν
Benoist Sand, IL	1,783	1,100	90	18,000	3.55	0.31
Cotton Valley, TX	9,835	4,000	260	27,500	6.35	0.17
Cotton Valley, LA	11,018	8,000	280	29,000	3.50	0.13
Cotton Valley, LA	11,031	8,000	280	39,000	7.00	0.22
Austin Chalk, TX	7,997	8,000	210	26,800	6.67	0.25

P_c = confining pressure.
C_o = ultimate compressive strength.
All samples have been cored along vertical direction.

TABLE 9.2 In Situ Rock Elastic Constants at 6,550 ft (Piedras Negras Field, Mexico) [8]

P_c (psi)	T (°F)	E (10^6 psi)	ν
(a) Vertical tests for C_o = 36,000 psi			
0	75	2.3	0.14
1,000	75	5.2	0.24
2,000	75	6.3	0.23
4,000	75	6.5	0.30
5,000	75	6.9	0.31
5,000	190	6.4	0.30
(b) Horizontal tests for C_o = 20,000 psi			
0	75	12.4	0.34
1,000	75	11.0	0.27
2,000	75	6.0	0.32
4,000	75	6.3	0.35
5,000	75	6.9	0.36
5,000	200	8.8	0.40

TABLE 9.3 Typical Rock Elastic Constants of Various Rocks [9]

Nature and Origin of Rock	E (bar)	ν	C_o (bar)	Tensile Strength (bar)	(10^{-6} 1^{c}b/bar)
Hassi-Messaoud sandstone	300,000–500,000	0.14–0.21	1,100–1,250	20–90	4–6
El Agreb sandstone	400,000–550,000	≈0.2	1,350–1,550	–	4
Zarzaitine sandstone	450,000	–	–	–	–
Fine Vosges sandstone	125,000	–	≈300	–	–
Coarse Vosges sandstone	235,000	–	400	30–50	–
Fontainebleau sandstone	300,000–400,000	0.15–0.25	600–1,900	≈50	4–6
Clayey sandstone (35% clay)	50,000–90,000	–	700–740	–	–
Bituminous sandstone	30,000–60,000	0.25–0.30	160–260	–	≈30
Saint-Maximin limestone	66,000–82,000	0.19–0.25	90–120	10–12	20–25
Rouffach cornstone	–	–	450–700	–	–
Marquise sandstone	775,000–950,000	0.28–0.33	1,100–1,500	100–140	1.5
Marl	≈80,000–100,000	0.41	–	–	6
Tersanne salt	≈50,000	0.36	150–200	–	15–20

The four elastic constants (9.20, 9.21, 9.22, 9.23) are not independent of each other, and if any two of these are known it is possible to derive the other two from the following expressions:

$$G = \frac{E}{2(1+v)} \tag{9.25}$$

$$K = \frac{E}{3(1-2v)} \tag{9.26}$$

$$E = \frac{9\,KG}{3\,K+G} \tag{9.27}$$

and

$$v = \frac{3K-2G}{2(3K+G)} \tag{9.28}$$

Example

A stress versus deformation curve, shown in Figure 9.9, was obtained by axially loading at 200 lb/s a core in a triaxial cell, and by measuring the total load applied after ~87 s and the resulting deformation in both the axial and lateral

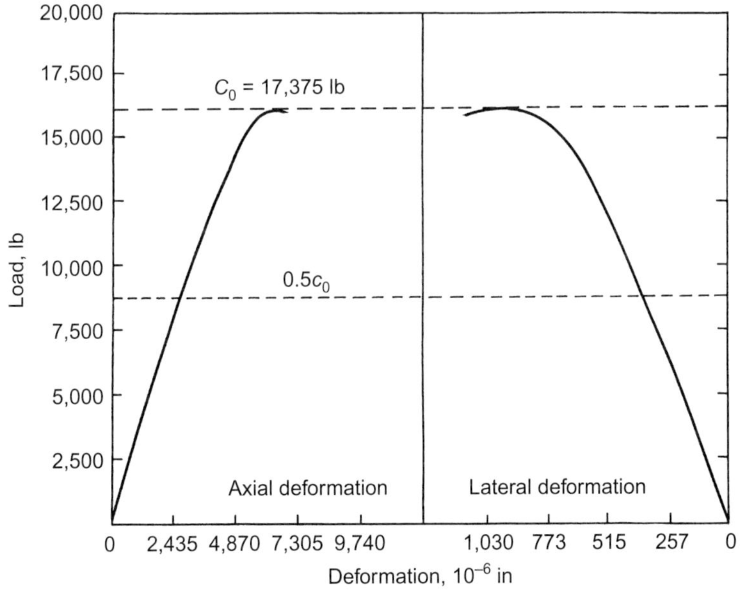

FIGURE 9.9 Load–displacement diagram [8].

directions. Inasmuch as no confining pressure was applied, this test is actually called a uniaxial test. The initial length and diameter of the core are 2.435 and 1.03 in., respectively. The change in diameter is 465×10^{-6} in. and the change in length is $3{,}288 \times 10^{-6}$ in. for a maximum load of 17,375 lb. Calculate:

1. The ultimate strength of the rock, that is, strength where rock fails.
2. Young's modulus, Poisson's ratio, modulus of rigidity, bulk modulus, and rock matrix compressibility.

Solution

The ultimate strength of this rock is:

$$\sigma = \frac{F}{A} = \frac{17{,}375}{\pi(1.03/2)^2} = 20{,}852 \text{ psi}$$

Young's modulus is calculated from Eq. (9.21), where the strain in the axial direction ε_{ax}, is:

$$\varepsilon_{ax} = \frac{\Delta L}{L_o} = \frac{3{,}288 \times 10^{-6}}{2.435} = 1.35 \times 10^{-3}$$

The axial and lateral deformations were measured by strain gauges, which are resistors mounted on the core. Thus

$$E = \frac{20{,}852}{1.35 \times 10^{-3}} = 15.4 \times 10^6 \text{ psi}$$

Poisson's ratio is given by Eq. (9.20), where the strain in the lateral direction is:

$$\varepsilon_{lat} = \frac{\Delta d}{d_0} = \frac{465 \times 10^{-6}}{1.03} = 0.451 \times 10^{-3} \qquad \nu = \frac{0.451}{1.35} = 0.33$$

The modulus of rigidity G is calculated from Eq. (9.25):

$$G = \frac{15.4 \times 10^6}{2(1 + 0.33)} = 5.77 \times 10^6 \text{ psi}$$

Equation (9.26) is used to determine the bulk modulus

$$K = \frac{15.4 \times 10^6}{3(1 - 2 \times 0.33)} = 15.1 \times 10^6 \text{ psi}$$

The rock bulk matrix (grain) compressibility is:

$$c_r = \frac{1}{15.1 \times 10^6} = 0.066 \times 10^{-6} \text{ psi}^{-1}$$

In order to standardize the values of elastic properties measured at different points in the formation, it is recommended that these properties be determined at the same reference stress point, for example, at the 50% point of the ultimate strength, which corresponds to the point of inflection on the stress versus deformation curves [8, 9]. In this case, the 50% compressive strength is:

$$C_{50} = 0.50 \times 20{,}852 = 10{,}426 \text{ psi}$$

$$\sigma_{50} = \frac{C_{50}}{A} = \frac{10{,}426}{0.8332} = 12{,}513 \text{ psi}$$

$$\varepsilon_{ax50} = \frac{\Delta L_{50}}{L_0} = \frac{2{,}800 \times 10^{-6}}{2.435} = 1{,}150 \times 10^{-6}$$

$$E_{50} = \frac{12{,}513}{1.150 \times 10^{-6}} = 10.88 \times 10^{6} \text{ psi}$$

The ΔL_{50} corresponds to the deformation obtained at C_{50}, as shown in Figure 9.9. Similarly:

$$v = \frac{360 \times 10^{-6}/1.03}{1,150 \times 10^{-6}} = 0.30$$

$$G = \frac{10.88 \times 10^{6}}{2(1+0.30)} = 8.17 \times 10^{6} \text{ psi}$$

$$K = \frac{10.88 \times 10^{6}}{3(1-2 \times 0.30)} = 9 \times 10^{6} \text{ psi}$$

and

$$c_r = \frac{10^{-6}}{9} = 0.11 \times 10^{-6} \text{ psi}^{-1}$$

Stress–strain curves are not generally linear, as shown in Figure 9.10. Consequently Young's modulus is not a simple constant, but is related to the level of stress applied.

The change of elastic limit from elastic to plastic deformation is known as the yield point or yield strength. If the stress on a material exceeds its elastic limit, then it is permanently strained, the latter being brought about by plastic flow. Plasticity is defined as time-independent, nonelastic, nonrecoverable, and stress-dependent deformation under uniform sustained load [10]. Although most rocks at room temperatures and pressures fail by rupture before attaining a stage of plastic deformation, at sufficiently high temperatures and confining pressures they deform plastically even in experiments lasting for a short time. Sometimes the term elastico-viscous flow is used to describe "creep" or slow continuous deformation, with the passage of time, which occurs in rocks within the field of plastic flow. Figure 9.11 shows a typical creep curve for rocks displaying four stages of deformation: (A) instantaneous elastic strain, (B) primary or transient creep strain, (C) secondary or steady-state creep, and (D) tertiary or accelerating creep. Primary creep occurs in the early stages of a long-term creep test or at low stresses. Secondary creep is observed in a long-term test or at intermediate stresses. Under continued increases in the stress, microfractures develop and propagate, causing the eventual failure (rupture) of the rock.

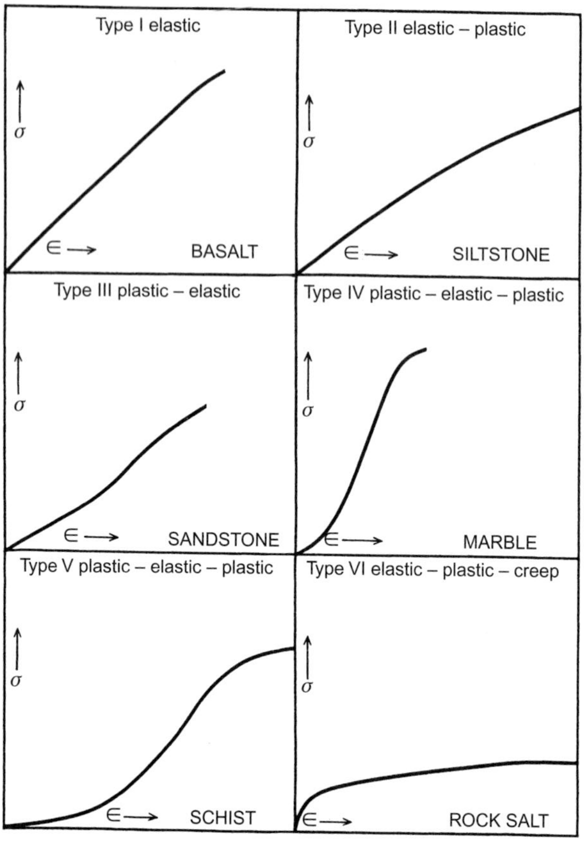

FIGURE 9.10 Typical stress–strain curves for various rocks in uniaxial compression tests [10].

If the rock ruptures before any significant plastic deformation occurs, the rock is described as brittle (Figure 9.12) [11]. Ductile rocks are those that undergo a large degree of plastic deformation before rupture. After the elastic limit has been exceeded, ductile rocks undergo a long interval of plastic deformation, and in some instances they may never rupture. Ruptures may be classified as either tension fractures or shear fractures. Tension fractures result from stresses that tend to pull the rock specimen apart, and when the rock finally breaks, the two walls may move away from each other. Shear fractures result from stresses that tend to slide one part of the specimen past an adjacent part, and when the sample finally breaks, the two walls may slide past one another. The arrangement and form of the fractures depend upon several factors, including homogeneity, isotropy, continuity, and fabric of the rock.

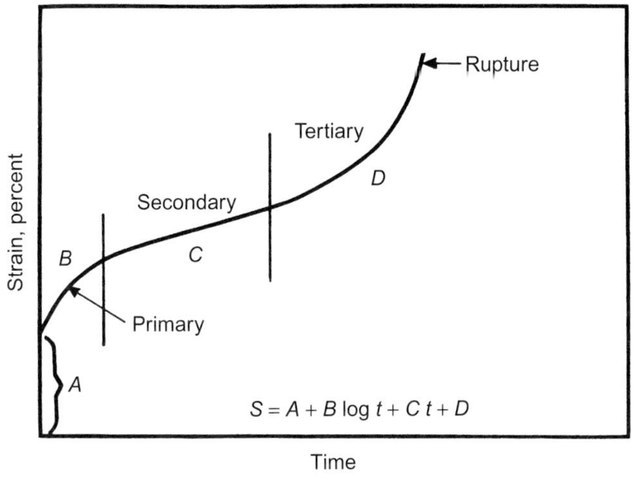

FIGURE 9.11 Ideal creep curves: A = instantaneous deformation; B, C, and D are primary, secondary, and tertiary creep, respectively; s = total strain; t = time.

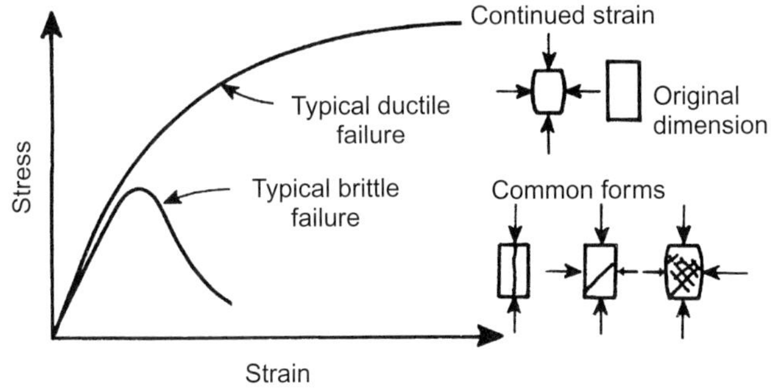

FIGURE 9.12 Typical stress–strain curves showing ductile and brittle failure [11].

STRESS–STRAIN DIAGRAMS

The relation between the stress and strain is commonly expressed in graphs known as stress–strain diagrams [12]. The rock in Figure 9.13 is under compression. With increasing stress the specimen becomes shorter, and the strain (deformation) is plotted in terms of the percentage of shortening of the rock sample. Curve A represents a typical behavior of a brittle rock, which deforms elastically up to a stress of ~20,000 psi (137.9 MPa), shortening 0.5% before rupture. Curve B describes an ideal plastic substance. First, it behaves elastically until reaching the proportional elastic limit, which is the point at which the curve departs from the straight line. Then the rock deforms continuously with any added stress. Curves C and D represent the more typical plastic behavior. Once the elastic limit is reached, rock sample

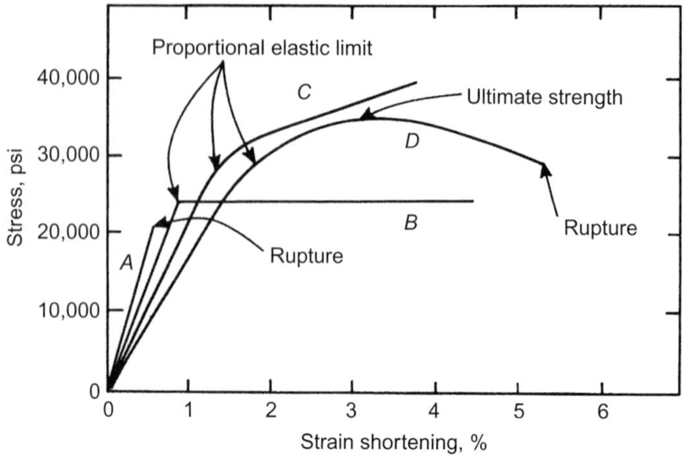

FIGURE 9.13 Stress–strain diagrams [12].

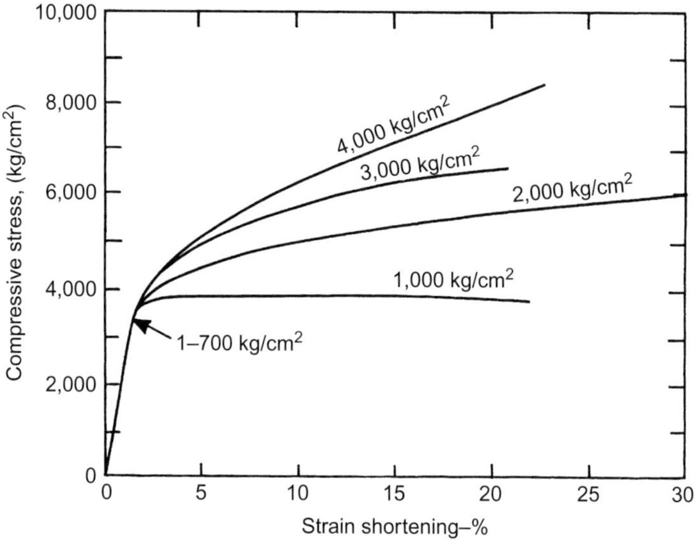

FIGURE 9.14 Effect of confining pressure on stress–strain relationship [13].

C becomes progressively more difficult to deform. With increased stress, rock sample D reaches its ultimate strength point, beyond which less stress is necessary to continue the deformation until rupture.

The mechanical behavior of rocks is controlled not only by their inherent properties such as mineralogy, grain size, porosity, width, and density of fractures, but also by properties including confining pressure, temperature, time, and interstitial fluids. Figure 9.14 illustrates the behavior of limestone

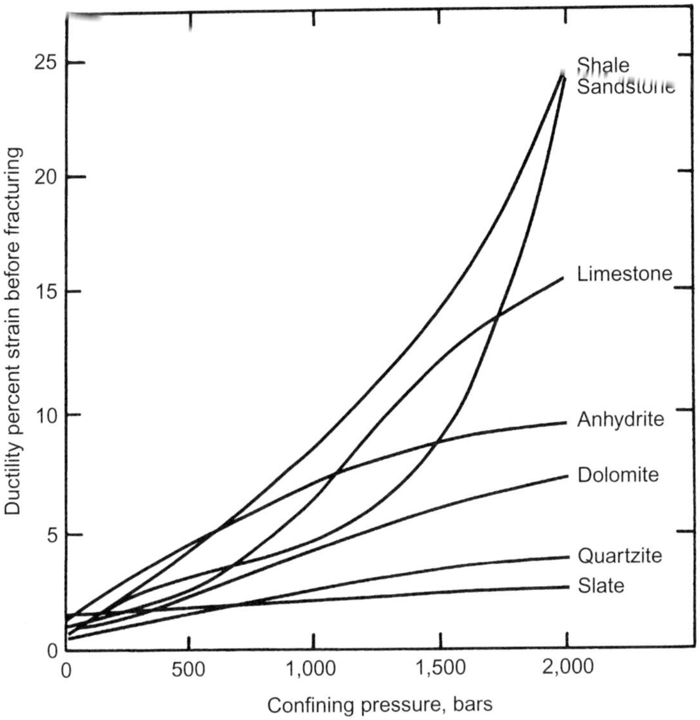

FIGURE 9.15 Effect of confining pressure on ductility of several common rocks [12].

under compression for different confining pressures, in a series of triaxial tests [13]. It is evident that the strength of the rock increases with confining pressure. Such experiments indicate that rocks exhibiting very little plastic deformation near the surface of the Earth may be more plastic under high confining pressure. Thus, under a confining pressure of 1,000 kg/cm^2 or greater, limestone will deform plastically. Figure 9.15 illustrates the effect of confining pressure on the breaking strength of several different rocks [12]. At atmospheric pressure, the rocks deform only a few percent before fracturing. Under a confining pressure of ~1,000 kg/cm^2, the sandstone and shale deform more than 5% before rupturing. Under a confining pressure of ~2,000 kg/cm^2, the limestone deforms nearly 15%, and shale and sandstone more than 20% before rupturing.

Heard showed that changes in temperature modify the strength of rocks [14]. The effects of temperature on sedimentary rocks, however, are of less consequence than the effects of pressure down to depths of 10 km. Nonetheless, with increasing temperatures there is a reduction in yield stress, and strain hardening decreases as shown in Figure 9.16 [15]. Heating particularly enhances the ductility—that is, the ability to deform permanently

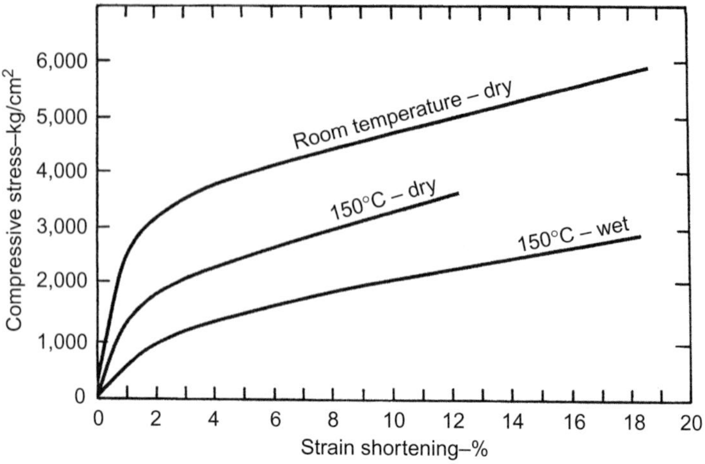

FIGURE 9.16 Effect of temperature on stress–strain curves [15].

without loss of cohesion—of calcareous and evaporitic rocks; however, it has little effect upon sandstones. Much rock deformation takes place while solutions capable of reacting chemically with the rock are present in the pore spaces. This is notably true of metamorphic rocks, in which extensive or complete recrystallization occurs.

The solutions dissolve minerals and precipitate new ones (neoformation). Under such conditions, the mechanical properties of rock are greatly modified. Colback and Wild showed that the compressive strengths of quartzitic shale and quartzitic sandstone under saturated conditions were approximately half what they were under dry conditions [16]. Bernaix showed that the water content reduced the strength of these two quartzitic materials by 30–45% [17].

It is important to emphasize that the discussion and tests described in the preceding sections assumed isotropic materials, that is, rocks whose mechanical properties were uniform in all directions. Reservoir rocks are not isotropic, and their deformation depends upon the orientation of the applied forces to the planar structures of the rock.

THE MOHR DIAGRAM

The relations between stress and rupture for many rocks may be determined graphically by Mohr's stress circles. Consider an imaginary plane through a cylindrical rock specimen inside a triaxial compression chamber (Figure 9.17) [7]. The confining pressure σ_3 is applied and the longitudinal load σ_1 is increased until failure occurs. Continued loading of the rock specimen will cause it to deform via microcracks which, as more loading is applied, extend

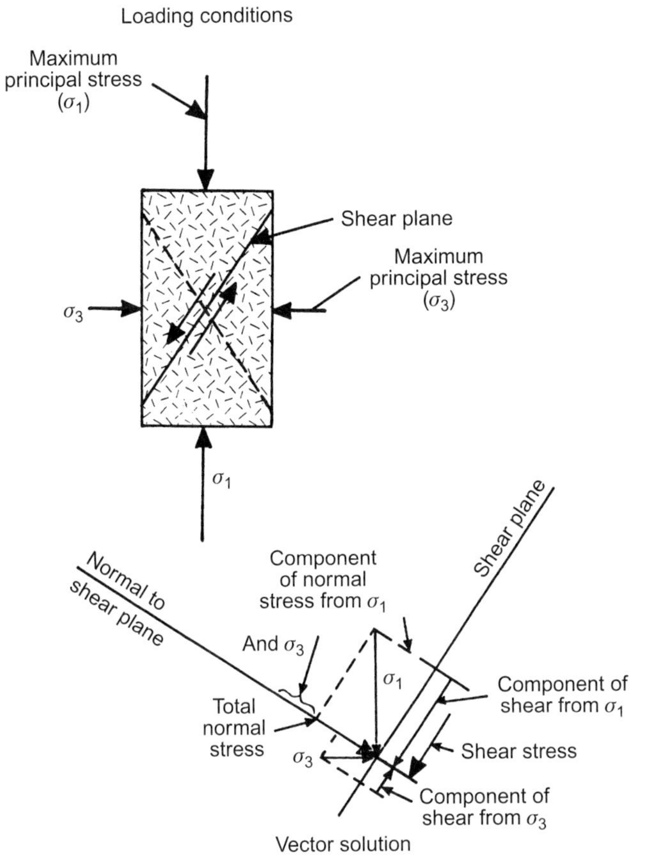

FIGURE 9.17 State of stress along any plane in a rock [7].

and ultimately join together to form a macro-weakness plane (shear plane) along which rupture will occur. At the peak load, the stress conditions are $\sigma_1 = F/A$ and $\sigma_3 = \text{p}$, where F is the highest load supportable parallel to the cylindrical axis and p is the pressure in the confining medium. The stress normal to the failure or crack plane, σ_n, is given by:

$$\sigma_n = \frac{1}{2}(\sigma_1 + \sigma_3) + \frac{1}{2}(\sigma_1 - \sigma_3)\cos 2\theta \tag{9.29}$$

The shear stress parallel to the crack plane, τ, is given by:

$$\tau = \frac{1}{2}(\sigma_1 - \sigma_3)\sin 2\theta \tag{9.30}$$

where θ is the angle between the failure plane and the direction of the minimum principal stress σ_3. Again, failure or rupture is caused by a critical

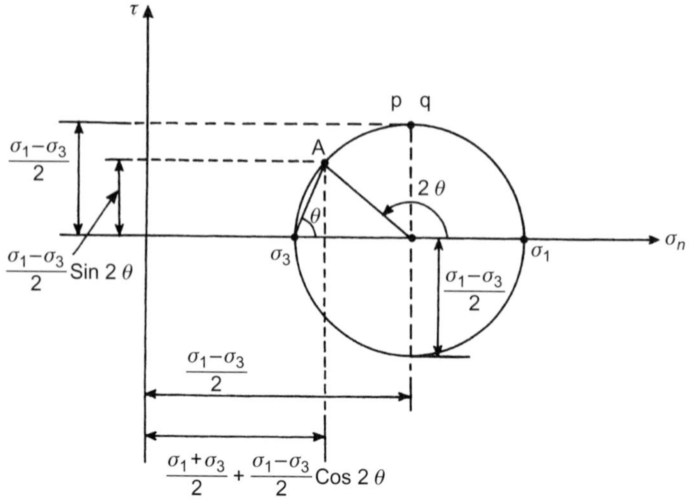

FIGURE 9.18 Mohr circle representation of stress on a plane [5].

combination of both shear and normal stresses. This state can be represented by a point in the plane of τ versus σ_n, known as Mohr's diagram. Figure 9.18 is a Mohr plot of a single run [5]. A circle is drawn through s3 and s1, with the center on the horizontal axis; the center of the circle is $(\sigma_1 + \sigma_3)/2$ and the radius is $(\sigma_1 - \sigma_3)/2$. Inasmuch as an increase in confining pressure will normally increase the strength of the rock specimen (i.e., as the normal stress σ_n increases, the shear stress τ increases), several triaxial tests at increasing confining pressures will lead to several Mohr's circles; each test must be run until rupture occurs. Figure 9.19 is the Mohr diagram for five runs with different stresses [12]. In the first experiment, the confining pressure was atmospheric. Each circle cuts the horizontal axis in two places. In each experiment, the left-hand intersection is the confining pressure whereas the right-hand intersection is the compressive stress causing rupture. The circles show that as the confining pressure is increased, the stress as well as the stress difference $(\sigma_1 - \sigma_3)$ must be increased to produce rupture.

A line drawn tangent to the circle is known as "Mohr's envelope." Stresses that fall within the envelope are below the point of failure, whereas outside the envelope, the stresses will cause failure. The angle that this envelope line makes with the horizontal axis of the diagram (σ_n) is the angle of internal friction ϕ_f. The intercept of the envelope line with the vertical axis, τ_o, is the cohesive strength of the rock. Evaluation of the results obtained from Mohr's stress circle normally assumes the validity of Coulomb's law, which determines the maximum shear stress at which a

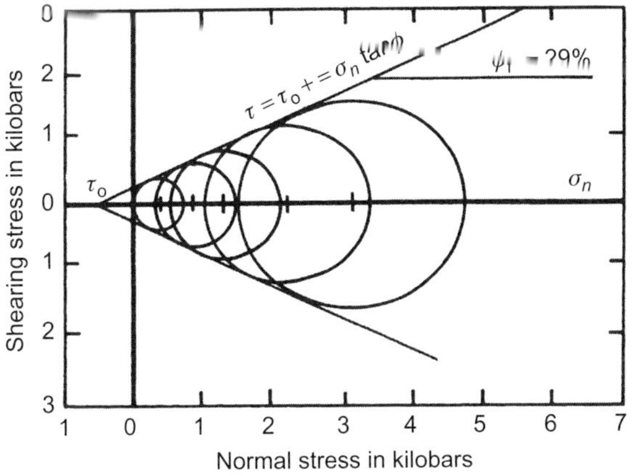

FIGURE 9.19 Mohr's stress envelope [12].

rupture will occur along a plane of weakness in a rock sample. This law can be expressed as:

$$\tau = \tau_o + \sigma_n \tan \phi_f \tag{9.31}$$

The angle that fractures theoretically should make with the greatest principal stress, σ_1, is obtained from:

$$\theta = 45 - 0.5\phi_f \tag{9.32}$$

In the experimental work, it is difficult to measure the fracture angles with great precision. Nevertheless, observations tend to confirm the fracture angles indicated by Mohr's envelopes. Figure 9.20 indicates that the angle of internal friction, ϕ_f, is 29°, the cohesive strength, τ_o, is about 0.35 kbars, and from Eq. (9.32) shear fractures should theoretically form at 31° [18]. One disadvantage of Mohr's method for determining the limits of failure of rocks is that it neglects the effect of the mean principal stress, σ_2, and, consequently, it yields answers that are not always consistent with the experimental results.

DYNAMIC ELASTIC PROPERTIES

Several methods have been used to determine the dynamic values of Young's modulus and Poisson's ratio. Hosking obtained the dynamic values of various elastic properties by determining the velocities of propagation in a rock using ultrasonic pulse, v_p, and the sound of resonance, v_r [19]. Young's modulus is obtained from:

$$E = \frac{\rho_b v_r^2}{12\,g} \tag{9.33}$$

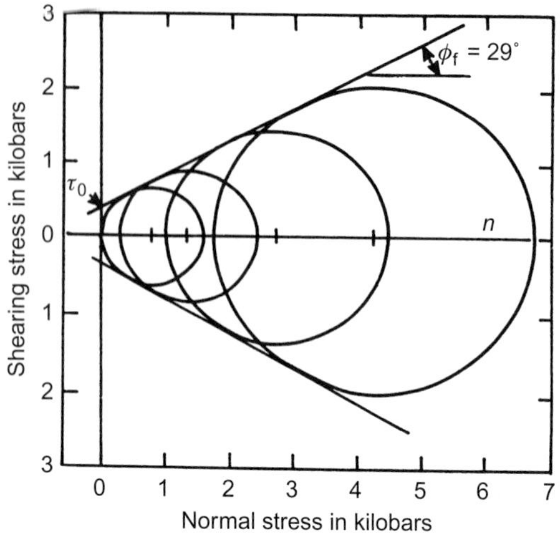

FIGURE 9.20 Mohr stress envelope for Berea sandstone at 24°C with confining pressures of 0, 0.05, 4, and 2 kilobars [12].

and Poisson's ratio from:

$$\frac{v_P}{v_r} = \frac{1 - v}{(1 + v)(1 - 2v)} \tag{9.34}$$

where ρ_b is the rock bulk density and g is the acceleration due to gravity. Deere and Miller used a similar method to derive the following relationship between compressional wave velocity (v_c), density of the rock (ρ_b), Young's modulus, and Poisson's ratio [20]:

$$v_c = \frac{\rho_b E(1 - v)}{(1 + v)(1 - 2v)} \tag{9.35}$$

Sonic logging and wave-form analysis provide the means for obtaining continuous measurements of compressional and shear velocities. These data, in conjunction with bulk density measurements, permit the in situ measurement and calculation of the mechanical properties of the rock [21]. Table 9.4 shows the elastic moduli relationships in terms of transit times and bulk density. The units applicable to the table are:

ρ_b = bulk density, g/cm^3
v_s = shear velocity, ft/s
Coefficient $a = 1.34 \times 10^{10}$ if ρ_b is in g/cm^3 and t is in μs/ft
v_c = compressional velocity, ft/s
t_s = shear transit time, μs/ft
t_c = compressional transit time, μs/ft

TABLE 9.4 Dynamic Elastic Properties [21]

Poisson's ratio, v	$\frac{\text{Lateral strain}}{\text{Longitudinal strain}}$	$\frac{1/2(t_s/t_c)^2-1}{(t_s/t_c)^2-1}$
Shear, G, psi	$\frac{\text{Applied stress}}{\text{Shear strain}}$	$\frac{\rho_b}{t_s^2} \times a$
Young's modulus, E, psi	$\frac{\text{Applied uniaxial stress}}{\text{Normal strain}}$	$2G(1+v)$
Bulk modulus, K_b, psi	$\frac{\text{Hydrostatic pressure}}{\text{Volumetric strain}}$	$\rho_b\left(\frac{1}{t_c^2}-\frac{4}{3t_s^2}\right) \times a$
Bulk compressibility (with porosity), C_b, psi^{-1}	$\frac{\text{Volumetric deformation}}{\text{Hydrostatic pressure}}$	$\frac{1}{K_b}$
Rock compressibility (zero porosity), C_r, psi^{-1}	$\frac{\text{Change in matrix volume}}{\text{Hydrostatic pressure}}$	$\rho_g\left(\frac{1}{t_{ma}^2}-\frac{4}{3t_{sma}^2}\right) \times a$
Biot elastic constant, α	Pore pressure proportionality	$1-\frac{C_r}{C_b}$

ε = strain, μin./in

σ = stress, psi

Example

A rock sample 2 in. in diameter, 5 in. long, and with a bulk density of 0.3 g/cm^3 is subjected to a static loading test. The load is 5,000 lb, the axial contraction $\Delta L = 0.01$ in., and diametrical expansion $\Delta d = 0.001$ in. From well log analysis, the same sample had a compressional wave travel time $t_c = 50$ μs/ft and a shear travel time $t_s = 80$ μs/ft. Calculate the static and dynamic values of Young's modulus and Poisson's ratio.

Solution

Using Eqs. (9.20) and (9.21), the static Young's modulus and Poisson's ratio are, respectively:

$$E = \frac{F/A}{\Delta L/L_o} = \frac{5,000/(\pi(2/2)^2)}{0.01/5} = 7.9 \times 10^5 \text{ psi}$$

$$\nu = -\frac{\varepsilon_{lat}}{\varepsilon_{ax}} = \frac{\Delta d/d_o}{\Delta L/L_o} = \frac{0.001/2}{0.01/5} = 0.25$$

The dynamic Young's modulus and Poisson's ratio are obtained from Table 9.4:

$$\upsilon = \frac{0.5(t_s/t_c)^2-1}{(t_s/t_c)^2-1} = \frac{0.5(80/50)^2-1}{(80/50)^2-1} = 0.18$$

$$E = 2G(1+\nu) = \frac{2a\rho_b}{t_s^2}(1+\nu) = \frac{2 \times 1.34 \times 10^{10} \times 0.3}{80^2}(1+0.18)$$

$$= 14.8 \times 10^5 \text{ psi}$$

DYNAMIC VERSUS STATIC MODULUS OF ELASTICITY

Generally, dynamic elastic constants are derived from the measurements of elastic wave velocities in rocks. Equipment such as geophones and seismographs can be used to measure in situ values of wave velocities. Inasmuch as the static moduli are required for design of most rock engineering projects, dynamic moduli measurements are not common. Also, equations used to calculate dynamic moduli assume ideal rock; that is, the rock is perfectly linear elastic, homogeneous, and isotropic. Reservoir rocks, of course, are not ideal. This causes the values of static and dynamic moduli to be different. Therefore, dynamic testing techniques will provide meaningful design data if dynamic moduli values can be converted into static values. Many studies [22] aimed at establishing a relation between dynamic and static moduli showed that the dynamic modulus of elasticity, E_s, tends to be greater than its static equivalent, E_s, and that ν_d is slightly smaller than ν_s. Using experimental data, the following correlation is obtained:

$$E_s = \exp\left[\frac{1.843 \ln E_d}{\sigma^{0.0724}} + 0.45 \ln(0.0266\sigma)\right] \tag{9.36}$$

where σ is the uniaxial compressive strength in MPa, and E_d and E_s are expressed in GPa. This correlation makes it possible to predict approximate values of the static Young's modulus from dynamic measurements for most reservoir pressures. Ten different types of rock materials were used to derive this correlation, including several sandstones, quartzites, and magnetite. If the dynamic value of Poisson's ratio ν_d is assumed to be equal to the static value ν_s, then the bulk modulus K and the modulus of rigidity G can be calculated from Eqs. (9.26) and (9.25), respectively.

ROCK STRENGTH AND HARDNESS

Strength is the ability of rock to resist stress without yielding or fracturing. It is influenced by the mineralogy of the rock particles and by the character of the particle contacts. These properties are the result of the various processes of deposition, diagenesis, and catagenesis that formed the rock, later modified by folding, faulting, fracturing, jointing, and weathering. Consequently, the strength of rocks reflects their geological history. Rock strength is estimated from two common laboratory techniques: uniaxial compressive strength tests and triaxial or confined compressive strength tests.

Uniaxial compressive strength tests are used to determine the ultimate strength of a rock, that is, the maximum value of stress attained before failure. The uniaxial strength is one of the simplest measures of strength to obtain. Its application is limited; however, it is generally used only when comparisons between rocks are needed. Uniaxial compression tests are influenced by several factors: size and shape of the test sample, rate of loading, amounts and types of fluid present in the rock sample, mineralogy, grain size, grain shape, grain sorting, and rate of loading.

The effects of these factors can be minimized by taking the following precautions [23]:

1. The length-to-diameter ratio, also called the slenderness ratio, of the rock sample should be approximately 2 to 1.
2. The ends of the sample should be parallel and ground flat to within 0.025 mm; otherwise, low values of compressive strength are obtained.
3. Size effects are considerable only if flaws exist in the rock sample: The larger the sample, the greater the probability of a flaw existing in the sample. Size effects can be reduced by testing a large number of samples with the same size and calculating the average, preferably the geometric mean, of compressive strength values.
4. Because fluid content could reduce the compressive strength, it is recommended to perform the uniaxial test under fluid saturations similar to those existing in the reservoir. Reduction in compressive strength due to the presence of fluids could occur in several ways. It is probable, however, that in many rocks the effect of pore pressure is the main cause of reduction in rock strength. The pore pressure could affect the intergranular contact stresses and cause instability along a weakness plane.
5. High rates of loading should be avoided, as they tend to yield abnormally high compressive strength values. Loading rates in the range of 0.5–3 MPa/s are considered normal and generally cause negligible change in compressive strength of rock samples.

Sometimes, only approximate compressive strength values are needed, in which case three testing techniques are available: point-load test, Protodyakonov test, and Brazilian test.

The point-load test is a widely accepted test for direct and quick evaluations of the rock strength of drill cores and irregular rock fragments. In the point-load test, the rock specimen of diameter d is placed between opposing cones and subjected to a compression load F_a, at a distance of at least $0.7d$ from either end (Figure 9.21). The load generates tensile stress normal to the axis of loading.

The determination of the direct tensile strength has proved complicated because of shape and size effect. The following empirical point-load equation, however, gives a good estimation of the tensile strength of rock [8]:

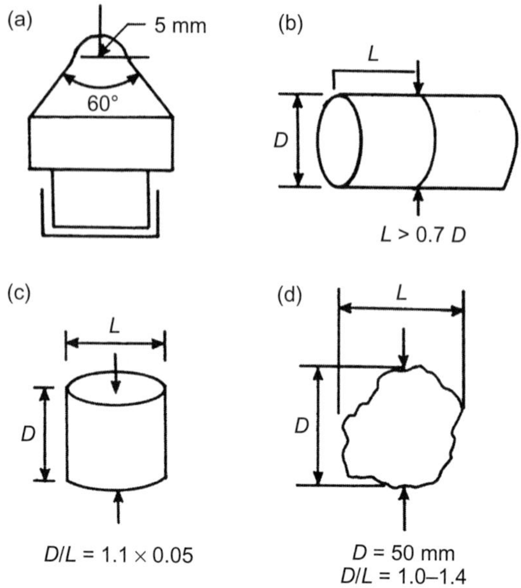

FIGURE 9.21 Definition of length of rock sample for the point-load test [23].

$$C_t = \frac{S_a F_a}{(L_s - 1.7F_a/C_u)^2} \tag{9.37}$$

where

C_t = tensile strength ($\approx F_a/L_s$), psi
S_a = shape factor [24], 0.70 L_s/d
F_a = applied load, psi
L_s = length as defined in Figure 9.21

C_u is the uniaxial, or unconfined, compressive strength, which can be estimated from the following equation:

$$C_u = S_s \frac{F_a}{L_s^2} = \left(\frac{S_s}{L_s}\right) C_t \tag{9.38a}$$

The size correction factor, S_s, can be estimated from the following correlation [25]:

$$S_s = 0.18d + 14 \tag{9.38b}$$

where d is in millimeters.

The Protodyakonov test is a widely accepted test for direct and quick evaluations of the rock strength of drill cores and irregular rock fragments.

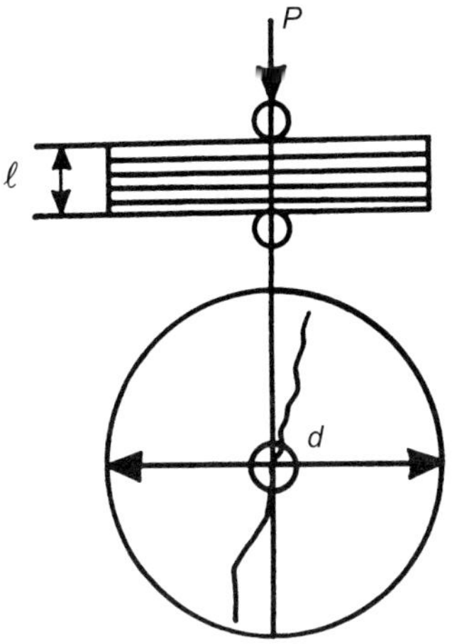

FIGURE 9.22 Protodyakonov test [23].

In the Protodyakonov test, the specimen is placed between opposing cones as shown in Figure 9.22 and subjected to compression, which generates tensile stress normal to the axis of loading.

The Brazilian test is another indirect method for obtaining the tensile strength of rocks. In this test, a core sample of length L and diameter d is subjected to a load F_a in a diametrical plane along its axis. The core sample generally ruptures along the line of diametrical loading, and the tensile strength C_t is calculated from:

$$C_t = \frac{2F_a}{\pi L_s d} \tag{9.39}$$

The Brazilian test is based on the fact that most rocks under biaxial stress fail in tension when the principal stress is compressive. This test is much more accurate for brittle rocks than ductile rocks. Rocks have a much higher compressive strength than tensile stress, about 8:1 in theory, but in practice it is generally 15:1 to 25:1. The uniaxial compressive strength can be also calculated from:

$$C_u = \left(\frac{2\cos\phi_f}{1-\sin\phi_f}\right)C_{ti} \tag{9.40}$$

where ϕ_f is the angle of friction in the Mohr–Coulomb failure model, generally set at about 29° or 30° and the initial tensile strength C_{ti} is measured from the following empirical equation:

$$C_{ti} = 2.6\times10^{-8}\frac{E}{C_b}[0.0035\,V_{sh} + 0.0045] \tag{9.41}$$

where V_{sh} is the shale fraction determined from well logs and C_b is the bulk compressibility of the rock sample.

Example

A core, 2.13 in. in diameter and 30 in. in length, has been recovered from a 5,000 ft deep sandstone. In order to run a point-load test, ten 3-in. long cores have been cut. The following failure loads (in psi) were obtained: 3,420; 3,150; 2,950; 3,280; 2,825; 3,410; 2,780; 3,050; 2,950; and 3,310. Calculate the following elastic properties: (1) compressive strength, (2) tensile strength, and (3) Young's modulus.

Solution

The compressive strength of the rock sample can be evaluated from Eq. (9.37), where the mean of failure load is equal to:

$$F_a = \frac{1}{10}[3{,}420 + 3{,}150 + \cdots + 3{,}310] = 3{,}112 \text{ lb}$$

The length of the core is the same as its diameter, that is, $L_s = 2.13$ in. The size correction factor, S_s, is estimated from Eq. (9.38b). For core diameter of 2.13 in. (54.1 mm), $S_s = 23.74$. Thus

$$C_u = S_S \frac{F_a}{L_s^2} = 23.74 \frac{3{,}112}{2.13^2} = 16{,}284 \text{ psi}$$

The tensile strength of the sandstone is calculated from Eq. (9.38), where the shape factor $S_a = 0.79$ for a disk-shaped core. Thus

$$C_t = \frac{S_a F_a}{(L_s - 1.7F_a/C_u)^2} = \frac{0.79 \times 3{,}112}{(2.13 - 1.7 \times 3{,}112/16{,}284)^2} = 754.4 \text{ psi}$$

Young's modulus can be estimated from Figure 9.23. Because the uniaxial compressive strength is about 16×10^3 psi and the core samples are sandstone (line 2 in Figure 9.23), Young's modulus E is 2×10^6 to 6×10^6 psi.

ROCK HARDNESS

Rock hardness is measured by the Brinell spherical indenter. Hardness determinations require only a small amount of core material. The Brinell rock hardness number, N_{Br}, is defined as the ratio of applied load F_a on the indenter (sphere of radius, r_s) to the indentation depth D_i:

$$N_{Br} = \frac{F_a}{2\pi r_s D_i} \tag{9.42}$$

or

$$N_{Br} = \frac{F_a}{\pi r_{is}^2} \tag{9.43a}$$

where r_{is} is the radius of the indentation circle. N_{Br} is a rock constant, but the sphere diameter can influence the measured hardness. This influence, however, is small in most cases and negligible in others.

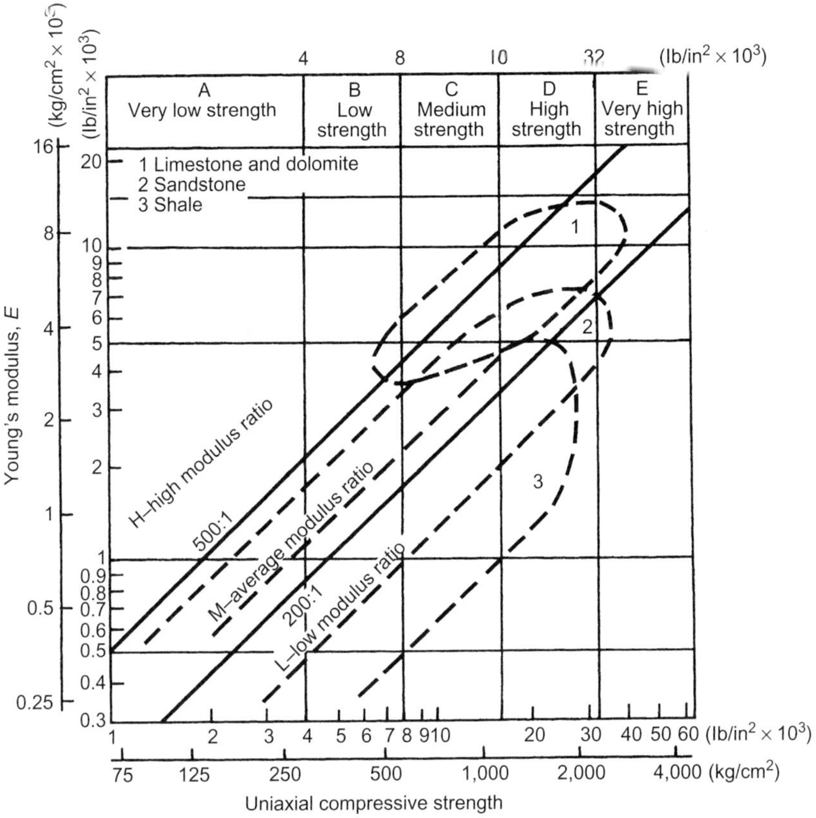

FIGURE 9.23 Young's modulus for three common reservoir rocks [23].

For well-consolidated rocks, N_{Br} may be estimated from the following empirical correlation:

$$N_{Br} = \frac{E}{77.25} \tag{9.43b}$$

The Brinell rock hardness number for the rock sample in the Example on page 588 is:

$$N_{Br} = \frac{14.8 \times 10^6}{77.25} = 1.9 \times 10^5 \text{ psi}$$

COMPRESSIBILITY OF POROUS ROCKS

Reservoir rocks are subjected to the internal stress exerted by fluids in the pores, and to external stress which is in part exerted by the overlying rocks.

The depletion of fluids from the reservoir rocks results in a change in the internal (hydrostatic) stress in the formation, thus causing the rock to be subjected to an increased and variable overburden load, and the result is the compaction of the rock structure due to an increase in the effective stress. This compaction results in changes in the grain, pore, and bulk volume of the rock. The fractional change in the volume of solid rock constituent (grains) per unit change in pressure is defined as the rock matrix compressibility. The fractional change in the total or bulk volume of the formation per unit change in the reservoir pressure is called the rock bulk compressibility. Of principal interest to the reservoir engineer is the pore compressibility, which is the fractional change in the pore volume per unit change in pressure. In areas where fluid withdrawal from underground reservoirs could induce subsidence that might result in the loss of wells, appreciable property damage, or earthquakes, the bulk compressibility is very important.

PORE COMPRESSIBILITY

Many researchers have recorded the changes in compressibility of reservoir rocks as a function of fluid pressure decline [26–45]. Biot published a theory of elastic deformations of porous materials and their influence on fluid displacement within the pores [26]. Geertsma was, however, the first engineer to develop a set of practical pressure–volume relationships explaining pore and rock bulk volume variations in petroleum reservoirs [39]. He derived the following general expressions:

$$\frac{dV_p}{V_p} = c_r\, dp_p + \frac{1}{\phi}(c_b - c_r)d(\sigma - p_p) \tag{9.44}$$

and

$$\frac{dV_b}{V_b} = (c_r - c_b)dp_p + c_b\, dp = c_r\, dp_p + c_b\, d(\sigma - p) \tag{9.45}$$

where

V_p = pore volume
V_b = bulk volume
c_r = compressibility of the rock matrix material
c_b = rock bulk compressibility of the porous structure
p_p = pore pressure (internal pressure)
σ = confining stress (external pressure)

Equation (9.45) can be simplified by maintaining a constant difference between the confining stress σ and the pore pressure p_p

during the triaxial, or hydrostatic, test, that is, $d(\sigma - p_p) = 0$. Solving for c_r one obtains:

$$c_r = \frac{1}{V_b}\left(\frac{dV_b}{dp_p}\right) \tag{9.46}$$

On the other hand, if the pore pressure is held constant during the triaxial test, that is, $dp_p = 0$, Eq. (9.45) gives:

$$c_b = \frac{1}{V_b}\left(\frac{dV_b}{d\sigma}\right) \tag{9.47}$$

For $dp_p = -d\sigma$, Eq. (9.44) yields an expression of the formation pore compressibility:

$$c_\gamma = \frac{1}{V_p}\left(\frac{dV_p}{dp_p}\right) \tag{9.48}$$

and for $dp_p = 0$, the pore compressibility (c_p) is:

$$c_p = \frac{c_b - c_r}{\phi} = \frac{1}{V_p}\frac{dV_p}{d\sigma} \tag{9.49}$$

Assuming the rock bulk compressibility c_b is much greater than the rock matrix compressibility c_r, such as is the case in many consolidated sandstone reservoirs with $\phi > 0.05$, Eq. (9.49) becomes:

$$\frac{c_b}{\phi} = \frac{1}{V_p}\frac{dV_p}{d\sigma} \tag{9.50}$$

Comparing Eq. (9.48) with Eq. (9.50) for the case where $dp = d\sigma$ during a hydrostatic test, one obtains:

$$c_p = \frac{c_b}{\phi} \tag{9.51}$$

Geertsma derived a similar expression by showing that, in petroleum reservoirs, only the vertical component of hydrostatic stress is constant and that the stress components in the horizontal plane are characterized by the boundary condition [39]. For these boundary conditions, he developed the following approximation for sandstones:

$$c_p = \frac{2}{v_p}\frac{dv_p}{dp_p} \tag{9.52}$$

Thus, it is important to remember that the pore compressibilities of sandstones obtained on using triaxial apparatus are about twice as high as those obtained in a uniaxial test.

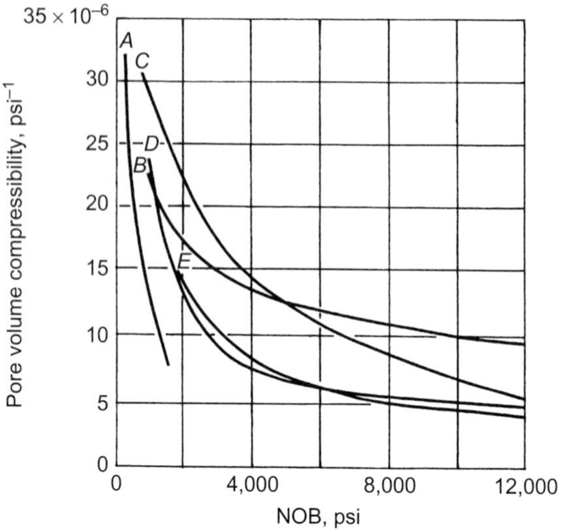

FIGURE 9.24 Influence of net overburden pressure on pore-volume compressibility for poorly sorted unconsolidated sand, curve A ($\phi = 0.36$) and for sandstone, curves B ($\phi = 0.13$), C ($\phi = 0.15$) and E and D ($\phi = 0.12$) [27].

Fatt reported results of experimental tests on a limited number of representative consolidated rock samples and sandpacks [27]. Figure 9.24 shows the pore-volume compressibilities as a function of net overburden pressure for sandstones containing poorly sorted grains, 20–45% cement, and intergranular detrital material. These compressibilities are higher than those given in Figure 9.25, which are for sandstones containing well-sorted grains and only 10–30% cement and intergranular detrital material.

Brandt defines the net overburden pressure as the external pressure, σ, minus 85% of the internal fluid pressure [28]. The constant 0.85 was introduced to take into account the fact that the internal fluid pressure does not really react against the external pressure. This constant depends on the structure of the rock, and ranges from 75% to 100% with an average of 85%.

Fatt's results showed the absence of correlation between the compressibility and porosity data [27]. This is contrary to the conclusion reached by Hall, who stated that, as the reservoir pressure declines, the pore compressibility of any reservoir rock is a result of two separate factors: expansion of the individual rock grains and the additional formation compaction brought about because the reservoir fluid becomes less effective in opposing the weight of the overburden [29]. Both of these factors, according to Hall, tend to decrease porosity as shown in Figure 9.26. Most limestone and sandstone formations have pore compressibilities of the order of 10^{-6} to 25×10^{-6} psi^{-1}.

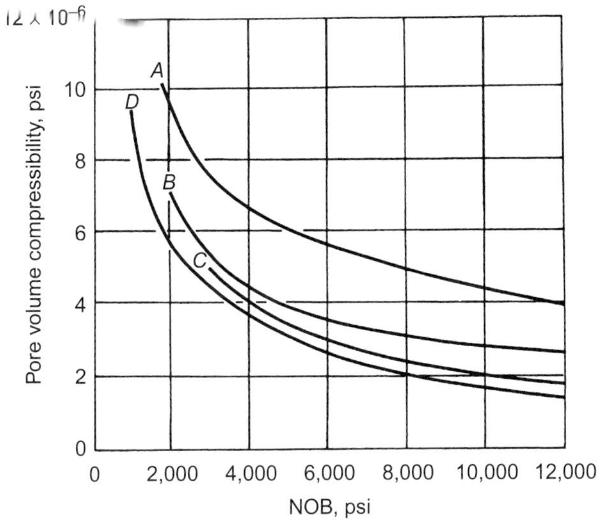

FIGURE 9.25 Influence of net overburden pressure on pore-volume compressibility for well-sorted sandstone grains [27].

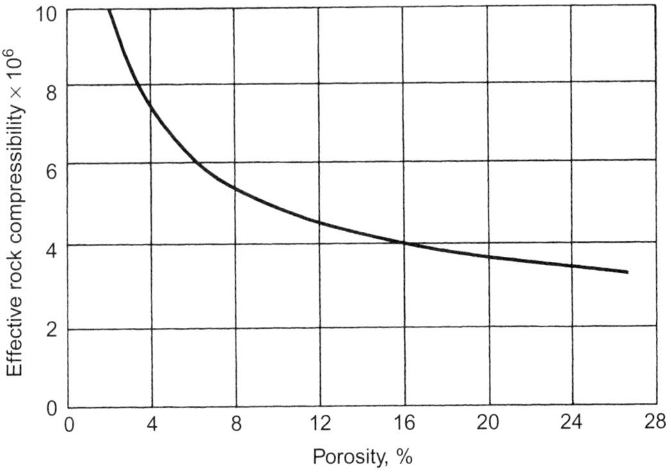

FIGURE 9.26 Relationship between effective rock compressibility and porosity [29].

The magnitude of the pore-pressure change, Δp_p, in the petroleum reservoirs for a given change in pore volume depends on the compressibility of the pore fluid, c_{fl}:

$$\Delta p_p = -\frac{1}{c_{fl}} \frac{\Delta v_p}{v_p} \tag{9.53}$$

The relative change in pore volume can be expressed from Eq. (9.44) as follows:

$$\frac{\Delta v_p}{v_p} = \frac{1}{\phi}[c_b - (1+\phi)c_r]\Delta p_p - \frac{1}{\phi}(c_b - c_r)\Delta\overline{\sigma} \tag{9.54a}$$

where $\overline{\sigma}$ is the mean effective stress, p_p is the pore pressure, and:

$$\overline{\sigma} = \frac{\sigma_1 + \sigma_2 + \sigma_3}{3} = \frac{\sigma_x + \sigma_y + \sigma_z}{3} \tag{9.54b}$$

where

$$\overline{\sigma} = \frac{(\sigma_1 + \sigma_2 + \sigma_3)}{3}$$

The change in pore pressure for a given change in the mean stress is obtained by combining Eq. (9.53) with Eq. (9.54a), and solving for Δp_p (fluid in the pores is assumed to be water, $c_{fl} = c_w$):

$$\Delta p_p = \frac{(c_b - c_r)\Delta\overline{\sigma}}{\phi c_w + [c_b - (1+\phi)c_r]} \tag{9.55}$$

Example

Calculate the change in pore pressure of a rock sample subjected to a hydrostatic test using the given data:

$\overline{\sigma} = 6{,}000$ psi	$c_r = 0.20 \times 10^{-6}$ psi^{-1}
$p_p = 4{,}700$ psi	$c_b = 2 \times 10^{-6}$ psi^{-1}
$\phi = 0.20$	$c_w = 2.75 \times 10^{-6}$ psi^{-1}

Solution

The mean effective stress change is given by:

$$\Delta\overline{\sigma} = 6{,}000 - 4{,}700 = 1{,}300 \text{ psi}$$

Change in the pore pressure is (from Eq. (9.55)):

$$\Delta p_p = \frac{(2 - 0.2) \times 10^{-6} \times 1{,}300}{0.2 \times 2.75 \times 10^{-6} + (2 \times 10^{-6} - 1.2 \times 0.2 \times 10^{-6})} = 1{,}014 \text{ psi}$$

EFFECTIVENESS OF PORE PRESSURE IN COUNTERING STRESS

The tests that measure stress–strain behavior of rocks described in the earlier sections assumed that pore pressure of the fluids developed during loading and might dissipate as the fluids are drained. Such tests are called drained tests, whereas the experiments in which drainage is prevented are called undrained tests [42]. Silt, sand, gravel, and other sediments and sedimentary

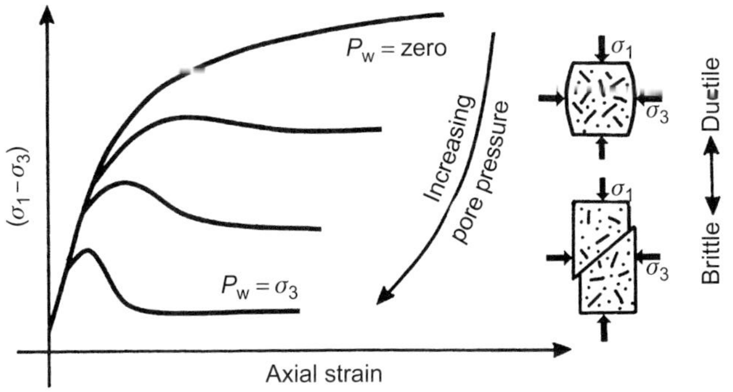

FIGURE 9.27 Transition from brittle to ductile failure in rock as a function of pore water pressure (p_w) [11].

rocks of relatively high permeability exhibit, when a load is applied, drained strength; that is, the rock is soft or ductile. This effect is more pronounced if the rock is saturated with an incompressible liquid than with a compressible gas [8]. Clay, shale, and many rocks of low permeability such as unweathered igneous rocks, exhibit undrained strength; when initially and instantaneously loaded, they become brittle and much weaker. Figure 9.27 illustrates the effect of pore pressure on the elastic behavior of rocks.

The effect of pore pressure on the mechanical properties of saturated rocks has been extensively investigated by using the concept of "effective stress," which Van Terzaghi defined as the stress controlling compression or shear in rocks, and is simply the difference between the applied overburden load or the total stress, σ_{OB}, and the pore pressure, p_p [31]:

$$\sigma_e = \sigma_{OB} - p_p \tag{9.56a}$$

or

$$p_e = p_t - p_p$$

Laboratory experiments and careful field observations of the deformation have led to the modification of this relationship as follows:

$$\sigma_e = \sigma_{OB} - \alpha p_p \tag{9.56b}$$

where α, according to Brandt, is a correction factor that measures the effectiveness of the pore pressure in counteracting the total applied load [28]. According to the Biot poroelasticity theory, the radial or horizontal Earth stress is:

$$\sigma_x = \left(\frac{\nu}{1-\nu}\right)\sigma_e + \alpha p_p \tag{9.57}$$

where the Earth's effective vertical or overburden stress is obtained from Eq. (9.56b).

According to Brandt, the value of α, which varies between 0 and 1, depends on the pore geometry and the physical properties of the constituents of the solid system. In the extreme cases when $\alpha = 0$, the pore pressure has no effect on the behavior of the rock, and when $\alpha = 1$ the pore pressure is 100% effective in counteracting the applied load. Equation (9.56a), when $\alpha = 1$, is used to evaluate the failure magnitude, and Eq. (9.56b) is used by some investigators to evaluate the deformation of the porous medium. Geertsma and Skempton respectively proposed the following expression for α [39, 40]:

$$\alpha = 1 - \frac{K}{K_r} = 1 - \frac{c_r}{c_b} \tag{9.58}$$

where

c_b = bulk compressibility, psi^{-1} [1/kPa]
c_r = compressibility of the rock matrix, psi^{-1} [1/kPa]
K = effective bulk modulus, psi [1/kPa]
K_r = bulk modulus of the rock solid only, psi [1/kPa]

The value of α, which is known as the Biot constant, in the previous example is given by:

$$\alpha = 1 - \frac{0.2 \times 10^{-6}}{2 \times 10^{-6}} = 0.90$$

Geertsma's derivation of Eq. (9.58) has been modified by Suklje to include porosity, which showed that [41]:

$$\alpha = 1 - (1 - \phi)\frac{K}{K_r} \tag{9.59}$$

where ϕ is the porosity. It is obvious from Eq. (9.59) that when $\alpha = 0$, that is, $\sigma = \sigma_1$, the K/K_r ratio is equivalent to:

$$\frac{K}{K_r} = \frac{1}{1 - \phi} \tag{9.60}$$

Equation (9.58) is, therefore, valid only in the ideal case where there is no porosity change under equal variation of pore pressure and confining pressure. The following expression is more general and accurate:

$$\alpha = \frac{3(v_u - v)}{B(1 - 2v)(1 + v_u)} \tag{9.61}$$

where

v = drained Poisson's ratio
v_u = undrained Poisson's ratio
B = Skempton's pore-pressure coefficient

In liquid-saturated compressible rocks, $B = 1.0$; when the porous space is partially saturated, $B < 1.0$; and $B = 0$ when the rock specimen is dry. Skempton showed that the value of B can be estimated from [42].

$$B = \frac{1}{(1 + c_r/\phi c_{fl})} \tag{9.62}$$

where c_r and c_{fl} are the compressibilities of the rock and the compressibility of the fluids (water) in the void space, respectively.

Example

Given the following data, calculate:

1. The bulk modulus of the rock sample
2. The correction factor α using Eqs. (9.58–9.61)

$\nu = 0.250$	$E = 2 \times 10^6$ psi	$c_w = 2.75 \times 10^{-6}$ psi^{-1}
$\nu_u = 0.362$	$K_r = 4.83 \times 10^6$ psi	$\phi = 0.20$

Solution

1. The bulk modulus of the rock sample is:

$$K = \frac{E}{3(1-2\nu)} = \frac{2 \times 10^6}{3(1 - 2 \times 0.25)} = 1.33 \times 10^6 \text{ psi}$$

2. The correction factor α

(a) Using Eq. (9.58) gives:

$$\alpha = 1 - \frac{1.33 \times 10^6}{4.83 \times 10^6} = 0.72$$

(b) Using Eq. (9.59) one obtains:

$$\alpha = 1 - (1 - 0.20)\frac{1.33 \times 10^6}{4.83 \times 10^6} = 0.78$$

(c) In order to use Eq. (9.61), one needs to calculate the rock compressibility from Eq. (9.24) and Skempton's pore-pressure coefficient B from Eq. (9.62):

$$c_r = \frac{1}{k_r} = \frac{1}{4.83 \times 10^6} = 0.207 \times 10^{-6}$$

$$B = \frac{1}{1 + 0.207/0.20 \times 2.75} = 0.726$$

Then, the value of α is given by:

$$\alpha = \frac{3(0.362 - 0.25)}{0.726(1 - 2 \times 0.25)(1 + 0.362)} = 0.68$$

There is no agreement between the three equations. Actually, Eq. (9.59) is the least accurate of the three equations.

ROCK COMPRESSIBILITY CORRELATIONS

The rock or formation compressibility for consolidated sandstones may be estimated from the following correlation by Newman [92]:

$$C_f = \frac{97.32 \times 10^{-6}}{1 + 55.8721\phi^{1.428586}} \tag{9.63a}$$

Newman used 79 cores with porosities ranging from 2% to 23%. He also developed the following correlation for limestone:

$$C_f = \frac{0.8535}{1 + 2.367 \times 10^6 \phi^{0.93023}} \tag{9.63b}$$

This correlation was developed for limestone porosities ranging from 2% to 33%. Porosity ϕ is expressed in fraction and compressibility C_f in psi^{-1} in both correlations.

EFFECT OF PORE COMPRESSIBILITY ON RESERVES CALCULATIONS

As a new oil reservoir begins production, the volume of oil-in-place is one of the first and most important parameters the reservoir engineer needs to determine. Basically, the method of estimating oil-in-place from the pressure decline data in an undersaturated reservoir above the bubble point, assuming volumetric conditions, is:

$$N = \frac{N_p}{c_e \, \Delta p} \tag{9.63c}$$

where

N = initial oil-in-place, bbl
N_p = oil production during the pressure decline Δp, bbl

$$\Delta p = p_i - p$$

where

p = reservoir pressure
c_e = effective compressibility of the reservoir, expressed as:

$$c_e = c_t / S_o \tag{9.64}$$

$$c_t = c_o S_o + c_g S_g + c_w S_w + c_f \tag{9.65}$$

where c_o, c_g, and c_w are the compressibilities of oil, gas, and water, and S_o, S_g, and S_w are the saturations of oil, gas, and connate water, respectively. c_f

is the formation compressibility, which is the same as pore compressibility, c_p, and c_t is the total compressibility. If N and N_p are expressed in stock tank barrels (STB), then Eq. (9.63c) becomes:

$$N = \frac{N_p}{c_e \Delta p} \frac{B_o}{B_{oi}} \tag{9.66}$$

where B_o and B_{oi} are the oil formation volume factors at the reservoir pressure p and p_i, respectively.

As the undersaturated oil reservoir is produced, the pore pressure declines, allowing the reservoir fluids to expand and provide energy for production. In addition to fluid expansion, the formation compacts as the net overburden pressure increases, providing additional energy to squeeze out the reservoir fluids. Hall showed that the magnitude of formation compressibility is such that, if neglected, in some cases calculated values for oil-in-place will be from 30% to 100% higher than the actual oil-in-place [29].

Example

A volumetric undersaturated oil reservoir has the following characteristics:

$p_i = 5{,}000$ psia	$c_o = 10.7 \times 10^{-6}$ psi^{-1}
$S_{wc} = 20\%$	$c_w = 3.6 \times 10^{-6}$ psi^{-1}
$\phi = 9\%$	$c_f = 5 \times 10^{-6}$ psi^{-1}
$B_{oi} = 1.354$ bbl/STB	

The cumulative oil produced is 1.25×10^6 STB and $B_o = 1.375$ bbl/STB when the reservoir pressure falls to 3,600 psia, with negligible water production. What is the effect of neglecting formation compressibility on the value of oil-in-place in this reservoir?

Solution

The effective compressibility of this undersaturated oil reservoir is obtained by Eq. (9.65a), where $c_g = 0$ above the bubble point:

$$c_e = \frac{0.80 \times 10.7 \times 10^{-6} + 0.20 \times 3.6 \times 10^{-6} + 5 \times 10^{-6}}{0.80}$$
$$= 17.86 \times 10^{-6} \text{ psi}^{-1}$$

Then the initial oil-in-place by using Eq. (9.66) is given by:

$$N = \frac{1.25 \times 10^6}{17.86 \times 10^{-6}(5{,}000 - 3{,}600)} \frac{1.375}{1.354} = 50.77 \times 10^6 \text{ STB}$$

If the formation compressibility is neglected, then the effective compressibility becomes:

$$c_e = \frac{1}{0.80}(0.80 \times 10.7 \times 10^{-6} + 0.20 \times 3.6 \times 10^{-6})$$
$$= 11.6 \times 10^{-6}\ \text{psi}^{-1}$$

and the initial oil-in-place is:

$$N = \frac{1.25 \times 10^6}{11.6 \times 10^{-6} \times 1{,}400}(1.0155) = 78.16 \times 10^6\ \text{STB}$$

Thus, neglecting formation compressibility results in the overestimated values of initial oil-in-place. In this case, N is overestimated by about 28×10^6 STB, that is, a nearly 50% error.

Below the bubble-point pressure, gas will be liberated from the oil, and a free-gas saturation will develop in the reservoir. As a first order of approximation, the gas compressibility, c_g, is given by $1/p$ and the effective compressibility becomes, owing to the much higher compressibility of the free-gas phase, of the order of $100 \times 10^{-6} - 500 \times 10^{-6}\ \text{psi}^{-1}$. Rock and water compressibilities usually are omitted from the calculations. Thus, for this example ($c_g = 1/3{,}000 = 333 \times 10^{-6}\ \text{psi}^{-1}$, which is obviously much higher than c_w and c_f), Eq. (9.66) cannot be used to estimate the initial oil-in-place.

CONVERTING LAB DATA TO RESERVOIR DATA

Although it is relatively easy to measure the pore-volume compressibility under a hydrostatic load, where the pressure is the same from all directions, this condition is not representative of the reservoir boundary conditions [39]. Consider, for instance, the case of a horizontal formation that is thin in comparison to both its depth of burial and lateral extent. As the reservoir fluid is depleted and pressure declines, the net overburden pressure increases, causing the formation to compact and the matrix to expand. The change in size will be in the vertical direction only, because the surrounding rock at the boundary of the reservoir will prevent any lateral expansion of the reservoir rock. In the laboratory, this condition is similar to the one observed during the uniaxial strain test, where a load is applied in the vertical direction and the confining stress in the lateral direction is adjusted to retain a zero lateral strain condition. Anderson developed a practical procedure for converting the hydrostatic-stress, pore-volume reduction test data into the uniaxial strain compressibility data, which simulate more accurately the reservoir conditions [43]. The three-step procedure involves the following:

1. Hydrostatic stress data are curve-fitted to the following power-law relationship:

$$\Delta V_p = a_h(\sigma_c^n - \sigma_o^n) \tag{9.67}$$

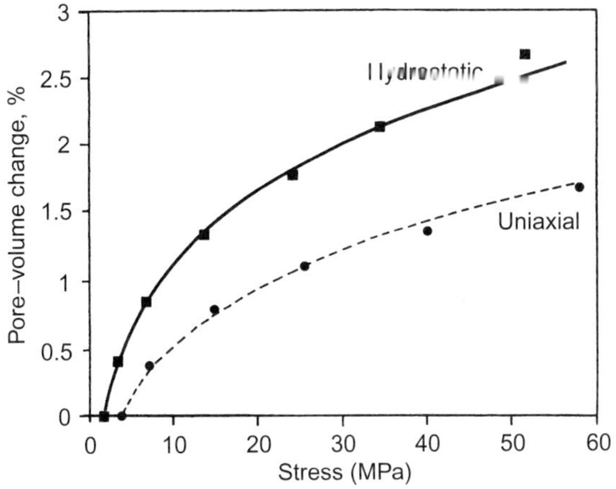

FIGURE 9.28 Power-law curve fit of hydrostatic-stress and uniaxial-strain pore-volume reduction data [43].

where

ΔV_p = change in the pore-volume as a function of stress
σ_e = net overburden stress (grain-to-grain stress or effective pressure, p_e)
σ_o = stress at the initial strain condition
a_h = hydrostatic coefficient of fit
n = power-law coefficient of fit

Figure 9.28 shows the curve-fits of pore-volume change data obtained from hydrostatic-stress and uniaxial-strain tests.

2. Calculate the "predicted" value of the change in pore volume using the parameters a_h and n from the curve fit:

$$\Delta V_p^* = R_{hu} a_h (\sigma_a^n - \sigma_o^n) \tag{9.68}$$

where σ_a is the axial stress and R_{hu} is the hydrostatic-to-uniaxial correction factor and a function of Poisson's ratio V:

$$R_{hu} = \frac{1}{3}\left(\frac{1+\nu}{1-\nu}\right) \tag{9.69}$$

For the Berea core tests, Anderson obtained a Poisson's ratio of about 0.40, which leads to R_{hu}, of 0.78 [43]. Figure 9.29 compares the predicted pore-volume change with that of the uniaxial test.

3. Calculate the predicted uniaxial strain pore-volume compressibility that is representative of reservoir conditions from the following equation:

$$c_p^* = c_f = \frac{n R_{hu} \alpha a_h \sigma_a^{n-1}}{1 - \Delta V_p^*} \tag{9.70}$$

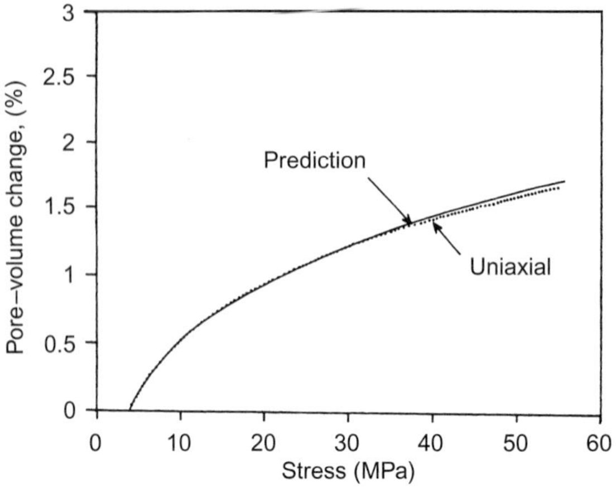

FIGURE 9.29 Comparison of measured to predicted pore-volume reduction as a function of stress [43].

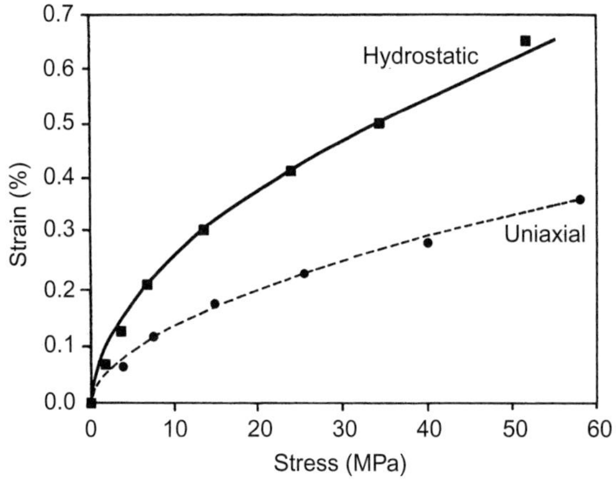

FIGURE 9.30 Power-law curve fit of uniaxial and hydrostatic stress−strain curves [43].

where c_p^* is the predicted pore compressibility and n is the power-law coefficient. Figure 9.30 shows that there is an excellent match between the power-law prediction and the calculation of uniaxial strain compressibility based on the experimental data.

Experimental data obtained by Nieto et al. indicate the following [46]:

1. Core compaction correction values between 0.87 and 0.91 are more representative than the 0.95 factor generally used when correcting ambient porosity to reservoir condition.

2. The rate of stress application appears to be the most significant factor in determining the end value of core compaction correction.
3. Different core compaction corrections observed in Jurassic and Triassic rocks having similar ambient porosity and permeability may indicate the significance of rock fabric and/or initial burial depth.
4. The use of uniaxial correction to porosity is incorrect. It is suggested that a hydrostatic total effective stress equivalent to reservoir conditions be applied slowly to allow the rock to creep back to a state representative of the reservoir. Stress application rate is a more significant factor in porosity reduction measurement than accurate simulation of reservoir stresses using biaxial strain.

EFFECT OF STRESS ON CORE DATA

The effect of overburden pressure on several petrophysical parameters of reservoir rocks, such as porosity, permeability, resistivity, and density, was extensively investigated respectively by Dobrynin and by Chierici et al. [44, 45].

In addition to the use of published experimental data, Dobrynin carried out experiments to investigate the main physical properties of sandstones under pressure. He concluded that the changes in these properties are determined to a large extent by the pore compressibility, which, in the range of 0–20,000 psi, can be characterized by the maximum pore compressibility and the net overburden pressure. He also developed several general equations that describe the behavior of the physical properties of sandstones under pressure.

Chierici et al. used a large number of samples to investigate, experimentally and theoretically, the influence of the overburden pressure on porosity, horizontal and vertical permeability, relative permeability to gas, formation resistivity factor, and capillary pressure curves [45]. From the experimental results they drew the following conclusions:

1. The effect of shale on pore compressibility must be taken into account when dealing with undersaturated oil reservoirs in shaly rocks.
2. The formation resistivity factor and permeability of clean sandstones are affected by stress. In the low-porosity rocks, however, the formation resistivity factor at reservoir conditions can be quite different from that measured at no-stress conditions.
3. Permeability anisotropy is only slightly affected by overburden pressure.
4. Capillary pressure curves are considerably affected by the stress tensor only at low capillary pressure values, whereas the irreducible water saturation is only slightly influenced by the overburden pressure.

EFFECT OF STRESS ON POROSITY

Well-cemented, elastic rocks do not undergo much change in volume upon release from their in situ environment and, as a result, porosity at ambient conditions approximates porosity under stress conditions. Unconsolidated and poorly consolidated sediments often increase in size when released from their natural reservoir state of stress. Porosity must therefore be measured on samples of this type under a confining pressure approximately equivalent to the in situ stress. Several approaches are commonly utilized to determine the approximate stress at which the laboratory tests are conducted. One such approach utilizes the concept of net overburden pressure. This approach requires that the reservoir pressure (or reservoir pressure gradient) is deducted from an appropriate overburden pressure (or gradient) to calculate the laboratory hydrostatic load. The data are subsequently translated to equivalent uniaxial loading conditions [47]. Another approach utilizes the concept of mean stress and requires knowledge of Poisson's ratio (ν) and elastic modulus. These parameters are generally determined from acoustic velocity tests at multiple stress conditions. A hydrostatic load is applied in the laboratory at the appropriate mean stress and the data are directly applied.

Since different approaches can be utilized to determine the net overburden pressure, that is, $P_{nob} = P_{ob} - P_{res}$, and since the net overburden pressure can vary as a function of time, measurements at a suite of pressure to define the parameter change as a function of pressure are generally recommended. The laboratory technique commonly employs hydraulic loading of the core, which differs from the reservoir loading and yields porosity values that are pessimistic. These values must be corrected to uniaxial loading prior to use.

Experimental data generated by Dobrynin on a large number of sandstone samples (Figure 9.31) show that between a certain minimum pressure p_m and

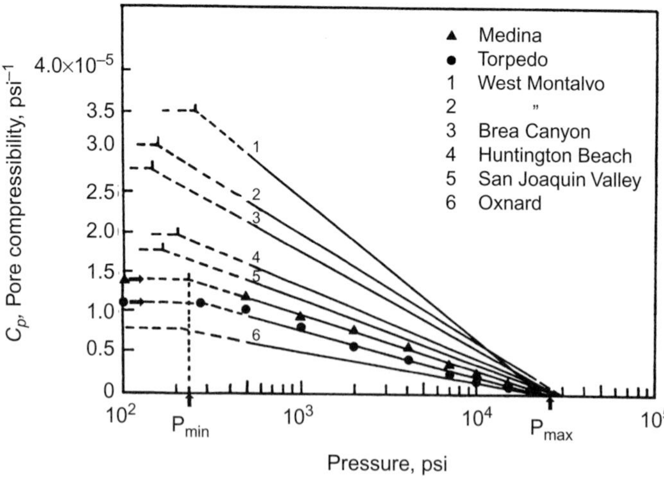

FIGURE 9.31 Pore compressibility as a function of net overburden pressure [44].

a certain maximum pressure p_M, the relation between pore compressibility and the logarithm of pore pressure can be approximated by a straight line, which can be expressed mathematically as follows [44]:

$$c_p = \frac{c_{pM}}{\log(p_M/p_m)} \log(p_M/p) \tag{9.71}$$

The maximum pore compressibility c_{pM} can be determined by extrapolating the experimental curves to zero pressure using Cartesian coordinates. The minimum pressure p_m is determined by assuming that there is essentially no change in pore compressibility within the small range of pressure from 0 to p_m. Actually, p_m corresponds to the first data point where the c_p curve begins to decline with increasing net overburden pressure as shown in Figure 9.31. The value of p_m is obtained by extrapolating the straight line to $c_p = 0$. For practical purposes, p_m is the pressure above which changes in pore compressibility are negligible. Dobrynin found that p_m is between 150 and 300 psi, and p_M is between 25,000 and 30,000 psi.

Combining the following relationship between c_b, c_r, and c_p:

$$c_b = \phi_{c_p} + (1 - \phi)c_r \tag{9.72}$$

and Eq. (9.52) yields:

$$c_b = \frac{c_{pM}}{\log(p_M/p_m)} \log(p_M/p) + (1 - \phi)c_r \tag{9.73}$$

It is assumed in this equation that the rock matrix compressibility is not affected by pressure in the range of 0–20,000 psi.

The relative changes of porosity under overburden pressure can be expressed as follows [44]:

$$\frac{\Delta\phi}{\phi} = 1 - \left[\left(1 - \frac{\Delta V_p}{V_p}\right) \Big/ \left(1 - \frac{\Delta V_b}{V_b}\right)\right] \tag{9.74}$$

where $\Delta V_p/V_p$ and $\Delta V_b/V_b$ are the relative changes of pore volume and bulk volume, respectively. Inasmuch as the rock matrix compressibility is assumed to be independent of pressure in the range of 0–20,000 psi, that is, $0 < p < p_m$, the relative change of bulk volume is essentially linear with respect to the relative change of pore volume:

$$\frac{\Delta V_b}{V_b} = \phi \frac{\Delta V_p}{V_p} \tag{9.75}$$

Combining Eqs. (9.74) and (9.75) yields:

$$\frac{\Delta\phi}{\phi} = 1 - \left[\frac{(1 - (\Delta V_p/V_p))}{(1 - \phi(\Delta V_p/V_p))}\right] \tag{9.76}$$

Within the range of pressure $0 < p < p_m$, the relative change of pore volume can be determined according to Eq. (9.48):

$$\frac{\Delta V_p}{V_p} = \int_0^{p_m} c_{p_m}\, dp + \int_{p_m}^{p} c_p\, dp \tag{9.77}$$

Substituting for c_p from Eq. (9.71), one obtains:

$$\frac{\Delta V_p}{V_p} = c_{pM} \int_0^{pm} dp + \frac{c_{pM}}{\log(p_M/p_m)} \int_{pm}^{p} \log\left(\frac{p_M}{p_p}\right) dp \tag{9.78}$$

which can be written, after integration, as follows:

$$\frac{\Delta V_p}{V_p} = c_{pM} D(p_p) \tag{9.79a}$$

where the Dobrynin pressure function $D(p_p)$ is:

$$D(p_p) = P_m + \frac{p_p}{\log(p_M/p_m)}\left[\log\left(\frac{p_M}{p_p}\right) + 0.434 - \frac{p_M}{p_p}\left(\log\left(\frac{p_M}{p_m}\right) + 0.434\right)\right] \tag{9.79b}$$

Substituting Eq. (9.79a) into Eq. (9.76) yields:

$$\frac{\Delta\phi}{\phi} = 1 - \frac{1 - c_{pM} D(p_p)}{1 - \phi c_{pM} D(p_p)} \tag{9.79c}$$

Figure 9.32 shows the experimental data that illustrate the effect of net overburden pressure on the $\Delta\phi/\phi$ ratio for five different representative values of c_{pM}, and porosity values of 5%, 10%, and 20%. It is evident that experimental data are in agreement with the calculated curves using Eq. (9.79c). The average values of p_m and p_M are 200 and 25,000 psi, respectively.

A practical method for determining the porosity and compressibility reduction from experiments in which rocks are hydrostatically stressed at a constant pore pressure was presented by Schutjens et al. [48]. Assuming (1) linear poroelasticity, (2) small strains, (3) compressible grains, and (4) rock deformation are caused by changes in total stress at constant pore pressure, they showed that, for small variations of the porosity, the porosity reduction during compaction can be accurately calculated from:

$$\Delta\phi = \phi - \phi_0 = C_S \frac{\Delta V_p}{V_{b0}} \tag{9.80a}$$

where V_{b0} and ϕ_0 are, respectively, the sample or reservoir volume and porosity at reference stress conditions. The compaction correction factor, C_S, which is typically in the range 0.6–0.9, is given by:

$$C_S = \frac{\alpha - \phi_0}{\alpha + (\Delta V_p/V_{b0})} \tag{9.80b}$$

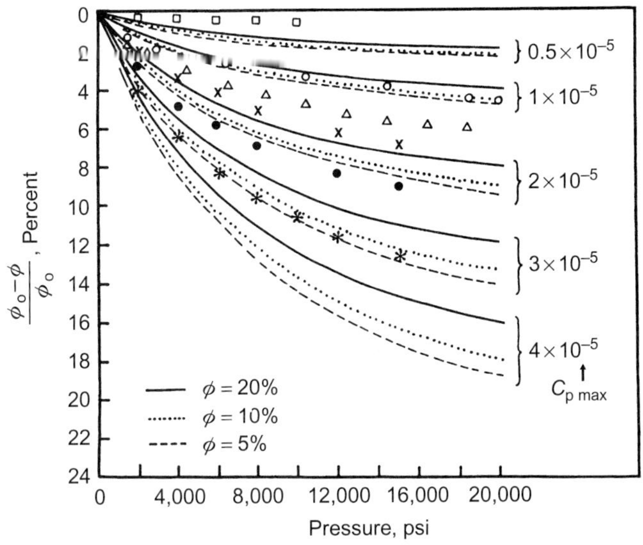

FIGURE 9.32 Influence of net overburden pressure on changes in porosity for five representative maximum pore compressibilities [44].

For incompressible or only slightly compressible grains $\alpha \approx 1$. For compressible grains $\alpha < 1$.

Combining Eqs. (9.80a) and (9.80b) and assuming α is much larger than $\Delta V_p/V_{b0}$ yields:

$$\Delta\phi = \left(1 - \frac{\phi_0}{\alpha}\right)\frac{\Delta V_p}{V_{b0}} \tag{9.80c}$$

In an experiment where the pore pressure is constant, $\Delta V_p = \alpha\,\Delta V_{b0}$ [34] and the sample (bulk) volumetric strain $\varepsilon_b = -\Delta V_b/\Delta V_{b0}$. Substituting these two terms into Eq. (9.80c) gives:

$$\Delta\phi = \frac{\varepsilon_b}{1-\varepsilon_b}(\phi_0 - \alpha) \tag{9.80d}$$

Solving explicitly for porosity yields:

$$\phi = \left(\frac{1}{1-\varepsilon_b}\right)\phi_0 - \left(\frac{\varepsilon_b}{1-\varepsilon_b}\right)\alpha \tag{9.80e}$$

The pore compressibility is obtained from the change in porosity as follows [48]:

$$C_{pc} = \frac{\alpha\,\Delta\phi}{\phi_0(\phi_0 + \Delta\phi - \alpha)}\left(\frac{1}{\Delta P_c}\right) \tag{9.81}$$

where ΔP_c is the change in confining pressure.

EFFECT OF STRESS ON PERMEABILITY

Assuming that changes in permeability due to changes in pore pressure depend mainly upon the contraction of the pore channels, Dobrynin derived the following semiempirical equation [44]:

$$\frac{\Delta k}{k} = 2(1 + f_{ps})c_{pM}D(p_p) \tag{9.82}$$

where f_{ps} is the pore shape factor. Figure 9.33 shows a comparison of experimental data with calculated curves using the more practical form of Eq. (9.82):

$$\frac{k_p}{k} = 1 - 2(f_{ps} + 1)c_{pM}D(p_p) \tag{9.83}$$

where k_p is the actual permeability under pressure, that is, $k - \Delta k$, and k is the permeability under zero pressure. The following approximate relationship between the pore shape coefficient f_{ps} and the maximum pore compressibility c_{pM} for sandstones with poor sorting is obtained from experimental data:

$$f_{ps} = 8 \times 10^{-5}(c_{pM}^{-0.9}) \tag{9.84}$$

For a uniform pore-size distribution or for a very high pore compressibility, $f_{ps} = 0.33$. It is obvious that the formation permeability decreases with increasing values of stress. This fact should be taken into account when interpreting the results of pressure transient tests, drawdown, or buildup. As an example, the stresses that are at their maximum effect in the vicinity of the well, due to the depletion of reservoir fluids and consequent rapid pressure decline and reduction of permeability in the same zone, can account for skin effects, which are not necessarily caused by mud-filtrate invasion [45].

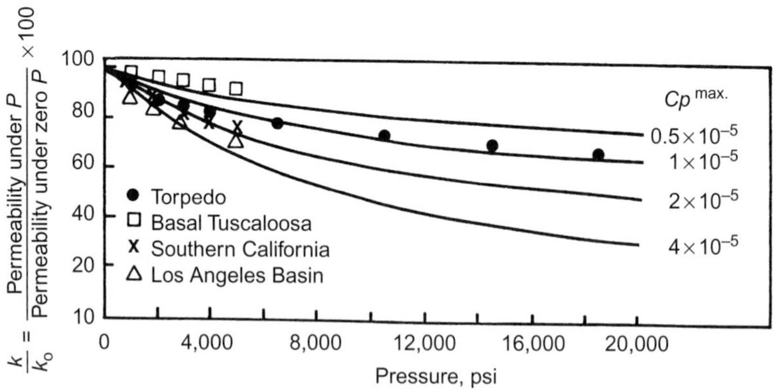

FIGURE 9.33 Comparison of experimental and calculated data showing changes in permeability of sandstones as a function of net overburden pressure [44].

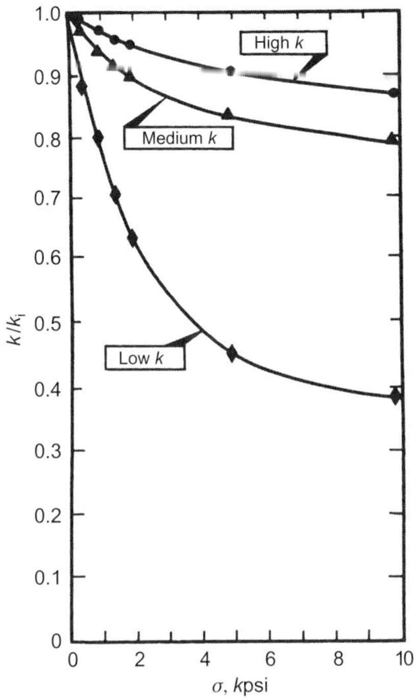

FIGURE 9.34 Curve-fit of permeability versus net stress data [50].

Figure 9.34 shows that the permeability decreases rapidly at low stress and stabilizes with increasing confining or overburden stress [49].

EFFECT OF STRESS ON RESISTIVITY

To investigate the relationship between resistivity and pressure, Dobrynin used Archie's equation relating the formation factor F and porosity ϕ (assuming $a = 1$):

$$F_{\mathrm{R}} = \frac{1}{\phi^{\mathrm{m}}} \tag{9.85}$$

where m is the cementation exponent. Assuming (1) changes in resistivity of porous rocks, when subjected to stress, are primarily dependent upon the shrinkage of the smaller pore channels, which are mostly filled with irreducible water and (2) fine materials are present in the small pores or channels, Eq. (9.85) becomes:

$$F_{\mathrm{RP}} = \frac{1}{(\phi - \Delta\phi)^{m + \Delta m}} \tag{9.86}$$

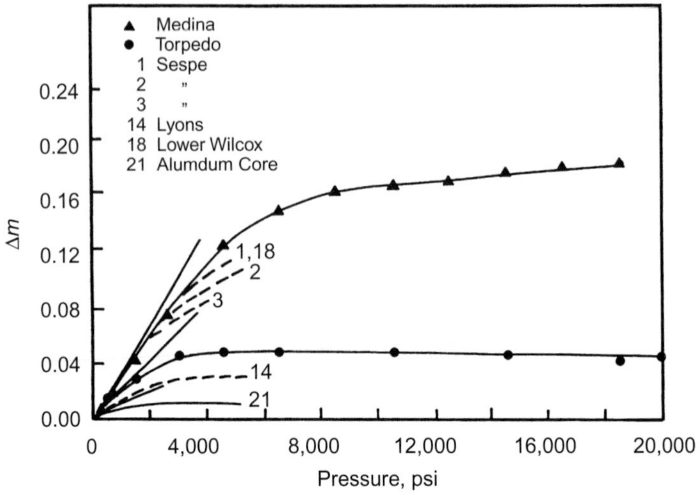

FIGURE 9.35 Effect of net overburden pressure on exponent m [44].

Dividing Eq. (9.86) by Eq. (9.85) and assuming that Δm and $\Delta\phi$ are very small such that $(\phi - \Delta\phi)^{\Delta m} = \phi^{\Delta m}$:

$$\frac{F_{RP}}{F_R} = \frac{1}{(\phi - \Delta\phi/\phi)^m \phi^{\Delta m}} \tag{9.87}$$

If the cementation exponent is ~2, such as in the low-porosity sands and limestones, which tend to be highly cemented, then $(\Delta\phi/\phi)^2 = 0$ and Eq. (9.87) becomes:

$$\frac{F_{RP}}{F_R} = \frac{1}{(1 - 2\,\Delta\phi/\phi)\phi^{\Delta m}} \tag{9.88}$$

Substituting Eq. (9.81) into Eq. (9.88) gives:

$$\frac{F_{RP}}{F_R} = \frac{1}{\{2[1 - c_{pM}D(p_P)]/[1 - \phi c_{pM}D(p_P)]\}\phi^{\Delta m}} \tag{9.89}$$

Figure 9.35 illustrates the effect of the net overburden pressure on Δm. It is evident from this figure that the curves are of similar character and, consequently, it is possible to distinguish them by the maximum change in Δm. Assuming (1) this maximum change, Δm_M, depends upon the number of flow channels and (2) clay content controls the percentage of flow channels, Dobrynin generated, experimentally, the two graphs shown in Figure 9.36 [44]. These graphs show a typical behavior of F_{RP}/F_R as a function of net overburden pressure, porosity, and relative clay content.

In conclusion, changes in the physical properties of sandstones under overburden pressure are determined by the pore compressibility, which can

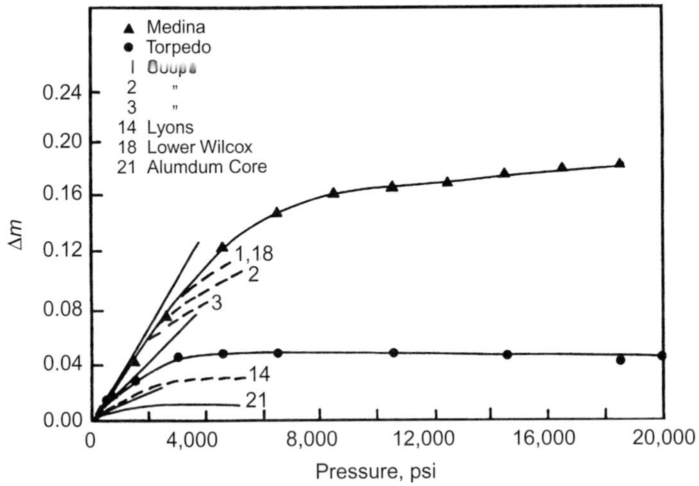

FIGURE 9.36 Relative formation factor as a function of net overburden pressure, porosity, and relative clay content [44].

be characterized by c_{pM}, and the net overburden pressure (p_p) in the range of 0–20,000 psi.

POROSITY–PERMEABILITY–STRESS RELATIONSHIP

When the compressibility of grains, c_r, is negligible compared with the change in porosity, and assuming that all the stress relief occurs as a result of utilization of pore space, the following relationship is applicable [50]:

$$d\phi = -c_b(1-\phi)\sigma \tag{9.90}$$

where c_b is the bulk compressibility which is defined by Eq. (9.72). For $c_r = 0$:

$$c_b = \phi c_p \tag{9.91}$$

Combining Eq. (9.90) and Eq. (9.91) and integrating over the stress range yields:

$$\int_{\phi_0}^{\phi} \frac{d\phi}{\phi(1-\phi)} = \int_{\sigma_0}^{\sigma} c_p \, d\sigma \tag{9.92}$$

where ϕ_0 is the porosity at initial or zero effective stress σ_0 (total overburden load).

Assuming pore compressibility declines exponentially, one can express c_p as follows:

$$c_p = c_{po}\, e^{-b\Delta\sigma} \tag{9.93}$$

where $\Delta\sigma = \sigma_0 - \sigma$. Defining the average pore compressibility as:

$$\overline{c}_p = \frac{1}{\Delta\sigma}\int_{\sigma_0}^{\sigma} c_p\, d\sigma \tag{9.94}$$

substituting Eq. (9.93) into Eq. (9.94) and integrating yields:

$$\overline{c}_p = \frac{c_{po}}{b\,\Delta\sigma}\left(1 - e^{-b\Delta\sigma}\right) \tag{9.95}$$

where b is the rate of decline of pore compressibility as the effective stress increases. Pore compressibility (c_p) is not constant and represents the average pore compressibility over a stress interval $\Delta\sigma$. Thus, as σ changes, c_p also will change over the sand interval. In some cases, however, assuming constant pore compressibility may give an excellent curve fit, as shown in Figure 9.37 for a coal core sample.

Combining Eqs. (9.93) and (9.92) gives:

$$\frac{\phi}{1-\phi} = \frac{1-\phi}{1-\phi_0}\, e^{-\overline{c}_p(\sigma_0 - \sigma)} \tag{9.96}$$

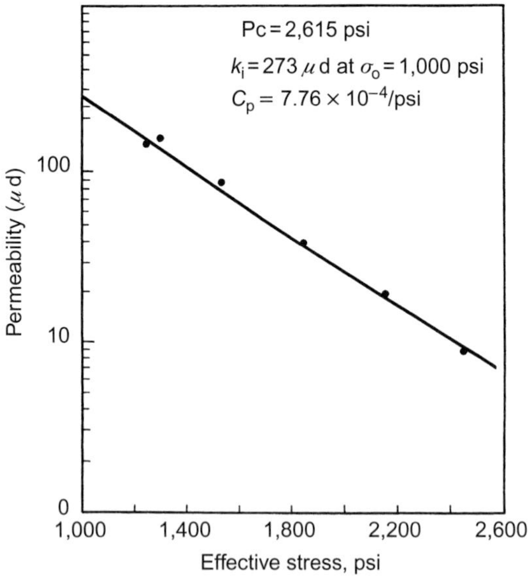

FIGURE 9.37 Laboratory-measured permeability versus effective stress for a coal sample from a depth of 2,766 ft [51].

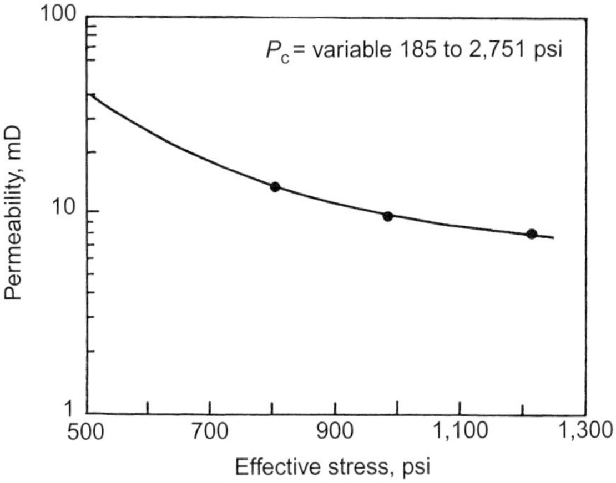

FIGURE 9.38 Laboratory-measured permeability versus effective stress and theoretical match using variable compressibility for a coal sample from a depth of 2,767 ft [51].

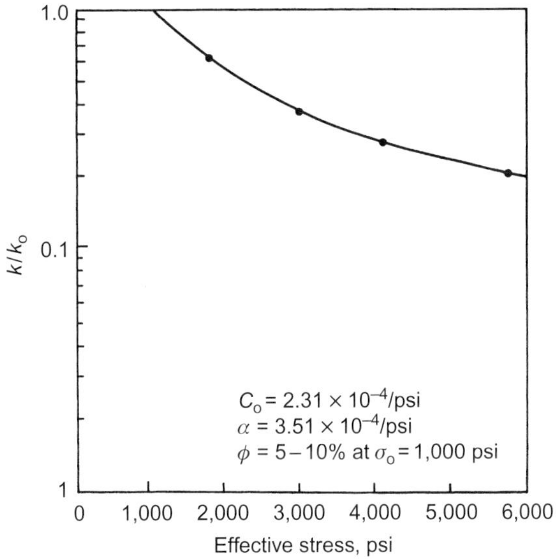

FIGURE 9.39 Laboratory-measured permeability versus effective stress and theoretical match using variable compressibility for a sandstone core sample [51].

Equation (9.96) shows that, for an average constant pore compressibility value, a semilog plot of the void ratio $\phi/(1-\phi)$ versus effective stress will result in a straight line. Using laboratory core tests on sandstone, coal, clay, and granite samples, McKee et al. found that the theoretical curves for permeability and porosity as functions of stress fit the experimental values well, as shown in Figures 9.38–9.40 [51]. Figure 9.40 shows a plot of laboratory-measured

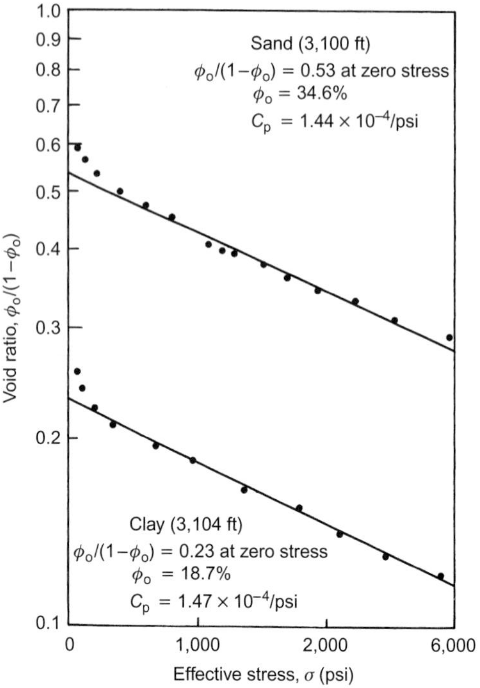

FIGURE 9.40 Laboratory-measured void ratio versus effective stress for sand and clay core samples [51].

void ratio versus effective stress data for adjacent sand and clay samples from a Venezuelan oil field. It is apparent from this figure that the theoretical curve matches the experimental data well within a practical range of accuracy, especially for the clay core sample, which was taken within 4 ft of the sand sample. The data points obtained at very low stress values (<200 psi) were not included in the least-squares fit. The void ratio for the sand sample decreased from ~0.60 (experimental point) to 0.30, which corresponds to a decline in porosity from 0.38 to 0.22, respectively, when the effective stress, σ_e, increased to about 4,400 psi.

Solving Eq. (9.96) for ϕ/ϕ_0, one obtains:

$$\frac{\phi}{\phi_0} = \frac{e^{-\bar{c}_p \Delta\sigma}}{1 - \phi_0(1 - e^{-\bar{c}_p \Delta\sigma})} \tag{9.97}$$

Using the Carman–Kozeny correlation (Eq. (3.26)) and assuming that the specific surface area per unit of grain volume of a porous material (S_{GV})

is independent of the net overburden stress and permeability is independent of pore (fluid) pressure, one can obtain the following formula:

$$\frac{k}{k_0} = \frac{\phi^3/(1-\phi)^2}{\phi_0^3/(1-\phi_0)^2} \tag{9.98}$$

Combining Eq. (9.96) with Eq. (9.98) gives:

$$\frac{k}{k_0} = \frac{\phi}{\phi_0} e^{-2\bar{c}_p\,\Delta\sigma} \tag{9.99}$$

Substituting Eq. (9.97) into Eq. (9.99) yields:

$$\frac{k}{k_0} = \frac{e^{-3\bar{c}_p\Delta\sigma}}{1-\phi_0(1-e^{-\bar{c}_p\Delta\sigma})} \tag{9.100}$$

If the permeability is dependent on fluid pressure, then from Eq. (9.57) the following is true:

$$d\sigma = -\alpha\, dp \tag{9.101}$$

where dp is the change in fluid pressure, or:

$$\Delta\sigma = \sigma_t - \sigma_0 = \alpha(p_p - p) = \alpha\,\Delta p \tag{9.102}$$

where σ_t is taken as constant caused by the overburden, and α is the rate of decline. Substituting Eq. (9.102) into Eq. (9.95) and solving for ϕ/ϕ_0 gives:

$$\frac{\phi}{\phi_0} = \frac{e^{-\alpha\bar{c}_p\,\Delta p}}{1-\phi_0(1-e^{-\alpha\bar{c}_p\Delta p})} \tag{9.103}$$

A combination of Eqs. (9.99) and (9.103) yields a new expression for the permeability ratio k/k_0:

$$\frac{k}{k_0} = \frac{e^{-3\alpha\bar{c}_p\,\Delta p}}{1-\phi_0(1-e^{-\alpha\bar{c}_p\Delta p})} \tag{9.104}$$

Figure 9.39 is a semilog plot of k/k_0 versus the effective strefs [σ_e(psi)], assuming $\alpha = 1$ for a sandstone sample from the Rocky Mountain region.

Permeability values were measured at 100 psi with initial confining pressure (σ_0). For these samples, the rate of decline a of the effective stress is 3.5×10^{-4} psi^{-1}; and the theoretical curve, based on Eq. (9.100) can be obtained with an initial porosity value ϕ_0 ranging from 0.05 to 010. It is important to emphasize that the above equations (Eqs. (9.90–9.104)) are derived with the assumption that the grain compressibility c_r is negligible.

McKee et al. also showed that the dependence of formation density on the effective stress is related to the change in pore compressibility [51]. If

$$\rho = \rho_g(1 - \phi) \tag{9.105}$$

where ρ and ρ_g are the specific gravities of the formation and grain material, respectively, substituting Eq. (9.97) into Eq. (9.105) for ϕ yields:

$$\rho = \frac{(1 - \phi_0)\rho_g}{1 - \phi_0(1 - e^{-\bar{c}_p \Delta\sigma})} \tag{9.106}$$

Figure 9.41 shows a considerable data scatter; however, there is a fair match of this expression to experimental data of density versus depth for 400 shale core samples from northern Oklahoma, where approximately the same amounts of sediment were deposited with no intervening unconformity. The grain density in this area was 2.65 g/cm^3, whereas the density at zero depth is 2.02 g/cm^3. The effective stress in psi at any depth D in feet for lithostatic conditions can be approximated by the following formula:

$$\sigma_e = 0.572\,D \tag{9.107}$$

where the constant 0.572 psi/ft is the effective stress gradient, g_{se}, and is obtained from:

$$g_{se} = g_s - g_w \tag{9.108}$$

where g_w is the water gradient (0.4335 psi/ft for freshwater) and g_s is the stress gradient:

$$g_s = \rho_g(1 - f) + \rho_w\phi \tag{9.109}$$

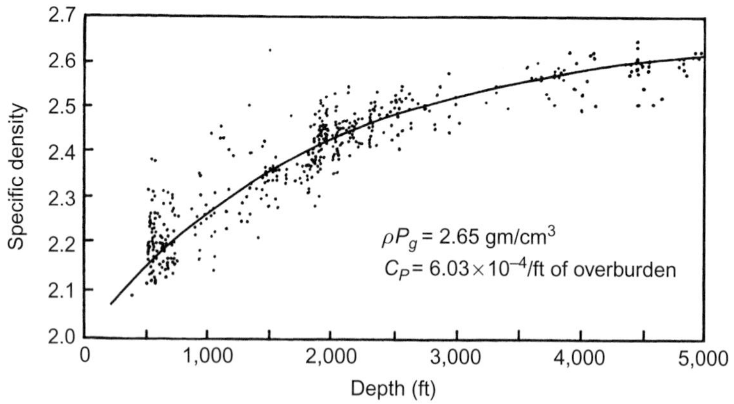

FIGURE 9.41 Laboratory-measured density versus depth for shale core samples [50].

Assuming $\rho_g = 2.65$ g/cm^3, $\rho_w = 1$ g/cm^3, $g_w = 0.433$ psi/ft, the stress gradient g_s is given by:

$$g_s = [2.65(1 - 0.20) + (1)(0.2)]62.4\ \text{lbm/ft}^3 \times 44\ \text{ft}^2/\text{in.}^2 = 1.005\ \text{psi/ft}$$

and the effective stress gradient g_{se} is given by:

$$g_{se} = 1.005 - 0.43 = 0.572\ \text{psi/ft}$$

Although this approach of using lithostatic conditions to calculate the effective stress gradient yields acceptable results, it is important to use the local effective stress gradients when applying this theory to field cases, especially in deep formations.

As natural gas reservoirs are found at deeper zones, understanding stress-dependent permeability is essential because, under large drawdown, reduced permeability can lower the production from a stress-sensitive reservoir [51].

Jones presented empirical equations that accurately fit permeability and porosity data versus net confining stress [49]. Each of these equations has four adjustable parameters and, with little loss of accuracy, two of the coefficients can arbitrarily be preset. Consequently, permeability and porosity measurements need to be made at only two confining stresses to quantify the effect of stress with a good accuracy. Figure 9.42 was fitted by the following equation:

$$\frac{k}{k_0} = \frac{\exp[a_k(1 - e^{-\sigma_e/\sigma^*})]}{1 + a_J\sigma} \tag{9.110}$$

where

k = slip-corrected (Klinkenberg) permeability at effective stress σ_e, mD
K_0 = slip-corrected permeability at zero effective stress, mD
a_k = slope of the straight line in Figure 9.42
a_J = Jones coefficient, 3×10^{-6} psi^{-1}
σ_e = effective stress, hydrostatic confining stress minus average pore pressure, psi
σ^* = Jones decay constant, 3,000 psi

Figure 9.42 shows that greater reductions occur with low-permeability samples than with high-permeability samples. A laboratory study of low-permeability gas sands revealed that a 10-fold permeability reduction is not uncommon [50]. Equation (9.110) can be expressed as follows:

$$\ln[k(1 + a_J\sigma_e)] = -a_k(1 - e^{-\sigma_e/\sigma^*}) + \ln k_0 \tag{9.111}$$

A semilog plot of this expression, that is, $k(a + a_J\sigma)$ versus $[1 - \exp(-\sigma_e/\sigma^*)]$, results in a straight line with intercept k_0 and a slope of $-a_k$, as shown in Figure 9.42. The theoretical curves fit the laboratory data of permeability well, when these data were assumed to decline exponentially as

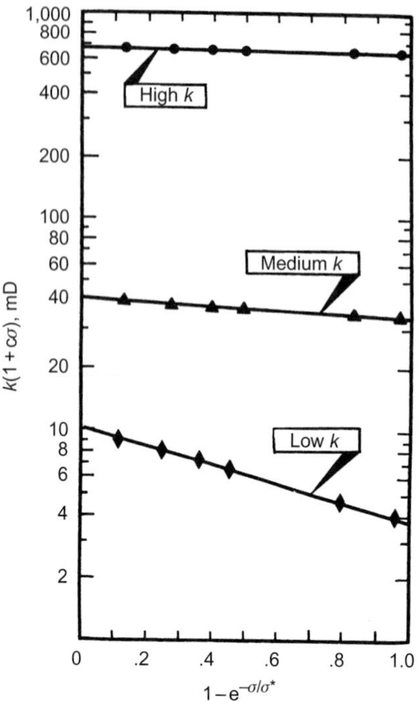

FIGURE 9.42 Straight-line relationship between permeability and stress [50].

a function of stress. Also, the abscissa ranges from 0, which corresponds to $\sigma = 0$, to 1, which corresponds to infinite stress. These correlations do not include the further loss of permeability when a rock sample is contacted by a brine, which can be especially severe in right samples [52].

Pore-volume reduction with increasing stress exhibits the same general behavior as the permeability reduction, as shown in Figure 9.43 for the same three core samples. The points on this figure were curve-fitted by the following expression:

$$\frac{V_p}{V_{po}} = \frac{\exp[a_v(e^{-\sigma_e/\sigma^*} - 1)]}{1 + a_J\sigma_e} \tag{9.112}$$

where V_p (in cm^3) is the pore volume of sample at net stress σ, and V_{p0} is the pore volume at zero net stress. All four of the adjustable coefficients in Eq. (9.112) (V_{po}, a_v, σ^*, and a_J) were fit using least-squares technique. Figure 9.44 is a semilog plot of the same data points in Figure 9.43 according to the following correlation:

$$\ln[V_p(1 + a_J\sigma)] = -a_v(e^{-\sigma_e/\sigma^*} - 1) + \ln V_{po} \tag{9.113}$$

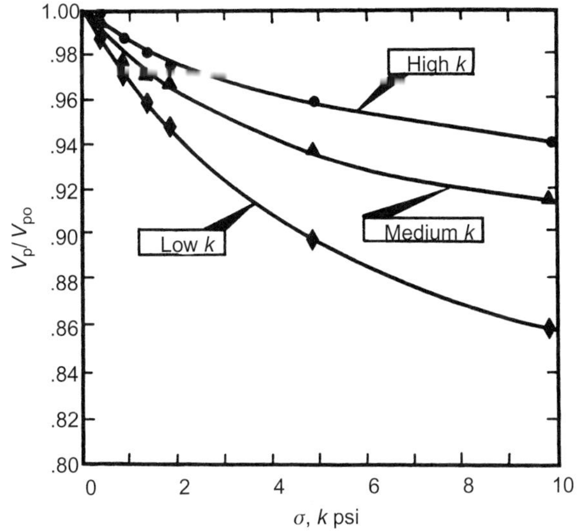

FIGURE 9.43 Effect of stress on pore volume [50].

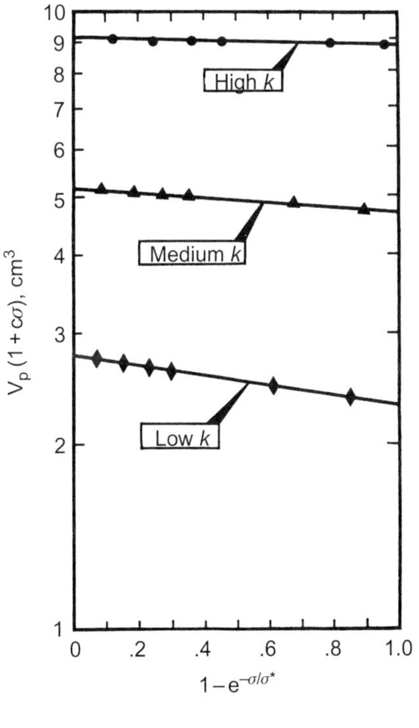

FIGURE 9.44 Straight-line relationship between pore volume and stress [50].

A majority of experimental data of k and V_p versus effective a revealed that the Jones decay constant σ^* (3,000 psi) and the coefficient a_J of 3×10^{-6} psi^{-1} yielded the best curve fits. Thus, if these two factors could be fixed a priori, then Eqs. (9.111) and (9.113), or their corresponding plots, suggest that only two stress tests are needed to describe the behavior of permeability and pore volume with respect to σ_e. The two stresses of choice, according to Jones, would be approximately 1,500 and 5,000 psi, because:

1. Stresses much higher than 5,000 psi result in reduced sleeve life in a Hassler-type core holder.
2. Stresses less than 1,500 psig are unreliable because a thick rubber sleeve does not completely conform to microscopic irregularities of the core surface at low stresses, which results in too high measurements of pore-volume compressibility, c_p.

By definition:

$$c_p = \frac{1}{V_p}\frac{dV_p}{d\sigma} = \frac{d\ln(V_p)}{d\sigma} \tag{9.114}$$

Differentiating Eq. (9.113) with respect to stress σ and substituting into Eq. (9.114) yields:

$$c_p = \left(\frac{a_v}{\sigma^*}\right)e^{-\sigma_e/\sigma^*} + \frac{a_j}{1 + a_J\sigma_e} \tag{9.115}$$

Figure 9.45 shows the behavior of c_p versus σ according to Eq. (9.115) for the same three core samples used in the previous figures for k and V_p.

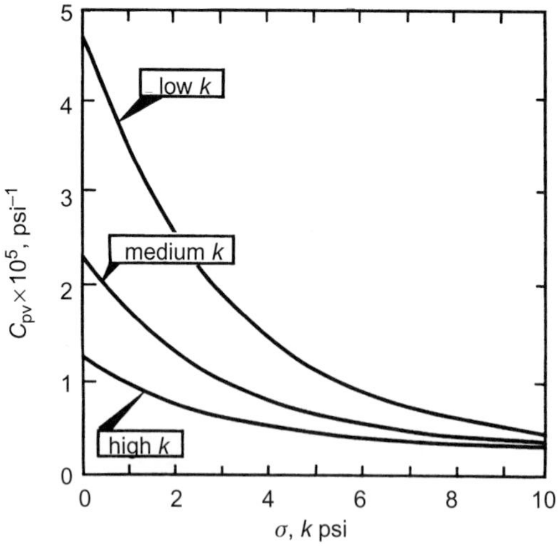

FIGURE 9.45 Effect of stress on compressibility [50].

The curve fit is obviously excellent. Because pore-volume compressibility is a derivative of pore-volume–stress data (Eq. (9.114)), several precautions need to be observed to ensure that the data are about an order of magnitude better quality than when the derivative is not required. These added precautions include the following:

1. The core sample must be cut such that it is free from surface irregularities.
2. The ends of the rock sample are square and do not have chips missing.
3. Additional stress tests are needed between the 1,500 and 5,000 psi measurements tests. Flaws in the rock samples cause low-stress compressibility measurements to be too high.

The decrease in porosity with increasing stress can be predicted similarly to that of permeability. Inasmuch as:

$$\phi = \frac{V_p}{V_p + V_g}$$

solving for the void ratio ϕ_R:

$$\phi_R = \frac{\phi}{1-\phi} = \frac{V_p}{V_g} \tag{9.116}$$

Assuming grain volume compressibility is negligible because it is much lower than the pore-volume compressibility, the void ratio ϕ_R is directly proportional to V_p. Thus, from Eq. (9.113) one obtains:

$$\frac{\phi/(1-\phi)}{\phi_o/(1-\phi_o)} = \frac{V_p}{V_{po}} = \frac{\exp[a_v(e^{-\sigma_e/\sigma^*} - 1)]}{1 + a_J\sigma_e}$$

or

$$\frac{\phi}{1-\phi} = \left(\frac{\phi_o}{1-\phi_o}\right)\frac{\exp[a_v(e^{-\sigma_e/\sigma^*} - 1)]}{1 + a_J\sigma_e} \tag{9.117}$$

and

$$\ln\left(\frac{\phi}{1-\phi}\right)(1 + a_J\sigma_e) = -a_v\left(1 - e^{-\sigma_e/\sigma^*}\right) + \ln\left(\frac{\phi_o}{1-\phi_o}\right) \tag{9.118}$$

Thus, a semilog plot of $[\phi/(1-\phi)] - (1 + a_J\sigma)$ versus $(1 - e^{\sigma_e/\sigma^*})$ results in a straight line with intercept $\phi_o/(1-\phi_o)$ and a slope of $-a_v$.

Figure 9.46 shows two-point fits of experimental porosity–stress data according to Eq. (9.118), where $\sigma^* = 3{,}000$ psi and $a_J = 3 \times 10^{-6}\ \text{psi}^{-1}$. Occasionally, the yield strength of a rock sample is between the preferred stress range of 1,500–5,000 psig, causing the sample to rupture. Two types of failure are recognized. The first type causes the sample to be crushed, but the interpretation of porosity–stress data obtained prior to crushing is

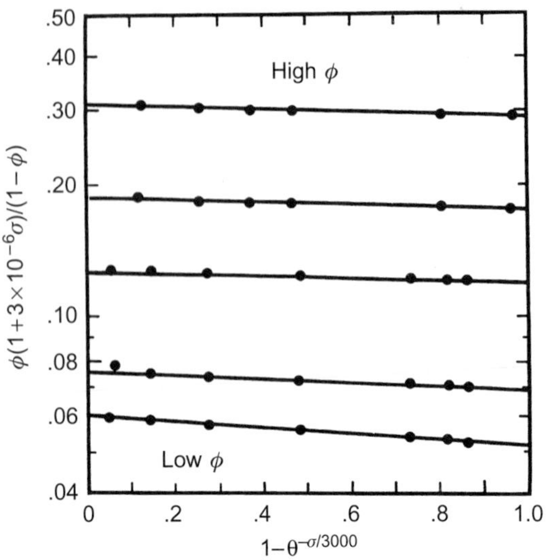

FIGURE 9.46 Straight-line relationship between porosity and stress [50].

unchanged. The second type of failure results in fracturing, slippage of grain, and irreversible compaction, causing changes in the permeability–stress and pore-volume–stress curves. Jones found that in this case semilog plots, such as shown in Figure 9.42 or 9.44, exhibit two straight-line portions, and the intersection points of the two segments correspond to the compressive yield strength of the rock sample. The second segment has a steeper slope than the first one. It is suggested that an intermediate test point at 3,000 psi be run for every few samples if irreversible compaction is suspected. If the three points, that is, 1,500, 3,000, and 5,000 psig, do not lie on the same straight line, the yield strength probably has been exceeded and additional points are needed. These additional points should be between 3,000 and 5,000 psig, because the change in slope usually does not occur at hydrostatic stresses less than approximately 4,000 psig.

Jones investigated the sensitivity of the curves of k versus σ to the correlation parameters σ^* and a_J and concluded the following:

1. The fits are appreciably affected by the parameter a_J only at stresses higher than 6,000 psig, as shown in Figure 9.47.
2. The theoretical fits for k, V_p, ϕ, or c_p are relatively insensitive to σ^* and a_J in the recommended range of stress, that is, 1,500–5,000 psig.
3. Extrapolations beyond this stress range may be made with good reliability, provided they are not too large.
4. Five to eight stress-test points for a sample are recommended for a new reservoir.

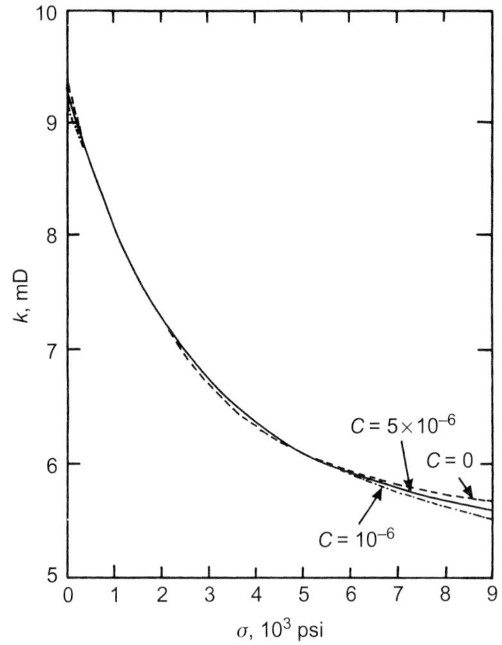

FIGURE 9.47 Influence of compressibility on permeability–stress relationship [50].

Harari et al. investigated the pore compressibility characteristics of carbonate reservoir rock samples under elevated net confining pressure conditions [38]. The rock samples were obtained from a limestone carbonate sequence found at a depth of 6,500 ft. The tested samples were separated into four groups on the basis of their lithological descriptions: grainstones, packstones, wackestones, and mudstones. Berea sandstone samples also were tested for comparison purposes. The stress tests consisted of saturating the rock samples with water and then subjecting them to differential pressures, ranging from 0 to 4,500 psi, by reducing the pore pressure, p_p, while maintaining a constant hydrostatic confining pressure, p_c. The resulting reduction in pore volume was measured, and the corresponding pore compressibility was calculated from Eq. (9.26). Harari et al. found that a log–log plot of the pore compressibility determined under variable pore pressure, c_p versus $(p_c - p_p)$, is a linear relation of the general form [38]:

$$c_p = a_1(p_c - p_p)^{b_1} \tag{9.119a}$$

where

c_p = pore compressibility under variable p_p, psi^{-1}
p_p = pore pressure, psi
a_1, b_1 = correlation constants (Table 9.5)

TABLE 9.5 Correlation Constants for Eqs. (9.119) and (9.120)

Rock Type	Correlation Constants			
	Equation (9.119)		Equation (9.120)	
	a_1	b_1	a_2	b_2
Grainstone	0.00107	−0.632	1.06	−0.01055
Packstone	0.000837	−0.608	1.05	−0.00946
Wackestone	0.000953	−0.644	1.06	−0.00999
Mudstone	0.000114	−0.505	1.02	−0.00329
Berea	0.0236	−0.998	1.09	−0.0152

For the same rock samples, Harari et al. also found that porosity versus $(p_c - p_p)$ is a linear relation of the general form:

$$\frac{\phi}{\phi_0} = a_2(p_c - p_p)^{b_2} \tag{9.119b}$$

where

ϕ_0 = original porosity, %
ϕ = porosity under variable p_p, %
a_2, b_2 = correlation constants

Butalov proposed the following correlation between pore volume (V_p) of a consolidated sandstone, and the external pressure (P_e) and the pore pressure (P_p):

$$\frac{V_p}{V_{po}} = aP^3 + bP^2 + cP + d \tag{9.120}$$

where pressure is expressed in 10^{-2} kg$_f$/cm^2, and

- For $P = P_p$

$$a = -39.46 \times 10^{-3}, \quad b = 0.32826, \quad c = -3.083, \quad d = 100.045$$

- For $P = P_e$

$$a = -27.77 \times 10^{-2}, \quad b = 3.15362, \quad c = -10.810, \quad d = 100.859$$

As with all correlations, Eqs. (9.119a), (9.119b), and (9.120) should be used only to obtain an order of magnitude of c_p, ϕ, and V_p/V_{po} for carbonate rocks under variable pore-pressure conditions.

EFFECT OF STRESS ON FRACTURING

The stress conditions at the bottom of the wellbore greatly influence the strength and ductility of the rock being drilled or fractured [53–61]. Knowledge of the stress redistribution that occurs on drilling a wellbore is important in understanding (1) the causes of the reduction in the rate of penetration, (2) fluid loss problems, and (3) borehole and perforation instability problems in friable clastic formations.

Howard and Fast defined hydraulic fracturing as the process of creating a fracture or fracture system in a porous medium by injecting a fluid under pressure through a wellbore to overcome native stresses and to cause material failure of the porous medium [62]. The basic purpose of a fracture treatment is to increase the conduction of reservoir fluid into the wellbore. Hydraulic fracturing may be beneficial in several cases: (1) low-permeability reservoirs, (2) damaged wellbore due to invasion of drilling mud, (3) secondary recovery, and (4) disposal of industrial waste. It should be remembered, however, that migration of natural gas to the surface can occur along these fractures. Several researchers have published theories of the mechanics of failure of rocks subjected to internal fluid pressure [30, 62, 63]. In this section, only the practical effects of some elastic constants on fracturing are discussed.

EFFECT OF POISSON'S RATIO ON FRACTURE GRADIENT

Hydraulic fracturing has been extensively used for more than 40 years to stimulate the production of oil and natural gas from many different reservoir rocks. The subject of many debates during these years has been the fracture gradient (FG) required to induce fracturing of subsurface formations. Failure to accurately predict the formation fracture pressure gradient has resulted in the unintentional fracturing or opening of natural fissures and, consequently, very expensive and disastrous lost circulation problems. Lost circulation, which is the loss of large quantities of drilling fluid from a formation during drilling or well completion, may occur at any depth and anywhere the total pressure against the formation exceeds the total pressure in the formation. This pressure differential causes the sandface to fracture and/or enhances natural fractures, which may then open and drain circulation fluid. Once a fracture has been created, the fluid lost to the fracture will wash out and widen the fracture. In many cases, even when the total pressure against the formation is reduced, the fracture may not close completely and the fluid loss will continue. This is why the FG is an essential parameter required in the design of wells and why it has received a great deal of attention in the literature.

Hubbert and Willis showed that the Poisson's ratio of rocks, reservoir pore pressure gradient, and overburden stress gradient are the main factors controlling fracture pressure gradient, according to the following expression [64]:

$$FG = \frac{1}{D}\left[(\sigma_{OB} - p_p)\frac{\nu}{1-\nu} + p_p\right] \tag{9.121}$$

where σ_{OB} is the overburden pressure, D is the depth, p_p is the reservoir pore pressure and, ν is Poisson's ratio. Assuming that the stress gradient $g_s = \sigma_{OB}/D = 1.0$ psi/ft and $\nu = 0.25$, Eq. (9.121) becomes:

$$FG = \frac{1}{3}\left(1 + \frac{2p_p}{D}\right) \tag{9.122}$$

This equation is still widely used even though it is known to predict values that are usually too low compared with the values from field data, especially in the U.S. Gulf Coast at shallow depths. Equation (9.121) should give accurate values of FG if in situ values of σ_{OB} and n are available. The reservoir fluid (or pore) pressure (p_p) can be determined with great accuracy from pressure transient tests.

Matthews and Kelly modified the Hubbert and Willis equation by introducing the empirical concept of matrix stress coefficient, R_{ms} [65]:

$$FG = \frac{R_{ms}}{D}(\sigma_e + p_p) \tag{9.123a}$$

where

$$R_{ms} = \frac{\nu}{1-\nu} \tag{9.123b}$$

The effective stress, σ_e, is obtained from:

$$\sigma_e = g_s D - p_p = \sigma_{OB} - p_p \tag{9.124}$$

assuming the overburden stress gradient is 1.0 psi/ft and R_{ms} is obtained from Figure 9.48, where D_i is the depth for which the matrix stress σ_e would be the normal value:

$$D_I - \frac{\sigma_e}{0.535} = 1.869(D - P_p) \tag{9.125}$$

The matrix stress coefficient values used in Figure 9.48 are correlated with the equivalent normal pressure depth and, thus, depend on both depth and pore pressure.

Eaton investigated the effect of Poisson's ratio on the FG and concluded the following [66]:

1. Poisson's ratio for rocks increases with depth, particularly in the Gulf Coast region (Figure 9.49).

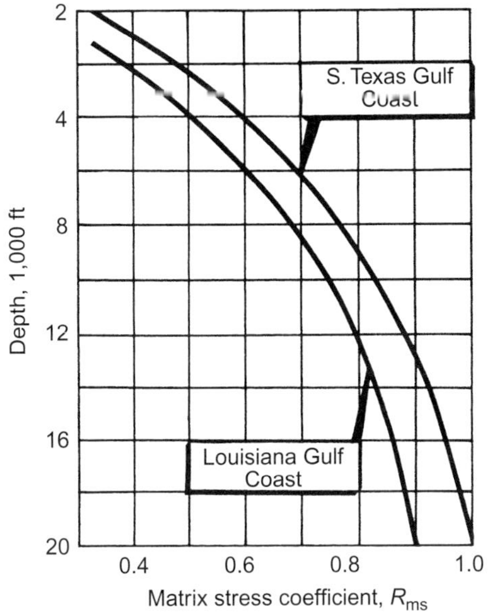

FIGURE 9.48 Matrix stress coefficient as a function of depth [66].

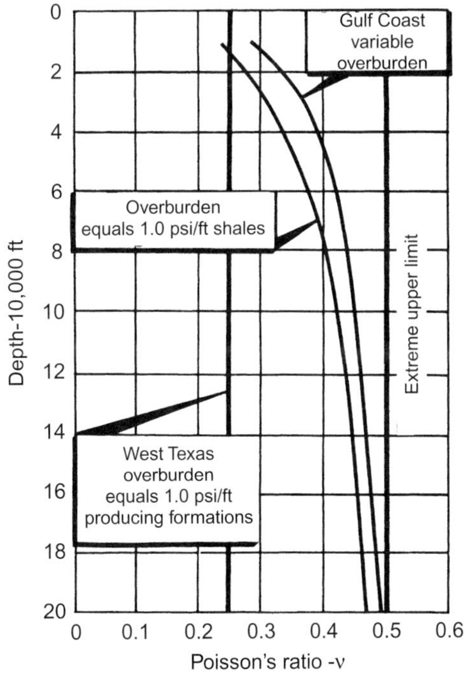

FIGURE 9.49 Poisson's ratio as a function of depth [66].

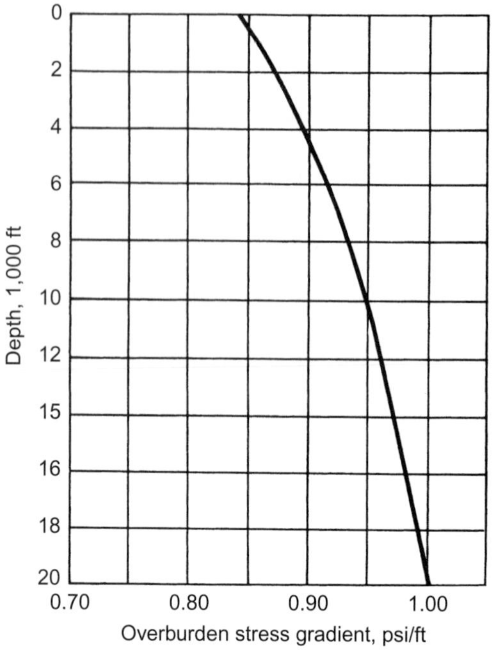

FIGURE 9.50 Overburden stress gradient as a function of depth for normally compacted Gulf Coast formations [66].

2. In the Gulf Coast region, the average overburden stress gradient is about 0.85 psi/ft near the surface and increases to 1.0 psi/ft at depths of about 20,000 ft (Figure 9.50).

A composite group of density logs from many Gulf Coast well logs was used to plot bulk density, σ_b, versus depth, as shown in Figure 9.51, and to generate overburden stress gradient data used to plot Figure 9.50. The curvature of the trend of depth versus Poisson's ratio in Figure 9.49 is caused by the sediments being younger and more compressible near the surface, but being less compressible and more plastic with depth. Eaton developed a monograph (Figure 9.52) for solving Eq. (9.121) to predict the fracture pressure gradients and suggested the following procedure for estimating the FGs for other tectonically relaxed areas of the Earth:

1. Obtain overburden stress gradient versus depth from bulk densities taken from well logs, seismic data, or shale density measurements.

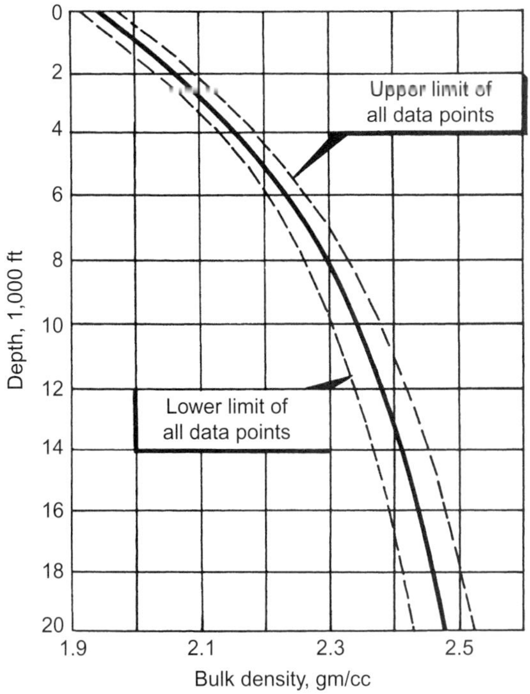

FIGURE 9.51 Bulk density as a function of depth for the Gulf Coast [66].

2. Convert this ρ_b versus D data into a plot of average overburden stress gradient versus depth similar to that shown in Figure 9.50. The values for ρ_b should be read at the midpoint of each 1,000-ft interval and averaged step-by-step downward to at least 20,000 ft of depth.
3. Obtain actual fracture pressure gradient, FG_a, for several depths from actual fracturing data, or lost-circulation or squeeze data.
4. Determine formation pressures that correspond to the same depths as in Step 3.
5. With these data and the following equation (Eq. (9.126)) Poisson's ratio curve can be back-calculated and plotted versus depth (similar to that shown in Figure 9.49):

$$\frac{\nu}{1-\nu} = \frac{FG_a - (p_p/D)}{(\sigma_{OB}/D) - (p_p/D)} \tag{9.126}$$

6. Combining these plots and Figure 9.52, accurate FG values can be predicted.

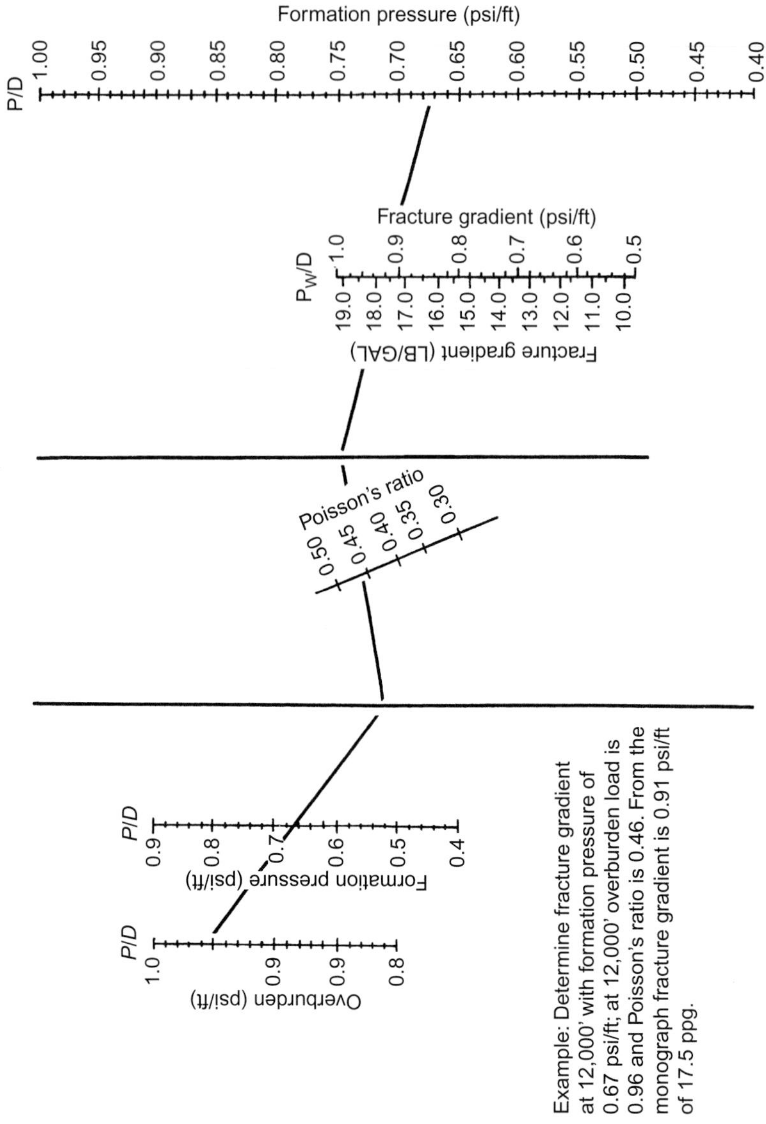

FIGURE 9.52 FG monograph [66].

7. Predicted values can be plotted as a function of depth as shown in Figure 9.53, and the resulting curves can be used in everyday operations: cementing, sand consolidation, matrix and fracture acidizing, hydraulic fracturing, and secondary recovery.

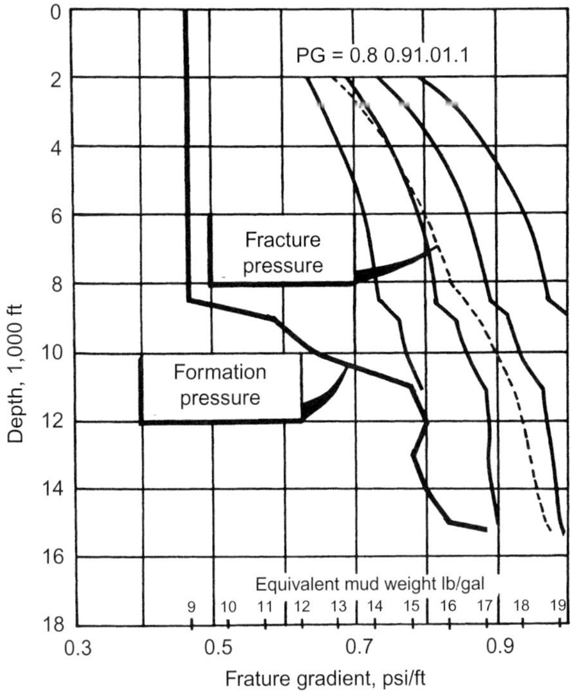

FIGURE 9.53 Fracture gradient as a function of depth for variable Poisson's ratio [66].

FG FOR OFFSHORE WELLS

Constant and Bourgoyne extended the correlations of Eaton and of Matthews and Kelly to offshore wells by including the effect of oceanwater depth on the overburden stress [67]. They developed the following equation for calculating the fracture pressure gradient:

$$\mathrm{FG} = \frac{p_\mathrm{f}}{(D_\mathrm{s} + D_\mathrm{W} + D_\mathrm{A})a_\mathrm{c}} \tag{9.127}$$

where D_s is the sediment depth, D_w is the water depth, D_A is the air gap from the rig to sea level, and a_c is the conversion constant. The fracture pressure, p_f, is the sum of the average horizontal matrix stress, σ_h, and the pore pressure p_p, where $\sigma_\mathrm{h} = R_\mathrm{ms}\sigma_\mathrm{e} = (v/(1-v))\sigma_\mathrm{e}$.

The vertical overburden stress, σ_OB, due to geostatic load at any depth, may be presented as follows:

$$\sigma_\mathrm{OB} = \int_\mathrm{O}^{D} \rho_\mathrm{b} g \, \mathrm{d}D \tag{9.128}$$

where g is the acceleration gravity, and the rock bulk density is given by:

$$\rho_b = \rho_g(1 - \phi) + \rho_{fl}\phi \tag{9.129}$$

Solving for the porosity ϕ, Eq. (9.129) yields:

$$\phi = \frac{\rho_g - \rho_b}{\rho_g - \rho_{fl}} \tag{9.130}$$

This equation is used to calculate porosity from density logs for a given ρ_{fl} and ρ_g. The trend of average porosity versus depth of sediment is assumed to be linear on a semilog plot, thus:

$$\phi = \phi_o \, e^{-bpD_s} \tag{9.131}$$

where b_p = porosity decline constant.

Inasmuch as, when moving further offshore, the sandstone layers gradually become much thinner, it is much more difficult to establish an accurate trend of porosity or to calculate with accuracy the values of ϕ_o and b_p, which are the intercept and slope, respectively, of the porosity decline curve on a semilog graph.

Substituting Eq. (9.129) into Eq. (9.128) yields:

$$\sigma_{OB} = \int_O^D [\rho g(1 - \phi) + \rho_{fl}\phi] dD \tag{9.132}$$

When an offshore well is considered, it is necessary to integrate Eq. (9.132) from the surface to the water line, D_w, and from the water line to the depth of interest, D_s. Hence:

$$\sigma_{OB} = \int_O^{D_W} \rho_{SW} \, dD + \int_O^{D_s} [\rho_g(1 - \phi) + \rho_{fl}\phi] dD \tag{9.133}$$

Substituting Eq. (9.131) for ϕ in Eq. (9.133) and integrating yields:

$$\sigma_{OB} = a_c \left[\rho_{SW} D_{SW} + \rho_m D_s - \frac{\phi_0}{b_p}(\rho_m - \rho_f)(1 - e^{-b_p D_s}) \right] \tag{9.134}$$

where D_{sw} is the seawater depth, ρ_{sw} is the seawater density, ρ_g is the grain density, ρ_f is the fluid density, and b_p is the porosity decline constant. If oil field units are used in Eq. (9.134) then $a_c = 0.052$, whereas for SI units $a_c = 9.81 \times 10^{-3}$.

The matrix stress ratio can be presented by an equation of the following general form:

$$R_{ms} = \frac{v}{1 - v} = 1 - ae^{bD_s} \tag{9.135}$$

The sediment depth, D_s, is used rather than the total depth so that the matrix stress starts at the mud line. The curve fit constants, a and b, are

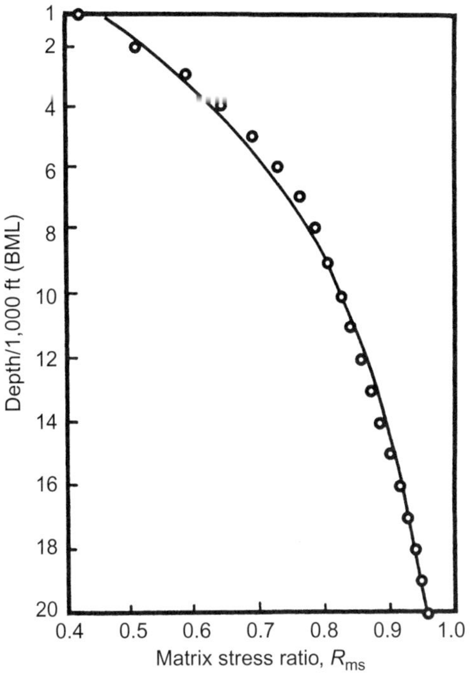

FIGURE 9.54 Curve fit of matrix stress ratio versus depth data [67].

dependent on local stress conditions. For instance, when Eaton's data for Poisson's ratio in the Gulf Coast was converted to R_{ms} and curve-fitted using Eq. (9.135), as shown in Figure 9.54, Constant and Bourgoyne found that $a = 0.629$ and $b = -1.28 \times 10^{-4}$ [67]. Thus, for stress conditions similar to those in the Gulf Coast, the matrix stress ratio as a function of depth of sediments can be estimated from the following equation:

$$R_{ms} = 1 - 0.629\, e^{-128 \times 10^{-4} D_s} \tag{9.136}$$

This correlation can be used in other regions until enough data are obtained to perform a curve fit and determine local values of a and b. Equation (9.136) tends to yield conservative fracture pressure gradient predictions for offshore wells.

Example

Data from a Louisiana Gulf Coast well in the Green Canyon area are summarized in Table 9.6. Calculate the predicted fracture pressure gradients at various depths and compare with the value obtained from leak-off test: 15.3 lb/gal = 0.795 psi/ft.

TABLE 9.6 Green Canyon Well No. 1 Data [67]

Water Depth = 1,223; Air Gap = 85 ft		
D_s, BML (ft)	Pore Press. Gradient (lb/gal)	Frac. Press. Gradient (lb/gal) (Leak-off Test)
4,100	9.8	13.9
6,692	11.6	15.3
8,692	13.2	16.0

Solution

The calculation procedure is presented for only one depth ($D_s = 6{,}692$ ft).
For Gulf Coast area:
$\rho_{sw} = 8.5$ lb/gal
$\rho_g = 21.6$ lb/gal (2.6 g/cm^3)
$\rho_{fl} = 8.95$ lb/gal (1.074 g/cm^3)
$b_p = 0.000085$ ft^{-1}
$\phi_o = 0.41$

1. The overburden stress at $D_s = 6{,}692$ ft is obtained from Eq. (9.134), where $a_c = 0.052$ for the units given in Table 9.6:

$$\begin{aligned}\sigma_{OB} &= 0.052(8.5 \times 1{,}223 + 21.66 \times 6{,}692) \\ &\quad - \frac{0.41}{8.5 \times 10^{-5}}(21.66 - 8.95)(1 - e^{-8.5 \times 10^{-5} \times 6{,}692}) \\ &= 6{,}694 \text{ psi}\end{aligned}$$

2. The effective overburden stress is $\sigma_{OB} - p_p$, where the pore pressure p_p is obtained from:

$$p_p = a_c g_p (D_s + D_w + D_A) \tag{9.137}$$

and g_p is the pore-pressure gradient; thus:

$$p_p = 0.052 \times 11.6(6{,}692 + 1{,}233 + 85) = 4{,}820 \text{ psi}$$
$$\sigma_e = 6{,}694 - 4{,}826 = 1{,}868 \text{ psi}$$

3. From Eq. (9.136), the matrix stress ratio is given by:

$$R_{ms} = 1 - 0629\, e^{-1.28 \times 10^{-4} \times 6{,}692} = 0.733$$

and Poisson's ratio is estimated from:

$$\nu = \frac{R_{ms}}{R_{ms} + 1}$$

thus

$$\nu = \frac{0.733}{1 + 0.733} = 0.42$$

4. The average horizontal rock stress is given by:

$$\sigma_h = 0.733 \times 1{,}868 = 1{,}369 \text{ psi}$$

5. The fracture pressure is given by:

$$P_f = 1{,}369 + 4{,}826 = 6{,}195 \text{ psi}$$

6. The fracture pressure gradient at the total depth of 6,692 + 1,233 + 85 = 8,040 ft is:

$$\text{FG} = \frac{6{,}195}{0.052 \times 8{,}000} = 14.9 \text{ lb/gal} = 0.775 \text{ psi/ft}$$

which is lower than the observed fracture pressure gradient value obtained from a leak-off test (15.3 lb/gal = 0.795 psi/ft). The difference may be due to at least three factors:

- Difficulty was encountered in obtaining porosity and density data from well logs at shallow depths; therefore, typical Gulf Coast values of ϕ and ρ were used.
- Presence of salt domes—when drilling on flanks of salt domes, lateral matrix stresses tend to be higher than expected and, in some cases, may be even greater than the vertical overburden stress.
- Presence of plastic shale domes, that is, alternating sandstone and shale layers. Plastic shale domes normally extend to depths in excess of 10,000 ft near the shoreline and gradually become much thinner further offshore. The predominantly plastic shale layers (15% or less sandstone) are reached at much shallower depths. It has long been recognized that FGs in sandstones are lower than in the plastic shales. The more plastically a formation behaves, the less tendency there will be for differences between the horizontal and vertical overburden stresses [67].

The difference between the calculated and observed FGs also may be due to the assumption that is implicit in the derivation of Eq. (9.134). Pilkington correlates the matrix stress ratio R_{ms} as follows [68]:

For $\sigma_{OB}/D \leq 0.94$:

$$R_{ms} = 3.9\left(\frac{\sigma_{OB}}{D_s}\right) - 2.88 \tag{9.138}$$

and, for $\sigma_{OB}/D > 0.94$:

$$R_{ms} = 3.2\left(\frac{\sigma_{OB}}{D_s}\right) - 2.224 \tag{9.139}$$

where σ_{OB} is the effective overburden stress, in psi, and D is the depth in ft. These statistical correlations are valid for both normally and abnormally pressured sandstones. Using these expressions, the equation for FG becomes:

$$\mathrm{FG} = \frac{1}{D_s}\left[R_{ms}\left(\frac{\sigma_e}{D_s}\right) + p_p\right] \tag{9.140}$$

This equation is valid for both normally and abnormally pressured formations in tectonically relaxed areas containing plastic shales with interbedded sands. It is not valid, however, for brittle or naturally fractured formations, including limestones and dolomites [68].

EFFECT OF POISSON'S RATIO ON FRACTURE DIMENSIONS

A vertical fracture formed by hydraulic fracturing extends (in length, height, and width) according to the so-called penny-shape crack theory, which assumes that the fracture height is constant along the length of the fracture [69]. In many cases, however, the fracture height is variable because shales are barriers possessing higher horizontal stress than sandstones. In naturally fractured reservoirs and shaly formations, it is very difficult to predict fracture height. Currently used field techniques to measure fracture height can be classified into two groups: (1) techniques that directly measure the fracture height, such as formation microscaner, borehole televiewer, and spinner survey and (2) techniques that are based on interpretation of well logs such as the temperature and gamma-ray logs. By comparing temperature logs run before and after fracturing, as shown in Figure 9.55, the zone cooled by injecting fracturing fluid can be identified and its height measured. Similarly, gamma-ray logs run before and after a stimulation treatment can be compared to locate intervals contaminated by the injected radioactive-tagged propping agent [30].

Labudovic determines hydraulic fracture height by calculating Poisson's ratio values from velocities of longitudinal, v_c, and transversal v_s, waves given by Pirson and by Rzhevsky and Novik [70–72]. These values are then tabulated versus depth. The area of lower values of Poisson's ratio ν is the area of fracture height extension, and higher values of ν represent shale barriers. This height is then compared, and corrected if necessary, with a log diagram for resistivity, R, and spontaneous potential, SP, from the same interval. According to Pirson:

$$v_c = \left[\frac{E}{\rho_b}\left(\frac{1-\nu}{(1-\nu-2\nu^2)}\right)\right]^{0.5} \tag{9.141}$$

$$v_c = \left[\frac{E}{\rho_b}\left(\frac{0.5}{(1-\nu)}\right)\right]^{0.5} \tag{9.142}$$

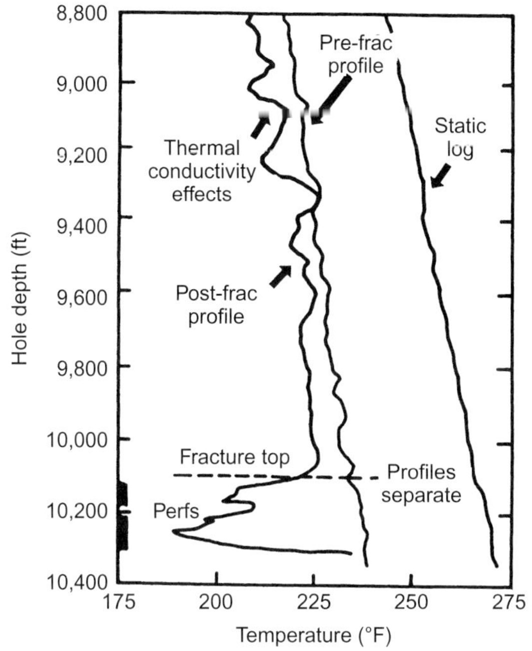

FIGURE 9.55 Temperature profile before and after fracturing [30].

where E is Young's modulus of elasticity and ρ_b is the bulk density of the rock.

Figure 9.56 shows the profile of Poisson's ratio of a fractured oil well located in a sandstone reservoir with thick deposits of shale on the bottom and shaly sandstone on the top. Values of Poisson's ratio are obtained from Eq. (9.142). The fracture was created with a hydrocarbon-based gel and propped with 0.8–1.2 mm sand, with continuous injection of radioactive tracer. Several days after hydraulic fracturing and casing decontamination, radioactive tracer logging was performed to establish the height of the fracture. It is evident from Figure 9.56 that the bottom tip of the fracture is located at a depth of 795 m, where Poisson's ratio increased from approximately 0.29 to 0.32. This increase in n corresponds to a definite change in SP, resistivity, and tracer profiles. The figure also shows that it is relatively difficult to locate the exact location of the top tip of the fracture, based on Poisson's ratio alone. This difficulty is due to the similarity between horizontal stresses in the top shaly sandstone deposit and stresses in the producing sandstone. The tracer and resistivity profiles, however, seem to indicate that the top tip, in this case, is at a depth of 784 m. Thus, the fracture height is 11 m. Labudovic found that the predicted fracture height using Eaton's method deviated considerably from the measured one, whereas the

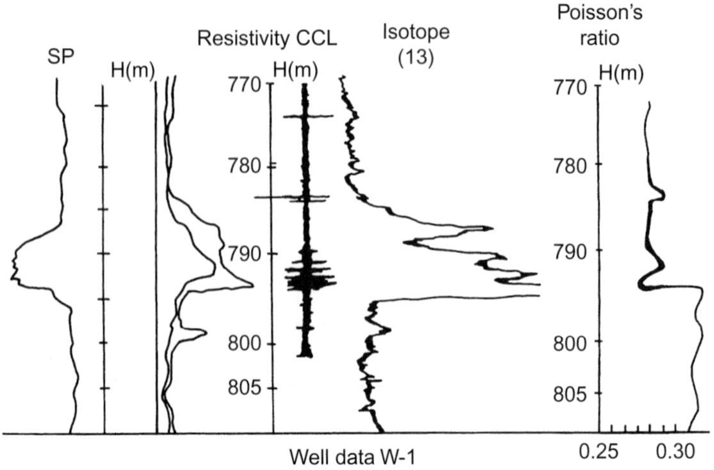

FIGURE 9.56 Poisson's ratio versus fracture height determined by using radioactive tracer isotope I-131 [70].

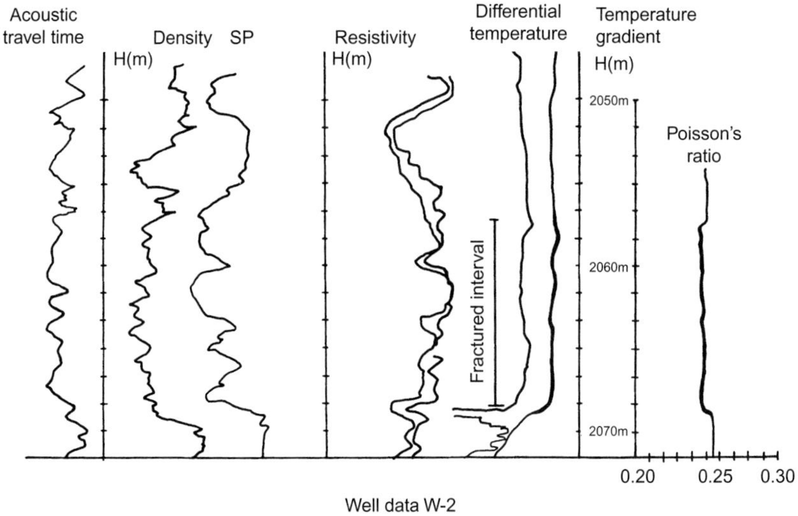

FIGURE 9.57 Poisson's ratio versus fracture height determined by using differential temperature log [70].

correlation of *n* values with SP and resistivity log values yielded only a slight deviation between the measured and predicted fracture heights [70].

Figure 9.57 shows how fracture height is determined from well logs, using differential temperature logs and Poisson's ratio values. The pay zone, in this figure, is a sandstone layer with shaly sandstone at the top and bottom. The predicted fracture height deviated slightly from the measured one. The well was fractured with a water-based gel, and the breccia–dolomite deposits were selectively acidized. Radioactive tracer I-131 was continuously added to the diverting agent (benzoic acid). In this case, the measured fracture height of the producing interval and the zone in which acid entered the formation coincided with the predicted interval. The reason for this accuracy is the relatively sharp difference between Poisson's ratio values of the top and bottom deposits, and that of the producing section.

The stress intensity factor, I_s, for an infinitely long and linear crack of height H_f, internally pressurized by a propping fluid and propagating through a homogeneous rock is given by [73]:

$$I_s = 1.25(p_{fl} - \sigma_h)\sqrt{H_f} = 1.25\ \Delta p\sqrt{H_f} \tag{9.143}$$

where p_{fl} is the fluid pressure inside the fracture, σ_h is the formation stress normal to the plane of fracture, and $\Delta p = P_{fl} - \sigma_h$. For a finite coin-shaped fracture of radius 0.5 H_f, Eq. (9.143) becomes:

$$I_s = 0.80\ \Delta p\sqrt{H_f} \tag{9.144}$$

It is believed that if I_s reaches a critical value, I_{sc}, the fracture will propagate. The critical stress intensity factor is the property commonly known as the fracture toughness. Measured values of I_{sc} are 950–1,650 psi in.$^{0.5}$ for silt stones, 400–1,600 psi in.$^{0.5}$ for sandstones, 400–950 psi in.$^{0.5}$ for limestones, and 300–1,200 psi in.$^{0.5}$ for shales. Knowing I_{sc}, one can estimate the width of the fracture, w_f, from the following equation:

$$w_f = \left(\frac{1-\nu}{G}\right)H_f\ \Delta P = \left(\frac{1-\nu}{G}\right)I_{sc}\sqrt{H_f} \tag{9.145}$$

where G is the shear modulus of the fractured formation.

Figure 9.58 shows a hydraulic fracture in a three-layer system, where G_1 is the shear modulus of the fractured producing zone of height h, G_2 is the

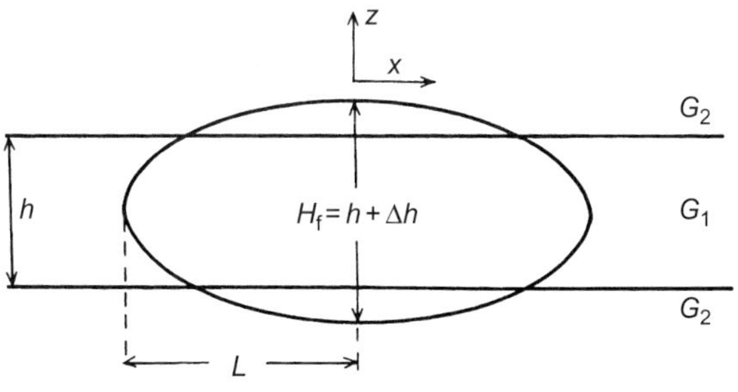

FIGURE 9.58 Vertical fracture in a three-layer system [73].

shear modulus of the top and bottom deposits, and the maximum width of the fracture is:

$$w_f = (1-v)\Delta P\left(\frac{h}{G_1} + \frac{\Delta h}{G_2}\right) = (1-v)\Delta P\left(\frac{h}{G_1} + \frac{H_f - h}{G_2}\right) \tag{9.146}$$

Equation 9.146 assumes implicitly that Poisson's ratio is the same in the three layers. If both G and n vary between the layers, Eq. (9.146) becomes:

$$w_f = \Delta p\left[\left(\frac{1-v_1}{G}\right)h + \left(\frac{1-v_2}{G}\right)(H_f - h)\right] \tag{9.147}$$

where $\Delta p = p_w - \sigma h$, and p_w is the fluid pressure in the wellbore. The half-length of the fracture can be estimated from:

$$L = \frac{h}{2}\left[1 + \frac{24\,G_2}{19\,G_1}\log\frac{H_f}{h}\right]^{0.5} \tag{9.148}$$

Whether or not the adjacent deposits at the top and bottom of the producing formation will act as a fracture barrier may depend on several factors: (1) differences of in situ stress, (2) elastic properties, (3) fracture toughness, (4) ductility, (5) permeability, and (6) type of bonding at the interface. Van Eekelen analyzed these factors with respect to their relative influence on fracture containment and concluded the following [74]:

1. The commonly used concept of stress intensity factor for predicting propagation or containment of fractures has only limited applicability.
2. In most cases, the fracture will penetrate into the layers adjoining the pay zone that is being fractured.
3. Contrasts in stiffness and in situ stress between the pay zone and adjoining layers tend to limit the penetration depth of the fracture into these layers.

Thus, Poisson's ratio and shear modulus have a considerable effect on fracture dimensions; however, differences of these properties between the reservoir rock and the bounding layers are not sufficient to contain the fracture. Warpinsky et al. analyzed the results of many hydraulic fracturing experiments and showed [75]: (1) the minimum in situ stress is the predominant influence on the propagation of fractures, (2) the orientation of the minimum in situ stress dictates the orientation of fractures, and (3) steep gradients and discontinuities in the magnitude of this stress can act as barriers to fracture propagations.

IN SITU STRESS DISTRIBUTION

The relationship between the strength and elastic properties as measured at laboratory conditions and those which exist at wellbore depths is still not well understood. However, several theories describing the induced stress distribution around a drilled hole are available. One of the early theories was proposed by Westergaard, who used the concept of effective stresses to show that, at great depths, a plastic state which relieves the stresses exists around the hole [76]. Later, Paslay and Cheatham investigated the rock stresses induced by producing reservoir fluid, assuming the rock behaved elastically [77]. Bradley developed a useful semiempirical approach for predicting the limit of elastic behavior in inclined boreholes [78].

Risnes et al. applied the theories of elasticity and plasticity to investigate the stresses in a poorly consolidated sand around a wellbore [79]. They showed that there is a plastically strained zone just around the wellbore of the order of magnitude of 1 m, and higher degree of consolidation yields smaller plastic zones. Breckels and van Eekelen used a large amount of field data to generate empirical equations to describe the trend of horizontal stress with depth for normally pressured formations in the U.S. Gulf Coast region, Venezuela, Brunei, the North Sea, and the Netherlands [80]. Coupled with regional correlations between the horizontal stress and pore pressure, these correlations enable horizontal stress levels to be estimated rather accurately for a given depth, provided the pore pressure is known.

The total vertical stress or overburden stress σ_{OB}, which is normally equivalent to the maximum principal in situ stress, can be determined from well logs. The in situ minimum principal (horizontal) stress, σ_{Hmin}, can be approximated, sometimes very accurately, by the instantaneous shut-in pressure recorded during or after a fracturing job. In regions without tectonic activities, the maximum total horizontal stress, σ_{Hmax}, is approximately equal to the minimum total horizontal stress, σ_{Hmin}. In the more general case ($\sigma_{Hmax} > \sigma_{Hmin}$), however, these approximations are invalid, and the correlations derived by Breckels and van Eekelen appear to be very useful [80]. Using a large number of basic data obtained from tests such as leak-off tests and casing-seat tests carried out to determine fracturing pressures or instantaneous shut-in pressures, Breckels and van Eekelen derived a relationship between the minimum horizontal stress, σ_{Hmin}, and depth, D, for various parts of the world.

Figure 9.59 shows a plot of σ_{Hmin} versus D for more than 300 data points for normally pressured formations in the U.S. Gulf Coast. These data are obtained from hydraulic fracturing tests and leak-off tests. The latter tests, however, can only give a range of pressures at which the formation starts taking fluid, that is, the fracturing pressure, p_f, which may range from σ_{Hmin} to $2\sigma_{Hmin} - p_p$, where p_p is the pore pressure. Consequently, the lower end of the range of values has been used to obtain an approximate trend for σ_{Hmin} (psi) as a function of depth (ft), as shown in Figure 9.59. The solid curve in this figure forms a lower limit to 93% of the data points. The following two correlations are valid for normally pressured sands:

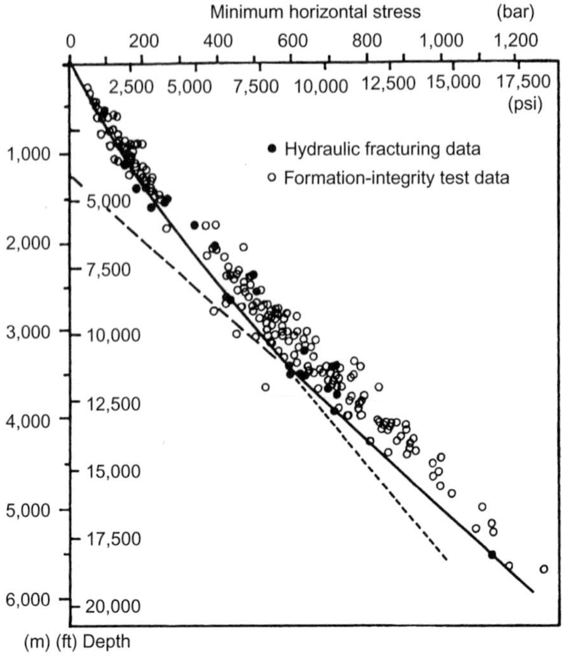

FIGURE 9.59 Minimum horizontal stress as a function of depth for Gulf Coast formations [80].

- For $D \leq 11{,}500$ ft:

$$\sigma_{\text{Hmin}} = 0.197\, D^{1.145} \tag{9.149}$$

- For $D > 11{,}500$ ft:

$$\sigma_{\text{Hmin}} = 1.167D - 4{,}596 \tag{9.150}$$

The lower curve represents a good correlation between the horizontal stress and depth for the normally pressured formations, because of the following:

1. Formation integrity test data, which are generally influenced by the hoop stress around the borehole, lead to an overestimation of σ_{Hmin}.
2. Only data points from normally pressured and overpressured formations, which normally have higher total horizontal stress than the underpressured formations, are included in Figure 9.59 [77].

The minimum horizontal stress in abnormally pressured formations in the Gulf Coast region can be estimated from the following correlations:

- For $D \leq 11{,}500$ ft:

$$\sigma_{\text{Hmin}} = 0.197D^{1.145} + 0.46(p_p - p_{pn}) \tag{9.151}$$

where p_{pn} is the normal pore pressure corresponding to a gradient value of 0.465 psi/ft, that is, $p_{pn} = 0.465D$.

- For $D > 11{,}500$ ft:

$$\sigma_{\text{Hmin}} = 1.167D - 4,596 + 0.46(p_p - p_n) \quad (9.152)$$

Although Eq. (9.151) is not supported by hydraulic fracturing, it can be used as a first estimate of the minimum in situ horizontal stress during the design of a fracturing job.

Using hydraulic fracturing data from Venezuela, particularly from Lake Maracaibo Block 1, and the original pore-pressure gradient of the Eocene formation of Block 1, which is 0.433 psi/ft, a relationship was derived between the σ_{Hmin} and depth for abnormally pressured formations:

$$\sigma_{\text{Hmin}} = 0.21\text{D}^{1.145} + 0.56(p_p - p_{pn}) \quad (9.153)$$

This correlation gives good results for $5{,}900 < D < 9{,}200$ ft. Currently, there is not enough data to support extrapolating this relationship beyond Lake Maracaibo. Normally, pressured formations in this area (Figure 9.60, Curve 1) are at a slightly higher stress level than the U.S. Gulf Coast sediments (Curve 2).

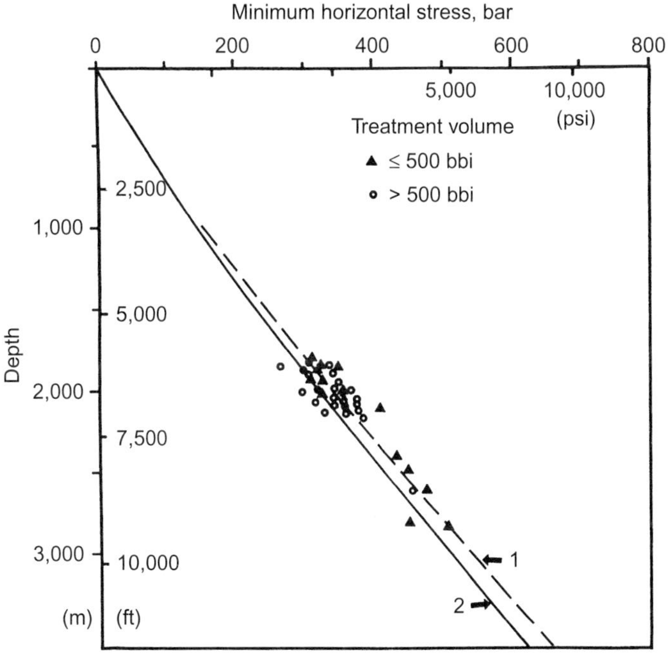

FIGURE 9.60 Minimum horizontal stress as a function of depth in Venezuela [80].

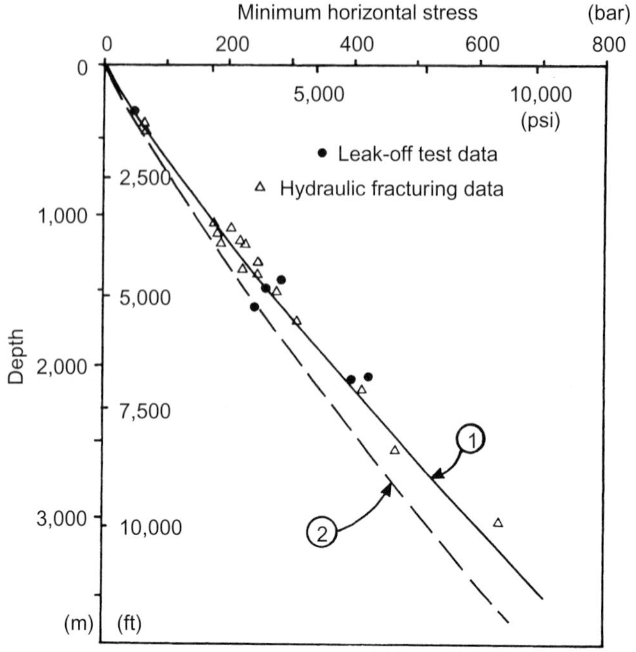

FIGURE 9.61 Minimum horizontal stress as a function of depth in Brunei [80].

An approach similar to that used in the U.S. Gulf Coast and Venezuela was adopted to derive the following correlation for abnormally pressured formations in Brunei for $D < 10{,}000$ ft using the leak-off test data and hydraulic fracturing data:

$$\sigma_{\text{Hmin}} = 0.227D^{1.145} + 0.49(p_{\text{p}} - p_{\text{pn}}) \tag{9.154}$$

The normal pore-pressure gradient is 0.433 psi/ft, that is, $p_{\text{pn}} = 0.433D$. Normally, pressured formations in the offshore Brunei appear to be subject to higher compressive stresses (Figure 9.61, Curve 1) than in the U.S. Gulf Coast (Curve 2). Comparing the leak-off with instantaneous shut-in pressure data obtained in Brunei, during the same test, it was found that the leak-off pressures exceed the instantaneous shut-in pressure values by 11%. Extending this percentage to the U.S. Gulf Coast data, the following best-fit correlation is obtained for $D \leq 11{,}500$ ft:

$$\text{Leak-off pressure} = 0.219D^{1.45} \tag{9.155}$$

Similarly, the best-fit curve for leak-off test data from the North Sea yields:

$$\text{Leak-off pressure} = 0.353D^{1.091} \tag{9.156}$$

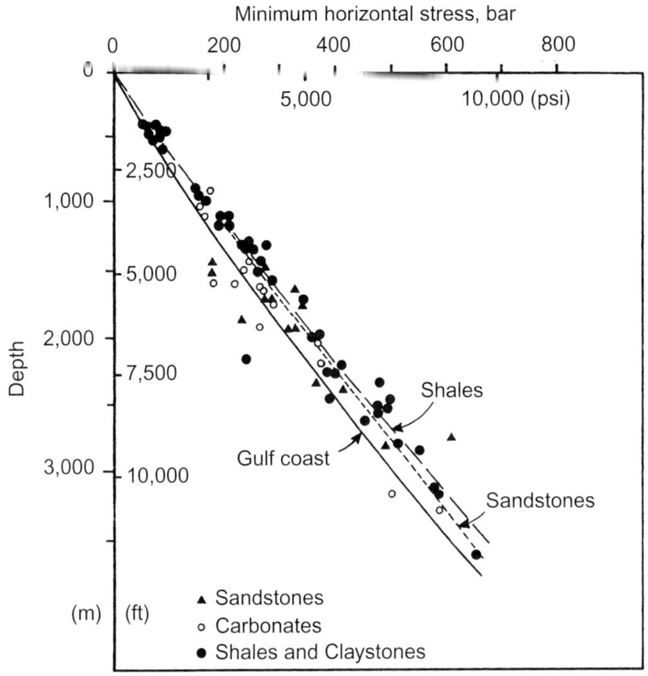

FIGURE 9.62 Minimum horizontal stress as a function of depth in the North Sea [80].

Breckels and van Eekelen analyzed leak-off test data from the North Sea (Figure 9.62, Curve 1) and concluded that the U.S. Gulf Coast correlations for normally pressured formations (Curve 2), that is, Eqs. (9.149) and (9.150), can be used with a fair degree of confidence to predict σ_{Hmin} as a function of depth. The same conclusion can be extended to normally pressured formations in Venezuela, Brunei, and other areas of the world.

Example

Estimate the minimum horizontal stress of an 8,000-ft deep well in the U.S. Gulf Coast region. The pore pressure of the sandstone reservoir is 4,200 psi.

Solution

The normal pore-pressure gradient for the U.S. Gulf Coast is 0.465 psi/ft. Thus, the normal pore pressure for this well is given by:

$$p_{pn} = 0.465 \times 8{,}000 = 3{,}720 \text{ psi}$$

Inasmuch as $p_{pn} < 4{,}200$ psi, this well is obviously abnormally pressured (overpressured). Using Eq. (9.151), one can determine σ_{Hmin}:

$$\begin{aligned}\sigma_{Hmin} &= 0.197(8{,}000)^{1.145} + 0.465(4{,}200 - 3{,}720)\\ &= 5{,}801 + 221 = 6{,}022 \text{ psi}\end{aligned}$$

Thus, the overpressure (4,200 −3,720 = 480 psi) causes the total horizontal stress to be higher (0.465 × 480 = 221 psi) than if the formation had normal pore pressure.

Lindner and Halpern, and Haimson have compiled from different sources, mostly from hydraulic fracturing tests, the in situ stress measurements from hundreds of locations in the North America and have plotted them on a map of the United States [81, 82]. Haimson showed that throughout the United States all three principal stresses are compressive, with the major (or maximum) horizontal principal compression direction generally toward the northeast between N45°E and N75°E; and a hydraulic fracture would generally be parallel to the maximum horizontal stress direction.

EFFECT OF STRESS CHANGE ON ROCK FAILURE

In reservoir rock, there are two primary stresses: (1) the effective stress (grain-to-grain stress) and (2) pore pressure. Overburden load is transmitted to the underlying layers through these two stresses. Under normal production conditions, reduction in pore pressure due to fluid withdrawal, grain-to-grain effective stress increases and is directly proportional to the decrease in pore pressure. Shear failure of the rock occurs when effective stress reaches the threshold value. The weak point is the first to fail and usually is the perforation cavity. This is also because the wellbore is the point of lowest pore pressure and maximum effective stress. In a homogeneous reservoir rock, the effective stress radially decreases with eye at the wellbore, as shown in Figure 9.63. However, in a reservoir rock with varying strength, the stress profile changes radically as we move away from the wellbore and depends on the shear strength of the rock, magnitude of drawdown pressure, reservoir

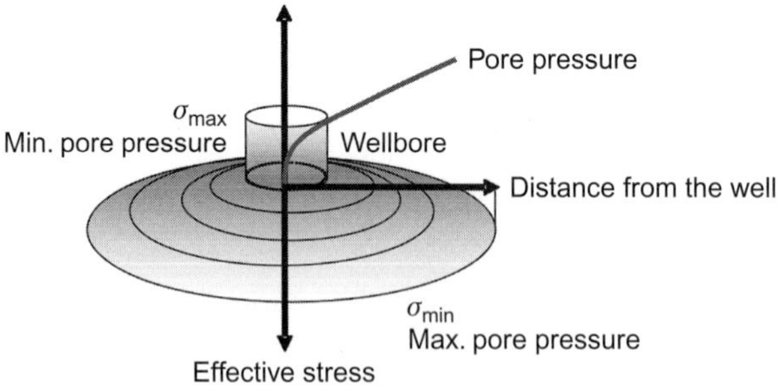

FIGURE 9.63 Effective stress radially decreasing from the wellbore in a homogeneous reservoir rock.

permeability and porosity, and fluid properties such as viscosity. The failing rock particles are then carried to the wellbore by the fluid, thereby plugging the wellbore tubular and equipment downhole. Crushed sand is very abrasive and erodes the metals with which it comes in contact.

Production history, fluid pressure, and uniaxial compressive strength are basic data in sand production evaluations. The two possible mechanisms of rock failure and sand production are tensile rupture and compressive rupture. Tensile rupture occurs under two conditions: (1) the fluid pressure gradient at the wellbore (sandface) is greater than the radial stress and (2) the tangential effective stress does not exceed the level of compressive failure of the rock ($\sigma_\theta < \sigma_{UCS}$). As fluid gradient and tangential compressive stress are linked through the equilibrium equation of the sand, conditions (1) and (2) impose an upper limit on the drawdown ($\Delta P_{max} = P_p - P_{wf}$) to avoid rock failure and sand production. ΔP_{max} is proportional to σ_{UCS} and various values of the ratio $\Delta P_{max}/\sigma_{UCS}$ can be found, depending on the drainage geometry and production history (influence of shut-in).

CHANGE IN STRESS FIELD DUE TO DEPLETION AND REPRESSURIZATION

Because of drilling operations, natural depletion, and injection of fluids, the stress distribution around the well shows changes. This change also affects the critical or threshold shear stress of the rock, the stress at which rock fails, and solids are dislodged. The principal stress is the overburden stress. As shown in Figure 9.64, the rock fails at the minimum horizontal stress.

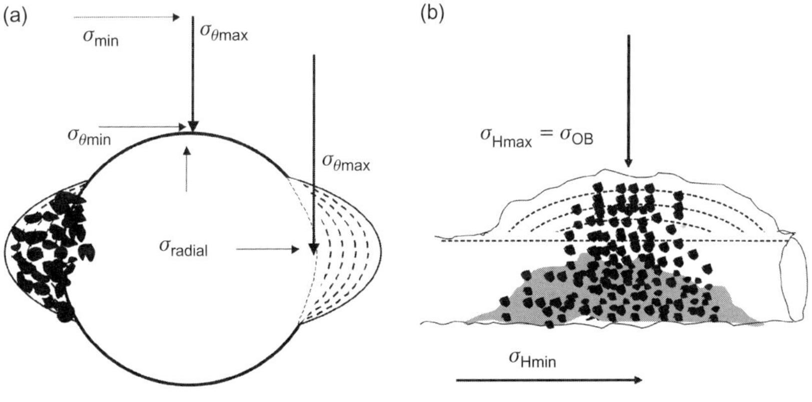

FIGURE 9.64 Schematic borehole breakout illustrates the relationship between far-field principal stresses, effective tangential stresses, shear failure of the wellbore wall, and solid in vertical wellbore and horizontal wellbore, respectively [83].

The knowledge of minimum horizontal and maximum horizontal stress is essential in order to determine borehole stability and the correct value of pressure drawdown.

The overburden stress does not change and is considered constant; however, for Biot coefficient equal to unity, the change in the effective stress is equal to the change in the pore pressure, in the opposite direction, caused by the depletion or injection. As shown in Figure 9.63, effective stress increases if the pore pressure decreases and vice versa. These changes in minimum and maximum horizontal stress are expressed as follows:

$$\Delta\sigma_{Hmax} = (1 - K_2)\Delta p_p \tag{9.157}$$

$$\Delta\sigma_{Hmin} = (1 - K_1)\Delta p_p \tag{9.158}$$

where K_1 and K_2 are the stress paths, defined as:

$$K_1 = \frac{\Delta\sigma_{Hmin}}{\Delta\sigma_{OB}} \tag{9.159}$$

$$K_2 = \frac{\Delta\sigma_{Hmax}}{\Delta\sigma_{OB}} \tag{9.160}$$

Equations (9.157) and (9.158) indicate that, for the value of $K < 1$, the total horizontal stresses decrease as the pore pressure is reduced.

STRESS RELATIONSHIP AT THE WELLBORE

The simple correlation which relates the maximum tangential stress to the principal stresses at the wellbore, is given as:

$$\sigma_{\theta max} = 3\sigma_{Hmax} - \sigma_{Hmin} - P_{bh} \tag{9.161}$$

The maximum tangential stress occurs in the direction of the least principal stress. The axial stress is estimated as:

$$\sigma_{axial} = \sigma_{OB} + 2\nu(\sigma_{Hmax} - \sigma_{Hmin}) \tag{9.162}$$

where ν is the static Poisson's ratio of the formation rock. Equations (9.161) and (9.162) are valid only if the pore pressure is constant throughout the formation. To incorporate the changing pore pressure profile and the pore elastic term, Haimson and Fairhurst [84] developed the following correlation:

$$\Delta\sigma = \alpha\left[\frac{(1-2\nu)}{(1-\nu)}(P_{wb} - P_p)\right] \tag{9.163}$$

where α is the Biot coefficient, and P_{wb} is the pore fluid pressure just behind the wellbore.

ESTIMATING CRITICAL BOREHOLE PRESSURE IN VERTICAL WELLS

Overbalanced Conditions

For an impermeable mud cake, the pore elastic effect does not exist. The maximum effective tangential stress is given by [83]:

$$\sigma_{\theta \max,\text{eff}} = 3\sigma_{\text{Hmax}} - \sigma_{\text{Hmin}} - P_{\text{bh}} - P_{\text{p}} \tag{9.164}$$

where P_{bh} is the wellbore pressure.

The radial effective stress at the wellbore wall is given by:

$$\sigma_{\theta\text{radial,eff}} = P_{\text{bh}} - P_{\text{p}} \tag{9.165}$$

The combination of Eqs. (9.164) and (9.165) and the Mohr–Coulomb shear failure criterion, Eq. (9.31), results in the critical bottomhole flowing pressure as shown by Rhett and Risnes [83]:

$$P_{\text{wfco}} = 0.5[3\sigma_{\text{Hmax}} - \sigma_{\text{Hmin}} - \sigma_{\text{UCS}}](1 - \sin\phi_{\text{f}}) + p_{\text{p}} \sin\phi_{\text{f}} \tag{9.166}$$

where ϕ_{f} is the internal friction angle and is a material property of the rock. It is obtained from the slope of the failure line on the Mohr–Coulomb figure. The most common overbalanced conditions are encountered during drilling operations and injection projects, such as waterflooding. If the calculated critical bottomhole flowing pressure is less than the pore pressure, then the assumption of overbalance is not valid.

The critical wellbore flowing pressure is a function of pore pressure, which continuously changes with depletion. The change in critical pressure with change in pore pressure is simply the derivative of Eq. (9.166), as shown by Rhett and Risnes [83]:

$$\Delta P_{\text{wfco}} = 0.5[(2 - 3K_2 + K_1) + \sin\phi_{\text{f}}(3K_2 - K_1)]\Delta P_{\text{p}} \tag{9.167}$$

Balanced and Underbalanced Conditions

The most common practice of underbalanced conditions is where wellbore flowing pressure is always lower than the reservoir pressure. Recently, drilling and perforation techniques have been frequently applied to maintain wellbore pressure in order to reduce damage to the formation. Such conditions, however, affect the effective stress at the wellbore. A useful correlation for underbalanced conditions is [83]:

$$p_{\text{wfcu}} = (1 - \nu)(3\sigma_{\text{Hmax}} - \sigma_{\text{Hmin}} - \sigma_{\text{UCS}}) - P_{\text{p}}(1 - 2\nu) \tag{9.168}$$

where σ_{UCS} is the unconfined compressive strength of the rock, psi. Equation (9.168) requires accurate knowledge of the downhole stresses. If the estimated value of critical borehole pressure (Eq. (9.168)) does not agree with the filed observations of the pressure at which first sand production was observed, the values of the stress can be adjusted to match the stress history of the formation.

The change in the critical pressure with a certain change in pore pressure during underbalanced conditions is given by:

$$\Delta P_{\mathrm{wfcu}} = [1 - (1 - \nu)(3K_2 - K_1)]\Delta P_{\mathrm{p}} \tag{9.169}$$

Thus, at any given stage of depletion, the net effective critical wellbore flowing pressure can be estimated as follows:

$$P_{\mathrm{wfcu}}(t) = (P_{\mathrm{wfcu}})_{\mathrm{i}} + (\Delta P_{\mathrm{wfcu}})_t \tag{9.170}$$

where $(P_{\mathrm{wfcu}})_{\mathrm{i}}$ is the critical wellbore pressure estimated at initial reservoir conditions and $(\Delta P_{\mathrm{wfcu}})_t$ is the additional stress caused by the depletion at time t.

CRITICAL BOREHOLE PRESSURE IN HORIZONTAL WELLS

Equations of vertical wellbore stresses can easily be applied to the horizontal well if the well orientation with respect to stress is known. Horizontal wells are drilled either parallel to maximum horizontal stress σ_{Hmax} or minimum horizontal stress σ_{Hmin}. For a horizontal well, $K_2 = 1$ and $\sigma_{\mathrm{max}} = \sigma_{\mathrm{OB}}$. All equations are transformed as shown below.

Overbalanced Conditions

1. Well drilled parallel to maximum horizontal stress:

$$P_{\mathrm{wfco}} = 0.5[3\sigma_{\mathrm{OB}} - \sigma_{\mathrm{Hmin}} - \sigma_{\mathrm{UCS}}](1 - \sin\phi_{\mathrm{f}}) + P_{\mathrm{p}}\sin\phi_{\mathrm{f}} \tag{9.171}$$

Additional stress due to change in pore pressure is:

$$\Delta P_{\mathrm{wfco}} = 0.5[(K_{1\mathrm{min}} - 1) + \sin\phi_{\mathrm{f}}(3 - K_1, \mathrm{min})]\Delta P_{\mathrm{p}} \tag{9.172}$$

where

$$K_{1\mathrm{min}} = \frac{\Delta\sigma_{\mathrm{Hmin}}}{\Delta\sigma_{\mathrm{OB}}} \tag{9.173}$$

2. Well drilled parallel to minimum horizontal stress:

$$P_{\mathrm{wfco}} = 0.5[3\sigma_{\mathrm{OB}} - \sigma_{\mathrm{Hmax}} - \sigma_{\mathrm{UCS}}](1 - \sin\phi f) + P_{\mathrm{p}}\sin\phi_{\mathrm{f}} \tag{9.174}$$

Additional stress due to change in pore pressure is:

$$\Delta P_{\mathrm{wfco}} = 0.5[(K_{1\mathrm{max}} - 1) + \sin\phi_{\mathrm{f}}(3 - K_{1\mathrm{max}})]\Delta P_{\mathrm{p}} \tag{9.175}$$

where

$$K_{1\mathrm{max}} = \frac{\Delta\sigma_{\mathrm{Hmax}}}{\Delta\sigma_{\mathrm{OB}}} \tag{9.176}$$

Underbalanced Conditions

1. Well drilled parallel to maximum horizontal stress:

$$P_{\mathrm{wfcu}} = (1 - \nu)[3\sigma_{\mathrm{OB}} - \sigma_{\mathrm{Hmin}} - \sigma_{\mathrm{UCS}}] - P_{\mathrm{p}}(1 - 2\nu) \tag{9.177}$$

Additional stress due to depletion is:

$$\Delta P_{\mathrm{wfcu}} = [1 - (1 - \nu)(3 - K_{1\mathrm{min}})]\Delta P_{\mathrm{p}} \tag{9.178}$$

where

$$K_{1\mathrm{min}} = \frac{\Delta\sigma_{\mathrm{Hmin}}}{\Delta\sigma_{\mathrm{OB}}} \tag{9.179}$$

2. Well drilled parallel to minimum horizontal stress:

$$P_{\mathrm{wfcu}} = (1 - \nu)[3\sigma_{\mathrm{OB}} - \sigma_{\mathrm{Hmax}} - \sigma_{\mathrm{UCS}}] - P_{\mathrm{p}}(1 - 2\nu) \tag{9.180}$$

Additional stress due to depletion is:

$$\Delta P_{\mathrm{wfcu}} = [1 - (1 - \nu)(3 - K_{1\mathrm{max}})]\Delta P_{\mathrm{p}} \tag{9.181}$$

where

$$K_{1\mathrm{max}} = \frac{\Delta\sigma_{\mathrm{Hmax}}}{\Delta\sigma_{\mathrm{OB}}} \tag{9.182}$$

Knowledge of K_1 and K_2 is essential in order to estimate critical pressure so as to avoid borehole stability problems. Maximum horizontal stress is usually determined from minifrac or leak-off tests, sonic logs, and borehole breakout analysis. Pressure at the wellbore is measured with gauges.

The critical flow rate in field units corresponding to critical pressure in a horizontal well can be estimated using the El-Sayed–Al-Sughayer correlation [85]:

$$q_{\mathrm{c}} = \frac{kh(P_{\mathrm{o}} - P_{\mathrm{wf}})}{141.2\mu_{\mathrm{o}}B_{\mathrm{o}}[\ln(r_{\mathrm{e}}/r_{\mathrm{w}}) + X_{\mathrm{h}}]} \tag{9.183}$$

$$X_{\mathrm{h}} = \ln\left(\frac{4r_{\mathrm{w}}}{\lambda L R_{\mathrm{K}}}\right) + \frac{h}{\lambda L}\left[\sqrt{\frac{Lh}{4r_{\mathrm{w}}}}\left(\frac{2\lambda R_{\mathrm{k}}}{1 + 1/\lambda}\right)\right] \tag{9.184}$$

$$R_{\mathrm{k}} = \sqrt{\frac{k_{\mathrm{h}}}{k_{\mathrm{v}}}} \tag{9.185}$$

$$\lambda = \sqrt{\cos^2\beta + \frac{1}{R_{\mathrm{k}}}\sin^2\beta} \tag{9.186}$$

where β is the inclination angle of the horizontal well from the vertical section. The effect of the inclination on wellbore pressure is high for an angle between 0° and 60° and negligible from 60° to 90°.

CRITICAL PORE PRESSURE

The only factor that reduces reservoir pressure is the primary depletion. Thus, initial effective stress, which is partially supported by the reservoir fluid, is progressively transferred to the rock matrix. The rock begins to fail if the increasing effective stress approaches the critical shear stress of the rock. Thus, any reservoir pressure value below the shear failure value of the rock matrix will cause crushing of the rock solids. Fluid movement further enhances the cavitations of the cementing material and dislodged particles are carried by the fluid toward the wellbore.

$$P_{pc} = \frac{(3\sigma_{Hmaxi} - \sigma_{Hmini} - \sigma_{UCS}) + P_{pi}(3K_2 - K_1 - 2)}{3K_2 - K_1} \tag{9.187}$$

For axial stress greater than the maximum tangential stress:

$$P_{pc} = \frac{\sigma_{OB} + 2\upsilon[\sigma_{Hmaxi} - \sigma_{Hmini} + (K_2 - K_1)P_{pi}] - \sigma_{UCS}}{1 + 2\upsilon(K_2 - K_1)} \tag{9.188}$$

Once the weight of the overburden is unloaded from the reservoir fluids and is supported by the grains, rock mass remains at the maximum shear stress evolved during the depletion [86, 87]. Thus, stress paths during the injection are different from the paths followed during the depletion. During the repressurization, critical pore pressure can be estimated by using the minimum values of the maximum and minimum horizontal stresses as follows [83]:

$$P_{pc} = (1 - \nu)(3\sigma_{Hmax-min} - \sigma_{Hmin-min} - \sigma_{UCS} - P_p) \tag{9.189}$$

where $\sigma_{Hmax-min}$ and $\sigma_{Hmin-min}$ are the lowest values of the maximum horizontal stress and minimum horizontal stress, respectively. Change in critical pore pressure with changing reservoir pressure at a given stage is then given by:

$$(\Delta P_{pc})_t = -(\Delta P_p)_t(1 - 2\nu) \tag{9.190}$$

where subscript t stands for any given time.

EXAMPLE OF A NORTH SEA RESERVOIR

A 14,150-ft deep North Sea reservoir having a porosity range of 5–12% with well-sorted uniform medium-grained quartz sandstone was analyzed with the sand prediction models discussed above [83]. The rock is stiff and strong. The elastic properties and strength of the rock were estimated using triaxial compression tests on 2.54- to 5-cm-long cylindrical plugs by applying an axial loading of 1,000 psi/h. Young's modulus ranged from 1.8×10^6 to 2.1×10^6 psi. Internal friction angle varied from 50° to 54°. The ranges of cohesive strength and unconfined compressive strength observed were 1,143–1,334 psi and 6,994–7,635 psi, respectively. Uniaxial strain compression tests indicated that the rock followed typical elastic rock behavior with $K_1 \approx 0.25$. The test samples of the rock were weakest when taken from the top of the reservoir at a true vertical depth of 14,150 ft with a friction angle of 53.8°, with a cohesive strength of 1,143 psi and a σ_{UCS} of 6,881 psi.

The following table summarizes the above-mentioned properties of the rock sample:

Total overburden stress	14,150 psi
Maximum horizontal stress	12,716 psi
Minimum horizontal stress	12,537 psi
Initial reservoir pore pressure	12,000 psi
Maximum horizontal stress path K_2	0.35
Minimum horizontal stress path K_1	0.25
Internal friction angle	53.8°
Cohesive strength	1,143 psi
Unconfined compressive strength, UCS	6,994 psi

Stress Field and Reservoir Stress Paths

No in situ stress measurements were available for this particular reservoir, so the stress boundary conditions had to be developed in the laboratory and from the geologic studies of the tectonic environment (Table 9.7). The total overburden stress gradient was assumed 1.0 psi/ft. This resulted in a total overburden stress of 14,150 psi. The initial reservoir pressure is 12,000 psi, resulting in a net effective stress of 2,150 psi (Figure 9.65).

There is no record of recent active normal faulting and, thus, it is assumed that the minimum horizontal stress is mainly controlled by the rock material properties. Accordingly the initial effective minimum horizontal stress was estimated from uniaxial stress–strain test path. It was determined to be ~25% of the effective overburden stress (538 psi).

No borehole breakouts were found in wireline logs, which led to the assumption that the maximum horizontal stress was larger than the minimum horizontal stress, or about 33% of the effective overburden stress (710 psi). The two horizontal stress paths K_1 and K_2 (from uniaxial strain stress paths) were estimated to be 0.25 and 0.35, respectively.

Analysis

The critical pore pressure at initial reservoir pressure is estimated using Eq. (9.187):

$$P_{pc} = \frac{(3\sigma_{Hmaxi} - \sigma_{Hmini} - \sigma_{UCS}) + P_{pi}(3K_2 - K_1 - 2)}{3K_2 - K_1}$$

$$P_{pc} = \frac{[3(12{,}716) - (12{,}537) - (6{,}991)] + 12{,}000[3(0.35) - 0.25 - 2]}{3(0.35) - 0.25}$$

$$= 5{,}275 \text{ psi}$$

TABLE 9.7 Stress Paths for Various Geologic and Tectonic Regions [83]

Regions		ν	K_1	K_2	Controlling Factors
Tectonically inactive basins lacking stress measurements	$K_1 \approx K_2 = \frac{\nu}{(1-\nu)}$	0.2	0.25		
Geologic setting of active normal faulting	$k_1 = \frac{(1 - \sin \phi_f)}{(1 + \sin \phi_f)} = \frac{1}{[(1 + \mu^2)^{1/2} + \mu]}$ where μ = coefficient of sliding friction	0.2	0.2 −0.23 (North Sea Area)		Friction on the fault and fractures
Regions with active thrust of reverse faulting			$\simeq$0.21	In the order of 4.7	Friction on the fault and fractures
Areas with strike and slip faults		>0.21	<4.7		

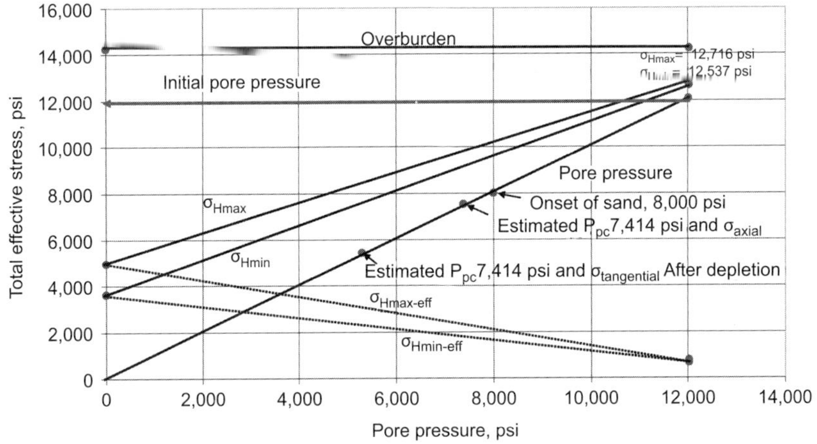

FIGURE 9.65 Depletion diagram indicating estimated critical pore pressure and stress change path [83].

The depletion causes an increase in the critical pore pressure and the axial stress replaces the maximum tangential stress as the greatest principal stress. Thus, critical reservoir pressure can be calculated from Eq. (9.188), using Poisson's ratio of 0.2:

$$P_{pc} = \frac{\sigma_{OB} + 2\upsilon[\sigma_{Hmaxi} - \sigma_{Hmini} + (K_2 - K_1)P_{pi}] - \sigma_{UCS}}{1 + 2\upsilon(K_2 - K_1)}$$

$$P_{pc} = \frac{14{,}150 + 2(0.2)[12{,}716 - 12{,}537 + (0.35 - 0.25)12{,}000] - 6{,}691}{1 + 2(0.2)(0.35 - 0.25)}$$

$$= 7{,}702 \text{ psi}$$

The field observations indicated the first sand production at a pore pressure of 8,000 psi, which is very close to the estimated value. At 8,000 psi, the maximum tangential stress is used to calculate the critical wellbore pressure, using Eq. (9.168):

$$P_{wfcu} = (1 - v)[3s_{Hmax} - s_{Hmin} - s_{UCS}] - P_p(1 - 2n)$$

$$p_{wcfu} = (1 - 0.2)[3(12{,}716) - (12{,}537) - (6{,}691)] - 7{,}703[1 - 2(0.2)]$$

Perhaps the major applications of rock shear modeling are borehole failure during drilling operations and sand production during depletion. Sand production is a major problem in many parts of the world. The main reason for sand production is the deteriorating rock strength with depletion. Rock strength varies from place to place within the same reservoir, ranging from loose sand to consolidated sandstone formations. Such formations are frequently encountered in the Gulf of Mexico, but can be encountered in oil-producing basins worldwide.

POROSITY AS STRENGTH INDICATOR TO EVALUATE SAND PRODUCTION

Petroleum production from poorly consolidated formations can be considerably hindered by the phenomenon of sand production. Gravel-packs can be used to prevent sand production, but they are generally harmful to well productivity and expensive. Sarda et al. proposed tensile rupture and compressive rupture as possible mechanisms of sand failure [88].

Field observations indicate that the formations possessing low porosity show significant rock strength. Thus, porosity can be used as a qualitative measure of rock strength and to predict sand production. Sand production can be expected if the product GK_b of two elastic parameters exceeds the threshold value 8×10^{11} psi^2 [89], where the shear modulus G and the bulk modulus K_b are derived from the interpretation of acoustic and density logs:

$$G = 1.34 \times 10^{10} \frac{\rho_b}{\Delta t_s^2} \tag{9.191a}$$

$$K_b = G\left(r^2 - \frac{4}{3}\right) \tag{9.191b}$$

where r is obtained from the acoustic log and is expressed as:

$$r = \frac{\Delta t_s}{\Delta t_c}$$

The ratio r is related to Poisson's ratios as follows:

$$v = \frac{r^2 - 1}{2r^2 - 1} \tag{9.192}$$

The bulk density and porosity are obtained from:

$$\rho_b = \phi\rho_f + (1 - \phi)\rho_{ma} \tag{9.193}$$

$$\phi = \frac{\Delta t_c - \Delta t_{ma}}{\Delta t_f - \Delta t_{ma}} \tag{9.194}$$

The product GK_b is actually the "Sand Production Indicator" (SPI):

$$\text{SPI} = G^2(r^2 - \tfrac{4}{3}) \tag{9.195}$$

Thus, using the suggested threshold [88]:

$\text{SPI} \leq 8 \times 10^{11}$ psi^2—the formation is stable; therefore, no sand production occurs.

SPI $\geq 8 \times 10^{11}$ psi^2—the formation is unstable; therefore, sand production could occur.

The sand strength limit can be estimated using the following correlation [57, 90]:

$$\sigma_{UCS} = 0.087 \times 10^{-6} E K_b [0.008 V_{sh} + 0.0054(1 - V_{sh})] \tag{9.196}$$

The internal friction angle, θ_f, of the sand is assumed to be equal to 30°. Parameters such as grain form, grain strength, grain size, and size distribution contribute to the frictional strength of the sand, whereas cementation, contact surface area, and pore fluid contribute to its cohesional strength.

Estimation of unconfined compressive rock strength from porosity data: In clean sandstones, the acoustic velocities depend on porosity. Unconfined compressive rock strength can be estimated using log-derived data as follows [91]:

$$\sigma_{UCS} = f(\theta_f) \frac{\rho_b^2}{\Delta t_s^2} \left[\Delta t_c^2 - \frac{4}{3} \frac{\Delta t_s^2}{} \right] g(V_{sh}) \tag{9.197}$$

where ρ_b, Δt_c, and Δt_s depend mainly on porosity. Equation (9.192) clearly introduces porosity as a basic and implicit variable. C_p represents some decompaction or some mechanical damage and θ_f is the frictional angle. Table 9.8 shows the various models of rock strength developed by Sarda et al. [88], which include porosity as an input variable.

TABLE 9.8 Unconfined Compressive Strength of Various Rocks [88]

Rock Type	Unconfined Compressive Strength	Remarks
Ceramics	$\sigma_{UCS} = \sigma_o \, e^{-\beta\phi}$	Use for 2–62% porosity range. $\beta = 8$ or 9 depending on the orientation of pores with respect to the loading direction
Undamaged rocks	**1.** σ_{UCS}(MPa) = 357 $e^{-10.8\phi}$ **2.** σ_{UCS} (MPa) = 258 $e^{-9\phi}$	**1.** Good for 0–7% porosity range. **2.** Use for porosity up to 30%. Uniaxial strength of the stones fits this correlation.
Damaged rocks	$\sigma_{UCS\text{-}min}$(MPa) = 111.5 $e^{-11.6\phi}$	Damage due to coring and plugging operations
Mechanical strength in the zone of high porosity	σ_{UCS} (MPa) = $\sigma(\phi - \phi_{max})^2$	If the porosity shows multiple trends

Example

Is a sand at 4,437 m, for which $\Delta t_s = 125$ μs/ft and $\Delta t_c = 74$ μs/ft, likely to break down under production conditions? The formation water has a density of 1.071 g/cm^3 [88].

Solution

The shear modulus can be estimated using the second equation in Table 9.4, with $a = 1.34 \times 10^{10}$; or Eq. (9.191a):

$$G = 1.34 \times 10^{10} \frac{\rho_b}{\Delta t_s^2}$$

Bulk modulus is given by the fourth equation in Table 9.4; or Eq. (9.191b):

$$K_b = 1.34 \times 10^{10} \left[\left(\frac{\rho_b}{\Delta t_s^2} \right) \left(r^2 - \frac{4}{3} \right) \right]$$

where r is estimated by:

$$r = \frac{\Delta t_s}{\Delta t_c} = \frac{125}{74} = 1.6893$$

Porosity has to be found in order to estimate the bulk density:

$$\phi = \frac{\Delta t_c - \Delta t_{ma}}{\Delta t_f - \Delta t_{ma}} = \frac{74 - 55.5}{185 - 55.5} = 14.28\%$$

Bulk density is solved from the density porosity equation as follows:

$$\phi = \frac{\rho_{ma} - \rho_b}{\rho_{ma} - \rho_f}$$

$$\rho_b = \phi \rho_f + (1 - \phi)\rho_{ma} = (0.1428)(1.071) + (1 - 0.1428)(2.65) = 2.424 \text{ g/cm}^3$$

Then

$$G = 1.34 \times 10^{10} \frac{2.424}{125^2} = 2{,}078{,}822.4$$

$$K_b = 1.34 \times 10^{10} \left[\left(\frac{2.424}{125^2} \right) \left(1.6892^2 - \frac{4}{3} \right) \right] = 3{,}159{,}941.651$$

$$\text{SPI} = GK_b = 6.554 \times 10^{12} \text{ psi}^2$$

Since the sand production index, SPI, is greater than the threshold value of 8×10^{11}, the formation is unstable and it is likely that sand production will occur.

Example

Estimate the unconfined compressive strength of the corresponding rocks using the porosity values given in Table 9.9:

1. Assuming undamaged rock.
2. Assuming damaged rock.

TABLE 9.9 Porosity Values for Example [88]

Depth (m)	ϕ (%)
891.4	32.48
892	7.01
892.4	33.07
895.7	32.5
896.3	24.85
896.5	30.36
897.1	19.02
897.5	22.11
897.7	23.67
898.5	22.08
904.1	19.95
904.5	16.13
904.9	21.28
914.9	22.38
915.1	29.47
915.7	33.56
916.5	34.81
Av.	25.98

3. Assuming that all the porosity values are from the same formation, estimate the unconfined strength in the regions of highest porosity values, assuming that the average porosity of this formation is 25%.

Solution

Since all of the porosity values are greater than 7%, the second equation in Table 9.8 can be used for undamaged rocks.

1. Undamaged rocks:

$$\sigma_{UCS} = 258\ e^{-9\phi}$$

TABLE 9.10 Example Results

σ_{UCS} (MPa) Undamaged	$\sigma_{UCS\text{-}min}$ (MPa) Damaged
13.87	2.58
137.29	49.45
13.15	2.41
13.85	2.57
27.56	6.24
16.79	3.29
46.58	12.28
35.27	8.58
30.65	7.16
35.37	8.61
42.84	11.02
60.42	17.17
38.01	9.45
34.42	8.31
18.19	3.65
12.59	2.27
11.25	1.97

$$\sigma_{UCS} = 258\ e^{-9(0.3248)} = 13.87\ \text{MPa}$$

2. Damaged rocks:

$$\sigma_{UCS-min} = 111.5\ e^{-11.6f}$$

$$\sigma_{UCS-min} = 111.5\ e^{-11.6(0.3248)} = 2.58\ \text{MPa}$$

Other results are shown in Tables 9.10 and 9.11, and Figure 9.66.

3. Unconfined strength in the highest porosity value regions at $\phi = 32.48\%$ (values lower than 25% are ignored):

$$\sigma_{UCS}(\text{MPa}) = s(\phi - \phi_{max})^2$$

TABLE 9.11 Example Results

ϕ	$\sigma_{UCS\text{-min}}$ (MPa)	σ_{UCS} (MPa)
	Undamaged	Damaged
0.3248	0.08	0.01
0.3307	0.09	0.02
0.325	0.08	0.01
0.3036	0.05	0.01
0.2947	0.04	0.01
0.3356	0.09	0.02
0.3481	0.11	0.02

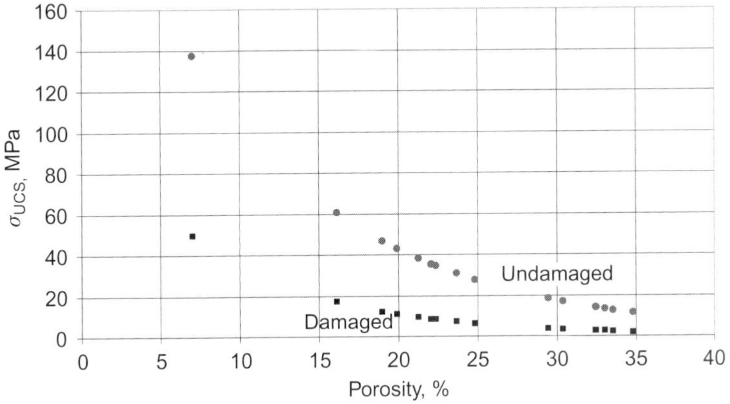

FIGURE 9.66 Unconfined compressive strength prediction in damaged and undamaged rocks—example results.

For undamaged rocks:

$$\sigma_{UCS} = 31.87(0.25 - 0.3248)^2 = 0.08 \text{ MPa}$$

For damaged rocks:

$$\sigma_{UCS} = 2.58(0.25 - 0.3248)^2 = 0.01 \text{ MPa}$$

TABLE 9.12 Uniaxial Compression Test

Core $L = 11.34$ cm, $d = 5.34$ cm		
Axial Load (kN)	Axial Deformation	Diametral Deformation (mm)
25	0.047	0.0045
50	0.075	0.0052
75	0.092	0.0095
100	0.115	0.0125
125	0.137	0.0150
150	0.158	0.0195
175	0.179	0.0235
200	0.208	0.025
225	0.23	0.038
250 (failure)	0.27	0.055
252	–	0.075

PROBLEMS

1. Consider a sandstone formation at a depth of 5,000 ft from which several 2-in. diameter core samples are available for laboratory investigations. The stress gradient is 1.1 psi/ft and the horizontal stress is 2,000 psi. Three compressional tests were run: a uniaxial test with an ultimate strength (σ_1) of 3,000 psi, and two triaxial tests with ultimate strengths of 6,000 and 8,500 psi for confining pressures (σ_3) of 1,000 and 2,000 psi, respectively.
   (a) Estimate the fracture angle.
   (b) Calculate the normal and shear stresses acting on the failure plane.
   (c) Comment on the stress field and implications if the formation stress field is such that $\sigma_1 = 750$ psi and $\sigma_3 = 6{,}250$ psi.
2. The results of four compression tests and other data are given in Tables 9.12–9.15. The four rock samples were obtained from a limestone formation. Calculate for each rock sample:
   (a) The ultimate compressive strength of the rock samples at (1) maximum load and (2) the same reference stress point at 50% of ultimate strength.
   (b) Young's modulus, Poisson's ratio, modulus of rigidity, bulk modulus, and rock compressibility.
   (c) Analyze the relation between stress and rupture of the limestone.

TABLE 9.13 Triaxial Test

Core $L = 11.11$ cm, $d = 5.32$ cm		
Minimum Principal Stress (σ_3) = 11.4 MPa		
Static Load = 25.3 kN		
Axial Minus Static	Axial Strain	Lateral Strain
Load (kN)	ε_a, 10^{-6}	ε_1 (10^{-6})
50	460	125
100	850	245
150	1,275	338
200	1,720	470
250	2,165	640
300	2,670	895
360 (failure)	3,950	1,500

ADDITIONAL SAMPLE DATA

Sample No.	σ_1 (MPa)	σ_3 (MPa)
1	250	0.0
2	385.3	11.4
3	511.1	22.8
4	667	57.0

3. Triaxial compression tests were made on 11 oriented cores at 60 MPa confining pressure. Table 9.16 shows the results of these static tests. All of the cores came from the Mesa Verde formation, Colorado, from depths of about 1,500–2,500 m. The average grain density is 2.4 g/cm^3. For each core:

(a) Calculate the rock compressibility.

(b) Estimate the dynamic values of the elastic and shear moduli.

(c) Determine the shear and compressional transit time.

4. A core sample was subjected to a hydrostatic test with the following stresses: $\sigma_1 = 5{,}800$ psi, $\sigma_2 = 2{,}200$ psi, and $\sigma_3 = 3{,}600$ psi. The pore pressure is 2,850 psi. The porosity of the core is 0.18. Knowing $c_r = 0.45 \times 10^{-6}$ psi^{-1}; $c_b = 2.3 \times 10^{-6}$ psi^{-1}; $c_w = 3.1 \times 10^{-6}$ psi^{-1}, determine the following:

TABLE 9.14 Triaxial Test

Core $L = 10.88$ cm, $d = 5.34$ cm		
Minimum Principal Stress (σ_3) = 22.8 MPa Static Load = 51.1 kN		
Axial Minus Static	Axial Strain	Lateral Strain
Load (kN)	ε_a (10^{-6})	ε_1 (10^{-6})
50	400	100
100	895	290
150	1,330	400
200	1,825	600
250	2,420	850
300	3,100	1,100
350	3,600	1,500
400	4,540	2,000
450	5,700	3,300
460 (failure)	6,200	4,100

(a) The relative change in pore volume.
(b) The change in pore pressure.
(c) The effectiveness (α) of the pore pressure in counteracting the total applied load, if the sample has a compressional wave travel time of 58 ms/ft and a shear travel time of 84 ms/ft. The bulk density of the rock sample is 0.42 g/cm^3. The undrained Poisson's ratio is 0.18.

5. A 15,265-ft deep sandstone formation has a stress gradient of 1 psi/ft. The formation is unconsolidated, with a pore pressure of 7,000 psi. The compressional and shear velocities are 10,836 and 7,462 ft/s, respectively. From laboratory measurements, the unconfined strength is 6,848 psi. The bulk density of this rock is 2.65 g/cm^3.

Calculate:

(a) Poisson's ratio, shear modulus, bulk modulus, and Young's modulus.
(b) Horizontal stress.
(c) Mean effective stress.
(d) Initial shear strength.

6. A stress-sensitive productive zone was cored from 8,000 to 8,050 ft. Core plugs were subsequently analyzed for porosity and permeability by an automated core measurement system (Core Laboratories CMS 300). This

TABLE 9.15 Triaxial Test

Core $L = 11.30$ cm, $d = 5.32$ cm		
Minimum Principal Stress (σ_3) = 57 MPa		
Static Load = 127 kN		
Axial Minus Static	Axial Strain	Lateral Strain
Load (kN)	ε_a (10^{-6})	ε_1 (10^{-6})
50	500	50
100	1,050	150
150	1,300	200
200	1,800	325
250	2,400	580
300	3,000	650
350	3,800	800
400	4,300	1,050
450	5,100	2,000
500	6,000	2,800
540 (failure)	6,800	3,850

system provides measurements of porosity and permeability as a function of overburden stress, as shown in Table 9.17. The stress gradient is 0.5 psi/ft.

7. Using Jones' method [49], determine porosity and Klinkenberg permeability at zero net stress.
   (a) Develop an empirical relationship between (1) the void ratio $\phi/(1-\phi)$ and stress, and (2) permeability and stress.
   (b) Estimate the porosity and permeability values at a depth of 8,025 ft.
   (c) A second well was drilled to a second zone at 10,500 ft. Assuming the two zones are similar, calculate the porosity and permeability of this zone.
   (d) Using data in Table 9.17 and the Jones and Owens [50] method, calculate the average pore compressibility over the stress intervals.
8. Given the following data of a sandstone formation,
   $\Delta t_s = 1{,}350$ μs/ft, $\Delta t_c = 75$ μs/ft, $\rho_w = 1.0$ g/cm^3, depth = 3,425 m
   Determine:
   (a) The likelihood of sand production using the threshold criteria of SPI.

TABLE 9.16 Elastic Properties of Oriented Cores Used in Stress Calculations

Depth (m)	Elastic Modulus (GPa)	Poisson's Ratio
1,500.8	25.1	0.13
1,507.2	26.9	0.18
1,507.5	27.4	0.17
1,508.8	25.1	0.15
1,509.4	25.2	0.15
1,995.5	29.6	0.21
1,998.0	30.7	0.20
2,404.3	39.6	0.23
2,406.4	40.4	0.22
2,474.7	35.6	0.18
2,475.2	33.9	0.17

TABLE 9.17 Automated Core Measurements

Net Stress (psi)	Porosity (%)	Klinkenberg Permeability (mD)
1,000	14.48	16.38
1,500	14.07	15.30
5,000	12.81	12.46
9,800	12.51	11.82

(b) Poisson's ratio.
(c) Pore pressure and overburden pressure.
(d) Horizontal stress of the Earth.
(e) Fracture gradient (FG). Use two methods for calculating the Biot constant and discuss its effect on the calculation of FG.

NOMENCLATURE

A area
B Skempton coefficient, formation volume factor

C compressibility
C compressional strength
D depth
D diameter
D(p) Dobrynin pressure function
E modulus of elasticity
f factor
F force, load, formation factor
F_R formation resistivity factor
FG fracture gradient
g gravity acceleration, gradient
G modulus of rigidity
H height
h power-law coefficient
I intensity factor
J stress invariant
L length
m cementation exponent
N oil-in-place
p pressure
R correction factor, ratio
S saturation
S specific surface area
S_s shape factor, saturation
t time
v velocity
V volume

SUBSCRIPTS

A air
ax axial
b bulk
bh bottomhole
c compressional
d dynamic, drained
e effective
f fracture, formation
fl fluid
gv grain value
H horizontal
h hydrostatic
hu hydrostatic to uniaxial
i initial
lat lateral
m matrix, minimum
M maximum

ms matrix stress
n normal
o original, oil
O overbalanced
OB overburden
p principal, propagation, pore
pc critical pore
pi initial pore
ps pore shape
r rock, resonance
s shear, static, stress
t tensile
T time, or any stage in depletion
θ tangential
u undrained, uniaxial, underbalanced
UCS unconfined compressive strength
UCS-min minimum unconfined compressive strength
w water
wb wellbore
wfco wellbore flowing critical at overbalanced conditions
wfcu wellbore flowing critical at underbalanced conditions

GREEK SYMBOLS

α correction factor
ε strain
ϕ_f angle of friction
γ shear strain
κ bulk modulus
ν Poisson's ratio
θ angle
ρ density
σ stress
τ shear stress

REFERENCES

1. Fairhurst C. *Rock mechanics*. London: Pergamon Press; 1963.
2. Gnirk PF. The mechanical behavior of uncased wellbores situated in elastic/plastic media under hydrostatic stress. *Trans AIME* 1972;**253**:49–59.
3. Holt RM, Ingsoy P, Mikkelsen M. Rock mechanical analysis of North Sea reservoir formations. Paper no. 16796. In: *Soc Pet Eng Annu Tech Conf*. Dallas, TX; September 27–30, 1987. 7 pp.
4. Geertsma J. Problems of rock mechanics in petroleum production engineering. In: *Proceedings of the first congress of the International Society of Rock Mechanics*, vol. I. Lisbon, Portugal: Nacional de Engenharia Civil; 1966. pp. 585–94.
5. Farmer I. *Engineering behavior of rocks*, 2nd ed. London: Chapman & Hall; 1983. 203 pp.

6. Poulos HG, Davis EH. *Elastic solutions for soil and rock mechanics*. New York: John Wiley & Sons; 1974. 411 pp.
7. Matthewson CC. *Engineering geology*. Columbus, OH: Merrile; 1981. 410 pp.
8. Ren, et al. *Rock mechanics tidbits*. Tulsa, OK: Compiled for Dowell; 1979–83.
9. LeTirant P, Gay L, Kerbouch P, Moulinier J, Veillons D. *Manuel de fracturation hydraulique*. Paris, France: ARTEP Editions Technip; 1972.
10. Bell FG. *Fundamentals of engineering geology*. London, England: Butterworth; 1983. 648 pp.
11. Blyth FGJ, de Freitas MH. *A geology for engineers*, 7th ed. New York: Elsevier; 1984. p. 325.
12. Billings MP. *Structural geology*, 3rd ed. Englewood Cliffs, NJ: Prentice Hall; 1972. 606 pp.
13. Robertson EC. Experimental study of the strength of rocks. *Geol Soc Am Bull* 1955;**66**:1294–314.
14. Heard HC. Effect of large changes in strain rate in the experimental deformation of yule marble. *J Geol* 1963;**71**:162–95.
15. Donath FA. Some information squeezed out of rocks. *Am Sci* 1970;**58**:54–72.
16. Colback PSB, Wild BL. Influence of moisture content on the compressive strength of rock. Canadian Department of Mines and Technical Survey Symposium, Ottawa; 1965. pp. 65–83.
17. Bernaix J. New laboratory methods of studying the mechanical properties of rocks. *Int J Rock Mech Min Sci* 1969;**6**:43–90.
18. Handin JW, Hager RV, Friedman M, Feather JN. Experimental deformation of sedimentary rocks under confining pressure: pore pressure tests. *Am Assoc Petrol Geol (AAPG) Bull* 1963;**47**:717–55.
19. Hosking JR. A comparison of tensile strength, crushing strength and elastic properties of roadmaking rocks. *Quarry Man J* 1955;**39**:200–12.
20. Deere DU, Miller RP. *Engineering classification and index properties for intact rock*. Technical Report No. AFWL-TR-65-115, Air Force Weapons Lab., Kirkland Air Base, New Mexico; 1966.
21. Schlumberger, Inc. *Log interpretation principles/application*. Schlumberger Educational Services, Houston, TX; 1987, 198 pp.
22. Savich AI. Generalized relations static and dynamic indices of rock deformability [translated from: *Gidrotekhnicheskoe Stroitel'stuo 8*]; 1984. pp. 50–4.
23. Reichmuth DR. Point load testing of brittle materials to determine tensile strength and relative brittleness. In: *Proceedings of the 9th Rock Mechanics Symposium*. University of Colorado, Boulder; 1967. pp. 134–59.
24. Touloukian YS, Ho CY. Physical properties of rocks and minerals. In: *McGraw-Hill/Cindas Data Series on Material Properties*, vol. II–2; 1981.
25. Bieniawski ZT. The point-load test in geotechnical practice. *Eng Geol* 1975;**9**:1–11.
26. Biot MA. Theory of elasticity and consolidation for a porous anisotropic solid. *J Appl Phys* 1955;**26**:182–97.
27. Fatt I. Pore volume compressibilities of sandstone reservoir rocks. *Soc Petrol Eng JPT* 1958;64–6.
28. Brandt H. A study of the speed of sound in porous granular media. *Trans ASME* 1955;**77**:479.
29. Hall HN. Compressibility of reservoir rocks. *Trans AIME* 1953;**198**:309–11.
30. Economides MJ, Nolte KG. *Reservoir stimulation*. Houston, TX: Schlumberger Educational Services; 1987.
31. van Terzaghi K. Die berechnung der durchlassigkeitsziffer des tones aus dem verlauf der hydrodynamischen spannungerscheinungen. *Sber Akad Wiss Wien* 1923;**132**:105–13.

32. Biot MA. General theory of three-dimensional consolidation. *J Appl Phys* 1941;**12**:155–64.
33. Biot MA. General solutions of the equations of elasticity and consolidation for a porous material. *J Appl Mech* 1953;**23**:91–6.
34. Zimmerman RW, Somerton WH, King MS. Compressibility of porous rocks. *J Geophys Res* 1986;**91**(B12):765–77.
35. Sulak RM, Danielsen J. Reservoir aspects of Ekofisk subsidence. *J Petrol Technol* 1989;709–16.
36. Ruddy I, et al. Rock compressibility, compaction, and subsidence in a high-porosity chalk reservoir—a case study of Valhall field. *J Petrol Technol* 1989;**41**:741–6.
37. Dean G, Hardy R, Eltvik P. Monitoring compaction and compressibility changes in offshore chalk reservoirs. *J SPE Form Eval* 1994;73–6.
38. Harari Z, Wang ST, Saner S. Pore-compressibility study of Arabian carbonate reservoir rocks. *J SPE Form Eval* 1995;207–14.
39. Geertsma J. The effect of fluid pressure decline on volumetric changes of porous rocks. *Trans AIME* 1953;331–40.
40. Skempton AW. Effective stress in soils, concrete, and rock. In: *Pore pressure and suction in soils*. Butterworth, London, 1960. pp. 4–16.
41. Skulje L. *Rheological aspects of soil mechanics*. 1969. 123 pp.
42. Skempton AW. *Selected papers on soil mechanics*. London: Thomas Telford; 1984. 65 pp.
43. Anderson MA. Predicting reservoir condition pore-volume compressibility from hydrostatic-stress laboratory data. In: *60th Annual SPE Technol Conf Soc Petrol Eng*. Paper No. 14213. Las Vegas, NV; September 22–25, 1985.
44. Dobrynin VM. Effect of overburden pressure on some properties of sandstones. *Soc Petrol Eng J* 1962;360–6.
45. Chierici GL, Ciucci GM, Eva F, Long G. Effect of the overburden pressure on some petrophysical parameters of reservoir rocks. In: *Proceedings of the 7th World Petroleum Conference*. Mexico City; 1967. pp. 309–30.
46. Nieto JA, Yale DP, Evans RJ. Improved methods for correcting core porosity to reservoir conditions. *Log Analyst* 1994;21–30.
47. Teeuw D. Prediction of formation compaction from laboratory compressibility data. *SPE* 1970;**2973**(October).
48. Schutjens PMTM, Hanssen TH, Hettema MHH, Merour J, de Bree P, Coremans JWA, et al. Compaction-induced porosity/permeability reduction in sandstone reservoirs: data and model for elasticity-dominated deformation. *SPERE J* 2004;202–15.
49. Jones SC. Two-point determinations of permeability and PV vs. net confining stress. *Soc Pet Eng Form Eval* 1988;235–41.
50. Jones FO, Owens WW. A laboratory study of low-permeability gas sands. *Soc Petrol Eng JPT* 1980;1631–40.
51. McKee CR, Bumb AC, Koening RA. Stress-dependent permeability and porosity of coal and other geologic formations. *Soc Pet Eng Form Eval* 1988;81–91.
52. Juhasz I. Conversion of routine air-permeability data into stressed brine-permeability data. In: *10th European Formation Evaluation Symposium*. Aberdeen; April, 1986. pp. 22–5.
53. Warren TM, Smith MB. Bottomhole stress factors affecting drilling rate at depth. *Soc Petrol Eng JPT* 1985;1523–33.
54. Maurer WC. Bit tooth penetration under simulated borehole conditions. *Soc Petrol Eng JPT* 1965;1433–42.

55. Cunningham RA, Eenink JG. Laboratory study of effect of overburden, formation and mud column pressure on drilling rate of permeable formations. *Soc Petrol Eng JPT* 1959;9–15.
56. Yang JH, Gray KE. Single-blow bit-tooth impact tests on saturated rocks under confining pressure–II. Elevated pore pressure. *Soc Petrol Eng J* 1967;389–408.
57. Coates GR, Denoo SA. Mechanical properties program using borehole analysis and Mohr's circle. *Soc Prof Well Log Anal 22nd Annu Log Symp* 1981;DD1–D16.
58. Hottman CE, Smith JH, Purcell WR. Relationship among Earth stresses, pore pressure, and drilling problems offshore Gulf of Alaska. *Soc Petrol Eng JPT* 1979;1477–84.
59. Geertsma J. Some rock-mechanical aspects of oil and gas well completions. *Soc Petrol Eng J* 1985;848–56.
60. Van der Vlis AC. Rock classification by a simple hardness test. In: *Intl Soc Rock Mech, 2nd Congr*; 1970. 2 pp.
61. Breckels IM, van Eekelen HAM. Relationship between horizontal stress and depth in sedimentary basins. *Soc Petrol Eng JPT* 1982;2191–9.
62. Howard GC, Fast CR. *Hydraulic fracturing*. Society Petroleum Engineering Monograph, vol. 2. Dallas, TX; 1970.
63. Craft BC, Holden WR, Graves ED. *Well design–drilling and production*. Englewood Cliffs, NJ: Prentice Hall; 1962.
64. Hubbert MK, Willis DG. Mechanics of hydraulic fracturing. *Trans AIME* 1957;**201**:153–66.
65. Matthews WR, Kelly J. How to predict formation pressure and fracture gradient. *Oil Gas J* 1967;**20**.
66. Eaton BA. Fracture gradient prediction and its application in oilfield operations. *Trans AIME* 1969;**246**:1353–60.
67. Constant WD, Bourgoyne AT, Jr. Fracture gradient prediction for off-shore wells. *Soc Petrol Engr* paper No. 15105, In: *Proceedings of the 56th California Regional Management*. Oakland, CA; April 1986. pp. 125–30.
68. Pilkington PE. Fracture gradient estimates in tertiary basins. *Petrol Eng Int* 1978;138–48.
69. Perkins TK, Kern LR. Width of hydraulic fractures. *Soc Petrol Eng JPT* 1961; September:937–49. Trans. AIME, vol. **222.**
70. Labudovic V. The effect of Poisson's ratio on fracture height. *Soc Petrol Eng JPT* 1984;287–90.
71. Pirson SJ. *Handbook of well log analysis*. Englewood Cliffs, NJ: Prentice Hall; 1963.
72. Rzhevsky V, Novik G. *The physics of rocks*. Moscow, USSR: MIR Publishers; 1971.
73. Sih GC. *Handbook of stress intensity factors*. Bethlehem, PA: Lehigh University; 1973.
74. van Eekelen HAM. Hydraulic fracture geometry: fracture containment in layered formations. *Soc Petrol Eng J* 1982;341–9.
75. Warpinsky NR, Schmidt RA, Northrop DA. In-situ stresses: the predominant influence on hydraulic fracture containment. *Soc Petrol Engr JPT* 1982;653–64.
76. Westergaard HM. Plastic state of stress around a deep well. *Boston Soc Civ Eng J* 1940;**27**:1–5.
77. Paslay PR, Cheatham JB. Rock stresses induced by flow of fluids into boreholes. *Soc Petrol Eng J* 1965;85–94.
78. Bradley WB. Failure of inclined boreholes. In: *ASME Energy Technology Conference*, Houston, TX; 1978. pp. 5–9.
79. Risnes R, Bratli RK, Horsrud P. Sand stresses around a wellbore. *Soc Petrol Eng J* 1982;883–98.
80. Breckels IM, van Eekelen HAM. Relationship between horizontal stress and depth in sedimentary basins. *Soc Petrol Eng J* 1982;2191–9.

81. Lindner EN, Halpern JA. In-situ stress in North American: a compilation. *Intl Rock Mech Min Sci Geochem J Abs* 1979;**15**:183–203.
82. Haimson BC. Crustal stress in the continental United States as derived from hydrofracturing tests. In: Heacock JC,, editor. *The Earth's AGU geophysics. Monograph services*, vol. 20. 1977. pp. 576–92.
83. Rhett DW, Risnes R. SPE/ISRM Paper 78150. In: *Predicting critical borehole pressure and critical reservoir pore pressure in pressure depleted and repressurized reservoirs*. Irving, Texas; October 20–23, 2002.
84. Haimson BC, Fairhurst C. Initiation and extension of hydraulic fractures in rock. *Soc Petrol Eng J* 1997;310–18.
85. El-Sayed A, Al-Sughayer AAA. Paper SPE 68134. In: *New concept to predict sand production from extended reach and horizontal wells.*. MEOS, Bahrain; March 17–20, 2001.
86. Teufel LW, Rhett DW. Failure of chalk during waterflooding of Ekofisk field. In: *SPE Annual Technical Conference*. Washington, DC; October 4–7, 1992.
87. Santarelli FJ, Tronvoll JJ, ORMIS, Svennekjaier M, Skeie H, Henriksen R, Bratli RK, Saga Petroleum. Reservoir stress path: the depletion and the rebound. In: *Proceedings of Eurock 98, PSE/ISRM Rock Mechanics in Petroleum Engineering*, Trondheim, Norway; July 10, 1998. pp. 203–9.
88. Sarda J-P, Kessler N, Wicquart E, Hannaford K, Deflandre J-P. Use of porosity as a strength indicator for sand production evaluation. Paper SPE 26454. In: *The 68th Annual Technical Conference and Exhibition of the SPE*. Houston, TX; October 3–6, 1993.
89. Tixier MP, Loveless GW, Anderson RA. Estimation of the formation strength from the mechanical properties log. *J Petrol Technol* 1975;**27**:283–93.
90. Bruce S. Mechanical stability log. Paper SPE 19942. In: *IADC/SPE Conference*. Houston, Texas; February 27–March 2, 1990.
91. Wyllie MR, Gregory AR, Gardner GHF. An experimental investigation of factors affecting elastic wave velocity in porous media. *Geophysics* 1958;**23**:459–93.
92. Newman GH. Pore volume compressibility of consolidated, friable, and unconsolidated reservoir rocks under hydrostatic loading. *J Petrol Technol* 1973;**25**(2):129–34.

Chapter 10

Reservoir Characterization

INTRODUCTION

The rock and fluid properties of a petroleum reservoir and their three-dimensional spatial variations are some of the most important criteria for overall economic production of the hydrocarbon resources. The locations and types of wells, and other surface equipment, should be based on the geometry of the reservoir, the petrophysical properties of the reservoir rocks, and the properties of the fluids to be produced. If required, a team approach should be used for the evaluation and development of newly discovered fields as well as for the re-evaluation and mature fields, or those near the point of abandonment as the economic limit of production is approached. The complete management of a field is a complex strategy as illustrated in the general sketch of the process in Table 10.1 showing that petrophysics holds a key role in all aspects for reservoir development and management.

Petrophysical, and fluid, properties are required as soon as possible for estimates of the original fluid volumes in place. Seismic and well-log data are used for initial development of isopach maps showing porosity and fluid saturation distributions. Production and fluid property data are used for estimates of the original hydrocarbons reserves by analysis of reservoir material balance, reservoir depletion/fluid-expansion, and pressure decline methods during the life of the reservoir. These analyses are refined as more wells are drilled and additional production and petrophysical data are acquired to yield reliable ultimate hydrocarbon recovery predictions (Table 10.2). The overall objectives are to minimize technical and financial risks, maximize the economic value of the reservoir, and constantly evaluate the economic value of the reservoir throughout its life. As production proceeds, reservoir characterization should be conducted during several stages in the life of the reservoir to review petrophysical data, redefine the long-term management of the reservoir, uncover conditions where well work-over can be beneficial, and to

TABLE 10.1 Overall Strategy for Field Development

Overall Field Management

- Coordination of team disciplines (job assignments)
- Overall evaluation and integration of data acquisition by team members
- Design of final plans for field development or modification
- Financial estimates
- Presentation to directors
- Execution of approved field engineering

Technology

- Seismic
  - Two- and three-dimensional
  - Cross-hole
  - Tomography
  - Profile
  - Reservoir description

- Drilling and completions
  - Coring and sample preservation
  - Fracture planning and execution
  - Logging

- Geology
  - Reservoir and overburden deposition description
  - Log analysis and core description
  - Maps
  - Porosity and permeability distributions
  - Reservoir heterogeneity

Engineering

- Well location
- Surface equipment (facilities)
- Pressure transient testing
- Integration of log analytic data
- Material balance
- Decline curve analysis
- Production management [primary (I°), secondary (II°), tertiary production (III°)]
- Overall data integration
- Reservoir simulation (I°, II°, III°, EOR, neural networks)
- Environmental concerns and precautions

Laboratory

- Core analyses
- Reservoir fluid analyses
- Thin sections
- Microscopic analyses
- Porosity; pore-size distribution
- Permeability; relative permeability

(*Continued*)

TABLE 10.1 (Continued)

- Capillary pressure; wettability
- Fluid flow performance
- Production limits (residual oil saturation)
- Enhanced oil recovery feasibility
- Compaction

TABLE 10.2 Data Acquisition for Reservoir Definition, Development, Management, and Consideration for Enhanced Oil Recovery

Seismic, Drilling, Cross-section Tomography
Maps (structure, faults/fractures, stratigraphy, lithology)
Original reservoir estimates
Initial formation volume factors
Initial well-log interpretations
Drilling, Logging, Coring, Fluids
Logs: depth, thickness, porosity, water saturation, fluid–hydrocarbon contacts, structure cross-section maps, location of perforations
Cores: porosity, permeability, capillary pressure, wettability, relative permeabilities, pore-size distributions, matrix compressibility, heterogeneity, lithology
Fluid samples: pressure–volume–temperature relationships (PVT data), formation volume factors, viscosities, densities, and fluid compressibilities
Well tests: reservoir pressure, average permeability, reservoir continuity (boundaries), productivity/injectivity indices
Primary Production
Fluid production rates, cumulative production with respect to time (decline curves), change of formation volume factors and viscosity with respect to time, reservoir pressure decline, change of fluid contacts with respect to time, water influx, gas-cap expansion, gravity segregation
Secondary Production
Cumulative production/injection rates with respect to time, change of reservoir and fluid properties with respect to time, prediction of ultimate production and abandonment

increase production potential from infill drilling, revision of the original gas and oil reserve estimates, and ultimate recovery [1–7].

RESERVOIR VOLUMES

Estimation of the volume of hydrocarbons in a reservoir requires a minimum of two distinct types of data: (1) physical description of the formation obtained from seismic data and well logs and (2) reservoir fluid properties and initial production behavior. Modern three-dimensional seismic surveys produce detailed descriptions of the physical dimensions of the reservoir that (together with well-log interpretations) lead to development of isopach maps that are the basis for determination of the volumes, locations (or distributions) of fluid saturations (gas, oil, and water) that are in place [8, 9]. The gross-thickness-isopach maps yield the overall volume of the reservoirs showing the limits of porous–permeable zones within fault blocks and facies changes where the porosity (ϕ)–permeability (K) capacity for containing reservoir fluid vanishes ($\phi < 0.05$; $K < 0.10$ mD) because of changes of lithology or rock properties.

In order to obtain realistic volumetric estimates, the gross-thickness-isopach map must be corrected by subtraction of the zones that do not contain mobile, productive hydrocarbons, or other inclusions that cannot contain hydrocarbons, such as shale within sand structures or calcite/chert deposits in carbonates. Well logs, core analyses, and well tests furnish the data required for preparation of the net-thickness-isopach maps. The basic methods for development of these maps are detailed by Amyx et al., Chapman, and Archer [10–12]. The net-thickness-isopach maps are used to determine the volumes of gas and oil in the reservoir using the average porosity and hydrocarbon saturation at ascending thickness of the reservoir, from the oil–water contact to the top.

Estimates of oil and gas (in terms of stock tank barrels and standard cubic feet) in a reservoir begin with three general equations that are used from the discovery of the field to the point of abandonment. Evaluations are updated as new field and laboratory data of the fluid and petrophysical properties are obtained (refer to section "Quantitative Use of Porosity" in Chapter 3). The equations are expressed in the general reservoir field units (feet, acres, barrels, etc.):

$$N = \frac{7,758\,Ah\phi S_o}{B_o} \tag{10.1}$$

$$G_s = NR_s \tag{10.2}$$

$$G = \frac{43,560\,Ah\phi S_g}{B_g} \tag{10.3}$$

Well-log and fluid data from the discovery well and bottom hole fluid samples yield the initial estimates of oil and gas volumes, expressed in terms of acre-feet. Subsequent wells and seismic data produce the maps from which estimates of the actual volumes in place are obtained. The well logs provide water saturations from the water/oil contact at the base of the hydrocarbon zone to the top of the oil zone, and the thickness of the gas zone above the oil. When originally discovered, reservoir fluids are at static fluid equilibrium since they have been undisturbed over long periods of time; therefore, the oil zone (at discovery only) does not contain free gas, and likewise the gas zone above the oil (if a gas-cap exists) does not contain oil. Thus, the initial oil and gas saturations are defined by:

$$1.0 = S_g + S_o + S_w \tag{10.4}$$

Example

A.

Estimate the original hydrocarbon volumes using well-log and fluid data from the initial discovery well where average porosities in the oil and gas zones are 0.15 and 0.21, respectively; thickness of the oil and gas zones are 18 and 16 ft; average water saturations for the oil and gas zones are 0.26 and 0.22.

Laboratory data are as follows: initial formation volume factors, $B_{oi} = 1.25$, $B_{gi} = 0.015$, and $R_{si} = 500$.

B.

If the reservoir area is 200 acres, determine the following: (a) the initial reservoir volume of original oil in place (OOIP); (b) the remaining oil in place at the end of primary recovery of 18% of the original oil; (c) the remaining oil saturation if the oil formation volume factor at the end of primary production is ($B_o = 1.12$); and (d) the remaining gas saturation if the water saturation at the end of primary production is $S_w = 0.31$.

Solution

A.

Initial volume of oil in place:

$$N_i = \frac{7,758(1 \text{ acre})(18 \text{ ft})(0.15)(1.0 - 0.26)}{1.25 \text{ RB/STB}} = 12,400 \text{ STB/acre}$$

Initial solution gas:

$$G_{si} = (12,400 \quad \text{STB/acre})(500 \quad \text{scf/STB}) = 6.2 \quad \text{MMscf/acre}$$

where Mscf = 10^3 scf; MMscf = 10^6 scf; Bscf = 10^9 scf

Initial volume of gas in the gas-cap:

$$G_i = \frac{43,560(1 \quad \text{acre})(16 \quad \text{ft})(0.12)(1.0 - 0.22)}{0.015 \quad \text{RB/scf}} = 1.36 \quad \text{MMscf/acre}$$

B.

a. $OOIP = 7758(200)(18)(0.15)(1.0 - 0.26) = 3{,}100{,}097$ Res. bbl
b. Remaining oil vol. $= (3{,}100{,}097)(1.0 - .18) = 2{,}542{,}080$ Res. bbl
c. Remaining oil sat. $= (1.0 - S_{wi})[1.0 - N_p(\%)]$
$= (2.0 - 0.26)(1.0 - 0.18)(1.12/1.25)\_ = 0.54$
d. Remaining gas sat. $= 1.0 - 0.54 - 0.31 = 0.15$

Correlations and general rules also can be used to estimate the reserves of a newly discovered well, where specific laboratory data are not available. Arp's rule states that the solution-gas/oil ratio (R_s) for a reservoir at hydrostatic pressure is equal to 10% of the depth (from the surface). In addition, an empirical equation has been developed from Standing's Correlation Chart [13] for estimation of the formation volume factor for a saturated liquid:

$$B = 0.972 + 0.000147C^{1.175}$$
$$C = R_s\left(\frac{\gamma_g}{\gamma_o}\right)^{0.5} + 1.25(T\ ^\circ F) \tag{10.5}$$
$$\gamma_o = \frac{141.5}{131.5 + ^\circ API}$$

Example

Estimate the original oil in place using Arp's rule and Standing's correlation where

$S_w = 0.38$, net reservoir thickness $(h) = 20$ ft, porosity $= 0.18$, temperature $=$ 166°F, °API $= 32$, gas specific gravity $(\gamma_g) = 0.68$, depth $= 7{,}200$ ft.

Solution

$$\gamma_o = \frac{141.5}{131.5 + 32} = 0.865$$
$$R_s = (0.10)(7{,}200) = 720$$
$$C = 720\left(\frac{0.68}{0.865}\right)^{0.5} + 1.25(166) = 638.4 + 207.5 = 845.9$$
$$B_o = 0.972 + (0.000147)(845.9)^{1.175} = 1.192$$
$$OOIP = \frac{7{,}758(0.18)(0.62)}{1.192} = 726\ \text{STB/acre–ft}$$
$$= 14{,}520\ \text{STB/acre}$$

After discovery, isopach maps (developed from seismic and well-log data) provide the first estimates of the gross rock volume and shape of the reservoir. The contour intervals (drawn at even intervals of 10–50 ft each) of the map representing subsea level depths are called isopachs. They are drawn by connecting points of equal elevation obtained from seismic and well-log subsurface data and indicate the external geometric configuration of the reservoir.

The water/oil and gas/oil contacts, determined from multiple wells, are indicated within the contours of the map; the elevation between the water/oil contact and the top of the structure provides the maximum height of the hydrocarbon column and estimates of the reservoir volume, as illustrated in Figure 10.1a. Figure 10.1b provides a profile of the cross section shown in Figure 10.1a.

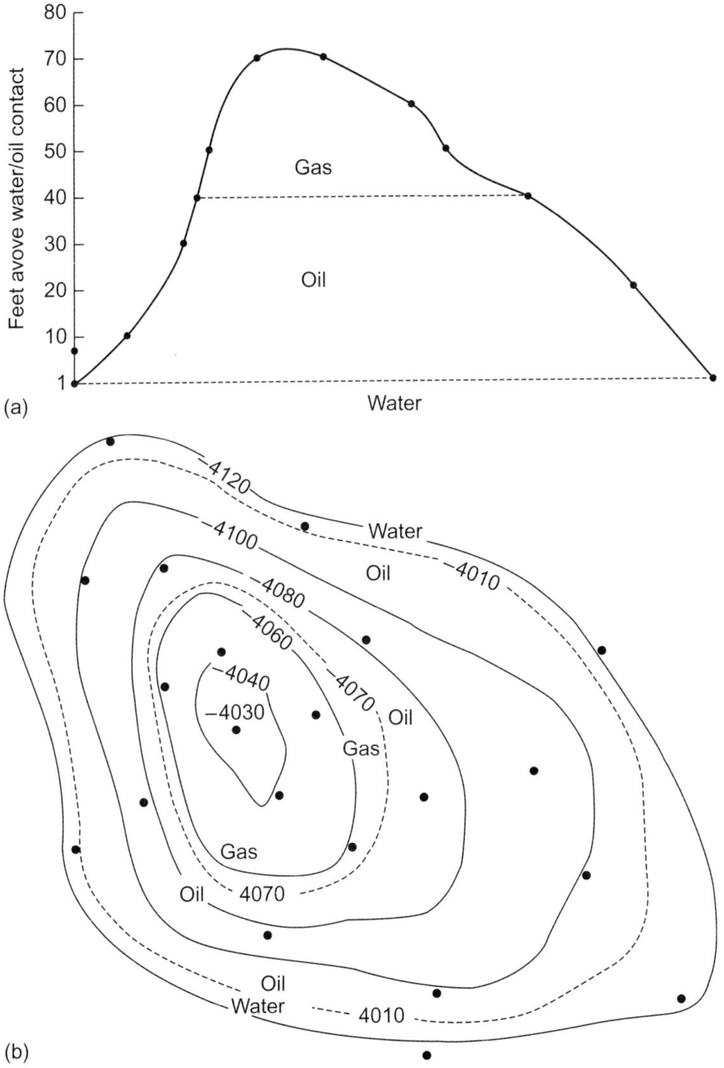

FIGURE 10.1 (a) Projection of the cross-section points on the contours to form an image of the reservoir with respect to height above the water/oil contact. (b) Relationship between the contours of the isopach map and a cross section with respect to the gas, oil, and water contacts.

Contours for every 10 ft in Figure 10.1b, from the water/oil contact to the top, were drawn at a scale that could be easily used with a planimeter, and the area of each contour was determined (Table 10.3). The volume of each 10-ft segment of the reservoir was then estimated by integrating each segment of a graph of the contours at 10-ft intervals versus the area of each contour (Figure 10.2). The volumes also may be estimated by dividing each segment into even increments and using the trapezoid rule:

$$V = \frac{h}{2}[A_0 + 2(A_1 + A_2 + \cdots + A_{n-1}) + A_n] + hA_n \tag{10.6}$$

Example

Calculate the overall gross volume of the reservoir using contour intervals of 10 ft in Figure 10.2b.

Solution

$$\begin{aligned} V &= \frac{10}{2}[38.6 + 2(26.7 + 24.0 + 21.3 + 8.5 + 6.4 + 5) + 3.5] + 6(3.5) \\ &= 1{,}150.5 \text{ acre-ft} \end{aligned}$$

After obtaining the gross volume of each segment of the reservoir, well logs are used to estimate the volumes of zones with porosities that are practically impermeable because of changes in lithology such as shale lenses in sandstones, and chert in limestone formations. This analysis is not exact but must be done to obtain more realistic estimates of the actual hydrocarbon volumes in place. The net volumes for the example reservoir are shown in Table 10.3. Water saturations are then used to obtain the average hydrocarbon volumes of each 10-ft segment. Finally, the volumes are converted to standard cubic feet (14.7 psi, 60°F) and stock tank barrels using Eqs. (10.1)–(10.3). The process is complicated, in actual practice, by fault zones that divide the reservoir into different compartments; each compartment must be analyzed individually.

The estimated saturations of gas and oil are average values determined from average porosity and water saturation data from well logs and core analyses. As more information is obtained from the field and laboratory, the estimates must continually be refined to yield realistic production predictions and to schedule changes of field management for optimization of the resource. Some of the early laboratory data from core analysis may be wettability and capillary pressure analyses from cores taken in various locations at different elevations in the reservoir. Wettability governs the fluid flow (production) behavior of the reservoir (refer to section "Effect of Wettability on Oil Recovery" in Chapter 6); capillary pressure is used to determine the saturation distribution within the reservoir, or sections of the reservoir (as divided by faults and vertical divisions due to permeability heterogeneity; refer to section "Statistical Zonation Technique" in Chapter 3). These data are important for the placement of wells and the well perforations because the water/oil producing ratio increases with respect to

TABLE 10.3 Areas from Planimeter Estimates for 10-ft Contours of Figure 10.2a, Average Values from Well Logs, and Estimation of the Gas (Top Portion of the Figure), and Oil Hydrocarbons in Standard Cubic Feet and Acre-Feet, respectively

Subsea Depth	Planimeter Area (acres)	Average Volume, 10-ft Segments (acre-ft)	Porosity (%)	Gross Volume (acre-ft)	Net Volume (acre-ft)	Average *K* (mD)	Water Saturation (%)	Hydrocarbon Volume (Gas: SCF, Oil: acre-ft)
−4,034	0							
		11.6	13.2	1.5	1.1	78	37.2	0.7
−4,040	3.5							
		44.0	20.2	8.9	4.3	206	24.4	3.3
−4,050	5.0							
		55.7	18.3	10.2	8.6	133	28.3	6.2
−4,060	6.4							
		72.3	19.8	14.3	13.9	186	26.4	10.2
−4,070	8.5							
		150.8	22.3	33.6	33.0	219	33.6	21.9
−4,080	21.3							
		229.2	23.6	54.1	51.4	233	44.3	28.6
−4,090	24.0							
		256.5	19.4	49.8	44.2	154	51.4	25.9
−4,100	26.7							
		330.4	21.2	70.0	58.6	167	68.2	21.7
−4,110	38.6							

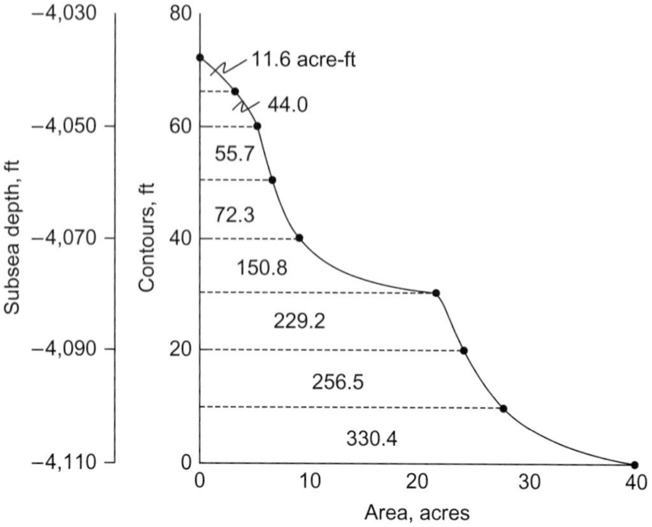

FIGURE 10.2 Volumes (acre-feet) of 10-ft segments of the reservoir from areas obtained by planimeter (Table 10.1).

depth as the water/oil contact is approached. Equation (5.63) is used to determine the saturation profile with respect to height above the water/oil contact (illustrated in Figure 5.22). If the reservoir is continuous through a large vertical height, capillary pressure will determine the height above the free water zone (100% water), the position of the water/oil contact, the vertical transition zone when water and oil are produced, and the zone above which 100% hydrocarbon is produced (where the water saturation is equal to the irreducible water saturation). The irreducible water saturation is generally the same in both the oil and gas zones and therefore the gas/oil contact (GOC) cannot be easily distinguished on an electric well log. Radioactivity logs, fluid flow testing, and core analyses showing the absence of residual oil are necessary factors to establish the GOC.

The other parameters of Eqs. (10.1)–(10.3) that are necessary for determination of the gas and oil volumes in the reservoir are discussed in detail in the chapters on porosity and the formation resistivity factor. The amount of gas (scf) and oil (STB) in the reservoir are an optimistic estimate of the economic value because the overall production recovery is influenced by several factors that control the areal- (horizontal) and vertical-sweep efficiencies of the hydrocarbons displaced to the wells by pressure maintenance and advancing gas and water drives (WDs). The areal-sweep efficiency (E_A) is principally governed by the mobility ratio between the displacing water and reservoir oil, well design, and the geometric properties of the overall reservoir. The vertical-sweep efficiency (E_V) is controlled by another set of conditions: the variations of petrophysical properties (especially permeability heterogeneity) within vertical zones (or layers).

Statistical correlations indicate ranges for E_A from 0.60 to 0.70, and E_V from 0.4 to 0.6 [11]. The mobile oil saturation is included for calculation of the initial overall recovery factor for the reservoir:

$$R_F = E_A \cdot E_V \cdot \frac{S_{oi} - S_{or}}{S_{oi}} \tag{10.7}$$

Arps [14] and Doscher [15] used statistical correlations to develop equations for estimating overall recovery factors for solution gas drive (SGD) and WD reservoirs as follows:

$$R_F^{SGD} = 41.815 \left[\frac{\phi(1 - S_{wi})}{B_{oB}}\right]^{0.1611} \left(\frac{K}{\mu_{oB}}\right)^{0.0979} (S_{wi})^{0.3722} \left(\frac{P_B}{P_A}\right)^{0.1741} \tag{10.8}$$

$$R_F^{WD} = 54.898 \left[\frac{\phi(1 - S_{wi})}{B_{oi}}\right]^{0.0422} \left(\frac{K\mu_{wi}}{\mu_{oi}}\right) (S_{wi})^{-0.1903} \left(\frac{P_i}{P_A}\right)^{-0.2159} \tag{10.9}$$

Using Eq. (10.1), the ultimate oil recovery (expressed as stock tank barrels (STB)) is calculated using the recovery factor (R_F):

$$RR = R_F \cdot \frac{7,758Ah\phi S_{oi}}{B_{oi}} \tag{10.10}$$

Example

Using the correlation values of areal- and vertical-sweep efficiencies, calculate the maximum, average, and minimum recoverable reserves for a reservoir with the following properties: $V = 1,000$ acre-ft, $\phi = 0.24$, $S_{oi} = 0.68$, $B_{oi} = 1.2$.

Solution

The maximum, average, and minimum recovery factors are as follows:

$$R_{F(max)} = (0.7)(0.6) = 0.42$$
$$R_{F(ave)} = (0.65)(0.5) = 0.325$$
$$R_{F(min)} = (0.6)(0.4) = 0.24$$

$$RR = R_F \cdot \frac{7,758(1,000)(0.24)(0.68)}{1.2} = 1.06 \text{ MMSTB} \times R_F$$

$$RR_{(max)} = 0.44 \text{ MMSTB}$$
$$RR_{(ave)} = 0.34 \text{ MMSTB}$$
$$RR_{(min)} = 0.25 \text{ MMSTB}$$

HYDRAULIC FLOW UNITS

One of the most useful methods for visually identifying zones with consistent petrophysical properties is based on self-potential/resistivity log correlations between wells (Figure 10.3). The constructions make it very clear

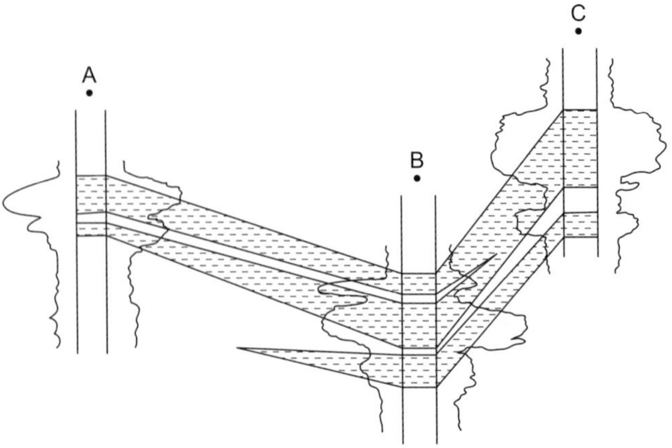

FIGURE 10.3 Structural cross sections between the wells based on the self-potential and resistivity logs. Two flow units at A and B merge into a single unit at C, and a third appears between B and C.

that there are apparently two zones at well **A**, which are still identifiable at well **B**, and they continue on to well **C** merging into a single zone as a shale stringer pinches out between **B** and **C**. A third flow unit apparently begins near **B** and is connected to **C**. Each of these sand bodies may have different fluid properties, porosities, permeabilities, wettability, changes of lithology, etc.; therefore, they can each have quite different fluid transmissibilities identified as unique flow units within the reservoir. There are many methods for characterizing these flow units based on their consistent fluid and petrophysical properties (refer to the section on flow units in Chapter 3).

MOBILITY RATIO

Darcy's law, the development of linear and radial flow equations, and applications are discussed in detail in Chapter 7. The radial flow equations for water, oil, and gas, used for flow rate calculations of injection and production wells (Eq. (7.57)), expressed in US Field Units are as follows:

$$\text{Water:} \quad q_w = \frac{0.00708 k_w h (P_e - P_w)}{B_w \mu_w Ln(r_e/r_w)} \tag{10.11}$$

$$\text{Oil:} \quad q_o = \frac{0.00708 k_o h (P_e - P_o)}{B_o \mu_o Ln(r_e/r_w)} \tag{10.12}$$

$$\text{Gas:} \quad q_g = \frac{0.03975 k_g h (P_e - P_w)}{B_g \mu_g Ln(r_e/r_w)} \tag{10.13}$$

When Eq. (10.11) is divided by Eq. (10.12), the ratio, q_w/q_o, is the producing water/oil ratio of the well, which is:

$$\text{WOR} = \left(\frac{k_w}{k_o}\right)\left(\frac{\mu_o}{\mu_w}\right)\left(\frac{B_o}{B_w}\right) = M\left(\frac{B_o}{B_w}\right) \tag{10.14}$$

Removing the volume factors from Eq. (10.14) (which are used to convert from reservoir volumes to volumes at surface conditions) yields the reservoir mobility ratio for water/oil, and similarly the gas/oil mobility is obtained; in the case of the gas/oil mobility, however, the amount of gas flowing in the reservoir is augmented by the solution gas evolving from the oil as the pressure declines toward the well-bore.

Within the reservoir, the water/oil ratio is equal to the *water/oil mobility ratio*; equally, the free-gas/oil ratio is the *free-gas/oil mobility ratio*:

$$M^{\text{gas/oil}} = \frac{k_g}{k_o}\frac{\mu_o}{\mu_g} \tag{10.15}$$

At the surface, the produced gas/oil ratio is equal to the free gas (G_F) augmented by solution gas ($G_s = q_s = R_s$) that evolved from the oil when the pressure and temperature were changed from reservoir conditions to surface conditions. Therefore, the producing gas/oil ratio of the well is the sum of the solution-gas/oil ratio (R_s) and the free-gas/oil ratio at surface conditions:

$$\begin{aligned} q_g &= G_S + G_F\frac{B_o}{B_g} \\ R_P &= R_S + \frac{k_g}{k_o}\frac{\mu_o}{\mu_g}\frac{B_o}{B_g} \end{aligned} \tag{10.16}$$

Mobility is determined for conditions existing before water breakthrough occurs at a well (unless other specifications are given) because the mobility increases with respect to water saturation. The mobility ratio determines the hydrocarbon displacement properties. The effective permeabilities incorporate the characteristics of the rock matrix (pore-size distribution and wettability) and determine the mobile hydrocarbon saturation and amount of oil left behind (the residual saturation) after the reservoir has been swept by the water. The viscosities influence the displacement properties and frictional drag on oil by displacing water in the reservoir. Actual displacement of the hydrocarbons deviates from a complete piston-like process as water travels through the pores displacing a portion of the oil in an advancing front, but leaving behind the oil that is moved by viscous drag of the water on the oil; this creates a saturation gradient extending from high values at the advancing oil displacement front to residual oil saturation at some distance behind the front. The fingering (or passage of water through the oil phase ahead of the advancing front) is greater and the oil-to-water viscosity ratio increases, as reservoir wettability changes from water-wet conditions toward more oil-wet conditions from one zone to another [16]. The fingering of water toward the producing wells is described mathematically as a series of parallel reservoir layers with different permeability properties (refer to the sections on layered reservoir flow in Chapter 7).

Enhanced recovery methods are simply explained by the mobility ratio. The overall hydrocarbon displacement characteristics are governed by the mobility ratio, M; therefore, methods (beyond those of just water injection) can be explained as influencing the mobility ratio, for example:

1. The addition of 10–100 ppm of sodium tripolyphosphate to the injection water reduces scale in the pipes and precipitation of sulfate deposits in the reservoir and around the wells, thus improving the relative permeabilities, resulting in a decrease of M.
2. The use of surfactants to decrease the oil/rock adhesion tension (produce water-wet conditions) and to create miscible phase displacement at the front by removing the viscous interface between water and oil (micellar/polymer flood) affects both the permeability and viscosity terms, resulting in a decrease of M.
3. The addition of water-soluble polymers to the injection water process increases water viscosity without change in the oil viscosity and thus produces a reduction of M and so an increase of displacement efficiency.
4. Thermal methods of enhanced oil recovery (EOR) cause a large decrease of the oil viscosity with a relatively negligible effect on the water viscosity, producing a large decrease in the value of M.

Thus, relative permeability and viscosity data received from the laboratory after discovery of a petroleum reservoir for various heights above the water/oil contact should be carefully evaluated to determine the fluid flow properties of the reservoir, possible vertical layering (produced by various depositional stages over geologic time), and the implementation of improved recovery methods, or isolation of "thief zones" (zone of higher-than-normal rates of water flow) by physical or chemical methods [16].

Example

(a) Calculate the mobility ratio for various points in the reservoir from relative permeability and fluid data.

(b) Discuss the effect of high mobility value zones on overall production.

Solution

(a)

$\mu_o/\mu_w = 12.0$; $B_o = 1.28$, $B_w = 1.02$

Relative permeability values taken at $S_w = 0.50$.

Depth (subsea)	k_{rw}	k_{ro}	M	WOR
3,066	0.08	0.40	2.4	3.0
3,074	0.12	0.18	8.0	10.1
3,087	0.19	0.11	20.8	26.2
3,102	0.13	0.22	7.1	8.9
3,109	0.10	0.21	5.7	7.2

(b)

The zone at 3,066 ft indicates very favorable production and therefore the thickness of this zone should be determined so that well completion can be modified to enhance the production rate from this zone.

The mobility ratio at 3,087 ft indicates that it will contribute to a high WOR and diminished production of oil. Therefore, it should be examined to determine the thickness of the zone so that it can be isolated. In addition, adjacent well data should be closely examined to determine if the zone extends beyond the current well. This extreme permeability variation in the middle of the oil reservoir could be caused by a change of lithology or wettability (a zonal change to a strongly oil-wet region).

J-FUNCTION

Layered hydraulic flow units are found within thick sand reservoirs that develop during sedimentation in response to changing environmental conditions: weathering, changing grain type and size, erosion, and factors that affect the permeability and porosity during diagenesis and compaction. Clastic carbonate reservoirs composed of shell fragments and limestone grains, sometimes mixed with sand, have performed well as productive petroleum reservoirs, similarly to sandstone reservoirs. Carbonate eolianites generally develop parallel to paleo-shorelines as discontinuous sequences and may have distinctive flow characteristics that can be characterized by permeability–porosity correlations. Other carbonate reservoirs exhibit considerable heterogeneity and low values of permeability and porosity (tight matrices), which can nevertheless be characterized by the correlations.

A variety of factors contribute to the extant factors that control the transmissibility of layered beds of sediments that have consistent petrophysical properties over long distances. One or more of the flow units can have very different directional transmissibilities that adversely affect the overall reservoir production. The mobility analyses provide one indication of this, as explained above.

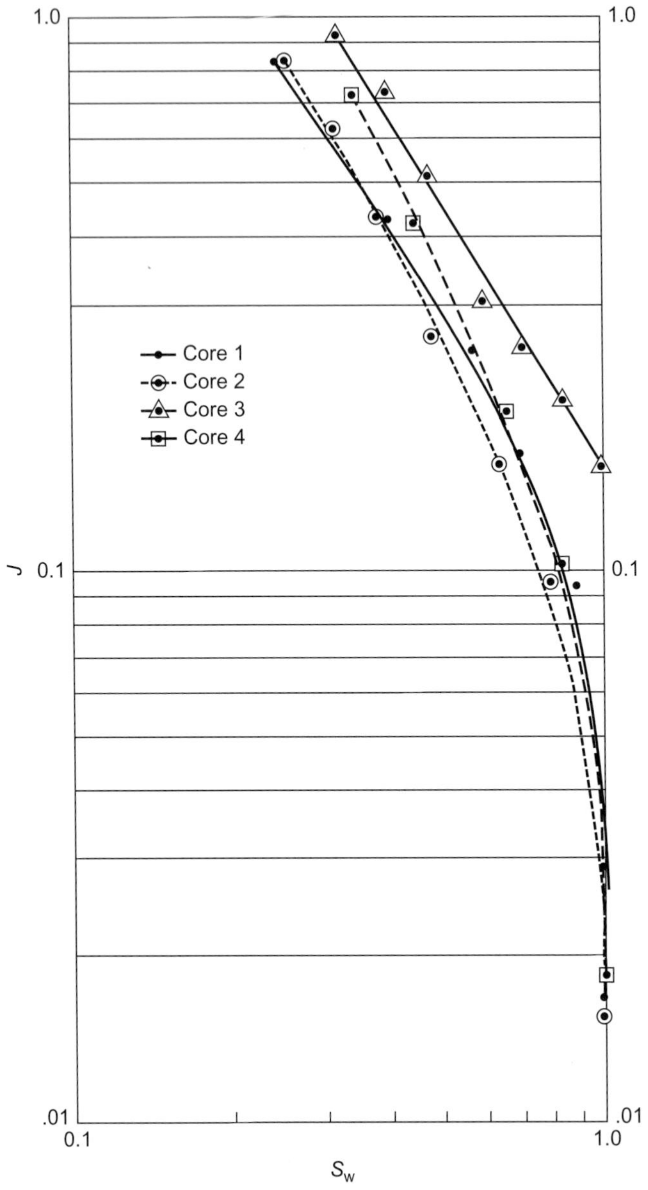

Two other correlations that can be used to define the flow units are the exponential permeability–porosity correlations (Figure 3.12) and the *J*-function, which is a function of water saturation, and includes the capillary pressure in a dimensionless correlation with permeability and porosity (Eq. (5.64)).

The *J*-function also shows characteristic differences between different reservoirs, and differences within a reservoir (when examined as a function

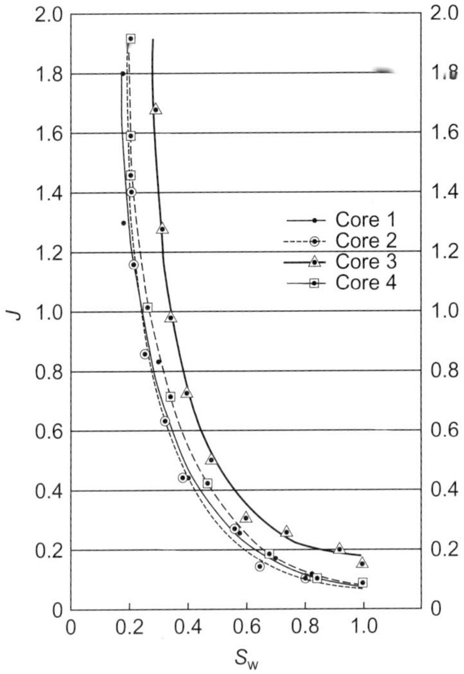

FIGURE 10.4 *J*-Function versus water saturation used to define flow units in a reservoir. *Source: Data from Table 5.3.*

of depth) and it is therefore used for correlation of flow units between wells. The *J*-function calculations in Table 5.3 (refer to section "Vertical Saturation Profile in a Reservoir" in Chapter 5) were taken from four samples in the same sandstone formation. When plotted, as *J* versus S_w, (see Figure 10.4) it becomes obvious that cores 1, 2, and 4 have very similar petrophysical properties (and hence flow properties), but core 3 exhibits completely different properties and defines a zone that is considerably more water wet. The log (*J*) versus log(S_w) is a straight line with an intercept (J_1 at $S_w = 1.0$) and $1.0/J_1^2$ is equal to the K_T parameter of the Kozeny–Carman equation, Eq. (3.27) (refer to the section "Mathematical Theory of Flow Units" in Chapter 3). The parameter K_T is a function of the grain properties and heterogeneity where low values of K_T ($J_1 > 0.1$) indicate high permeability, uniformity in pore size, and connectivity; whereas high values of K_T indicate low permeability and heterogeneity.

Example

Use the values of *J* and S_w in Table 5.3 (Figure 10.5) to make a plot of log(*J*) versus log(S_w) by plotting them on log–log graph paper. Compute the values of K_T for each core and discuss the results.

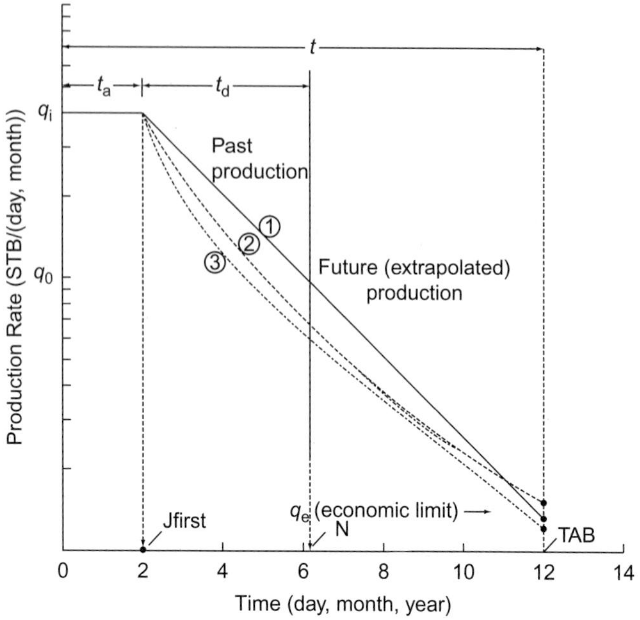

FIGURE 10.5 Generalized charts of production rate versus time (decline curves): (1) exponential decline, (2) hyperbolic decline, and (3) harmonic decline.

Solution

The log–log plot shows that the data for Core 3 deviates from that of the other three cores, indicating clearly that Core 3 is from a zone that can be considered a distinctly different flow unit in the reservoir.

Values of K_T are: Core 1 = 210; Core 2 = 308; Core 3 = 44; Core 4 = 179.

The values of K_T are even more revealing, showing that Core 3 must exhibit higher permeability and greater uniformity of pores and considerably better pore connectivity.

Saner et al. [17] developed a permeability–porosity correlation for a carbonate reservoir using 75 cores that had a correlation coefficient of 0.81. This could (theoretically) be used throughout the reservoir to obtain the permeability, where porosity was known similar to the correlations shown in Figure 3.12 for granular matrices.

$$K = 0.007873 \times 10^{0.166\phi} \tag{10.17}$$

An exponential correlation was used by Shedid and Almehaideb [18] for a different carbonate reservoir to develop the overall relationship between K and ϕ, where

$$K = 0.2112 \times e^{0.4505\phi} \tag{10.18}$$

A comparison of the results for the two different reservoirs showed that the reservoir examined by Shedid and Almehaideb [18] exhibited much greater permeabilities (47–183 times greater) for the same values of porosity, thus demonstrating the great difference in petrophysical properties that can occur between carbonates because of differences in pore geometry and connectivity.

An examination of the exponential K versus ϕ, as a function of depth, at different wells can indicate differences within flow units.

Improvements in the J-function were made by Amaefule et al. [19], who derived a term that was labeled the Reservoir Quality Index (RQI) which is plotted on log–log graph paper versus the ratio of the pore volume to the grain volume (ϕ_z) of the specimens (refer to section on flow units, and example calculation in Chapter 3). This produces points with linear trends for separate flow units:

$$\mathrm{RQI} = 0.0324\sqrt{\frac{K}{\phi}} \tag{10.19}$$

Shedid and Almehaideb [18] later developed another correlation that is more useful for identifying multiple flow units in a reservoir from well to well. They used dimensionless analysis to develop another equation (labeled the characterization number, CN) that incorporates more fluid and rock properties and is plotted versus the $\sqrt{K/\phi}$ term. Distinctly different flow units are evident by separation of lines with slightly different slopes:

$$\mathrm{CN} = \left(\frac{\rho_o \sigma_{w-o}}{\mu_o^2 \cos\theta}\right)\left(\frac{k_{ro(\text{cross-over})}}{k_{rw(\max S_w)}}\right)\sqrt{\frac{K}{\theta}} \tag{10.20}$$

where $K = \mathrm{m}^2$; $\rho = \mathrm{kg/m^3}$; $\sigma = \mathrm{N/m}$; $\mu = \mathrm{kg/(m\ s)}$

Another method for defining flow units is to determine the foot-by-foot flow capacity (Kh) versus the storage capacity (ϕh) for a productive zone. The sums of the values for the zone are taken as the total flow capacity and storage capacity of the zone. A plot of the percentage flow capacity for each foot, or interval of depth, versus the percentage storage capacity is used to define the flow units based on inflection points of the graph [20–24].

ANALYSIS OF RESERVOIR PERFORMANCE

Primary performance of oil and gas reservoirs is principally analyzed using the petrophysical, fluid and production properties through decline curves, material balance calculation, and reservoir simulation. Several specific reservoir oil-drive mechanisms of primary recovery have been recognized, which can be treated with good accuracy; however, reservoirs with combination drives (e.g., WD plus gravity segregation) are difficult because of the large number of interacting variables affecting production. Analyses made on the assumption of single-drive mechanisms, however, have been very successful and useful for reservoir performance prediction, reserve estimation, and diagnosis of insipient chemical/mechanical problems.

The mechanisms that influence production performance and are subject to accurate analyses are as follows:

- Reservoir rock and liquid expansion: *Reservoir pressure* declines continuously and very rapidly when the initial reservoir pressure (P_i) is greater than the bubble-point pressure (P_{BP}). *Gas–oil ratio* is low and remains constant. *Brine production* is negligible unless the reservoir water saturation is high ($S_w > 0.45$). *Production efficiency* is very low, averaging between 3% and 5%.
- Solution-gas drive: *Reservoir pressure* declines continuously and rapidly. *Gas–oil ratio* is quite variable; it is low in the beginning and at some point begins to rise to a maximum before declining. *Brine production* is negligible except for reservoirs with high water saturation. *Production efficiency* varies widely between 10% and 40% with an average somewhere around 20%; in addition it becomes necessary to begin pumping early in the life of the reservoir. In the later stages of production, the SGD may vanish merging into a gravity-drainage mechanism.
- Gas-cap drive: Recognized by gas breakthrough in down-dip wells. *Reservoir pressure* declines slowly at close to a constant rate. *Gas–oil ratio* increases continuously. *Brine production* is absent or negligible. *Production efficiency* varies from approximately 15% to 40% with an average close to 25%.
- Gravity drainage: *Reservoir pressure* declines rapidly at an almost constant rate. *Gas–oil ratio* is constant in down-dip wells but high in up-dip wells. *Brine production* is absent or negligible. *Production efficiency* is high, averaging 60% in most cases.
- Water drive: *Reservoir pressure* generally remains high and responds rapidly, decreasing or increasing as the fluid production rates are increased or decreased, respectively. *Gas–oil ratio* will be low if the reservoir pressure is maintained at a high value. *Brine production* increases significantly with respect to production. *Production efficiency* is between 35% and 80%, with a general average for this type of reservoir of approximately 50% [25].

DECLINE CURVES

The analysis generally consists of plots of the *logarithm of the flow rate* of a well (or group of wells) *with respect to time*, or *cumulative oil produced*, using semilog graph paper. As primary, or secondary, production declines, the data are extrapolated to the designated economic limit of production for estimates of production rates at specific future dates, ultimate economic production, and oil reserves (initial and remaining). Oil flow rate is expressed in STB/day, month, or year; and gas-flow rate as scf/day, month, or year.

The log(q) versus time plots (Figure 10.5) describe three distinct types of curves: (1) an *exponential* curve, (2) a *hyperbolic* curve, and (3) a *harmonic* decline curve; the exponential and harmonic curves, however, are no more than special cases of the hyperbolic curve. Therefore, the least-squares solution of the data described by a hyperbolic equation can be used to represent the decline curves. The three-constant hyperbolic equation (Eq. (5.44)) is explained and an example of its use for least-squares analysis is presented in Table 5.2 for spread-sheet use. A FORTRAN program for analyses of decline curves together with the subroutine HYPER and input/output DATA files for an example problem are listed at the end of the chapter.

The graphs shown in Figure 10.5 present hypothetical curves for typical production facilities where the declining rate of production follows an initial restricted (or allowable) production rate (q_i). The first time period, t_a, is a period of erratic production that may be experienced when a well, or group of wells, is first brought into production the second period, t_d, encompasses *past production* (the time period of production; from JFIRST to the end of the production/time data) and *future production* (the period of forecasted production based on the decline data). TAB is the time (days, months, years) at which abandonment of the project is anticipated.

The slope (D) of the exponential curve (when the constant C in the hyperbola equation is equal to zero) is governed by the logarithmic decline, thus:

$$
\begin{aligned}
q_o &= q_i e^{-Dt} \\
D &= \frac{Ln(q_i/q_o)}{t} \\
N_p &= \frac{q_i - q_o}{D}
\end{aligned}
\tag{10.21}
$$

Determination of the cumulative production (N_p), at any time (t), for the other production versus time relationships is accomplished using the hyperbolic equation.

When sufficient production data are available to forecast the behavior of the future decline curve, any change of shape, or slope, of the curve should be immediately investigated to determine the cause. This could foretell a large number of developing conditions that are detrimental and can thus cause a severe loss of future production such as abnormal increase of water coning, adverse plugging of the production wells by precipitates, particle movement, scale, etc.

The log(q) versus cumulative production may also be used for decline analysis with the program DECLINE.FOR. In addition, the increasing water/oil ratio may be analyzed with the program whenever the economic limit of production is to be governed by the amount, or cost, of water production, treatment, and disposal. In fact, subroutine HYPER is a very versatile tool for analysis of any curves (increasing, constant, or declining) that one may encounter.

The example of decline curve analysis using DECLINE.FOR is listed at the end of the chapter because of its unusual length.

MATERIAL BALANCE

Optimum exploitation of an oil reservoir requires classification of the reservoir according to one of four dominant oil-drive mechanisms: (1) SGD, (2) gas-cap drive, (3) WD, or (4) gravity drainage. This is necessary for estimation of reserves, placement of wells, completion intervals, and development of primary/secondary/EOR oil recovery schemes.

Movement of oil toward a production well is governed by several of the drive mechanisms simultaneously, but for most reservoirs there exists a dominant drive mechanism whose energy exceeds the other mechanisms, and the interpretation and predictions of production can be based on the dominant drive mechanism with good accuracy. In fact, secondary recovery processes are often initiated to enhance the dominant drive mechanisms.

SOLUTION-GAS DRIVE

When a reservoir does not have a gas-cap or WD from an adjacent aquifer, and is initially at a pressure greater than the bubble-point pressure (P_{BP}), the oil production drive mechanism is a *solution-gas drive*. The pressure decline and gas/oil ratio (GOR) exhibit characteristic behavior with respect to time or the cumulative oil produced (Figure 10.6a).

1. Production begins at a pressure above the bubble point; then the pressure declines at a constant rate and the GOR remains constant until the bubble-point pressure is attained.
2. The pressure decline slows when the P_{BP} is exceeded as gas is released from solution. The pressure decline accelerates once more when the GOR increases, while the oil volume shrinks as expanded gas is removed.

   At the P_{BP}, gas begins to evolve from the oil (l, Figure 10.6a) and the GOR decreases until a critical gas saturation (or residual gas saturation, similar to S_{or}) is exceeded (2, Figure 10.6a) and gas begins to flow.

   The rapid rise of GOR occurs in all of the wells. If the rapid increase of GOR occurs in wells completed up-dip (high within the reservoir), it is an indication of a combined expanding gas-cap and an SGD mechanism.

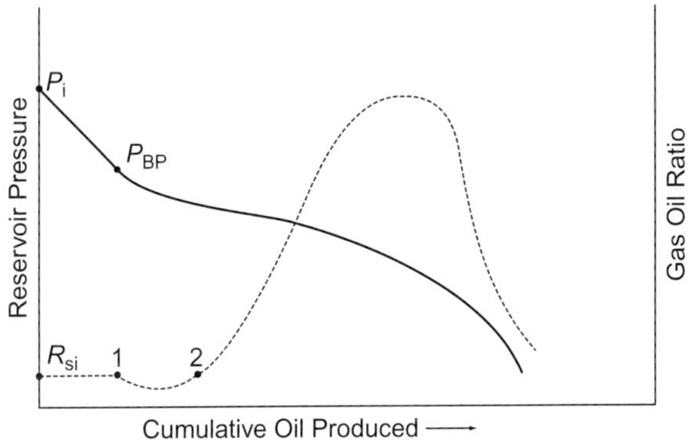

FIGURE 10.6a Idealized pressure and gas–oil ratio behavior with respect to cumulative oil produced.

3. No water will be produced initially if the connate water saturation is less than the irreducible water saturation (S_{wi}); however, as pressure declines, water will expand and all of the wells will produce some water. If there is an active WD, production of water would begin first in wells that are completed down-dip (low within the reservoir).

GAS-CAP DRIVE

When a gas-cap is present, gas expands and invades the oil zone as a fairly uniform, vertically advancing front. The stability of the front and efficiency of oil displacement are functions of the gas and oil vertical mobilities; using Darcy's equation:

$$u_g \uparrow = \left(\frac{K_g}{\mu_g}\right)\left(\frac{\Delta P}{L}\right), \qquad u_o \downarrow = \left(\frac{K_o}{\mu_o}\right)\left(\frac{\Delta P}{L}\right)$$
$$M = \left(\frac{K_g}{K_o}\right)\left(\frac{\mu_o}{\mu_g}\right) \tag{10.22}$$

Relatively low oil viscosity will enhance the downward counterflow of oil displacement. On the other hand, an unfavorable, or high, viscosity ratio promotes greater gas mobility that translates into (1) greater instability (gas fingering) at the advancing front, (2) increased by-passing of oil; hence, reduced displacement efficiency (greater S_{or}). Gas is the nonwetting phase; therefore, it preferentially enters, and occupies, the larger pores surrounding and trapping oil in smaller pores; this phenomenon is enhanced by increasing velocity of the movement of gas, and (3) lower gravity-assist for vertical oil drainage; greater vertical permeability aids the movement of the

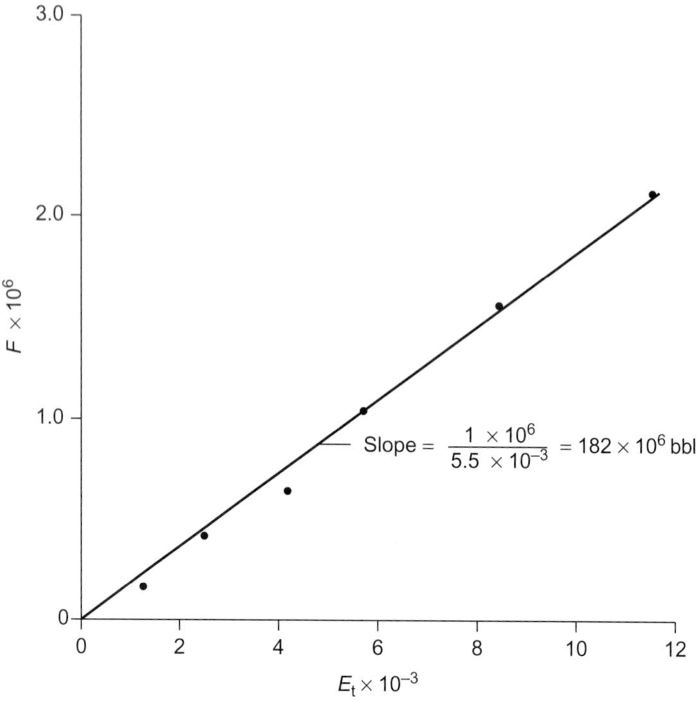

FIGURE 10.6b Graphical solution for MBE (no WDR and no gas cap).

expanding gas downward and facilitates gravity drainage and improved oil displacement.

As the gas-cap expands, the GOC and bubble-point pressure move downward (Figure 10.7a and b), and free gas develops at the top of the oil zone. The original solution gas evolves from oil as the level of the bubble-point pressure moves downward and the oil volume in this zone shrinks as a result. Hence, the drive mechanism in the upper oil saturated (with gas) zone becomes an SGD. Eventually the gas will move downward and completely encompass the oil zone.

Gas expansion is the source of the oil-drive energy; therefore, production of gas should be minimized as much as possible. This means that wells producing higher in the reservoir (that began production of gas when the bubble-point level arrives at the well completion depth) must be taken out of production to conserve gas in the reservoir for greater overall oil displacement efficiency.

The relationship between the reservoir pressure and the gas/oil producing ratio, with respect to cumulative oil produced, is illustrated in Figure 10.8. As production of oil increases, pressure declines slowly at first while the gas-cap expands into the oil zone and the GOR increases

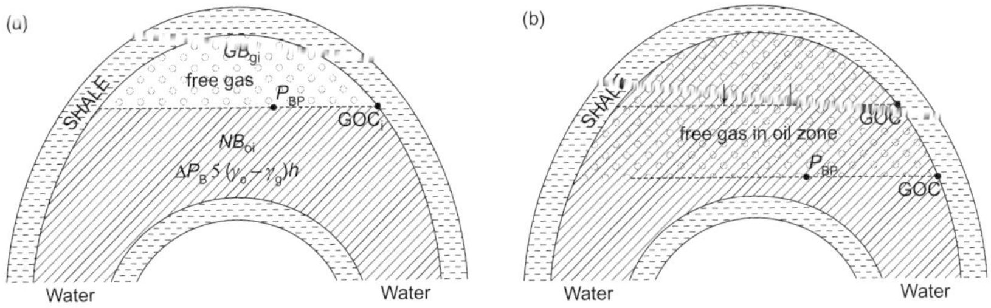

FIGURE 10.7. (a) Gas-cap drive reservoir. (b) Expansion of gas downward as oil is produced.

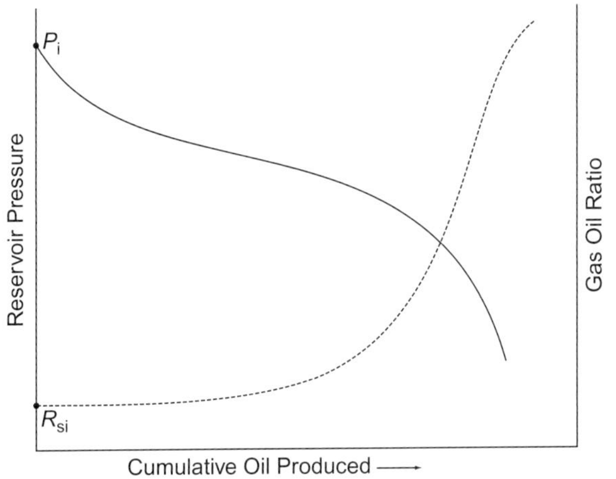

FIGURE 10.8 Behavior of reservoir pressure and gas–oil ratio as a function of the cumulative oil produced for a gas-cap oil drive mechanism.

gradually. When the gas-cap expands downward causing greater production of gas, the pressure begins to decline sharply in tandem with the ever-increasing GOR.

The rate of oil production is one of the major considerations for a gas-cap drive reservoir. The pressure drawdown in the vicinity of the well-bore is very sensitive to the rate of production. A high rate of oil production will create a considerable pressure difference between the well-bore and its radius of influence, which will result in severe increase of gas production. Excessive gas production will (1) decrease the oil displacement efficiency and (2) decrease the ultimate recovery. The drawdown pressure at any point from the well-bore is equivalent to the slope of the curve (Figure 10.10). The slope imparts a horizontal pressure, toward the well on the gas, and the buoyant pressure acts vertically. Thus, the gas moves diagonally toward the well because the resultant pressure acting on the gas becomes much greater as the slope of the drawdown curve increases.

Example

Calculate the resultant pressure per foot of height for (1) a drawdown pressure with a slope equal to 0.4 and (2) a greater drawdown pressure with a slope equal to 1.0.

Solution (refer to example Figure 10.14a)

Slope = 0.4

$$\Delta P_H = \frac{P_2 - P_1}{L} = \frac{27.5 - 20.0}{1.0} = 7.5 \text{ psi/ft}$$

$$\Delta P_B = (\gamma_o - \gamma_g)h = \left[\frac{\text{lb}}{\text{ft}^3} \times \frac{\text{ft}}{144 \text{ in}^2/\text{ft}^2}\right] = \text{psi/ft of height}$$

$$\Delta P_B = (0.2)(0.4/144) = 5.6 \times 10^{-4} \text{ psi/ft}$$

$$R = [7.5^2 + (5.6 \times 10^{-4})^2]^{1/2} \approx 7.5 \text{ psi/ft}$$

Slope = 1.0

$$\Delta P_H = (39 - 29)/1.0 = 10 \text{ psi/ft}$$

$$\Delta P_B = (0.2)\left(\frac{1.0}{144}\right) = 1.4 \times 10^{-3}$$

$$R \approx 10 \text{ psi/ft}$$

The net resultant force moving the gas toward the well is 2.5 psi greater for the steeper drawdown at a specific point away from the well-bore. The buoyant force, per foot of height, is very small, and hence it is essentially constant in the reservoir; it becomes a significant factor as the height of the column increases [26,27].

The forces acting on the gas beyond a critical radius from the well-bore will have sufficient buoyance to move gas upward rather than toward the well. Consequently, there is a critical production rate for each well (in a gas-driving oil reservoir) beyond which gas production will occur, resulting in lower overall recovery.

WATER DRIVE

There are several descriptive conditions that apply to the classification of a WD reservoir. If a gas-cap is not present (or is so small that its influence is not significant), gas dissolved in the oil will act as an SGD as part of the production mechanisms, but it does not have the dominant role. Hence, the producing gas/oil ratio rises slowly (Figure 10.9) with respect to cumulative production of oil because the strong pressure maintenance effect from the adjoining aquifer moderates the release of solution gas in the reservoir.

There are four general types of WD reservoirs, illustrated schematically in Figure 10.10. In all of these, the aquifer attached to the oil reservoir must be very large to perform a WD mechanism because the influx of water, as production of oil decreases the reservoir pressure, results in

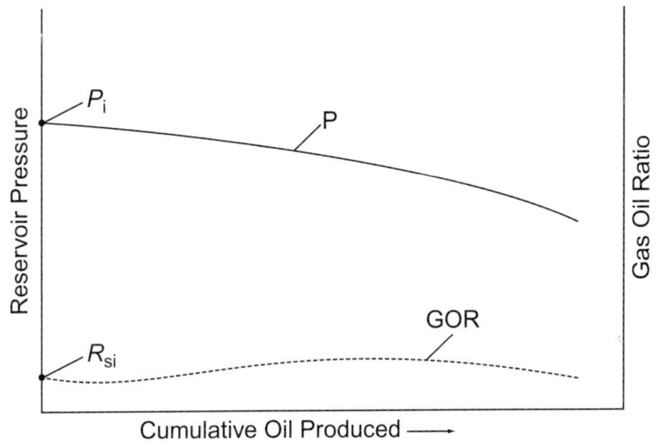

FIGURE 10.9 Behavior of reservoir pressure and gas–oil ratio as a function of cumulative oil produced for a gravity drainage mechanism.

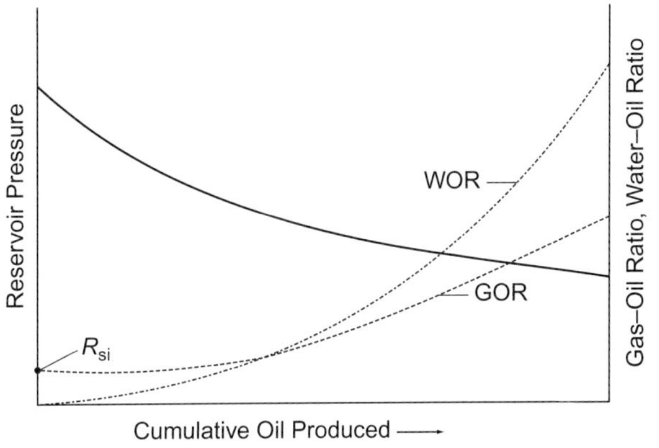

FIGURE 10.10 Behavior of the pressure, gas–oil ratio, and water–oil ratio of a water drive reservoir as a function of the cumulative oil produced.

expansion of water into the reservoir. A constant pressure drop develops at the interface between the oil and water while the aquifer behaves like an infinite system: the pressure remains constant at the outer boundary of the aquifer. Consequently, the reservoir pressure decline is gradual throughout the productive life of the reservoir, as shown in Figure 10.9.

The water/oil ratio increases steadily since water is constantly entering from the aquifer. Down-dip wells experience the water influx first and must be shut in when the WOR becomes excessive. There is a strong tendency toward water coning and therefore a WD reservoir is very sensitive to the production rate. The resultant water coning can bring about premature abandonment of the field, especially if there is a large difference between the oil and water viscosities.

A drawdown pressure should be determined (using reservoir simulation) for a production rate that will control water influx and pressure maintenance in the reservoir. In cases where the pressure decline is too steep, secondary recovery will be initiated early to augment the natural water drive mechanism. The oil drive is a frontal displacement mechanism and is especially effective with a favorable mobility ratio. The ultimate recovery of a carefully managed WD reservoir ranges as high as 60%. Recovery efficiency is governed by the balance between production rate and pressure maintenance because abandonment is tied directly to the quantity and cost of processing the ever-increasing water production.

GRAVITY DRAINAGE

By definition, a gravity drainage reservoir is one in which the other drive mechanisms are absent or very weak (little, or no, solution gas, no gas-cap or water influx). Heavy oil (°API<20) reservoirs generally meet these conditions, but the high oil viscosity seriously inhibits the process of gravity segregation within the reservoir.

The mechanism is one of vertical counterflow of gas and oil and is significant where low production rates allow sufficient time for gravity segregation of gas and oil to take place. A SGD transforms to one of gravity segregation near the end of the productive life of the field when the solution gas has been substantially depleted and the pressure has declined to a point where it is no longer an effective drive component.

Figure 10.11 illustrates the characteristic behavior of pressure and gas/oil ratio for a gravity drainage reservoir. The initial pressure is relatively low and its decline during production occurs at a very low gradient, which is almost constant throughout the life of the field. Production of gas must be controlled at a low rate to perpetuate the drive mechanism. Consequently, the gas–oil ratio also remains low and almost constant, only showing a slight rise near the center of the productive life of the field where up-dip wells begin greater gas production before being removed from production.

Water production is relatively low because there is little, or no, water influx in a gravity drive reservoir; however, the low overall rate of production generally dictates the initiation of a water injection program to enhance productivity. Although the gravity segregation process is slow, the ultimate recovery of a gravity drainage reservoir, when produced as such, can be as high as 60–70%.

DEVELOPMENT OF THE GENERAL MATERIAL BALANCE EQUATION

The material balance equation is used for calculations of the original volumes of hydrocarbons in the reservoir and for estimation of future

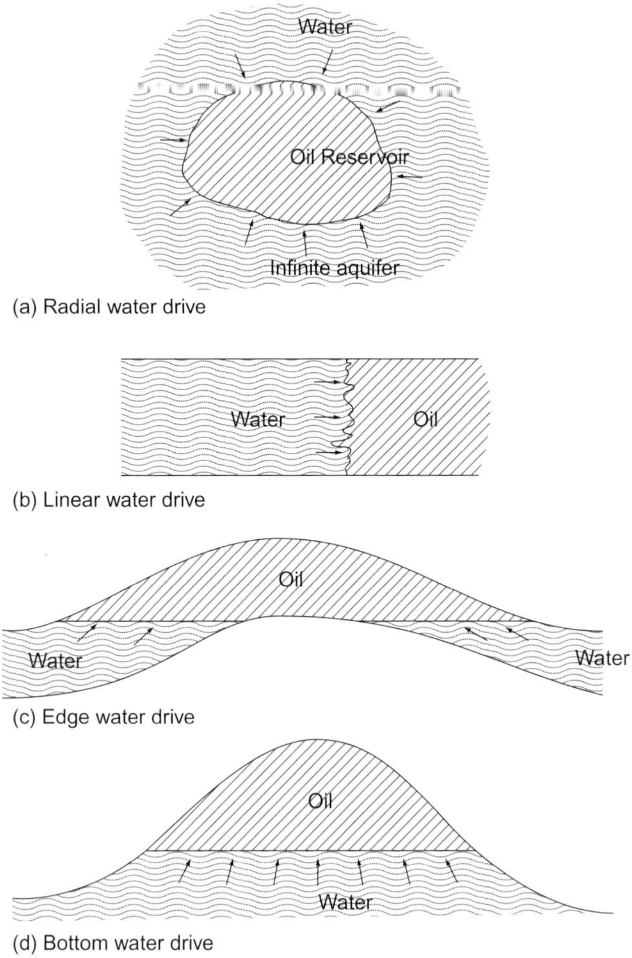

FIGURE 10.11 (a–d) Four general types of water-drive reservoirs.

performance. The equation is based on the premise that the reservoir volume remains constant throughout the life of the reservoir. Thus the volume is unaffected by changes of pressure, expansions, contractions, and movement of fluids into, and out of, the reservoir volume. Thus the sum of the volumetric changes in gas, oil, and water that take place during production of fluids is equal to zero.

By assuming that equilibrium between solution gas and reservoir oil exists within the reservoir at all times, a general volumetric balance equation can be developed to account for the fluids withdrawn, injected, and encroached, and for the expansion of fluids and rock. The analyses of the reservoir changes lead to (1) determination of reservoir parameters, gas-cap size, OOIP, and amount of water influx, (2) prediction of reservoir performance (after a period of production yields sufficient data for analysis), (3) future gas and oil production volumes, and (4) the pressure decline trend for the reservoir.

The data requirements for material balance analysis are (1) initial reservoir pressure and pressure decline as a function of time, (2) production data for gas, oil, and water as a function of time, (3) formation volume factors for gas, oil, and water, (4) ratio of the initial gas-cap volume to the initial reservoir oil volume (estimated from well logs, cores, and well-completion data), (5) quantity, or rate, or water influx (for simplicity it is assumed that water influx is a steady-state phenomenon in which the rate of water influx is proportional to the reservoir pressure decline: $dW_e/dt = k(p_i - p)$), and (6) water and rock compressibility.

GAS RESERVOIR

The original gas volume is equal to the gas present originally in the reservoir (G) and the amount of gas produced (G_p). If a water drive is not present (W_e and $W_p = 0$), the gas will expand into its original volume. If, however, the reservoir is connected to an aquifer, the original volume occupied by the gas (GB_{gi}) will contain the encroached water volume minus water than was produced along with the gas ($W_e - W_pB_w$), Figure 10.12. Thus encroachment of water reduces the gas pore volume and tends to maintain reservoir pressure.

$$GB_{gi} = (G - G_p)B_g + W_e - W_pB_w \tag{10.23}$$

$$\text{Gas production} = \text{Gas expansion} + \text{Water influx} - \text{Water produced}$$

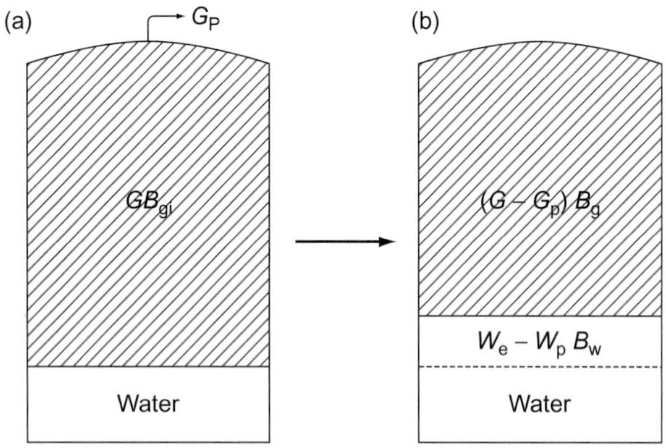

FIGURE 10.12 Gas produced from a reservoir in contact with an aquifer. The gas produced is equal to the expanded gas remaining in (b) plus (water influx − water produced).

Example

A gas reservoir has an original volume equal to 300 MMscf with the following data at some time after production commenced: calculate the volume of water invaded into the original volume:

$$G_p = 800 \text{ MMscf} \qquad W_p = 16{,}500 \text{ bbl}$$
$$B_{gi} = 0.0055 \qquad B_g = 0.0060$$
$$B_w = 1.028 \qquad \phi = 0.18$$
$$S_{wc} = 0.20$$

Solution

$$\text{OOIP} = \frac{(300 \times 10^6)(0.18)(1.0 - 0.20)}{0.0055} = 7{,}855 \times 10^6 \text{ scf}$$

Rearranging Eq. (10.23): $W_e = G_p B_g - G(B_g - B_{gi}) + W_p B_w$

$$W_e = (800 \times 10^6)(0.0060) - (7{,}855 \times 10^6)(0.006 - 0.0055) + (16{,}500)(1.028)(5.615)$$
$$= 957{,}000 \text{ ft}^3$$

The definitions of the terms to be used in the material balance equations are repeated here with more explicit definition than in the Nomenclature because it is *very important that each term be clearly understood* in order to work with the equations. In addition, the general behavior of the formation volume factors with respect to decreasing reservoir pressure, or increasing time, are indicated in Figure 10.13. Note that gas volumes are expressed as either scf or bbl; therefore care must be taken for consistent units.

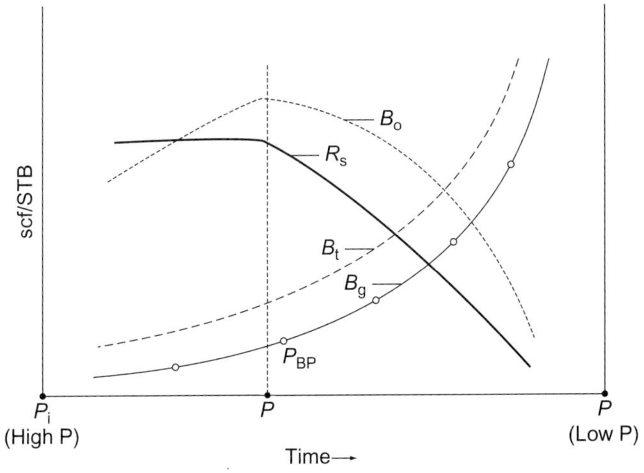

FIGURE 10.13 General behavior of formation volume factors as a function of declining reservoir pressure.

B_g gas formation volume factor: reservoir volume/STB (scf/STB or [bbl/STB = scf/5.61 STB])
B_o oil formation volume factor: reservoir barrels/STB (scf or bbl)
B_t total formation volume factor (volume (bbl) that 1 STB of oil *plus* its initial dissolved gas occupies at reservoir conditions ($B_t = B_o + B_g(R_{si} - R_s)$ (bbl/STB))
C_f formation, or rock, compressibility (average value = 6×10^{-6} psi^{-1})
C_w water compressibility (average value = 3×10^{-6} psi^{-1})
E_o $B_t - B_{ti}$ (expansion of oil plus original gas in solution)
E_g $B_g/B_{gi} - 1$ (expansion of the gas-cap)
E_{gw} $\left(\frac{NB_{ti}}{1 - S_{wc}}\right)[(1 + m)(S_{wc}C_w + C_f)]\,\Delta P$ (expansion of water and rocks)
F $N_p[B_t + (R_p - R_{si})B_g] - (W_{in}B_w + W_e + G_{in}B_g)$
G original free gas volume in place (as in a gas-cap) (scf or bbl)
G_{in} volume of gas injected (scf or bbl)
G_p gas produced (scf or bbl)
m (initial reservoir free gas volume)/(initial reservoir oil volume): $[GB_{gi}/NB_{oi}]$
N original oil volume in place (STB)
N_p oil produced (STB)
R_p producing gas–oil ratio: $R_p = R_s + \frac{k_g}{k_o}\frac{\mu_o}{\mu_g}\frac{B_o}{B_g}$
R_s solution-gas–oil ratio at a specific time, or specified pressure (scf/STB or bbl/STB)
R_{si} initial solution-gas–oil ratio (scf/STB or bbl/STB)
S_w water saturation, fraction
S_{wc} connate water saturation (initial reservoir water saturation at discovery)
W_e volume of water entering a reservoir from an adjacent aquifer (bbl)
W_{in} water injected (bbl)
W_p water produced (bbl)

When production is initiated, the reservoir and its fluids undergo changes that are the results of (1) *voidage*: withdrawal (production) of gas, oil, and water, (2) *influx*: water encroachment (influx from adjoining aquifers), and injection of gas and water for pressure maintenance or waterflood, (3) *expansion*: expansion of gas, oil, water, and rock.

1. Voidage (production in terms of reservoir barrels) [bbl gas = scf/(5.61 × STB)]

$$N_p(R_p - R_{si})B_g = \text{gas-cap production}\ \left[\text{STB}_{oil}\left(\frac{\text{bbl}_{gas}}{\text{STB}_{oil}} - \frac{\text{bbl}}{\text{STB}}\right)\frac{\text{bbl}_{gas}}{\text{STB}_{oil}} = \text{bbl}_{gas}\right]$$

$$N_pB_t = \text{produced oil with associated dissolved gas}$$

$$W_pB_w = \text{produced water}$$

2. Influx

$$W_e = \text{water influx from adjoining aquifer}$$

$$G_{in}B_g = \text{gas injected}$$

$$W_{in}B_w = \text{water injected}$$

3. Expansion
   a. Gas-cap.

$$\frac{NmB_{ti}(B_g - B_{gi})}{B_{gi}} = mNB_{ti}\left(\frac{B_g}{B_{gi}} - 1\right)$$

$$= (\text{Gas-cap volume at time } t) - (\text{The original gas-cap volume})$$

b. Connate water in the gas-cap and the oil zone

$$\left(\frac{mNB_{ti}}{1-S_{wc}}\cdot S_{wc}+\frac{NB_{ti}}{1-S_{wc}}\cdot S_{wc}\right)(C_w)(\Delta P)=NB_{ti}(1+m)\left(\frac{S_{wc}}{1-S_{wc}}\right)(C_w)(\Delta P)$$

c. Oil expansion with its associated solution gas

$$NB_t - NB_{ti} = N(B_t - B_{ti}) = (\text{Initial oil volume}) - (\text{Oil volume at time } t)$$

d. Rock expansion

$$\left(\frac{mNB_{ti}}{1-S_{wc}}+\frac{NB_{ti}}{1-S_{wc}}\right)(C_w)(\Delta P)=\left(\frac{NB_{ti}}{1-S_{wc}}\right)(1-m)(C_f)(\Delta P)$$

The general material balance equation results from equating the total voidage volume to the sum of the fluid and rock expansions, thus:

$$\sum_{i=1}^{j}(\text{voidage})_i=\sum_{i=1}^{k}(\text{expansion})_i \tag{10.24}$$

$$N_pB_t+N_p(R_p-R_{si})+W_pB_{wt}-(W_{in}+W_e+G_{in}B_g)=N(B_t-B_{ti})+mNB_{ti}\left(\frac{B_g}{B_{gi}}-1\right)+\left(\frac{NB_{ti}}{1-S_{wc}}\right)[(1+m)(S_wC_w+C_f)](\Delta P) \tag{10.25}$$

(10.25)

{Oil zone production + Gas − cap production + Water produced − Water injected − Water influx − Gas injected} = {Oil zone expansion + Gas − cap expansion + Water and rock expansion}

Equation (10.23) applies to a reservoir that is producing with a gas-cap drive, dissolved gas drive and WD simultaneously. Consequently, conditions that do not apply to a specific reservoir result in simplifications:

1. In a reservoir without a gas-cap, $m=0$.
2. Where gas injection is not in use, $G_{in}=0$.
3. Where water production is negligible, $W_p=0$.
4. If the reservoir is not hydraulically connected to a large aquifer, $W_e=0$.

STRAIGHT LINE SOLUTIONS OF THE MATERIAL BALANCE EQUATION

Terms in the general material balance equation (Eq. (10.23)) that do not apply are removed and the balance of the terms then arranged to yield a straight line ($Y=A+mX$) with G, N, and W_e as the unknowns. Extrapolation of the straight line is then used for prediction of future performance. Any deviations from the line indicate that one, or more, of the assumptions about the reservoir are in error and must be investigated in greater detail.

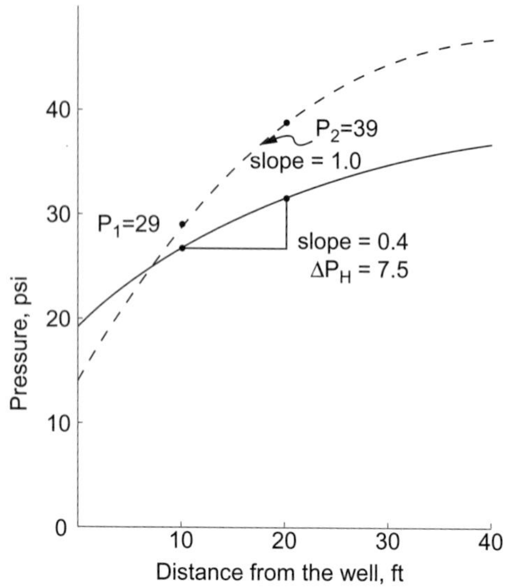

FIGURE 10.14a Graphical solution for Example on p. 692.

Gas Reservoir (Eq. (10.21))

(a) No WD (W_e and $W_p = 0$)

$$G_pB_g = GE_g + E_{fw} = GE_t \quad (Y = mX) \tag{10.24}$$

A plot of G_pB_g versus E_g is a straight line with G equal to the slope.

(b) With WD (Figure 10.14)

Equation (10.21) is rearranged to obtain a linear equation that can be solved by trial-and-error estimation of W_e to produce a straight line (assuming $E_{fw} = 0$). Deviations of the straight line indicate that the estimated value of W_e is either too small or too large (Figure 10.14b).

$$\frac{G_pB_g + W_pB_{io}}{B_g - B_{gi}} = \frac{W_e}{B_g - B_{gi}} + G \quad (Y = X + G) \tag{10.25}$$

This solution is deceptively simple because it is complicated by the production characteristics of a gas well with water production. Water in the production tubing restricts the production of gas, which in turn distorts the production rate of gas (as a function of cumulative gas produced). Consequently, there can exist vast differences in the production performance of different wells in the field with respect to time, thus reducing the overall production that would be obtained in the absence of water production.

A condensate gas reservoir adds to the complications because of changing relative permeability and gas composition with respect to time. Therefore, a

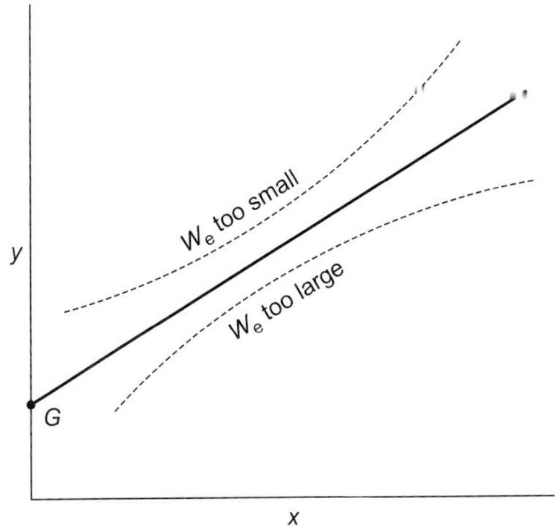

FIGURE 10.14b Linear solution of material balance equation for a WD gas reservoir using trial-and-error estimation of the amount of water influx.

more complex analysis based on variable composition is required for more accurate calculations of the gas in place and water influx [5,9,16].

Oil Reservoirs

Using the definitions of the expansion terms, the general equation is reduced to:

$$F = N\left[E_o + \left(\frac{mB_{ti}}{B_{gi}}\right)E_g + E_{fw}\right] + W_e = NE_t + W_e \qquad (10.26)$$

This general equation is then fit to the conditions that are most applicable to the specific reservoir and arranged to represent a linear equation whose slope and intercept are interpreted to obtain the original oil, and/or gas, in place and water influx. Adjusting the linear equation to represent different drive mechanisms gives excellent insight to the type of energetic system which is governing production of the reservoir. Figure 10.15a–d illustrates four types of applications of the material balance concept where the reservoir is considered to be composed of a single, constant, control volume where different fluids enter and leave maintaining the control volume completely filled with fluids at all times.

In addition, the reservoir is considered to have uniform pressure throughout (no pressure drawdown at wells) and that changes in reservoir pressure with respect to time occur evenly throughout the reservoir. Pressure variations and multiple tank models are used for more complex computer simulation, but the simple, single-tank model, nevertheless accurately represents most reservoirs and facilitates the overall development of the general material balance equation concept.

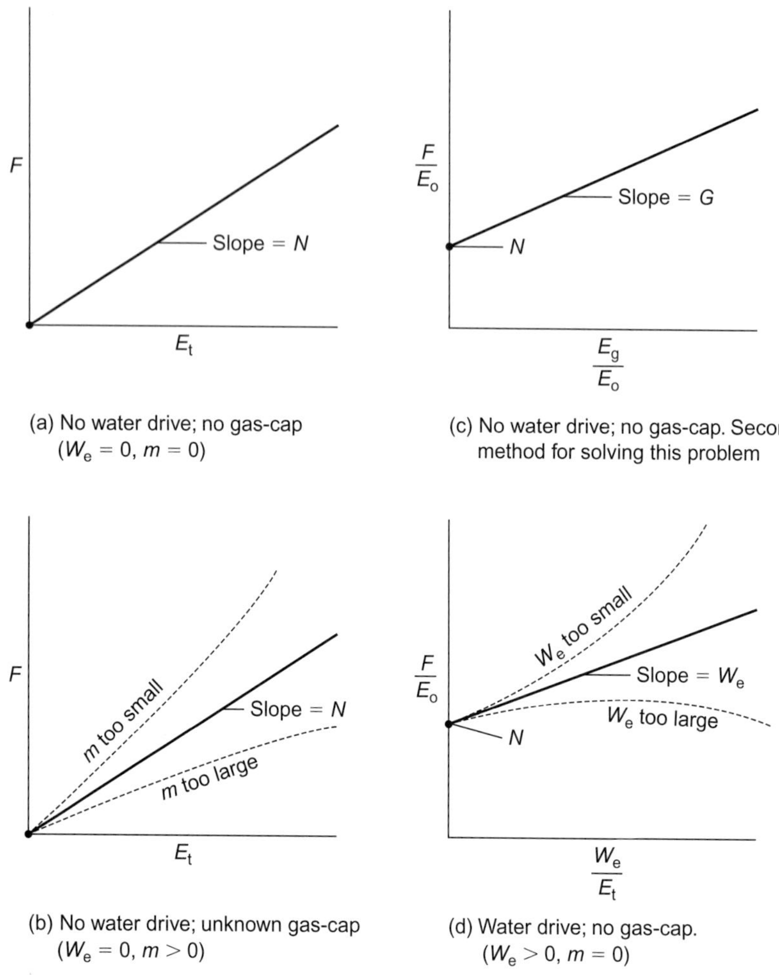

FIGURE 10.15 (a–d) Linear solutions of the general material balance equation.

Figure 10.15a. It is assumed that a WD and a gas-cap are nonexistent. The slope of the curve of F versus E_t is then the amount of the OOIP (as shown in Figure 10.6b, p.690). If the intercept does not occur at zero, then one or more assumptions about the drive mechanism of the reservoir are incorrect.

Figure 10.15b. It is assumed that there is no active WD, but the oil reservoir is connected to a gas-cap of unknown size. The problem is solved by trial-and-error iteration on the value of the ratio of the gas-cap size to the oil reservoir size (m). When the assumed values of m cause deviation of the line as a curve with an increasing positive slope, the values of m are too small and vice versa. When values that fit the curve are obtained, G can be calculated from $G = mNB_{oi}/B_{gi}$.

Figure 10.15c. Another solution to the conditions assumed in (b) is to divide the general equation by E_o, thus:

$$\frac{F}{E_o} = N + \left(m\frac{B_{ti}}{B_{gi}}\right)\left(\frac{E_g}{E_o}\right)$$

In this case, N is the intercept and G is the slope of the line.

Figure 10.15d. It is assumed that there is an active WD, but not gas-cap. In this case, the problem is solved by trial-and-error estimations of the value of W_e, and N is once more the intercept while W_e is the slope of the line.

Example

Oil and water production, and average reservoir pressure, data are listed in the table below for a reservoir that is undersaturated at its original reservoir pressure and does not have a WD mechanism.

Select an appropriate straight-line solution and determine the OOIP.

Solution

No SGD; therefore, $m = 0$, and $R_{si} = 0$

No WD; therefore, $W_e = 0$

$$F = N(E_o + E_{f,w,o}) \quad \text{(reduction of Eq. (10.26))}$$

$$F = N_p B_o + W_p \quad \text{(reduction of } F\text{... definition in Nomenclature)}$$

$$E_o = B_t - B_{ti} = B_o - B_o = 0 \quad \text{(reduction of } E_o\text{)}$$

$$F = NE_{f,w,o}$$

The slope of the plot of F versus $E_{f,w,o}$ shows that N (OOIP) = 182 × 10^6 bbl of oil

$$N_p B_o = NB_o\left(\frac{S_o C_o + S_{wc} C_w + C_f}{1 - S_{wc}}\right)\Delta P$$

$$P_i = 2{,}997 \text{ psi}$$

$N_p \times 10^3$	$W_p \times 10^3$	P_{ave}	ΔP	$F \times 10^3$	$E_t \times 10^{-6}$
20	0	2,997	6	26	231
35	0	2,987	10	46	286
80	0	2,981	16	104	617
100	0	2,977	20	130	771
200	1	2,962	35	261	1,350
350	2	2,932	65	457	2,507
550	3	2,887	110	718	4,243
800	5	2,347	150	1,045	5,786
1,250	7	2,777	220	1,632	8,485
1,700	9	2,697	300	2,219	11,571

EXAMPLE FORTRAN PROGRAM ANALYSIS OF A DECLINE CURVE

Oil production rates for a hypothetical well are listed in the input data file DECIN.DAT. The first three numbers in the file are as follows: *N*, the number of data point in the time/production data; JFIRST, the time after an initial period of unstable production, when the decline curve begins (this can be one (1), or any other number); and QABAN, the final production rate at which the well, or field, is abandoned.

The computed results are listed in output file DECOUT.DAT with self-explanatory labels. The slope of the decline curve is constant; therefore, the curve is classified as an exponential curve: the hyperbolic equation constant, *C*, is approximately equal to zero and log(q) versus time is a straight line. The slope is printed at time intervals of 5 that may be changed to any number desired by removing, or changing, the number 5 at the end of the *WRITE* statement (WRITE 12, 908).

DECLINE.FOR

```
C
C       PRODUCTION DECLINE CURVE ANALYSIS. Subroutine HYPER is required
C
C       Data files DECIN and DECOUT are used for data input and output: N
C        = total number of production-rate versus time points;JFIRST
C        = Time of the initial point of the decline curve, see Fig. 10.6:
C       C QABAN (final production rate at which the well, or field, must be
C       abandoned.
C
C       A,B,C = constant of the LnQR vs TIME decline curve: JFIRST =
C       time at which the decline curve begins. FUTUR = cumulative
C       production estimate extrapolated from the end of QR vs T data to
C       QR abandonment (see Fig. 10.6). PAST = cumulative production from
C       beginning (JFIRST) to current time – (end of production data).
C       TOTAL = Total production from Time = 0 to Time = abandonment.
C
        REAL LNQR(200),QR(200),QR1(200),QR2(200),QR3(200),SLOPE(200)
        REAL T(200),T2(200),TSLOPE(200),CUMNP(200),LNQR2(200),NP(200)
        REAL NP1, NP2
C
        OPEN (UNIT = 10, FILE = 'DECIN.DAT', STATUS = 'OLD')
        READ (10, *) N, JFIRST, QABAN
        READ (10, *) (T(I), QR(I), I = 1,N)
        CLOSE (10)
C
        JF = JFIRST-1
        DO 10 I = JFIRST,N
                LNQR(I) = ALOG10(QR(I))
                J = I-JF
                LNQR2(J) = LNQR(I)
```

```
                    T2(J) = T(I)
10       CONTINUE
         KK = N-JF
         CALL HYPER (KK, T2, LNQR2, P, Q, R)
                    A = P
                    B = Q
                    C = R
C
C        Slope of the Decline Curve
         CON = B-A*C
         DO 12 J = 1,KK
         SLOPE(J) = CON/(1.0 - C*T2(J))
         TSLOPE(J) = T2(J) + JF
12       CONTINUE
C
C        Initial Production
         SUM1 = 0.0
         DO 14 I = 1,JFIRST
            SUM1 = SUM1 + QR(I)
            CUMNP(I) = SUM1
14       CONTINUE
         NP1 = CUMNP(JFIRST)
         QR1AVE = SUM1/JFIRST
         DO 16 I = 1, JFIRST
                    QR1(I) = QR1AVE
16       CONTINUE
C
C        Cumulative production using hyperbolic Eq. (T(1) -> N)
         SUM2 = 0
         DO 18 I = JFIRST + 1, N
                    AA = (A + B*T(I))/(1.0 + C*T(I))**2
                    QR2(I) = 10**AA
                    SUM2 = SUM2 + QR2(I)
              CUMNP(I) = SUM2 + SUM1
                    QR1(I) = QR2(I)
18       CONTINUE
         PAST = CUMNP(N)
C
C        Estimated Time of Abandonment (TAB)
         TAB = (A - ALOG10(QABAN))/( - B + C*ALOG10(QABAN))
C
C        Estimated future production (NFUTUR)
         SUM3 = 0
         NUM = N + 1
         DO 20 I = N + 1, TAB
                    T(I) = NUM
                    AAA = (A + B*T(I))/(1.0 + C*T(I))
                    QR3(I) = 10**AAA
                    SUM3 = SUM3 + QR3(I)
                    CUMNP(I) = SUM3 + PAST
                    QR1(I) = QR3(I)
                    NUM = NUM + 1
```

```
20      CONTINUE
        FUTUR = SUM3
        TOTAL = CUMNP(TAB)
C
        OPEN(UNIT = 12, FILE = 'DECOUT.DAT', STATUS = 'OLD')
        WRITE (12,900)
900     FORMAT (5X, 'PRODUCTION DECLINE CURVE ANALYSIS', //)
C
        WRITE (12,902) A, B, C
902     FORMAT ('Hyberbolic Eq. Constants',3X, 'A = ',F8.3,3X, 'B = ',F8.3,3X, &'C
        = ',F8.3,//)
C
        WRITE (12,904)
904     FORMAT ('SLOPE of the LnQR-versus-T curve at several increments & of Time')
        WRITE (12,906)
906     FORMAT ('TIME', 4X, 'SLOPE')
C
        WRITE(12,908) (T2(J), SLOPE(J), J = 1,KK,5)
908     FORMAT(F5.1, 3X, F5.3)
C
        WRITE(12,910) TAB,QABAN
910     FORMAT (//, 'Time at abandonment = ',F6.0,3X, 'Production Rate at
        &abandonment = ',F10.0,//)
        WRITE(12,912) QR1AVE
912     FORMAT ('Initial value of the production rate (QR at T = JFIRST)' &,F8.0,/)
        WRITE (12,914) NP1
914     FORMAT ('Initial Production (From T = 1 -> T = JFIRST)',F8.0,/)
        WRITE (12,916) PAST
916     FORMAT ('Cumulative Production from Beginning to current time & (T = 1 ->
        T = N)',F10.0,/)
        WRITE (12,918) FUTUR
918     FORMAT('Cumulative Production from Current-Time to abandonment & (T = N
        -> T = TAB)',F10.0,/)
        WRITE (12,920) TOTAL
920     FORMAT ('Total Estimated Production (T = 1 -> T = TAB)',F10.0,//)
C
        WRITE (12,922)
922     FORMAT ('TIME' 4X , 'PROD-RATE',4X, 'CUM-PROD')
        WRITE (12,924) (T(I), QR1(I), CUMNP(I), I = 1,TAB)
924     FORMAT (F5.0,2X,F7.2,8X,F6.2)
        STOP
        END
C       Subroutine HYPER: Least Squares curve fit of data to a hyperbolic function. X =
        (P + Q*X)/(1 + R*X). The coefficients P, Q and R are returned.
C
C       The derivative dY/dX = (Q - P*R)/(1 + R*X)^2. The integral is: (Q*X)/R + ((P*R
        - Q)/R^2)Ln(1 + R*X) from b to a.
C
        SUBROUTINE HYPER (N,X,Y,P,Q,R)
C
        REAL X(200), Y(200), NUM1, NUM2, NUM3
```

```
C
      A = 0.0
      B = 0.0
      C = 0.0
      D = 0.0
      E = 0.0
      F = 0.0
      G = 0.0
      H = 0.0
C
      DO 10, I = 1,N
            A = A + X(I)
            B = B + Y(I)
            C = C + X(I)*Y(I)
            D = D + X(I)**2
            E = E + Y(I)**2
            F = F + X(I)*Y(I)**2
            G = G + (X(I)**2)*Y(I)
            H = H + (X(I)**2)*(Y(I)**2)
10    CONTINUE
C
C
      NUM1 = D*(C*F-B*H) + C*(A*H-C*G) + G*(B*G-A*F)
      NUM2 = N*(G*F-C*H) + A*(B*H-C*F) + C*(C**2-B*G)
      NUM3 = N*(D*F-C*G) + A*(B*G-A*F) + C*(A*C-B*D)
      DNOM = N*(G**2-D*H) + A*(A*H-C*G) + C*(C*D-A*G)
C
      P = NUM1/DNOM
      Q = NUM2/DNOM
      R = NUM3/DNOM
C
      RETURN
      END
```

DECIN.DAT

```
24, 5, 20

1     172.1
2     160.3
3     283.1
4     213.5
5     206.9
6     198.2
7     177.3
8     155.8
9     142.2
10    133.1
11    119.1
12    106.7
13    95.8
14    89.2
```

15	76.5
16	73.1
17	67.1
18	55.9
19	54.0
20	47.8
21	45.2
22	37.4
23	35.9
24	31.9

DECOUT.DAT

PRODUCTION DECLINE CURVE ANALYSIS

Hyberbolic Eq. Constants A = 2.527 B = − 0.047 C = − 0.003

SLOPE of the LnQR-versus-T curve at several increments of Time

TIME	SLOPE
5.0	−.039
10.0	−.039
15.0	−.038
20.0	−.038

Time at abandonment = 28. Production Rate at abandonment = 20.

Initial value of the production rate (QR at T = JFIRST) 207.

Initial Production (From T = 1 –> T = JFIRST) 1036.

Cumulative Production from Beginning to current time (T = 1 –> T = N) 3058.

Cumulative Production from Current-Time to abandonment (T = N –> T = TAB) 98.

Total Estimated Production (T = 1 –> T = TAB) 3156.

TIME	PROD-RATE	CUM-PROD
1.	207.	172.
2.	207.	332.
3.	207.	616.
4.	207.	829.
5.	207.	1036.
6.	209.	1245.
7.	193.	1438.
8.	177.	1615.
9.	163.	1777.
10.	149.	1927.
11.	137.	2064.
12.	125.	2189.
13.	115.	2304.

14.	105.	2409.
15.	96.	2505.
16.	87.	2592.
17.	80.	2672.
18.	72.	2744.
19.	66.	2810.
20.	60.	2870.
21.	54.	2924.
22.	49.	2973.
23.	44.	3018.
24.	40.	3058.
25.	29.	3086.
26.	26.	3112.
27.	23.	3135.
28.	21.	3156.

PROBLEMS

10.1 Assume that you have been designated as the leader for assessment and development of a newly discovered oil field. If you are limited to three engineers of different disciplines (specialties), develop an integrated team approach for management of this task and discuss the responsibilities of each person assigned to a specific discipline.

10.2 Initial estimates for a reservoir are listed below. Calculate the original gas in place (G, scf), OOIP (N, STB/acre-ft), and original solution gas in place (G_s, scf/acre-ft).

$$A_{\text{oil-zone}} = 144 \text{ acres}, \quad \phi_{\text{oil-zone}} = 0.18, \quad S_{\text{wc(oil)}} = 0.32, \quad S_{\text{wc(gas)}} = 0.26,$$
$$B_{oi} = 1.32 \text{ RB/STB}, \quad R_{si} = 430 \text{ scf/STB-oil}, \quad B_{gi} = 30.3\text{RB/scf},$$
$$A_{\text{gas-cap}} = 105 \text{ acres}, \quad \phi_{\text{gas-cap}} = 0.14$$

10.3 Assume that an oil reservoir is discovered under a gas-cap containing 350 MMscf of gas. The decision is made to allow the gas to expand into the oil reservoir to establish a gas-drive mechanism for oil production (no production of gas from the gas-cap). The initial gas pressure and gas formation factor are 4,000 psi and 0.0015, respectively. When the gas-cap pressure has declined to 3,500 psi, how much gas will have expanded into the oil zone if the gas formation factor at 3,500 psi is 0.0020?

10.4 The following production data for an oil reservoir described a decline curve. Use computer program DECLINE.FOR to determine the type of decline curve, the change in slope of the curve with respect to time, and forecast the development of the field.

The number of points, $N = 16$; JFIRST = 3; and production rate at abandonment = 0.4 Mbbl/day

Time (years)	Time (*N*)	Production Rate Mbbl/0.5 year
0.0	0	0.0
0.5	1	10.0
1.0	2	10.5
1.5	3	9.8
2.0	4	8.5
2.5	5	6.9
3.0	6	5.7
3.5	7	5.0
4.0	8	4.0
4.5	9	3.7
5.0	10	2.6
5.5	11	2.3
6.0	12	2.0
6.5	13	1.8
7.0	14	1.5
7.5	15	1.5
8.0	16	1.4

10.5 Use a straight-line plot of the general material balance equation to determine (1) N (original oil in place), (2) G (size of the gas-cap), and (3) W_e (water influx). The list of data includes all of the fluid, reservoir, and production data available to the engineers. Plot the graph.

Rock volume (oil zone) = 1,258,000 acre-ft
Weighted average porosity = 4.79%
Initial reservoir pressure = 3,950 psi
Saturation pressure = 3,950 psi
Reservoir temperature = 188°F
Initial gas–oil ratio at 115 psi separator pressure = 936 scf/STB
Initial formation volume factor at 115 psi separator pressure = 1.5328 RB/STB
Oil viscosity = 0.595 cp at 188°F
Formation compressibility = 6.6×10^6/psi
Water compressibility = 3×10^6/psi
Water viscosity = 0.46 cp at 188°F
Initial water saturation (oil zone) = 65.0%
Initial reservoir gas saturation = 0.4%
Water displaced residual oil saturation = 23.5%
Gravity drainage residual oil saturation = 13.0%
Oil gravity (°API) = 21
$N_p = \Sigma N_{pi}$ $G_p = \Sigma G_{pi}$

$N_{p1,MM\ bbl}$	N_{p2}	N_{p3}	$G_{p1,MMMscf}$	G_{p2}	G_{p3}	W_p	W_i	R_p	B_t	B_g
1.464	1.421	0	1.257	1.221	0	10.2	0	884	1.550	7.70
4.814	4.771	0	4.176	4.140	0	83.3	25.7	892	1.555	7.79
6.648	5.124	1.391	5.278	4.546	1.147	184.7	118.4	879	1.557	7.82
8.024	5.214	2.767	6.916	4.546	2.335	302.3	222.3	898	1.557	7.84
9.560	5.214	4.303	8.258	4.546	3.673	471.5	375.0	906	1.548	7.87
11.037	5.214	5.780	9.381	4.546	5.010	468.1	561.4	921	1.314	7.90
12.467	5.214	7.210	10.785	4.546	6.204	809.0	788.9	914	1.502	7.91
12.695	5.214	8.438	11.865	4.546	7.284	1,139.0	1,016.0	917	1.565	7.92
15.239	5.214	9.082	13.096	4.546	8.515	1,371.0	1,243.0	916	1.552	7.93
16.674	5.214	11.417	14.462	4.546	9.881	1,614.0	1,486.0	919	1.555	7.97

NOMENCLATURE

A area, acres
B_g gas formation volume factor, RB/scf
B_o oil formation volume factor, RB/STB
B_{oB} oil formation factor at the bubble-point pressure (P_{BP})
B_t total formation volume factor [$B_t = B_o + B_g(R_{si} - R_s)$ bbl/STB]
Constant $7,758 = \frac{43,560\ \text{ft}^2/\text{acre}}{5.514\ \text{ft}^2/\text{bbl}}$
C_f formation, or rock, compressibility
C_w water compressibility
E_A area (horizontal) sweep efficiency
E_V vertical-sweep efficiency
E_o $B_t - B_{ti}$ (expansion of oil plus original gas in solution)
E_g $B_g/B_{gi} - 1$ (expansion of the gas-cap)
E_{gw} $\left(\frac{NB_{ti}}{1-S_{wc}}\right)[(1+m)(S_{wc}C_w + C_f)]\Delta P$ (water and rock expansion)
F $N_p[B_t + (R_p - R_{si})B_g] - (W_{in}B_w + W_e + G_{in}B_g)$
G free gas in place (as in a gas-cap), scf
G_{in} volume of gas injected
G_p volume of gas produced
G_s solution gas in place (SGIP), scf
h thickness of the hydrocarbon interval, ft
k_i effective permeabilities, mD (water, oil, or gas)
K absolute permeability (mD, Darcy, m^2 as defined in the text)
K_T constant in the Kozeny–Carman equation (Eq. (3.27))
m (initial gas volume)/(initial oil volume)
M mobility ratio
N oil in place, STB
N_p oil produced
P_A pressure at abandonment, psi
P_B pressure at the bubble point, psi
P_e static res. pressure, psi (shut-in pressure)
P_i initial reservoir pressure

P_w flowing (bottom hole) pressure, psi
q_g gas production rate at the surface (60°F, 14.7 psi), Mscf/day
q_o oil flow rate, STB/day
q_w water flow rate, sur. bbl/day
r_e drainage radius of the well, ft
r_w well-bore radius, ft
R_F recovery factor
R_p producing gas–oil ratio
R_s solution gas/oil ratio, scf/STB
R_{si} initial solution-gas–oil ratio
S_o oil saturation
S_g gas saturation
S_w water saturation
S_{wc} connate water saturation
W_e water influx
W_{in} water injected
W_p water produced
GOC gas–oil contact
GOR gas–oil ratio
RB reservoir barrel, volume, bbls, at reservoir conditions
RR recoverable reserves
scf standard cubic feet (14.7 psi, 60°F)
SGD solution-gas drive
SGIP solution gas in place
STB stock tank barrels (42 U.S. gal, 14.7 psi, 60°F)
WD water drive
ΔP_B pressure difference due to buoyant pressure
ΔP_H pressure difference due to vertical distance
μ_i fluid viscosity (water, oil, or gas) (cP or kg/(m s), as defined)
μ_o oil viscosity at the bubble point
μ_{oi} oil viscosity at initial reservoir conditions
μ_{wi} water viscosity at initial reservoir conditions
ϕ porosity
σ interfacial tension, mN/m

REFERENCES

1. Chorn LG, Croft M. Resolving reservoir uncertainty to create value. SPE 49094. In: *Annual Technical Conference*. New Orleans, LA; September 1998.
2. Al-Hussainy R, Humphreys N. Reservoir management: principles and practices. SPE 30144. In: *PetroVietnam Conference*. Ho Chi Minh City, Vietnam; March 1996.
3. Al-Bazzaz WH, Gupta A. Reservoir characterization and data integration of complex Mauddud-Burgan carbonate reservoir. SPE 100318. In: *SPE Europec/EAGE Annual Conference*. Vienna, Austria; June 2006.
4. Meng HZ, Godbey KS, Gilman JR, Uland MJ. Integrated reservoir characterization and simulation for reservoir management using a web-based collaborative technical workflow

manager. SPE 77673. In: *Annual Technical Conference*. San Antonio, TX; September–October 2002.
5. Zellou AM, Hartley LJ, Hoogerduijn-Strating EH, Al-Dhahab SHH, Boom W, Hadrami F. Integrated workflow applied to the characterization of a carbonate fractured reservoir: Qarn Alam Field. SPE 81579. In: *13th Middle East Oil Show and Conference*. Bahrain; April 2003.
6. Ehllg-Economides CA. Engineering application for integrated reservoir characterization. SPE 29994. In: *International Conference on Petroleum Engineering*. PR China; November 1995.
7. Norris SO, Reinhardt BK. Improved reservoir characterization and management of a mature oil field via integrated team approach. In: *Annual Technical Conference*. Dallas, TX; October 1995.
8. Robinson JW, McCabe PJ. Sandstone-body and shale-body dimensions in a braided fluvial system: salt wash sandstone member (Morrison Formation) Garfield County, Utah. *AAPG Bull* 1997;**81**(8):1267–91 (SPE Reprint Series No. 60. *Advances in Reservoir Characterization*).
9. Sullivan M, Jensen G, Goulding F, Jennette D, Foreman L, Stern D. Architectural analysis of deep-water outcrops: implications for exploration and development of the Diana sub-basin, Western Gulf of Mexico. *GCSSEPM Foundation, 20th Annual Research Conference on Deep-Water Reservoirs of the World*; December 3–6. p. 1010–31 (SPE Reprint Series No. 60. *Advances in reservoir characterization*) 2000.
10. Amyx JW, Bass DM, Whiting RL. *Petroleum reservoir engineering*. New York: McGraw-Hill; 1960. 610 p.
11. Chapman RE. *Petroleum geology*. Amsterdam: Elsevier Science; 1983. 415 p.
12. Archer JS, Wall CG. *Petroleum engineering: principles and practice*. London: Graham & Trotman; 1986. 362 p.
13. Lyons WC, Plisga GJ, editors. *Standard handbook of petroleum and natural gas engineering*. 2nd ed Amsterdam, New York: Elsevier Science; 2005.
14. Arps JJ, et al. A statistical study of recovery efficiency. *API Bull* 1967.
15. Doscher TM, et al. Statistical analysis of crude oil recovery and recovery efficiency. *API Bull* 1984;**D14**:5–46.
16. Donaldson EC, Alam W. *Wettability*. Houston, TX: Gulf Publishing Company; 2008. 336 p.
17. Saner S, Kissami M, Al-Nufaili S. Estimation of permeability from well logs using resisitivity and saturation data. *SPE Form Eval* 1997;**12**:27–31.
18. Shedid SA, Almehaideb RA. Robust reservoir characterization of UAE heterogeneous carbonate reservoir. SPE 81580. In: *13th Middle East Oil Conference*; April 2003.
19. Amaefule JO, Atunbay M, Tiab D, Kersey DG, Keelan D. Enhanced recovery description: using core and log data to identify hydraulic (flow) units and predict permeability in uncored intervals/wells. SPE 26436. In: *66th Annual Conference*. Houston, TX; October 1993.
20. Aminian K, Thomas B, Ameri S, Bilgesu HI. A new approach for reservoir characterization. SPE 78710. In: *Eastern Regional Meeting*. Lexington, Kentucky; October 2002.
21. Holtz MH, Jackson JA, Jackson KG. Petrophysical characterization of Permian shallow-water dolostone. SPE 75214. In: *SPE/DOE Improved Oil Recovery Symposium*. Tulsa, OK; April 2002.
22. Gunter GW, Finneran JM, Hartmann DJ, Miller JD. Early determination of reservoir flow units using an integrated petrophysical method. SPE 38679. In: *Annual Technical Conference*. San Antonio, TX; October 1997.

23. Aminian K, Ameri S, Bilgesu HI, Alla V, Mustafa R. Characterization of a heterogeneous reservoir in West Virginia. SPE 84830. In: *Eastern Regional/AAPG Eastern Section Joint Meeting*. Pittsburgh, PA; September 2003.
24. Hosseini MS, Haystdavoudi A. Reservoir characterization of Tuscaloosa sand mineralogical and petrophysical data. *Form Eval* 1986;**December**:584–94.
25. Craft BC, Hawkins MF. *Petroleum reservoir engineering*. Englewood Cliffs, NJ: Prentice-Hall; 1950. 437 pp.
26. Chilingar GV, Serebryakov VA, Robertson Jr JO. *Origin and prediction of abnormal formation pressures*. Amsterdam, New York: Elsevier; 2002. 373 pp.
27. Dake LP. *Fundamentals of reservoir engineering*. Amsterdam, New York: Elsevier Scientific Publishing Co.; 1978. 443 pp.

Chapter 11

Fluid–Rock Interactions

INTRODUCTION

Laboratory and field studies indicate that rock properties, especially permeability, are altered or damaged during almost every field operation: drilling, cementing, perforation, completion and workover, production, stimulation, and injection of water and chemicals for enhanced oil recovery. Fine solid particles introduced from well fluids during any of these operations or generated in situ by the interaction of invading fluids with rock minerals and/or formation fluids are the main cause of formation damage. Regardless of their origin, these particles can concentrate at pore restrictions, causing severe plugging and large reduction in near-wellbore permeability. This zone of reduced permeability, commonly referred to as a "skin" and ranging from a few inches to a few feet, can reduce well productivity to only a fraction of its potential value. Numerous studies have been conducted to determine the magnitude of skin effect under various formation damage conditions. Other studies investigated the composition, physical characteristics, and other factors controlling the migration and deposition of fine particles in porous media. Because repair of permeability damage is generally difficult and expensive, all of these studies emphasize the importance of preventing damage.

IMPORTANCE OF NEAR-WELLBORE PERMEABILITY

Krueger and Amaefule and Kersey showed that although the thickness of the skin zone is only a few inches to a few feet, whereas the drainage radius may be several hundred feet, the effective permeability in the skin zone has an extremely disproportionate effect on well productivity, as illustrated in Figure 11.1 [1,2]. This figure also shows the effect of improving (by acidizing or fracturing) permeability of the formation rock in the vicinity of the wellbore. The impact of the skin zone on well productivity can be evaluated

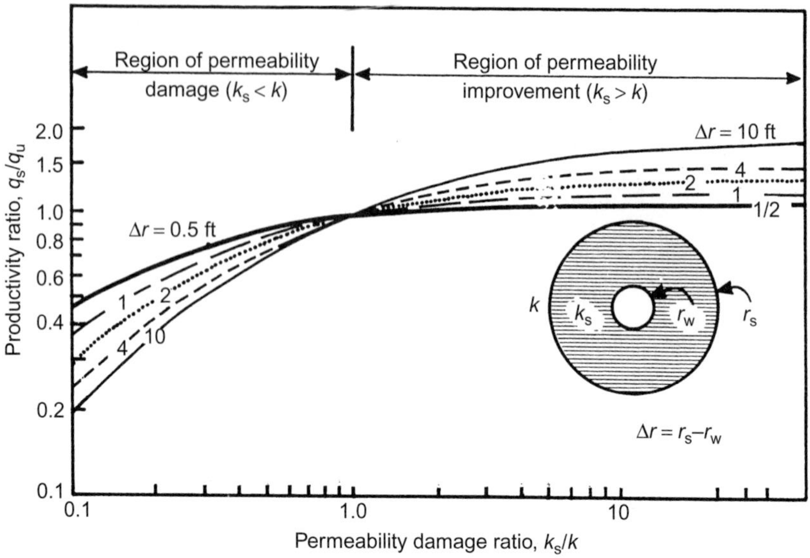

FIGURE 11.1 Effect of permeability damage on productivity ratio [1].

by calculating the annual revenue loss due to skin damage, or gain due to skin improvement, from the following equation:

$$\Delta\$ = 365 q_u (DF) \$_o \tag{11.1}$$

where $\$_o$ is the price of oil in U.S. dollars per STB, q_u is the production rate without skin effect in STB/D, and DF is the damage factor expressed as:

$$DF = 1 - PR = 1 - \frac{q_s}{q_u} \tag{11.2}$$

where q_s is the production rate with skin effect and PR is the productivity ratio (see Figure 11.1). Another common factor used to express the effect of skin is the damage ratio, which is the inverse of the productivity ratio:

$$DR = \frac{1}{PR} = \frac{q_u}{q_s} \tag{11.3}$$

The following example illustrates the relative importance of the formation condition in the vicinity of the wellbore.

Example

A pressure drawdown test yielded a flow efficiency of 60%, indicating a skin damage. The production rate of this well is 252 STB/D and the price of oil is 91 \$/STB. From core analysis, the depth of the skin zone is approximately 2 ft. The undamaged formation permeability is 55 mD. The wellbore radius is 0.25 ft. After an acid stimulation treatment, the productivity ratio of this well has doubled. Calculate:

1. Skin permeability, skin factors, and annual revenue loss before stimulation
2. Skin permeability and annual revenue gain after stimulation.

Solution

1. Before stimulation, i.e., damaged well: From Figure 11.1, the permeability damage ratio for $\Delta r_s = 2$ ft and PR = 0.60 is approximately 0.2. Therefore, the skin permeability of the skin zone before stimulation (k_{sb}) is given by:

$$k_{sb} = 0.20k_u = 0.20 \times 55 = 11 \text{ mD}$$

Knowing the radius, r_s, and the permeability, k_s, of the skin zone, the skin factor, s, may be calculated from the following expression:

$$s = \left(\frac{k}{k_s} - 1\right) \ln\left(\frac{r_s}{r_w}\right) \tag{11.4}$$

where k is the permeability of the undamaged zone of the formation and r_w is the wellbore radius. The skin factor before stimulation is then:

$$s = \left(\frac{55}{11} - 1\right) \ln\left(\frac{2}{0.25}\right) = 8.3$$

Thus, the permeability of the formation rock near the wellbore has been damaged to 20% of its original value by some field operation to a depth of 2 ft. The undamaged production rate is obtained from Eq. (11.3):

$$q_u = \frac{q_{sb}}{PR} = \frac{252}{0.60} = 420 \text{ STB/D}$$

The annual revenue from this damaged well is:

$$\Delta\$_b = 365 q_{sb}\$_o = 365 \times 256 \times 91 = \$8.5 \times 10^6$$

If the well is not treated for removal of permeability damage, the annual revenue loss from this well is obtained from Eq. (11.1):

$$\Delta\$_L = 365 q_u(1 - PR)\$_o = 365\$_o(q_u - q_{sb}) = 365 \times 420(1 - 0.60)91 = \$5.58 \times 10^6$$

2. Inasmuch as the productivity ratio is now 120%, the permeability in the skin zone after stimulation (k_{sa}) is given by:

$$k_{sa} = 1.20k_u = 1.20 \times 55 = 66 \text{ mD}$$

and the new production rate is:

$$q_{sa} = 1.20q_u = 1.20 \times 420 = 504 \text{ STB/D}$$

The annual revenue gain due to stimulation is:

$$\Delta\$_G = 365 \times \$_o(q_{sa} - q_{sb}) = 365 \times 91(504 - 252) = \$8.37 \times 10^6$$

The annual revenue from this stimulated well is given by:

$$\Delta\$_a = 365 q_{sa}\$_o = 365 \times 504 \times 91 = \$16.74 \times 10^6$$

which is double the annual revenue before stimulation. This example clearly illustrates the impact of near-wellbore permeability damage or stimulation on the well productivity, which is essential to the profitable development of new reserves.

NATURE OF PERMEABILITY DAMAGE

Permeability in the vicinity of wellbore may be damaged during any operation between drilling injection and injection. The primary and probably the most important cause of permeability damage is associated with:

1. the introduction of solid particles into the formation from wellbore fluids during any of these operations, or
2. movement of formation fines and chemical reactions in the pore channels resulting from the interaction invading fluids with rock minerals and formation fluids.

Not all reductions in well productivity are due to the impairment of rock permeability. Damages of mechanical origin, which are called "pseudo-skins," can result from partial completions, slanted wells, low perforation density, short perforations, and high production rates, which can cause turbulent flow and stress changes near the wellbore. Productivity decline can also result from the alteration of reservoir fluid viscosity, which typically occurs during the invasion of drilling and completion filtrates, and often leads to the formation of emulsion blocks in the vicinity of the wellbore.

Permeability damage can occur anywhere along the flow path, from the formation to perforations and into the wellbore as shown in Figure 11.2. Thus, when designing a remedial treatment, both the type and the location of permeability damage must be considered.

ORIGIN OF PERMEABILITY DAMAGE

Krueger, Amaefule and Kersey, and Economides and Nolte provided an extensive analysis of formation damage problems [1–3]. They all recognize that from the time the drill bit enters the formation and until the well is put on production, invasion of mud filtrate and solids and migration of formation fines are the major causes of permeability damage.

Formation damage during drilling is practically unavoidable because the mud preferred by the reservoir engineer may not be what the drilling supervisor needs. For instance, low-filtrate mud may be necessary to combat formation damage and differential pressure sticking, whereas high-filtrate mud minimizes cuttings hold-down and, therefore, provides for a high penetration rate. Also, inhibition of shale swelling and dispersion of clay solids are extremely important for borehole stability and the prevention of formation

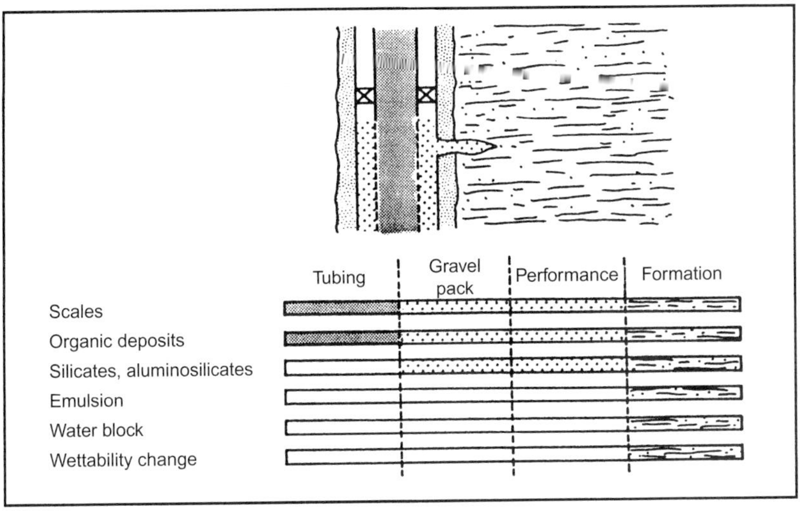

FIGURE 11.2 Location of various types of permeability damage near the wellbore [3].

TABLE 11.1 Depth of Invasion of Drilling Filtrate into the Formation Near Wellbore [4]

Time (days)	Oil Mud	Depth of Invasion (in.) (Low-Colloid Oil Mud)	Water Mud
1	1.2	3.3	7.7
5	4.6	11	12
15	10	21	23
20	12	23	27
25	14	29	31
30	16	32	34

damage, but dispersion of clay shale solids helps control the viscosity, gel strength, and filtration of water-based muds [4]. Filtrate damage may extend from a few inches to several feet as shown in Table 11.1, depending on the sensitivity of the formation rock and the type of filtrate [5]. High permeability of the filter cake, high wellbore overpressure—i.e., wellbore pressure higher than formation pressure—and long exposure of the formation to drilling fluid are some of the most important factors that enhance formation damage by filtrate invasion [1,3]. Potassium-based drilling fluids, which cause minimal permeability damage, are widely used. However, when drilling through highly shaly (clayey) formations, which are extremely sensitive to water, oil-based muds are preferred, even though they contain more solid particles than water-based muds and, therefore, have the potential to cause a severe permeability impairment by solids invasion.

Invasion of the drilling mud solids—such as clay particles, cuttings, and weighting and lost-circulation agents—into the formation is usually shallower (2–4 in.) than the filtrate invasion (up to 15 ft), but the resultant permeability reduction can be as high as 90% [3]. Formation damage from drilling fluid solids is dependent on the pore and pore throat size distribution of the formation rock (large pore and throat sizes favor solids invasion), particle size distribution in the drilling mud, existence of fissures and natural fractures in the vicinity of the wellbore, and wellbore overpressures. For minimizing formation invasion by mud solids, some operators use brines devoid of solids and fluid-loss agents.

Table 11.2 gives a complete list of potential formation damage problems during various well operations. A discussion of all sources of formation damage is beyond the scope of this work. Table 11.3 summarizes the origin, mechanisms, and types of formation damage problems, whereas Table 11.4 ranks the damage severity for various stages of wellbore development and reservoir exploitation.

TYPES OF PERMEABILITY DAMAGE

Keelan and Koepf identified four main types of near-wellbore permeability damage that can be evaluated by core analysis [6]:

1. Plugging of pores and pore throats by solids introduced during drilling, completion, workover, or improved recovery operations
2. Clay hydration and swelling clay particle dispersion and their movement with produced or injected water
3. Water blockage or increased water saturation near the wellbore caused by extraneous water introduced into the formation during various operations
4. Caving a subsequent flow of unconsolidated sands, causing loss of well productivity

Figure 11.3 illustrates the effect of water blockage on relative permeability curves. Points A and a correspond to the condition of irreducible water saturation in the pay zone. A 14% increase in water saturation (from a to d) causes a 60% decline of relative oil permeability (from A to D). Points C and c represent a situation where water blockage near the wellbore is so severe that the relative oil permeability is essentially zero. A formation water block occurs by invasion of water-based filtrate during drilling and completion operations, or through fingering or coning of connate water during production [3]. Therefore, to minimize or prevent water blocks in oil wells, low

TABLE 11.2 Potential Formation Damage Problems during Various Well Operations [2]

1. **Drilling**

   Mud Solids and Particle Invasion
   - Pore throat plugging
   - Particle movement

   Mud Filtrate Invasion
   - Clay swelling, flocculation dispersion, and migration
   - Fines movement and plugging of pore throats
   - Adverse fluid–fluid interaction resulting in either emulsion/water block, or inorganic scaling
   - Alteration of pore structure near wellbore through drill bit action
2. **Casing and Cementing**
   - Blockage of pore channels by cement or mud solids pushed ahead of the cement
   - Adverse interaction between chemicals (spacers) pumped ahead of cement and reservoir minerals fluids
   - Cement filtrate invasion with resulting scaling, clay slaking, fines migration, and silica dissolution
3. **Completion**
   - Excessive hydrostatic pressure can force both solids and fluids into the formation
   - Incompatibility between circulating fluids and the formation with resultant pore plugging
   - Invasion of perforating fluid solids and explosives debris into the formation with resultant pore plugging
   - Crushing and compaction of near-wellbore formation by explosives during perforation
   - Plugging of perforation of extraneous debris (mill scale, thread dope, and dirt)
   - Wettability alteration from completion fluid additives
4. **Well Servicing**
   - Problems similar to those that can occur during completion
   - Formation plugging by solids in unfiltered fluids during well killing
   - Adverse fluid–fluid and fluid–rock interaction between invading kill fluid and reservoir minerals
   - Damage to clays from dumping of packer fluids
5. **Well Stimulation**
   - Potential plugging of perforations, formation pores, and fractures from solids in the well kill fluid
   - Invasion of circulating fluid filtrate into the formation with resultant adverse interaction
   - Potential release of fines and collapse of the formation during acidizing
   - Precipitation of iron reaction products
   - Plugging of pores and fractures by dirty fracture liquids
   - Inadequate breakers for high viscosity fracture fluids may cause blockage of propped fracture
   - Fluid loss or diverting agents may cause plugging of the perforations, formation pores, or propped fractures
   - Crushed proppants may behave like migratory fines to plug the fracture

(Continued)

TABLE 11.2 (Continued)

- Inorganic/organic scales in the wellbore along with remnant debris may plug perforations, pores, or etched fractures
- Fracture conductivity decline due to proppant embedment

6. **Production**
   - Initiation of fines movement during initial DST by using excessive drawdown pressures
   - Inorganic/organic scaling through abrupt shift in thermodynamic conditions
   - Sand production in unconsolidated formations triggered by water encroachment into producing zones
   - Screens of gravel packs can be plugged by produced silt, clay, mud, scale, etc.
   - Sand-consolidated wells may be plugged by debris, and sand-consolidating material may reduce reservoir permeability
7. **Secondary Recovery Operations—Injection Wells**
   - Formation wettability alteration from surface-active contaminants in the injection water
   - Impairment of injectivity due to suspended solids (clays, scale, oil, and bacteria) in the injection water
   - Formation plugging by iron corrosion products
   - Inorganic scaling due to incompatibility of injected and formation waters
   - In pressure maintenance with gas injection, formation may be plugged by compressor lubricants that may also alter wettability
   - Reduced well injectivity from injected corrosion inhibitors in gas zones
8. **Enhanced Oil Recovery**
   - Fines migration, clay swelling, and silica dissolution initiated by contact of high pH steam generator effluents (condensates) with the formation rock during thermal recovery
   - Dissolution of gravel packs and increased sanding during thermal recovery
   - Inorganic scaling due to changes in thermodynamic conditions during steam injection
   - Plugging due to carbonates deposition during CO_2 injection
   - Deposition of asphaltenes with CO_2 contacts asphaltic crude oils
   - Potential emulsion formation during CO_2 wag process
   - Fines movement due to hydrodynamic conditions of velocity and viscosity during chemical EOR process with surfactants and polymers

fluid-loss muds should be used when drilling or coring with water-based muds. Oil-based or inverted oil-emulsion muds should be used during completion operations. In gas wells, air is recommended for drilling and coring. Using oil-based mud in gas wells may cause severe permeability damage [6]. Once a water block has occurred, core testing is the best available tool for determining the extent of the water block and finding the best remedial treatment. In oil wells, water blocks are usually treated by surface tension-reducing chemicals such as surfactants and alcohols. In gas wells, various alcoholic acid solutions are injected into the formation to vaporize the liquid block. Water blocks form in oil-wet rocks, whereas oil blocks form in water-wet rocks.

TABLE 11.3 Mechanisms and Types of Formation Damage Problems [2]

Origin	Mechanism	Types of Formation Damage
A. **Solids Invasion**		
A-1. *Solid Types*		Plugging of effective flow path
1. Drill solids (sand, silt, clays, and colloids)	Physical	
2. Weighting materials (barite, bentonite)		
3. Lost circulation materials		
4. Fluid-loss additives		
5. Solid precipitates		
6. Living organisms (bacteria)		
7. Suspended solids (silt, clays, and oil)		
8. Perforation debris (pulverized rock and charge debris)		
9. Crushed proppants		
A-2. *Solids Origin*		
1. Drilling fluids		
2. Completion fluids		
3. Workover fluids		
4. Stimulation fluids		
5. Supplemental injected fluids (water, steam, and chemical)		
B. **Fluid Invasion**		
B-1. *Fluid Types*		Alteration of fluid saturation distribution
1. Water	Chemical	Changes in capillary pressure
2. Chemicals	Physical	Destabilization of resident minerals
3. Oil	Biological	Clay swelling
		Fines migration
		Mica alteration

(*Continued*)

TABLE 11.3 (Continued)

Origin	Mechanism	Types of Formation Damage
B-2. *Fluid Origin*		Wettability alteration
1. Drilling mud filtrates		Reduction of hydrocarbon relative permeability
2. Cement filtrates and spacers		Emulsion blockade of pores
3. Completion and workover fluids pores		Inorganic scales
4. Stimulation fluid		Organic scales
5. Chemical additives		Mineral transformation
6. Supplemental injected fluids		Sand movement
C. **Thermodynamic (Pressure and Temperature) and Stress Changes**		
C-1. *Origin*		Inorganic scales (induced/natural)
1. Production (pressure drawdown)	Chemical/ Physical	Organic scales (induced/natural)
2. Drilling/completion/ workover fluids		Ion exchange with resultant destabilization of resident minerals
3. Stimulation fluids (acids/ fractures)		Permeability decay with pressure drawdown
4. Supplemental injected fluids (water, thermal, and gas)		Secondary mineral precipitation
D. **Operating Conditions**		
D-1. *Parameter*		Wellbore erosion
1. Wellbore pressure (over- or underbalanced)	Chemical	Destabilization of resident minerals
2. Operating time	Physical	Clay/fines movement
3. Production rates		Fluid/solids invasion
4. Injection rates		Weakening of rock integrity
D-2. *Origin*		
1. Drilling		
2. Completion (perforations)		

(*Continued*)

TABLE 11.3 (Continued)

Origin	Mechanism	Types of Formation Damage
D-2. *Origin*		
1. Stimulation (acidizing/ fracturing)		
2. Drill stem testing		
3. Production		
4. Supplemental fluid injection		
E. **Types of Materials**		
1. Salt types	Chemical	Salt precipitation
2. Additives (surfactants, corrosion inhibitors, etc.)	Physical	Destabilization of resident minerals
3. Gelling materials (viscosifiers)		Wettability alteration
4. Acids (HCl, HF)		Inorganic scales
5. Alkalinity control		Organic scales
		Emulsion formation

When formation clays come in contact with aqueous fluids used in drilling or completion operations, clay swelling occurs and usually causes plugging of flow channels and reduction in well productivity. Permeability damage by the dispersion and subsequent migration of various clay particles, resulting in the plugging of pore channels, is more prevalent than originally suspected, and is now the subject of intense experimental and theoretical investigations [7–22], some of which are presented in this chapter.

A wellbore in an unconsolidated formation is likely to be unstable and usually leads to sand production and consequent reduction in well productivity. This instability is aggravated when fluid systems with poor filtration properties or reactive fluids are used during drilling or completion operations. Highly reactive fluids that dissolve the cementing material may cause the wellbore to collapse [23]. Sand production in weakly consolidated formations also can be triggered by the onset of water production. Muecke showed that production of formation waters promotes the movement of formation fines, which eventually develop bridges at pore restrictions (throats) near the wellbore [7]. Severe bridging increases the pressure drawdown necessary to maintain production, which in turn leads to the movement and production of

TABLE 11.4 Formation Damage Severity Scale for Various Stages of Well Development and Reservoir Exploitation [2]

Type of Problem	Well Development				Reservoir exploitation		
	Drilling and Cementing	Well Completion	Workover	Stimulation	DST	Primary Production	Supplemental Fluid Injection
Mud solids plugging	****	**	***	—	*	—	—
Fines migration	***	****	***	****	****	***	****
Clay swelling	****	**	***	—	—	—	**
Emulsion/Water block	***	****	**	****	*	****	****
Wettability alteration	**	***	***	****	—		****
Reduced relative permeability	***	***	****	***	—	**	—
Organic scaling	*	*	***	****	—	****	—
Inorganic scaling	**	***	****	*	—	****	***
Injected particulate plugging	—	****	***	***	—	—	****
Secondary mineral precipitation	—	—	—	****	—	—	***
Bacteria plugging	**	**	**	—	—	**	****
Sanding	—	***	*	****	—	***	**

(****) Very severe.
(***) Severe.
(**) Less severe.
(*) Not severe.
(—) Negligible.

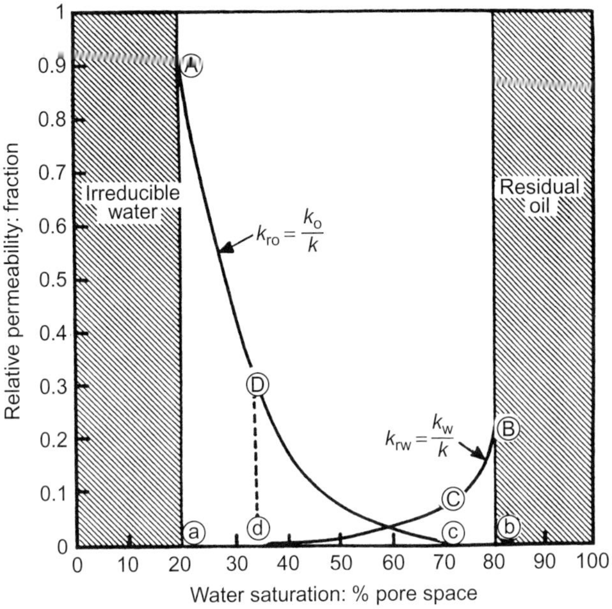

FIGURE 11.3 Effect of water blocking on relative permeability curve [6].

formation sand. High drawdowns also can cause premature compaction due to sudden changes in the stress state near the well, which invariably result in the caving and subsequent flow of sand into the wellbore. Oil-based gravel packs, particularly in open holes, wire-wrapped screens, and prepacked liners are some of the most commonly used sand-control techniques.

Other common types of formation damage are emulsion formation, wettability change, scale deposition, and organic deposits. The normal locations of these damages are shown in Figure 11.2. Emulsions may result from the intermixing of either (1) water-based fluids and reservoir oil or (2) oil-based fluids and formation brines during the invasion of drilling and completion filtrates. Adsorption of surface-active agents (surfactants) from oil-based fluid is the major cause of wettability change in water-wet oil reservoirs [3]. Wettability changes cause formation of water blocks and, therefore, reduction in well productivity. The drop in temperature and pressure in or near the wellbore during production causes minerals to precipitate and deposit either within the formation pores or inside the wellbore. These inorganic deposits, called scales, are a common source of severe well plugging. Organic deposits, such as paraffins and asphaltenes, are also an important source of formation damage. As with mineral deposits (scales), organic deposits (waxes) form during production because of pressure or temperature drop and may accumulate either in the formation near the rock face or in the tubing. Figure 11.4 shows various products and methods available for treating several types of formation damage.

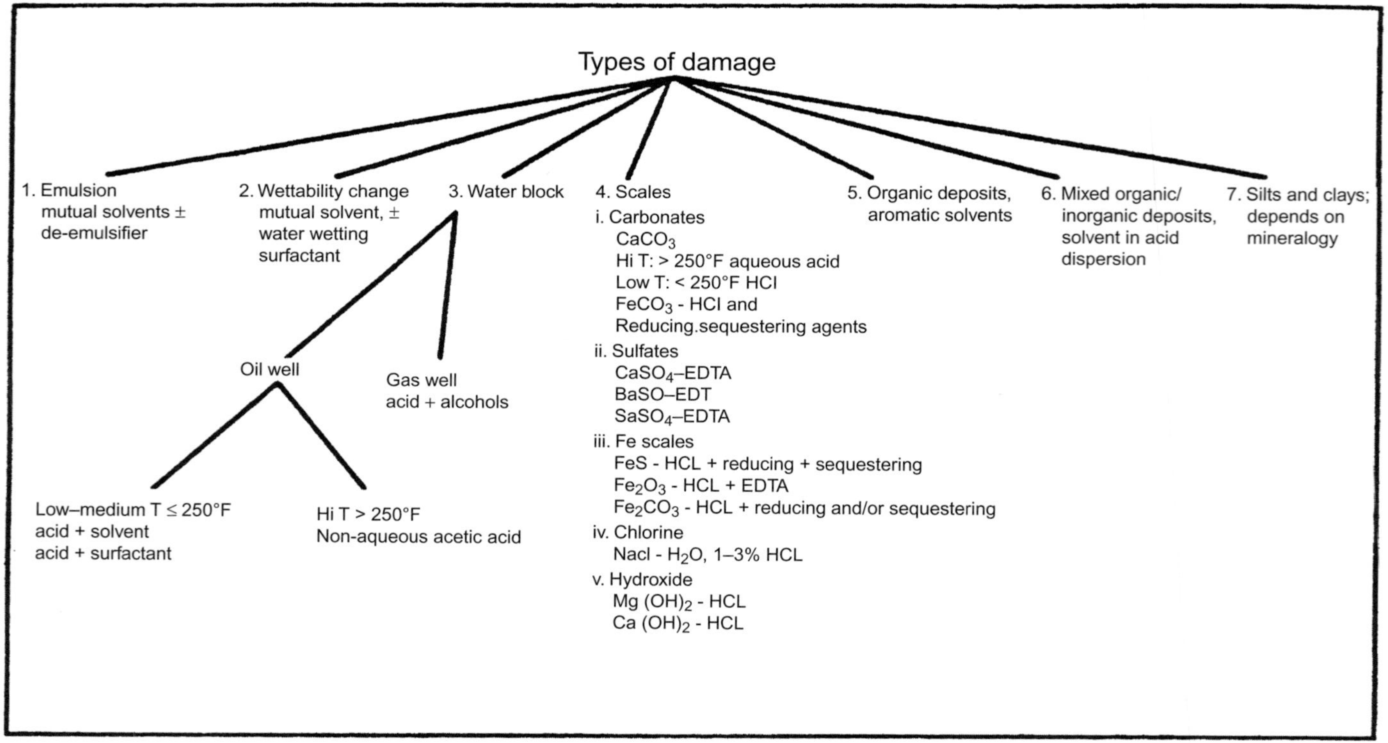

FIGURE 11.4 Basic types of formation damage and treatment selection [3].

Formation damage during stimulation treatments, such as wellbore cleanup and acidizing, and during injection of fluids for improved recovery is as common as during production. Most problems causing productivity decline can occur during wellbore cleanup, acidizing, waterflooding, chemical flooding, and steam injection operations. If the treatment is well designed and properly executed, however, the net result will be improved productivity.

EFFECT OF FINES MIGRATION ON PERMEABILITY

Clay swelling and dispersion, and the subsequent movement and entrapment of clay fines in the formation pores and pore throats, are probably the most important causes of permeability damage. Therefore, a fundamental knowledge of clay properties is necessary to diagnose the damage problem and to design an effective remedial treatment.

TYPES AND SIZES OF FINES

There are two types of clays in most sandstone formations [21]:

1. Detrital or allogenic clays, which were introduced into a sandstone by physical processes at the time the sediment bed was deposited or by biogenic processes shortly after deposition
2. Authigenic clays, which were developed by direct precipitation from formation waters or were formed by the interaction of formation waters with preexisting clay minerals

Detrital clays normally form an integral part of the supporting rock matrix and, therefore, are not mobile. Figure 11.5 shows the various modes of occurrence of clays in sandstones shortly after deposition. Inasmuch as detrital clays in most sandstone beds are altered after burial to form regenerated authigenic clays, most clay minerals in ancient depositional environments are authigenic. Authigenic minerals, which typically fill, line, or bridge pore systems, have the greatest potential to damage a formation through migration or adverse chemical reaction with invading drilling and well completion fluids because of their morphology, sensitive locations within pore systems, and very high surface area-to-volume ratios [2]. Smectite and vermiculite have the highest potential for adverse chemical reactions (Table 11.5). Figure 11.6a–c shows scanning electron microscope (SEM) microphotographs of several examples of authigenic clays typical of those present on the surfaces of formation sand grains. Authigenic clays are generally classified as cementing agents, and the frequency of their occurrence in decreasing order is presented in Table 11.6.

The major characteristics of five authigenic clay groups commonly found in sandstones are presented in Table 11.7. Amaefule and Kersey, Wilson and

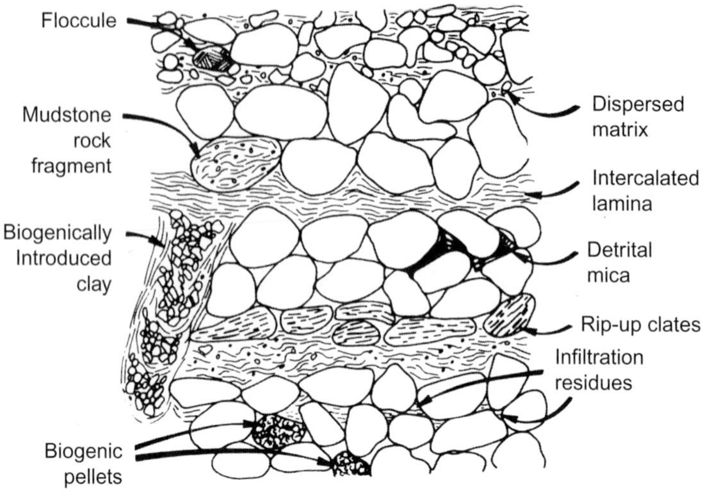

FIGURE 11.5 Modes of occurrence of clays in sandstone [24].

TABLE 11.5 Surface Area of Clay Minerals [2]

Clay	Internal Surface Area (m^2/g)	External Surface Area (m^2/g)	Total (m^2/g)
Smectite	750	50	800
Vermiculite	750	<1	750
Chlorite	0	15	15
Kaolinite	0	15	15
Illite	5	25	30
Quartz	0	3	3

Pittman, and Eslinger and Pevear summarized many of the important concepts explaining the role of clay minerals in permeability damage [2,24,25].

It is important to emphasize that not all the fines that cause permeability damage are clay minerals. Reed showed that when micaceous sands are leached with neutral sodium chloride (NaCl) or calcium chloride ($CaCl_2$) solutions, mica is altered by exchange of interlayer potassium by sodium or calcium ions Figure 11.7 [26]. This causes the edges of mica particles to break off, migrate downstream, and plug pore throats, which decreases permeability as shown in Table 11.8. He also showed that when large volumes of

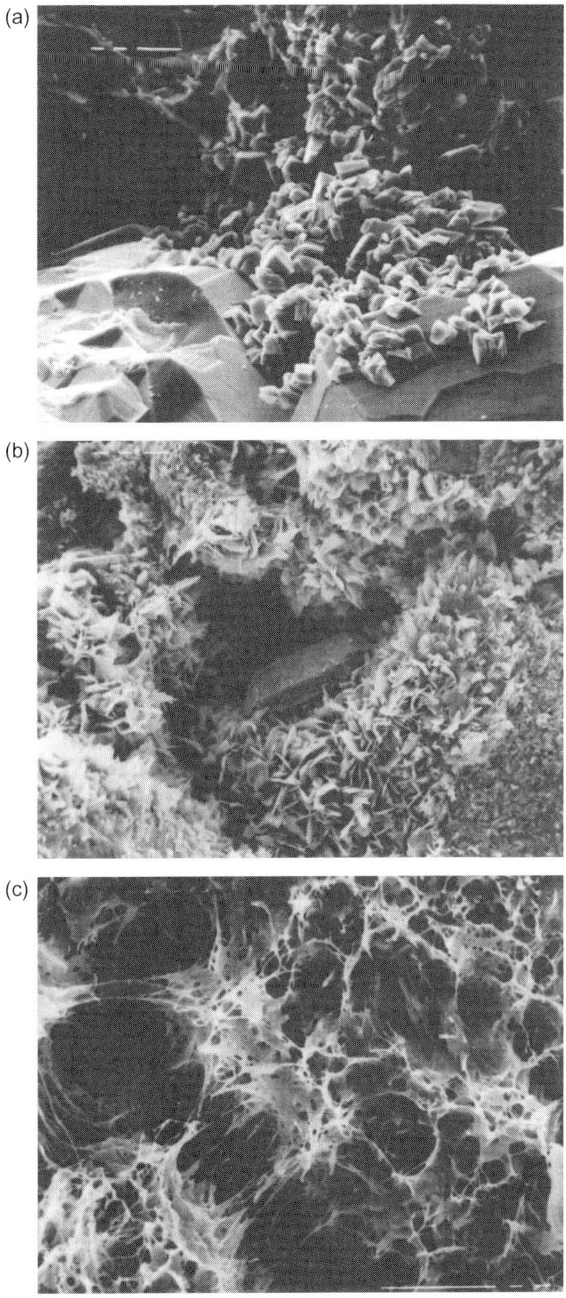

FIGURE 11.6 (a) SEM microphotographs of pore-filling authigenic kaolinite. Note the smooth quartz overgrowths on the framework grains (×1,000 magnification). (b) SEM microphotographs of euhedral authigenic feldspar in a chlorite-lined pore (×1,000 magnification). (c) SEM microphotograph of delicate fibers of illite bridge pores. Under high flow rates, the fibers may break off and migrate, becoming lodged in pore throats (×3,000 magnification).

TABLE 11.6 Frequency of Occurrence of Cements in Decreasing Order [2]

1. Quartz	6. Muscovite	11. Hematite
2. Calcite	7. Kaolinite	12. Halite
3. Dolomite	8. Chlorite	13. Barite
4. Siderite	9. Orthoclase	14. Celestite
5. Anhydrite	10. Albite	15. Zeolites

leaching solution pass through poorly consolidated micaceous sands, significant amounts of carbonate cement are removed causing mineral particles to move and plug flow channels, thus reducing permeability as shown in Figure 11.8 and Table 11.9. Reed's findings are based on reservoir samples obtained from several Southern California oilfields. The minerological composition of these samples is provided in Table 11.10

Regardless of their origin and type, it is ultimately the sizes and amounts of solid particles that mostly dictate the extent of permeability damage. To define these two important factors, Muecke analyzed a large number of samples of unconsolidated sandstones in different U.S. Gulf Coast wells. He conducted three types of analyses on these samples [7]:

1. SEM microphotographs, such as those shown in Figures 11.6 and 11.9, were used to determine the size and shape of fines present in the sandstones. Arbitrarily defining formation fines as particles small enough to pass through the smallest mesh screen available (400-mesh or a 37-mm opening), Muecke found that these particles varied widely in size and ranged from 37 mm to considerably less than 1 mm in all samples analyzed. Examination of SEM microphotographs also revealed that the concentration of fine particles located on surfaces of sand grains was high and variable.
2. Standard dry-sieve analyses confirmed the range of these fines and showed that their amount varied from 2 to 15 wt%. The SEM examination of several sieved sandstone samples, however, revealed that large concentrations of fines were still present on the surfaces of sand grains. Thus, the range of 2–15 wt% is a conservative estimate of the amount of fines present in sandstone samples. Even wet sieving, which gives more accurate size distribution of particles, did not remove all fines from the surfaces of sand grains.
3. X-ray diffraction analyses of formation fines to determine mineralogical content showed that not all formation fines are clay minerals. The results of these analyses are presented in Table 11.11 and Figure 11.10. As expected, quartz was the dominant species (39 wt%), followed by

TABLE 11.7 Characteristics of Authigenic Clays [24]

	Morphology of Individual Flakes	Form of Aggregates	Relationship to Sand Size Detrital Grains	Thickness of Coating or Long Dimension of Aggregates (μm)	Special Features
Kaolinite and dickite	Pseudohexagonal	Stacked plates (book)	Pore filling	2–2,500 (generally 2–20)	Flakes notched or embayed (twinned?)
	Pseudohexagonal	Vermicule	Pore filling	10–2,500 (generally 20–200)	Flakes notched or embayed (twinned?)
	Pseudohexagonal	Sheet	Pore filling	0.1–1	
Chlorite	Pseudohexagonal	Plates (2-D cardhouse)	Pore filling	2–10	
	Curled equidimensional with rounded edges	Honeycomb	Pore lining	2–10	
	Equidimensional with angular or lobate edges	Rosette or fan	Pore lining and pore filling	4–150 (generally 4–20)	
	Fan-shaped fibrous bundles	Cabbage head	Pore lining and pore filling	8–40	
Illite	Irregular with elongate spines	Sheet	Pore lining	0.1–10	Bridging between sand grains
Smectite	Not recognizable	Wrinkled sheet or honeycomb	Pore lining	2–12	Bridging between sand grains
Mixed-layer smectite/illite	Subsequent with stubby spines	Imbricate sheet to ragged honeycomb	Pore lining	2–12	Bridging between sand grains

amorphous material (32 wt%) and other minerals (18 wt%), such as feldspars, muscovite, calcite, dolomite, and barite. Clays represented only 11 wt% of the total fines. Similar analyses of the mineralogical composition of several Southern California oil reservoir sands by Reed (Table 11.10) gave almost identical average mineralogical contents of

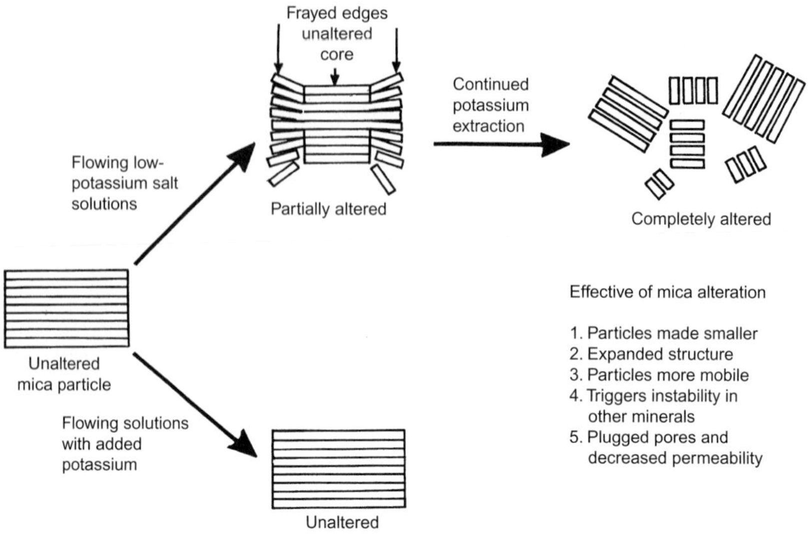

FIGURE 11.7 Damage mechanisms by mica alteration [26].

fine particles [26]. Reed's and Muecke's studies showed that formation fines are present in all reservoir sands in sufficient amounts to cause serious formation damage, and any remedial treatment to remove this damage must be capable of treating not only clay minerals but also all other types of substances (minerals and nonminerals).

FINES MIGRATION

Migration of fines in petroleum reservoirs has been typically investigated from two different perspectives:

1. Chemical interactions of drilling and completion fluids with the reservoir rock and fluids
2. Mechanical flow forces in the portion of formation near the wellbore

Veley reported that as freshwater contacts a clay-containing formation, clays swell, disperse, migrate, and plug, causing a rapid and severe decline in permeability [27]. Swelling and dispersion of clay particles is a function of the amount of water absorbed, which in turn is a function of the crystal structure and the cations present on the mineral surfaces. The smectite group of minerals has the greatest water sensitivity as shown in Table 11.12. Kersey indicated that smectite minerals, which have a crystal structure that

TABLE 11.8 Permeability Damage in Micaceous Sands [26]

Well	Initial Permeability (mD)	Volume Throughput (ml)	Permeability Decline (%)
A—Permeability damage in field cores during flow of a 3% $NaCl_2$			
1030	1,010	3,000	59
1030	1,313	1,600	63
1030	813	4,000	59
1030	952	3,400	66
4290	1,662	2,500	69
4290	827	2,500	71
4290	635	2,450	93
			Mean 69
B—Permeability damage in field cores during flow of a 3% $CaCl_2$			
4290	1,188	3,000	38
4290	480	3,000	81
4290	411	2,500	5
4290	644	2,500	71
			Mean 49
C—Permeability damage in field cores during flow of a 3.7% KCl			
1030	1,867	3,900	70
1030	1,399	3,900	15
1030	1,446	3,000	56
1030	2,165	4,000	8
1030	408	4,000	50
4290	645	2,450	4
4290	1,424	2,500	0
			Mean 29

favors absorption of water and some organic molecules, such as polymers, between unit layers, will not swell unless they are contacted by a drilling or completion fluid with different salinity and chemical composition than the formation water [29]. He maintained that smectite minerals can swell to

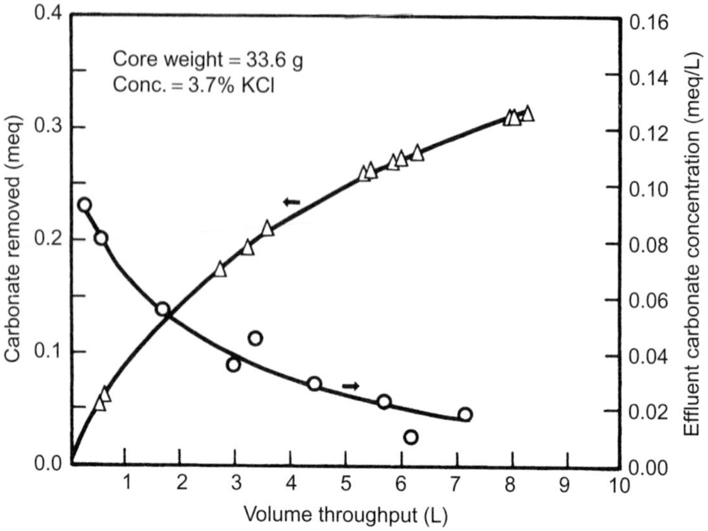

FIGURE 11.8 Carbonate leached from a field core by flowing 3.7% KCl solution [26].

TABLE 11.9 Permeability Damage in Field Cores during a Flow of 3.7% KCl Solution Saturated with $CaSO_3$ [26]

Initial Permeability (mD)	Volume Throughput (mL)	Permeability Decline (%)
150	2,200	7
78	1,900	9
518	1,600	14
977	3,500	−7
1,690	2,500	−20
140	2,475	0
73	2,500	0
1,585	2,550	−29
221	2,475	−2
278	2,500	2
351	2,500	−2
200	2,500	2
		Mean −4

TABLE 11.10 Mineralogical Composition of Several Southern California Oil Reservoir Sands [26]

	Composition (wt%)				
	Field A	Field B	Field C	Field D	Field E
Quartz	50.7	42.7	35.8	34.7	29.8
K-feldspar	10.5	16.6	15.4	33.8	18.2
Plagioclase	25.8	28.8	28.7	17.8	30.9
Calcite	0.0	0.0	0.0	0.0	2.4
Dolomite	Tr	4.1	8.5	0.0	0.0
Siderite	0.0	0.0	0.0	0.0	0.0
Anhydrite	0.0	0.0	0.0	0.0	0.0
Pyrite	0.8	0.7	0.0	0.2	0.6
Kaolinite	1.1	0.0	2.7	1.8	4.2
Mica	7.8	6.0	7.6	7.9	11.6
Chlorite	0.3	0.0	0.0	0.9	1.2
Mixed layer	0.0	1.0	0.0	0.0	0.0
Montmorillonite	3.1	0.0	0.0	0.5	0.4

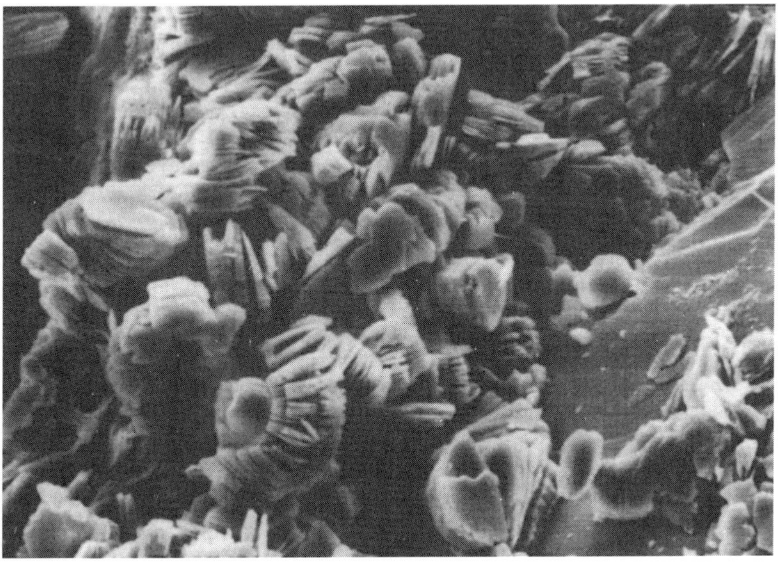

FIGURE 11.9 SEM microphotograph of pore-filling kaolinite plates. Source: *Courtesy of Core Laboratories.*

TABLE 11.11 Results of X-ray Analysis of Formation Fines* in wt% [7]

	Well A	Well B	Well C	Well D	Well E
Clays					
Montmorillonite	5.5	13.4	2.2	1.4	—
Illite	6.2	9.1	3.0	1.7	—
Kaolinite	0.8	4.2	1.3	0.7	—
Chlorite	3.9	—	—	—	—
Quartz	36.7	24.0	17.3	17.0	6.83
Other minerals					
Feldspar	8.6	5.7	9.1	5.4	11.4
Muscovite	1.6	—	1.6	1.0	—
Sodium chloride	1.1	1.3	7.8	5.0	1.5
Calcite	—	1.6	—	—	1.5
Dolomite	—	—	1.8	2.8	—
Barite	—	—	—	22.1	—
Amorphous materials	35.6	40.7	25.9	42.9	17.3
Total	100.0	100.0	100.0	100.0	100.0

**Passed 400-mesh screen.*

1,000% (20 times) of their original volume if the invading fluid is appreciably incompatible with the formation water. Table 11.13 shows the importance of clay content and type of clay mineral in determining the mechanism of permeability damage. It is evident from this table that, regardless of the clay content, water sensitivity is of major importance in formations containing grain-coating authigenic minerals (e.g., smectite, illite, and chlorite), whereas migration of fines is prominent in pore-filling authigenic minerals (e.g., silicates, kaolinite, and illite). Because of their very large surface areas and loose attachment to sand grain surfaces, kaolinite and illite clay particles are very susceptible to migration [29].

Chemical Damage

Khilar et al. investigated the sensitivity of sandstones to freshwater and determined the existence of a critical rate of salinity decrease at which a

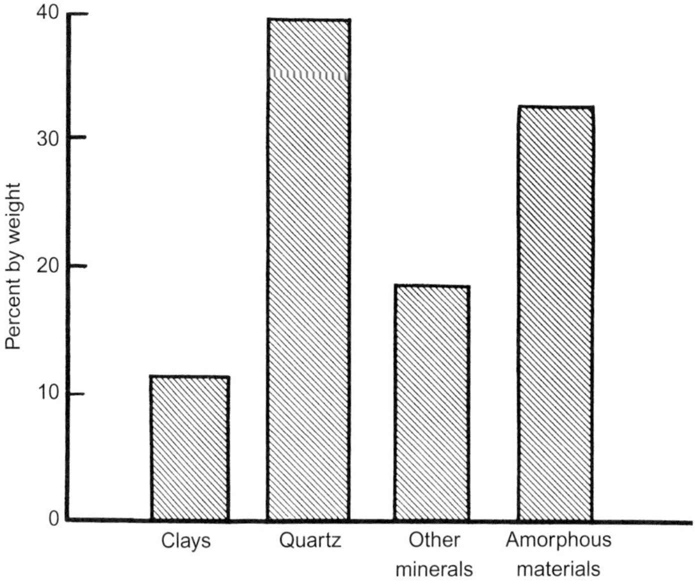

FIGURE 11.10 Typical average mineralogical content of fine particles present in U.S. Gulf Coast formations [7].

rapid and drastic reduction in permeability occurs [30]. This rate is reached when the saltwater present in the sandstone is replaced by the injected freshwater. Figure 11.11 shows a typical behavior of the permeability damage ratio k/k_o during a standard core flood test to demonstrate the water sensitivity of clayey sandstones. As a few pore volumes (2.5) of freshwater were injected into a core containing 0.51 M NaCl (1 ppm = 10^{-5} m), the permeability ratio dropped from 100% (no damage) to approximately 1%, i.e., the final core permeability of $0.01k_o$. When freshwater was injected in the opposite direction, the permeability increased only temporarily; however, when saltwater was injected in the same direction as freshwater, no change in permeability was observed. The last portion of Figure 11.11 corresponds to the case where saltwater was injected in the same direction at the start of the flow test. In this case, the permeability ratio increased and then stabilized at approximately 0.85. These permeability trends indicate that mobilization and dispersion of clay fines are occurring, and the primary cause of permeability reduction is pore plugging [31,32]. These permeability trends also led Khilar to the deduction that the permeability reduction is related to the rate at which the salt concentration is decreased [31]. To determine this relationship, Khilar conducted a large number of flood tests in Berea sandstone core samples of 1 in. diameter and 1 in. length [31]. These cores ($\phi = 19\%$) typically contain approximately 8 wt% of dispersible clays, such as kaolinite and

TABLE 11.12 Mineralogical Sensitivity to Formation Damage [7]

Water sensitive	Chlorite/Smectite	Illite/Smectite
	Illite	Smectite (Montmorillonite)
Acid sensitive (HCl)	Chamosite	Glauconite
	Chlorite (iron-rich)	Hematite
	Chlorite/Smectite	Pyrite
	Dolomite (iron-rich)	Siderite
Acid sensitive (HF)	Calcite	Silicate minerals
	Dolomite	
Scale	Anhydrite	Halite
	Barite	Hematite
	Brucite	Magnetite
	Calcite	Siderite
	Celestite gypsum	Trolite
Migration of "fines"	Illite	Silicate minerals
	Kaolinite	
"Sand" production	Rock-forming minerals	

illite. The experimental apparatus outlined in Figure 11.12 was used to carry out several experimental runs in which the salt concentration was decreased exponentially with time by using a continuously stirred mixer. Table 11.14 shows the results of 15 such experimental runs, where the superficial velocity is equivalent to the Darcy velocity and the space velocity is the ratio of the flow rate of the stream to the volume of the mixer (q/V). The space velocity was used to characterize the rate of salinity decrease.

Figures 11.13 and 11.14 are typical plots showing the behavior of the permeability damage ratio k/k_o and salt concentration ratio C_o/C, where C and C_o are the effluent and original salt concentrations, respectively. Figure 11.13 shows the case where the salt concentration is decreased slowly as indicated by a high residence time of 8.0 h or, inversely, a low space velocity of 0.125 h^{-1}. Figure 11.14 corresponds to the case of a high rate of decrease as indicated by a higher space velocity of 1.316 h^{-1}. Comparison of these two figures shows that a high rate of salinity decrease causes severe and abrupt permeability reduction. Consequently, there is a critical rate of salinity decrease below which negligible permeability damage occurs and

TABLE 11.13 Mechanism of Formation Damage as a Function of Clay Content and Type [2]

	Silicate and/or Carbonate Cement		Clay Cement	
Type	**Clay content**			
Distribution	<10%	>10%	<10%	>10%
Clay Minerals	**Mechanism of Formation Damage**			
Detrital/ Laminae	Freshwater	Freshwater	Freshwater	Freshwater
	Acid (HCl)	Acid (HCl)	Acid (HCl)	Acid (HCl)
	ACID (HF)	ACID (HF)	Acid (HF)	Acid (HF)
		Migration of "Fines"	"SAND" PRODUCTION	Migration of "Fines"
				"SAND" PRODUCTION
Detrital/ Bioturbated	Freshwater	Freshwater	Freshwater	Freshwater
	Acid (HCl)	Acid (HCl)	Acid (HCl)	Acid (HCl)
	Acid (HF)	ACID (HF)	Acid (HF)	Acid (HF)
	Migration of"Fines"	MIGRATION OF "FINES"	Migration of "Fines"	Migration of "Fines"
			"SAND" PRODUCTION	"SAND" PRODUCTION
Authigenic/ Grain Coating	FRESHWATER	FRESHWATER	FRESHWATER	FRESHWATER
	ACID (HCl)	ACID (HCl)	ACID (HCl)	ACID (HCl)
	ACID (HF)	ACID (HF)	Acid (HF)	Acid (HF)
			"SAND" PRODUCTION	"SAND" PRODUCTION
Authigenic/ Pore Filling	Acid (HCl)	Acid (HCl)	Acid (HCl)	Acid (HCl)
	Acid (HF)	Acid (HF)	Acid (HF)	Acid (HF)
	MIGRATION OF "FINES"	MIGRATION OF "FINES"	MIGRATION OF "FINES"	MIGRATION OF "FINES"
			"SAND" PRODUCTION	"SAND" PRODUCTION

1. The categories of formation damage are based on the assumption that normal rock-forming minerals are present (including calcite and pyrite).
2. Scale can form in each category if incompatible fluids are mixed.
3. Capitalized words indicate major importance.

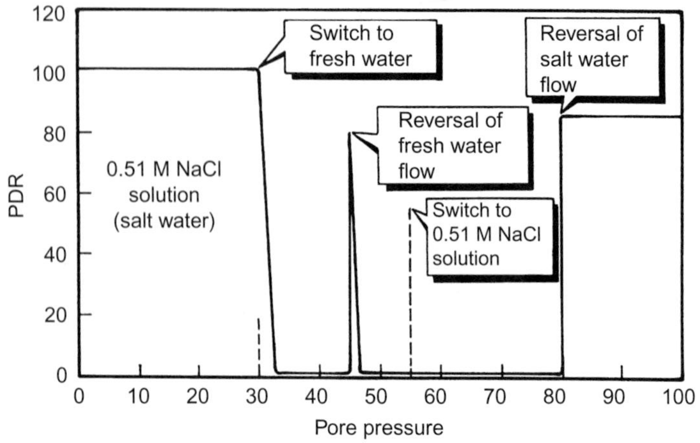

FIGURE 11.11 Behavior of permeability damage ratio during a standard core flood test [30].

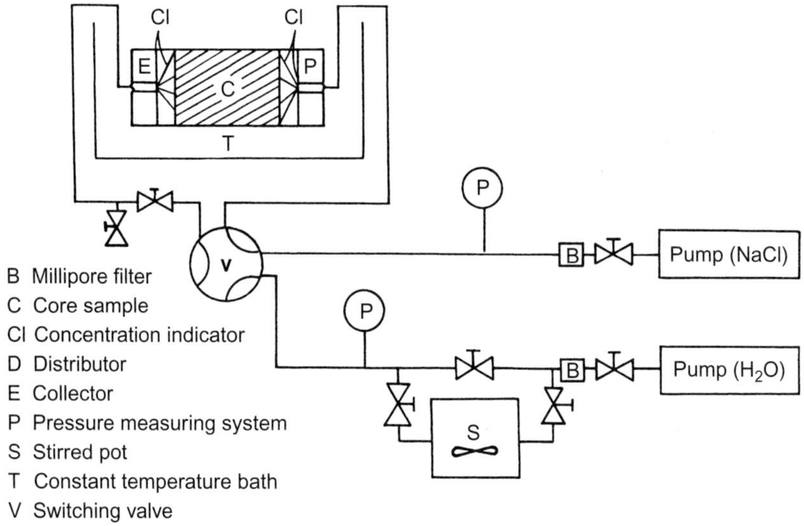

FIGURE 11.12 Schematic diagram of experimental apparatus used to investigate chemical damage [31].

beyond which permeability reduction is rapid and drastic. This critical rate can be obtained by plotting the final permeability reductions from Table 11.14 versus space velocities, as shown in Figure 11.15. For $v = 25.6$ cm/h, the critical rate corresponds to a space velocity of approximately 0.26 h^{-1} and a permeability damage ratio of 0.62 (Figure 11.15). For

TABLE 11.14 Results of Chemical Damage Tests [31]

Experimental Run #	Flowrate (q) (cc/s)	Superficial Velocity (cm/h)	Volume (V) of Mixer (cc)	Space ($^{-1}$) Velocity h^{-1}	k/k_o
1	1.67×10^{-3}	1.2	150	0.040	0.45
2	8.89×10^{-3}	6.3	150	0.222	0.67
3	8.89×10^{-3}	6.3	150	0.222	0.56
4	2.22×10^{-2}	15.8	36	2.22	0.008
5	3.61×10^{-2}	25.6	1,600	0.083	0.70
6	3.61×10^{-2}	25.6	1,080	0.125	0.67
7	3.61×10^{-2}	25.6	530	0.263	0.62
8	3.61×10^{-2}	25.6	385	0.357	0.16
9	3.61×10^{-2}	25.6	255	0.555	0.01
10	3.61×10^{-2}	25.6	100	1.320	0.008
11	3.61×10^{-2}	25.6	36	3.85	0.008
12	3.33×10^{-3}	2.4	148	0.082	0.36
13	3.33×10^{-3}	2.4	65	0.212	0.11
14	3.33×10^{-3}	2.4	36	0.345	0.02
15	3.33×10^{-3}	2.4	24	0.526	0.008

a lower superficial velocity of 2.4 cm/h, the critical rate of salinity corresponds to 0.15 h^{-1} and k/k_o of approximately 0.30.

Samples of the effluent from several runs were analyzed both microscopically and chemically. Scanning electron micrographs revealed that clay particles are released regardless of the rate of salinity decrease. To quantify this observation, clay particles were dissolved in hydrofluoric acid solution to yield mostly silicon ions (Si^{4+}) and aluminum ions (Al^{3+}). Assuming that most migrating clay particles are kaolinite, the concentrations of aluminum ions were converted to concentrations of clay particles (X) and plotted as shown in Figures 11.16–11.18.

From these figures, Khilar et al. concluded [30] the following:

1. The effluent is free of clay particles until the critical rate of salinity decrease is attained.
2. The clay concentration in the effluent stream is directly proportional to the space velocity and inversely proportional to the superficial velocity.

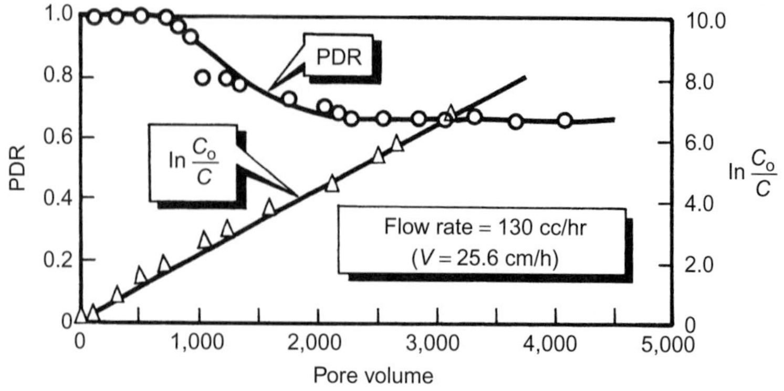

FIGURE 11.13 Permeability damage ratio and salt concentration ratio at a space velocity of $0.125\ h^{-1}$ [31].

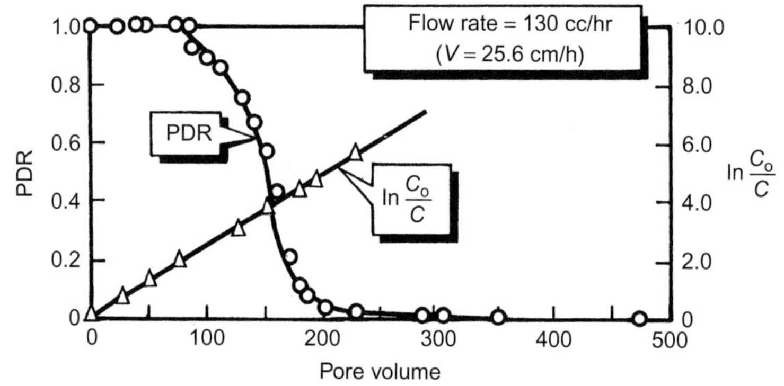

FIGURE 11.14 Permeability damage ratio and salt concentration ratio at a space velocity of $1.316\ h^{-1}$ [31].

3. Approximately the same amount or mass of clay particles is released for a given drop in the salt concentration below the critical one, irrespective of the flow rate and the rate of salinity decrease (Table 11.15 and Figure 11.19). This figure shows the amount of clay particles released as a function of salt concentration. The amount of clay particles released can be estimated by calculating the area under the clay concentration curve *X*.

Based on chemical and microscopic analyses, the dependence of the permeability damage ratio k/k_o on the rate of salinity change can be explained by the so-called logjam mechanism. For instance, Figure 11.16 shows that an abrupt decrease in the salt concentration causes a sudden and rapid increase in clay concentration (*X*). As more particles are released, they start arriving

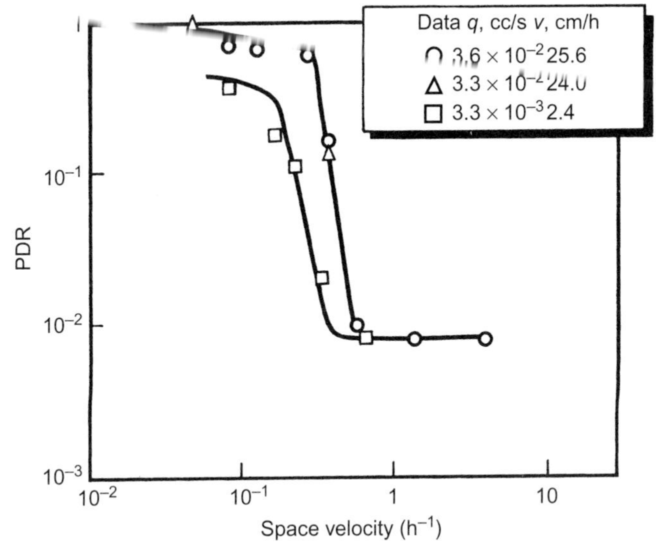

FIGURE 11.15 Critical rate of salinity decrease in water sensitivity tests of Berea core [31].

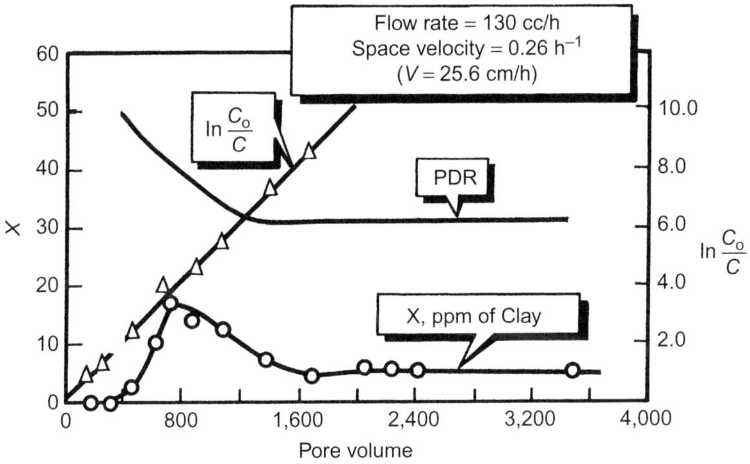

FIGURE 11.16 Behavior of clay concentration, X, and permeability damage ratio, k/k_o, at slow decrease in salt concentration, C, and high flow rate [30].

at pore throats at the same time, causing logjams or bottlenecks and consequently a significant reduction in the permeability. Continued release and capture of clay particles in the same flow direction can result in total plugging of the sandstone pore throats. Conversely, if the salt concentration is reduced slowly, then the clay particles are released slowly and in low

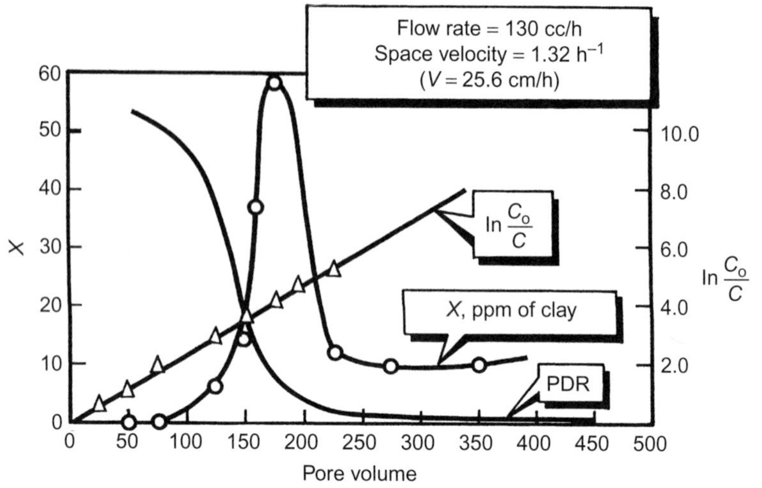

FIGURE 11.17 Behavior of clay concentration, X, and permeability damage ratio, k/k_o at fast decrease in salt concentration, C, and high flow rate [30].

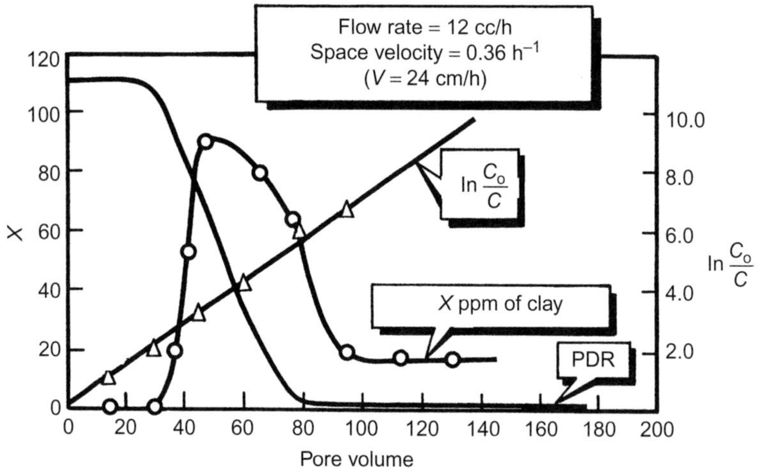

FIGURE 11.18 Behavior of clay concentration, X, and permeability damage ratio, k/k_o, at fast decrease in salt concentration, C, and low flow rate [30].

concentration, thereby avoiding the logjam effect. Based on these experimental results, a theoretical model was developed to predict the critical rate of salinity decrease. The predictions appear to be in reasonable agreement with these experimental measurements.

TABLE 11.15 Mass of Clay Particles Collected at Different Velocities [30]

Space Velocity (h)	Superficial Velocity (cm/h)	(ΔM) (g)
0.125	25.6	0.025
0.263	25.6	0.022
9.34	2.4	0.020

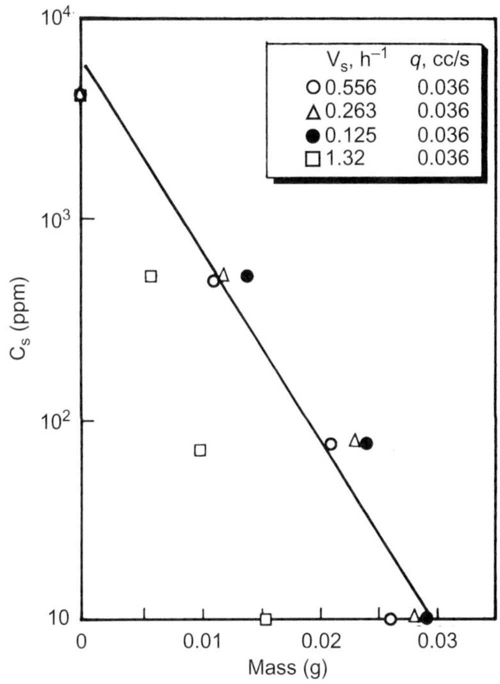

FIGURE 11.19 Mass of clay particles collected as a function of salt concentration at various space velocities [30].

Mechanical Damage

Permeability damage by fines migration resulting from mechanical flow forces has been systematically investigated first by Muecke, and by Gruesbeck and Collins [7, 9]. Earlier studies, which were mostly based on analyses of core flooding experiments, yielded only qualitative results with limited practical applications [33–38]. Gray and Rex found that nonswelling

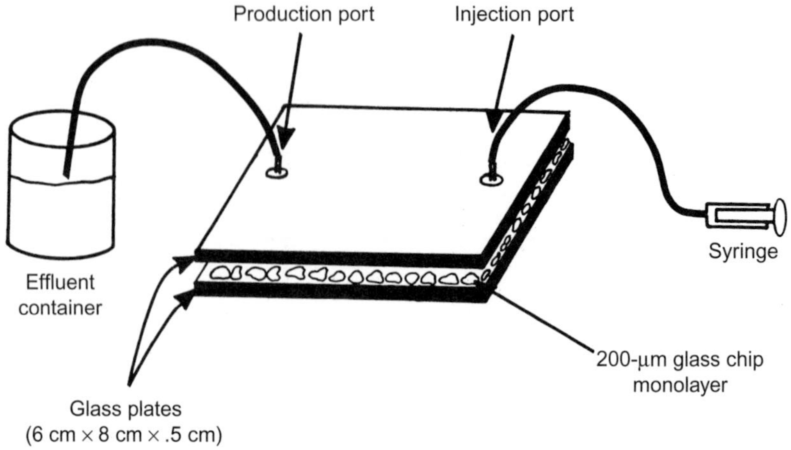

FIGURE 11.20 Schematic diagram of experimental apparatus used to investigate fines movement in porous media [7].

mica crystals, partially mixed-layer minerals, and kaolinite crystals constituted the principal sources of permeability damage [33]. The electrostatic forces bonding these clay particles and keeping them in equilibrium are weak and may be altered by any change in cation concentration, causing these particles to dislodge from the walls of sandstone pore channels.

Muecke designed a micromodel of porous media, shown in Figure 11.20, to investigate the many factors controlling the migration of formation fines in porous media [7]. This micromodel allows direct visual observation of fines movement during fluid flow through an optical microscope using transmitted light at various magnifications. Precipitated calcium carbonate ($CaCO_3$) particles ranging in size from 2 to 15 μm were used as fines and introduced into the micromodel as a suspension. Water, oil, and various common solvents were used, separately and in various combinations, as carrier fluids. When a single-liquid phase is present, Muecke observed the following:

1. The fines move freely through the porous system, unless they mechanically bridge at pore restrictions as depicted in Figure 11.21.
2. The tendency to bridge is directly proportional to the concentration of fines; consequently, high concentrations of fines in sandstone formations can lead to severe permeability reduction by the bridging or plugging mechanism.
3. Bridges formed at high flow rates or velocities are much more stable to flow reversals or pressure disturbances than those formed at low velocities; therefore, pulsating flow may cause less permeability damage than continuous flow.

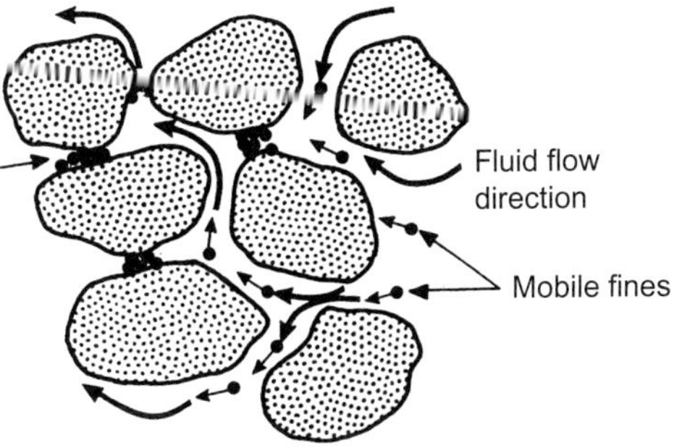

FIGURE 11.21 Fines migration during single-phase flow [7].

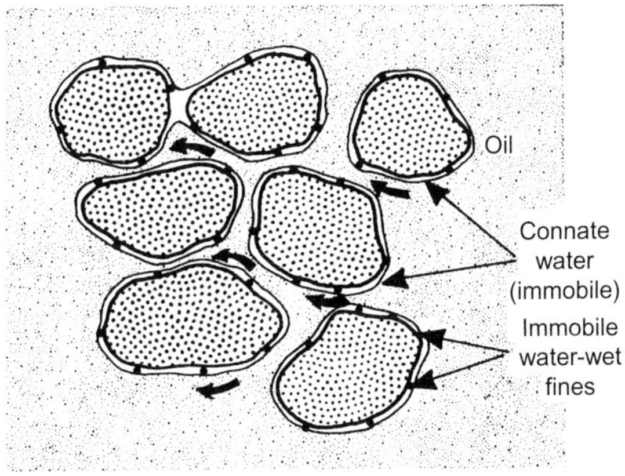

FIGURE 11.22 Water-wet fines are immobile when the connate water is immobile [7].

4. There is an equilibrium-bridged condition reached at some constant flow rate when the movement of fines practically ceases. If the flow rate is changed or the flow direction is reversed, however, the bridges become unstable, causing the fines to move again.

Using the same micromodel, Muecke investigated fines migration in porous media containing two or more immiscible fluids and concluded the following:

1. Because fine particles have a high surface area-to-mass ratio, fines wettability and surface–interfacial forces play a dominant role in their movement.
2. Formation fines become mobile only if the liquid phase that wets them becomes mobile, as shown in Figures 11.22 and 11.23, for a water-wet fines case.

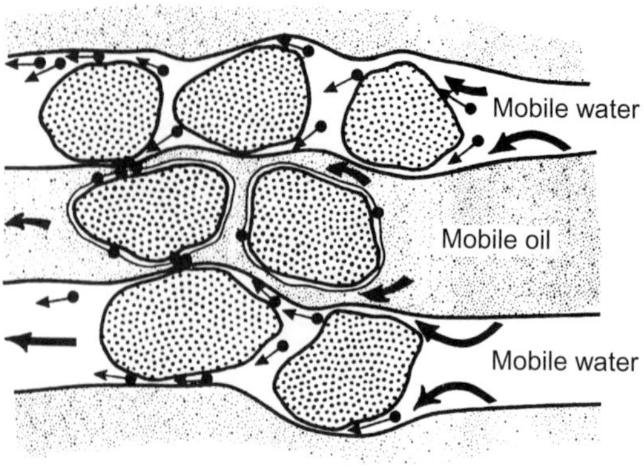

FIGURE 11.23 Fines migration during two-phase flow [7].

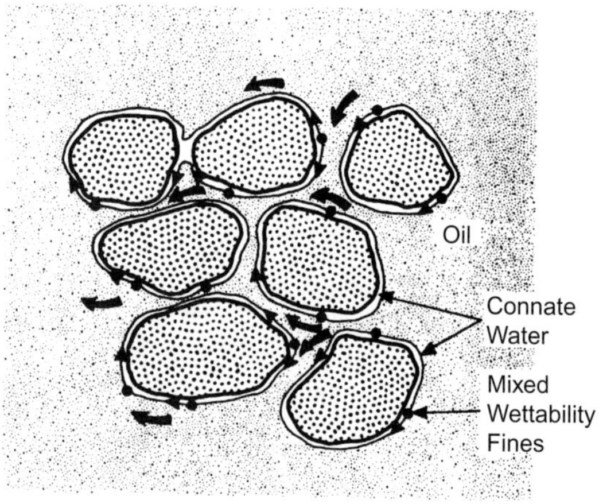

FIGURE 11.24 Fines of mixed wettability migrate along the water/oil interface when the connate water is immobile [7].

3. Simultaneous flow of oil and water causes considerable fines migration because the localized pressure disturbances at the oil–water interfaces keep the fine particles agitated.
4. Continued water flow at residual oil saturation, i.e., the oil phase is immobile, rapidly establishes an equilibrium-bridged condition.
5. Fines of mixed wettability, i.e., located at the interface between the oil and water, tend to move only along the oil–water interface (Figure 11.24). If one of the wetting phases is immobile, interfacial forces confine the fines movement to only a few grain sizes.
6. Injection of mutual solvent or surfactant solution releases formation fines held by wetting and interfacial forces, causing them to migrate at high

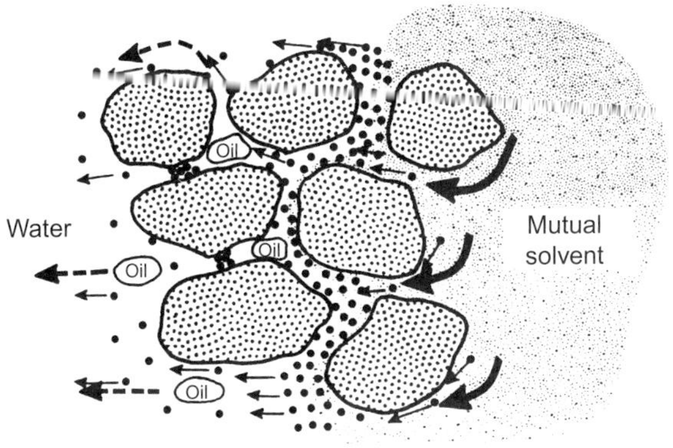

FIGURE 11.25 Fines migration during injection of chemicals for improved oil recovery [7].

TABLE 11.16 Typical Mineralogical Composition of Berea Cores [21]

	Bulk Core	Fines (<5 m)
Quartz	86%	51%
Feldspar	5	11
Dolomite	1	—
Siderite	1	1
Illite	4	15
Kaolinite	3	20
Chlorite	—	2
Total	100%	100%

concentrations as shown in Figure 11.25. This suggests that permeability damage by fines migration is likely to occur during improved oil recovery processes. Muecke reached the same conclusions when he used a large-scale linear flow cell (4 ft long and 1.5 in. in diameter) to study the movement of fines under conditions that are more representative of those present in the reservoir rocks.

Gabriel and Inamdar investigated the simultaneous effects of both the chemical and mechanical mechanisms on fines migration and subsequent permeability damage [20]. The flow tests were performed mostly on Berea sandstone core samples (10.2 cm long and 2.54 cm in diameter). Table 11.16 shows the mineralogical composition of these samples. This experimental investigation essentially confirmed the findings of Muecke and Khilar [7,31].

The primary contribution of this study is the estimation of critical velocity. For Berea sandstone core samples, which typically have an average permeability of approximately 150 mD, permeability damage occurred at velocities greater than 0.007 cm/s. Generally, however, the extent of permeability damage is a function of flow area, direction, original core permeability, and wettability.

MIGRATION OF FOREIGN SOLIDS

Clays or other solid particles often are included in the formulation of drilling, completion, or workover fluids to provide the required density, rheology, and filtration. These solid particles may be carried into the formation with the filtrate from the fluid system. Once these solids are inside the pore channels of the formation, they behave similarly to formation fines. Permeability impairment by foreign solids is strongly dependent on their size distribution and the pore throat size distributions of the formation. Penetration of these solids is generally shallow (5–10 cm); however, the resultant permeability damage ratio, k/k_o, can be as low as 10% [3].

CRITICAL VELOCITY CONCEPT

Gruesbeck and Collins used a parallel-pathway model of a porous medium, as shown in Figure 11.26, to determine the local laws of mobilization and deposition of formation fines [8]. In this model, it is assumed that flow channels have two parallel branches: (1) one consists of large pore sizes in which only surface nonplugging deposition occurs, and (2) the other consists of smaller pore sizes in which plug-type deposition of fines may occur. The

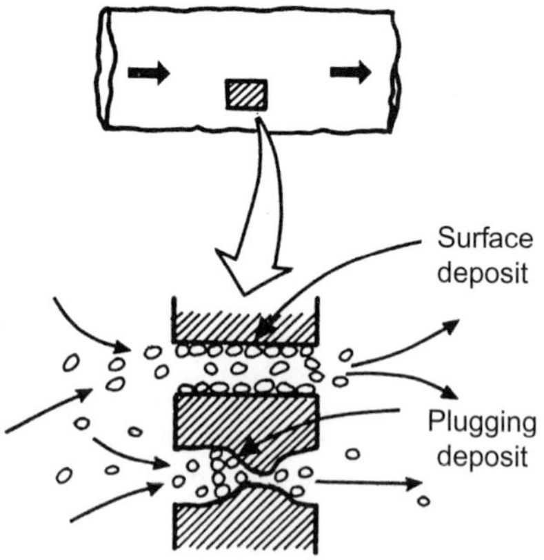

FIGURE 11.26 Parallel pathway model of fines migration and deposition [8].

fraction of flow channels that are plug-type pathways is determined by the size of fines relative to the size of pores.

ENTRAINMENT AND SURFACE DEPOSITION

Assuming no plugging deposition occurs when very small fine particles are entrained through a porous medium having large pores, Gruesbeck and Collins performed a set of experiments using clean sandpacks with grain diameters ranging from 840 to 2,000 mm and a suspension of $CaCO_3$ particles with a mean diameter of 8 mm [8]. The carrier fluid was a 2% KCl solution with pH adjusted to 8 with sodium hydroxide. Effluents were collected and analyzed for fines concentration and particle size distribution with a turbidimeter and a Coulter Counter. Figure 11.27 shows a typical plot of C_e/C_i versus PV throughput for different interstitial velocities, v/ϕ_i, where

C_e = fines concentration in effluent, cm^3/cm^3
C_i = fines concentration in inlet fluid, cm^3/cm^3
u = volume flux density (q/A), cm/s
ϕ_i = initial porosity, fraction
q = flow rate, cm^3/s
A = cross-sectional area, cm^2

Figure 11.27 shows that at each change in flow rate, there exists a constant value of C_e/C_i, indicating steady-state conditions. Thus, from a material balance on fines at any point x in the sandpack, the following equation is obtained:

$$\frac{\partial}{\partial t}(\phi C + \phi_i V_{fp}) + u\frac{\partial C}{\partial x} = 0 \tag{11.5}$$

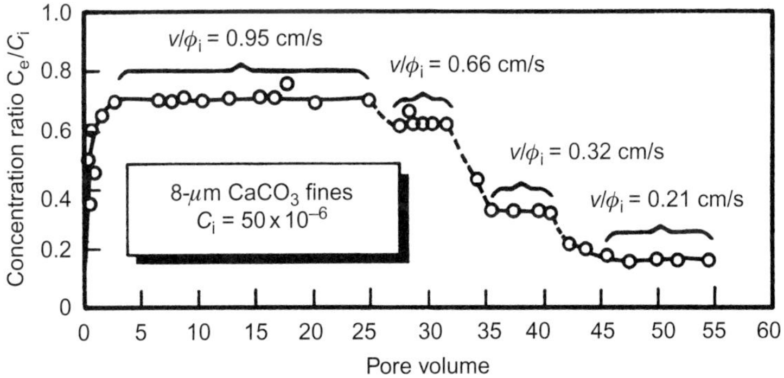

FIGURE 11.27 Deposition of fines in a porous medium (sandpack) for different interstitial velocities [8].

where

ϕC = volume of fines in suspension per unit bulk volume of porous medium, cm^3
$\phi_i V_{fp}$ = volume of fines deposited per unit bulk volume of porous medium
ϕ = remaining porosity after deposition, fraction
C = fines concentration in fluid, cm^3/cm^3
x = distance along the core, cm
V_{fp} = volume of fine particles deposited per unit initial PVV, cm^3/cm^3

At steady-state conditions, the volume of fines in suspension, ϕC, is constant and $\partial(\phi C)/\partial t = 0$. Thus, Eq. (11.5) becomes:

$$\frac{\partial V_{fp}}{\partial t} = \frac{u}{\phi_i}\frac{C}{x} \tag{11.6}$$

Assuming the rate of fines deposition, $\partial V_{fp}/\partial t$, is proportional to fines concentration C, i.e.:

$$\frac{\partial V_{fp}}{\partial t} = \alpha_1 C \tag{11.7}$$

where α_1 = constant of proportionality, s^{-1}.

Substituting $\alpha_1 C$ for $\partial V_{fp}/\partial t$ in Eq. (11.6) and integrating yields:

$$\alpha_1 = \frac{-u}{\phi_i L}\ln\frac{C_e}{C_i} \tag{11.8}$$

where C_i is the inlet fines concentration and C_e is the fines concentration at the outlet of the core.

Thus, the surface-type deposition occurs according to the following equation:

$$\frac{\partial V_{fp}}{\partial t} = \frac{-uC}{\phi_i L}\ln\frac{C_e}{C_i} \tag{11.9}$$

Using the data in Figure 11.27 and Eq. (11.8), it was found that $\alpha_1 = 0.01\ s^{-1}$ for each flow rate, which indicates that the surface deposition constant α_1 is independent of flow rate.

Based on another *set* of data shown in Figure 11.28, which was obtained from flowing a clean fluid through a dirty sandpack—i.e., the 840- to 2,000-μm-grain diameter sand was mixed and packed wet with a fines suspension—Gruesbeck and Collins derived an equation for calculating the average rate of entrainment. Assuming steady-state conditions and integrating Eq. (11.6) over the length of the sandpack with $C = 0$ at $x = 0$, and $C = C_e$ at $x = L$ gives:

$$\left(\frac{\partial V_{fp}}{\partial t}\right)_{avg} = -\frac{uC_e}{\phi L} \tag{11.10}$$

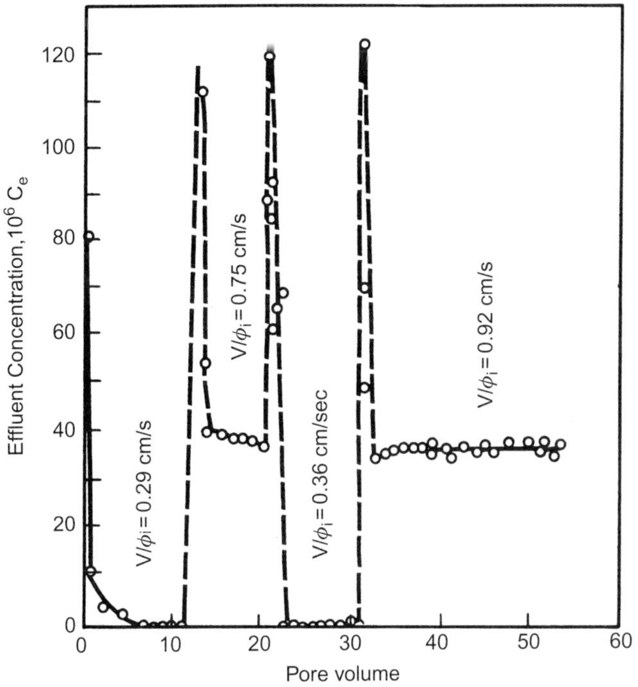

FIGURE 11.28 Concentration of fines in the effluent for different interstitial velocities [8].

This equation is used to calculate the average rate of entrainment within the sandpack from the data in Figure 11.28, which is then plotted against the interstitial velocity as shown in Figure 11.29 (curve a). This figure also shows the result of experiments using the same fines, sand, and solution, but with a polymer added to increase the fluid viscosity (curve b). These plots reveal the existence of a "critical velocity," or flow rate, below which entrainment of fine particles does not occur, i.e., $\partial V_{fp}/\partial t = 0$, and above which the rate of mobilization, entrainment, and deposition of formation fines increases linearly with the flow rate according to the following expression:

$$\frac{\partial V_{fp}}{\partial t} = \alpha_1 C - \alpha_2 V_{fp}(u - u_c), \quad u > u_c \tag{11.11}$$

where

u_c = critical volume flux density, cm/s
α_2 = constant (determined experimentally), cm^{-1}

Inasmuch as the inlet concentration is zero, the term $\alpha_1 C$ should be negligibly small throughout the sandpack for the brief duration of these type of tests and, therefore, may be dropped from Eq. (11.8).

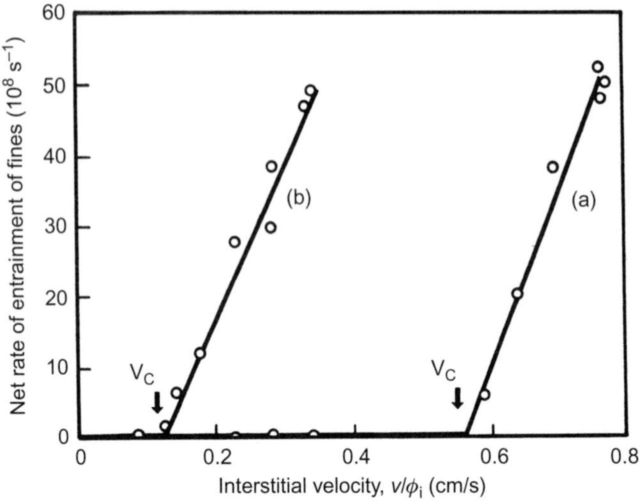

FIGURE 11.29 Net rate of entrainment of fines in a porous medium as a function of interstitial velocity [8].

From Figure 11.29, it appears that the only effect of increasing the viscosity of flowing fluid on the rate of entrainment of fines is the increase in the value of the critical velocity. This increase may be justified by postulating the existence of a force that binds fine particles on the sand grain surface. If this force is exceeded by the viscous drag force of the flowing fluid, the release and entrainment of fine particles would occur. The large fluctuations observed in the effluent concentration at each change in flow rate are caused by turbulent flow (Figure 11.28). Turbulence enhances the viscous drag force of the flowing fluid and, consequently, increases the rate of entrainment of formation fines.

ENTRAINMENT AND PLUGGING

Experiments in which both surface and plug-type deposition occur were made by using flowing suspensions of 5- to 10-μm glass beads through a clean pack of 250- to 297-mm-diameter sand grains. The results of these tests are shown in Figure 11.30, which reveals the existence of a limiting condition of zero deposition rate at some equilibrium value of fines deposition. This equilibrium value is dependent on the interstitial velocity of the flowing fluid. Using the parallel pathway model in Figure 11.26 and assuming, within any elemental volume of the porous medium, a fraction f of the

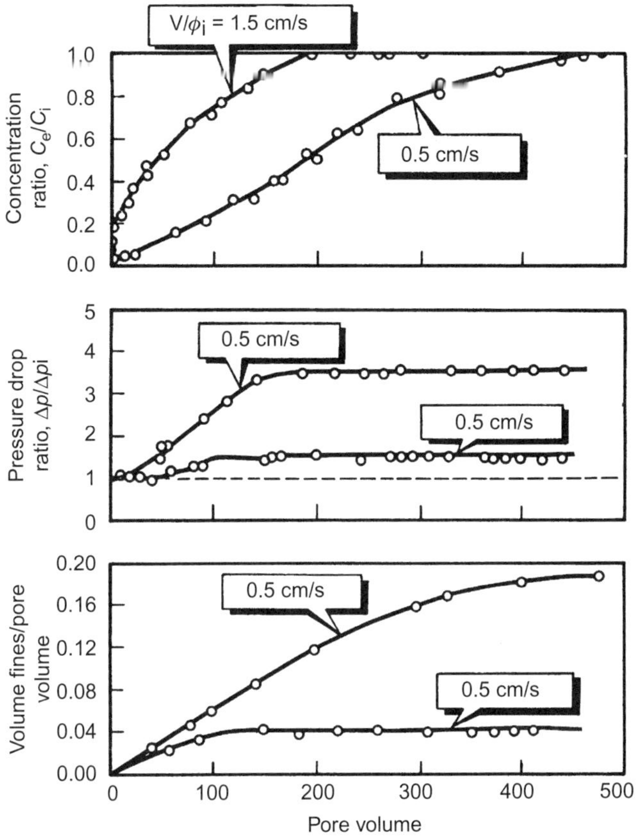

FIGURE 11.30 Effect of interstitial velocity on entrainment and deposition of formation fines [8].

element made up of pluggable pathways and a fraction $(1-f)$ of nonpluggable pathways, the total flux density is given by [8]:

$$u = fu_{\mathrm{p}} + (1-f)u_{\mathrm{np}} \tag{11.12}$$

where the subscripts "p" and "np" stand for pluggable and nonpluggable, respectively. The total volume of deposits per unit bulk volume of porous medium is expressed as:

$$\phi_i V_{\mathrm{fp}} = f\phi_{\mathrm{i}}(V_{\mathrm{fp}})_{\mathrm{p}} + (1-f)\phi_{\mathrm{i}}(V_{\mathrm{fp}})_{\mathrm{np}} \tag{11.13}$$

Nonpluggable pathways are assumed to be sufficiently large compared with the size of the suspended fines so that total plugging never occurs. However, in pluggable pathways, which are composed of narrower pore openings, the formation permeability and consequently the flux density u_{p} can be reduced to zero by fines deposition. Based on these assumptions and

Eqs. (11.5), (11.12), and (11.13), Gruesbeck and Collins postulated that the rate of entrainment and deposition in the pluggable pathways behaves according to the following expression [8]:

$$\frac{\partial (V_{fp})_p}{\partial t} = \left[\alpha_3 + \alpha_4 (V_{fp})_p\right] \frac{u_p}{\phi_i} C \tag{11.14}$$

where α_3 and α_4 are constants (in cm^{-1}) that can only be determined experimentally. In nonplugging pathways, Eq. (11.11) applies by changing the terms u and V_{fp} to u_{np} and $(V_{fp})_{np}$, respectively:

$$\frac{\partial (V_{fp})_{np}}{\partial t} = \alpha_1 C - \alpha_2 (u_{np} - u_c)(V_{fp})_{np} \tag{11.15}$$

It is evident from Eqs. (11.12) and (11.13) that as the volume of fine particles deposited in the plugging pathway, $(V_{fp})_p$, increases, the velocity, u_p, decreases, causing the fluid velocity through the nonplugging pathways, u_{np}, to increase; i.e., fluid flow is diverted from plugging pathways to nonplugging pathways. Hence, entrainment in nonplugging pathways increases until $\partial(V_{fp})_{np}/\partial t$ approaches zero. Setting $\partial(V_{fp})_{np}/\partial t$ equal to zero in Eq. (11.11) or (11.15) gives the equilibrium value, V^*_{fp}, observed in Figure 11.30:

$$V^*_{fp} = \frac{\alpha_1 C}{\alpha_2 (u - u_c)} \tag{11.16}$$

where it is assumed that, during equilibrium, all pluggable pathways are plugged, and consequently $u_{np} = u$. As a test of the general validity of the critical concept theory, Gruesbeck and Collins carried out numerical integration of these equations and generated data (plotted in Figure 11.31). It is evident that these theoretical data match, almost perfectly, the data generated experimentally (Figure 11.30). The simple theory of critical velocity, therefore, is an excellent representation of deposition and entrainment processes in porous media. For the purpose of numerical integration, the permeability is assumed to change as deposits accumulate according to the following expressions:

$$\frac{k_p}{k_{pi}} = \exp\left[-\alpha_3 (v_{fp})_p^4\right] \tag{11.17}$$

and

$$\frac{k_{np}}{k_{npi}} = \frac{1}{1 + \alpha_4 (v_{fp})_{np}} \tag{11.18}$$

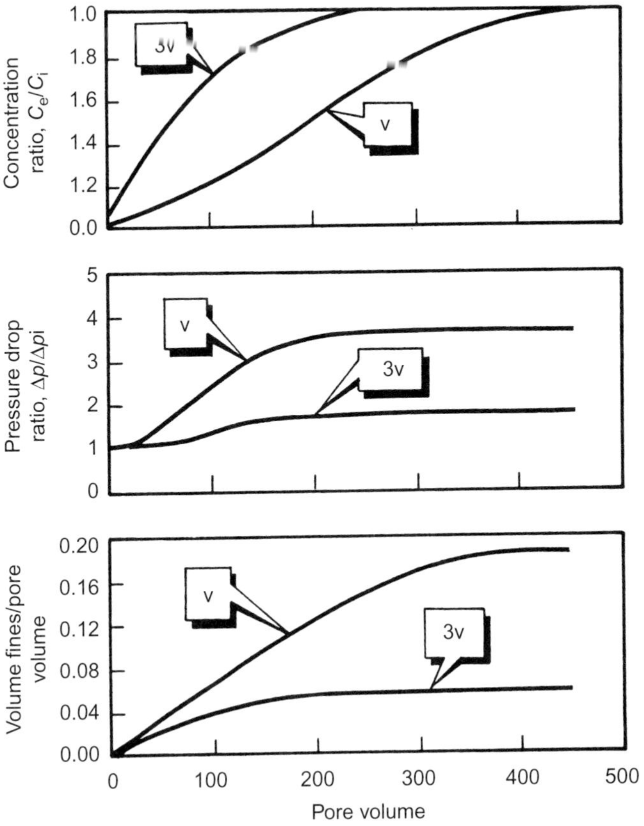

FIGURE 11.31 Numerical simulation of the effect of interstitial velocity on fines migration [7].

where α_3 and α_4 are dimensionless constants that may be determined experimentally. Also, the fraction of flow through pluggable pathways is assumed to change according to the following equation:

$$\left(\frac{u_p}{u_{np}}\right) = \frac{k_p(V_{fp})_p}{k_p(V_{fp})_p + k_{np}(V_{fp})_{np}} \tag{11.19}$$

The experimental tests used to develop the concept of critical velocity were run at a constant flow rate, even though in producing wells flow occurs more nearly at a constant pressure differential, especially in the vicinity of wellbore where formation damage occurs. Figure 11.32 shows the results of deposition and entrainment tests run at a constant pressure drop across the sandpack for two types of sand. The permeability damage ratio k/k_o for finer sands (curve a) approaches zero rather rapidly because the deposits tend to concentrate near the inlet end of the sandpack. In the coarser sand (curve b),

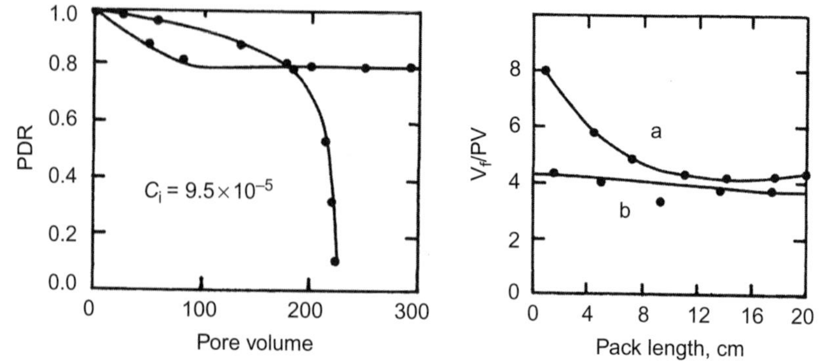

FIGURE 11.32 Entrainment and deposition of formation fines under constant Δp of 5- to 10-mm-diameter fines; curve a: 177- to 210-mm-diameter sandpack, $\Delta p/L = 900$ kPa/m; curve b: 250- to 297-mm-diameter sandpack, $\Delta p/L = 450$ kPa/m [8].

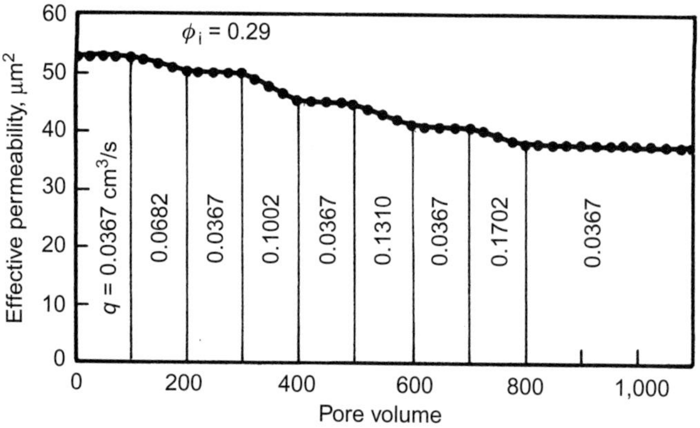

FIGURE 11.33 Effect of flow rate on the entrainment and redeposition of fines in a Berea core during flow of a 2% KCl solution [8].

deposits are more uniform throughout the sandpack and, consequently, after a brief decline, the permeability damage ratio becomes constant. Gruesbeck and Collins observed that more deposition results, for a given pore volume throughput, under constant Δp than for a constant flow rate q.

Figures 11.33 and 11.34 show permeability changes versus PV and interstitial velocity, respectively, when a Berea sandstone core is exposed to a deposition and entrainment test. Data in Figure 11.33 were obtained by maintaining each indicated flow rate for a throughput of 100PV, while Δp was measured continuously. For the flow rate of 0.0367 cm^3/s, the permeability of the core became constant. Figure 11.34 is a plot of k/k_o versus u/ϕ_i for the data in Figure 11.33 and three other similar data sets obtained using different flowing fluids. The curves plotted in Figure 11.34 have the same

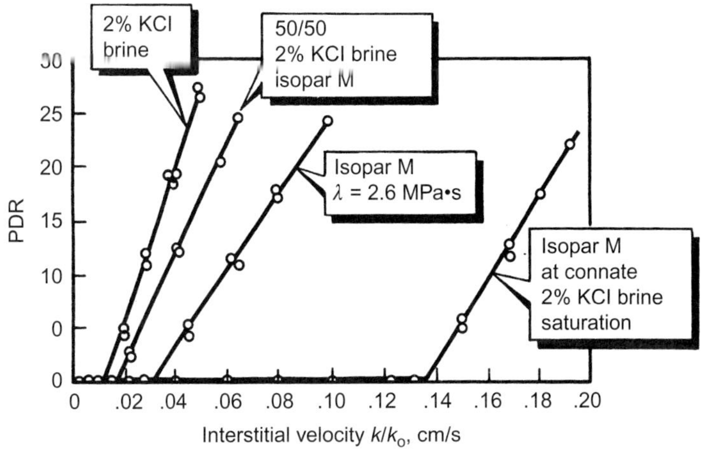

FIGURE 11.34 Effect of interstitial velocity of the permeability damage ratio k/k_o for different flowing solutions [8].

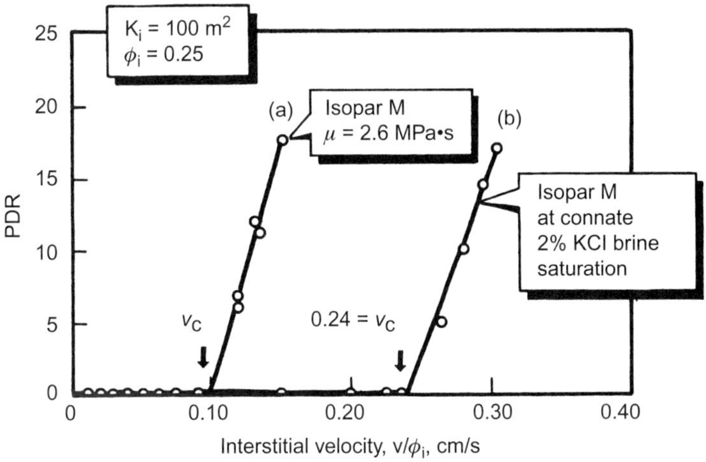

FIGURE 11.35 Permeability damage ratio as a function of interstitial velocity in field core [8].

characteristic behavior as those shown in Figure 11.29. This characteristic behavior also can be observed in Figure 11.35, which shows the results of similar experiments on actual cores from a field where some wells have exhibited abnormal productivity decline. These results confirm the role played by naturally occurring fines in formation damage and consequent well productivity decline. Hence, the critical velocity theory is an excellent representation of deposition and entrainment processes and is applicable to consolidated as well as unconsolidated porous media.

IDENTIFICATION OF PERMEABILITY DAMAGE MECHANISMS

Assuming that, at certain times, only one mechanism of permeability damage is dominant, Wojtanowicz et al. developed a simple and practical technique for identifying the prevailing mechanism [18]. This technique is based on the "systems analysis" approach in which the experimental data of permeability changes versus the flowing time are used to infer quantitative values of selected factors involved in the fluid–rock interaction. For practical applications, this approach can be used to recognize the damage mechanisms by analyzing Cartesian plots of the permeability damage ratio, k/k_o, as a function of time. The type of damage mechanism occurring is identified by the existence of a straight line of a specific slope. The general assumptions used in the development of this technique are as follows:

1. The concentration of formation fines exponentially decreases with time.
2. The concentration of solids in completion fluids invading the formation is constant.
3. The migration of these solids behaves according to the constant-rate filtration process.
4. The concentration of solids is low, so that the volume reduction due to particles capture is negligible.
5. The filter cake is incompressible.
6. The formation is homogeneous and pore geometry is regular.
7. The flow is linear and laminar.

PERMEABILITY DAMAGE FROM FOREIGN SOLIDS

There are three mechanisms of foreign solids transport in porous rocks by which permeability damage can occur—gradual pore blocking, single-pore blocking or screening, and cake forming or cake straining.

Gruesbeck and Collins defined gradual pore blocking as a surface-type deposition and showed that during this mechanism the rate of deposition or capture is directly proportional to the solids concentration in the flow stream [9]. Using a first-order particles capturing model, Wojtanowicz et al. derived the following equation describing the behavior of formation permeability during gradual blocking [18]:

$$k = (\sqrt{k_o} - m_g t)^2 \tag{11.20}$$

or

$$\sqrt{\frac{k}{k_o}} = -m_g t + 1 \tag{11.21}$$

which is an equation of a straight line with a negative slope m_g and intercept 1.00. The slope is a function of the capture factor f_c, solids concentration

C_s (g/cc), average length of the flow path L (cm), density of solids ρ_s (g/cc), and original flow area A_{fo} (cm^2).

$$m_g = \frac{f_c c_s}{L \rho_s A_{fo}} \quad (11.22)$$

The physical meaning of the capture factor, f_c, is analogous to the cake-to-filter ratio used in deep-bed filtration theory and can be estimated from the solids' capture equation [36]:

$$f_c = -\frac{1}{C_s}\frac{\Delta M}{\Delta t} \quad (11.23)$$

where M is the mass of foreign solids captured by the rock over a period of time, Δt. Single-pore blocking occurs when solid particles of critical size, i.e., of size close to the pore size, instantly block individual pores.

The permeability damage function describing this mechanism is:

$$\sqrt{\frac{k}{k_o}} = -m_s t + 1 \quad (11.24)$$

which is also an equation of a straight line of slope $-m_s$ and intercept 1.00. The slope is given by:

$$m_s = -\frac{6 q C_s C_{sc} A_p}{\pi d_c^3 \rho_s A_{fo}} \quad (11.25)$$

where

q = flow rate, cc/min
C_{sc} = concentration of solid particles of critical diameter, g/cc
d_c = critical diameter of particles, cm
A_p = area of single pore, cm^2

During the straining mechanism, a filtration cake builds up at or near the formation face. Cake forming can be initiated by solid particles larger than the pore size or by a high concentration of solids smaller than the pore size. The permeability change during this mechanism is:

$$\frac{k_o}{k} = m_c t + 1 \quad (11.26)$$

This equation is a straight line of positive slope m_c and intercept 1.00. The slope m_c is approximated by the following equation:

$$m_c = \frac{\alpha q C_s}{A_c R_r} \quad (11.27)$$

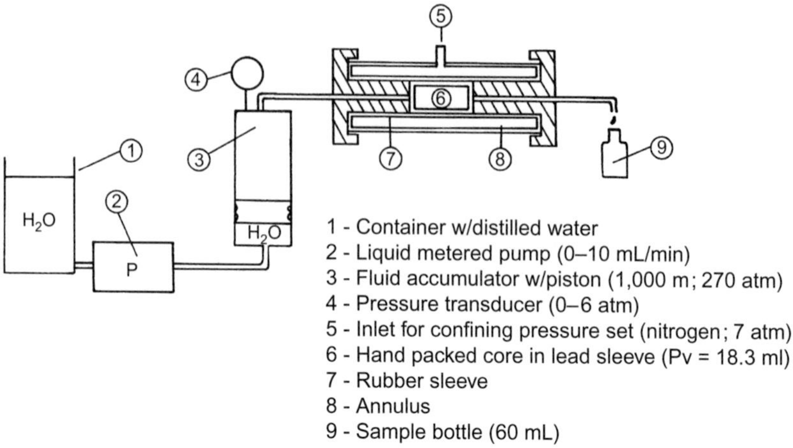

FIGURE 11.36 Schematic diagram of experimental apparatus used to investigate the damage mechanisms [18].

where

α = average flow resistance, cm/g ($= A_c R_c / M$)
A_c = area of cake, cm^2
R_r = rock resistance to filtration, cm^{-1}
R_c = cake resistance to filtration, cm^{-1}
L_r = actual length of the rock, cm
A_r = rock area, cm^2

Using the laboratory setup shown in Figure 11.36, Wojtanowicz et al. performed a series of experiments by flowing completion fluids with different levels of mud contamination through core samples, and generated Figures 11.37–11.40 [18]. Tables 11.17 and 11.18 show the completion fluids and the mineral composition of the reservoir rock samples used in these tests. Figure 11.37 clearly shows fundamental qualitative change in permeability for different fractions of contamination. Figure 11.38 is a plot of $\sqrt{k/k_o}$ versus time, which shows the existence of an early straight line for three different levels of contamination (0.2%, 0.5%, and 1.0%). In all three cases, the intercept is approximately equal to 1 and the slope is negative. Thus, according to Eq. (11.37), the gradual pore-blocking mechanism is dominant during the early part of the tests. As shown in Figure 11.38, the duration of this mechanism is a function of the level of contamination, and the higher the level of contamination, the shorter the duration of gradual pore-blocking mechanism. At the end of this mechanism, the curve is declining rapidly, i.e., there is severe permeability damage, suggesting the existence of a single-pore blockage (screening) mechanism. Figure 11.39 is a Cartesian plot of k/k_o versus time for the same data used to plot Figure 11.38. The two

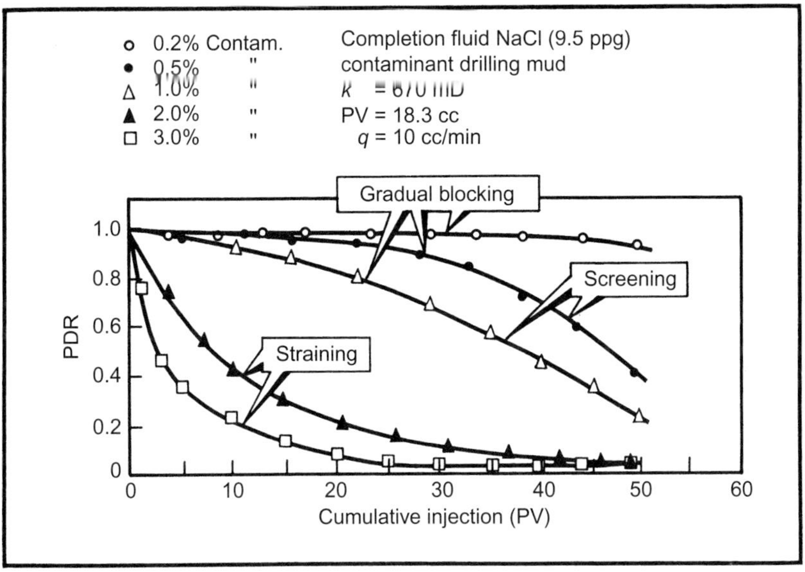

FIGURE 11.37 Behavior of PDR is indicative of the damage mechanism [18].

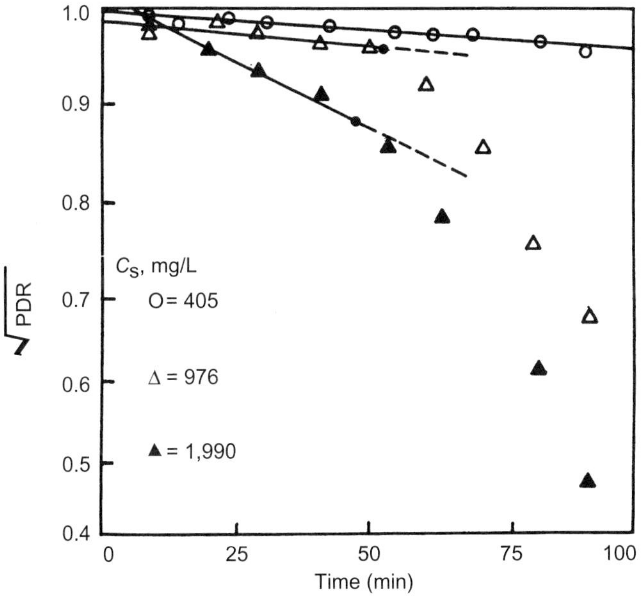

FIGURE 11.38 Early-time straight line is indicative of a "gradual pore blockage" damage mechanism during foreign solids invasion [18].

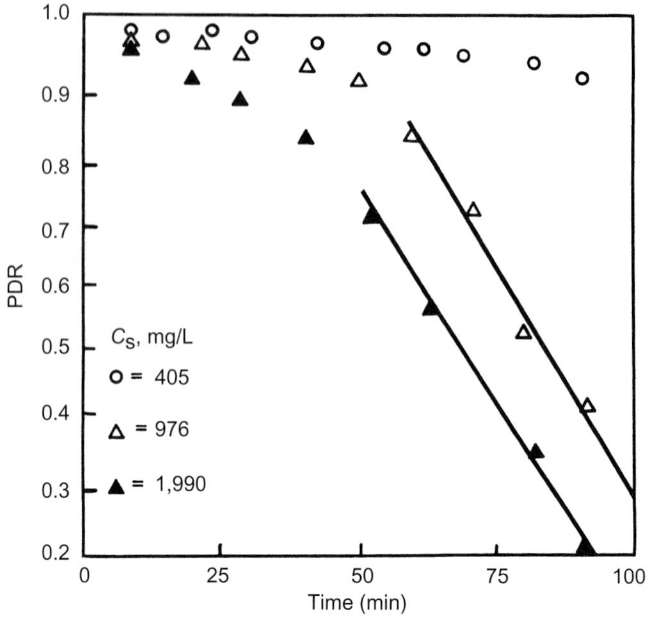

FIGURE 11.39 Late-time straight line is indicative of "single-pore blockage or screening" damage mechanism during foreign solids invasion [18].

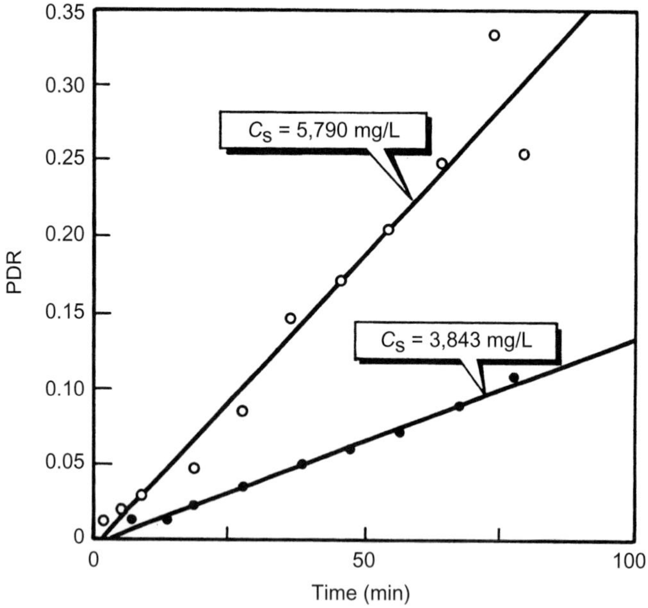

FIGURE 11.40 Diagnostic plot of "cake-forming damage mechanism" during foreign solids invasion [18].

TABLE 11.17 Completion Fluids Tested [18]

Fluid	TSS (ppm)	CST (s)	Salt Concentration (g/l)	Density (ppg)	Viscosity (cP)	pH
Calcium chloride	1.0	9.3	190	9.5	1.5	8.6
Sodium chloride	0.0	8.9	215	9.5	1.9	8.1
Ammonium nitrate	0.2	9.8	260	9.5	1.7	3.2
Ammonium nitrate + 20% methanol	0.3	10.0	320	9.5	1.9	3.1

Filtered through 0.4-μm filter.

TABLE 11.18 Mineral Composition of Core Samples [18]

Mineral	Concentration (%, w/w)
Quartz	56–68
Feldspar	11–23
Calcite	5–8
Dolomite	4–11
Illite/mica	3–7
Chlorite	3–5

straight lines corresponding to 0.5% and 1.0% contamination, which behave according to Eq. (11.24), confirm the existence of single-pore blockage mechanism. Figure 11.40 is a plot of k_o/k versus time of the same data corresponding to the two "straining" curves in Figure 11.37. Thus for very high levels of contamination (3% or more), the cake-forming mechanism is dominant and behaves according to Eq. (11.26). Figure 11.41 describes the effect of solids size and concentration on the gradual pore-blocking mechanism. Large particle sizes and concentrations shorten considerably the duration of this mechanism and, therefore, allow the more damaging single-pore blocking mechanism to dominate.

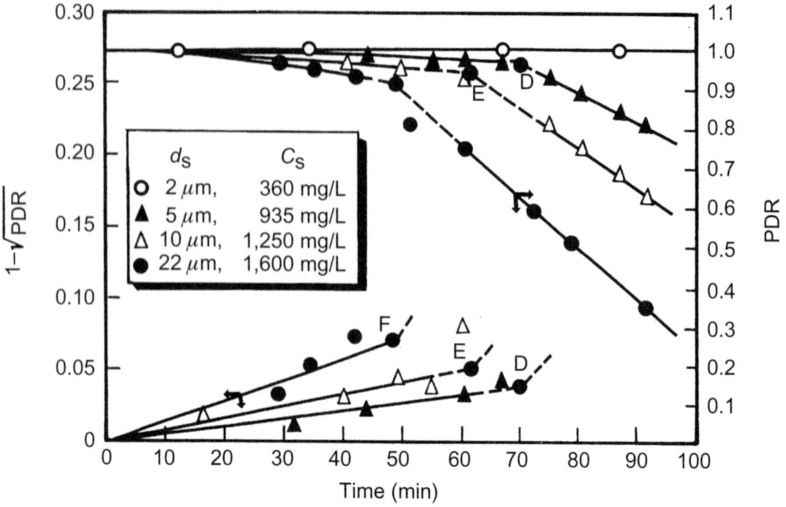

FIGURE 11.41 Effect of solids size and concentration on PDR [18].

Based on these results and other tests not included here, Wojtanowicz et al. made the following observations [18]:

1. Inasmuch as a solids-free completion fluid is virtually impossible to achieve, the gradual blocking mechanism is always present. This mechanism is indicated as follows:
   a. Parabolic-type permeability reduction.
   b. Deep invasion of particles.
   c. Steady decrease in the size of invading solids. The duration of this mechanism is a function of size and concentration of solid particles in the completion fluid.
2. A single-pore blockage or screening mechanism, which may produce irreversible permeability damage, is associated with the following factors:
   a. Steep and linear reduction in the formation permeability.
   b. Deep invasion of solid particles into the rock near the wellbore.
3. A cake-forming or cake-straining mechanism has the following three characteristics:
   a. Steep and linear reduction in permeability.
   b. Deep penetration of invading solids into the formation.
   c. Small solids passing through the damaged zone.

PERMEABILITY DAMAGE FROM FORMATION FINES

Formation fines may be either generated as chemical precipitates from the chemical reaction between the completion fluid and formation water, or released from the surface rock as the completion fluid interacts with the various clay and nonclay minerals present in the formation rock. Once mobilized, these fines can damage the permeability in a similar manner as foreign solids. The equations describing the permeability response to damage, however, are different. Assuming the solids are mobilized exponentially, by analogy to the decay equation and chemical reaction kinetics, Wojtanowicz et al. showed that when the size of mobilized fines is much smaller than the size of pores, a simultaneous gradual blocking and sweeping mechanism occurs [18]. Assuming the initial mass of mobile fines at the pore throats is negligible when compared to the mobile fines on the pore walls, the permeability response to this mechanism can be described by the following equation:

$$\sqrt{\frac{k}{k_o}} = 1 - a_{gs} t\, e^{-f_r t} \tag{11.28}$$

where

$$a_{gs} = \frac{f_c f_r M_{pi}}{q L_t A_{fo}} \tag{11.29}$$

where

M_{pi} = initial mass of fines on the pore surface, g
L_t = average length of the pore throat, cm
f_r = release coefficient, min^{-1}

Equation 11.28 can be written as follows:

$$\ln\left(\frac{1-\sqrt{k/k_o}}{t}\right) = -f_r t - \ln a_{gs} \tag{11.30}$$

Thus, a semilog plot of $(1-\sqrt{k/k_o})/t$ versus time should yield a straight line of slope $-f_r$. Such a plot generally yields two straight portions, as shown in Figure 11.42. The first straight line corresponds to the case where the concurrent mechanism is dominated by gradual pore blocking. The second straight line is for the case where the damage mechanism is dominated by pore sweeping. The slope of the second straight line is approximately five to six times the slope of the first straight line. The minimum permeability is reached at time $t = 1/f_r$.

The gradual pore-blocking and pore-sweeping mechanisms also can be analyzed separately in case the previous approach does not yield definite results. The general form of permeability response to these mechanisms is as follows:

$$\sqrt{\frac{k}{k_o}} = 1 + a_m(1 - e^{-f_r t}) \tag{11.31}$$

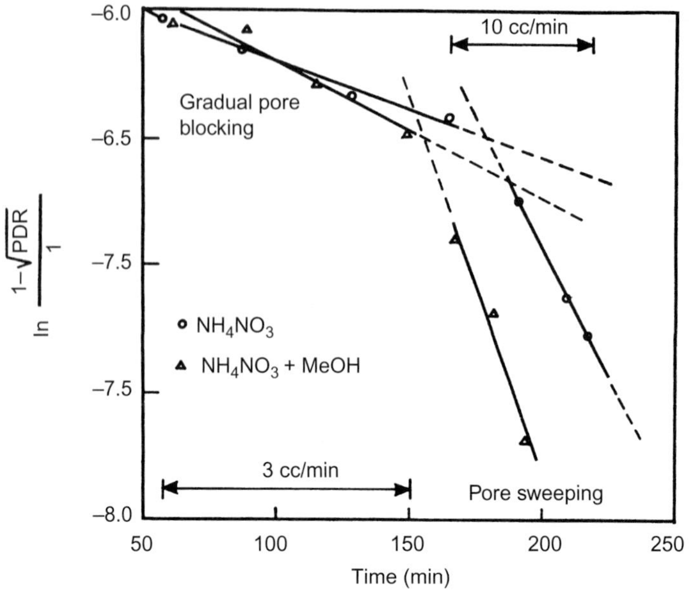

FIGURE 11.42 Combined effects of gradual pore-blocking and pore-sweeping mechanisms [18].

where, for gradual pore blocking, the constant a_m is given by:

$$a_m = -\frac{f_c M_{pi}}{A_{fo}\rho L_t} \tag{11.32}$$

and for pore sweeping:

$$a_m = \frac{M_{ti}}{A_{fo}\rho L_t} \tag{11.33}$$

M_{ti} is the initial mass of solids at the pore throat. Equation (11.30) can be written as:

$$\ln\left(1 + \frac{1 - \sqrt{k/k_o}}{a_m}\right) = -f_r t \tag{11.34}$$

Thus, a semilog plot of $[1 + (1 - \sqrt{k}/k_o)/a_m]$ versus time should yield a straight line of slope $-f_r$.

When the mobilized fines resulting from interaction of completion fluid and formation are within the range of the pore throat size, the single-pore blocking mechanism takes place and causes a reduction in permeability. When the size of mobilized fines is considerably greater than the pore throat size, however, the plugging or cake-forming mechanism occurs, which also causes permeability damage [18]. The permeability decline during single-pore blocking is:

$$\frac{k}{k_o} = 1 + a_{sb}(1 - e^{-f_r t}) \tag{11.35}$$

where a_{sb} is a group of measurable parameters:

$$a_{sb} = \frac{6C_{sc}M_{pi}d_p^4}{\pi d_c^3 \rho_s (32A_r C_s)^2} \tag{11.36}$$

Rearranging and taking the logarithm, Eq. (11.35) becomes:

$$\ln\left(1 + \frac{1 - k/k_o}{a_{sb}}\right) = -f_r t \tag{11.37}$$

The permeability response to plugging or cake forming due to the mobilization of fines is given by:

$$\frac{k_o}{k} = b_c + a_c(1 - e^{-f_r t}) \tag{11.38}$$

or

$$\ln\left(1 + \frac{b_c - k_o/k}{a_c}\right) = -f_r t \tag{11.39}$$

where

$$a_c = \frac{\alpha A_r K_o M_{pi}}{A_c^2 L_r} \tag{11.40}$$

$$b_c = \frac{A_r k_o R_r}{A_c L_r} \tag{11.41}$$

Figures 11.43 and 11.44 illustrate the permeability response to the interaction of four completion fluids, described in Table 11.17, and reservoir rock samples. Based on this study, Wojtanowicz et al. made the following observations [18]:

1. The amount of fines available for mobilization is constant for a particular formation rock, and not dependent on the type of completion fluids.
2. The mechanism by which fines are mobilized and captured is dependent on the type of completion fluid.
3. The level of compatibility of the completion fluid with a formation rock can be quantified by values of the release and capture coefficients f_r and f_c. Khilar and Fogler investigated the mechanism of water sensitivity of

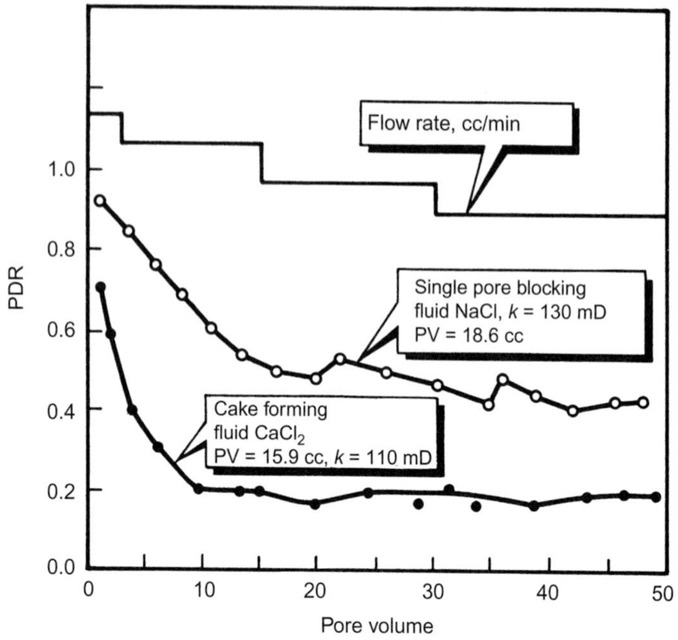

FIGURE 11.43 Behavior of PDR during the mobilization of fines in reservoir rock samples by NaCl and $CaCl_2$ [18].

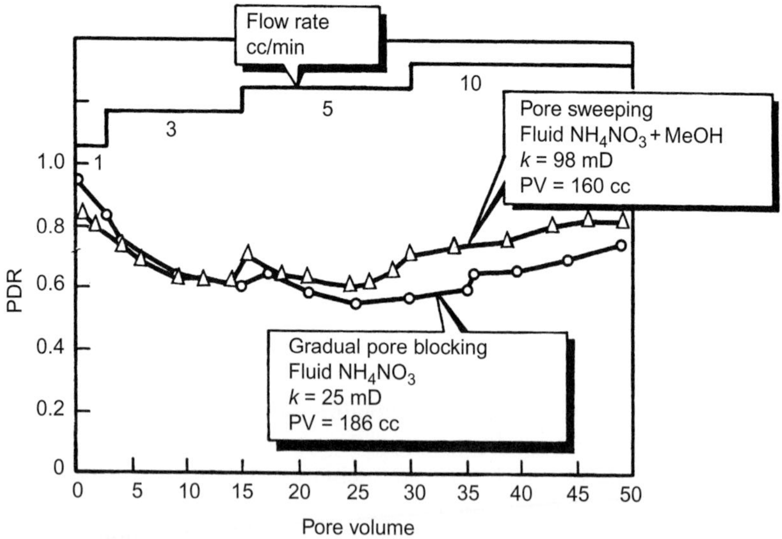

FIGURE 11.44 Behavior of PDR during the mobilization of fines in reservoir rock samples using two different well completion fluids [18].

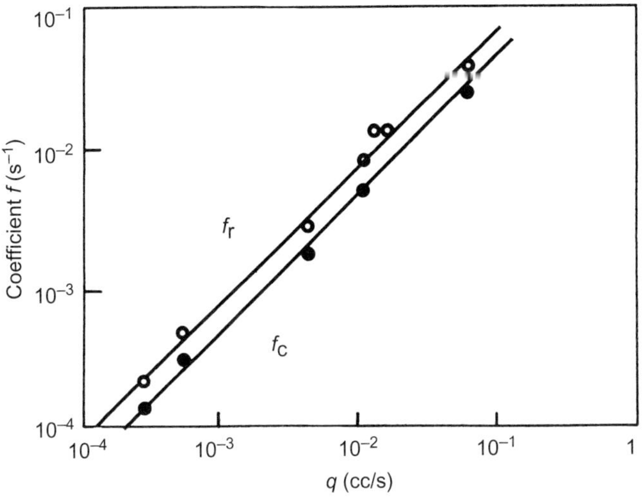

FIGURE 11.45 Variation of release and capture coefficients with flow rate [32].

Berea cores and found that these two coefficients were proportional to the fluid flow rate as shown in Figure 11.45 [32]. A curve fit of these data yields:

$$f_c = 0.385q^{0.965} \tag{11.42}$$

and

$$f_r = 0.640q^{0.970} \tag{11.43}$$

where the flow rate q is expressed in cm^3/s and the two coefficients in s^{-1}. The temperature affects the release coefficient and, consequently, the permeability damage ratio k/k_o. According to Figure 11.46, decreasing the temperature delays the onset of permeability damage and reduces the rate of decline of k/k_o. Figure 11.47 shows that the release coefficient f_r can be related to the temperature by an Arrhenius-type relationship. The capture coefficient f_c is independent of temperature [32].

EFFECT OF WATER QUALITY ON PERMEABILITY

The quality of water injected into a sandstone formation during a water flood or a water disposal project is affected by various types of contaminants, including suspended silts, clays, scale, oil, and bacteria, which can be a source of severe permeability damage and subsequent injectivity decline. In considering the effects of suspended solid particles on the rate of injectivity decline, Barkman and Davidson introduced a measure of "water quality"

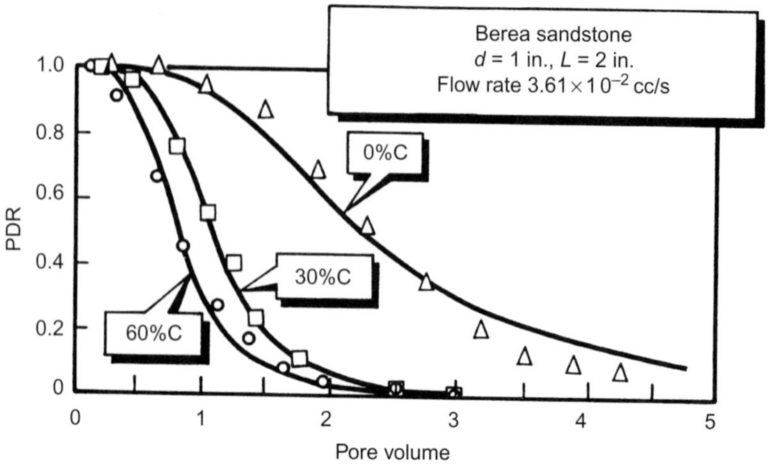

FIGURE 11.46 Effect of temperature on PDR [32].

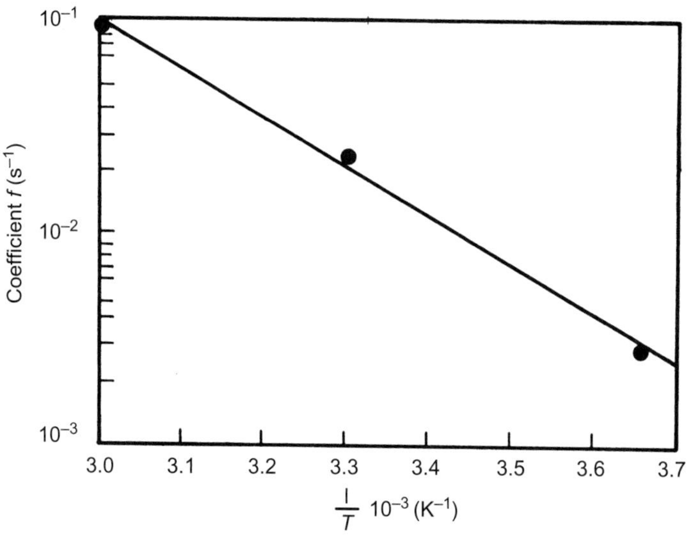

FIGURE 11.47 Effect of temperature on the release coefficient [32].

[22]. They defined it as the ratio of the concentration of suspended solids to the permeability of the filter cake formed by those solids. It can be obtained directly from the membrane or core filtration experiments. Filtration usually can reduce the concentration of suspended solids, but it is unlikely that sufficient water treatment can be achieved over long periods of time to prevent

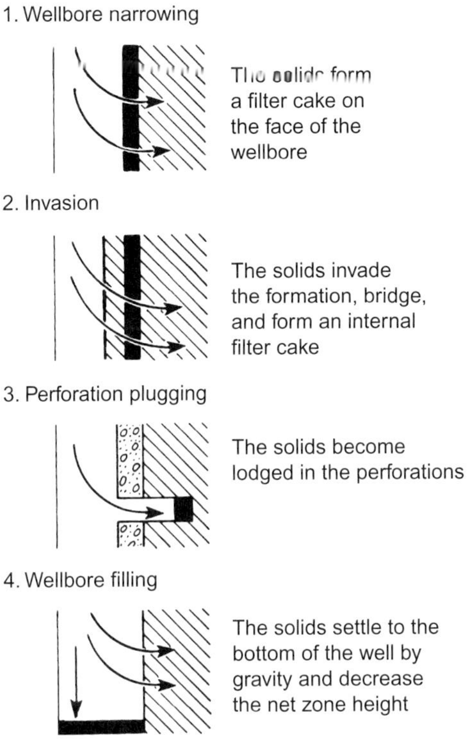

FIGURE 11.48 Types of formation damage caused by solids near the wellbore [22].

injectivity decline. A convenient way to determine how long an injector can be used before well stimulation or replacement is required is to calculate its "half-life," which is simply the time required for the injection rate to decrease to half of its initial value, and to estimate the rate and location of permeability damage. Barkman and Davidson investigated four mechanisms by which impairment of permeability can occur [22]:

1. Wellbore narrowing caused by the formation of a filter cake by solids on the face of the wellbore
2. Invasion by solids of the formation rock in the vicinity of the wellbore or skin zone, causing bridging and buildup of an internal filter cake
3. Perforation plugging as the solid particles lodge in the perforations
4. Wellbore fill-up as the solids settle to the bottom of the well by gravity, causing a decrease in the net zone thickness

Figure 11.48 gives a schematic representation of these types of permeability damage. Assuming that, in all four damage mechanisms, fluid is injected at a constant pressure, the flow state is a semi-steady state, and the flowing fluid is incompressible, Barkman and Davidson showed that, for each mechanism, the time required for the injection rate to decline to some fraction of its initial value can be expressed as follows:

$$t_\alpha = t_T G_{BD} \tag{11.44}$$

where t_T is the total time required to completely fill the wellbore volume with solids at the initial water injection rate i_{wi}, and is independent of the impairment mechanism:

$$t_T = \frac{\pi r_w^2 h \rho_c}{i_{wi} C_{sw} \rho_w} \tag{11.45}$$

where

r_w = wellbore radius
h = injection interval
C_{sw} = concentration of solids in water
ρ_c and ρ_w = densities of filter cake and water, respectively

The Barkman–Davidson function, G_{BD}, adjusts t_T according to which

EXTERNAL FILTER-CAKE BUILDUP

Assuming that:

1. the solids are trapped in the wellbore as in Figure 11.48,
2. there is a resistance to flow across the filter cake and across the formation, and
3. the filter cake resistance is a function of time as the cake builds, then the total pressure drop across the formation and filter cake is given by:

$$\Delta p = \frac{i_w \mu}{2\pi k_c h} \ln\left(\frac{r_{eD}^{k_c/k}}{r_{cD}}\right) \tag{11.46}$$

where

i_w = water injection rate
k_c = filter cake permeability
k = formation permeability
r_e = external radius of the formation or injection pattern
r_c = radial distance to the face of filter cake
$r_{eD} = r_e/r_w$
$r_{cD} = r_c/r_w$

From a material balance on the solid particles in the wellbore, one obtains:

$$i_w = \left(\frac{2\pi r_w^2 h \rho_c}{C_{sw} \rho_w}\right) r_{eD} \frac{dr_{cD}}{dt} \tag{11.47}$$

Substituting Eq. (11.47) into Eq. (11.46) and integrating over time gives the fractional life of a well with an external filter cake:

$$t_{E\alpha} = t_T G_{BDE} \tag{11.48}$$

where

$$G_{BDE} = 1 + \frac{0.5}{\ln\theta} - \left[\frac{1}{\alpha} + \frac{0.5}{\ln\theta}\right]\theta^{2(1-1/\alpha)} \tag{11.49}$$

$$\alpha = \frac{i_w}{i_{wi}} \tag{11.50}$$

and

$$\theta = r_{eD}^{k_c/k} \tag{11.51}$$

At half-life, i.e., the time at which the injection rate is 50% of its initial value or $\alpha = 0.5$, the function G_{BDE} can be approximated by 1 for $k_c/k > 0.05$ and, therefore, $t_E = t_T$. For $k_c/k < 0.05$:

$$G_{BDE} = 3\left(\frac{k_c}{k}\right)\ln\frac{r_e}{r_w} \tag{11.52}$$

and the corresponding half-life is given by:

$$t_{E1/2} = 3t_T\left(\frac{k_c}{k}\right)\ln\left(\frac{r_e}{r_w}\right) \tag{11.53}$$

For $k_c/k > 0.05$, the half-life of a typical injection well, where $r_e/r_w = 1{,}800$ and $r_w = 4$ in., and in which impairment from an external filter cake is dominant, is expressed as:

$$t_{E1/2} = \frac{340}{(i_{wi}/h)C_{sw}} \tag{11.54}$$

and for $k_c/k < 0.05$:

$$t_{E1/2} = \frac{7{,}600}{(C_{sw}/k_c)(i_{wi}/h)k} \tag{11.55}$$

where C_{sw} is expressed in ppm; k and k_c in mD; i_{wi} in bbl/d; h in ft; and time t in years.

A key finding of Barkman and Davidson is that the water quality ratio (WQR), C_{sw}/k_c, can be used to calculate the rate of formation impairment. For instance, if the permeability of the deposited cake, k_c, is small compared with the formation permeability, k, the WQR, from Eq. (11.55), is given by:

$$\text{WQR} = \frac{C_{sw}}{k_c} = \left(\frac{7{,}600}{t_E}\right)\frac{1}{(i_{wi}/h)k} \tag{11.56}$$

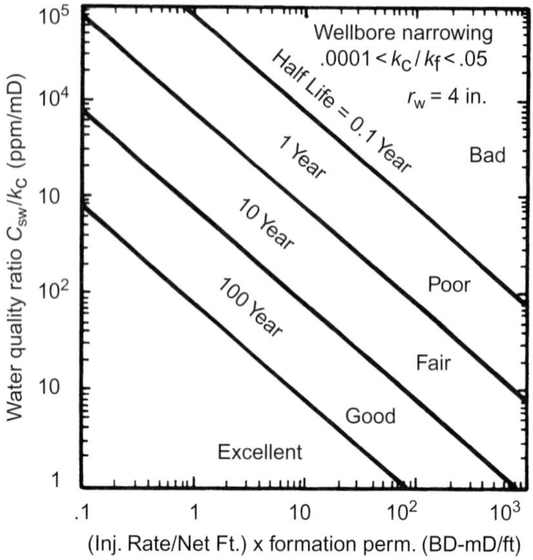

FIGURE 11.49 Qualitative scale of water quality for the case of wellbore narrowing with a low k_c assuming $r_e/r_w = 1{,}800$ [22].

which can be written as:

$$\log(\mathrm{WQR}) = -\log\left(\frac{i_{\mathrm{wi}}k}{h}\right) + \log\left(\frac{7{,}600}{t_{\mathrm{E}}}\right) \tag{11.57}$$

Thus, a plot of WQR versus $i_{wi}k/h$ on a log–log graph should yield a straight line of slope -1, as shown in Figure 11.49. This figure shows a qualitative scale of water quality (good, fair, poor, and bad) based on the order of magnitude of the typical injector half-life. A quantitative measure of water quality can be obtained by taking the derivative of Eq. (11.56) with respect to either (i_{wi}/h) or t_E. Figure 11.50 is a plot of C_{ws} in ppm versus i_{wi}/h for the case where k_c is larger than k (Eq. (11.54)).

INTERNAL FILTER-CAKE BUILDUP

Impairment from invasion occurs in two main stages:

1. Solid fines are transported and deposited, because of gravity and surface forces, causing restriction of the pore throats.
2. Bridging occurs, which results in a buildup of an internal filter cake within the formation in the vicinity of the wellbore.

Assuming that the depth of invasion and, consequently, the extent of formation impairment depends only on fluid velocity, pore throat sizes of the

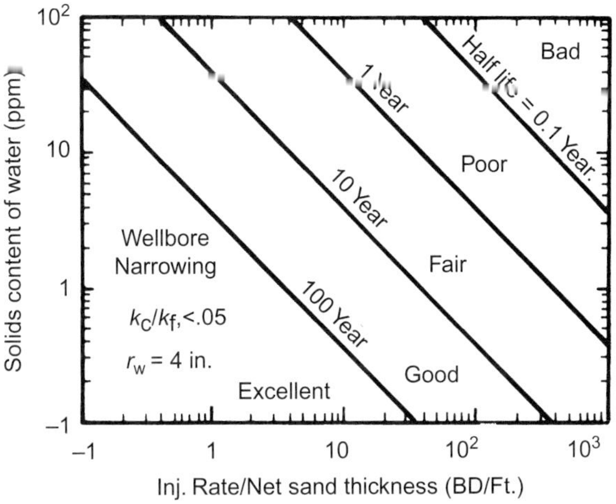

FIGURE 11.50 Half-life of a typical injection well ($r_e/r_w = 1{,}800$) for the case of wellbore narrowing with high k_c [22].

rock, and size of invading fines, Barkman and Davidson found the following:

1. In systems that have a broad distribution of pore and solid particle sizes, invasion may last for a long time before a filter cake forms on the rock face.
2. Bridging will occur unless $r_s/r_p < 0.1$, where r_s is the mean radius of the solid particles and r_p is the mean pore radius. This corresponds to a permeability ratio of $k_c/k < 10^{-4}$, assuming Kozeny's equation is valid.

Inasmuch as formation damage from invasion is caused by the growth of internal filter cakes, the total pressure drop across the formation and filter cake can be expressed as:

$$\Delta p_t = \frac{i_w \mu}{2\pi k_c h}\left[\frac{k_c \phi}{k}\ln\frac{r_e}{r_w} - s\right] \tag{11.58}$$

where the skin factor s is given by:

$$s = \left(1 - \frac{k_c}{k}\right)\ln\frac{r_c}{r_b} \tag{11.59}$$

and r_b is the radius at which solid particles bridge pores and form internal filter cakes. The permeability in the damaged zone is approximated by the product ϕk_c. This approximation is valid only if the formation pore channels are blocked by the filter cake, which is not often the case. This approximation, however, should be useful when only a qualitative estimate of permeability damage is necessary. A material balance on the solids yields:

$$i_w = \left(\frac{2\pi r_b^2 \phi h \rho_c}{C_{sw}\rho_w}\right) r_{cD} \frac{dr_{cD}}{dt} \tag{11.60}$$

where $r_{cD} = r_c/r_b$. Combining Eqs. (11.58) and (11.60) and integrating in time, one obtains an equation of the following general form:

$$t_{I\alpha} = t_T G_{BDI} \tag{11.61}$$

where G_{BDI} is the Barkman–Davidson function for impairment from invasion and subsequent formation of an internal filter cake:

$$G_{BDI} = \left(\frac{\phi r_b}{r_w}\right) 2\left[1 + \frac{1 - k_c/k}{2\ln\theta} - \left(\frac{1}{\alpha} + \frac{1 - k_c/k}{2\ln\theta}\right)\theta^\beta\right] \tag{11.62}$$

where

$$\beta = \frac{2(\alpha - 1)}{\alpha(1 - k_c/k)} \tag{11.63}$$

At half-life and for small filter cake permeability such that $k_c/k < 0.10$, the function G_{BDI} can be approximated by:

$$G_{BDI\,1/2} = 3\left(\frac{\phi r_b}{r_w}\right)^2 \frac{k_c}{k} \ln\frac{r_e}{r_w} \tag{11.64}$$

Thus, the half-life of formation impairment from invasion is:

$$t_{I\,1/2} = 3t_T\left(\frac{\phi r_b}{r_w}\right)^2 \frac{k_c}{k} \ln\frac{r_e}{r_w} \tag{11.65}$$

Formation damage from invasion can be related to that from wellbore narrowing, at half-life time, by combining Eqs. (11.53) and (11.65) for $k_c/k < 0.05$:

$$t_{I\,1/2} = \left(\frac{\phi r_b}{r_w}\right)^2 t_{EI/2} \tag{11.66}$$

The predicted half-lives of the two impairment mechanisms will be the same when:

$$\frac{\phi r_b}{r_w} = 10 \tag{11.67}$$

Because filtrate invasion can be as deep as 8 ft or more, as shown in Figure 11.51, the half-life of impairment from invasion, $t_{I\,1/2}$, generally will be greater than the half-life of impairment from wellbore narrowing,

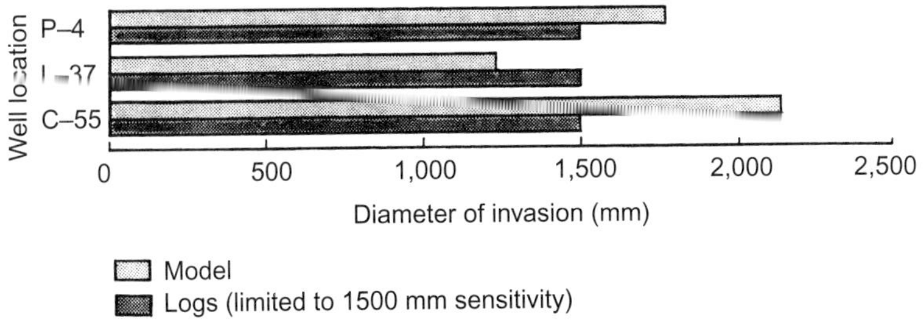

FIGURE 11.51 Depth of invasion for McKenzie Delta wells [39].

especially in high-permeability clean sandstones [39]. For instance, if $\phi = 0.20$, $r_w = 0.5$ ft, and $r_b = 10$ ft:

$$t_{I1/2} = \left(\frac{0.20 \times 10}{0.5}\right)^2 t_{EI/2} = 16 t_{EI/2}$$

The severity of formation damage from invasion depends on the sensitivity of the formation to the filtrate. Although high-permeability clean sandstones, i.e., not containing clays that can be dispersed and/or swollen, undergo more invasion than the low-permeability ones, they usually are not too severely damaged (40% or less permeability reduction) when their connate water is chemically compatible with the invading fluid [38].

INJECTIVITY DECLINE FROM PLUGGING OF PERFORATIONS

Impairment from perforation plugging occurs when the solid particles become lodged in the perforations, as shown in Figure 11.52. The extent of this impairment mechanism is a function of particle size. Plugging of perforations will occur unless the injected particles are so small that they can be transported by suspension through the rock matrix. The existence of near-wellbore fractures minimizes this impairment mechanism because they allow the solid particles to bypass the perforations [22]. Assuming that the filter cake forms within the perforation, Barkman and Davidson found that the total pressure drop, Δp, is constant and equal to the sum of the pressure drop across the filter cake and the pressure drop across the formation, thus:

$$\Delta p = \frac{i_w \mu}{2\pi h k_c}\left[\frac{k_c}{k} \ln \frac{r_e}{r_w} + \frac{8 L_p}{n_p d_p}\right] \tag{11.68}$$

where n_p is the number of perforations per unit length of sand interval, d_p is the diameter of perforation, and L_p is the filter-cake thickness in perforation.

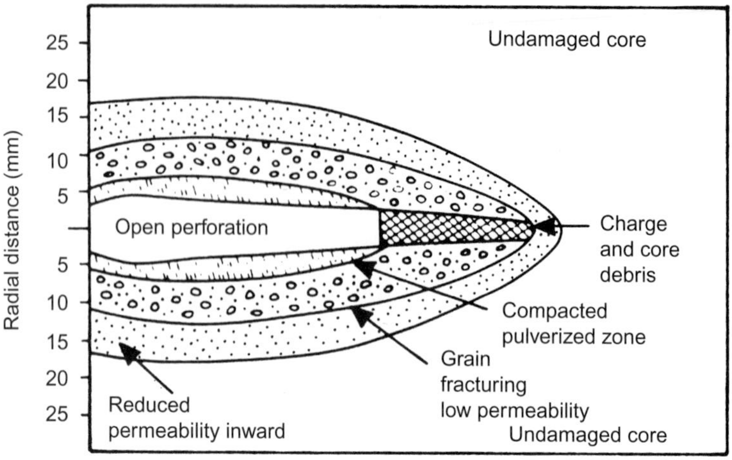

FIGURE 11.52 Types of formation damage near perforations [3].

A material balance on solid particles along the filter-cake thickness in perforation yields:

$$i_w = \left(\frac{\pi n_p h \rho_c d_p^2}{4C_{sw}\rho_w}\right)\frac{dL_p}{dt} \tag{11.69}$$

Combining Eqs. (11.68) and (11.69) and integrating with respect to time yields the time at which injection rate is a fraction of the form:

$$t_{PP\alpha} = t_= G_{BDPP} \tag{11.70}$$

where the Barkman–Davidson impairment function for perforation plugging is expressed as:

$$G_{BDPP} = \left(\frac{1}{\alpha^2} - 1\right)\frac{k_c}{k}\left(\frac{n_p d_p^2}{8r_w}\right)\ln\frac{r_e}{r_w} \tag{11.71}$$

The half-life approximation of G_{BDPP} is:

$$G_{BDPP1/2} = \frac{3}{64}\left(\frac{k_c}{k}\right)\left(\frac{n_p d_p^2}{r_e}\right)^2 \ln\frac{r_e}{r_w} \tag{11.72}$$

Thus, the half-life of formation damage from perforation plugging is:

$$t_{PP1/2} = \frac{3t_T}{64}\left(\frac{k_c}{k}\right)\left(\frac{n_p d_p^2}{r_w}\right)^2 \ln\frac{r_e}{r_w} \tag{11.73}$$

Assuming average parameters, Barkman and Davidson found that, because of the smaller injection surface area offered by perforations, formation damage in perforated completion occurs at a faster rate than the formation damage from wellbore narrowing in the open-hole completion. Consequently, the half-life of perforated completion is considerably smaller than that of an open-hole completion. This can be demonstrated by comparing Eqs. (11.53) and (11.73):

$$t_{\mathrm{PP}1/2} = \left(\frac{n_{\mathrm{p}} d_{\mathrm{p}}^2}{8 r_{\mathrm{w}}}\right)^2 t_{\mathrm{E}1/2} \tag{11.74}$$

For instance, if $n_{\mathrm{p}} = 4$ shots/ft, $d_{\mathrm{p}} = 0.4$ in., and $r_{\mathrm{w}} = 5$ in., then:

$$\left(\frac{n_{\mathrm{p}} d_{\mathrm{p}}^2}{8 r_{\mathrm{w}}}\right)^2 = \left[\frac{4 \times (0.4/12)^2}{8 \times (5/12)}\right]^2 = 1.77 \times 10^{-6}$$

thus

$$t_{\mathrm{PP}1/2} = 1.77 \times 10^{-6} t_{\mathrm{E}1/2}$$

When injecting into a perforated well, the water quality requirements must be more stringent. Figure 11.53 illustrates the effect of poor water quality (high solids content) on the formation permeability in perforated wells [22].

IMPAIRMENT FROM WELLBORE FILL-UP

Formation damage from wellbore fill-up occurs when large solid particles settle to the bottom of the well by gravity and reduce the injection interval, as illustrated in Figure 11.48. Assuming all solid particles fall to the bottom of the well and fluid flow through the settlement of solid particles is negligible because of its negligible permeability, the total pressure drop across the formation is given by the following equation:

$$\Delta p = \frac{i_{\mathrm{w}} \mu}{2\pi k h} \ln \frac{r_{\mathrm{e}}}{r_{\mathrm{w}}} \tag{11.75}$$

A material balance on the solids along the injection interval h gives:

$$i_{\mathrm{w}} = \left(\frac{\pi r_{\mathrm{w}}^2 \rho_{\mathrm{c}}}{C_{\mathrm{sw}} \rho_{\mathrm{w}}}\right) \frac{\mathrm{d}h}{\mathrm{d}t} \tag{11.76}$$

Equating Eqs. (11.75) and (11.76) and integrating in time yields:

$$t_{\mathrm{WF}\alpha} = t_{\mathrm{T}} G_{\mathrm{BDWF}} \tag{11.77}$$

where

$$G_{\mathrm{BDWF}} = \ln \frac{1}{\alpha} \tag{11.78}$$

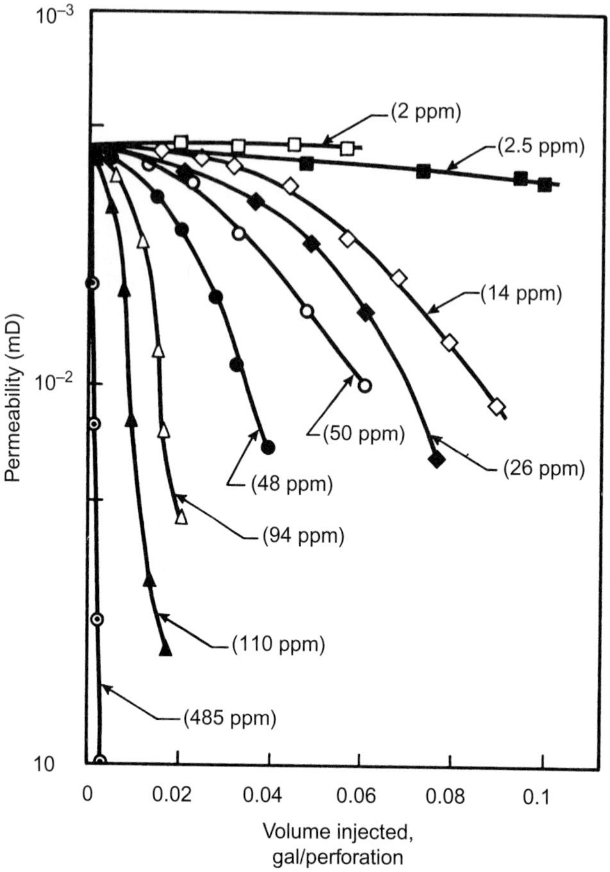

FIGURE 11.53 Effect of poor water quality on permeability in perforated wells [38].

G_{BDWF} is the Barkman–Davidson function for wellbore fill-up at half-life time, i.e., $\alpha = 0.5$ and $G_{BDWF} = 0.69$. Thus, the injector half-life during wellbore fill-up is given by:

$$t_{WF1/2} = 0.69 t_{T1/2} \tag{11.79}$$

Impairment from wellbore fill-up can be reduced by injecting water at a low rate, because from Eq. (11.45), t_T is inversely proportional to i_w.

MEMBRANE FILTRATION TESTS

Barkman and Davidson extended the surface filtration theory of McCabe and Smith to include the effect of solid particles invasion and derived a simple equation for calculating the WQR (C_{sw}/k_c) [40]. During a linear filtration

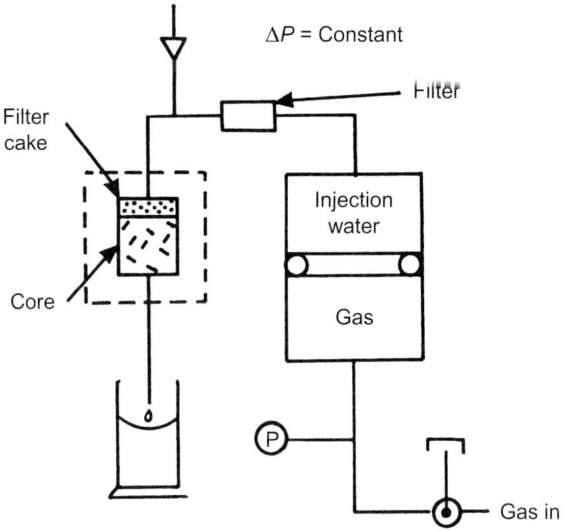

FIGURE 11.54 Schematic diagram of experimental apparatus used in standard linear filtration tests [2].

test, as shown in Figure 11.54, the total pressure drop is the sum of the pressure drop across the filter cake and that across the filter membrane [22]:

$$\Delta p = q\mu\left[\frac{L_c}{k_c A_c} + \frac{L_m}{k_m A_m}\right] \tag{11.80}$$

where L, k, and A are length, permeability, and area, respectively, of the filter cake (subscript c) and filter medium (subscript m; core or membrane). The flow rate q is obtained from the following equation:

$$q = \frac{A_c \rho_c}{C_{sw}\rho_w}\frac{dL_c}{dt} \tag{11.81}$$

Substituting Eq. (11.81) into Eq. (11.80) and integrating (between 0 and L_c, and t and t_B), assuming a constant pressure drop, one obtains:

$$\Delta p = \frac{A_c \mu \rho_c}{C_{sw}\rho_w}\left[\frac{L_m L_c}{k_m A_m} + \frac{L_c^2}{2k_c A_c}\right]\frac{1}{t - t_B} \tag{11.82}$$

where t_B is the time at which bridging occurs. The relationship between the cumulative volume of filtrate after bridging occurs, V_B, and the thickness of the cake, L_c, is:

$$V_B = V - \left(\frac{A_c \rho_c}{C_{sw}\rho_w}\right)L_c \tag{11.83}$$

where V is the cumulative throughput volume of filtered solutions. Solving this equation for L_c and substituting into Eq. (11.82) yields the following relationship between V and the total testing time t:

$$V = [D_1^2 + D_2(t - t_B)]^{0.5} + (V_B - D_1) \tag{11.84}$$

where

$$D_1 = \left(\frac{k_c}{C_{sw}}\right)\left(\frac{A_c^2}{A_m}\right)\left(\frac{\rho_c}{\rho_w}\right)\frac{L_m}{k_m} \tag{11.85}$$

and

$$D_2 = \left(\frac{k_c}{C_{sw}}\right)\left(\frac{\rho_c}{\rho_w}\right)\frac{2A_c^2\,\Delta p}{\mu} \tag{11.86}$$

Even though invasion of solid particles takes place during the early part of the test, bridging will occur if V is large enough [22]. When the testing time, t, is much larger than the bridging time, t_B, and $D_2 t \gg D_1^2$, Eq. (11.84) becomes:

$$V = m\sqrt{t} + b \tag{11.87}$$

where

$$m = \sqrt{D_2} \tag{11.88}$$

$$b = V_B - \frac{m^2}{2q_B} \tag{11.89}$$

Equation (11.87) indicates that a plot of the cumulative throughput volume V versus $\sqrt{t}$ should yield a straight line portion, when $t \gg t_B$, of slope m and intercept b. Knowing the slope, the WQR, (C_{sw}/k_c), can be calculated by the following equation:

$$\text{WQR} = \frac{C_{sw}}{k_c} = \frac{1}{m^2}\left(\frac{2\rho_w A_c^2\,\Delta p}{\mu\rho_c}\right) \tag{11.90}$$

Once the WQR is obtained and C_{sw} is determined by weighting the deposited solids, the filter cake permeability, k_c, can be calculated. The intercept b is used to determine whether or not invasion has occurred: (1) if $b < 0$, there is no invasion, as shown in Figure 11.55; and (2) if $b > 0$, such as in Figures 11.56 and 11.57, then solids invasion has occurred. In this case, the cumulative volume and filtration rate at bridging time, V and q_B, can be calculated by trial and error from the plot of V versus $\sqrt{t}$ and Eq. (11.87), which, at bridging time, becomes:

$$V_B = m\sqrt{t_B} + b \tag{11.91}$$

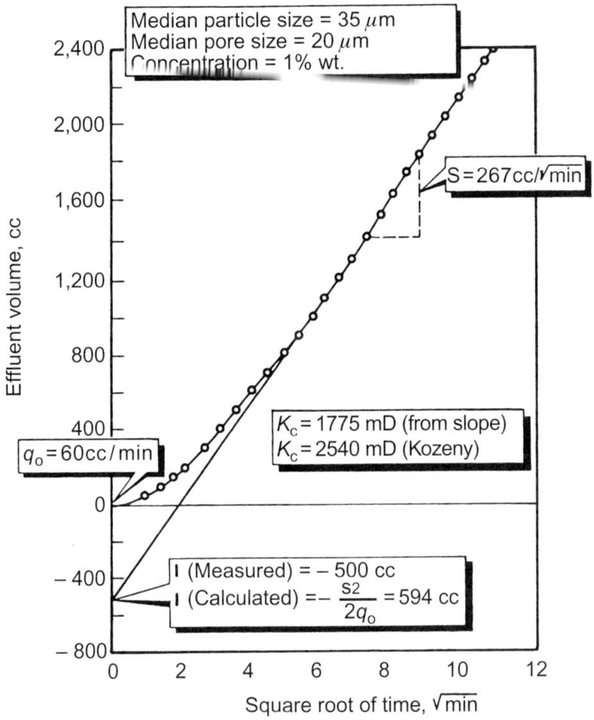

FIGURE 11.55 Typical filtration curve without invasion in Berea cores [22].

Then, q_B is obtained from Eq. (11.89). For routine field tests, Barkman and Davidson suggested the use of a membrane filter of 47 mm diameter and using a differential test pressure of 20 psi. Then, V (in cm^3) is plotted versus $\sqrt{t}$, where time is in minutes, and the slope of the straight line portion is determined. This slope is used in Figure 11.58 to calculate the WQR, C_{sw}/k_c.

The shape of the V versus $\sqrt{t}$ depends on the properties of the suspended solid particles and of the filter medium (core or membrane).

There are three possible shapes:

1. Figure 11.55 is a typical filtration curve without invasion, as indicated by a negative intercept. This may indicate that the median particle size is greater than the median pore size.
2. Figure 11.56 is an example of filtration curve with invasion, as indicated by a positive intercept. The median particle size in this case is smaller than the median pore size.
3. Figure 11.57 is a typical S-shaped filtration curve with invasion. Figures 11.58 and 11.59 illustrate the effect of membrane pore sizes and pressure tests on the shape of filtration curves. Figures 11.55–11.57

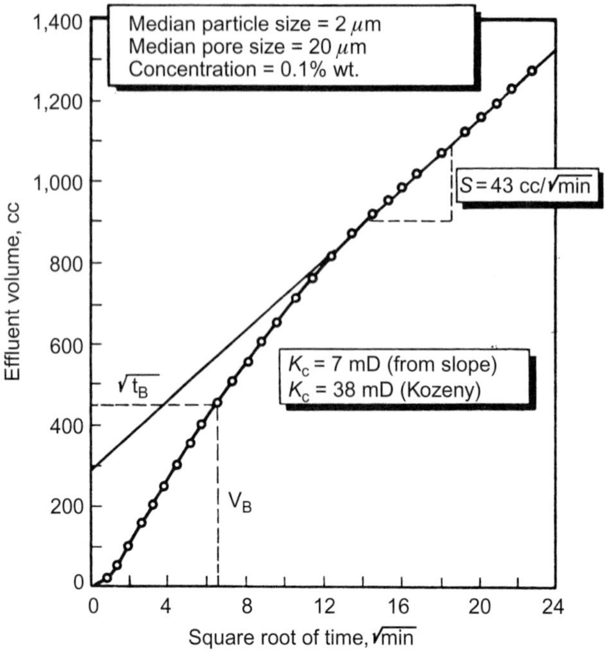

FIGURE 11.56 Typical filtration curve with invasion in Berea cores [22].

indicate the importance of running the filtration test long enough to reach the proper straight line. To achieve this, the test should be run until $\sqrt{t} \geq 2\sqrt{t_B}$. If short tests are required, then the test should be run at pressures higher than 20 psi [22].

CORE FILTRATION TESTS

Membrane filtration testing is an attractive way of describing phenomena occurring in cake filtration because of its simplicity and repeatability. Vetter et al. and Ershagi et al., however, investigated particle invasion and related injectivity decline and concluded that core flow tests yielded results that are more representative of field conditions than membrane filtration tests [42,43]. Todd et al. investigated the influence of core plug preparation on particle invasion and found that the decline of permeability in the case of sawn-faced core is generally sharper than that of broken-faced core of similar characteristics [44]. Broken-faced core can be obtained by fracturing a core at a plane perpendicular to the longitudinal axis of the core. The length of the core is measured from the main plane of the broken-face (inlet) to the outlet (sawn or broken face). The broken-faced core, which represents more

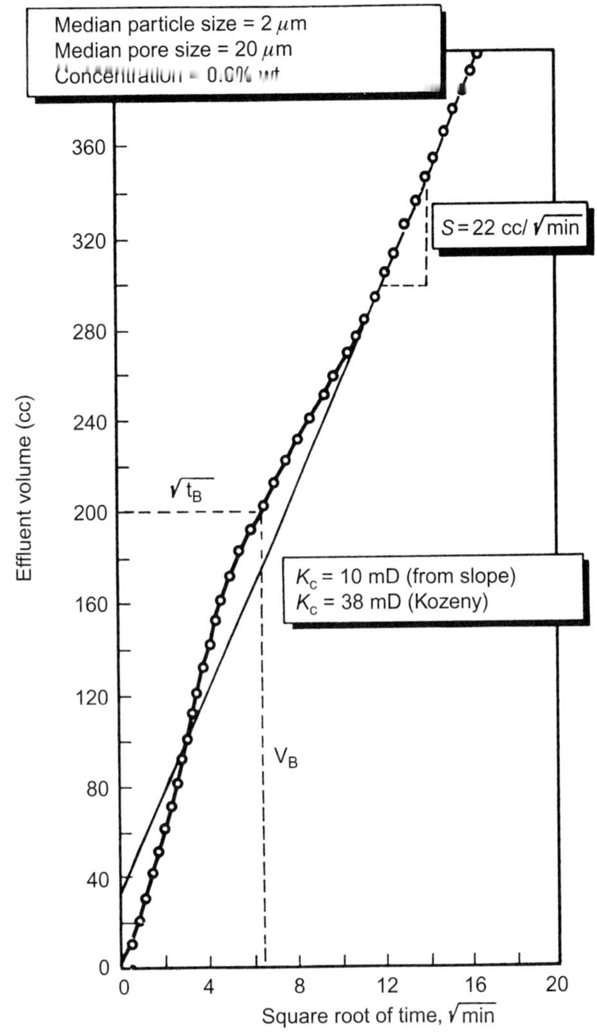

FIGURE 11.57 Typical S-shaped filtration curve with invasion in Berea cores [22].

accurately the well sandface, generally showed little or no external accumulation of filter cake. After running numerous core tests, Todd et al. concluded there is a very definite influence of the nature of the inlet face on the entrainment and deposition of solid particles [44]. It is possible, however, that fracturing a core to obtain a broken face may introduce microfractures parallel to the plane of fracturing, which could explain the lack of external filter cake.

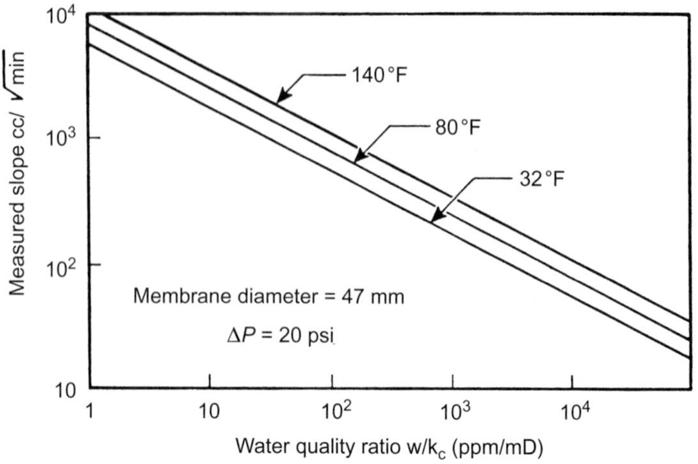

FIGURE 11.58 Finding WQR directly from the slope of a standard filtration test using a membrane [22].

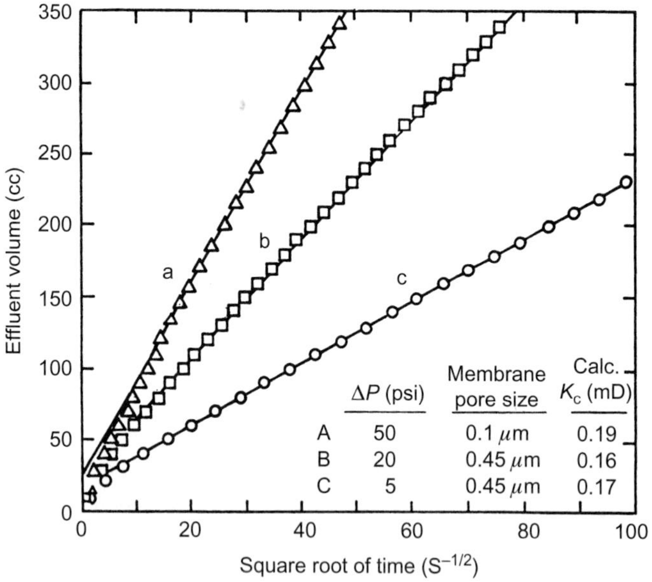

FIGURE 11.59 Effects of membrane pore sizes and pressure on k_c value [22].

External Filter-Cake Formation

Eylander used the results of laboratory core flood tests and Coulter Counter analysis of the injection water to modify Barkman and Davidson's fractional life expressions [45]. A material balance between the solids in the injection water and the cake yields:

$$AL_c\rho_s(1-\phi_c) = (AL_c\rho_c + V)\rho_w\left(\frac{C_{sw}}{1-C_{sw}}\right) \tag{11.92}$$

where V is the cumulative injected volume, ρ_s is the density of solids, and A is the cross-sectional area of the core or formation subjected to water injection. Solving Eq. (11.92) for L_c and substituting it into Eq. (11.80) (assuming $A = A_c = A_m$), and then taking the derivative of V with respect to time yields:

$$\frac{\Delta p}{dV/dt} = m_eV + b_e \tag{11.93}$$

where

$$m_e = \frac{\mu\rho_w}{A^2\rho_s(1-\phi_c)}\left(\frac{C_{sw}}{k_c}\right) \tag{11.94}$$

and

$$b_e = \frac{\mu L_m}{Ak} \tag{11.95}$$

Assuming constant Δp, m_e, and b_e and integrating Eq. (11.93) over time and volume until bridging occurs, i.e., $t = t_B$ and $V = V_B$, will give an equation similar to Eq. (11.84), which is then approximated by Eq. (11.87). The latter, in turn, is used to calculate the WQR. Inasmuch as the pressure difference, Δp, generally fluctuates during the filtration tests, Eylander found that direct use of Eq. (11.93) gives more representative results [45]. A Cartesian plot of $\Delta p/(dV/dt)$ versus V should yield a straight line having a slope m_e and intercept b_e, as shown in Figure 11.60. The intercept is used to calculate the matrix permeability, k. The slope is used to calculate the filter-cake porosity, ϕ_c, from Eq. (11.94) by substituting the Carman–Kozeny equation for permeability for the parameter k_c, i.e.:

$$k_c = \frac{\phi_c^3}{5S_{GV}^2(1-\phi_c)^2} \tag{11.96}$$

where the specific surface area of solids, S_{GV} per unit grain volume, is obtained from Coulter counter analysis of the injected water:

$$S_{GV} = (36\pi)^{1/3}\frac{\sum n_i(V_i^{2/3})}{\sum n_i V_i} \tag{11.97}$$

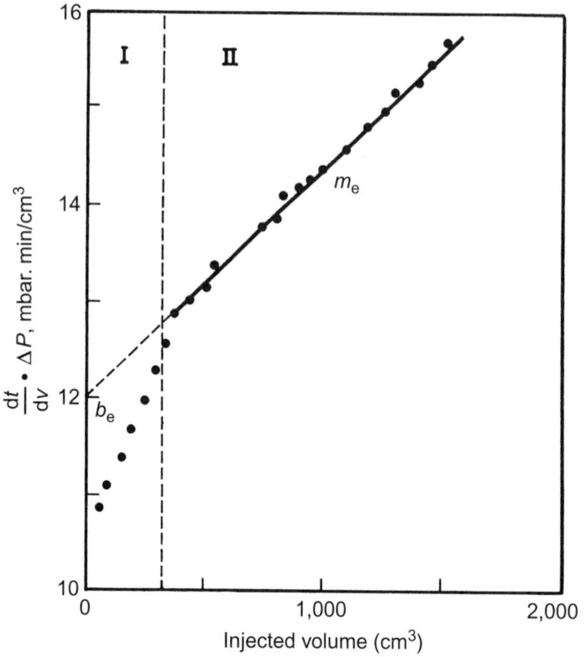

FIGURE 11.60 Core filtration test showing formation of external filter cake [43].

where n_i and V_i are the number and volume of particles counted in the ith channel of the Coulter counter. Combining Eqs. (11.94) and (11.96) gives:

$$\phi_c^3 + M\phi_c - M = 0 \tag{11.98}$$

where

$$M = \frac{\mu\rho_w C_{sw}}{A^2\rho_s}\left(\frac{5S_{GV}^2}{m_e}\right) \tag{11.99}$$

Equation (11.98) has only one real solution (the other two are imaginary):

$$\phi_c = \left[\frac{M}{2} + \left(\frac{M^3}{9} + \frac{M^2}{4}\right)^{0.5}\right]^{1/3} + \left[\frac{M}{2} - \left(\frac{M^3}{9} + \frac{M^2}{4}\right)^{0.5}\right]^{1/3} \tag{11.100}$$

Knowing ϕ_c, the filter-cake permeability can be calculated from Eq. (11.96).

Assuming:

1. the total bottomhole pressure drop is maintained constant during the injection period,
2. the flowing fluid is incompressible,
3. the injection zone is homogeneous, and

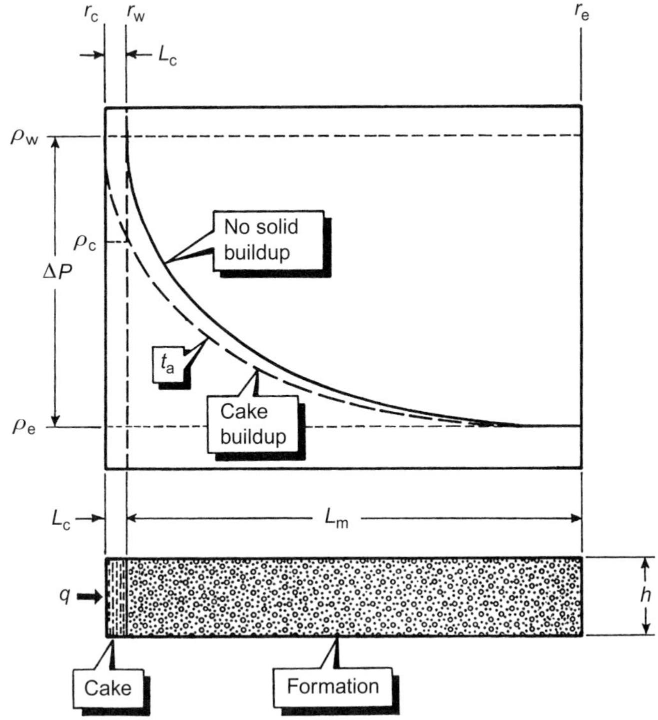

FIGURE 11.61 Effect of external filter cake on reservoir pressure [43].

4. k_c and ϕ_c are constant during the injection period, Eylander showed that the fractional life of an injector with an external filter cake, $t_{E\alpha}$, is given by:

$$t_{E\alpha} = \frac{\pi r_w^2 h(1-\phi_c)}{i_{wi} C_{sw}} \left(\frac{k_c}{k}\right)\left(\frac{1}{\alpha^2} - 1\right) \ln \frac{r_e}{r_w} \tag{11.101}$$

where C_{sw} is the fractional solids concentration in water. The cumulative injected volume at time $t_{E\alpha}$, can be expressed as follows:

$$V_{E\alpha} = \left(\frac{2i_{wi}}{(1/\alpha)+1}\right) t_{E\alpha} \tag{11.102}$$

At half-life, i.e., $\alpha = i_w/i_{wi} = 0.5$, Eq. (11.102) becomes:

$$V_{E1/2} = \left(\frac{2i_{wi}}{3}\right) t_{E1/2} \tag{11.103}$$

Figure 11.61 shows that external filter-cake formation results in a rapid pressure decline and, consequently, a rapid reduction of well injectivity.

Internal Filter-Cake Formation

Equating the total pressure drop across the porous sample to the sum of the pressure drops across the particle deposition zone, the filter cake, and the uncontaminated matrix, and assuming that the filter-cake thickness, L_c, can be approximated by:

$$L_c = \frac{VC_{sw}\rho_w}{A\rho_s(1-\phi_c)\phi} \tag{11.104}$$

where ϕ is the porosity of uncontaminated portion of the porous medium, the following expressions can be derived from a material balance:

$$\frac{\Delta p}{dV/dt} = m_i V + b_i \tag{11.105}$$

where

$$b_i = \frac{\mu L_m}{A k_m} \tag{11.106}$$

and

$$m_i = \frac{\mu\rho_w C_{sw}}{A^2\rho_s(1-\phi_c)\phi}\left(\frac{1}{k_d} - \frac{1}{k_m}\right) \tag{11.107}$$

By analogy of electric flow through parallel elements, the permeability of the damaged zone, k_d, is expressed as [43]:

$$k_d = f_d\phi k_c + (1-f_d)(\phi)k_m \tag{11.108}$$

where f_d is the formation-damage intensity factor when $0 < f_d < 1$. The lowest value of k_d, such as during complete plugging, is obtained at $f_d = 1$:

$$k_d = \phi k_c \tag{11.109}$$

The highest value of k_d as a result of invasion corresponds to $f_d = 0$; thus:

$$k_d = \phi k_m \tag{11.110}$$

Generally, however, $\phi k_c < k_d < \phi k_m$. To investigate the worst-case situation, substitute Eq. (11.109) into Eq. (11.107) and solve Eq. (11.105) (similar to Eq. (11.93)) for the WQR, C_{sw}/k_c ratio, the internal filter-cake permeability, k_c, and the porous medium permeability, k_m.

Figure 11.62 shows a typical plot of $\Delta p(dV/dt)$ versus the cumulative injected volume V, showing the formation of an internal filter cake. In contrast to the case of external filter cake, the intercept b_i in Eq. (11.106) is obtained from the plot by extrapolating the filtration line (phase II in Figure 11.62), not to $V = 0$, but to the "displaced" origin, which corresponds to the end of phase I where $V = V_I$. According to the deep-bed filtration theory, during phase I, i.e., at early stages of the filtration process, the permeability of the porous

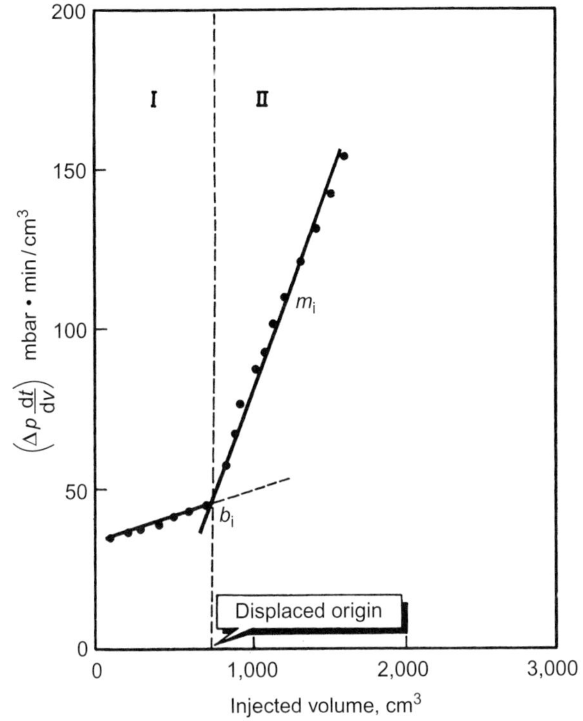

FIGURE 11.62 Core filtration test showing formation of internal filter cake [43].

medium is changing. Consequently, at the onset and during the semi-steady state filtration performance line, the permeability k_m is less than the pre-phase I value of k. The change in porosity during phase I is negligible. The data points during this phase also fall on a straight line. The two straight lines intersect at $V = V_I$ and $\Delta p(dV/dt) = b_i$.

The fractional life of an injection well, in which impairment of the formation by invasion is dominant, can be predicated by:

$$t_{I\alpha} = \frac{\pi r_d^2 h\phi(1-\phi_c)}{i_{wi}C_{sw}} \frac{1}{((k/\phi k_c)-1)} \left(\frac{1}{\alpha^2} - 1\right) \ln\frac{r_e}{r_w} \tag{11.111}$$

where r_d is the radius of filtrate invasion, which can be estimated from the core flood tests conducted at reservoir conditions of pressure and temperature [2]:

$$r_d^2 = r_w^2 + \frac{V_{LT}}{\pi\phi h(1 - S_{orf})} \tag{11.112}$$

where S_{orf} is the residual oil saturation after water-based filtrate invasion. If filtrate is from the oil-based drilling fluid, S_{orf} is replaced by S_{wif}, which is the irreducible water saturation after invasion of water-based mud. Assuming filtrate loss per unit surface area in the laboratory test is equivalent to that in the field, the total fluid loss, V_{LT}, can be obtained from the following equation:

$$V_{LT} = \frac{8r_w h}{d_c^2}(V_{LD}t_D + V_{LS}t_s) \tag{11.113}$$

where d_c is the diameter of the core sample, V_{LD} is the 30-min fluid loss during dynamic circulation in the laboratory, V_{LS} is the 30-min fluid loss during static condition in the laboratory, t_D is the mud circulation time across the interval of interest in the well, and t_s is the time the mud was left stagnant in the open hole across the interval of interest. Eq. (11.112) is obtained by scaling standard laboratory filtrate data to field conditions. Any convenient system of units can be used here. For instance, if r_d is expressed in feet, Eq. (11.112) becomes:

$$r_d = \left[r_w^2 + \frac{4.316\times 10^{-3} r_w(V_{LD}t_D + V_{LS}t_s)}{\phi d_c^2(1 - S_{orf})}\right]^{0.5} \tag{11.114}$$

where r_w and d_c are in feet, V_{LD} and V_{LS} are in cc/30 min, and t_D and t_s are in days. Once r_d is determined and if the true skin factor s is calculated from a well pressure test, then the permeability of the damaged portion of the formation, k_d, can be estimated from:

$$k_d = \frac{k}{[s/\ln(r_d/r_w)] + 1} \tag{11.115}$$

The cumulative injected volume, $V_{I\alpha}$, at time $t_{I\alpha}$ is given by:

$$V_{I\alpha} = \left(\frac{2i_{wi}}{(1/\alpha) + 1}\right)t_{I\alpha} \tag{11.116}$$

Comparing Eqs. (11.102) and (11.116) gives:

$$\frac{V_{E\alpha}}{t_{E\alpha}} = \frac{V_{I\alpha}}{t_{I\alpha}} = \frac{2i_{wi}}{1/(\alpha + 1)} \tag{11.117}$$

and comparing Eqs. (11.101) and (11.111) yields:

$$\frac{t_{I\alpha}}{t_{E\alpha}} = \frac{\phi^2}{1 - \phi k_c/k}\left(\frac{r_d}{r_w}\right)^2 \tag{11.118}$$

Equations (11.117) and (11.118) are useful in relating external and internal filter cakes. There are relatively fewer uncertainties in analyzing the effect of an external cake on the performance of an injector than that of an internal cake. The principal uncertainty in analyzing the performance of an injection well impaired by invasion is the value of r_d, which is determined in the laboratory at fixed testing times (30-min fluid loss), when, actually, r_d varies with time [15,16]. Nevertheless, if representative core samples are flooded at reservoir conditions and the test is run for a long time, i.e., until the permeability damage ratio becomes constant, the previous procedures [22,45] can give good results.

PROBLEMS

1. If cores from a well show that the original formation permeability is 120 mD, and a well test 5 years later indicates that the permeability has been reduced to 60 mD:
   (a) What percent reduction of production rate has taken place?
   (b) What is the value of the damage factor?
   (c) If the initial production rate was 60 STB/d, what is the revenue loss assuming the price of oil is $35.00 per barrel?
2. If the formation damage presented in Problem 1 occurred gradually over the 5-year period:
   (a) What is the most probable cause and what could be done to avoid this in the future?
   (b) Discuss three other causes of formation damage that may be active.
3. **(a)** Illustrate the changes of the relative permeability curves for water and oil that will occur when the formation near the wellbore is damaged by a water block.
   (b) Explain how this could occur.
4. **(a)** What are some common organic deposits that cause formation damage in the vicinity of production wells?
   (b) Why are these organic deposits localized near the production wells?
5. Define allogenic and authigenic clays, and explain their effects on permeability.
6. **(a)** List three water-sensitive clays and discuss how this property produces formation damage at injection and production wells.
   (b) What well treatments can be used to abate this problem at production wells?

NOMENCLATURE

A area
a constant
B breakthrough
C concentration
DF damage factor
f fraction

h thickness of formation
k permeability
L length
LD dynamic loss
LS static loss
LT total loss
M mass
m membrane
m slope
p pressure
PDR permeability damage ratio, k/k_o
PR productivity ratio
q production rate
r radius
R resistance
S specific surface area; saturation
sw solids in water
t time
u mass velocity
v interstitial velocity
V volume
WQR water quality ratio, C_{sw}/k_c

SUBSCRIPTS

1/2 half-life
B bridge
BD Barkman–Davidson
c capture, cake, critical
d damage
e effluent, external
E external
fp fine particles
G function
i injection rate, inlet, initial
I invasion
np nonpluggable
o oil, time zero
p pluggable, pore, perforation
PP perforation plugging
s skin, solid
sa after stimulation
sb before stimulation
T total
u undamaged
w wellbore
WF wellbore fill-up

SYMBOLS

α fraction i_w/i_{wi}, constant
ϕ porosity
μ viscosity
ρ density

REFERENCES

1. Krueger RF. An overview of formation damage and well productivity in oilfield operations. *Soc Petrol Eng JPT* 1986;**38**:131–52.
2. Amaefule JO, Kersey DG. *Advances in formation damage assessment and control strategies.* Houston, TX: Core Laboratories, Division of Western Atlas International; November, 1988.
3. Economides MJ, Nolte KG. *Reservoir stimulation.* Houston, TX: Schlumberger Educational Services; 1987.
4. Simpson JP. The drilling mud dilemma—recent examples. *Soc Petrol Eng JPT* 1985;**37**:201–6.
5. Simpson JP. Paper No. 4779. *Drilling fluid filtration under simulated downhole conditions.* Oklahoma City, OK: Society of Petroleum Engineers; 1974.
6. Keelan DK, Koepf EH. The role of cores and core analysis in evaluation of formation damage. *Soc Petrol Eng JPT* 1977;**29**:482–90.
7. Muecke TW. Formation fines and factors controlling their movement in porous media. *Soc Petrol Eng JPT* 1979;**31**:144–50.
8. Gruesbeck C, Collins RE. Entrainment and deposition of fine particles in porous media. *Soc Petrol Eng J* 1982;**22**:847–55.
9. Gruesbeck C, Collins RE. Particle transport through perforations. *Soc Petrol Eng J* 1982;**22**:857–65.
10. Khilar KC, Fogler HS. Water sensitivity of sandstones. *Soc Petrol Eng J* 1983;**23**(1):55–64.
11. Khilar KC, Fogler HS. The existence of a critical salt concentration for particle release. *J Coll Int Sci* 1984;**101**(1):214–24.
12. Porter KE. An overview of formation damage. *Soc Petrol Eng JPT* 1989;**41**:780–6.
13. Sharma MM, Yortsos Y. Transport of particulate suspensions in porous media: model formulation. *AIChE J* 1987;**33**(10):1636–43.
14. Sharma MM, Yortsos Y. Fines migration in porous media. *AIChE J* 1987;**33**(10):1654–62.
15. Chamoun H, Schechter RS, Sharma MM. The hydrodynamic forces necessary to release non-Brownian particles attached to a surface. In: *Symposium on Advances in Oilfield Chemistry.* Toronto, OH: American Chemical Society; June 1988. pp. 5–11.
16. Sharma MM, Yortsos YC, Handy LL. Release and deposition of clays in sandstones. Paper No. 13562. In: *International Symposium on Oilfield and Geothermal Chemistry.* Phoenix, AR: Society of Petroleum Engineers; April 9–11, 1985. pp. 125–135.
17. Vitthal S, Gupta A, Sharma MM. A rule based system for estimating clay distribution, morphology, and distribution in reservoir rocks. Paper No. 16870. In: *Annual Technical Conference and Exhibition.* Dallas, TX: Society of Petroleum Engineers; October 9–11, 1987.
18. Wojtanowicz AK, Krilov Z, Langlinais JP. Study on the effect of pore blocking mechanisms on formation damage. Paper No. 16233. In: *Production Operations Symposium*, Oklahoma City, OK: Society of Petroleum Engineers; March 8–10, 1987.

19. Civan F, Knapp RM. Effect of clay swelling and fines migration on formation permeability. Paper No. 16235. In: *Production Operations Symposium*. Oklahoma City, OK: Society of Petroleum Engineers; March 8–10, 1987.
20. Gabriel GA, Inamdar GR. An experimental investigation of fines migration in porous media. Paper No. 12168. In: *58th SPE Annual Meeting*. San Francisco, CA: Society of Petroleum Engineers; October, 1983.
21. Egbogah EO. *An effective mechanism for fines movement control in petroleum reservoirs*. Paper No. 84-35-16. Calgary, AB: Canadian Institute of Mining; June, 1984. pp. 269–82.
22. Barkman HJ, Davidson DH. Measuring water quality and predicting well impairment. *Soc Petrol Eng JPT* 1972;**24**:865–73.
23. Methven NE, Kemick JG. Drilling and gravel packing with oil base fluid system. *Soc Petrol Eng JPT* 1969;**21**:671–8.
24. Wilson MD, Pittman ED. Authigenic clays in sandstones: recognition and influence on reservoir properties and paleoenvironmental analysis. *J Sed Petrol* 1977;**47**:3–31.
25. Eslinger E, Pevear D. *Clay mineralogy for petroleum geology and engineering*. SEPM Short Course. Notes No. 22. Tulsa, OK: SEPM; 1988.
26. Reed MG. Formation permeability damage by mica alteration and carbonate dissolution. *Soc Petrol Eng JPT* 1977;**29**:1056–60.
27. Veley CD. How hydrolyzable metal ions react with clays to control formation water sensitivity. *Soc Petrol Eng JPT* 1969;**21**:1111–18
28. Reed MG. Stabilization of formation clays with hydroxy-aluminum solutions. *Soc Petrol Eng JPT* 1972;**24**:860–4.
29. Kersey DG. The role of petrographic analyses in the design of nondamaging drilling, completion, and stimulation programs. Paper No. 14089. In: *International Meeting on Petroleum Engineering*. Beijing, China: Society of Petroleum Engineers; March 17–20, 1986.
30. Khilar KC, Fogler HS, Ahluwalia JS. Sandstone water sensitivity: existence of a critical rate of salinity decrease for particle capture. *Chem Eng Sci* 1983;**38**(5):789–800.
31. Khilar KC. *The water sensitivity of sandstones*. Ph.D. thesis, University of Michigan, Ann Arbor, MI; 1981.
32. Khilar KC, Fogler HS. Water sensitivity of sandstones. *Soc Petrol Eng J* 1983;**23**:55–64.
33. Gray DH, Rex R. Formation damage in sandstones caused by clay dispersion and migration. In: *Proceedings of the 14th National Conference on Clay and Clay Minerals*. New York; 1966. pp. 355–66.
34. Abrahm A. Mud design to minimize rock impairment due to particle invasion. *Soc Petrol Eng JPT* 1977;**29**:586–92.
35. Donaldson EC, Baker, BA. Particle transport in sandstones. Paper No. 6905. *Annual Fall Technical Conference*. Denver, CO: Society of Petroleum Engineers; October 9–12, 1977.
36. Ives KJ. Deep bed filtration. In: Svarovsky L, editor. *Solid–liquid separation*. London: Butterworths; 1981.
37. Vitthal S, Sharma MM, Sepehrnoori K. A one-dimensional formation damage simulator for damage due to fines migration. Paper No. 17146. In: *Formation Damage Control Symposium*. Bakersfield, CA: Society of Petroleum Engineers; February 8–9, 1988.
38. Peden JM, et al. The analysis of the dynamic filtration and permeability impairment characteristics of inhibited water based muds. SPE Paper No. 10655. In: *SPE Formation Damage Symposium*. Lafayette, LA: Society of Petroleum Engineers; March 24–25, 1982.

39. Hassen BR. New technique estimates drilling filtrate invasion. Paper No. 8791. In: *4th Symposium on Formation Damage Control.* Bartlesville CA: Society of Petroleum Engineers; January, 1980.
40. McCabe WL, Smith JC. *Unit operations of chemical engineering.* New York: McGraw-Hill; 1956.
41. Tuttle RN, Barkman JH. New nondamaging and acid-degradable drilling and completion fluids. *Soc Petrol Eng JPT* 1974;**26**:1221–6.
42. Vetter OJ, et al. Particle invasion into porous medium and related injectivity problems. Paper No. 16255. In: *International Symposium on Oilfield Chemical.* San Antonio, TX: Society of Petroleum Engineers; February 4–6, 1987.
43. Ershagi I, et al. Injectivity losses under particle cake buildup and particle invasion. Paper No. 15073. In: *California Regional Meeting.* Oakland, CA: Society of Petroleum Engineers; April 2–4, 1986.
44. Todd AC, Somerville JE, Scott G. The application of depth of formation damage measurements in predicting water injectivity decline. Paper No. 12498. In: *Symposium on Formation Damage Control.* Bakersfield, CA: Society of Petroleum Engineers; February 13–14, 1984.
45. Eylander JGR. Suspended solids specifications for water injection from coreflood tests. Paper No. 16256. In: *International Symposium on Oil Field Chemistry.* San Antonio, TX: Society of Petroleum Engineers. February 4–6, 1987.
46. Todd AC, Kumar T, Mohammadi S. The value and analysis of core-based water quality experiments as related to water injection schemes. Paper No. 17148. In: *Symposium on Formation Damage Control.* Bakersfield, CA: Society of Petroleum Engineers. February 8–9, 1988.

Chapter 12

Basic Well-Log Interpretation

INTRODUCTION

A new era of well-log interpretation began when Archie (in 1942) published his equation for the quantitative evaluation of resistivity well logs. A correction for the influence of shale, V_{sh} (volume of shale, %), is added as a reminder that the terms within the equation should have corrections for the amount of shale in the zones of interest (water zones or hydrocarbon zones).

$$S_w = \left(\frac{FR_w}{R_t}\right)^{1/n} - V_{sh} \tag{12.1}$$

This text, and the accompanying computer program, has been organized to follow the logical sequence for determination of the parameters that make up Eq. (12.1). Therefore, it begins with the self-potential log (SP log) that is used to determine the formation water resistivity (R_w), the shale content (V_{sh}), and indications of lithology and the original sedimentary environment. Next, the gamma ray log (GR log) is introduced because of its use in characterization of lithology and shale content. The porosity logs (acoustic and density/neutron) follow for determination of the formation resistivity factor (F), lithology, and the presence of gas. The mini log (16 in normal) and the two focused micrologs (microlatero and proximity) are considered for calculations of porosity, formation water saturation, and the delineation of porous zone boundaries (and hence thickness of the formation of interest). These data, together with indication of the state of wettability of the formation, are used to select the value of the saturation exponent (n).

The planet Earth is a continually changing restless body. However, most of the changes are so slow that they are imperceptible during the ordinary life span of humans. Forces within the planet raise mountains in regions that were once flat plains while the relentless processes of erosion reduce mountains to plains, depositing the fine production of erosion in low areas and subsiding zones that we recognize as sedimentary basins. Erosion produces much of the direct evidence that internal forces have been operating throughout geologic time. Interior forces acting on a global scale are very slow, changing the geography of the planet by moving the continental masses around. Mountain-range building, erosion, and plate tectonics are slow acting events that are not obvious unless they are closely examined. Some cataclysmic events, however, make rapid changes that demonstrate the tremendous forces acting on the surface earth.

Volcanic eruptions are capable of building a mountain in a few years and regional movements associated with earthquakes can produce scarps instantaneously; and the gradual movement of the western coastal area of California in a northerly direction with respect to the continent as slippage occurs along the San Andreas fault is clearly evident by the dislocation of surface features such as highways and streams.

The motions of the earth's crust deform and break layers of sediments that were originally deposited in horizontal attitudes, leading to folding, faulting, and production of steeply dipping sedimentary structures. Some of this kneading of the sedimentary layers produces stratigraphic and structural porous traps that are sealed by other impermeable layers of sediments, preventing further migration of the hydrocarbons. The gravitational field and capillary phenomena produce a saturation gradient of fluids within the trap due to the differences in density among gas, oil, and water.

Structural traps develop from folding, faulting, and intrusion of salt and igneous rocks. Stratigraphic traps develop from tilting and exposure of a porous bed to the surface where erosion removes part of the structure, and later deposition of impermeable formation on top produces a seal at the top that is known as an unconformity. Accumulation of oil originating from a source rock generally takes place after formation of the trap. Migration of oil into a structure that is exposed to the surface results in evaporation of the less dense components, polymerization of polar compounds in the crude oil by oxygen, and decomposition of paraffinic components by bacteria; the combination of processes leads to the development of viscous crude oil, known as heavy oil (API gravity $<20^\circ$, viscosity $<10{,}000$ cP) or a bitumen (viscosity $>10{,}000$ cP). A second type of structural trap is formed by a facies change, such as a pinch-out, when the porous formation gradually changes to an impermeable structure. These are common in (1) deltaic environments where changes from sand to shale to carbonate as a function of depth occur frequently, (2) alluvial structures that are produced by meandering streams, and (3) eolian deposits that change from one type of sediment to another as wind currents shift, changing the source and particle size of the sediments.

The complete discussions of the genesis or oil and origin of sedimentary structures are extensive subjects that are vital to a complete understanding of well-log interpretation.

There are several attributes of hydrocarbon zones that are very important to well-log interpretation: (1) the free water level (FWL) occurs at a capillary pressure, (P_c), equal to zero; (2) the oil–water contact (OWC) occurs above the FWL in a water-wet reservoir; (3) in an oil-wet reservoir, the FWL and the water–oil contact occur at the same location; (4) the water saturation at which residual oil occurs is slightly above the OWC (this is the limiting saturation for oil production in the reservoir; (5) the irreducible water saturation (S_{wi}) occurs at a point where the capillary pressure continually rises without further displacement of water; and (6) the connate water saturation (water born with the oil), which is established at the time of oil accumulation, may be greater than or equal to the irreducible water saturation, but the connate water saturation cannot be less than S_{wi}.

GEOTHERMAL GRADIENT

The temperature in the subsurface increases with depth in accordance with the local geothermal gradient and the mean annual surface temperature of the region. The relationship is linear, except where abnormally high-temperature zones are encountered due to unique geological strata that cause this effect.

The average geothermal gradient is an increase in temperature (1.0°F/100 ft of depth). When other information is not available for calculation of the local geothermal gradient, a very good estimation of the temperature of the subsurface formation is expressed as:

$$T_{for} = T_{sur} + GG\left(\frac{D}{100}\right) \quad (^\circ F) \tag{12.2}$$

The temperature of the formation is required for interpretation of most logs. If the bottom hole temperature (BHT) is given on the well log, use it and the mean surface temperature of the region (60–80°F) to determine the geothermal gradient, GG. If the BHT is not given on the well log, use the average geothermal gradient and the mean surface temperature to estimate the temperature of the zone of interest.

SELF-POTENTIAL LOG

The SP log records the electrical potentials (millivolts) that develop in the borehole because of the presence of a conductive drilling mud and the contrasting formation water conductivity. The voltage between an electrode moving in the borehole and a surface reference electrode is recorded as the SP log (Figure 12.1). Shales are very similar; thus, they exhibit a very constant low voltage, which is defined as the *shale line* (the dashed line on the right side). Examination of the SP log in Figure 12.1, from the top, shows a large deep deflection to the left that indicates a clean sand formation. The top and bottom bed boundaries occur at the points where the deflection begins at the top and ends at another shale bed. The following small deflection (at 4,820 ft) to the left results from a limestone between the shale beds; the limestone also exhibits high resistivity. Beginning at 4,850 ft, a shaley water sand is indicated by a shallow deflection to the left.

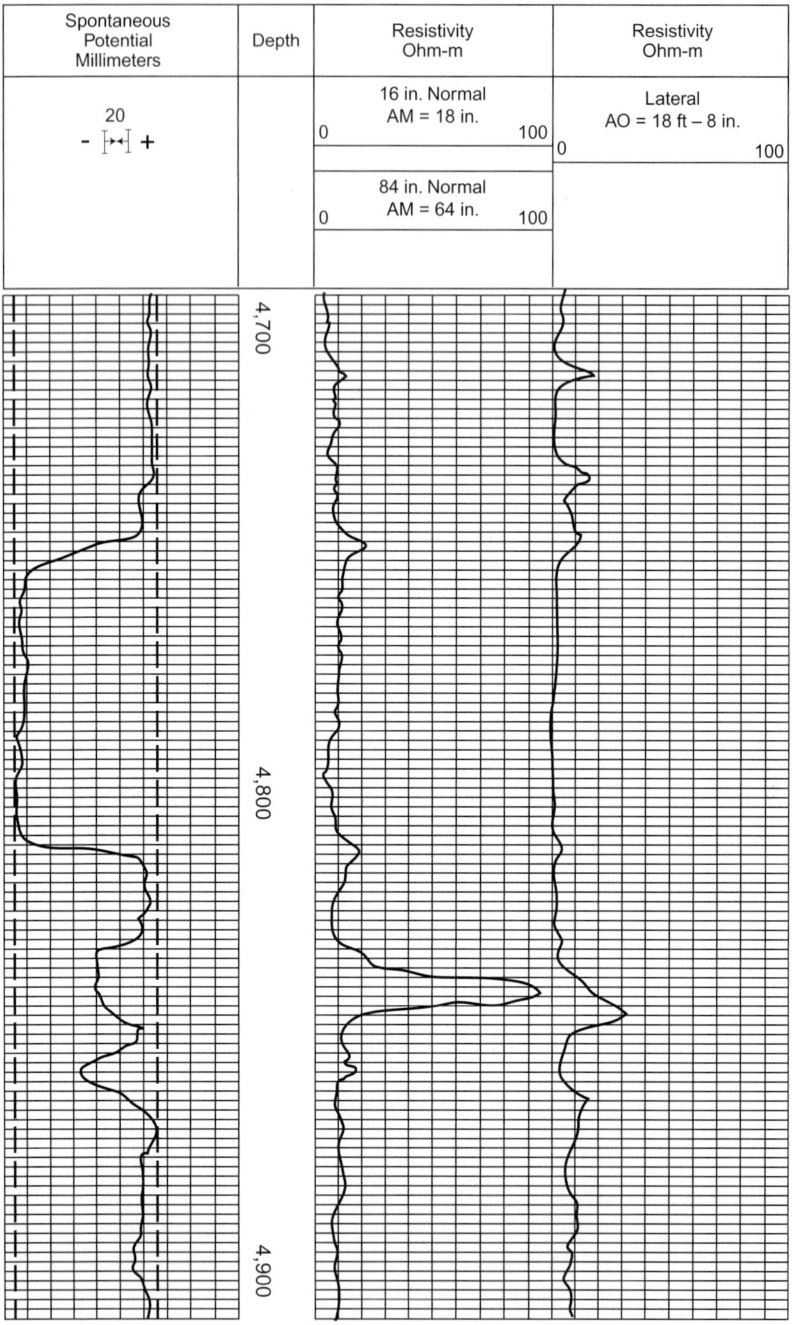

FIGURE 12.1 Behavior of the SP log, mini log, and microlaterolog.

The SP is a measure of the mud resistivity adjacent to the porous formation, but the complete circuit (the static self-potential)) is the total potential that the tool will measure if it is held in a static position adjacent to the formation; the SSP is the sum of contributions of resistances from the sand, shale, and mud:

$$\mathrm{SSP} = I(r_{\mathrm{sd}} + r_{\mathrm{sh}} + r_{\mathrm{m}}) \tag{12.3}$$

Referring to Figure 12.1, in very thick, clean (no shale), the SP will give a value that can be considered equal to the SSP, which is denoted by a dashed line on the left side of the SP log.

The potential is proportional to the two solutions (borehole mud fluid and formation water) and is expressed by:

$$E_{\mathrm{sh}} = -59.1 \log\left(\frac{R_{\mathrm{mf}}}{R_{\mathrm{w}}}\right) \tag{12.4}$$

This is the SSP that can be expressed to include the formation temperature and the equivalent resistivity of the formation fluid:

$$\mathrm{SSP} = -(60 + 0.133T_{\mathrm{f}})\log\left(\frac{R_{\mathrm{mf}}}{R_{\mathrm{we}}}\right) \tag{12.5}$$

$$\log\left(\frac{R_{\mathrm{we}}}{R_{\mathrm{mfz}}}\right) = \left(\frac{\mathrm{SSP}}{60 + 0.133T_{\mathrm{z}}}\right) = +y$$

$$\frac{R_{\mathrm{we}}}{R_{\mathrm{mfz}}} = 10^{y} \tag{12.6}$$

The equivalent water resistivity (R_{we}) is equal to the true formation water resistivity ($R_{\mathrm{w}}SP$) determined from the SP log when R_{we} is greater than 0.08 (Figure 12.2). Where R_{we} is less than 0.08, charts for correction of the equivalent water resistivity to the true formation water resistivity can be used; however, the equation relating the two values is presented as Eq. (12.7):

$$\text{If } R_{\mathrm{we}} > 0.08, \quad R_{\mathrm{w}}SP = R_{\mathrm{we}}$$
$$\text{If } R_{\mathrm{we}} < 0.08, \quad R_{\mathrm{we}} \text{ must be corrected}$$

$$R_{\mathrm{w}}SP = \frac{R_{\mathrm{we}} + 0.131 \times 10^{[1/\log(T_{\mathrm{f}}/19.9)]} - 2}{-0.5R_{\mathrm{we}} + 10^{[0.426/\log(T_{\mathrm{f}}/50.8)}} \tag{12.7}$$

The SSP assumes a clean, non-shaley permeable sand, where the only factor developing the SSP is the differential mobility of the ions in the formation waters (formation water and mud fluid). Shale mixed in the sand (or clay) reduces the SP by an amount that is proportional to the percentage of shale in the bed. Using the clean sand as the SSP, the following relation

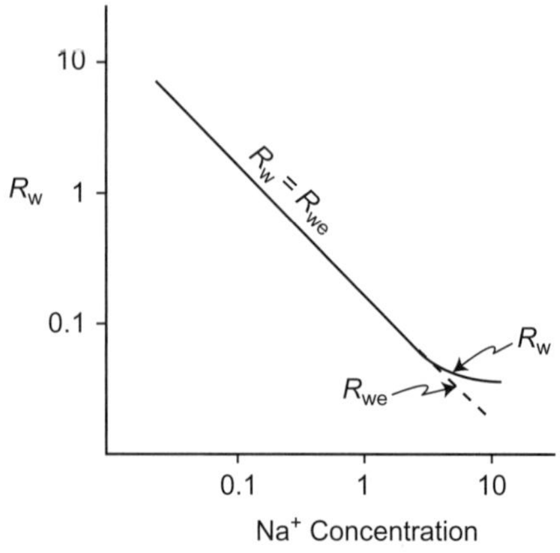

FIGURE 12.2 Devitation of water resistivity at values less than 0.08 ohm-meters.

was developed to measure the volume of shale (or the fractional percentage) of shale in the bed (V_{sh}).

Thus, the SP log is used to determine the thickness of the formation of interest ($DELH_{sp}$), the true resistivity of the formation water that, in turn, is used in Archie's equation (Eq. (12.1)), and the percentage of shale that is used for corrections to other log values.

GAMMA RAY LOG

Gamma rays of sufficient intensity to be easily detected are emitted by three elements that accumulate principally in clays. These common elements are potassium, thorium, and uranium. The gamma rays emitted by these three form the basis of the natural gamma ray well log. Shales are composed of about 60 clay minerals mixed with other clastic materials and sometimes organic matter. Thus, the shales emit much more gamma rays than do sands, limestones, and anhydrites. Carbonates result from the deposition of marine organisms, and because living bodies tend to eliminate radioactive elements, the limestones have very little, if any, radioactivity. Dolomites usually exhibit a small amount of radioactivity because they are formed with percolating waters that contain small amounts of radioactive nuclei as cations. Clean quartz sands (no clay or shale content) have no radioactivity because retention of the radioactive elements is a function of the cation exchange capacity of the sediments, and crystalline quartz sands have no cation exchange capacity. Therefore, the GR log basically distinguishes between shales and other types of formations.

The GR log is recorded so that the curve deflects to the right as the radioactivity increases, and thus the curve is very similar to the SP log. The GR log has many advantages over the SP log because it can be recorded in open holes with any type of fluid in the hole, or in cased wells. The gamma rays are attenuated to some degree by the cement and steel casing, but the penetrating power is so high that sufficient energy remains to allow unambiguous detection. The log is used where the SP log cannot be employed, such as in boreholes filled with nonconductive muds, air-filled holes, cased holes, and when the mud filtrate resistivity is almost equal to the resistivity of the formation water. Used alone with a casing collar locator, the GR log permits direct correlation of depth measurements with open-hole logs. This is used to ensure accurate depth control for cased-hole wireline services, such as perforating, formation testing, fracturing, location of leaks, channeling behind casing, and many other applications where a small amount of radioactive element can be added to a fluid for subsequent detection downhole.

The API unit is adopted for standard calibration of the GR logs, where one API unit represents about 0.07 μg of radium equivalent per ton of formation. A test pit is maintained at the University of Houston that contains a standard radioactive cement sandwiched between two nonradioactive cements. The API unit is defined as 1/200 of the difference in deflection between the two types of cements. The average mid-continent shale will record at about 100 units.

Shales are distinguished from other formations by their relatively high deflection to the right on the log, just as the SP log responds to shale beds by deflection to the right. The GR log, which can be run with any other type of log, is recorded on the left of Figure 12.3 with the compensated densilog survey and caliper log on the right. At a depth of 5,500 ft, at the top to 5,508 ft, indicates a shale bed followed by a sand bed containing shale extending to 5,632 ft; this is followed by another porous zone between 5,638 and 5,656 ft. The strong deflections to the right at 5,570–5,584 ft and again at 5,621–5,633 ft indicate 100% shale beds.

One of the important uses of the GR log is the evaluation of the fraction of shale in porous formations. The fraction of shale present in the formation is required for corrections of the porosity logs for quantitative evaluation of formation water saturation using Archie's equation.

The gamma ray shale index is defined as:

$$I_{GR} = \frac{GR_z - GR_{cs}}{GR_{sh} - GR_{cs}} \tag{12.8}$$

The I_{GR} is empirically correlated to the fraction of shale in formations. The linear scaling provides an upper limit of shale content in any formation where $V_{sh} = I_{GR}$. Equation (12.9) is used for analyses of pre-Tertiary rocks and is applicable to highly consolidated Mesozoic rock formations, and Eq. (12.10) is used for Tertiary clastics (unconsolidated sands).

$$V_{sh} = 0.33(2^{(2.0 I_{GR})} - 1) \tag{12.9}$$

$$V_{sh} = 0.083(2^{(3.7 I_{GR})} - 1) \tag{12.10}$$

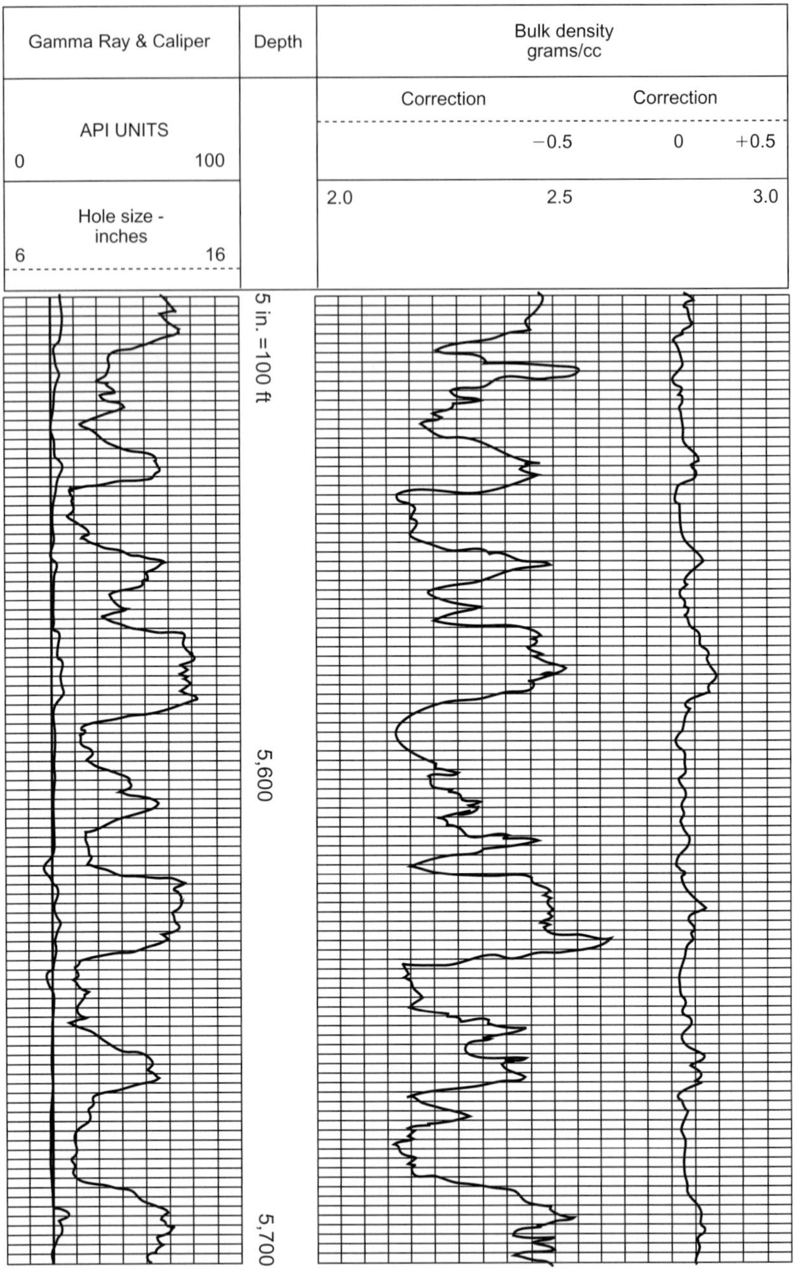

FIGURE 12.3 Gamma ray and bulk density logs.

ACOUSTIC LOG

The acoustic log was introduced in boreholes to make measurements similar to those that had been developed for seismic exploration. Acoustic data were soon found to be valuable for the determination of porosity and lithology correlation. Therefore, it soon became a standard formation evaluation tool.

An acoustic pulse signal (20 kHz) is transmitted into the formation and the time required for the signal to arrive at a receiver at a specific distance from the transmitter is recorded as Δt (the travel, or transit, time in μs/ft). The recordings in the borehole are generally within 40–200 μs/ft. This transit time is a function of the density of the formation and its elastic properties, and thus contains information about the porosity (as a function of the density) and lithology (a function of the elasticity).

The relationship defining the porosity in terms of transit times is quite accurate for a broad range of conditions. The equation expresses porosity in terms of the recorded transit times, with correction for the degree of compaction of the formation (C_p) and $V_{sh.}$

$$\phi_{ac} = \left(\frac{\Delta t_z - \Delta t_m}{\Delta t_f - \Delta t_m}\right)\left(\frac{100}{C_p}\right) - V_{sh}\left(\frac{\Delta t_{sh} - \Delta t_m}{\Delta t_f - \Delta t_m}\right) \tag{12.11}$$

The fluid transit time, Δt_f, for 20% NaCl is 189 μs/ft and for pure water 218 μs/ft. Oil has a general value of 238 μs/ft and methane has a very high value equal to 626 μs/ft. Values for some of the types of matrices are as follows: unconsolidated sands = 58.8, consolidated sands = 52.6, limestone = 47.6, dolomite = 43.5, and shale = 167 μs/ft. Of course, these are average values and the matrix transit time may vary from one area to another.

The shale compaction factor (C_p) is estimated as $100.0/\Delta t_{sh}$.

In general, the equations for oil and gas zones are as follows:

$$\text{Oil zone: } \Delta t_f = 190(1 - S_{or}) + 238[1.0 - (1 - S_{or})] \tag{12.12}$$

$$\text{Gas zone: } \Delta t_f = 190(1 - S_{gr}) + 626[1.0 - (1 - S_{gr})] \tag{12.13}$$

DENSITY LOG

The density log is based on gamma ray scattering as a function of the bulk density of the irradiated matrix. The bulk density is the overall density of the matrix and the fluids (water, oil, gas) within the pores. A gamma ray source irradiates a stream of gamma rays into the formation, some of which are adsorbed, some passed on through the matrix, and some scattered. The ability of the matrix to attenuate the gamma rays is recorded as the intensity of scattered gamma rays arriving at two fixed distances from the gamma ray source. Bulk density is determined by a correlation between the gamma ray intensity at the detectors and data used for calibration of the tool. The gamma ray intensity arriving at the detectors is an inverse function of the bulk density.

The density log tool is sometimes referred to as a compensated density log that measures the matrix bulk density with compensation for effects due to the thickness of the mud cake and borehole irregularities. The recorded log has a linear scale of bulk density (g/cm^3). A second curve is included that shows the degree of compensation that was applied, and a caliper log is included with the density log survey (Figure 12.2).

The total density (bulk density) of the formation is an average of densities of the matrix and fluids in the pores of the mud-fluid flushed zone, and the porosity is affected by the presence of shale. The density of shale varies between 2.20 and 2.85 g/cm^3, depending on the types of clay minerals in the shale.

The density log is used to estimate the effective porosity of shaley sands, assuming the shale density is approximately 2.65 g/cm^3. Nevertheless, the expression for the bulk density of a shaley, water-bearing formation is:

$$\phi_{\mathrm{d}} = \frac{\rho_{\mathrm{m}} - \rho_{\mathrm{z}}}{\rho_{\mathrm{m}} - \rho_{\mathrm{f}}} - V_{\mathrm{sh}}\left(\frac{\rho_{\mathrm{m}} - \rho_{\mathrm{sh}}}{\rho_{\mathrm{m}} - \rho_{\mathrm{f}}}\right) \tag{12.14}$$

When hydrocarbons are present in the formation, they lower the density of the residual hydrocarbons in the mud-fluid flushed zone. The density of the hydrocarbons (and their residual saturation) must be known, along with the mud-fluid resistivity (R_{mf}) and the resistivity of the flushed zone (R_{xo}) that is obtained from a microlog. The expression for this complex relationship is:

$$\phi_{\mathrm{d}} = \frac{(\rho_{\mathrm{m}} - \rho_{\mathrm{z}}) + 1.97\sqrt{R_{\mathrm{mf}}/R_{\mathrm{xo}}}(1.11 - 1.24\rho_{\mathrm{h}})}{(\rho_{\mathrm{m}} - 1) + 1.07(1.11 - 1.24\rho_{\mathrm{h}})} \tag{12.15}$$

When the resistivity of the flushed hydrocarbon zone is not available, the porosity may be calculated from values of the resistivities of the formation water, the hydrocarbon formation:

$$\phi_{\mathrm{d}} = \left(\frac{\rho_{\mathrm{m}} + \sqrt{R_{\mathrm{w}}/R_{\mathrm{t}}}}{\rho_{\mathrm{z}} - \rho_{\mathrm{f}}}\right) - V_{\mathrm{sh}}\left(\frac{\rho_{\mathrm{m}} - \rho_{\mathrm{sh}}}{\rho_{\mathrm{m}} - \rho_{\mathrm{f}}}\right) \tag{12.16}$$

When a formation contains gas, the density log values are high and the neutron log values are low; but when the two logs are used in combination, they compensate for the differences and yield an accurate value of the porosity. Hence, as a general rule, the density log is used in combination with the neutron log for accurate evaluation of porosity.

NEUTRON LOG

The details of the theory behind the neutron log require considerable study of the principles of nuclear radiation behavior, which is beyond the scope of this chapter. A neutron source (beryllium/radium or beryllium/plutonium) on the tool emits a continuous flux of energetic neutrons. The neutrons lose energy as they migrate spherically from the source across the wellbore and into the formation. The neutrons are absorbed by the nuclei of atoms in the wellbore and the formation, and by a radiation detector spaced from 11 to 18 in. from the source of radiation. The penetration of the neutrons is governed by the hydrogen atoms on the fluids (water or hydrocarbons) and the porosity of the formation, and the measurement of neutrons by the detector indicates the relative amount of hydrogen in the formation. When the principal source of hydrogen in the formation is from the liquids in the pores, the measured neutrons are correlated to the quantity of liquid present in the pores. In high-porosity zones, only a small number of neutrons reach the detector, and conversely in low-porosity zones, there will be a high count of neutrons. Thus, the neutron log can be correlated to the porosity of the formation where the pore spaces are filled with water or oil (the ratio of hydrogen in water and oil is almost equal). Thus, the neutron log is correlated to yield a log of porosity of the subsurface formations traversed by the log. The presence of shale in the formation results in high values of the porosity; therefore, a V_{sh} correction is required; the correction adds to the neutron log measured porosity because the measured porosity is lower when shale is present. By assuming that the false porosity measurement of a shale adjacent to the zone of interest is the same as the response to the shale content within the porous bed of interest:

$$\phi_n = \phi_z + V_{sh}\phi_{sh} \tag{12.17}$$

The neutron log yields too low a porosity when gas is present in the zone. This can often be used to detect gas zones. If the formation has uniform porosity, the gas/liquid contact can be found from the neutron log.

A density log and neutron log combination provides a clear indication of the gas/liquid interface as well as more accurate porosity values, using the following relationship:

$$\phi_{true} = \sqrt{\frac{(\phi_d^2 + \phi_n^2)}{2.0}} \tag{12.18}$$

MINI LOG

During drilling, the pressure of the drilling mud causes the mud fluid to penetrate porous zones and displace the formation fluids while leaving a cake of mud adjacent to the wellbore wall. The mud in the borehole has a resistivity R_m, whereas the mud cake on the wall has a resistivity equal to R_{mc}, which is usually slightly higher than R_m; and the flushed zone where the formation fluid has been replaced by mud filtrate (resistivity equal to R_{mf}) has a resistivity designated R_{xo}. The mini log, microlaterolog, and proximity log are designed to measure the resistivity a few inches into the flushed zone.

A small electrode at the lower part of an insulated pad that is braced against the wall of the borehole is kept at a constant potential. Two button electrodes (small spherical electrodes), above the current electrode, are spaced 1 in. apart. The closest measuring electrode to the electrical current input electrode theoretically measures the resistivity at a depth of 1.5 in. (R_1), and the second electrode (2 in. distant from the current electrode) measures the resistivity at a depth of 4 in. (R_2).

The values of the two resistivity curves provide a good estimate of the porosity of the formation. The general relationship of the resistivities and the porosity of the formation is given in Archie's equation:

$$\left(\frac{R_o}{R_w}\right) = F = \left(\frac{1.0}{\phi^m}\right) \tag{12.19}$$

where

R_o = resistivity of the formation filled with water
R_w = resistivity of the water in the pores
m = cementation exponent that depends on the type of matrix
ϕ = porosity

The relationship generally used between the formation resistivity factor and porosity is:

$$F = \frac{0.62}{\phi^{2.15}} \tag{12.20}$$

For the conditions within the flushed zone, the equation may be written as:

$$\frac{R_{xo}}{R_{mf}} = F = \frac{0.62}{\phi^{2.15}} \tag{12.21}$$

To account for the presence of residual hydrocarbons in the flushed zone, Eq. (12.21) becomes:

$$\phi_{mi} = \left[\frac{0.62(R_{mf}/R_{xo})}{(1 - S_{rh})^2}\right] \tag{12.22}$$

Equation (12.19) when applied to the water saturation of the undisturbed formation (beyond the flushed zone) yields:

$$S_w = \left(\frac{FR_w}{R_t}\right)^{1/2} \tag{12.23}$$

and when applied to the saturation of mud fluid in the flushed zone (S_{xo}) yields:

$$S_{xo} = \left(\frac{FR_{mf}}{R_{xo}}\right)^{1/2} \tag{12.24}$$

and

$$\frac{S_w}{S_{xo}} = \left(\frac{R_{xo}/R_t}{R_{mf}/R_w}\right)^{1/2} \tag{12.25}$$

Substituting the empirical relationship that was developed with extensive laboratory analyses ($S_{xo} = S_w^{1/5}$) into Eq. (12.24) yields:

$$S_w = \left(\frac{R_{xo}/R_t}{R_{mf}/R_w}\right)^{0.625} \tag{12.26}$$

MICROLATERO AND PROXIMITY LOGS

The microlog has a central current electrode surrounded by another electrode to focus an electrical beam into the formation with the focusing or guard electrode. The potential difference between the guard electrode and the central beam electrode is continuously monitored to maintain it at zero. This causes the current leaving the center (beam) electrode to flow through the mud cake and into the formation as a narrow beam; the mud cake causes some reduction of the current beam potential. The beam current spreads into the formation as a function of the distance from the wall of the borehole, and thus a decrease in potential occurs between the beam electrode and the detector electrode that is placed at a remote distance from the focusing electrode pad. The measured resistivity is the apparent resistivity of the formation within about 3 in. from the borehole wall.

The proximity log has two focusing electrodes surrounding the central beam electrode. The extra focusing electrode (referred to as the shield electrode) eliminates the scatter of the beam current by the mud cake. Thus, the proximity log responds to about 6 in. into the formation. Equation (12.22) is used with any of the R_{xo} reading micro tools to yield the porosity of the formation of interest and Eq. (12.26) is used for determination of the water saturation where R_w is determined from the SP log or from analysis of the formation-produced water.

Any of the deep-reading tools (64-in. normal log, induction log, and laterolog) furnishes the value of R_t, which is used in Eqs. (12.22) and (12.26).

MEASUREMENT OF OIL AND GAS

The Simandoux equation is accepted as one of the best for calculation of S_w for the widest range of conditions. Water saturation, along with a correction for V_{sh}, is expressed as a quadratic equation as follows:

$$\frac{1}{R_t} = \left(\frac{V_{sh}}{R_{sh}}\right) S_w + \left(\frac{\phi^m}{aR_w}\right) S_w^2 \tag{12.27}$$

When $V_{sh} = 0$, Eq. (12.26) reduces to Archie's equation; thus:

$$\frac{1}{R_t} = \left(\frac{\phi^m}{aR_w}\right) S_w^2 \quad \text{or} \quad S_w = \left(\frac{aR_w}{\phi^m R_t}\right)^{1/2} \tag{12.28}$$

The formation factor, F, is used below in the Simandoux equation, allowing the chances required for selection of different types of lithology (Table 12.1); thus:

$$\left(\frac{1}{FR_w}\right) S_w^2 + \left(\frac{V_{sh}}{R_{sh}}\right) S_w - \frac{1}{R_t} = 0$$

$$(a) \qquad (b) \qquad (c)$$

$$S_w = \frac{-b + \sqrt{b^2 - 4ac}}{2a} \tag{12.29}$$

$$S_w = \left(\frac{FR_w}{2}\right) \sqrt{\left(\frac{V_{sh}}{R_{sh}}\right)^2 \left(\frac{4}{FR_w R_t}\right)} - \left(\frac{V_{sh}}{R_{sh}}\right)\left(\frac{FR_w}{2}\right)$$

TABLE 12.1 Selection of Well-Log Interpretation Parameters: Matrix Designation, Fluid Transit Time for Various Types of Lithology, Density of Various Types of Lithology, and the Formation Resistivity Factor, *F*

Matrix	Δt_f, μs/ft	Type of Formation	ρ_m, g/cm^3	F
1	48	Carbonate (limestone)	2.71	$1.0/\phi^2$
2	50	Calcareous sand	2.69	$1.45/\phi^2$
3	53	Consolidated sand	2.65	$0.81/\phi^2$
4	59	Unconsolidated sand	2.60	$1.0/\phi^{1.5}$
5	65	Shaley sand	2.60	$1.65/\phi^{1.33}$
6	44	Dolomite	2.88	$1.0/\phi^2$

When the SP log is not available for calculation of S_w, the resistivity of a water zone (bottom water, or an adjacent porous formation containing 100% water) may be used in the Simandoux equation to calculate the water saturation of the hydrocarbon zone with very good accuracy.

Thus, there are three possible calculations of S_w that can be used to obtain the quantities of hydrocarbons in the reservoirs:

1. Water saturation derived from the SP log using the value for water resistivity ($R_w SP$),
2. The water saturation obtained from a microlog
3. Using the resistivity of an adjacent water zone in the Simandoux equation in place of R_wSP.

When the water saturation of a hydrocarbon zone is determined, the quantity of oil or gas is determined from:

$$
\begin{aligned}
\text{BBL}-\text{Res/acre}-\text{ft} &= 7,758\,\phi\,(1.0 - S_w)\,\Delta H_{av} \\
\text{STB/acre}-\text{ft} &= 7,758\,\phi\,(1.0 - S_w)\left(\frac{\Delta H_{av}}{B_o}\right) \\
\text{CF/acre}-\text{ft} &= 43,560\,\phi\,(1.0 - S_w)\,\Delta H_{av} \\
\text{SCF/acre}-\text{ft} &= 43,560\,\phi\,(1.0 - S_w)\left(\frac{\Delta H_{av}}{B_g}\right)
\end{aligned}
\tag{12.30}
$$

Each of the well log programs is set individually. Data are entered as indicated in the PRINT/READ statements of each program.

SP Log requires SUBROUTINE SPLOG
GAMMA RAY requires SUBROUTINE GAMMA
ACOUSTIC LOG requires SUBROUTINE ACOUSTIC
DENSITY/NEUTRON requires SUBROUTINE DENEUT

```
C     SP LOG PROGRAM MAIN....requires the SPLOG SUBROUTINE
C     Enter the data as shown in the PRINT/READ statements
C     Refer to the NOMENCLATURE for the meaning of symbols
C
      PRINT *, 'Tsur,BHT,BHD,Rmf,Tmf,Dzone,SSP,SPzone'
      READ *, Tsur,BHT,BHD,Rmf,Tmf,Dzone,SSP,SPzone
C
      CALL SPLOG (Tsur,BHT,BHD,Rmf,Tmf,Dzone,SSP,SPzone,RwSP,Rmfz, & VshSP,
      Vsh)
C
      PRINT 12, RwSP, Rmfz, VshSP, Vsh
```

```
12    FORMAT(1X, 'RwSP = ', F6.4, 3X, 'Rmfz = ', F6.4, 3X, 'VshSP = ', & F6.4, 3X,
      'Vsh = ', F6.4,/)
C
      STOP
      END
```

```
C     Subroutine SPLOG uses data from the well log to calculate RwSP and VshSP
C
      SUBROUTINE SPLOG (Tsur, BHT, BHD, Rmf, Tmf, Dzone, SSP, SPzone, $ RwSP,
      Rmfz, VshSP, Vsh)
C
C     Calculate Tzone, RwSP, and VshSP
C
      Tzone = Tsur + (BHT - Tsur)*(Dzone/BHD)
      Rmfz = Rmf*((Tsur + 7.0)/(Tzone + 7.0))
C
      A = SSP/(60 + 0.133*Tzone)
      Y = 10**A
      Rwe = Rmfz/Y
C     IF(Rwe .LT. 0.08) GO TO 2
      RwSP = Rwe
      GO TO 4
C
2     A = (1/ALOG10(Tzone/19.9)) - 2.0
      AA = 10**A
      B = Rwe + 0.131*AA
C
      C = 0.0426/ALOG10(Tzone/50.8)
      CC = 10**C
      D = -0.5*Rwe + CC
C
      RwSP = B/D
C
4     VshSP = 1.0 - (SPzone/SSP)
      Vsh = VshSP
C
      RETURN
      END
```

```
C     GAMMA-RAY LOG
C
      PRINT *, 'N1,GRzone, GRcs, GRsh, ZONEG'
      READ *, N1,GRzone, GRcs, GRsh, ZONEG
C
      IF(N1 .GT. 1) GO TO 2
      GO TO 3
C
2     CALL GAMMA (GRzone, GRcs, GRsh, ZONEG, VshGR)
```

```
C
      PRINT 14, VshGR
14    FORMAT (1X, 'VshGR = ', F6.4, /)
C
3     CONTINUE
      STOP
      END
```

Subroutine GAMMA

```
C
C   Calculates VshGR when Gamma-Ray Log is available. N1 = 0 if there is no GR-
    Log.
C
    SUBROUTINE GAMMA (GRzone, GRcs, GRsh, ZONEG, VshGR)
c
    V1 = (GRzone - GRcs)/(GRsh - GRcs)
    IF(ZONEG .eq. 1) GO TO 2
    GO TO 4
C
2   TERM = 3.7
    GO TO 6
C
4   TERM = 2.0
    GO TO 6
C
6   A = V1*TERM
    B = 2**A - 1.0
    C = 2**TERM - 1.0
    VshGR = B/C
C
    RETURN
    END
```

```
C     ACOUSTIC LOG
C
      PRINT *, 'N2, DELTz, DELTsh, ZONEA, MATRIX, VshSP, VshGR'
      READ *, N2, DELTz, DELTsh, ZONEA, MATRIX, VshSP, VshGR
c
      PRINT *, 'Bo, Bg, Joil, Jgas, Rw, Rt,Rsh, DELHav'
      READ *, Bo, Bg, Joil, Jgas, Rw, Rt,Rsh, DELHav
C
      IF(N2 .GT. 1) GO TO 4
      GO TO 30
C
4     CALL ACOUSTIC (DELTz,DELTsh,ZONEA,MATRIX,VshSP,VshGR,Vsh,PHIac)
c
      IF(MATRIX .EQ. 1) THEN
      Fac = 1.0/PHIac**2
```

```
C
      ELSE IF(MATRIX .EQ. 2) THEN
      Fac = 1.45/PHIac**2
C
      ELSE IF(MATRIX .EQ. 3) THEN
      Fac = 0.81/PHIac**2
C
      ELSE IF(MATRIX .EQ. 4) THEN
      Fac = 1.0/PHIac**1.5
C
      ELSE IF(MATRIX .EQ. 5) THEN
      Fac = 1.65/PHIac**1.33
C
      ELSE IF(MATRIX .EQ. 6) THEN
      Fac = 1.0/PHIac**2
      END IF
C
      CONTINUE
C
      IF(MATRIX .EQ. 5) GO TO 26
      Swac = SQRT(Fac*(Rw/Rt))
      STBac = Joil*7758*PHIac*(1.0-Swac)*DELHav/Bo
      SCFac = Jgas*43560*PHIac*(1.0-Swac)*DELHav/Bg
      GO TO 28
C
26    CONTINUE
C     Use the Simandoux Shaley Sand Equation to compute Sw for MATRIX 5
      TE = SQRT((5*PHIac**2/Rw*Rt) + (Vsh/Rsh)**2)
      Swac =0.42*Rw/PHIac**2*TE - Vsh/Rsh
C
      STBac = 7758*PHIac*(1.0 - Swac)*DELHav/Bo
      SCFac = 43560*PHIac*(1.0 - Swac)*DELHav/Bg
      GO TO 28
C
28    PRINT 900, Vsh, PHIac
900   FORMAT(1X, 'Vsh = ', F6.4, 3X, 'PHIac = ', F6.4,/)
C
      PRINT 902, STBac, SCFac
902   FORMAT(1X, 'STBac = ', E12.4,3X, 'SCFac = ', E12.4,/)
C
30    CONTINUE
      STOP
      END
```

```
C     Subroutine ACOUSTIC calculates the porosity from data on the Acoustic Log,
C     if the Acoustic Log is present.
C     N2 = O if there is no Acoustic Log
C     PHIac is calculated for a water zone, oil zone, or gas zone.
C
      SUBROUTINE ACOUSTIC (DELTz, DELTsh, ZONEA, MATRIX, VshSP, $ VshGR,
      Vsh, PHIac)
```

```
C
C     Select the value of DELTm based on the type of MATRIX
C
      IF(MATRIX .EQ. 1) THEN
      DELTm = 48
C
      ELSE IF(MATRIX .EQ. 2) THEN
      DELTm = 50
C
      ELSE IF(MATRIX .EQ. 3) THEN
      DELTm = 53
C
      ELSE IF(MATRIX .EQ. 4) THEN
      DELTm = 59
C
      ELSE IF(MATRIX .EQ. 5) THEN
      DELTm = 65
C
      ELSE IF(MATRIX .EQ. 6) THEN
      DELTm = 44
      END IF
C
C     Select the proper zone of interest and cal. PHIac
C
      IF(ZONEA .EQ. 1) GO TO 6
      IF(ZONEA .EQ. 2) GO TO 8
      IF(ZONEA .EQ. 3) GO TO 10
C
6     DELTf = 189.0
      GO TO 11
C
8     DELTf = 107
      GO TO 11
C
10    DELTf = 98
C
C     Select Vsh from the Gamma Ray Log if it is available for
      porosity correction.
C     Otherwise, use VshSP for Vsh.
C
11    IF(VshGR .GT. 0) GO TO 12
      Vsh = VshSP
      GO TO 14
C
12    Vsh = VshGR
14    CONTINUE
C
      IF(DELTsh .GT. 100) GO TO 16
      A = ((DELTz - DELTm)/(DELTf - DELTm))
      B = Vsh*((DELTsh - DELTm)/(DELTf - DELTm))
      PHIac = A - B
      GO TO 18
```

```
C
16    Cp = 100.0/DELTsh
      A = ((DELTz - DELTm)/(DELTf - DELTm))
      B = Vsh*((DELTsh - DELTm)/(DELTf - DELTm))
      AA = A*1.0/Cp
      PHIac = AA - B
C
18    CONTINUE
C
      RETURN
      END



C      DENSITY/NEUTRON (Compensated Density/Neutron Logs, therefore they
       indicate porosity directly on the logs)
C
       PRINT *, 'N3,PHIdz, PHIdsh, PHInz, PHInsh, MATRIX,Vsh'
       READ *, N3,PHIdz, PHIdsh, PHInz, PHInsh,MATRIX, Vsh
C
       PRINT *, 'Bo, Bg, Joil, Jgas, Rw,Rt, Rsh, DELHav'
       READ *, Bo, Bg, Joil, Jgas, Rw,Rt, Rsh, DELHav
C
       IF(N3 .GT. 1) GO TO 7
       GO TO 30
C
7      CALL DENEUT (PHIdz, PHIdsh, PHInz, PHInsh, Vsh, PHIdn)
C
       IF(MATRIX .EQ. 1) THEN
       Fdn = 1.0/PHIdn**2
C
       ELSE IF(MATRIX .EQ. 2) THEN
       Fdn = 1.45/PHIdn**2
C
       ELSE IF(MATRIX .EQ. 3) THEN
       Fdn = 0.81/PHIdn**2
C
       ELSE IF(MATRIX .EQ. 4) THEN
       Fdn = 1.0/PHIdn**1.5
C
       ELSE IF(MATRIX .EQ. 5) THEN
       Fdn = 1.65/PHIdn**1.33
C
       ELSE IF(MATRIX .EQ. 6) THEN
       Fdn = 1.0/PHIdn**2
       END IF
C
       CONTINUE
C
       IF(MATRIX .EQ. 5) GO TO 26
       Swdn = SQRT(Fdn*(Rw/Rt))
```

```
      STBdn = Joil*7758*PHIdn*(1.0-Swdn)*DELHav/Bo
      SCFdn = Jgas*43560*PHIdn*(1.0-Swdn)*DELHav/Bg
      GO TO 28
C
26    CONTINUE
C     Use the Simandoux Shaley Sand Equation to compute Sw for MATRIX 5
      TF = SQRT((5*PHIdn**2/Rw*Rt) + (Vsh/Rsh)**2)
      Swdn = 0.42*Rw/PHIdn**2*TF - Vsh/Rsh
C
      STBdn = 7758*PHIdn*(1.0 - Swdn)*DELHav/Bo
      SCFdn = 43560*PHIdn*(1.0 - Swdn)*DELHav/Bg
C
28    PRINT 908, PHIdn
908   FORMAT (1X, 'PHIdn = ', F6.4,/)
C
      PRINT 910, STBdn, SCFdn
910   FORMAT (1X, 'STBdn = ', E12.4,3X, 'SCFdn = ', E12.4,/)
c
30    CONTINUE
      STOP
      END
```

```
      Subroutine DENEUT calculates the Density/Neutron Log porosity
C     This porosity is generally considered the most accurate.
C
      SUBROUTINE DENEUT(PHIdz, PHIdsh, PHInz, PHInsh,Vsh,PHIdn)
C
      PHId = PHIdz - Vsh*PHIdsh
      PHIn = PHInz + Vsh*PHInsh
C
      A = PHId**2 + PHIn**2
      PHIdn = SQRT(A/2)
C
      RETURN
      END
```

```
C     MINI-LOG, MICRO-LOG, or PROXIMITY LOG
C
      PRINT *, 'N4, Rmf, Rxo,Rw,Rt,Rsh,Vsh, DELHav,MATRIX'
      READ *, N4, Rmf, Rxo,Rw,Rt,Rsh,Vsh, DELHav,MATRIX
C
      PRINT *, 'Tsur,BHT,BHD,Dzone'
      READ *, Tsur,BHT,BHD,Dzone
C
      PRINT *, 'Bo, Bg, Joil, Jgas'
      READ *, Bo, Bg, Joil, Jgas
```

```
C
      Tzone = Tsur + (BHT - Tsur)*(Dzone/BHD)
      Rmfz = Rmf*((Tsur + 7.0)/(Tzone + 7.0))
C
      IF(N4 .GT. 1) GO TO 10
      GO TO 30
C
10    TERM = 0.62*(Rmfz/Rxo)
      PHImi = TERM**0.465
C
      IF(MATRIX .EQ. 1) THEN
      Fmi = 1.0/PHImi**2
C
      ELSE IF(MATRIX .EQ. 2) THEN
      Fmi = 1.45/PHImi**2
C
      ELSE IF(MATRIX .EQ. 3) THEN
      Fmi = 0.81/PHImi**2
C
      ELSE IF(MATRIX .EQ. 4) THEN
      Fmi = 1.0/PHImi**1.5
C
      ELSE IF(MATRIX .EQ. 5) THEN
      Fmi = 1.65/PHImi**1.33
C
      ELSE IF(MATRIX .EQ. 6) THEN
      Fmi = 1.0/PHImi**2
      END IF
C
18    CONTINUE
C
      IF(MATRIX .EQ. 5) GO TO 26
      Swmi = SQRT(Fmi*(Rw/Rt))
      STBmi = Joil*7758*PHImi*(1.0-Swmi)*DELHav/Bo
      SCFmi = Jgas*43560*PHImi*(1.0-Swmi)*DELHav/Bg
      GO TO 28
C
26    CONTINUE
C     Use the Simandoux Shaley Sand Equation to compute Sw for MATRIX 5
      TG = SQRT((5*PHImi**2/Rw*Rt) + (Vsh/Rsh)**2)
      Swmi =0.42*Rw/PHImi**2*TG - Vsh/Rsh
C
      STBmi = Joil*7758*PHImi*(1.0-Swmi)*DELHav/Bo
      SCFmi = Jgas*43560*PHImi*(1.0-Swmi)*DELHav/Bg
      GO TO 28
C
28    PRINT 930, PHImi
930   FORMAT(1X, 'PHImi = ', F6.4,/)
C
      PRINT 932, STBmi, SCFmi
932   FORMAT(1X,'STBmi = ', E12.4, 3X, 'SFCmi = ', E12.4,/)
```

```
C
30     CONTINUE
       STOP
       END
```

DATA for SP Log Example

Tsur = 70, BHT = 142, BHD = 4700, Rmf = 1.126, Tmf = 74, Dzone = 4660, SSP = 108, SP = 88

ANSWERS: RwSP = 0.036, Rmfz = 0.584, VshSP = 0.185, Vsh = 0.185

DATA for Gamma Ray Log Example

N1 = 2, GRzone = 75, GRcs = 57, GRsh = 148, ZONEG = 2 (consolidated sand)

ANSWER: VshGR = 0.1052

DATA for the ACOUSTIC Log Example

N2 = 2, DELTz = 64, DELTsh = 127, ZONEA = 2, MATRIX = 1, VshSP = 0.12, VshGR = 0.18
Bo = 1.4, Bg = 0.006, Joil = 1, Jgas = 0, Rw = 0.036, Rt = 52, Rsh = 8, DELHav = 10

ANSWERS: Vsh = 0.18, PHIac = 0.1034, STBac/acre = 0.8372 E + 04

DATA for the DENSITY/NEUTRON Log Example at 4418 feet depth

N3 = 2, PHIdz = 0.19, PHIdsh = 0.07, PHInz = 0.21, PHInsh = 0.24, MATRIX = 3
Vsh = 0.18
Bo = 1.4, Bg = 0.006, Joil = 1, Jgas = 0, Rw = 0.036, Rt = 12, Rsh = 9, DELHav = 8

ANSWERS: PHIdn = 0.2186, STBdn/acre = 0..1471 E + 05

DATA for the MICRO Log Example

N4 = 2, Rmf = 1.268, Rxo = 7.0, Rw = 0.03, Rt = 8, Rsh = 5, Vsh = 0.07, DELHav = 30
MATRIX = 3
Tsur = 75, BHT = 151, BHD = 9417, Dzone = 5720
Bo = 1.4, Bg = 0.006, Joil = 1, Jgas = 0

ANSWERS: PHImi = 0.294, STBmi/acre = 0.7782 E + 05

NOMENCLATURE

B_g gas formation volume factor, Res-CF/SCF
B_o oil formation volume factor, bbl/STB
BHD bottom hole depth, ft
BHT bottom hole temperature, °F

CF cubic feet
DELHav average formation thickness
DELHsp thickness of the zone from the SP log
DELHgr thickness of the zone from the gamma ray log
$DELT_f$ transit time of the formation fluid (acoustic log)
$DELT_m$ transit time of the matrix (acoustic log)
$DELT_z$ transit time in the zone of interest (acoustic log)
D_z depth (ft) of the zone of interest
F formation resistivity factor (Archie's equation: a/ϕ^2)
GR_{cs} gamma ray value in clean sand (minimum)
GR_{sh} gamma ray value in shale (maximum)
GR_z gamma ray value in the zone of interest
I electrical potential
I_{GR} gamma ray index
J_{gas} no gas zone = 0; gas zone = 1
J_{oil} no oil zone = 0; oil zone = 1
MATRIX type of formation for calculation of F (listed in the Table 12.1)
PHI_{ac} porosity from the acoustic log
PHI_d porosity from the density log
PHI_{dsh} porosity from the density log in a shale zone
PHI_{dz} porosity from the density log in the zone of interest
PHI_n porosity from the neutron log
PHI_{nsh} porosity from the neutron log in a shale zone
PHI_{nz} porosity from the neutron log in the zone of interest
PHI_{mi} porosity from a microlog
r_m electrical resistance of drilling mud
r_{sd} electrical resistance of sand
r_{sh} electrical resistance of shale
$R_{mf\text{-}sur}$ resistivity of the mud fluid at surface temperature
R_{mfz} resistivity of the zone of interest
R_{sh} resistivity of a shale zone
R_t resistivity of the formation from a 64-in. normal, induction, or laterolog
R_{xo} resistivity of the flushed zone (from a microlog)
R_w formation water resistivity
R_{we} equivalent water resistivity
R_wSP water resistivity determined by the SP log
S_{gr} residual gas saturation
S_{or} residual oil saturation
SP_z SP log value in the zone of interest
SSP static self-potential
SCF standard cubic feet of gas
STB stock tank barrels
S_{rhc} saturation of residual hydrocarbon
S_{wi} irreducible water saturation
S_w water saturation
T_f temperature of the formation
T_{ma} temperature of the formation matrix

T_{mf} temperature at which R_{mf} was measured
T_{sur} average surface temperature (60–80°F)
V_{sh} fraction of shale in the formation
$V_{sh\text{-}av}$ average value from the SP log and GR log
$V_{sh}GR$ fraction of shale determined from the gamma ray log
$V_{sh}SP$ fraction of shale determined from the SP log
ZONE = 1 unconsolidated formation
ZONE = 2 consolidated formation

GREEK SYMBOLS

ΔH_{av} average formation thickness
ΔH_{GR} formation thickness from gamma ray log (ft)
ΔH_{SP} formation thickness from the SP log
Δt_f transit time of formation fluid, acoustic log
Δt_m transit time of the formation matrix
Δt_{sh} transit time in a shale zone
Δt_z transit time in the zone of interest
ϕ_{ac} porosity determined by the acoustic log
ϕ_d porosity determined by the density log
ρ_f density of the formation fluid
ρ_m density of the formation matrix
ρ_{sh} density of a shale zone
ρ_z density of the formation of interest

Appendix

Measurement of Rock and Fluid Properties

Experiment 1 Fluid Content of Rocks by the Retort Method

Introduction

The theories of the formation of oil reservoirs consider that oil traps (structural or stratigraphic) originally were filled with water of marine origin. The oil and/or gas is believed to have entered the trap, displacing the water to some original reservoir saturation (the connate water saturation). Thus, a petroleum reservoir normally contains both petroleum hydrocarbons and water occupying the same, or adjacent, pores. Quantitative evaluation of the fluids is necessary for reservoir characterization.

The retort distillation is divided into two parts: (1) as the rock is first heated (to approximately 400°F or 204°C), water and all but the heaviest fraction of the oil present in the sample are vaporized; and (2) in the second stage of heating, the temperature is raised to about 1,100°F (593°C) and the hydrocarbons remaining in the sample are vaporized or cracked by the heat and removed as a vapor. Part of this vapor is condensable and part is not. Generally, the process of cracking leaves a carbon residue within the core. Therefore, the amount of oil recovered by retort distillation is less than the amount of oil in the core. Thus, the intense heat removes water of crystallization from the clays and other hydrated minerals present in the core. The amount of water obtained is slightly greater than the amount of free water in the pores because of the added water of crystallization. Empirical correction factors (obtained from retorting cores containing known amounts of oil) are used to correct for the retorting errors. The correction factors are defined as follows:

C_o = fraction of oil left in the rock as coke, with respect to the total oil recovered,

C_w = amount of excess water recovered due to dehydration (removal of water of crystallization) with respect to the dry mass of the sand.

The fluids produced by the retort method are collected in the centrifuge tubes, which then may be centrifuged to separate the oil and water for accurate volumetric measurements.

Because the oil is less dense than the water, it will separate to the top of the liquid column in the centrifuge tubes. An emulsion (a fine mixture of oil in water) will form in the centrifuge tubes between the oil and water that cannot be separated by the centrifuge. As a first approximation, one can assume that the emulsion is composed of 80% water and 20% oil, by volume. If better results are obtained by changing the water/oil ratio of the emulsion, one should do so.

Equipment and Procedures

Equipment

Retort assembly
Graduated centrifuge tubes
Analytical balance
Oil injection porosimeter
Oven-dried sandstone
Saturated sandstone

Retort Specimen

1. Obtain core samples from the instructor.
2. Determine the bulk volume of the core sample by measurement with a caliper or using an oil or mercury porosimeter (Experiment 8).
3. Obtain the mass of the core sample.
4. Place the sample in the retort cylinder and screw the lid on (hand-tight only).

Retort Calibration

1. Crush part of the dried sandstone to half-centimeter fragments.
2. Place the fragments into the retort receptacle to within 1/2 in. of the top and pack them in tightly.
3. Using a pipette, slowly drip 6 ml of oil and 4 ml of water over the crushed rock. The centrifuge tube should be under the retort to catch any flow.

Retorting

1. Start circulating water through the condensers.
2. Plug in both retorts.
3. After 45 minutes, prop the circulating hoses upright. Fill them with water and close the water valve.
4. Thirty (30) minutes later, drain all of the water from the condensers.
5. Fifteen (15) minutes later, unplug the retorts.
6. Cool the retorted samples. DO NOT PLACE THE RETORTED SAMPLES ON WAXED PAPER. Record the dry mass of the samples.
7. Centrifuge the tubes that contain the fluids and record the total fluid volumes obtained in (1) the calibration retort and (2) the sample retort.

Sample Calculations

Retort Calibration

Data

Volume of oil introduced = 6.0 ml
Volume of water introduced = 4.0 ml
Volume of oil recovered = 5.7 ml
Volume of water recovered = 4.2 ml

Mass of dry retort sample = 75.0 g
$C_o = (V_{\text{oil introduced}} - V_{\text{oil recovered}})/V_{\text{oil recovered}}$
$C_o = (6.0 - 5.7)/5.7 = 0.0526$
$C_w = (V_{\text{water recovered}} - V_{\text{water introduced}})$/mass of dry calibration retort sample
$C_w = (4.2 - 4.0)/75.0 = 0.002667$

Results From the Test Specimen

Data

Volume of oil recovered = 4.1 ml
Volume of water recovered = 2.1 ml
Mass of dry retort sample = 69.8 g
Correct volume of oil in sample = $(1.0 + C_o) \times$ oil recovered = $(1.0 + 0.0526) \times 4.1 = 4.32$ ml
Correct volume of water in sample = (volume of water recovered) − (mass of the dry specimen sample multiplied by C_w) = $2.1 - (69.8 \times 0.002667) = 1.91$ ml

Saturation Calculations

Data

Saturated mass of sample = 75.2 g
Mass of dry retort sample = 69.8 g
Bulk volume of sample = 34.98 ml
Saturated density of sample = 75.2/34.98 = 2.15 g/ml
Sand grain volume of sample = (mass of dry specimen retort sample)/[mineral grain density (2.65 g/ml)] = 69.8/2.65 = 26.34 ml
Sample pore volume = (sample bulk volume) − (sample sand grain volume) = 34.98 − 26.34 = 8.64 ml
Porosity = sample pore volume/sample bulk volume = 8.64/34.98 = 0.247
Stock tank oil saturation S_o = corrected volume of oil in sample/sample pore volume = $(4.32 \times 100)/8.64 = 50\%$
Connate water saturation S_w = corrected volume of water in sample/sample pore volume = $(1.91 \times 100)/8.64 = 22.1\%$
Gas saturation $S_g = 1.00 - S_o - S_w = 1.00 - 0.50 - 0.22 = 28\%$

Questions and Problems

1. Discuss the adverse effects that can result from penetration of drilling mud fluid into the cores used for laboratory analysis.
2. If the water saturation around the wellbore is at irreducible water saturation (S_{iw}), will water flow into the wellbore? Why?
3. Why is it that part of the vapors from the retort vessel cannot be condensed?
4. Why is it that the amount of water obtained from the retort is greater than the amount of free water in the core?

References

Anderson G. Coring and core analysis handbook. Tulsa, OK: PennWell Books; 1975, chapter 6.

Chilingarian GV, Robertson Jr. JO, Kumar S. Surface operations in petroleum production, II. Amsterdam, The Netherlands: Elsevier Science; 1989, Appendix A.

Experiment 2 Measurement of Saturation by Solvent Extraction

Introduction

A hydrocarbon solvent that is insoluble in water is used to leach the fluids from a saturated rock. A special graduated receiving tube may be used to collect the extracted water, thus providing a direct measurement of the water saturation (Figure A2.1). Once the water saturation is known, the oil saturation may be calculated from indirect gravimetric measurement of the rock. The gas saturation is then obtained indirectly by difference because:

$$S_g + S_w + S_o = 1.0 \tag{A2.1}$$

After extraction of the fluids and drying, the cleaned core may be used for measurement of other petrophysical properties and quantities (permeability, porosity, sieve analysis, mineral grain density, etc).

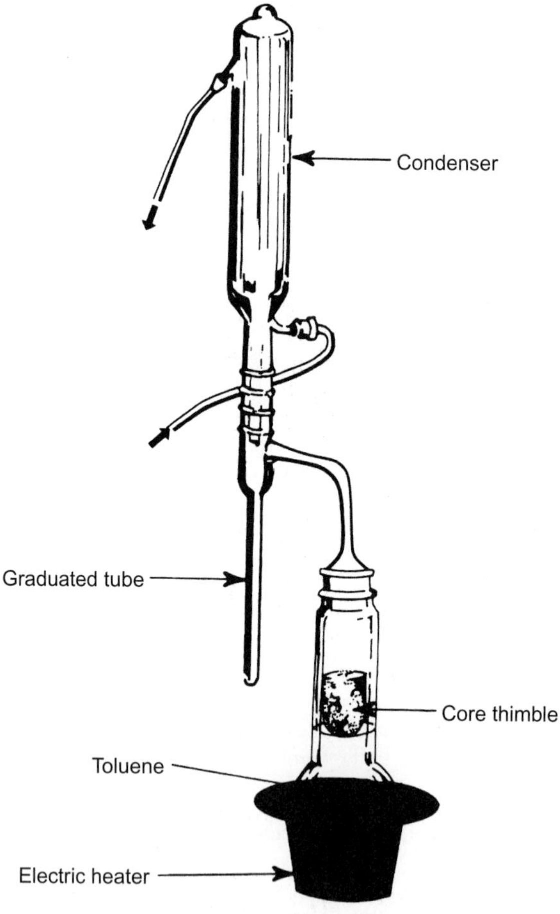

FIGURE A2.1 Equipment for measurement of water saturation by solvent extraction.

The boiling point of the hydrocarbon fluid used for the extraction must be higher than the boiling point of water and preferably lower than the boiling point of the hydrocarbons in the core (toluene is the most frequently used solvent: BP = 110.6°C). When the boiling point of the solvent is higher than 212°F (100°C), water vaporizes from the core and travels upward within the rising solvent vapors. The vapors (solvent plus water) are condensed and collected in a side receiver where the water is trapped at the bottom of the tube because its density is higher than the density of the solvent. The excess solvent drains back into the boiling pot where it is reheated to continue the process of extraction.

The crude oil that was originally a part of the core fluids will dissolve in the solvent. If the boiling point of the crude oil is higher than the boiling point of the solvent, the crude oil will remain in the boiling pot.

Equipment and Procedures

1. Pour toluene into the boiling flask until the flask is three-fourths full of toluene (or whatever hydrocarbon solvent is being used).
2. Obtain the mass of the saturated core sample.
3. Determine the bulk volume of the saturated core using a pycnometer.
4. Assemble the extraction equipment, which will consist of a heating mantle, boiling pot, extraction/collection tube, and a condenser.
5. Determine the density of the crude oil.
6. Set the heater control at a point where no gas or liquid can be seen escaping from the top of the condenser during condensation.
7. Record the volume of water collected in the graduated tube every 30 minutes. When no more water is collected after two readings, turn off the heater and record the volume of water collected.
8. When the solvent cools, remove the core and dry it in an oven equipped with an exhaust vent connected to a hood. If this type of oven is not available, allow the solvent to evaporate from the core by placing it in a hood overnight, and then dry the core in an oven.
9. Obtain the mass of the dried core.
10. Calculate the fluid saturation of the core from the data collected.

Sample Calculations

Data

$$\text{Porosity} = 0.20$$

$$\text{Bulk volume} = 25.0 \text{ ml}$$

$$\text{Density of the oil} = 0.88 \text{ g/ml}$$

$$\text{Mass of saturated core} = 57.0 \text{ g}$$

$$\text{Mass of dried core} = 53.0 \text{ g}$$

$$\text{Volume of water collected} = 1.4 \text{ ml}$$

$$\text{Pore volume} = \text{porosity} \times \text{bulk volume} = 0.20 \times 25.0 = 5.0 \text{ ml}$$

$$\text{Water saturation } S_w = \text{water volume/pore volume} = 1.4/5.0 = 0.28 = 28\%$$

Oil saturation S_o = oil volume/pore volume

= mass of oil/density of oil × 1.0/pore volume

= (mass saturated core − mass dry core − mass of water)/(oil density × pore volume)

= (57.0 − 53.0 − 1.4)/(0.88 × 5.0)

= 0.59 = 59%

Gas saturation $S_g = 1.0 - S_o - S_w = 1.8 - 8.28 - 0.59$

= 0.13 = 13%

Questions and Problems

1. Compare the accuracy of the "Measurement of Saturation by Solvent Extraction" (Experiment 2) to the "Fluid Content of Rocks by the Retort Method" (Experiment 1).
2. Which method of saturation measurement (solvent extraction or retort method) would you prefer to use if you were an engineer sending cores to a laboratory to be tested? Explain why you would choose one method over another.
3. List as many advantages and disadvantages as you can for both methods: (1) solvent extraction and (2) retort method.

Reference

Amyx JW, Bass Jr. DM, Whiting RL. Petroleum reservoir engineering. New York: McGraw-Hill; 1960, chapter 2.

Experiment 3 Density, Specific Gravity, and API Gravity

Introduction

The density of a substance is the ratio of its mass to its volume. It is always necessary to state the units of density because it may be expressed in a number of different mass and volume units.

Specific gravity is the ratio of the mass of a volume of a substance to the mass of an equal volume of some substance taken as a standard. For convenience, the following standards are generally used: for water, 39.2°F (4°C) and 60°F (15°C), and for gases, the standard is dried air at the same temperature and pressure as the gas for which the specific gravity is sought.

API hydrometers were developed for crude oils and are defined at 60°F. The specific gravity (SG) also was defined at 60°F as the density of a fluid at 60°F referred to the density of water at 60°F. Thus, direct conversion between the API gravity and specific gravity is possible.

$$\text{Specific gravity} = \frac{141.5}{(\text{API gravity} + 131.5)} \tag{A3.1}$$

A complete set of charts has been prepared for the conversion of hydrometer readings at temperatures from 0°F to 150°F (−18°C to 66°C) for the range of API gravities from 0 to 99; these charts are published in the *Petroleum Production*

Handbook (McGraw-Hill). For example, using the API charts, assume that an API hydrometer reading is 12 at an observed temperature of 85°F (29°C). Enter the 10–19° API chart at a temperature of 85°F, move to the column under the observed value of 12, and find that the API gravity at 60°F is 10.7. Using Eq. (A3.2), the specific gravity is 0.9951.

The hydrometer design is based on Archimedes' principle. A body floating on a liquid will displace a volume of liquid equal to the mass of a floating body divided by its density. Hydrometers of different mass will float at different depths in liquids, and the depth to which each floats is read from a scale on the stem of the hydrometer, which is scaled to read the API gravity directly. The API scale is an arbitrary one related to the specific gravity as shown in Eq. (A3.1).

Equipment and Procedures

Measurement of API Gravity

1. Transfer a sample of oil to a cylinder that is at least 25 mm greater in diameter than the bulb of the hydrometer.
2. Place the hydrometer carefully into the liquid: do not wet the entire stem of the hydrometer because the mass of the liquid on the stem will adversely affect the measurement. Spin the hydrometer lightly and let it come to rest.
3. Record the scale reading on the stem where it encounters the surface of the liquid, and record the temperature of the liquid using the thermometer in the stem thermometer. Use a separate thermometer if one is not provided in the stem of the hydrometer.
4. Convert the API gravity measured at the liquid temperature to API gravity at 60°F using the API tables (*Petroleum Production Handbook*). Convert the API gravity to specific gravity using Eq. (A. 31). The specific gravity at 60°F may be converted to the specific gravity at any temperature using the K_{SG} multipliers (given in Table A3.1 and Eq. (A3.2)).

$$\text{Specific gravity at } T°\text{F} = \text{SG at } 60°\text{F} - K_{SG} \times (T°\text{F} - 60) \quad \text{(A3.2)}$$

$$\text{Density}(g/\text{ml}) \text{ at } T°\text{F} = \text{specific gravity at } T°\text{F} \times \text{density of water at } T°\text{F} \quad \text{(A3.3)}$$

Westphal Balance

The Westphal balance, the design of which is based on Archimedes' principle, provides a method for direct measurement of the specific gravity (Figure A3.1).

TABLE A3.1 Multiplier Constant for Conversion of Specific Gravity at 60°F to Specific Gravity at Other Temperatures

Specific Gravity at 60°F	K_{SG}
0.90	0.00035
0.85	0.00037
0.80	0.00040
0.75	0.00043
0.70	0.00048

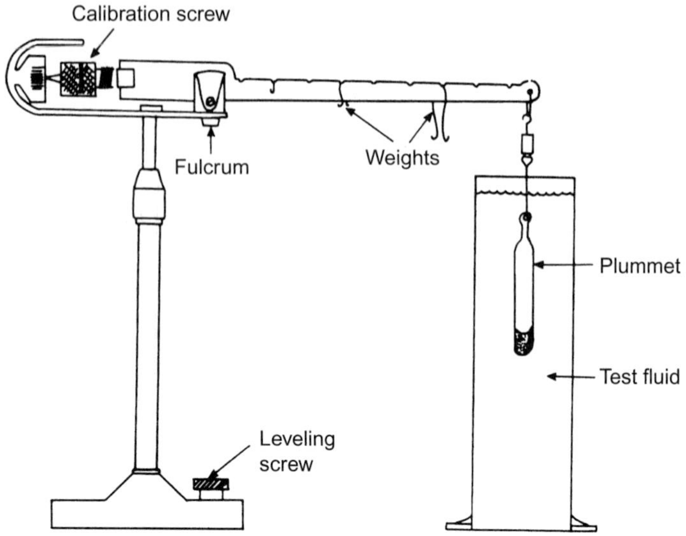

FIGURE A3.1 Westphal balance for measurement of specific gravity.

A plummet is attached to the end of the balance beam and immersed in a liquid. The beam is then balanced by adding or subtracting mass from the beam.

1. The instrument is first calibrated using deaerated distilled water.
2. The plummet is submerged in the water and the beam is balanced with a unit mass placed on mark 10 of the beam. Balancing is accomplished by turning the threaded counterweight at the end of the beam until the two pointers coincide. The temperature of the water is noted.
3. The plummet is removed and dried, and then submerged to about the same depth in the unknown liquid. The masses are adjusted to balance the beam. The masses are in ratios of 10: 1.0/0.1/0.01. The specific gravity of the unknown fluid, at the specified temperature, is equal to the total of the masses. If the temperature is 77°F (25°C) and the unit mass was placed on mark 8 of the balance beam, the 10th mass on mark 5, and the 100th mass on mark 2, the specific gravity is 0.852.

Questions and Problems

1. Convert the following °API to specific gravity at 60°/60°F: 50, 40, 30, 20, 10, and 1. Explain the meaning of 60°/60°F (refer to *Petroleum Production Handbook*).
2. Using the theory of additive gravities, compute °API for a crude oil–kerosene mixture as follows: 60% crude oil of °API = 32 plus 40% kerosene of °API = 48.
3. How would you measure the specific gravity of a viscous crude oil using a Westphal balance when the viscosity of the fluid hampers free movement of the plummet?

References

Amyx J W, Bass Jr DM, Whiting RL. Petroleum reservoir engineering. New York: McGraw-Hill; 1960, chapter 2.

Archer J S, Wall CG. Petroleum engineering. London: Graham & Trotman; 1986, chapter 2.

Fisher Scientific Co. Fisher-tag manual for inspectors of petroleum. 28th ed. New York; 1954. p. 218.

Frick TC, editor. Petroleum production handbook. New York: McGraw-Hill; 1962.

Experiment 4 Specific Gravity of Gases

Introduction

The Schilling specific gravity analyzer for gases is based on the principle that less dense gas molecules move with greater velocity than do heavier molecules at the same temperature and pressure. Therefore, a gas with the lowest specific gravity escaping from a container through a small orifice will escape in the shortest time.

Equipment and Procedures

The Schilling instrument is a glass tube (1 in in diameter and 10 in long) with the bottom end open and suspended in a tube about 4 in in diameter. The smaller tube has a mark just below its midpoint. Water is added to the larger tube to within about 1 in from the top.

Air pressure is applied at the top of the water in the larger tube, forcing water to rise into the smaller tube until it reaches the mark. The gas entry valve is then closed.

A valve connected to a small platinum orifice is opened, and the time required for the water level in the inside tube to return to its original level is recorded.

Gas, the specific gravity of which is to be determined, is introduced into the instrument, filling the smaller inner tube and the top of the larger tube. The gas should be allowed to bubble through the water for a short period to remove dissolved air from the water. The experiment is then repeated for the gas, and the specific gravity of the gas is calculated as follows:

$$SG = \frac{(t_{gas})^2}{(t_{air})^2} \tag{A4.1}$$

where

SG = specific gravity

t_{gas} = effusion time for the gas

t_{air} = effusion time for the air

The Edwards gas density balance consists of a beam mounted on a fulcrum with a large bulb attached to one end of the beam. The bulb is large enough so that the forces acting on the beam and counterweight may be neglected. The bulb furnishes a volume on which the buoyant force of a gas in the chamber can act.

The beam is mounted inside the chamber, which is surrounded by a water jacket to maintain constant temperature during measurements. A manometer attached to the gas chamber measures the difference between the pressure inside the chamber and atmospheric pressure. The pressure must be converted to absolute pressure by adding the barometric pressure.

A drying tube (containing anhydrous calcium chloride) is attached to the inlet gas line to remove moisture from the gas. Small pet-cocks are used to introduce the gas gradually into (or to let gas out of) the chamber.

The buoyant force of an unknown gas acting on the bulb is equalized to the buoyant force of air in the balance, and the specific gravity of the unknown gas is calculated as follows:

$$SG = \frac{P_{air}}{P_{gas}} \quad \text{(A4.2)}$$

where

SG = specific gravity
P_{air} = absolute pressure of air that balances the beam
P_{gas} = absolute pressure of the unknown gas that balances the beam.

1. Flush the chamber with dry air.
2. Carefully raise the pressure of air in the chamber until a definite pressure is attained, as shown by the manometer.
3. Balance the beam to a steady zero reading and record the pressure shown by the manometer.
4. Flush the system with the gas to be measured. Be sure to pass all gases entering the chamber through the drying tube.
5. Raise the pressure of the gas in the chamber until the beam is balanced once more.
6. Record the pressure on the manometer and the temperature of the water bath.
7. Calculate the results using Eq. A4.2

Sample Calculations

Pair = 51.1 mmHg
$P_{propane} = 33.6$ mmHg
$T_{ambient} = 79°F$
$SG_{79°F} = 51.1/33.6 = 1.521$

Questions and Problems

1. Measure the specific gravities of several known gases (nitrogen, helium, etc.) and compare your measurement with handbook (e.g. Perry's Chemical Engineer's Handbook, McGraw-Hill, New York) values. When you have developed sufficient skill to make accurate measurements, measure the specific gravity of unknown samples.
2. Discuss the sources of error inherent in the two methods: Schilling versus Edwards. Which is more accurate? Which is simplest in execution?
3. What is the mass, in kg, of 500 m^3 of air at 3 atm of pressure and 361°F?

Reference

Amyx JW, Bass Jr. DM, Whiting RL. Petroleum reservoir engineering. New York: McGraw-Hill; 1960, chapters 1 and 4.

Experiment 5 Viscosity

Introduction

Viscosity is the measure of a fluid's inherent resistance to flow. Fluid viscosity is very sensitive to changes in temperature; therefore, in this experiment, the viscosities of fluids are determined at several temperatures, and a plot of viscosity versus temperature for oil samples is made on semilog paper.

The terms normally used for viscosity are "dynamic" or "absolute" viscosity, which imply that the fluid is in motion. These terms are distinguished from the "kinematic viscosity," which is defined as the dynamic viscosity divided by the density of the fluid.

A poise (g/cm $\times$ s in CGS units; kg/m $\times$ s in SI units) is the fluid viscosity that requires a shearing force of 1 dyne to move a 1-cm^2 area plate (parallel to another plate) through the liquid with a velocity of 1 cm/s. A stoke (cm^2/s in CGS units; m^2/s in SI units) is defined as a poise divided by the density.

Equipment and Procedures

Several instruments have been developed for measurement of viscosity; they depend on the pressure of the liquid being tested to provide the force to drive the liquid through the instruments, thus providing a measure of the kinematic viscosity, which is expressed in centistokes (cS = cm^2/100 s). The Cannon-Fenske viscometer equipped with a clear, liquid-controlled temperature bath provides a method for determining viscosity at several temperatures (Figure A5.1). The Saybolt viscometer contains brass viscometer tubes set in a temperature-controlled oil bath with a calibrated orifice placed at the outlet of the viscometer tube.

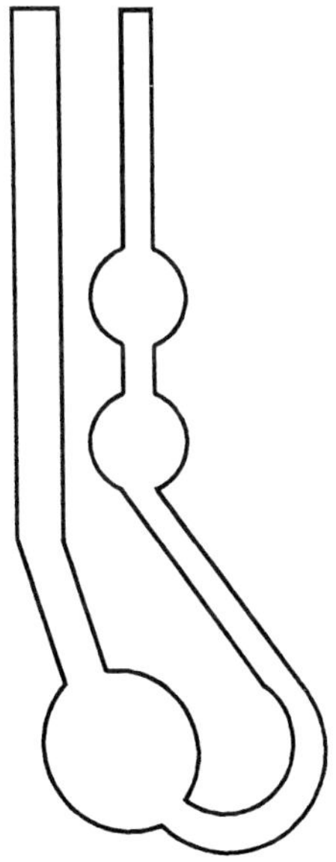

FIGURE A5.1 Cannon-Fenske viscometer.

The time required for 60 ml of the liquid being tested to flow from the viscometer is a measure of the viscosity in Saybolt universal seconds.

Saybolt universal viscosity, determined at various temperatures, may be converted to kinematic viscosity using the proper relationship, for example:

$$\begin{aligned}\text{Saybolt seconds} &= \text{cS} \times 4.628 \text{ at } 100°\text{F}\\ \text{Saybolt seconds} &= \text{cS} \times 4.629 \text{ at } 130°\text{F}\\ \text{Saybolt seconds} &= \text{cS} \times 4.652 \text{ at } 218°\text{F}\end{aligned} \tag{A5.1}$$

Cannon-Fenske Viscometer

1. Invert the viscometer, immerse the small diameter tube in the liquid to be measured, and using suction on the large-diameter tube, draw the liquid into the viscometer filling the two small bulbs (Figure A5.1). It is important that the proper amount of fluid be used in order to correspond to the calibration conditions that were used.
2. Return the viscometer to its normal, upright position and place it in a holder in the cool water bath. The large bulb at the bottom must be vertically below the two smaller bulbs, and the capillary tube must be slightly inclined from the vertical as shown in Figure A5.1.
3. Allow the liquid to drain into the larger bulb at the bottom. Then draw the liquid back up by suction on the smaller tube, until the top of the liquid is about 0.5 cm above the mark between the two small bulbs.
4. Allow the liquid to drain by gravity and record the time required for the top of the liquid to move from the top etch-mark to the bottom etch-mark under the bottom small bulb. Then record the temperature of the liquid bath. Repeat the measurement once more.
5. Place the viscometer, with its test liquid, in a warmer temperature bath and allow at least 10 minutes to achieve temperature equilibrium. Always move from the cooler to the warmer baths. Record the temperature and time for two readings at each bath temperature.
6. Using the capillary number of the Cannon-Fenske viscometer that was used for the test, obtain the calibration constant, K_{vis}, from Table A5.1 below. Table A5.1 may also be used to select an appropriate viscometer if an estimate of the viscosity is known.

Saybolt Viscometer

1. Clean the viscometer tube with a light paraffin solvent.
2. Before making the test, completely fill the viscometer tube with the oil to be tested. Allow this oil to flow through the orifice into a beaker in order to wet the wall of the tube.
3. Insert the cork stopper in the opening in the base of the tube below the orifice and fill the tube until it overflows slightly. The excess oil will collect in the annular tray at the top of the tube.
4. Gently stir the sample in the tube with the thermometer until a constant temperature is attained.
5. Place a receiving flask under the orifice; snap the cork from the opening and simultaneously start recording the time with a stopwatch.
6. Stop the watch when the tube empties, or when 60 ml is collected.

TABLE A5.1 Calibration Constants (K_{vis}) and Approximate Viscosity Ranges to be Used for Specific Cannon-Fenske Viscometers

Capillary Number	Capillary (mm)	Constant K_{vis} (cS/s)	Measuring Range (Approximate cS)
25	0.30	0.002	0.4–1.6
50	0.44	0.004	0.8–3.2
75	0.54	0.008	1.6–6.4
100	0.63	0.015	3–15
150	0.77	0.035	7–35
200	1.01	0.10	20–100
300	1.26	0.25	50–200
350	1.52	0.5	100–500
400	1.92	1.2	240–1,200
450	2.30	2.5	500–2,500
500	3.20	8	1,600–8,000
600	4.10	20	4,000–20,000

7. Check the measurement by running the test three times.
8. Convert the Saybolt universal seconds to kinematic viscosity (cS) using Eq. (A5.1). Obtain the density of the oil from the API gravity measurement using Eq. (A3.1), and convert the kinematic viscosity to absolute viscosity (cP) by multiplying the kinematic viscosity by the density.

Questions and Problems

1. How do your values of viscosity compare with literature values for the same fluids?
2. Explain Stokes' law.
3. How does the viscosity of a liquid vary with temperature? How does the viscosity of a gas vary with temperature?
4. Describe two methods for measuring viscosity.

Reference

Amyx JW, Bass Jr. DM, Whiting RL. Petroleum reservoir engineering. New York: McGraw-Hill; 1960, chapter 1.

Experiment 6 Fluorescence

Introduction

Modern well-logging methods locate petroleum-bearing formations without the aid of odor, taste, and visual observations of cores, or drill cutting; however, these methods are occasionally useful.

Rock samples are analyzed using a petrographic binocular microscope to qualitatively provide a description of the core materials along with any indication of oil or gas.

Fresh samples can indicate the presence of hydrocarbons by a distinctive aromatic odor, suggesting that more careful analyses of other samples or logs from the zone should be made.

Crushed samples of the cores, or cuttings, can be extracted with a light solvent such as pentane, carbon tetrachloride, or ether. The extract is then examined visually for change of color and for fluorescence to provide another qualitative indication of the presence of petroleum in the formation.

Fluorescence of the samples, or a vial of the solvent extract of samples, is examined in a box under ultraviolet light. The aromatic ring compounds in crude oils fluoresce with a white to yellow glow. This property has been proved to be a very sensitive indicator of the presence of crude oil in rocks (fresh and aged in storage), muds, and solvents. It has also been used for the qualitative analyses of lighter fractions of oils and lubricants.

Ultraviolet light has a higher frequency than visible light and, therefore, is just outside of the visible light spectrum. Under ultraviolet light, however, many molecules absorb some of the ultraviolet light energy, causing their electrons to jump to a higher energy-level orbit. In the case of molecules, the higher energy level produces strain between molecular bonds. The excited state exists for a very short period (about 10^{-8} s), after which the electron returns to the lower orbit accompanied by the emission of a small amount of light in the visible region. The continued excitation and relaxation of the electrons produces sufficient light to observe fluorescence. Small molecules, which cannot store much of the energy as vibrational energy, fluoresce at shorter wavelengths, producing a blue light. Larger molecules, which are capable of absorbing more energy, fluoresce at longer wavelengths, yielding fluorescence that appears yellow, orange, or brown.

Dilution of liquid samples sometimes may increase fluorescence. In concentrated solutions, dimerization (reaction of excited molecules with unexcited molecules) can take place. The gain in energy is distributed throughout the new double molecule, and when the dimer splits (after a very short duration, about 10^{-3} s), the energy of excitation is emitted as heat and, consequently, no fluorescence occurs. If the sample is diluted, however, the opportunity for molecular collision is diminished, and fluorescence will increase considerably.

Equipment and Procedures

The equipment used for fluorescence analysis is a sealed box with a door at one end for the introduction of samples, and viewing ports for observation. An ultraviolet light is located on the inside top, out of the viewing range. Samples, and comparative rocks or solvents, are placed on the floor of the box, and observations are recorded as positive or negative.

Considerable practice in the analyses of samples by this method is needed to gain sufficient skill for discrimination between blanks and samples containing very small amounts of aromatic, or multiple-bond, compounds.

1. Test a small sample (5–10 cm^3) of oil by observing it first under "white" light and then under ultraviolet light. Describe the differences noted in the observations.

2. Add about 0.5 cm^3 of carbon tetrachloride to the oil in the beaker, and compare it to another beaker containing the oil alone under the ultraviolet light. Describe the differences noted in the appearance of the diluted and undiluted samples under ultraviolet light.
3. Obtain samples of cores that do not contain crude oil and others that contain a small amount of residual crude oil. Compare the fluorescence of these samples under the ultraviolet light. Describe the differences observed.
4. Obtain some drilling fluid that has not been used and a sample of drilling fluid from a drilling rig. Place small amounts of each in beakers and observe them under the ultraviolet light. Add a small amount of crude oil (2–3 drops) to the fresh drilling mud, stir, and observe the sample under the ultraviolet light. Present as detailed a description as possible of the observations.

Questions and Problems

1. Describe the types of compounds in crude oils that are fluorescent under ultraviolet light.
2. What properties of the fluorescent hydrocarbons are responsible for fluorescence?
3. If the fluorescence of a sandstone rock is suspected to be due to the presence of minerals, such as calcite, how can the test be modified to ensure correct analysis?
4. Define the fluorescence "show number."

Reference

Helander DP. Fundamentals of formation evaluation. Tulsa, OK: Oil & Gas Consultants Publications; 1983, chapter 1.

Experiment 7 Absolute and Effective Porosity

Introduction

The most basic property of reservoir rocks is porosity (ϕ), which allows the rock to store fluids (gas, oil, and water). The total volume of fluid that can be stored in a given *bulk volume* (V_b) of rock is its *pore volume* (V_p). The solids *volume* (or grain volume), (V_s), is the portion of the rock consisted only of solid matter. The total exterior or bulk volume of the rock is, therefore, the sum of the grain and pore volumes. The total or *absolute porosity* is the pore volume divided by the bulk volume (expressed in percentage, or fraction, of the bulk volume). The *void ratio* (e) is the pore volume divided by the solids volume.

The complex internal structure of pores and solution vugs in rocks frequently results in the formation of isolated pores. These isolated pores contribute to the overall (or absolute) porosity of the rock but are not interconnected to the main body of pores and, therefore, are not involved in the flow of fluids through the rock. The interconnected pores that support the flow of fluids make up the *effective porosity*, which is numerically less than the absolute porosity, that is, the intercommunicating porosity *excluding* the pores containing irreducible fluid saturation. The small pores are occupied by the irreducible fluid (water or oil), and cracks and dead-end pores are not involved in the flow process.

For example, the mass (M) of a sample of rock with a bulk volume (V_b) of 15 cm^3 is 30.3 g; it was found to contain an absolute grain volume (V_s) of 11.25 cm^3, an interconnected pore volume (V_p) of 3.0 cm^3, an unconnected (isolated) pore space (V_i) of 0.75 cm^3, and an irreducible water saturation (S_{iw}) equal to 6%. The rock sample has the following characteristics:

Bulk density $(\rho_b = M/V_b) = 30.3/15 = 2.02\ \text{g/cm}^3$

Solids volume $(V_s = V_b - V_p - V_i) = 15 - 3 - 0.75 = 11.25\ \text{cm}^3$

Effective pore volume $(V_p) = 3\ \text{cm}^3$

Isolated pore volume $(V_i) = 0.75\ \text{cm}^3$

Absolute pore volume $(V_t = V_p + V_i) = 3 + 0.75 = 3.75\ \text{cm}^3$

Grain density $(\rho_g = \text{mass}/V_s) = 30.3/11.23 = 2.70\ \text{g/cm}^3$

Bulk density $(\rho_b = \text{mass}/V_b) = 30.3/15 = 2.02\ \text{g/cm}^3$

Effective porosity $(\phi_e = V_p/V_b - S_{iw}) = 3/15 - 0.06 = 0.14 = 14\%$

Isolated porosity $(\phi_i = V_i/V_b) = 0.75/15 = 0.05 = 5\%$

Absolute porosity $(\phi_t = V_t/V_b) = 3.75/15 = 0.25 = 25\%$

Void ratio $(e = V_t/V_s) = 3.75/11.25 = 0.33 = 33\%$

Void ratio $[e = \phi_t/(1 - \phi_a)] = 0.25/0.75 = 0.33 = 33\%$

Equipment and Procedures

Absolute Porosity from Grain Volume

As shown here, there are two types of rock porosity: the absolute (or total) porosity, which is composed of both isolated and interconnected pore spaces, and the effective porosity, which includes only the interconnected pores containing mobile fluid. The effective porosity is of greater interest to petroleum engineers because fluids saturating a rock can move only through the effective pore spaces.

Three basic parameters determine the porosity: (1) bulk volume, (2) pore volume, and (3) rock matrix, or solids, volume ($V_b = V_p + V_s$).

To determine the absolute porosity, the density of the solid portion of the rock (its grain density) is first determined from a sample of the rock that has been crushed using an impact crusher (not a grinder). An appropriate size pycnometer, whose volume is known, is dried and weighed, and then the volume and mass of a portion of the sand grains are determined using the pycnometer.

1. Fill the pycnometer with a liquid (hydrocarbon or water) and obtain its mass ($M_{pyc} + M_1$).
2. Empty and dry the pycnometer.
3. Place a sample of crushed rock in the pycnometer (about one-half the volume of the pycnometer) and determine the mass ($M_{pyc} + M_{grains}$).
4. Fill the pycnometer (containing the sand grains), then with the liquid used in step 1 above, and determine the mass ($M_{pyc} + M_{grains} + M_1$).

5. The sand grain density is calculated from the data as follows:
Sample Calculation

$$V_{pyc}(\text{known volume}) = 10.0 \text{ cm}^3$$
$$M_{pyc}(\text{measured}) = 16.57 \text{ g}$$
$$M_{pyc} + M_l(\text{measured}) = 26.58 \text{ g}$$
$$M_1 = \left[(M_{pyc} + M_1) - M_{pyc}\right] = 26.58 - 16.57$$
$$= 10.01 \text{ g}$$
$$\rho_1 = \left(\frac{M_1}{V_{pyc}}\right) = \frac{10.01}{10.0} = 0.01 \text{ g/cm}^3$$
$$M_{pyc} + M_{grains}(\text{measured}) = 20.59 \text{ g}$$
$$M_{pyc} + M_{grains} + M_1(\text{added}) = 29.175 \text{ g}$$
$$M_{grains} = (M_{pyc} + M_{grains}) - M_{pyc} = 20.59 - 16.57$$
$$= 4.02$$
$$M_1(\text{added}) = \left[M_{pyc} + M_{grains} + M_1(\text{added})\right] - (M_{pyc} + M_{grains})$$
$$= 29.175 - 20.59 = 8.585$$
$$V_1(\text{added}) = \left(\frac{M_1(\text{added})}{\rho_l}\right) = \frac{8.585}{1.01} = 8.5 \text{ cm}^3$$
$$V_{grains} = \left[V_{pyc} - V_1(\text{added})\right] = 10.0 - 8.50 = 8.5 \text{ cm}^3$$
$$\rho_{grains} = \left(\frac{M_{grains}}{V_{grains}}\right) = \frac{4.02}{1.50} = 2.68 \text{ g/cm}^3$$

Grain and Bulk Volumes

The next step is to determine the grain and bulk volumes of a rock sample (core):

1. A core of convenient size is dried and its mass is determined (M_{core}).
2. The core is then saturated with liquid by first evacuating the core to remove air and then admitting the liquid into the vacuum flask containing the core to fill the pores with the liquid.
3. The bulk volume (V_b) of the core may be determined accurately using a beam balance. Tie the core to a fine wire, attach it to the beam balance, and then immerse the core in a beaker with the same liquid used for the measurements above. The core should be immersed until it is just below the surface of the liquid, and it should not touch the sides or the bottom. Obtain the mass of immersed core (M_{im}).
4. Remove the beaker of liquid, carefully wipe the excess liquid from the surface of the core, and obtain the mass of the saturated core ($M_{core\text{-}sat}$).
Sample Calculation

$$M_{core} = 39.522 \text{ g}$$
$$M_{im} = 24.393 \text{ g}$$
$$M_{core-sat} = 43.797 \text{ g}$$
$$V_{grains} = \frac{M_{core}}{\rho_{grains}} = \frac{39.522}{2.68} = 14.747 \text{ cm}^3$$
$$V_b = \left[\frac{(M_{core-sat} - M_{im})}{\rho_1}\right]$$
$$= \frac{(43.797 - 24.393)}{1.01} = 19.212 \text{ cm}^3$$

$$\text{Porosity(absolute)} = \left(\frac{1 - V_{\text{grains}}}{V_b}\right) = 1 - \frac{(14.747)}{19.212} = 0.232$$

Effective Porosity

The effective porosity is the ratio of the interconnected pore space in the rock to the bulk volume minus the irreducible saturation. For highly cemented sandstones and carbonate rocks, there can be a significant difference between the absolute and effective porosities.

Sample Calculation

$$\text{Effective pore volume of core } (V_p) = \left[\frac{(M_{\text{core-sat}} - M_{\text{core}})}{\rho_{\text{liq}}}\right]$$

$$= \frac{(43.797 - 39.522)}{1.01} = 4.233 \text{ cm}^3$$

$$\text{Effective porosity}\left(\frac{V_p}{V_b - S_{iw}}\right) = \frac{4.233}{19.212} - 0.06 = 0.16$$

Summary

$$\text{Absolute porosity} = \frac{1.0 - V_{\text{grains}}}{V_b} = 0.232$$

$$\text{Pore volume} = \frac{(M_{\text{core-sat}} - M_{\text{core}})}{\rho_1} = 4.225 \text{ cm}^3$$

$$\text{Effective porosity} = \frac{V_{\text{pore}}}{V_b - S_{iw}} = 0.160 \text{ or } 16\%$$

Porosity Measurement by Mercury Injection

The mercury porosimeter is designed to yield bulk volume as well as pore volume (Chapter 5 text and Figure 5.11). The porosimeter consists of a hand-screw mercury pump attached to a pycnometer. The hand-screw has a precision-measuring screw that measures the volumetric displacement of the piston, which is indicated on a scale and on a micrometer dial.

1. Open the pycnometer valve. Displace mercury into the pycnometer by turning the hand-crank until mercury appears in the pycnometer valve seat.
2. Set the scale reading to zero by adjusting the dial on the hand-crank without turning.
3. Withdraw 15 ml of mercury from the pycnometer.
4. Place the core in the pycnometer. Lock the pycnometer lid and leave the valve open.
5. Inject mercury into the pycnometer until it appears in the pycnometer valve seat. The bulk volume of the core is read directly from the scale on the hand-crank wheel dial.
6. Adjust the scales to zero without turning the hand-crank.
7. Close the pycnometer valve and increase the pressure to about 750 psig. The volume scale on the instrument gives the apparent pore volume (V_{p-app}), which must be corrected for instrument volumetric deviation when the pressure is raised from atmospheric pressure to 750 psig. The correction factor (F_c) is obtained as described in steps 8–10.
8. Remove the core from the pycnometer and displace mercury into the pycnometer until it appears in the pycnometer valve seat.

9. Set the scales to zero without turning the hand-crank.
10. Close the pycnometer valve and increase the pressure to 750 psig. The volumetric correction factor (F_c) is the volume of mercury displaced by the piston when the pressure in the mercury-filled instrument is raised from zero to the reference pressure (750 psig).

Sample Calculation

$$V_p = (V_{p-app} - F_c) \times 1.02 \tag{A7.1}$$

where multiplication by 1.02 is used to account for compression of air trapped in the core.

$$\text{Porosity} = \frac{V_p}{V_b} \tag{A7.2}$$

Porosity Measurement by Gas Compression/Expansion

The method consists of placing a dry core sample in a sealed chamber of known volume (V_1), which is then filled with gas to a convenient pressure (p_1) (Figure A7.1). The second chamber, connected by a closed valve, is either evacuated or filled with gas at a different pressure (p_2). The valve between the two chambers is then opened and the final pressure, common to both chambers (p_f) is observed.

The grain volume of the core sample can then be calculated using the gas law: $pV = nRT$, where p is the pressure, V is the volume of the fluid, n is the number of moles of the fluid, R is the gas constant, and T is the absolute temperature.

The moles of gas in each chamber, after the initial pressure has been established, are:

$$n_1 = p_1 \times \frac{V_1}{R} \times T \text{ and } n_2 = p_2 \times \frac{V_2}{RT} \tag{A7.3}$$

where the temperature, T, is assumed to be constant throughout the experiment.

After the valve between the chambers is opened and the pressures have come to equilibrium, the total moles of gas in the system are calculated.

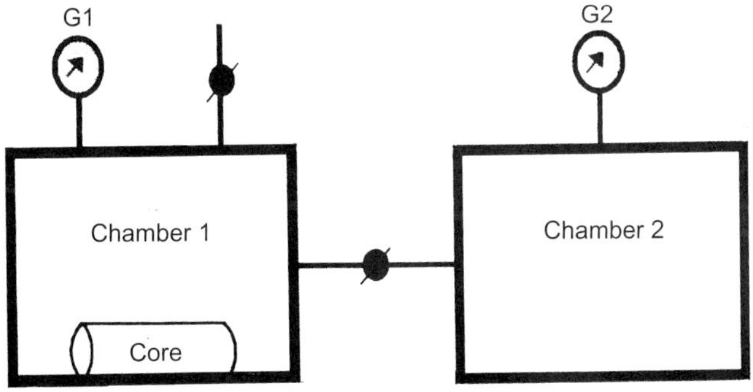

FIGURE A7.1 Equipment for measurement of porosity by gas compression/expansion.

Sample Calculation

$$n_{total} = n_1 + n_2$$
$$= p_f \times \frac{V_f}{RT} = p_1 \times \frac{V_1}{RT} = p_2 \times \frac{V_2}{RT} \tag{A7.4}$$
$$p_f \times V_f = p_f(V_1 + V_2) = p_1 V_1 + p_2 \times V_2$$

When a core is placed in chamber 1, the actual volume of gas contained in the chamber is not the volume of the original chamber because the solids volume from the core sample (V_S) occupies some of the space; hence, the gas volume of chamber 1 is actually ($V_1 - V_2$), and this term must, therefore, be substituted into Eq. (A7.4):

$$p_f(V_1 - V_s + V_2) = p_1(V_1 - V_s) + p_2 V_2 \tag{A7.5}$$

Equation (A7.5) may be rearranged to yield the solids, or grain, volume of the sample as follows.

$$V_s = \frac{[V_1(p_f - p_1) + V_2(p_f - p_2)]}{(p_f - p_1)} \tag{A7.6}$$

The bulk volume is then determined by fluid displacement, or by careful measurement of the core dimensions, and porosity is calculated as before:

$$\text{Porosity} = \frac{(1.0 - V_s)}{V_b} \tag{A7.7}$$

Sample Calculations

First, use Eqs. (A7.4) and (A7.5) to determine the volumes of the two chambers (V_1 and V_2) using a steel cylinder of known volume, V_c.

1. Close the valve between the two chambers and open chamber 1 (V_1) to the atmosphere ($p_1 = 0.0$ kPa).
2. Admit gas to chamber 2 (V_2) to a pressure of 413.7 kPa ($p_2 = 413.7$ kPa).
3. Close chamber 1 (V_l) so that it is no longer open to the atmosphere; then open the isolation valve between the chambers and record ($p_{f(1)} =$ 183.869 kPa).
4. Place a steel cylinder of known volume ($V_c = 2.54$ cm in diameter × 5.08 cm long = 25.741 cm^3) into chamber 1 at atmospheric pressure; close the isolation valve and chamber 1 valve to the atmosphere.
5. Admit pressure to chamber 2 to a convenient value: for example, $p_2 = 354.541$ kPa.
6. Open the isolation valve and record $p_{f(2)}$ ($p_{f(2)} = 354.541$ kPa).
7. Calculate the volumes V_1 and V_2 by simultaneous solution of Eqs. (A7.4) and (A7.5) (where V_c is substituted for V_g in Eq. (A7.5) and the appropriate p_f is used):

   $183.869 \times (V_1 + V_2) = (0 + 413.700) \times V_2$
   $183.860 \times (V_1 - 25.741 + V_2) = (0 + 354.541) \times V_2$
   $V_1 = 100$ cm^3
   $V_2 = 80$ cm^3

   Next, use the following procedure and Eq. (A7.5) to measure the porosity.
8. Place the dry core in chamber 1 and close the isolation valve.
9. Admit gas to chamber 2, where ($p_2 = 413.700$ kPa).

10. Open the isolation valve and record p_f, which is equal to 199.783 kPa.

11. Calculate the solids volume of the core using Eq. (A7.5):

$99.783 \times (100 - V_s + 80) = (0 + 413.700) \times 80$

$V_s = 14.338\ \text{cm}^3$

12. Obtain the bulk volume from the core dimensions (1 in. diameter × in. long = 18.382 cm^3)

Porosity $= 1.0 - V_s/V_b = 1.0 - 14.338/18.382 = 0.22$ or 22%

The mass of the core saturated with liquid is greater than the mass of the saturated core immersed in the liquid due to the buoyancy of the sample in the liquid. This result was used to determine the mass of liquid displaced by the saturated core.

The grain density of the crushed rock is equivalent to the density of the dry core. This result was used to obtain the grain volume of the core.

The absolute porosity is greater than the effective porosity, which indicates that there are some pores that are not connected to the main flow channels in the rock. This occurs because of changes in the rock's internal structure due to compaction during burial and cementation of the grains by precipitation of various minerals, e.g., calcite, dolomite, silica, clays, and ferruginous materials.

Statistical Evaluation of Porosity Data

The vertical and areal porosity distributions of reservoirs approximate normal distributions. The simplest method for testing this, and for initiation of statistical evaluation of porosity data, is to plot the porosity versus the cumulative frequency on arithmetic probability paper. If the data can be approximated by a straight line on such a plot, the data can be described by a normal distribution, which leads to convenient statistical evaluations.

- *Mode* is the value of the porosity that occurs most frequently and can be estimated from the peak of the histogram of frequency versus porosity, about 15 in Figure A7.2.
- *Mean* or average value varies slightly depending on how the data are analyzed. The most familiar arithmetic mean is the sum of individual values divided by the total number of values:

$$\phi_{ave} = \frac{(\sum \phi_i)}{N} \tag{A7.8}$$

When the data are classified in ranges (Table A7.1), the arithmetic mean is expressed as:

$$\phi_{ave} = \frac{(\sum \phi_i F_i)}{\sum F_i} \tag{A7.9}$$

where

ϕ_i = class mark (porosity value at the midpoint of the interval) of the *i*th interval

N = number of class intervals

F_i = frequency of the *i*th class interval

- *Standard Deviation:* The standard deviation gives a measure of the dispersion of the data from the arithmetic mean. For unclassified data:

$$S_D = [\Sigma(\phi_i - (\phi_{mean})^2 \times N]^{0.5} \tag{A7.10}$$

where S_D = standard deviation.

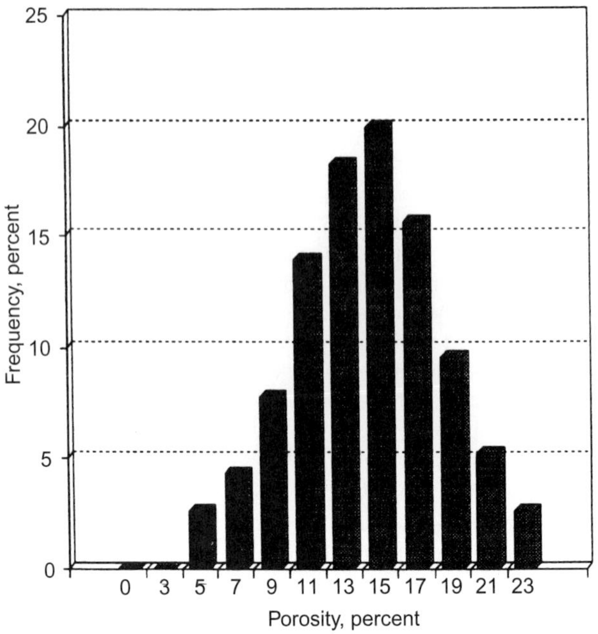

FIGURE A7.2 Histogram of frequency versus porosity.

TABLE A7.1 Statistical Evaluation of a Set of Porosity Data

(1) Porosity Interval	(2) Class Mark	(3) Number of Samples	(4) Frequency %	(5) Cumulative Frequency
4–6	5	3	2.61	2.61
6–8	7	5	4.35	6.96
8–10	9	9	7.83	14.79
10–12	11	16	13.91	28.70
12–14	13	21	18.26	46.96
14–16	15	23	20.00	66.96
16–18	17	18	15.65	82.61
18–20	19	11	9.57	92.18
20–22	21	6	5.22	97.40
22–24	23	3	2.61	100.01

A histogram of the frequency (*F*, col. 4) versus the porosity (*f*, col. 2) is illustrated in Figure A7.2 and the cumulative frequency (col. 5) is plotted as a function of porosity in Figure A7.3. Mean = 14.22, standard deviation = 4.21, and coefficient of variance = 0.30.

For classified data (Table A7.1), the standard deviation is calculated as follows:

$$S_D = [\Sigma(\phi_i - (\phi_{mean})^2 \times F_i]^{0.5} \tag{A7.11}$$

- *Normal curve:* The normal curve, represented by the straight line drawn through the points plotted on probability paper, may be represented analytically by an expression that is defined by the mean and the standard deviation:

$$f(\phi) = \left[\frac{1}{S_D} \times (2 \times \pi)^{0.5}\right] \times \left[\exp -0.5\left[\frac{(\phi_i - \phi_{mean})}{S_D}\right]^2\right. \tag{A7.12}$$

where $f(\phi)$ = function of porosity.

- *Variance:* The degree of heterogeneity can be expressed as a function of the coefficient of variation (or dispersion) of the data. The coefficient of variance is the standard deviation divided by the mean. A coefficient value of zero means that there is zero porosity variation and, consequently, the rock is completely homogeneous. A coefficient value of 1.0 indicates that the rock is completely heterogeneous.
- *Median:* The median is the central point of the distribution. Values are equally divided on each side of the median that occurs at the 50 percentile mark on a cumulative frequency curve (Figure A7.3). The cumulative frequency curve differs from the frequency distribution histogram because each point represents the frequency of each class as well as the sum of all percentages of the

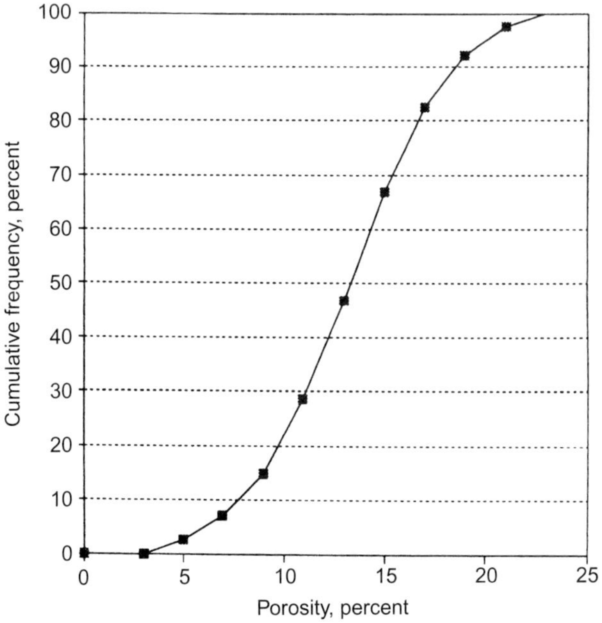

FIGURE A7.3 Cumulative frequency versus porosity. Each point represents the frequency of each class and the sum of all preceding size classes.

preceding size classes. Various vertical scales may be used for cumulative frequency curves: arithmetic, probability, and logarithmic.

Questions and Problems

1. Which porosity (absolute or effective) is measured by the gas compression/expansion method? Which porosity is measured by the mercury injection method?
2. How accurate are laboratory measurements of porosity from core samples when applied to an entire reservoir?
3. In the calculation of pore volume by the mercury injection method, why is it necessary to multiply by 1.02?
4. Why is the measurement of porosity so important to petroleum engineering?

References

Amyx JW, Bass Jr. DM, Whiting RL. Petroleum reservoir engineering. New York: McGraw-Hill; 1960, chapter 2.

Anderson G. Coring and core analysis handbook. Tulsa, OK: PennWell Books; 1975, chapter 2.

Archer JS, Wall CG. Petroleum engineering. London: Graham & Trotman; 1986, chapter 5.

Experiment 8 Particle Size Distribution

Introduction

One of the methods used for determining grain size distribution is sieve analysis. After the grain size distribution has been determined, the depositional history of the rock may be inferred from graphical analysis of the grain size distribution. The distribution of sizes of the grains in sediments is related to (1) the availability of different sizes of particles in the parent matter from which the grains are derived and (2) the processes operating where the sediments were deposited, particularly the competency of fluid flow (in other words, the history of sedimentary processes).

Measurement of the grain size distribution will yield a plot of the cumulative mass percent (frequency) of ranges of grain sizes versus the phi scale used for particles size notation, the surface area of the rock per unit of pore volume and bulk volume, and the surface area per unit of grain volume of the sediment.

Equipment and Procedures

In general, the procedure is to carefully crush the rock with an impact crusher (not a grinder) to obtain individual grains. A set of sieve trays is assembled with the finest screens at the bottom (Figure A8.1). In sieve analysis, it is assumed that the grains caught on the individual screens have sizes smaller than the openings of the screen above and larger than the screen they are resting on. The amount of sand caught on each screen in the stack is weighed and, as a first approximation, the average grain diameter is assumed to be equal to the average of the screen opening sizes between which it was trapped. The second

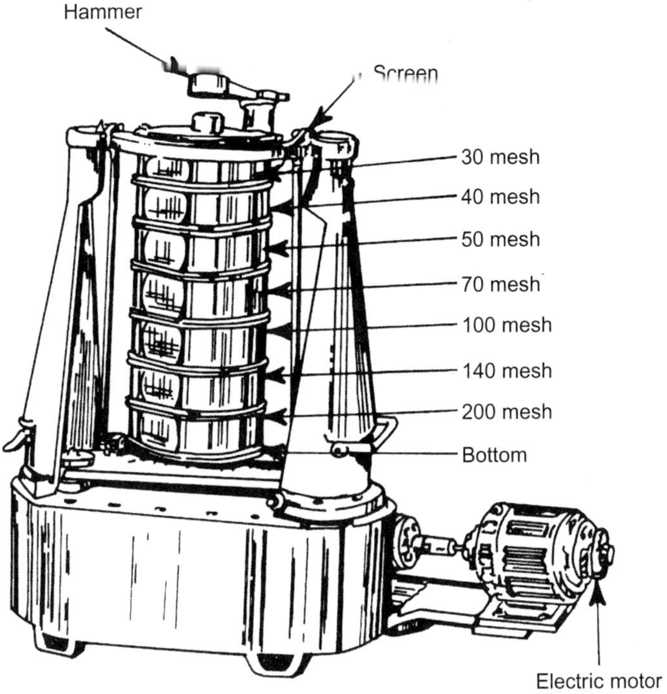

FIGURE A8.1 Assembly of sieve trays (finest screens at the bottom) for analysis of the grain size distribution of sediments.

assumption applied is that the mineral grains are spherical in shape. With these assumptions, several statistical analyses can be made.

The relationship between the surface area and volume of a sphere is as follows:

$$\frac{S_{A(\text{sphere})}}{V_{(\text{sphere})}} = \frac{\pi d^2}{\pi d^3/6} = \frac{6}{d} \qquad \text{(A8.1)}$$

where

$S_{\text{A(sphere)}}$ = surface area of a sphere

$V_{\text{(sphere)}}$ = volume of a sphere

The average surface of the grains is taken as the sum of the surface areas determined for the sand caught on each screen size. This surface area is based on the average grain diameter. Summing up the calculated surface area for each of the sieve screens is performed as follows:

$$S_{total} = V_{total}\left(\frac{S_{total}}{V_{total}}\right) = \sum V_{ith\text{SC}} \times \frac{6}{d_{ith\text{SC}}} = \sum \frac{M_{ith\text{SC}}}{\rho_{sd}} \times \frac{6}{d_{ith\text{SC}}} \qquad \text{(A8.2)}$$

where

$d_{ith\text{SC}}$ = diameter of the ith screen

ρ_{sd} = density of the sand

S_{total} = total surface area of the sample
V_{ithSC} = volume collected on the ith screen
V_{total} = total grain volume of the sample
M_{ithSC} = mass (grams) collected on the ith screen

The surface area of a porous medium does not quantify the value; instead, the surface area must be a function of the volume of material in which it was contained (cm^2/cm^3). Smaller particles packed in a unit volume have much more surface area than do larger particles filling the same space. On the basis of the three rock volumes (pore volume, grain volume, and bulk volume), the surface area of a porous medium can be expressed in three ways. The surface areas are defined as follows:

S_{BV} = surface area per unit bulk volume of the porous medium
S_{MG} = surface area per unit volume of the mineral grains comprising the porous medium
S_{PV} = surface area per unit volume of pore space within the porous medium.

The three surface area expressions are related through the definition of porosity (the ratio of the pore volume to the bulk volume), thus:

$$S_{BV} = (1 - f) \times S_{MG} = f \times S_{PV} \quad \text{(A8.3)}$$

The expression for calculating the surface area per unit volume of mineral grains (S_{MG}) from sieve analysis data may be derived as follows:

$$S_{MG} = \frac{S_{total}}{V_{total}} = \sum \left(\frac{M_{ithSC}}{\rho_{sd}} \times \frac{6}{d_{ithSC}} \right) = 6 \sum \frac{M_{ithSC}/d_{ithSC}}{\sum M_{ithSC}} \quad \text{(A8.4)}$$

The standard size classifications for sedimentary particles are listed in Table A8.2. The size classes are based on a geometric scale in which the adjacent orders within the scale differ by a factor of 2. Thus, going from large to small, the ratio is one half: $1, \frac{1}{2}, \frac{1}{4}, \frac{1}{8}$, 8 mm, etc. In the other direction, the ratio is 2 : 1, 2, 4, 8 mm, etc. The sizes also are expressed as the negative logarithm of these dimensions to base 2, which is known as the phi scale:

$$\text{Phi}(\phi) = -\log_2 d = -3.322 \times \log_{10} d \quad \text{(A8.5)}$$

The principal advantage of the phi scale is that it allows plotting of the particle size distributions on linear graph paper, and the calculation of the various statistical parameters is simplified. It also simplifies the geologic practice of plotting the larger sizes on the left and smaller sizes on the right.

Sieve Analysis Procedure

Equipment:

Analytical balance
Rock crushing instrument
Mortar and pestle
Sieves
Mechanical shaker for stacked sieves

1. Using a sledgehammer, break a clastic rock sample into pieces approximately 1 in. in diameter.
2. Crush the 1-in. pieces in the rock crusher and collect the crushed rock in a mortar.

3. Crush the sample into grains using the mortar and pestle. Do not crush the individual grains by using a rotary motion of the pestle. Separate the grains by using a back-and-forth motion.
4. Obtain a crushed sample of about 200 g.
5. Clean each of the sieves using a bristle brush (brush from the bottom).
6. Strike the sieve on the outside of the rim 2–3 times to loosen grains that are tightly trapped between the screen openings. It may not be possible to remove all of them.
7. Assemble the sieve on the shaker with the pan on the bottom and the coarsest of the screens on the top.
8. Pour the weighed, crushed sample of sand into the top sieve (be careful not to spill any of the sample).
9. Place the cover on the top sieve, tighten the stack of sieves onto the shaker, and shake the assembly for 5 minutes. Place any grains left on the top sieve into the mortar and recrush them. (Consult the lab instructor about this.) Return the crushed grains to the top sieve.
10. Shake the assembly for 15 minutes. A modification of the procedure at this point is to separate the sample into two new series of sieves for thorough sorting with additional shaking for 20 minutes. If the sample is not to be separated, continue shaking for an additional 10 minutes and then proceed with the weighing.
11. Weigh the contents of each sieve individually on an analytical balance and complete the sample calculation as follows:

Sample Calculations

$$S_{MG} = \frac{(6\Sigma M_i/d_i)}{\Sigma M_i} = \frac{6 \times 1{,}508.7}{49.24} = 183.8 \text{ in.}^2/\text{in.}^3 = 15.32 \text{ ft}^2/\text{ft}^3 \text{ mineral grains}$$

Assume a porosity of 10%, then:

$$S_{BV} = S_{MG} \times (1 - \phi) = 15.32 \times 0.90 = 13.8 \text{ ft}^2/\text{ft}^3$$

$$S_{PV} = \frac{S_{MG} \times (1 - \phi)}{\phi} = \frac{15.32 \times 0.90}{0.10} = 137.9 \text{ ft}^2/\text{ft}^3$$

TABLE A8.1 Surface Area

Sieve Size (in.)	Mass of Sand Trapped on Sieve, M_i(g)	Average Diameter d_i (in.)	M_i/d_i
0.12	0.00	–	
0.08	8.07	0.01	80.7
0.04	20.82	0.06	347.0
0.02	14.31	0.01	477.0
0.00	6.04	0.01	604.0
Total	49.24		1,508.7

TABLE A8.2 Data Graphs

Sieve Size (mm)	Sand Trapped on Sieve (g)	Average Diameter (mm)	Phi(ϕ) Size	Mass (%)	Cumulative Mass (%)
11	0	–	–	–	–
5	20	8	−3	0.10	0.10
3	30	4	−2	0.15	0.25
1	50	2	−1	0.25	0.50
pan	100	0.5	+1	0.05	1.00

Questions and Problems

1. Plot the cumulative weight percent versus phi size on linear coordinate paper and on probability-linear paper. These plots indicate the depositional environments of sedimentary particles. What is the probable depositional environment of the sample measured for this experiment?
2. Discuss the relationship of sorting and particle size to the overall surface area of clastic rocks.
3. How can knowledge of the grain size distribution aid in the evaluation of a petroleum reservoir?
4. How can grain size distributions be correlated to the shapes of the self-potential and gamma ray well logs?

References

Chilingarian GV, Wolf KH. Compaction of coarse-grained sediments, I. Amsterdam, The Netherlands: Elsevier Science; 1975, chapter 1.

Friedman GM, Sanders JE. Principles of sedimentology. New York: John Wiley & Sons; 1978, chapter 3.

Experiment 9 Surface Area of Sediments

Introduction

The theory of monolayer adsorption developed by Brunauer et al. (1938) is used for surface area measurement of sediments and other materials. Molecular adsorption is produced by van der Waals forces, which cause condensation of liquids and without which no adsorption would occur. Adsorption occurs at the most active site of the solid surface and then progresses to areas of lesser activity; finally, pore-filling will occur as gas condenses in the pores of the solid. Normally, only a monolayer of gas molecules covers the surface area of the solid, which is exposed to the gas at the temperature of condensation. When only a monolayer of gas is adsorbed on the surface, one can calculate the surface area that was covered by the gas molecules from the amount of gas that was adsorbed. The adsorption isotherm of nitrogen (volume adsorbed versus the relative pressure at constant temperature) is generally used for surface area measurement of sediments, grains, and powders. The volume of gas physically adsorbed as a unimolecular layer is determined by arranging the adsorption parameters in the form of a straight line:

$$\frac{P}{V(P_{N2} - P)} = \frac{1}{V_m C} + \frac{(C-1)P}{V_m C P_{N2}} \tag{A9.1}$$

where

P = equilibrium vapor pressure

P_{N2} = pressure of saturated nitrogen vapor at the test temperature

V = volume of nitrogen adsorbed at pressure P and the test temperature

V_m = volume of gas required to form a monolayer

C = dimensionless constant related to the heart of adsorption

A plot of $P/(P_{N2} - P)$ versus P/P_{N2} yields a straight line with intercept $1/V_m C$ and slope $(C-1)/V_m C$ from which V_m is determined by simultaneous solution of the equations. The volume gas (V_m) is then converted into molecules of nitrogen adsorbed on the sample. Hexagonal packing of the nitrogen molecules in the monolayer on the surface is assumed to determine the average area occupied by each molecule. This leads to the average area of the liquid nitrogen molecule 16.24 $Å^2$(1.624×10^{-15} cm^2). The surface area is calculated from the number of molecules adsorbed on the surface.

Helium is used to measure the volume of the system and sample cell (containing the degassed sample) because it does not adsorb on surfaces at ambient temperature and low pressure.

Equipment and Procedure

The equipment consists of (1) a gas burette, (2) sample cell, (3) Dewar flask containing liquid nitrogen, (4) vacuum pump, and (5) mercury manometer, as illustrated in Figure A9.1.

The gas lines and burette are flushed once with helium to remove air from the system. Then the burette is filled with helium, the remainder of the equipment is evacuated, and helium is expanded into the system and sample cell to determine their volumes. The system and sample cell are then flushed with nitrogen, and the amount of nitrogen adsorbed on the sample is determined for incremental increases in nitrogen pressure.

1. *Sample preparation:* Determine the mass of the sample (79.871 g for this example) and place it in the sample cell. Evacuate the sample and sample cell while heating to about 60°C (140°F) to degas the sample.
2. Cool the sample cell, which contains the sample, to ambient temperature, and connect it to the system.
3. *Evacuation of the system and sample cell.* Close three-way Valve 2 to the system, open Valves 3 and 4, and close Valve 5. Evacuate the system and sample cell to the lowest possible pressure and then close Valve 4.
4. *Fill the burette with helium:* With Valve 2 closed to the evacuated system, fill the burette with mercury. Connect the burette to a bottle of helium and draw some helium into the burette; then, expel this helium from the burette by disconnecting it from the helium bottle and filling the burette with mercury once more. (This step is used to remove the small amount of air from the line.)

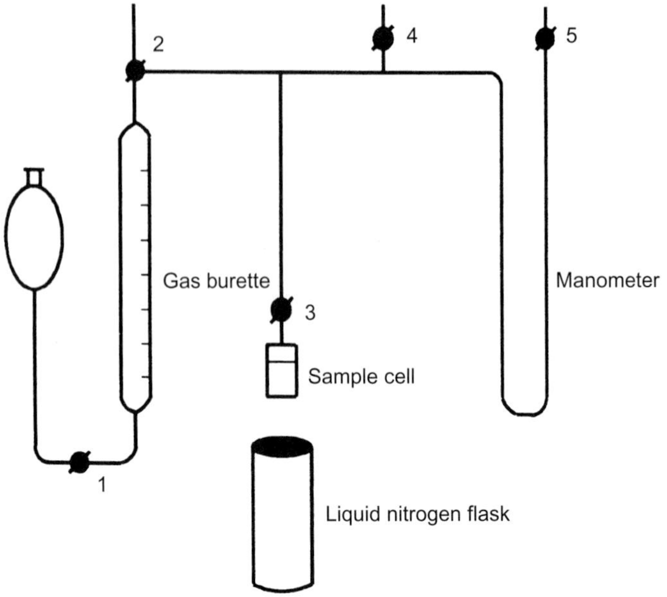

FIGURE A9.1 Equipment of measurement of the surface area of sediments by adsorption of nitrogen.

Reconnect to the helium bottle and fill the burette to the bottom mark with helium by withdrawing mercury from the burette.

5. *Determine the volume of the system (V_s), exclusive of the sample cell:* Close valve 3 and open three-way Valve 2 to admit helium from the burette into the system. Open Valve 5 to activate the manometer. Displace helium into the system to pressure P ($P = 14.55$ cm Hg for this example). Record the amount of helium displaced into the system at pressure P and at ambient temperature T_a (96 ml at P and $T_a = 293.5$ K for this example).
6. Determine the volume of the sample cell and sample (V_{sc}):
   - **(a)** Open Valve 3 and displace helium into the system from the burette until pressure P ($P = 14.55$ cm Hg) is attained once more, and record the volume of helium displaced as V_1 ($V_1 = 47.0$ ml at P and T_a).
   - **(b)** Immerse the sample cell in liquid nitrogen to a predetermined depth. Displace helium from the burette until pressure P is once more attained. Record the volume of helium displaced as V_2 ($V_2 = 145.0$ ml at P and T_{N2}).
   - **(c)** Adjust V_2 to ambient temperature:

$$V_{2(p,Ta)} = \frac{145.0 \times 77.4}{293.5} = 38.2 \text{ ml} \quad \text{(A9.2)}$$

   - **(d)** Calculate the volume of the sample cell at P and T_a:

$$V_{sc} = V_1 + V_2 = 47.0 + 38.2 = 85.2 \text{ ml}$$

(e) Correct for nonideality of nitrogen used for the adsorption isotherm. Factor to account for nonideal behavior of nitrogen at liquid nitrogen temperature, I_{N2}, is equal to 0.05, therefore:

$$V_{sc} = 85.2(1.00 + 0.05) = 89.5 \text{ ml}$$

7. Measurement of nitrogen adsorption:
   (a) Remove the liquid nitrogen from the sample cell.
   (b) Close three-way Valve 2 to the system, open Valves 3 and 4 and close Valve 5.
   (c) Evaluate the system.
   (d) Flush the burette and line with nitrogen from a nitrogen bottle and fill the burette with nitrogen.
   (e) Immerse the sample cell in liquid nitrogen to the predetermined depth. Close Valve 4, open the three-way Valve 2 to the system, and open Valve 5. Record the initial pressure registered by the manometer, P_i ($P_i = -0.9$ cm Hg for this example). This is a correction that will be necessary for all pressure measurements during the balance of the experiment (Table A9.1).
   (f) Displace nitrogen into the system and sample cell until a small positive pressure is attained. Record the pressure P, the amount of nitrogen displaced into the system (V_D), and the amount of nitrogen remaining in the burette at equilibrium (V_R).
   (g) Continue successively displacing increments of nitrogen into the system and sample cell until the limits of the equipment are attained (Table A9.1). Refill the burette with nitrogen whenever necessary by opening three-way Valve 2 to the nitrogen line (closed to the system); then, close Valve 2 to the nitrogen line, which opens it to the system once more.

Sample Calculations

Refer to Table A9.1 for a list of data and results of calculations.

1. Correct all pressure measurements by adding the initial pressure (P_i) in the system and sample cell to the incremental pressure P:
   $P_{corr,0} = P - P_i = 1.75 - (-0.9) = 2.65$ cm Hg
   $P_{corr,l} = 0.45 + 0.9 = 1.35$ cm Hg
   $P_{corr,2} = 1.10 + 0.9 = 2.00$ cm Hg
2. Add the measured volumes (column G):
   $V_{D,0} + V_{s,0} + V_{sc,0} = 0$
   $V_{D,\,1} + V_{s,1} + V_{sc,1} = 500 + 96 + 0 = 596$ ml
   [The initial sample cell volume, $V_{sc,1} = 0$]
   $V_{D,2} + V_{s,2} + V_{sc,2} = 300 + 96 + 89.5 = 485.5$ ml
3. Calculate the term in column H (G refers to the values in column G):
   $G_0 \times P_{1,0}/P_{2,0} = 0$
   $G_l \times P_{1,1}/P_{2,1} = 596 \times (2.65/1.35) = 1{,}167$
   $G_2 \times P_{1,2}/P_{2,2} = 485.5 \times (1.35/2.00) = 327.7$
4. Calculate the term in column I, Table A9.1:
   $G_0 - \Delta V_0 = 0$
   $G_1 - \Delta V_1 = 596.0\text{–}200 = 396.0$
   $G_2 - \Delta V_2 = 485.5\text{–}200 = 285.5$

TABLE A9.1 Data and Calculations for Surface Area Measurement (Greer et al., 1962)

A Run No.	B P cm Hg	C P_{corr} cm Hg	D V_D ml	E V_R ml	F ΔV ml	G $V_D + V_S + V_{SC}$ ml	H $G^* P_1/P_2$	I $G - \Delta V$	J $H - I$ V_a @P_2, T_a	K V_a @ STP	L V_a (accum)	M P/P_{N2}	N $(P/[V_a(P_{N2} - P)]) \times 10^3$
0	1.75	2.65	–	–	–	–	–	–	–	–	–	–	–
1	0.45	1.35	500	300	200	596	1,167.0	396.0	681.5	11.27	11.27	0.0177	1.60
2	1.10	2.00	300	100	200	485.5	327.1	285.5	42.2	1.03	12.30	0.0262	2.18
3	1.70	2.60	500	300	200	685.5	527.7	485.5	41.8	1.34	13.64	0.0340	2.58
4	3.95	4.85	300	20	280	485.5	260.3	205.5	64.8	3.25	16.89	0.0635	4.01
5	5.10	6.00	500	350	150	685.5	554.1	535.5	18.6	1.37	18.26	0.0785	4.66
6	7.75	8.65	350	163	187	535.5	371.6	348.5	23.1	2.45	20.71	0.1130	6.16
7	11.60	12.50	163	40	123	348.5	242.2	225.5	16.7	2.55	23.26	0.1635	8.38
8	14.45	15.35	500	383	117	685.5	560.0	551.5	8.5	1.60	24.86	0.2010	10.10
9	18.60	19.50	383	130	253	568.5	447.1	438.5	8.6	2.05	26.91	0.2550	13.20
10	23.80	24.70	253	152	101	438.5	346.1	337.5	8.6	2.60	29.51	0.3230	16.18
11	30.50	31.40	152	78	74	286.5	224.8	212.5	12.3	4.72	34.23	0.4110	20.35

5. Calculate the term in column J, Table A9.1:
   $H_0 - I_0 = 0$
   $H_1 - I_1 = 1,167 - 396.0 = 771$
   $H_2 - I_2 = 327.7 - 285.5 = 42.2$
6. Calculate the term in column K (STP = 273.2 K, 76 cm Hg), Table A9.1:
   $V_{a,0} = 0$
   $V_{a,1} = (273.2 \times 1.35 \times 771)/(76 \times 293.5) = 12.75$
   $V_{a,2} = (273.2 \times 2.00 \times 42.2)/(76 \times 293.5) = 1.03$
7. Calculate the term in column L, Table A9.1:
   $V_{a,0(accum)} = 0$
   $V_{a,1(accum)} = V_{a,0} + V_{a,1} = 0 + 11.27 = 11.27$
   $V_{a,2(accum)} = V_{a,1} + V_{a,2} = 11.27 + 1.03 = 12.30$
8. Calculate the term in column M (system pressure relative to the pressure of saturated nitrogen vapor at 77.4 K):
   $P_0/P_{N2} = 2.65/76.6 = 0.0346$
   $P_1/P_{N2} = 1.35/76.5 = 0.0176$
   $P_2/P_{N2} = 2.00/76.5 = 0.0261$
9. Calculate the term in column N, Table A9.1:
   $N_0 = 0$
   $N_1 = (P_1 \times 10^3)/[V_{a,1}(P_{N2} - P_1)]$
   $= (1.35 \times 10^3)/(12.30 \times 75.15) = 1.46$
   $N_2 = (P_2 \times 10^3)/[V_{a,2}(P_{N2} - P_2)]$
   $= (2.00 \times 10^3)/(12.30 \times 74.5) = 2.18$
10. Determine the slope $(C-1)/(V_m C)$ and intercept $1/(V_m C)$ from a plot of column N versus column M (Table A9.1), or by a least-squares solution of the data. Use simultaneous solution to obtain the values of C and V_m:
    From a least-squares solution of the data plotted in Figure A9.2:
    Slope $(C-1)/(V_m C) = 47.33$
    Intercept $(1/V_m C) = 0.87$
    From simultaneous solution of the slope and intercept:
    $C = 42.23$
    $V_m = 27.18$
11. Calculate the surface area of the sample and the surface area per gram and per cm^3, assuming the density of the sedimentary sample, ρ, is equal to 2.51:

$$S/cm^3 = \frac{A \times N}{V} = \frac{(1.624 \times 10^{-15})(6.03 \times 10^{23})(10^{-4}\ m^2\ cm^2)}{22,400\ ml}$$

$$= 4.372$$

$$S/cm^3 = 4.372$$

$$S_a \text{ of sample} = 118.989\ m^2$$

$$S_a/g = 118.989/79.871 = 1.49\ m^2/g$$

$$S_a/cm = S_a/g \times \rho = 1.49 \times 2.51 = 3.79\ m^2/cm^3$$

where

S = surface area of a cubic centimeter of nitrogen molecules
S_a = surface area
V_s = volume of the system

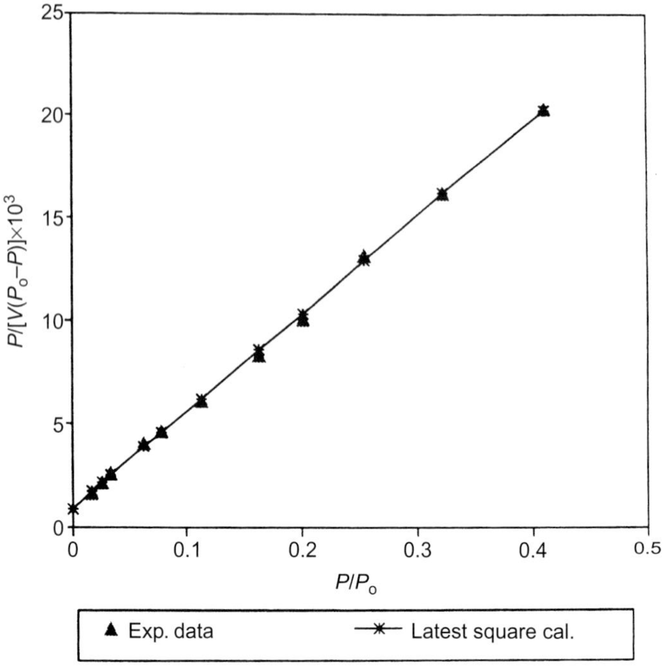

FIGURE A9.2 Experimental data and least squares analysis to determine the slope and intercept of the best straight line fit of the data.

V_{sc} = volume of the sample cell

Questions and Problems

1. Explain the difference between Langmuir and Freundlich adsorption isotherms.
2. What is the physical meaning of the intercept of the line in Figure A9.1 (the point at which $P/P_0 = 0$)?

References

Brooks CS, Purcell WR. Surface area measurement of sedimentary rocks. Soc Pet Eng Trans 1952;195:289–96.

Brunauer S, Emmett PH, Teller E. Adsorption of gases n multimolecular layers. J Am Chem Soc 1938;60:300–19.

Donaldson EC, Kendall RF, Baker BA, Manning FS. Surface area measurement of geologic materials. Soc Pet Eng J 1975;15(2):111–6. April.

Dullien FAL. Porous media fluid transport and pore structure. New York: Academic Press; 1979. pp. 68–72.

Greer FC, Carrol M, Beeson RM, Chilingar GV. Determination of surface areas of sediments. J Sed Petrol 1962;32(1):140–6. March.

Experiment 10 Absolute Permeability

Introduction

Permeability is a measure of the ease with which a fluid can flow through a porous medium; it is the inverse of resistance to flow. The permeability of a sample is, therefore, determined by measuring the rate at which a liquid will flow through a porous medium of specific dimensions with a given pressure gradient across the length of the porous medium. A permeability equal to one darcy is obtained when a liquid, having a viscosity of 1 cP, flows at a rate of 1 cm^3/s through a sample with a cross-sectional area of 1 m^2 under a pressure gradient of 1 atm/cm:

$$Q(\mathrm{cm}^3/\mathrm{s}) = \frac{k(\mathrm{darcy}) \times A(\mathrm{cm}^2)}{\mu(\mathrm{cP})} \times \frac{\Delta p(\mathrm{atm})}{L(\mathrm{cm})} \tag{A10.1}$$

Generally, permeability is expressed in mD because the darcy is a fairly large unit and mD is more convenient to use. In SI units, the darcy is expressed as micrometers squared (μm^2):

$$\mathrm{darcy} = 0.987 \times 10^{-8}\ \mathrm{cm}^2 = 0.987\ \mu\mathrm{m}^2 \tag{A10.2}$$

Equipment and Procedures

Absolute Permeability Using a Liquid

The liquid permeability measurement is made by determining the time required for a fixed volume of liquid, at constant temperature, to pass through a core at a specific pressure gradient. The Ruska liquid permeameter has a core holder, thermometer well, cutoff valve, a burette for liquid volume measurement, and a pressure regulator. The upstream pressure is indicated on a calibrated pressure gauge.

1. Core samples are cut to the specific size required by the core holder. The cores are washed gently to remove cuttings and dried in an oven at 110°C. If an outcrop core is used, it should be cleaned in a steam bath before use to remove humus materials that are generally present in outcrop core samples. (Steam cleaning does not disturb water-sensitive clays.)
2. A dried core is then placed in a vacuum chamber, evacuated to remove air, and saturated with a salt solution by closing off the vacuum pump. A deaerated salt solution is then admitted to the vacuum flask. A salt solution (approximately 20,000 ppm) is used to avoid clay swelling and particle transport during the test. To avoid these problems with very sensitive rocks, one may use a hydrocarbon solvent, such as Soltrol from Phillips Petroleum Co., in place of a water solution.
3. Insert the core in the core holder. Figure A10.1 illustrates the high-pressure Hassler sleeve core holder. The burette is filled above the zero level with the salt solution and connected to the air pressure source at the top.
4. The inlet pressure is set at the desired value and the discharge valve is opened. Start the timing when the water level reaches the zero mark on the burette. Stop the timer when the burette empties. Record the inlet pressure (p_1), the volume of fluid passed through the core (V), and the time required for the volume of fluid to pass through the core, t. Record the temperature of the fluid.

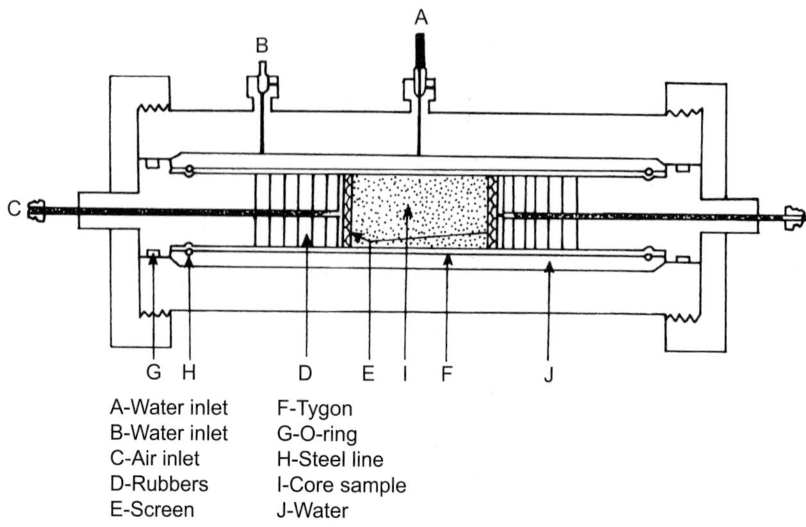

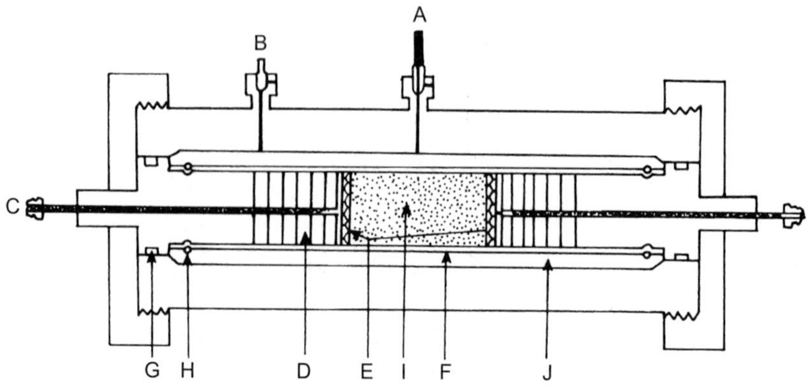

FIGURE A10.1 Hassler sleeve core holder used for fluid flow experiments (fractions of oil and water flowing, absolute permeability, and relative permeability).

5. Calculate the absolute permeability as follows:

$$k = \frac{\mu \times V \times L}{A \times \Delta p \times t} \tag{A10.3}$$

where

k = permeability, darcies
μ = fluid viscosity, cP, at the observed temperature
V = volume of fluid, cm^3.
L = length of the core sample, cm
A = cross-sectional area of the core, cm^2
Δp = pressure gradient across the core ($p_1 - p_2$), atmospheres
P_a = atmospheric pressure
t = time required for passage of $V\ cm^3$ of fluid, s

6. Example using a hydrocarbon fluid:

$\mu = 0.895$ cP at 25°C.
$V = 10.0\ cm^3$.
$L = 1.90$ cm.
$A = 2.83\ cm^2$.
$p = 2.0$ atm.
$t = 30$ s. $k = (0.895 \times 10.0 \times 1.90)(2.83 \times 2.0 \times 30) = 0.10$ darcy
$= 100$ mD
$= 98.7\ \mu m^2$.

Absolute Permeability Using a Gas

It is often more convenient to determine the permeability of core samples using a gas as the test fluid (Figure A10.2). When a gas is used, however, one must

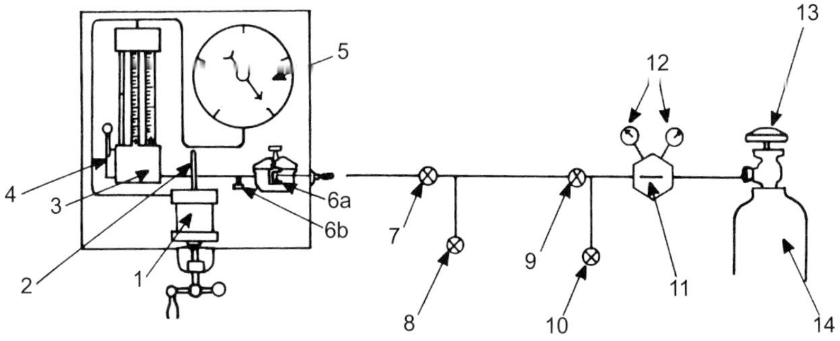

1-Coreholder
2-Thermometer, built into coreholder
3-Three flowmeters, small, medium, large
4-Flowmeter selector switch
5-Pressure guage
6-Pressure regulator
7-Second flowline valve
8-Bleed off valve (to other equipment)
9-First flowline valve
10-Bleed off valve
11-Flowline regulator
12-Nitrogen pressure gauge
13-Nitrogen tank valve
14-Nitrogen

FIGURE A10.2 Ruska gas permeameter.

account for expansion of the gas along the length of the core as the pressure decreases with respect to length. In addition, the "Klinkenberg effect" accelerates the flow of gas when the free mean path of the gas is greater than the pore diameter, because under these conditions some of the random kinetic energy of the gas is transferred to movement of the gas molecule through the pore, or slippage at the pore walls.

To account for the expansion of gas, the average rate of flow and the average pressure are used to calculate the permeability. Referring to the figure of the core holder, Figure A10.1, Q_1 and Q_2 (cm^3/s) are the entrance and exit gas flow rates, respectively. Because of expansion, Q_1 is less than Q_2. Using the ideal gas law at isothermal conditions:

$$p_1 \times Q_1 = p_2 \times Q_2 = p_{av} \times Q_{av} \tag{A10.4}$$

$$p_{av} = \frac{(p_1 + p_2)}{2}; \quad Q_{av} = \frac{(p_2 \times Q_2)}{p_{av}} \tag{A10.5}$$

In making the measurements, the time required to collect a specific volume of gas at the exit and the two pressures, p_1 and p_2, are measured. The average flow rate, Q_{av}, is equal to the volume collected divided by the time. Solving Darcy's equation for permeability:

$$k(\text{darcy}) = \frac{Q_{av}\mu L}{A(p_2 - p_1)} \tag{A10.6}$$

Due to the Klinkenberg effect, the measured gas permeabilities are greater than the absolute permeabilities of the core. Klinkenberg discovered that if the permeability to gas were measured at several pressures and plotted against the reciprocal of the average pressure, the points would lie on a straight line. When the line is extrapolated to $1/p_{av} = 0$ (infinite pressure), the intercept represents the absolute permeability because all gases become liquids at infinite pressure. Thus, the procedure for measurement of absolute permeability using a gas is to make three or more measurements at different pressure gradients and then to extrapolate the results to infinite pressure.

Effect of Overburden Pressure on Absolute Permeability

If a core is placed in a high-pressure Hassler sleeve core holder, which is capable of exerting very high overburden pressure (10,000 psi or more), the effect of increasing overburden pressure can be obtained by measuring the gas permeability of a dry core after each incremental increase in overburden pressure. After each increase in pressure, the core should be allowed to equilibrate for about 24 hours before the permeability is determined again.

Permeability is a measure of the ease with which a fluid flows through a porous medium; hence, it is related to the pore and pore throat size distributions, shape and continuity of the pores, and tortuosity. Compression of the rock changes the pore and pore throat size distributions; shape changes of the pores may increase tortuosity and close some of the fluid flow paths.

The overburden pressure (p_{OB}) of a subsurface sand is supported by the grain-to-grain pressure (p_G) of the sand and the interstitial fluid pressure (p_F), as shown in Figure A10.3, where:

$$p_{OB} = p_G + p_F \tag{A10.7}$$

When the overburden pressure is equal to the sum of the grain-to-grain pressure and the fluid pressure, a static condition exists at the boundary. However, if the compacting overburden pressure is increased, or the reservoir fluid pressure is reduced by fluid withdrawal, compaction of the sand grains will occur. The compaction is accompanied by reduction of porosity and permeability, and an increase in tortuosity in the compacting zone.

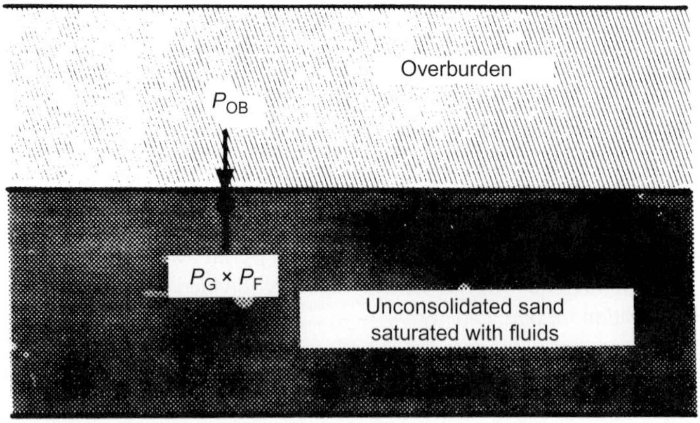

FIGURE A10.3 Relationships between the overburden pressure and the fluid and grain pressures in a subsurface environment.

Sample Calculations

1. Measure a specific volume of gas exiting a core at atmospheric pressure ($V_2 = 800\ \text{cm}^3$).
2. Time required to collect $V_2 = 500$ s.
3. $p_1 = 1.5$ atm; $p_2 = 1.0$ atm: $\Delta p = 0.5$ atm.
4. Core diameter $d = 2.5$ cm; core length $L = 4$ cm; gas viscosity $\mu = 0.02$ cP.

Run 1

$$
\begin{aligned}
p_{av} &= (1.5 + 1.0)/2 = 1.25\ \text{atm} \\
Q_2 &= V_2 Q_2 / p_{av} = 1.60 \\
Q_{av} &= p_2 Q_2 / p_{av} = (1.0 \times 1.6)/1.25 = 1.28\ \text{cm}^3/\text{s} \\
A &= (\pi/4)\, d_2 = (\pi/4)(2.5)^2 = 4.9\ \text{cm}^2 \\
k_{\text{gas at 125 atm}} &= 1.28 \times 0.02 \times (4.0/4.9) \times 0.5 = 0.0418\ \text{darcy} \\
&= 41.8\ \text{mD} = 0.0412\ \mu\text{m}^2
\end{aligned}
$$

Run 2

$$
\begin{aligned}
&p_1 = 2.333;\ p_2 = 1.00;\ \Delta p_{av} = 1.666 \\
&V_2 = 1,470\ \text{cm}^3;\ t = 300\ \text{s};\ Q_2 = 4.9\ \text{cm}^3/\text{s};\ Q_{av} = 2.94\ \text{cm}^3/\text{s} \\
&k_{\text{gas at 166 atm}} = 2.94 \times 0.02 \times (4.0/4.9) \times 1.666 = 0.036\ \text{darcy} \\
&\qquad = 36\ \text{mD} = 0.335\ \mu\text{m}^2
\end{aligned}
$$

Plot $k_{\text{gas at av } p}$ versus $1.0/p_{av}$ and extrapolate the straight line to intercept the Y-axis (permeability axis) to obtain the absolute permeability, which is equal to 0.020 darcy (0.0197 μm^2) for the example problem illustrated above.

Areal permeability distributions of reservoir generally fit exponential curves, unlike porosity, which most frequently falls into the category of normal distributions. When the permeabilities of individual samples, or sample intervals, for a field are plotted on semilog paper against the cumulative number of samples, one or more straight lines will be obtained. Each straight line segment represents a different statistical distribution of permeability, which is described by:

$$\log(k) = mN + B \tag{A10.8}$$

where m and B are the slope and intercept, respectively.

Mean: Because permeability is best described by an exponential curve, the geometric mean (rather than the arithmetic mean) is used as the permeability descriptor.

For unclassified data:

$$\log(k_{\text{mean}-g}) = \Sigma \log(k_i)/N \tag{A10.9}$$

For classified data (Table A10.2)

$$\log(k_{\text{mean}-g}) = \Sigma F_i \log(k_{\text{mean}-a})^i \tag{A10.10}$$

where

F_i = cumulative frequency of the ith interval
k_i = permeability of the ith sample
k_{mean} = arithmetic average permeability

k_{mean-a} = arithmetic average permeability of a logarithmic class interval
N = total number of samples

Sample Calculation 1

Assume that the following list of 50 permeability values was obtained from a reservoir and the data are arranged in ascending order (Table A10.1):

A semilog plot of the permeability data (versus chronological number of data) reveals two lines with the break between samples number 21 and 22. Therefore, two statistical analyses results are evident because the data show two distinct distributions of permeability.

Sample Calculation 2

Using Table A10.1 (Samples 1–21) and Eq. (A10.9):

$$\log(k_{mean-g}) = \frac{21.6528}{21} = 1.0311$$

$$k_{mean-g} = 10.7$$

Using Table A10.2 (Samples 1–21) and Eq. (A10.10):

$$\log(k_{mean-g}) = 1.0393$$
$$k_{mean-g} = 10.9$$

Taking the arithmetic mean for Samples 1–21 (Table A10.1):

$$k_{mean} = \frac{333}{21} = 15.9 \text{ mD}$$

The analyses of the data show that the two distributions of permeability have geometric mean permeabilities of 10.9 and 143.7 mD and arithmetic means of 15.8 and 172.2 mD, respectively. Characteristically, the geometric mean is less than the arithmetic mean.

Questions and Problems

1. Explain the meaning of the Klinkenberg effect.

TABLE A10.1 List of Permeability Values Arranged in Ascending Order

K(mD)	K(mD)	K(mD)	K(mD)	K(mD)
2	11	46	109	220
3	13	53	111	225
3	16	59	111	230
4	17	60	120	245
4	19	69	148	295
5	24	70	150	330
6	27	72	150	335
7	31	81	180	340
8	36	92	181	350
9	42	92	195	420

TABLE A10.2 Classification of the Data from Table A10.1 and Determination of the Geometric Mean for Samples 1–21 (First Set) and Samples 22–50 (Second Set) Using Eq. (A10.10)

Permeability Range (mD)	Average Permeability	Cumulative Log (k_{mean-g})	Frequency	$F_1 \times \log (k)_{mean-g}$
0–5	3.5	0.00	0.286	0.1556
6–10	7.5	0.8751	0.190	0.1663
11–15	13.0	1.1139	0.095	0.1058
16–20	17.3	1.280	0.143	0.1770
21–30	25.5	1.4065	0.095	0.1336
31–40	33.5	1.5250	0.095	0.1449
41–50	44.0	1.6435	0.095	0.1561
				1.0393
51–60	60.7	1.7832	0.103	0.1837
61–80	70.3	1.8470	0.103	0.1902
81–110	88.3	1.9460	0.103	0.2000
111–120	112.8	2.0523	0.138	0.2832
121–160	149.3	2.1741	0.103	0.2239
161–200	185.3	2.2679	0.103	0.2336
201–250	230.0	2.3617	0.138	0.3259
251–300	262.5	2.4191	0.069	0.1669
301–350	341.7	2.5336	0.103	0.2610
351–450	420.0	2.6232	0.034	0.0892
				2.1576

For Group 1: $k_{mean-g} = 10.9$ and $k_{mean} = 15.9$.
For Group 2: $k_{mean-g} = 143.7$ and $k_{mean} = 172.2$.

2. Why is nitrogen passed through a drying tube before being allowed to enter the core?
3. Why measure permeability with a gas, rather than with water?
4. Plot the permeability data in Table A10.1 versus the cumulative number of samples and observe the two distinct distributions of permeability.
5. Discuss the meaning of two or more permeability distributions with respect to the behavior of fluid flow in a reservoir.

References

Amyx JW, Bass Jr. DM, Whiting RL. Petroleum reservoir engineering. New York: McGraw-Hill; 1960, chapters 2 and 7.

Archer JS, Wall CG. Petroleum engineering. London: Graham and Trotman; 1986, chapter 5.

Chilingarian GV, Wolf KH. Compaction of coarse-grained sediments, I. Amsterdam, the Netherlands: Elsevier Science; 1975, chapter 1.

Experiment 11 Verification of the Klinkenberg Effect

Introduction

Refer to the section titled *Absolute Permeability Using a Gas* in Experiment 10.

When the mean free path of the measuring gas is greater than the diameter of the capillary through which it is traveling, the random kinetic energy of the gas is transferred to movement of the gas molecule through the capillary, or slippage of the molecules occurs at the pore walls. This "slippage" causes the molecules of the gas to travel at a higher velocity in the direction of transfer. This phenomenon, known as the "Klinkenberg effect," causes the measured permeability of a gas to be greater than the absolute permeability of the sample.

Equipment and Procedures

The mean free path of the gas decreases as the pressure increases and vanishes when the gas becomes liquid at infinite pressure. Therefore, extrapolation of the measured permeability to $1.0/p_{av} = 0$ (or infinite pressure) yields the absolute permeability, which would be obtained if the measurement had been made with a liquid. Measurements of absolute permeability are generally made more conveniently with a gas than with a liquid; therefore, this method is in common use.

The relationship between the measured gas permeability (k_g) and $1/p_{av}$ is a straight line expressed as:

$$k_g = k_a + \frac{k_a B}{p_{av}} \tag{A11.1}$$

where

k_g = measured permeability to gas
k_a = absolute permeability
B = Klinkenberg constant that is a function of the gas being used
p_{av} = mean (or average) flow pressure

The intercept of the line, at $1/p_{av} = 0$, is the absolute permeability (k_a). The slope of the line in Eq. (A11.1) is equal to $k_a B$; therefore, after k_a has been determined, the Klinkenberg constant for the gas (B) can be obtained. Klinkenberg reported that $B = (0.777 k_a - 0.39)$ for any gas.

Sample Calculations

Using air, measure the permeability of a core having diameter 2.54 cm and length 2.54 cm:

Experimental Data

Core diameter = 2.54 cm	Core length (L) = 2.54 cm
Core area (A) = 5.07 cm^2	Air viscosity (m) = 0.018 cP

Effluent absolute pressure (p_2) = 1.0 atm

V_2, cm^3	Time, s	p_1 (gauge), atm
100	917.2282	0.144737
250	330.3390	0.671053
500	107.8261	2.039474

Calculations are based on Eq. (A10.6):

$$k_g = \frac{Q_{av}\mu L}{A(p_2 - p_1)} \tag{A11.2}$$

Figure A11.1 shows the extrapolation to infinite pressure using linear regression.

$$k_{absolute} = 4.07 \text{ mD}$$
$$B_{Klink} = \frac{\text{slope}}{k_a} = \frac{2.03}{4.07} = 0.5$$

atm	p_{av}	$1/p_{av}$	$p_1 - p_2$	V_{av}	Q_{av}	k_g	k_g, mD
1.0	1.072	0.933	0.145	92	0.102	0.0059	5.91
1.0	1.336	0.749	0.671	187	0.567	0.0057	5.70
1.0	2.020	0.495	2.039	248	2.300	0.0050	5.04
							4.07

Useful equations:

$$p_{av} = \frac{p_1 + p_2}{2} \text{(absolute pressure)}$$

$$p_{abs} = p_{gauge} + p_{atm}$$

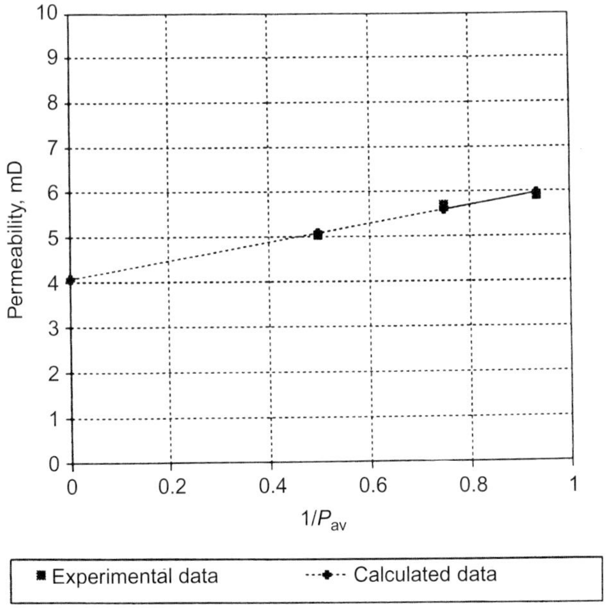

FIGURE A11.1 Extrapolation of permeability (measured with a gas) versus average pressure for several measurements. The absolute permeability is the value at the intercept (infinite pressure, where $1/p_{av} = 0$).

$$V_{av} = \frac{p_2 V_2}{p_{av}}$$

$$Q_{av} = \frac{V_{av}}{t}$$

Questions and Problems

1. Select at least two cores from different sources and determine the absolute permeability of each core with at least two different gases.
2. Calculate the Klinkenberg B-constant and verify it with the equation given in this experiment.
3. Explain why the B-constant is different for each gas.

References

Amyx JW, Bass Jr. DM, Whiting RL. Petroleum reservoir engineering. New York: McGraw-Hill; 1960, chapter 2.

Archer JS, Wall CG. Petroleum engineering. London: Graham and Trotman; 1986, chapter 5.

Experiment 12 Relative Permeability

Introduction

The absolute permeability is a characteristic of a rock and is determined from measurements of the flow rate of a single fluid through the rock. Relative permeability, however, is a function of the rock chemical and physical properties, and the fluid chemical and physical properties. Relative permeability, therefore, is sensitive to temperature and the relative wetting characteristics of the rock and fluids. As the temperature of the system increases, the water–oil–rock system becomes more water-wet and hence this results in the temperature effect.

Three fluids, gas, oil, and water, may be present and mobile in a rock at certain saturations. In this case, three-phase relative permeabilities must be considered; however, no general method has been established for three-phase relative permeability measurements. Several papers in the literature discuss this, but they do not agree with respect to data or procedures. However, methods for the determination of relative permeabilities for two phases (gas–oil, gas–water, water–oil) have been well established. Two methods are used: (1) the steady-state method in which the two fluids are made to flow into and out of the core at steady flow rates at various saturations and (2) the unsteady-state method, which is conducted by displacement of two fluids saturating a core with either gas or water.

Relative permeability is defined as:

$$k_{rg} = \frac{k_{eg}}{k}; \quad k_{rw} = \frac{k_{ew}}{k}; \quad k_{ro} = \frac{k_{eo}}{k} \tag{A12.1}$$

where

k = absolute permeability

k_{eg}, k_{ew}, and k_{eo} = effective permeabilities to gas, water, and oil at specific saturations, respectively

k_{rg}, k_{rw}, and k_{ro} = relative permeabilities to gas, water, and oil at specific saturations, respectively

Steady-State Method

To determine the relative permeabilities of water and oil, a core saturated to some predetermined saturation is placed in a Hassler sleeve core holder and a pressure transducer is connected to the inlet. If a back pressure regulator is used to maintain a high pore pressure, a pressure transducer is connected also to the outlet end of the core to allow measurement of the pressure difference. Metering pumps are used to pump the water and oil at steady flow rates into a small mixing cell, which transfers the fluids to the face of the core. The individual flow rates of the fluids are controlled by adjusting the pump rates.

Water and oil are injected at predetermined flow rates and the effluent rates are monitored until they are equal to the influent rates. At this point, it is assumed that steady state (constant saturation throughout the core) has been attained. The flow rates and pressure drop are recorded, and the core is removed and weighed. The saturation is calculated from the core mass, the pore volume, and the density of the two fluids. The relative permeability at that specific saturation is calculated as shown in Eq. (A12.2). Then the core is reassembled in the sleeve and the water/oil ratio is adjusted to another value to change the fluid saturation in the core. The procedure is repeated until sufficient data are obtained to describe a complete set of relative permeabilities as a function of the saturation of one of the fluids. For a water–oil system:

$$k_{ew} = \frac{Q_w \mu_w L}{A \Delta p_w} = f(S_w)$$
$$k_{eo} = \frac{Q_o \mu_o L}{A \Delta p_o} = f(S_w) \qquad \text{(A12.2)}$$

where k_{rw} and k_{ro} are calculated from Eq. (A12.1) for the specific saturations, assuming the capillary pressure between the phases is negligible; then $p_w = p_o$. The saturation (S_w) is calculated using the following relationship:

$$\text{Total mass } (M), \text{ core, and fluids} = (D_M) + S_w \times V_P \times \rho + (1 - S_w) \times V_P \times \rho_o$$

$$S_w = \frac{(M - D_M - V_p \times \rho_o)}{[V_p \times (\rho_w - \rho_o)]} \qquad \text{(A12.3)}$$

where

D_M = dry mass of the core, g
V_p = pore volume of the core, ml
M = total mass of the core and interstitial fluids, g
ρ = density, g/cm^3

Unsteady-State Method

Three methods for calculating the relative permeabilities are available. All the methods are based on the assumption that the core is homogeneous, and that capillary pressure and gravity may be neglected.

1. The *alternate method* yields the relative permeabilities as a function of the average fluid saturation of the core. The calculations for the alternate method are simplified because it is only necessary to apply Darcy's law to the

displacement method and plot the calculated relative permeabilities as a function of the average, rather than the terminal, saturation of the core.

2. The *Johnson–Bossler–Naumann (JBN) method* is used to calculate the relative permeabilities of the cores as a function of the effluent (terminal) fluid saturation for fluid displacement at constant injection rate. The experiment must be carried out at high flow rates to avoid capillary end effects (abnormally high wetting phase saturation at the end of the core).
3. The *Toth et al. method* is more general than the alternate and JBN methods because it is applicable to constant injection rate displacement and constant pressure. The method allows direct calculation from the displacement data and, consequently, offers greater computational accuracy than the other two methods.

Equipment and Procedures

1. Saturate a core 100% with water and then displace the water to S_{iw} by pumping the oil into the core until no more water is produced. Centrifuge the collected water and oil in a graduated tube (or tubes) and measure the amount of water displaced by the oil. Calculate the initial water saturation.
2. Allow the core and fluids to adjust to capillary equilibrium overnight.
3. For constant injection rate, adjust a metering pump to deliver the displacing fluid into the core at a convenient constant flow rate. For constant injection pressure, adjust a gas-driven piston pump to deliver the displacement fluid at a convenient constant pressure.
4. When the water displacement pump is turned on, start recording the time using a stopwatch.
5. If water and crude oil are used, collect the effluent fluids in graduated centrifuge flasks and record the time when each fraction is taken. If a gas and liquid are used, direct the effluent gas through a gas meter and collect the liquid in a graduated cylinder or burette.
6. Stop the displacement test when oil stops flowing, or at some predetermined volume of water to be injected (3 pore volumes, for example).
7. When crude oil is used, centrifuge the samples to separate the water and oil. Record the water and oil displaced, and the time when the fraction was taken.
8. Compute the cumulative fractions of water and oil collected. Plot the cumulative oil produced as a function of pore volumes of water injected.
9. At each time period, determine the rates of water and oil flowing, relative permeabilities, and the average water saturation, $S_{w(av)}$, as follows:

$$S_{w(av)} = \frac{S_{iw} + V_{oil(produced)}}{V_p} \tag{A12.4}$$

Sample Calculations

1. Alternate method (see Table A12.1):

   $L/A = 0.899$; $V_p = 4.012\ cm^3$; $k_{absolute} = 0.312$ darcy
   Q_w(water injection rate – constant) $= 0.0168\ cm^3/s$
   $S_{iw} = 0.261$; $k_o(\text{at } S_{iw}) = 0.233$ darcy
   $T = 73°F$; $\mu_o = 25.7$ cP; $\mu_w = 0.938$ cP

TABLE A12.1 Alternate Method (Loomis and Crowell, 1962)

(1) Time (s)	(2) W_i (cum)	(3) V_o (cum)	(4) Q_o	(5) S_w (av)	(6) Δp	(7) k_{ro}	(8) k_{rw}
60.7	1.02	0.84	0.000733	0.47	1.41	0.389	0.033
128.6	2.16	1.09	0.00264	0.53	0.78	0.254	0.059
255.4	4.29	1.31	0.00081	0.59	0.57	0.106	0.081
799.4	13.43	1.61	0.00024	0.66	0.42	0.043	0.109
336	51.01	1.91	0.000083	0.74	0.34	0.018	0.135
6,335	106.42	2.05	0.000031	0.77	0.20	0.008	0.153

(1): Time at which each fraction was taken, t, s.
(2): Cumulative pore volume of water injected, $W_{i\text{-}pv}$, ml.
(3): Cumulative volume of oil produced, V_o, cm^3.
(4): Volumetric flow rate of oil, Q_o, at the specific point obtained from the slope of the line of Vp_o versus time, cm^3.
(5): Average water saturation, $S_{w(av)} = S_{iw} + (3)/V_p$.
(6): Pressure difference between inlet and outlet $(p_1 - p_2)$, atmospheres.
(7): Relative permeability to oil (Eqs. (A12.1) and (A12.2)) to be plotted versus $S_{w(av)}$.
(8): Relative permeability to water (Eqs. (A12.1) and (A12.2)) to be plotted versus $S_{w(av)}$.

Example using the first data point for the water injection rate (0.0168 cm^3/s):

$$k_{ew} = \frac{0.0168 \times 0.983 \times 0.899}{1.41} = -0.185$$

$$k_{rw} = \frac{0.0185}{0.312} = 0.059$$

$$k_{eo} = \frac{0.00733 \times 25.72 \times 0.899}{1.41} = 0.120$$

$$k_{ro} = \frac{0.120}{0.312} = 0.389$$

$$S_{w(av)} = 0.261 + \frac{0.84}{4.012} = 0.17$$

2. JBN method: see Tables A12.2 and A12.3.
3. Toth et al. method:

Toth et al. developed formulas for calculation of relative permeabilities from unsteady-state fluid displacement data taken from constant rate and constant pressure experiments:

$$k_{rd} = m_d \frac{q_d}{q_i} Y$$
$$k_{rk} = m_k \frac{q_k}{q_i} Y \quad \text{(A12.5)}$$

The term Y is obtained from the cumulative injected fluid volume V_i, which is expressed as follows:

$$V_i = a_2 t^{b_2} \quad \text{(A12.6)}$$

TABLE A12.2 Johnson–Bossler–Naumann Method (Table 5 of Loomis and Crowell, 1962)

(1) $V_o - p_v$	(2) $W_i - p_v$	(3) f_o	(4) S_2	(5) Δp	(6) $1/I_r$	(7) Slope	(8) k_{ro}	(9) k_{rw}/k_{ro}	(10) k_{rw}
0.209	0.254	0.533	0.330	1.41	0.847	0.5786	0.320	0.0295	0.00944
0.271	0.539	0.171	0.440	0.783	0.471	1.351	0.231	0.176	0.0407
0.327	1.071	0.0875	0.494	0.570	0.343	2.112	0.185	3.80	0.0703
0.401	3.349	0.0234	0.584	0.417	0.251	3.299	0.0772	1.52	0.117
0.475	12.72	0.00563	0.664	0.336	0.202	4.284	0.0241	6.43	0.155
0.511	26.54	0.00265	0.702	0.296	0.178	4.701	0.0125	13.7	0.298

(1): Cumulative pore volumes of oil produced, $V_o - p_v$, ml.
(2): Cumulative pore volumes of water produced, $W_i - p_v$, ml.
(3): Fraction of oil flowing.
(4): Terminal water saturation $= S_{w(av)} - f_o W_i$.
(5): Pressure difference between inlet (1) and outlet (2), atm.
(6): Reciprocal of the relative injectivity, $1/I_r = k_{eo}\Delta p/[\mu_o(L/A)Q_w]$.
(7): Slope of the curve of $1/W_i$ plotted versus $1/W_i \times I_r$ (Figure A12.1). Plot on log–log paper and read across one decade of the $1/W_i$ axis for values of $\log(1/W_i \times I_r)2/[(1/W_i \times I_r)1] = \log(\text{term})$. Then calculate the slope at each point from slope $= I_r/\log_{(\text{term})}$. The slopes also may be obtained by determining the equation of the line on log–log paper ($y = Bx^n$) and determining the derivative at specified points ($dy/dx = B \times n \times x^{n-1}$), where $x = 1/(W_i \times I_r)$.
(8): Relative permeability to oil at the terminal end of the core (to be plotted versus $S_{w2} = f_o$ slope (Figure A12.2)).
(9): Relative permeability to water divided by the relative permeability to oil. $k_{rw}/k_{rw} = (\mu_w/\mu_o)\ [f_w/(1-f_w)]$, where f_w = fraction of water flowing.
(10): Relative permeability to water at the terminal end of the core (to be plotted versus S_{w2}), column 8 multiplied by column 9 (Figure A12.2).

TABLE A12.3 Calculation of K_{rw} and K_{ro} Using a Power Function, $Y = BX^n$

$1/W_i$	$1/W_i \times I_r$	Slope	$k_{ro} = f_o \times$ Slope	k_{rw}/k_{ro}	k_{rw}
3.937	3.335	1.006	0.536	0.032	0.017
1.855	0.874	1.515	0.259	0.177	0.046
0.934	0.320	2.059	0.180	0.380	0.069
0.299	0.075	3.209	0.075	1.523	0.114
0.079	0.016	5.156	0.029	6.444	0.187
0.038	0.007	6.709	0.018	13.731	0.244

Use a least-squares solution to obtain an expression for $1/W_i = f(1/W_i \times I_r)$; this is then used to obtain the slope [slope $= Bnx^{n-1}$]. Refer to Figures A12.1 and A12.2.

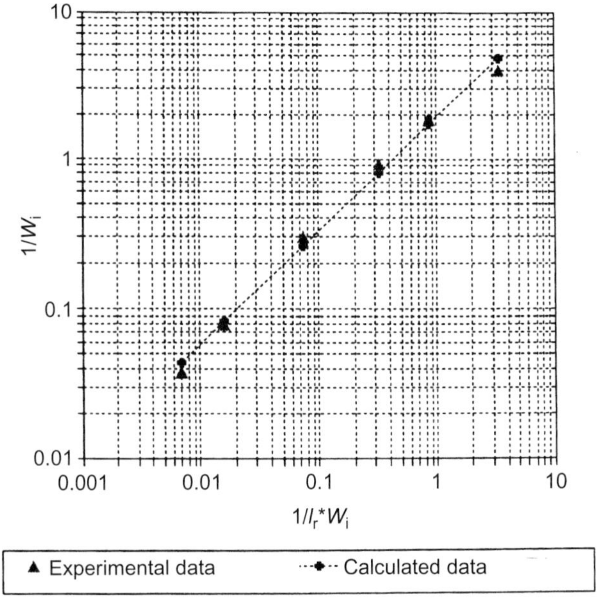

FIGURE A12.1 Log–log plot of the reciprocal of cumulative water injected ($1/W_i$) versus the reciprocal of the term relative injectivity times cumulative water injected $[1/(I_r^* W_i)]$.

For constant fluid displacement, Y is calculated from Eq. (A12.7) as follows:

$$Y = \frac{La_2 b_2^2 (V_p/a_2)^{(1-1/b_2)}}{\Delta p \times KA(2b_2 - 1)} \left(\frac{V_i}{V_p}\right)^{(1-1/b_2)} \quad \text{(A12.7)}$$

where

A = core cross-sectional area, cm^2
a = correlation constant for cumulative volume injected
b = correlation constant for cumulative volume injected

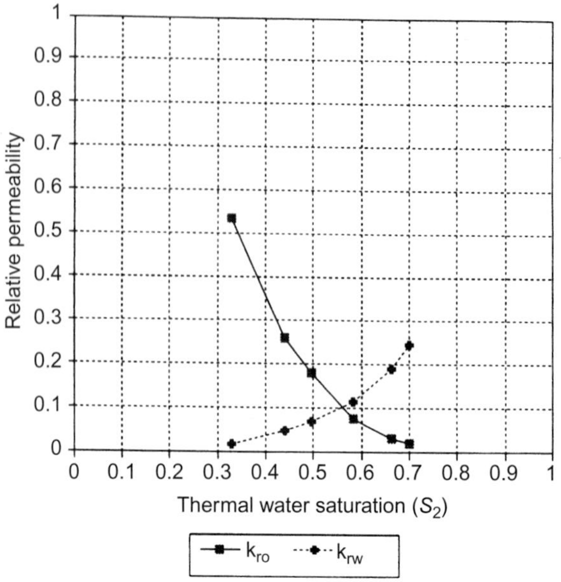

FIGURE A12.2 Relative permeability calculated using the Johnson–Bossler–Naumann method plotted as a function of the terminal water saturation.

K = absolute permeability
k_r = relative permeability
L = core length, cm
p = pressure, bar
q = flow rate, cm^3/s
t = time, s
V_i = cumulative volume injected (function of time), cm^3
Y = function of the displacing fluid saturation at the outlet of the core
μ = viscosity, cP

Subscripts
d = displacing fluid
i = injected fluid
k = displaced fluid
1 = inlet face of the core
2 = outlet face of the core

Questions and Problems

1. What are relative permeability curves used for?
2. What are pseudorelative permeability curves?
3. What is the effect of wettability on relative permeability curves?
4. What is the effect of temperature increase on relative permeability curves?
   Use a least-squares calculation of the function $y = bx^n$. The derivative = $dy/dx = Bnx^{n-1}$; the integral = $[B/(n+1)]x^{n+1}$ (see Table A12.4).

TABLE A12.4

x	y	log(x)	log(y)	(log$x$)^2	(log$x$) (log$y$)
3.335	3.937	0.523	0.595	0.274	0.311
0.874	1.855	−0.0509	0.268	0.003	−0.016
0.320	0.934	−0.494	−0.030	0.245	0.015
0.075	0.299	−1.125	−0.525	1.266	0.591
0.016	0.079	−1.799	−1.104	3.237	1.987
0.007	0.0368	−2.173	−1.424	4.724	3.095
		Sum(x) −5.128	Sum(y) −2.220	Sum(x^2) 9.749	Sum(xy) 5.983

Constant $B = 10^{[\mathrm{sum}(y) - n \times \mathrm{sum}(x)]/N}$.
Constant $n = [\mathrm{sum}(x) \times \mathrm{sum}(y) - N \times \mathrm{sum}(xy)]/[\mathrm{sum}(x^2) - N \times \mathrm{sum}(x^2)]$.
N (number of data points) = 6, $B = 1.909192$, $n = 0.761449$.

5. Plot the cumulative volume injection as a function of time on log–log paper. Determine the constant for Eq. (A12.6), and calculate the Toth et al. relative permeabilities. Compare the Toth et al. calculated values to those of the alternate and JBN methods and explain the differences observed.

References

Amyx JS, Bass Jr. DM, Whiting RL. Petroleum reservoir engineering. New York: McGraw-Hill; 1960, chapter 2.

Archer JS, Wall CG. Petroleum engineering. London: Graham and Trotman; 1986, chapter 7.

Johnson EF, Bossler DP, Naumann VO. Calculation of relative permeability from displacement experiments. Trans AIME 1959;216:370–2.

Loomis AG, Crowell DC. Relative permeability studies: gas–oil and water–oil systems. *US Bureau of Mines Bulletin*, Vol. 599, Washington, DC: US Government Printing Office; 1962; p.39.

Merle CM. Multiphase flow in porous media. Houston, TX: Gulf Publishing Company; 1981, chapter 3.

Toth J, Bodi T, Szucs P, Civan F. Convenient formulae for determination of relative permeability from unsteady-state fluid displacements in core plugs. J Pet Sci Eng 2002;36:33–44 October.

Experiment 13 Basic Well-Log Petrophysical Parameters

Introduction

The principal objectives of well-log interpretation are the identification of porous zones containing hydrocarbons and the determination of the water saturation using Archie's equation:

$$S_w^n = \frac{FR_w}{R_t} = \frac{a}{\phi^m} \times \frac{R_w}{R_t} \qquad \text{(A13.1)}$$

where

n = Archie's saturation exponent
F = formation resistivity factor ($F = R_w/R_o$)
R_o = resistivity of core 100% saturated with formation water
R_w = resistivity of the formation water
R_t = resistivity of the zone of interest
a = structural parameter
ϕ = porosity of the zone of interest
m = cementation factor

The average value of the saturation exponent for water-wet rocks is 2.0, which is generally used in well-log interpretation; however, n is a function of the wettability of the system, whose value increases as the oil–water–rock system becomes progressively more oil-wet. The value of n has been reported to range from as low as 1.4 to exceeding 10.

The formation resistivity factor is a function of the formation lithology, grain size, grain shape, grain distribution, grain packing, internal structure or tortuosity (τ), the porosity, and the degree of cementation of the rock. Therefore, the determination of F requires some knowledge of the lithology; several average formulas used for different types of formations are listed in Table A13.1.

The resistivity of the formation water (R_w) may be determined directly from the formation water. If the measurement is made at room temperature, it must be corrected to the value of the subsurface zone temperature. It also may be calculated from the value of the self-potential log (SP log) in the zone of interest. Knowledge of the theory and behavior of the SP log as well as practical experience must be developed for accurate evaluation of R_w from well logs; therefore, this aspect is not discussed as a part of these laboratory procedures.

The resistivity of the formation of interest is obtained from logging tools that read deep into the formation: the deep laterolog, the deep induction log, the 64-in normal log, or the 18-ft lateral log. Knowledge of the theory and experience are required for this; therefore, it is beyond the scope of these procedures.

The structural parameter is principally a function of the tortuosity of the capillary paths in the rock. Values have been developed from comparisons of log

TABLE A13.1 General Parameters Used for Calculation of the Formation Resistivity Factor (*F*)

Equation	Criteria for Equation Selection
$F = 1.0/\phi^2$	Carbonates (limestones, dolomites)
$F = 1.45/\phi^{1.70}$	Calcareous sands
$F = 0.81/\phi^2$	Consolidated sandstones
$F = 0.62/\phi^{2.15}$	Humble formula for consolidated sandstones
$F = 1.0/\phi^{1.5}$	Unconsolidated sands
$F = 1.97/\phi^{1.29}$	Unconsolidated Miocene sands, US Gulf Coast
$F = 1.65/\phi^{1.33}$	Shaly sands

interpretation results of core analyses and from experience (for example, the value used for the Miocene sands along the Gulf Coast, Table A13.1).

Porosity may be obtained from core analyses or from special porosity logs, discussion of which is beyond the scope of these procedures.

The cementation parameter (m) is generally assumed to be 2.0, unless a specific equation for F is being used, or it may be determined from laboratory experiments with cores presented herein.

Equipment and Procedures

Resistivity of the Formation Water

Place about 100 ml of brine (or reconstituted formation water), to be used for saturation of the rock cores, in a beaker and insert the water resistivity probe (dip cell). Make sure that the brine fills the small glass tube containing the electrodes. Record the resistance registered on the ohmmeter and multiply by the dip cell constant (0.001 m) to obtain R_w. Also record the temperature of the brine: for example, assume that the temperature of the brine is 77°F, and $R_w = 176$ ohm (recorded on the ohmmeter). The resistivity R_w is then $176 \times 0.001 = 0.176$ ohm-m^2/m.

Assuming the formation temperature is 155°F, convert the resistivity measured at laboratory temperature to the resistivity at formation temperature using Arp's equation:

$$R_2 = R_1\left(\frac{T_1 + 6.77}{T_2 + 6.77}\right)$$

$$R_{w(155^\circ)} = 0.176 \times \left(\frac{77 + 6.77}{155 + 6.777}\right) = 0.0091 \text{ ohm-m}^2/\text{m} \tag{A13.2}$$

Rock Resistivity Saturated with Brine

1. Review the instruction manual for operating the resistivity matching unit. Using that, and any additional instructions given by the laboratory instructor, familiarize yourself with the instrument.
2. Cut a core to the size required by the instrument and completely saturate it with the brine previously tested. Wrap the core with a piece of paper saturated with the brine and cut it so that it is about 3 mm from the end of the core. This process is done to avoid evaporation of water from the core while the measurement is being made.
3. With the unit in the test position, check the calibration value by performing a resistivity match by setting the meter to zero using the output control knob. Set the variable resistor to 300 ohms and throw the potential-match-switch. Then note the direction in which the needle moves. Set the variable resistor to 700 ohms and throw the potential-match-switch, and once more note the direction in which the needle moves. Finally, set the variable resistor to 500 ohms and throw the potential-match-switch. If the needle is in the same position both before and after the potential-match-switch is thrown, the resistance being measured and the resistance showing on the variable resistor are the same.
4. Measurement of R_w now can be made using this instrument if it was not done with an ohmmeter as described above. Connect the dip cell to the

resistivity matching unit and immerse it in the brine. Make sure that the brine fills the small glass tube containing the electrodes. Turn the select switch to measure and determine the brine resistance (r_w). TURN THE SELECT SWITCH BACK TO TEST. Record the temperature of the water after measuring it with the thermometer. Multiply r_w by the dip cell constant (0.001 m) to obtain the brine resistivity (R_w).

5. In order to take core resistance measurements, attach the core holder to the resistivity unit. Soak two small pieces of paper towel with brine and stick them to the faces of the electrodes. This is to ensure good electrical contact between the electrodes and the core. Keep the paper wet during the experiment.
6. Place the core (100% saturated with brine) in the core holder, ensuring a firm connection exists between the core and the electrodes, and then CLOSE THE LID of the core holder.
7. Turn the select switch to measure and find the core resistance, r_o. Turn the select switch to test and remove the core. Using a caliper, find the diameter D and the length L of the core.
8. Calculate the area of the core face A and then the resistivity of the core 100% saturated with brine using the following equation:

$$R_o = \frac{(r_o A)}{L} \tag{A13.3}$$

9. The formation resistivity factor may now be calculated from Archie's equation (Eq. (A13.1)) because $S_w = 1.0$ for a brine-saturated core:

$$F = \frac{R_o}{R_w} \tag{A13.4}$$

10. Assuming the core is a sandstone, and selecting the appropriate equation from Table A13.1, calculate the porosity. For example:

$$\phi = \left(\frac{0.81}{F}\right)^{0.5} \tag{A13.5}$$

11. Calculate the tortuosity, τ, which is a measure of the heterogeneity of the rock pores:

$$\tau = (F\phi)^2 \tag{A13.6}$$

12. Displace the brine in the core with oil using a centrifuge, or by pumping oil into the core until the irreducible water saturation (S_{iw}) is attained, and then measure the resistivity of the water and oil-saturated core containing irreducible water saturation, R_t. Calculate S_{iw} using Archie's equation and compare it to the S_{iw} estimated from the water displacement test:

$$S_w = \left(\frac{FR_w}{R_t}\right)^{0.5} = \left(\frac{R_o}{R_t}\right)^{0.5} \tag{A13.7}$$

13. Assuming the porosity of the core being used was determined independently, as described under porosity measurement, report and compare the results of independent measurement of porosity with the porosity calculated using Eq. (A13.5).

Sample Calculations

$$r_w = 176 \text{ ohms}$$
$$R_w = 176 \times 0.001 = 0.176 \text{ ohm} \cdot \text{m}$$
$$L = 1.95 \text{ cm} = 0.0195 \text{ m}$$
$$D = 1.83 \text{ cm} = 0.0183 \text{ m}$$
$$A = 0.000263 \text{ m}^2$$
$$r_o = 422 \text{ ohms (resistance of } 100\% \text{ brine-saturated core)}$$
$$R_o = \frac{422 \times 0.000263}{0.0195} = 5.692 \text{ ohm} \cdot \text{m}$$
$$F = \frac{5.692}{0.176} = 32.34$$
$$f = \left(\frac{0.81}{32.34}\right)2 = 0.158 = 15.8\%$$
$$t = (0.158 \times 32.34)2 = 26.109$$
$$R_t = 47.82 \text{ ohm} \cdot \text{m (for core saturated with oil where brine is at } S_{wi})$$
$$S_{iw} = \left[\frac{(32.34 \times 0.176)}{47.28}\right] 0.5 = 0.345 = 34.5\%$$
$$m = \frac{[\log(R_w/R_t) - 2 \times \log(S_w)]}{\log(f)}, \text{ when}$$
$$S_w = 1.0; \quad m = \frac{\log(R_w/R_t)}{\log(f)}$$

Assume that $R_w = 3.00$; $R_t = 7.31$; $f = 0.224$; and $S_w = 1.0$; then

$$m = \frac{\log(0.300/7.31)}{\log(0.224)} = 2.134$$

Questions and Problems

1. Explain the influence of water–oil–rock relative wetting on the Archie saturation exponent (n).
2. Compare the formation resistivity factor for generally unconsolidated sand and the US Gulf Coast Miocene Age sands using a sand porosity of 0.280. Why is there a difference between these values?
3. Describe the relationship between water resistivity and temperature. If the brine resistivity measured at 80°F is 0.12 ohm · m, what is the resistivity in the subsurface formation, which is at a temperature of 180°F?

References

Donaldson EC, Chernoglazov V. Characterization of drilling mud fluid invasion. J Pet Sci Eng 1987;1:3–13. August.

Donaldson EC, Madjidi A, White L. Conductivity mapping to determine interwell fluid saturation. J Pet Sci Eng 1991;5:247–59. April.

Donaldson EC, Siddiqui TK. Relationship between the Archie saturation exponent and wettability. SPE Formation Evaluation 1989;**4**:359–62. September.

Helander DP. Fundamentals of formation evaluation. Tulsa, OK: Oil & Gas Consultants International; 1983, chapter 4.

Experiment 14 Surface and Interfacial Tensions

Introduction

Attraction among molecules of a liquid at the surface is subjected to a net force directed toward the bulk liquid. This results in the formation of a film-like structure, or membrane, at the surface that resists change; therefore, work must be done to create a new surface and the liquid surface will tend to adjust itself in a way so as to minimize its surface area. The surface tension is the force per unit length required to create a new surface.

The term "surface tension" refers to the tension of a liquid surface in contact with a gas (usually air). "Interfacial tension," which means the same thing, is generally used when referring to the tension of an interface between two liquids.

The SI units of interfacial tension are millinewtons per meter (mN/m), which is exactly equal to the now-obsolete unit dynes per centimeter.

Equipment and Procedures

Several methods for measurement of interfacial tension have been developed. The two methods selected for this experiment are (a) the du Noüy ring method that employs a beam analytic balance to measure the force required to pull the ring through the interface and (b) the capillary rise method. Absolute cleanliness is essential for these experiments.

du Noüy Ring Method

1. Select a platinum du Noüy ring and measure its diameter accurately. Hold the ring in a flame until it is red in color in order to burn off any residual organic matter that may have remained from its last use. When the ring is cooled, place it on the hook at the left end of the beam analytic balance lever.
2. Place the fluid whose air–liquid interfacial tension (surface tension) is to be measured in a beaker about 2 in. in diameter in order to avoid wall effects.
3. Immerse the platinum ring in the liquid until it is just under the surface and adjust the balance to equilibrium.
4. Slowly lower the beaker so that the ring gradually breaks the surface, and record the grams-force required to just break through the surface. Repeat the experiment a few times to gain skill and improve accuracy.

Calculate the interfacial tension (IFT) using Eq. (A14.1):

$$\text{IFT} = \frac{g \times g_c}{2\pi d} \tag{A14.1}$$

where

IFT = interfacial tension ($N \times 10^{-3}$/m)
g = grams-force measured with the analytic balance
g_c = gravitational constant (980 cm/s^2)
d = diameter of the ring

There are instruments available that measure the IFT directly as the ring is pulled (sometimes automatically) through the interface of the liquid. They are all based on the same principle and use the relationship expressed by Eq. (A14.1).

The interfacial tension between two liquids is measured by the same procedure, but certain precautions must be taken to avoid contamination of the ring.

The denser liquid is transferred to the glass vessel first, and the ring is then immersed below the surface (<1/8 in). Then the lighter liquid is carefully poured on top of the denser liquid until it is covered by the lighter liquid with a thickness of about 1 cm. At this point, the procedure detailed in steps 1–4 is followed.

Capillary Rise Method

The height of the rise of a liquid in a capillary tube is expressed by:

$$h = \frac{2 \times \sigma \times \cos\theta}{r \times \rho \times g_c} \tag{A14.2}$$

where

h = distance to which the liquid rises in the capillary tube above the open surface of the liquid, cm

σ = interfacial tension, $N \times 10^{-3}$/m

$\cos\theta$ = cosine of the contact angle of the surface inside the capillary to the capillary wall

r = radius of the capillary, cm

ρ = density of the more dense liquid, g/cm^3

g_c = gravitational constant (980 cm/s^2)

1. Insert the capillary tube into the beaker of distilled water and measure the height of the rise of water in the tube. Record the temperature of the water, and obtain the interfacial tension of water/air and the density for that temperature from Table A14.1 (interpolate from values of temperatures not listed in the table). Assuming $\cos\theta = 1.0$, calculate the radius of the capillary from Eq. (A14.2). This step may be omitted if the manufacturer of the capillary tube has given the exact diameter of the tube.
2. Dry the capillary tube in an oven and then place it in the liquid whose air/liquid interfacial tension is to be measured, and measure the height of the liquid rise in the tube. Assuming $\cos\theta = 1.0$, calculate the interfacial tension from Eq. (A14.2).
3. If the interfacial tension between two liquids is to be measured, the same precautions used for the du Noüy method must be employed with the capillary rise method:
   a. Stopper the top of the capillary: a finger may be used or a small cork may be fitted over it.

TABLE A14.1 Surface Tension (Air/Water) and Density of Water at Several Temperatures

Temperature (°C)	Surface Tension ($N \times 10^{-3}$/m)	Density (g/cm^3)
15	73.49	0.9991
18	73.05	0.9986
20	72.75	0.9982
25	71.97	0.9971
30	71.18	0.9957

b. Pour the denser liquid into a beaker, place the stoppered capillary tube into the liquid, and then pour the lighter liquid on top of the denser liquid until it is covered by about 2 cm of the lighter liquid.
c. Remove the stopper from the top of the capillary and allow the denser liquid to rise in the tube. Measure the height of the rise and record the temperature of the liquids.
d. Use the density of the denser liquid (from a handbook or from previous measurement) to calculate the interfacial tension from Eq. (A14.1).

Sample Calculations

du Noüy ring:

Diameter of the du Noüy ring = 1.75 cm

Force required to pull the ring through the interface

$= 0.8075$ grams–forceIFT $= (0.8075 \times 980)/(2 \times 3.1416 \times 1.75)$

$= 71.97\ \text{N} \times 10^{-3}/\text{m}$ (from Eq. (A14.1))

Capillary rise:

Height of rise of distilled water in the capillary = 0.5 cm

Temperature of the water = 25°C

Density of water at 25°C = 0.9971

$\cos\theta = 1.0$

$r = 2 \times 71.97/(0.5 \times 0.9971 \times 980) = 0.1473$ cm (from Eq. (A13.2))

Surface tension of kerosene:

Height of rise in the capillary tube = 0.6 cm

Density of kerosene at 25°C = 0.9917

$\sigma = hrdg_c/2 = 0.6 \times 0.1473 \times 0.8817 \times 980/2 = 38.18\ \text{N} \times 10^{-3}/\text{m}$

Questions and Problems

1. Suppose that a small amount of residual detergent contaminated water is used to determine the radius of the capillary tube.
   a. What would this do to the height to which the water would rise in the capillary? What would it do to the surface tension of the water?
   b. Why is the surface tension of a hydrocarbon solvent less than the surface tension of water at any specific temperature?

References

Amyx JW, Bass Jr. DM, Whiting RL. Petroleum reservoir engineering. New York: McGraw-Hill; 1960, chapter 3.

Dullien FAL. Porous media. New York: Academic Press; 1979, chapter 2.

Experiment 15 Capillary Pressure

Introduction

The pore openings of most porous rocks are only a few microns (μm) in size. When the rock contains two immiscible fluids, one of the fluids tends to wet the rock surfaces preferentially and is labeled the "wetting" fluid. The relative wetting characteristics of the system are functions of the chemical properties of the fluids and the rock surfaces. Capillary pressure is the pressure difference between the nonwetting phase and the wetting phase:

$$P_c = p_{nonwet} - p_{wet} = \rho_{wet} g_c h = 2\sigma\cos\theta/r \quad (A15.1)$$

where

$\cos\theta$ = angle of contact of the wetting phase to the capillary wall (degrees)
g_c = gravitational constant (980 cm/s^2)
h = height of capillary rise of the wetting fluid (cm)
P_c = capillary pressure (pascals, N/m^2); function of water saturation
p_{nonwet} = pressure of the nonwetting fluid
p_{wet} = pressure of the wetting phase
r = radius of the capillary (m)
r_{wet} = density of the wetting fluid (g/cm^3)
s = interfacial tension (N $\times$ 10^{-3}/m)

Equipment and Procedures

There are three common methods used for measuring capillary pressure as a function of water saturation in rocks. The mercury injection method is the most rapid, but the core cannot be used for other tests after injection of mercury because mercury cannot be removed from the rock. In addition, the mercury injection curve does not yield data on capillary hysteresis. Nevertheless, because it offers a rapid method for obtaining capillary pressure data on irregular-shaped samples (any shape can be used), it remains as one of the standard petrophysical tests.

A second method is the use of a porous diaphragm, which relies on the selection of a suitable porous disk to provide a barrier that excludes the passage of the nonwetting fluid and permits the passage of the wetting fluid. Porcelains can be made with low permeability (<5 mD) and very uniform pore openings; therefore, a porcelain is the most frequently used material for the porous diaphragm. When water-wet systems are used, the diaphragm is saturated with water and a core is placed on it in good capillary contact. Then, the pressure exerted in the nonwetting phase (gas or oil) is recorded for incremental displacements of water. The displacement also may be reversed by saturating the disk with the nonwetting phase and recording the pressure required to displace incremental volumes of the nonwetting fluid. The drawback of this method is the long period of time required for completion of the test (sometimes several weeks) because one must wait for completion of the incremental fluid displacement after each increase in pressure.

The third method is the centrifuge method. Cores are placed in specially designed holders equipped to collect either water or oil in a calibrated portion of the core holder. A centrifuge is then used to displace one of the fluids by centrifugal force. The angular velocity of the centrifuge (revolutions per minute) is increased in increments, and the amount of fluid displaced at each incremental velocity is measured. The capillary pressure is calculated from the centrifugal force.

Core Preparation for Capillary Pressure Measurement

1. Cut the core and make sure that the ends of the core are smooth and parallel.

2. Extract the hydrocarbons from the core by solvent extraction.
3. Dry the core and measure its permeability using gas flow.
4. Measure the porosity of the core by any convenient method and calculate its pore volume V_p.
5. Weigh the dry core, saturate with the fluid to be displaced, and weigh again. Using the density of the fluid, calculate the pore volume and compare it with the pore volume of step 4.

Mercury Injection Method

Calibrations are required before the mercury injection instrument is ready for use. First, the volume of the sample chamber is measured by evacuating the system and then injecting mercury to the lower reference mark. The reading scale is then set to zero, and the amount of mercury required to reach the upper mark is recorded as the sample chamber volume. The second calibration is to run a blank (with a core in the sample chamber) in order to obtain an instrument expansion calibration as a function of pressure.

A dry core sample is placed in the sample chamber of the mercury injection apparatus. The pump piston is withdrawn and the system is evacuated. The vacuum valve is closed, and mercury is injected into the sample chamber. When the mercury level reaches the lower reference mark, the reading scale and the vernier are set to read the sample chamber volume. Mercury is then injected into the sample chamber until it reaches the upper reference mark. The bulk volume of the sample is read directly from the movable pump scale and hand-wheel dial numbers. The hand-wheel dial and pump scale are adjusted to read exactly 0.000 cm^3. The vacuum valve is then closed by opening the bleed valve, and nitrogen gas is admitted to the system until the mercury level is 4–5 mm below the upper reference mark. The pressure of the system is noted, and by operating the pump, mercury is injected into the sample chamber until it reaches the upper reference mark. The volume of mercury injected into the core is noted. After atmospheric pressure is attained in the system, pressure is applied using the compressed nitrogen gas. After each incremental injection of mercury to the upper reference mark, a waiting period of 2–5 minutes is imposed to allow the system to reach equilibrium. The procedure is repeated for a number of intervals until a pressure of about 800 psi is attained. The volume of mercury injected corresponds to the nonwetting phase volume because mercury is a nonwetting fluid, whereas the mercury vapor corresponds to a wetting phase. Each incremental pressure increase (P_c) is plotted versus the corresponding wetting phase saturation.

Porous Diaphragm Method

1. A schematic diagram of the Ruska diaphragm pressure cell is presented in Figure A15.1. The base, together with the diaphragm assembly, is detached and the pipette (19) is withdrawn from the base.
2. The diaphragm is dried in the oven with the core to be tested and then the diaphragm and core are evacuated for at least 5 hours. The fluid to be used as the saturating fluid is also evacuated for a brief period to remove dissolved gases prior to contact with the core and diaphragm.
3. While the vacuum is maintained on the system, the saturating fluid is introduced slowly into the evacuated diaphragm and core until they are

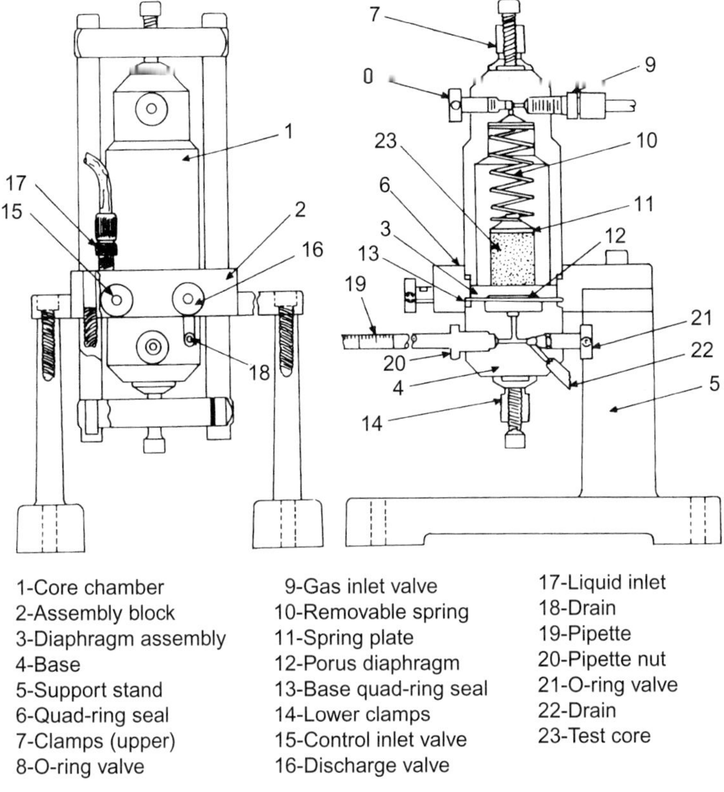

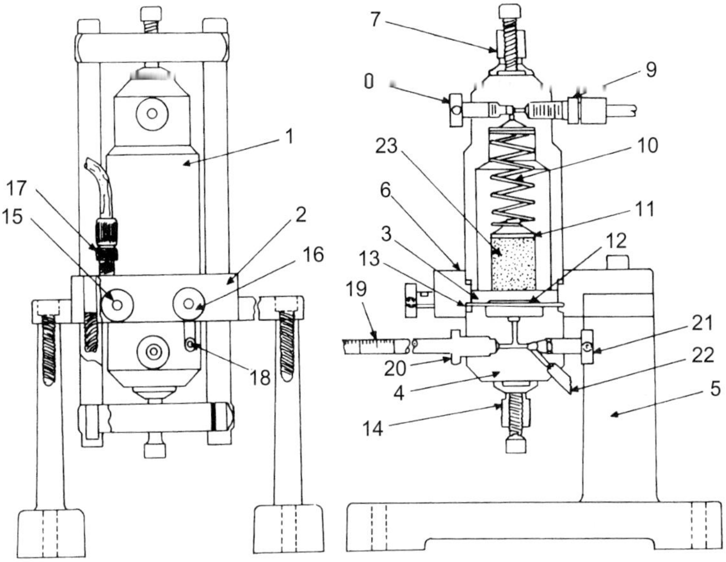

1-Core chamber
2-Assembly block
3-Diaphragm assembly
4-Base
5-Support stand
6-Quad-ring seal
7-Clamps (upper)
8-O-ring valve
9-Gas inlet valve
10-Removable spring
11-Spring plate
12-Porus diaphragm
13-Base quad-ring seal
14-Lower clamps
15-Control inlet valve
16-Discharge valve
17-Liquid inlet
18-Drain
19-Pipette
20-Pipette nut
21-O-ring valve
22-Drain
23-Test core

FIGURE A15.1 Ruska diaphragm pressure cell for measurement of capillary pressure.

immersed at least 1/2 in. After 1 hour, air is slowly admitted to the system until atmospheric pressure is attained.

4. The base valve (21) is closed, and while capping the pipette nut outlet with a finger, the diaphragm assembly cavity is filled with liquid. Air is displaced from the base flow channels through the pipette outlet and drained by opening them slightly.
5. Grease the quad-ring (6) and position it properly, then insert the diaphragm assembly into the cavity and hold it firmly in place.
6. A pipette of proper size (estimate the volume of fluid to be displaced for selection of the pipette) is inserted and the nut is tightened.
7. Clamp the complete base and diaphragm assembly to the underside of the assembly block. Place a piece of soft tissue paper on the diaphragm and saturate it with liquid. Now place the core on the tissue paper. Drain the pipette to the reading of zero, and the cell is now ready to measure capillary pressure.
8. Apply gas (air or nitrogen) pressure to the upper chamber through the gas inlet valve (9) in increments of about 2 psi. After each increase in pressure, wait until fluid collection in the pipette stops. Record the pressure and the

amount of fluid displaced, and then increase the pressure once more. Repeat this procedure until no more fluid is displaced from the core, or until the gas breaks through the porous diaphragm.

9. Calculate the average saturations of the core after each step increase in pressure and plot the pressure as a function of core saturation (% of pore volume, V_p) to obtain the capillary pressure curve. Extrapolate the curve to 100% fluid saturation to obtain an estimate of the threshold pressure (pressure required to enter the largest pore of the sample). The irreducible saturation (S_{iw}) of the fluid is the limiting saturation where no more fluid can be displaced for continued pressure increase.

$$S_{iw}(fraction) = \frac{V_p - cumulative\ volume\ of\ fluid\ displaced}{V_p}$$

Centrifuge Method

Several different types of centrifuges are available for the centrifuge method. However, the bowl temperature must be controlled within a few degrees because oil–water–rock systems become more water-wet due to increase in temperature. Therefore, the temperature at which the measurement is made should always be specified.

Only the initial displacement of fluid to irreducible saturation was discussed for the preceding tests (capillary pressure curve I, Figure A15.2). Mercury injection may be used for displacement of air, in which case the capillary pressure is generally plotted as a function of mercury saturation; or it may be used for mercury displacement of a wetting fluid such as water, and in this case, the capillary pressure is plotted against the wetting phase saturation. The porous diaphragm method may be used for displacement of a wetting phase by a gas, or the core (saturated with the wetting fluid) may be covered with a nonwetting phase and gas pressure used to cause displacement of the wetting phase by the nonwetting phase. All of these yield P_c curve I.

The centrifuge may be used to obtain three capillary pressure displacement curves (Figure A15.2):

a. P_c-I: initial displacement of a fluid to irreducible saturation. A wetting or a nonwetting fluid may be used for the initial saturation. For ease of understanding, however, it is assumed that the core is initially saturated with water, and oil is used to displace the water to irreducible saturation: (P_c-I, oil $\gg$ water, from $S_w = 1.0$ to S_{iw})

b. P_c-II: displacement of the oil with water to the water saturation equivalent to the residual oil saturation ($S_{wor} = 1 - S_{or}$). This displacement is equivalent to a waterflood where water is displacing oil from a core that was initially at saturation S_{iw}. S_{or} = residual oil saturation.

c. P_c-III: displacement of the water to irreducible water saturation (S_{iw}) once more by starting with the core saturated with water and oil at saturation S_{wor}. (P_c-III, oil $\gg$ water, from S_{wor} to S_{iw})

Centrifuge Procedure

1. Small core plugs for centrifuge measurement (2.5 cm in diameter by 2.5 cm long) should be jacketed with a thin Viton innertube surrounded by heat-shrinkable Teflon tubing. The reduction of exposed surface area is 67%.

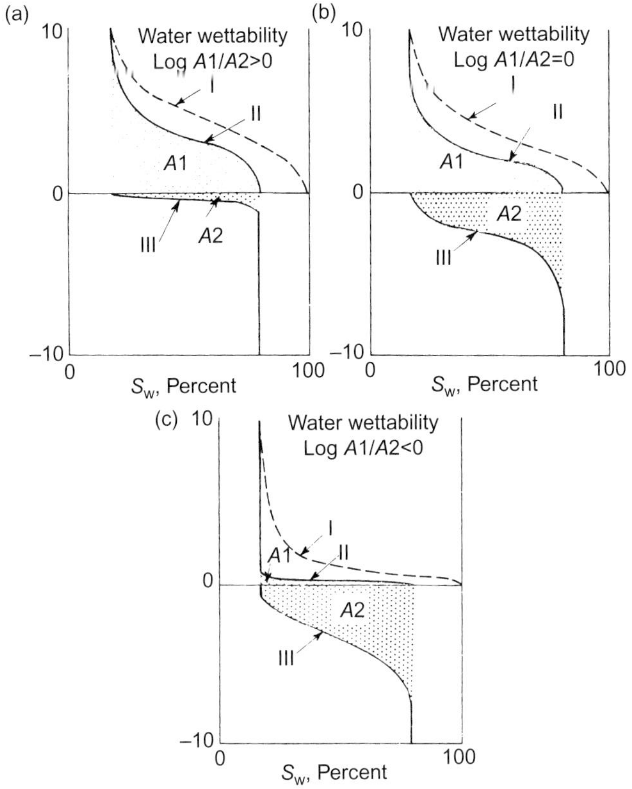

FIGURE A15.2 Capillary pressure curves showing the three curves that can be obtained from displacements using a centrifuge and calculation of wettability by the USBM method. For water–oil capillary pressure, the figure shows curve I for displacement of water from a 100% water-saturated core with oil to the irreducible water saturation, curve II represents displacement of oil by water to the residual oil saturation, and curve III represents displacement water (from the core saturated with water and oil at residual oil saturation) to irreducible water saturation.

Without the jacket, it is difficult to control the outside fluid films when changing core holders at the endpoint of runs, and large errors of saturation measurement may occur.

(P_c-I): At least two saturated cores are placed in graduated centrifuge core-tubes, and the tubes are filled with oil (Figure A15.3). The tubes are then placed in the metal shields, and the weights of the assemblies are adjusted with small weights until they are equal.

2. The assemblies are placed in the trunnion rings across from each other in the centrifuge.
3. In order to complete the USBM wettability test in a single day, the centrifuge speed may be increased until the maximum speed is attained. It is operated at the maximum speed until water displacement stops, as observed using a strobe light. Water displacement will usually stop within 15 minutes for a

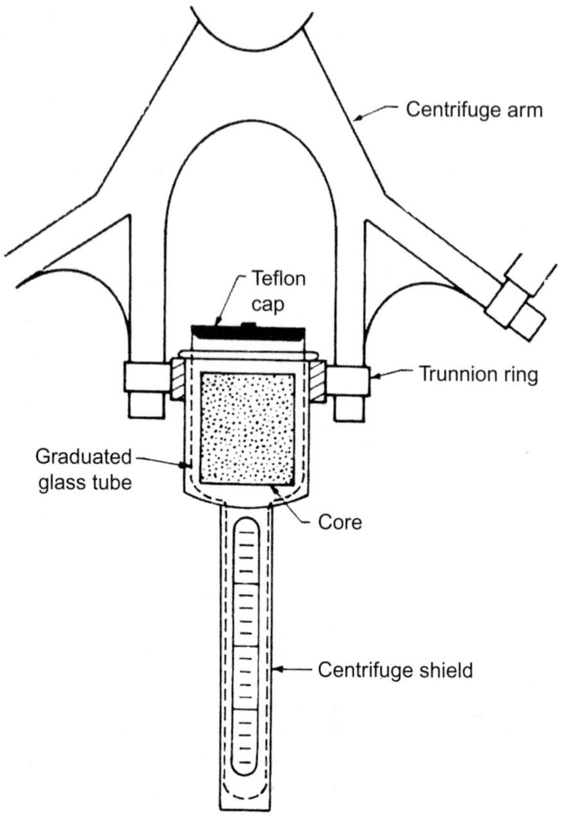

FIGURE A15.3 Centrifuge tube core holder assembled for displacement of water by oil or air.

light oil (API gravity >20) and within 30 minutes for a heavy oil. The water displacement is used to calculate the irreducible water saturation, which is the beginning point for the USBM wettability test.

If the initial oil-displacing water capillary pressure curve (P_c-I) is required, the centrifuge is started at low rpm (200 or less). Using a strobe light, the volumes of water collected in the graduated end of the core holder tubes are recorded.

The capillary pressure at the top end of the core, P_{c1}, is calculated using Eq. (A15.2) and plotted against the corresponding water saturation at the top of the core, S_{w1}, using Eq. (A15.3) or (A15.4). Occasionally, the slope of the curve S_a versus P_c (dS_a/dP_c) is so great that these corrections cannot be applied; in such cases, the average saturation versus the capillary pressure is used to determine the wettability index.

4. (P_c-II): The cores (containing water at S_{iw} and oil) are removed from the glass core holder tubes and placed in a second pair of glass core tube. The tubes are then filled completely with water and stoppered. Caution is taken to

minimize the time spent in transferring the core from one glass centrifuge core tube to another to prevent loss of fluid from the core due to evaporation. The core tubes are then inverted and placed in metal centrifuge shields, and the weights of the two assemblies are adjusted to zero (Figure A15.4).

5. The assemblies are placed in trunnion rings and centrifuged at incremental centrifugal velocities as before. The volume of oil displaced to the top, into the graduated end of the core tubes, is measured at each increment of rpm.
6. The capillary pressure and water saturation (P_c-II) are calculated using Eqs. (A15.2) and (A15.3). The density difference is reversed, yielding a negative capillary pressure value ($\rho_o - \rho_w$) for the displacement from S_{iw} to S_{wor}. The radius of the centrifuge arm for this displacement is different from the one used in the first displacement (Table A15.1).
7. (P_c-III): The cores are removed from the glass centrifuge core tubes and placed in a new set. The cores are immersed in oil and the core tubes are inserted in the appropriate centrifuge shield. Calculate the capillary pressure and saturation at the top of the core using Eqs. (A15.2) and (A15.3).

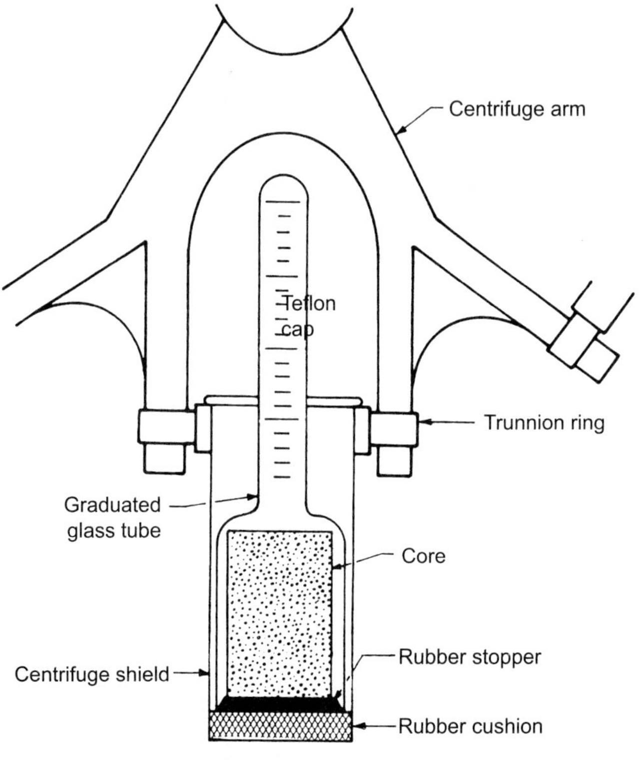

FIGURE A15.4 Centrifuge tube core holder assembled for displacement of oil by water or air.

TABLE A15.1 P_c **Curve II: Brine Displacing Crude Oil from** S_{iw} **to** S_{wor}

rpm	Oil Disp. (cm^3)	S_A	P_{C1} (psi)
0	0.000	0.221	0.000
300	0.600	1.416	−0.040
750	0.650	0.432	−0.253
1,200	1.175	0.063	−0.647
1,650	1.400	0.676	−1.223
2,210	1.575	0.732	−2.194
2,600	1.625	0.749	−3.036
3,280	1.650	0.757	−4.832
3,650	1.750	0.789	−5.984
4,090	1.700	0.773	−7.513
4,610	1.750	0.789	−9.545

HYPER constants for $P_{c1} = f(S_a)$:
$A = -0.0605$; $B = -1.1824$; $C = -19016$
$V_p = 3.08$; Porosity $= 0.238$; $K = 3{,}422$; $L = 2.55$; $R_2 = 16.68$; brine density $= 1.01$; oil density $= 0.938$; $P_c = 1.588 \times 10^{-7} \times 0.072 \times (R_2 - L/2) \times L \times \text{RPM}^2$.

8. The assemblies are placed in trunnion rings and centrifuged once more at incremental centrifugal velocities to the limit of the centrifuge, or to some preset limit. The stabilized volumes displaced at each incremental rotational velocity are recorded and the capillary pressure is calculated using Eq. (A15.1).

Calculation of Capillary Pressure and Saturation

The cores, which are saturated with one or two phases and surrounded by the displacing fluid phase, are rotated in the centrifuge at incremental speeds. The expelled fluid is measured at each incremental speed with the aid of a strobe light, which yields the average fluid saturation of the core. The saturation along the core, however, varies from a low value at the inlet to 100% at the outlet, whereas the capillary pressure exhibits a maximum at the inlet and zero at the outlet. The maximum capillary pressure at the inlet of the core is calculated from Eq. (A15.2). The saturation at the inlet end of the core (corresponding to the maximum capillary pressure saturation) is calculated from the following equations:

$$P_{c1} = 1.096 \times 10^{-6} \times (\rho_w - \rho_o)\left(R_2 - \frac{L}{2}\right)L \tag{A15.2}$$

$$\begin{aligned} S_1 = S_{AV} &+ \left[\frac{2R}{1+R}\right] \times P_{c1}\left[\frac{dS_{AV}}{dP_{c1}}\right] \\ &+ \left[\frac{R}{1+R}\right]\int_0^{P_{c1}} \left(\frac{1-[1-(P_c/P_{c1})(1-R^2)]^{0.5}}{1-[(P_c/P_{c1})(1-R^2)]^{0.5}}\right) \\ &\times \frac{dS_{AV}}{dP_{c1}} \times dP_c \end{aligned} \tag{A15.3}$$

$$S_1 = \frac{d}{dP_{c1}}[P_{c1} \times S_{AV}] = S_{AV} + P_{c1} \times \frac{dS_V}{dP_{c1}} \quad \text{(A15.4)}$$

where

L = core length, cm
N = centrifuge speed, revolutions per minute (rpm)
P_c = capillary pressure at any point in the core
$P_{c(AV)}$ = capillary pressure at the center of the core, kPa
P_{c1} = capillary pressure at the inlet end of the core, kPa
$R = R_1/R_2$
R_1 = centrifuge arm radius to the inlet end of the core, cm
R_2 = centrifuge arm radius to the outlet end of the core, cm
S = brine saturation at any point in the core
S_{av} = average brine saturation
S_1 = saturation at the inlet end of the core
ρ_o = oil density, g/cm^3
ρ_w = water (brine) density, g/cm^3

Hassler and Brunner (1945) proposed a solution (Eq. (A15.4)), which is a simplification made by neglecting the centrifugal gravity gradient and assuming that $R_1/R_2 = 1.0$. This approximation is frequently used when $R_1/R_2 > 0.7$; however, for values less than 0.7, errors will result.

USBM Wettability Index

The USBM wettability index (I_u) is defined as the logarithm of the ratio of (1) the area under P_c-III from $P_c = 0$ to P_c at $S_w = S_{iw}$ divided by (2) the area under P_c-II from $P_c = 0$ to P_c at $S_w = S_{wor}$, (Figure A15.2). The area under the curve may be determined by first matching the curve with a least-squares solution of the hyperbola (Eq. (A15.5)) and then by integrating the analytic expression, by using Simpson's rule, or by simply using a planimeter.

$$P_c = \frac{1.0 + A \times S_w}{B + C \times S_w} \quad \text{(A15.5)}$$

$$I_u = \log \frac{A_1}{A_2} \quad \text{(A15.6)}$$

Neutral, or fifty-fifty, relative wetting of the two phases occurs at $I_u = 0.0$; increasing positive values represent increasing water-wet conditions, whereas increasing negative values represent increasing oil-wet conditions. The example problem below exhibits a wettability of −0.085, which means that this water–oil–rock system is slightly oil-wet.

Energy Required for Fluid Displacement

The area under the capillary pressure curve represents the thermodynamic energy required for displacement of the volume of fluid from the core:

$$dF = P_c \times d(S_w \times V_p) \text{ between the limits of saturation change} \quad \text{(A15.7)}$$

where

F = free energy of displacement, joules (Nm)/volume displaced
P_c = capillary pressure, pascals (N/m^2)
V_p = pore volume, m^3

Sample Calculations

Capillary pressure curve I was not determined (Figure A15.2). Crude oil was used to displace the brine to irreducible water saturation at high centrifuge speed. Curves II and III, which are used to determine the USBM wettability index, were obtained as explained previously. The curves were corrected using the simpler Hassler–Brunner correction with the aid of the least-squares fit of capillary pressure curves using program HYPER. All of the data and results are presented in Tables A15.1–A15.7 and are illustrated in Figure A15.5.

Program HYPER is first used to fit $S_a = f(P_{c1})$ and to obtain the slope (dS_a/dP_c) for the Hassler–Brunner correction (constants A_1, B_1, and C_1). It is then used a second time (for constants A_2, B_2, and C_2) to obtain a fit for $S_{w1} = f(P_{c1})$; this curve is used to obtain the area for the wettability index by integration between the saturation limits (S_{iw} to S_{wor}).

Capillary pressure was expressed in psi; consequently, the area under the curves have the units of psi because the P_c-axis is expressed in psi and the S_w-axis is dimensionless. The displacement energy per volume of fluid displaced is the area multiplied by the pore volume, which yields the units of psi $\times$ cm^3. This is then converted into SI units by changing the psi to pascals and then cm^3 to m^3 as follows:

$$(\text{psi} \times \text{cm}^3) \times (6{,}985 \ \text{Pa/psi}) \times (\text{m}^3/\text{cm}^3 \times 10^{-6}) = \text{Pa} \times \text{m}^3 = (\text{N/m}^2) \times \text{m}^3 = \text{joules}$$

Questions and Problems

1. What are some applications of capillary pressure curves?
2. How are mercury capillary pressure curves made equivalent to water capillary pressure curves?
3. What is the capillary pressure J-function? What is it used for?
4. What is the value of threshold pressure for an oil-wet core?

TABLE A15.2 Calculation of S_{w1} Using Hassler–Brunner Method

S_{av}	P_{c1}	dS/dP_c	$P_c \times dS/dP_c$	S_{w1}
0.221	0.000	−0.259	0.000	0.221
0.416	−0.040	−0.034	1.001	0.417
0.432	−0.253	−0.024	0.006	0.438
0.603	−0.647	−0.016	0.011	0.613
0.676	−1.223	−0.062	0.076	0.752
0.732	−2.194	−0.119	0.261	0.883
0.749	−3.036	−0.138	0.420	1.168
0.757	−4.832	−0.148	0.718	1.474
0.789	−5.932	−0.193	1.156	1.945
0.773	−7.513	−0.170	1.278	2.051
0.789	−9.545	−0.193	1.844	2.633

$S_{w1} = S_{av} + P_{HP} \times dS_{av}/dP_{HP}$; $dS_{av}/dP_c = (1 + C \times S_{av})/(B - AC)$, brine saturation increasing; therefore, $S_{av} < S_{w1}$ (where S_{w1} = inlet water saturation; S_{av} = average water saturation; P_{HP} = capillary pressure from HYPER).

TABLE A15.3 P_{c1} **at Even Increments of** $S_{w(1)}$

S_{w1}	P_{c1}
0.228	0.007
0.250	−0.008
0.300	−0.044
0.350	−0.086
0.400	−0.136
0.450	−0.197
0.500	−0.273
0.550	−0.369
0.600	−0.495
0.650	−0.666
0.700	−0.915
0.726	−1.095

HYPER fit of valid corrected values of S_a to S_{w1} [from S_{iw} (0.2213) to S_{wor} (0.7891)]. $A_2 = 0.1141$; $B_2 = -0.4783$; $C_2 = -1.0841$.

TABLE A15.4 P_c**-Curve III**

rpm	Brine Disp.	S_{av}	P_{c1} (psi)
750	0.250	0.789	0.121
1,500	0.525	0.619	0.486
1,950	0.625	0.586	0.821
2,400	1.025	0.457	1.244
2,960	1.300	0.367	1.892
3,350	1.475	0.311	2.423
4,030	1.600	0.270	3.506
4,400	1.650	0.254	4.180
4,840	1.675	0.246	5.058
5,360	1.725	0.229	6.203
5,830	1.750	0.221	7.033

HYPER constants for $P_{c1} = f(S_{av})$: $A_1 = -2.6932$; $B_1 = -4.8378$; $C_1 = -6.4630$. Oil displacing brine from S_{wor} to S_{iw} centrifuge radius $(R_1) = 8.68$.

5. What is the capillary number? What is it used for?
6. What is the meaning of wettability?
7. a. Calculate the capillary pressure P_c, at the inlet (Eq. (A15.2)), and the saturation S_w, at the inlet (Eq. (A15.3)), and plot P_c versus S_w. Label this as curve A.

TABLE A15.5 Calculation of S_{w1} Using Hassler–Brunner Method

S_{av}	P_{c1} (HYPER)	dS/dP_c	$P \times dS/dP_c$	S_{w1}
0.789	0.121	−0.851	−0.10336	0.686
0.619	0.486	−0.452	−0.21939	0.399
0.586	0.821	−0.390	−0.32000	0.266
0.457	1.244	−0.188	−0.23366	0.223
0.367	1.892	−0.091	−0.17280	0.194
0.311	2.423	−0.048	−0.11593	0.195
0.270	3.506	−0.025	−0.08875	0.181
0.254	4.180	−0.018	−0.07645	0.177
0.246	5.058	−0.015	−0.0769	0.169
0.229	6.203	−0.010	−0.06135	0.168
0.221	7.338	−0.008	−0.05621	0.165

Brine saturation decreasing; therefore, $S_{av} > S_{w1}$.

TABLE A15.6 P_{c1} at Even Increments of S_{w1}

S_{w1}	P_{c1}
0.789	0.098
0.750	0.112
0.700	0.131
0.650	1.154
0.600	0.182
0.550	0.216
0.500	0.260
0.450	0.317
0.400	0.395
0.350	0.507
0.300	0.685
0.250	1.005
0.221	1.340

HYPER fit of valid corrected values of S_{av} to S_{w1} (from S_{wor} to S_{iw}). $A_2 = -1.2112$, $B_2 = 0.8773$, $C_2 = -7.9473$.

b. Calculate the capillary pressure at the inlet (Eq. (A15.2)) and the saturation at the inlet using the approximation expressed in Eq. (A15.4). Plot P_c versus S_w on the same graph used in (a) and label it as curve B.

c. Calculate the average capillary pressure (Eq. (A15.5)) and plot it on the same graph versus the average saturation. Label this curve C.

d. Compare the three curves and discuss the deviations of curves B and C from curve A.

TABLE A15.7 Wettability and Displacement Energy Calculation Using the Areas under P_c-Curves II and III (Tables A15.3 and A15.6, refer to Figure A15.2)

A_2 = area under NEGATIVE curve, Pc-II = 0.250206
A_1 = area under POSITIVE curve, Pc-III = 0.205808
Wettability index = log (A_1/A_2) = 0.08483
Displacement energy for oil displacing brine from S_{w1} to S_{wor} = 0.65 J/cm^3 = 0.031 BTU/bbl
Displacement energy for water displacing brine from S_{w1} to S_{wor} = 0.236 J/cm^3 = 0.028 BTU/bbl

Using the HYPER curves, Area = $(B_2 \times S_w/C) + [(A_2 \times C_2 - B_2)/C_2^2] \times \ln(1 + C_2 \times S_w)$.

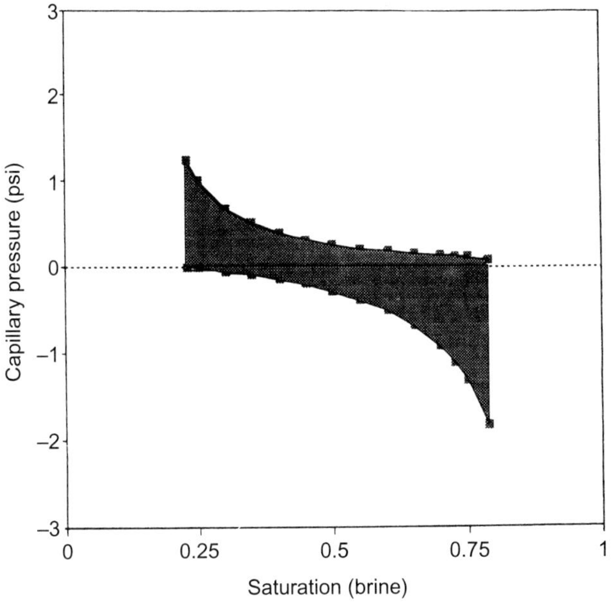

FIGURE A15.5 Capillary pressure curves II and III (Figure A15.2) matched with a least-squares solution using program HYPER for determination of the areas under the curves, wettability, and the energy required for fluid displacement.

References

Amyx JW, Bass Jr. DM, Whiting RL. Petroleum reservoir engineering. New York: McGraw-Hill; 1960, chapter 3.

Archer JS, Wall CG. Petroleum engineering. London: Graham and Trotman; 1986, chapter 6.

Donaldson EC, Ewall N, Singh B. Characteristics of capillary pressure curves. J Pet Sci Eng 1991;16:249–61 November.

Donaldson EC, Kendall RF, Pavelka EA, Crocker ME. Equipment and procedures for fluid flow and wettability tests of geological materials. DOE/BETC/IC 7915. Springfield, NTIS, VA; 1980: 40.

Hassler GL, Brunner E. Measurement of capillary pressure in small core samples. Trans AIME 1945;160:114–23.

Rajans RR. Theoretically correct analytic solution for calculating capillary pressure-saturation from centrifuge experiments. In: Proc. SPWLA annual logging symposium; June 9–13, 1986. pp. 1–18.

Slobod RL, Chambers A, Prehn Jr, WL. Use of centrifuge for determining connate water, residual oil, and capillary pressure curves of small core samples. Trans AIME 1951;192:127–235.

Experiment 16 Pore Size Distribution

Introduction

An air–water capillary pressure curve may be used to obtain a distribution of the average pore entry radii of a porous medium. Air is nonwetting; therefore, the contact angle can be considered to be 90°, which makes $\cos\theta$ in Eq. (A15.1) equal to 1.0. Thus, the equation to measure the average pore entry radius is as follows:

$$R = \frac{2\sigma}{P_c} \tag{A16.1}$$

$$R = \frac{144}{P_c} \tag{A16.2}$$

where

P_c = capillary pressure, pascals (N/m^2)
R = average pore entry size (microns, μm)
σ = interfacial tension (mN/m) (72 for air–water).

The centrifuge yields $S_{av} = f(P_{c1})$, the average saturation as a function of capillary pressure at the inlet end of the core (explained and illustrated in Experiment 15). The average saturation is corrected to the core inlet saturation using either the Rajan method or the Hassler–Brunner method; but in some cases, the slope (dS_w/dP_c) required by the correction equations (for air displacing water) is too great to allow the correction to be made. In such cases, the average saturation is used to obtain an estimate of the pore size distribution.

Equipment and Procedures

The core analysis centrifuge used for the wettability analysis (Experiment 15) is also used for determination of the air-displacing brine capillary pressure curves for pore size distribution measurement.

1. Cut several cores to appropriate size for the core analysis centrifuge. Clean them with solvents, and steam and dry the cores.
2. Obtain the mass of cores as accurately as possible, saturate with brine of known density, and obtain the mass of the saturated cores.

3. Place the cores in centrifuge tubes for collection of brine, balance the pairs of assemblies, and place the balanced pairs in opposition in the centrifuge.
4. Starting at the lowest rpm that is permissible with the centrifuge, obtain the amount of brine displaced as a function of step increases in centrifuge speed. Stop the experiment when no more brine is displaced in two successive runs (Table A16.1).

Sample Calculations

The volume of water displaced at incremental speeds of the centrifuge is used to calculate capillary pressure as a function of average saturation as shown in Table A16.1.

The air–water capillary pressure curve is fit with program HYPER (Chapter 5), and the HYPER constants are used to obtain capillary pressure as a function of even increments of average saturation (Table A16.2). The pore radii and pore size distribution corresponding to the even increments of saturation are then calculated as illustrated in Table A16.2 and are plotted in Figure A16.1 where $m_2 = D(R_i)$.

Questions and Problems

1. Explain the meaning of the distribution function D_{ri}.
2. Discuss and illustrate how measurement of pore entry size distributions can be used to evaluate formation damage.

TABLE A16.1 Pore Size Distribution of a Berea Sandstone Core

N (rpm)	V_{water}	S_{av}	P_c (psi)	P_c (kPa)
1,300	0.30	0.8266	4.1351	28.51
1,410	0.40	0.7688	4.8645	33.54
1,550	0.50	0.7110	5.8780	40.53
1,700	0.60	0.6532	7.0713	48.76
1,840	0.70	0.5954	8.2840	57.12
2,010	0.75	0.5665	9.8854	68.16
2,200	0.80	0.5376	11.8426	81.66
2,500	0.90	0.4798	15.2927	105.44
2,740	1.00	0.4220	18.3698	126.66
3,120	1.05	0.3931	23.8184	164.23
3,810	1.10	0.3642	35.5184	244.90
4,510	1.20	0.3064	49.7687	343.16
5,690	1.25	0.2775	79.2188	546.21

N = centrifuge speed, rpm.
V_{water} = brine displaced, cm^2.
S_{av} = average brine saturation $(V_p - V_{water})/V_p$.
P_c (psi) $= 1.588 \times 10^{-7} \times 0.9988 \times (8.68 - L/2) \times L \times N^2$.
P (kPa) $= 6.895 \times$ i (psi).
Calculation of the capillary pressure as a function of the average water saturation. ($V_p = 173$; $\phi = 0.170$; $K = 144$; $L = 2.01$; $S_{iw} = 0.278$; $p_{air} = 0.0012$). Centrifuge measurement of air displacing water from $S_w = 1.0$ to $S_w = S_{iw}$.

TABLE A16.2 Pore Size Distribution of a Berea Sandstone Core

S_{av}	P(HYPER) (psi)	P(HYPER) (kPa)	R_i	$D(R_i)$
1.000	2.258	15.570	9.249	0.367
0.950	2.682	18.490	7.788	0.454
0.900	3.167	21.840	6.593	0.549
0.850	3.731	25.724	5.598	0.654
0.800	4.391	30.279	4.756	0.767
0.750	5.177	35.697	4.034	0.890
0.700	6.127	42.249	3.408	1.022
0.650	7.300	50.331	2.861	1.162
0.600	8.782	60.550	2.378	1.312
0.550	10.716	73.885	1.949	1.471
0.500	13.345	92.014	1.565	1.639
0.450	17.127	118.092	1.219	1.817
0.400	23.033	158.812	0.907	2.003
0.350	33.549	231.323	0.623	2.198
0.300	57.533	396.688	0.363	2.403
0.278	81.631	562.846	0.256	2.496

$R_i = 144/P(\text{kPa})$.
$D(R_i) = (P \times V_p/R_i) \times (dS/dP)$.
$dS/dP = -(1 + C \times S)^2/(B - A \times C)$.
Calculation of the pore size distribution as a function of even increments of the average brine saturation. Program HYPER was used to smooth the data and to obtain the capillary pressure for even increments of average saturation. HYPER constant: $A = -15.5296$, $B = 17.6118$, and $C = -4.5064$.

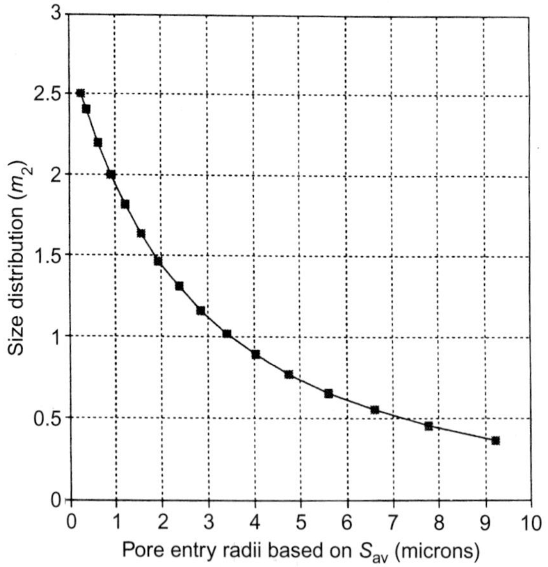

FIGURE A16.1 Pore entry size distribution of a core determined from an air–water capillary pressure curve.

References

Amyx JW, Bass Jr. DM, Whiting RL. Petroleum reservoir engineering. New York: McGraw-Hill; 1960, chapter 3.

Burdine NT, Gourney LS, Reichertz PP. Pore size distribution of petroleum reservoir rocks. Trans AIME 1950;189:195–204.

Donaldson EC. Use of capillary pressure curves for analysis of production well formation damage. In: Proc. SPE production operations symposium, SPE Paper No. 13809, Oklahoma City; March 10–12, 1985. p. 7.

Dullien FAL. Porous media. New York: Academic Press; 1979, chapter 3.

Hassler GL, Brunner E. Measurement of capillary pressure in small core samples. Trans AIME 1945;160:114–23.

Rajan RR. Theoretically correct analytic solution for calculating capillary pressure-saturation from centrifuge experiments. In: Proc. SPWLA annual logging symposium, June 9–13, 1986. pp. 1–18.

Torbati HM, Raiders RA, Donaldson EC, McInerney MJ, Jenneman GE, Knapp RM. Effect of microbial growth on pore entrance-size distribution in sandstone cores. J Ind Microbiol 1986;1:227–34.

Experiment 17 Determination of Z-Factors for Imperfect Gases

Introduction

The object of this experiment is to quantify the pressure–volume–temperature (PVT) behavior of imperfect gases. A diagram of required equipment is shown in Figure A17.1: pressure vessels, vacuum pump, water bath, and pressure gauges.

The Ideal, or Perfect, Gas Law expresses the pressure–volume–temperature relationship of an ideal gas and is a combination of Charles', Boyle's, and Avogadro's laws, which are special cases of the Ideal Gas Law:

$$PV = nRT \tag{A17.1}$$

where

P = pressure

V = volume

n = number of moles (mass/molecular mass)

R = gas constant that depends on the system of units used in Eq. (A17.10), as listed in Table A17.1

T = temperature

The Ideal Gas Law contains several inherent assumptions that result in deviation from the behavior of real gases: (1) the gas molecules do not have interactive attractive forces, (2) they themselves have zero volume, and (3) the only energy possessed by the molecules is kinetic energy, which is proportional to the absolute temperature. The real gases approach ideal behavior only under special conditions, such as elevated temperature where the molecular attractive forces are diminished by the rapid movements of the gas molecules and at low pressures where the intermolecular forces are diminished because the molecules

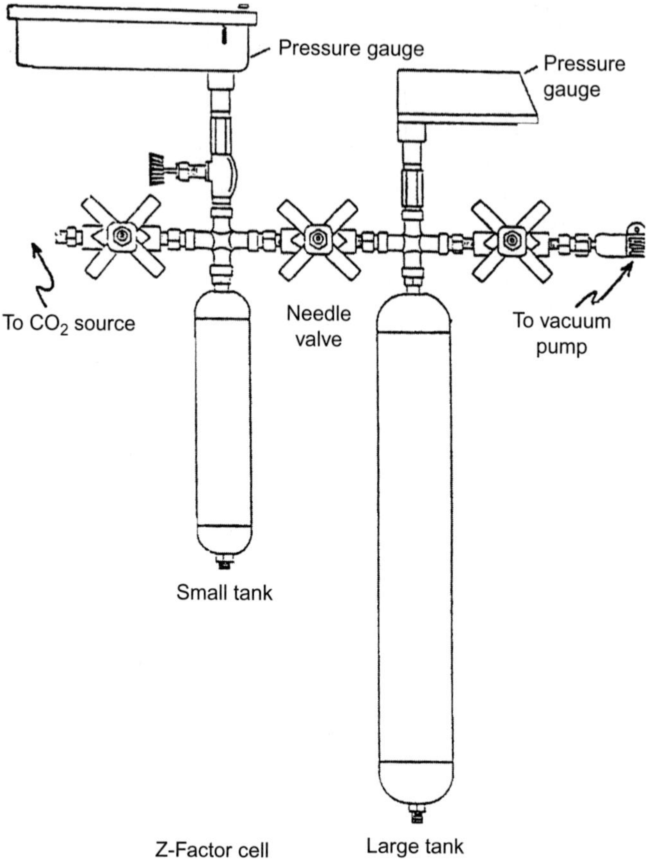

FIGURE A17.1 Assembly of equipment for pressure–volume–temperature (PVT) analyses of imperfect gases.

TABLE A17.1 Numerical Value of the Gas Law Constant as Related to the Units Used in the Gas Law Equations

P	*V*	*T*	*R*	*n*
psi	ft^3	°R	10.7	Pound-moles
lb/ft^2	ft^3	°R	1,545.0	Pound-moles
atmospheres	ft^3	°R	0.730	Pound-moles
atmospheres	cm^3	K	82.1	Gram-moles
atmospheres	liters	K	0.0321	Gram-moles
mm mercury	cm^3	K	62,369.0	Gram-moles
g/cm^3	cm^3	K	8.315	Gram-moles

are far apart. Since nonideal behavior is obvious outside of the two specific conditions mentioned, two methods to add corrections to the ideal gas law were developed to account for the nonideal behavior of real gases.

van der Waals Equation

The general van der Waals' correction for *n* moles of a gas is:

$$\left(\frac{P + n^2 \times a}{V^2}\right) \times (V - nb) = nRT, \text{ or in cubic form:}$$

$$n^3\left(\frac{ab}{V^2}\right) - n^2\left(\frac{a}{V}\right) + n(pb + RT) - PV = 0 \qquad \text{(A17.2)}$$

where

a = compensation for the attractive forces of the molecules

b = compensation for the volume of molecules

The van der Waals' constants are different for each gas, and tables of these constants are available in *Handbook of Chemistry and Physics*. The constants for a few common gases are listed in Table A17.2.

The second method for correcting for nonideal behavior is to add a parameter (Z) to the ideal gas law and determine this parameter for any gas of interest. Tables and charts of the compressibility, or Z-factors, for a wide variety of gases are available in numerous handbooks and textbooks. The nonideal (or imperfect) gas law is:

$$PV = ZnRT \qquad \text{(A17.3)}$$

Equipment And Procedures

1. Place the pressure vessels in a water bath and raise the temperature of the bath to 100°F.
2. Connect the vessels to the vacuum pump and evacuate both cells.
3. Close the valve between the two cells and introduce a test gas into the small cell; final pressure should be about 500 psi.
4. Make several readings of the pressure in the smaller cell until two of them are the same. The contents of the cells are now at thermal equilibrium.
5. Slightly open the valve between the cells and bleed gas from the smaller cell (at high pressure) into the larger (under vacuum) until the pressure in

TABLE A17.2 van der Waals' Constants for a Few Gases[a]

Gas	1	2
CO_2	3.592	0.04267
CH_4	2.253	0.04278
C_2H_2	4.390	0.05136
C_2H_4	4.471	0.05714
C_2H_6	5.489	0.06380

[a]*These apply only when the variables are expressed in the following units:* P *in atmospheres,* V *in liters,* T *in K,* n *in gram-moles, and* R = *0.08205. Using these units, the constants are expressed as (1) atm-liters*^{2} *and (2) liters.*

the larger cell reaches approximately atmospheric pressure; then close the valve.

6. After the cells reach thermal equilibrium (constant pressure), record the pressure readings of both cells.
7. Calculate the number of moles of gas bled into the larger cell using the ideal gas law ($Z = 1.0$).
8. Evacuate the larger tank, bleed gas into it until it reaches approximately atmospheric pressure once more, and record both pressures, as before.
9. Repeat the procedure until the pressure in the smaller cell reaches atmospheric pressure, and then calculate the moles of gas that remain in the system.
10. The total number of moles of gas introduced into the small cell in the beginning is the sum of the moles bled out of the system and the moles remaining in both cells at the end.
11. Use the nonideal gas law to determine Z at each step of the procedure (subtract the moles bled off at each step from the initial moles in the system to keep track of the moles in the system at each pressure decrement).

Questions and Problems

1. List all calculations for each step of the procedure. Plot Z as a function of P (at the constant temperature of the test) for the test gas.
2. Obtain the van der Waals' constants for the test gas and calculate the moles of test gas that were initially introduced into the smaller cell. Compare this to the calculations made using the Ideal Gas Law. Explain the difference.
3. A 30-ft^3 tank containing carbon dioxide is pressured up to 600 psi at 85°F. How many pounds of carbon dioxide were placed in the tank?
4. Calculate the temperature of 0.02 lb moles of ethane contained in a 6.6-ft^3 vessel at a pressure of 20 psia using the Ideal Gas Law and van der Waals' equation. Compare the results, and explain the difference.

References

Craft BC, Hawkins MF. Petroleum reservoir engineering. Englewood Cliffs, NJ: Prentice-Hall, Inc.; 1959, chapter 1.

Dake LP. Fundamentals of reservoir engineering. Amsterdam, the Netherlands: Elsevier Science; 1978, chapter 1.

Experiment 18 Basic Sediment and Water (BS&W)

Introduction

The basic sediment and water test is designed to determine the amount of water and sediments, or sludge, that is present in a stock-tank crude oil. Using a centrifuge, the water and sludge are accumulated and measured in a conical centrifuge tube.

The test is described in the ASTM manuals as D96-52T. It should be noted that this test does not remove all of the water contained in the oil. Some water remains as emulsified micelles that cannot be removed by the low-speed centrifuge generally used for this test. Pear-shaped centrifuge tubes containing a small-diameter, graduated tube at the bottom should be used for samples that contain small amounts of BS&W.

Equipment and Procedures

1. Place 50 cm^3 of benzene (industrial grade or better) into the centrifuge tubes and add 50 cm^3 of the oil to be tested; mix thoroughly. Place stoppers (lightly) at the top of the tubes and heat the tubes in a water bath to 120°F (49°C) for approximately 10 minutes. If the crude oil has a large quantity of waxy material, preheat the oil–benzene mixture to 140°F (60°C) before placing the tubes in the centrifuge.
2. Place the centrifuge tubes in holders, or shields, provided for the type of tube being used, and use a balance to match pairs of centrifuge tubes by adding, or subtracting, small amount of fluids from the tubes. Place the pairs of tubes across from each other in the trunnion rings. The samples are centrifuged at sufficient rpm to produce a minimum relative centrifugal force (RCF) of 600 at the tips of the tubes. The RCF is calculated from the following equation:

$$\mathrm{RPM} = 265 \times \left(\frac{\mathrm{RCF}}{d}\right)^{0.5} \qquad \text{(A18.1)}$$

where

RCF = relative centrifugal force

d = diameter of rotation (inches). Measure the radius from the center of the rotating arm to the tip of the tubes and multiply by 2.0.

3. Centrifuge for 10 minutes and record the amount of BS&W collected at the bottom of the tubes. Re-heat in the water bath and centrifuge once more for another 10 minutes. Record the second measurement. If the measurements are not equal, repeat the heating and centrifuging. Four repetitions should be the maximum number required to obtain stabilized measurements.
4. The combined measurement (water and sediment) in the tube is reported as percent BS&W.

Questions and Problems

1. Test several crude oils for basic sediment and water (BS&W).
2. What are the forces acting on the mixture of fluids and sediments that cause their separation?
3. Why is a centrifuge preferred for this test?

Reference

ASTM Test D96-52T. The American Society for Testing and Materials, West Conçhohochen, Pa., USA.

Experiment 19 Point-Load Strength Test

Introduction

The point-load test is designed to provide a rapid, portable index test for rock strength. A sample is compressed between solid steel cones that generate tensile stress normal to the axis of loading. The point-load strength index (I_s) is defined as the load (pressure) measured at the point of failure of the rock sample (F_a) divided by the distance between the conical platens at the moment of failure (L_s).

Empirical testing has revealed a direct correlation of the I_s index to the uniaxial compressive rock strength when the test is performed using a cylindrical core (the length of the core is greater than 1.4 times the diameter of the core) and the diametral test is performed.

Equipment and Procedures

The point-load test equipment is illustrated in the text (Figures 9.21 and 9.22). It consists of a hydraulic pump connected to a maximum indicating pressure gauge that measures the applied load (F_a, psi) and a movable conical platen. It is also equipped with a distance-measuring system to indicate the distance, L_s, between the platen contact points when failure occurs.

The loading system should be adjusted to accept and test available rock specimens, for example, in the size range of 25–100 mm for a maximum loading capacity of 50 kN.

When saturated with brine, samples should normally be tested to simulate natural strength conditions; however, oven-dried samples may also be tested. Ambient humidity weakens the rock; therefore, oven drying and testing immediately after cooling is necessary for consistent results.

(a) *Diametral test:* Prepare a core specimen with a length/diameter ratio equal to or greater than 1.4. The specimen is inserted in the test machine and the platens are closed to make contact along the core diameter at the center. The distance, L_s, is recorded and the load is increased to failure. The failure load, F_a, and the distance traveled by the platens, L_s, are recorded.

(b) *Axial test:* Prepare core specimens with length/diameter ratios of 1.1±0.5. The specimen is inserted in the test machine and the platens are closed to make contact along the core axis. The distance, L_s, is recorded and the load is increased to failure. The failure load, F_a, and the distance traveled by the platens, L_s, are recorded.

(c) *Irregular lump test:* Rock lumps with a diameter of 50 mm and the ratio of the longest to shortest diameter between 1.0 and 1.4 are prepared. The specimens are inserted in the test machine with contact along the longest diameter of the lump. The load is applied and F_a and L_s are recorded.

Sample Calculations

The point-load strength index, i_s, is defined by the ratio:

$$I_s = \frac{F_a}{L_s^2} \tag{A19.1}$$

For standard classification, the I_s index should be corrected to the equivalent value of 50 mm samples (when other sizes are used), using the correction chart (Figure A8.1). For example, the measured value of a 90-mm core is 0.9 mN/m^2 for a medium-hard sample. The I_s (50) value is 1.1 mN/m^2.

Empirical correlation tests between the uniaxial compressive strength and the point-load test index have shown a direct ratio of 23.7 (ratios of uniaxial strength to point-load strength), therefore:

$$D_{\text{uniaxial}} = 23.7 \times I_s(50)$$

Questions and Problems

1. Define uniaxial compressive strength. How is it measured?
2. What are the advantages and disadvantages of the point-load strength test?
3. What is the Brazilian point-load strength test?

Reference

Broch E, Franklin JA. The point-load strength test. Int J Rock Mech Min Sci 1972;9:669–97.

UTILITIES

This section is added to explain a few useful laboratory procedures that were not developed into experiments.

PRESERVATION OF CORES

1. Original wettability of cores can be preserved for long periods by immersing the core in degassed formation crude oil as soon as possible after it is obtained (at the wellsite) and then storing it in a "cold room" at about 2°C (36°F).
2. Humic materials and small amounts of residual polynuclear aromatic compounds (such as asphaltenes and resins) can be removed from cores by cleaning them with steam (Figure A19.1). Steam does not disturb water-sensitive clays (Donaldson et al., 1991).
3. Cores used in waterflood experiments, etc., with oils and water may be returned to original wettability using the following procedure:
   - **(a)** Clean with toluene to remove crude oils.
   - **(b)** Clean with steam to remove residual toluene and some heavy ends that cannot be removed with toluene alone (Figure A19.1). Steam will not disturb the clays.
   - **(c)** Dry, evacuate, and saturate the core with brine.
   - **(d)** Displace the brine to S_{iw} with crude oil from the formation.
   - **(e)** Place the core under the crude oil in a sealed steel vessel and heat at 65°C (150°F) for at least 100 hours. The core supposedly will be restored to original wettability.
4. Poorly consolidated sands have to be protected from mechanical disturbance as much as possible. Cores of friable sands can be secured.

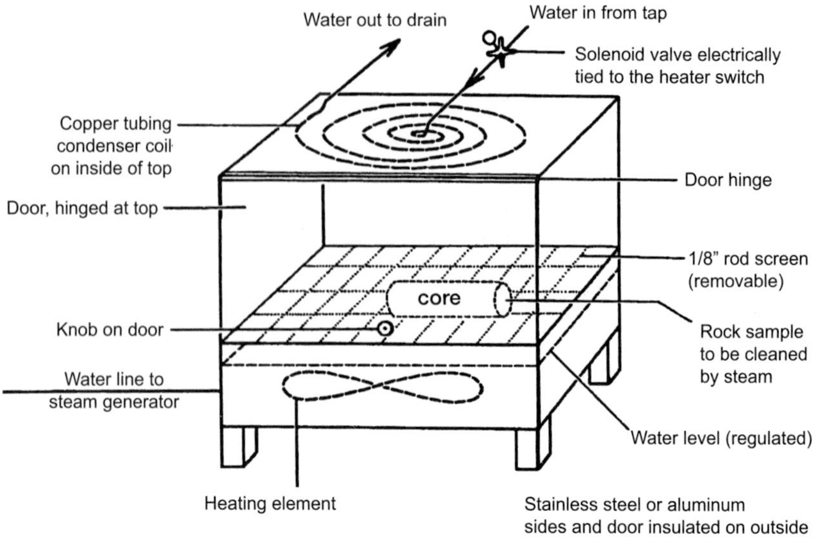

FIGURE A19.1 Design of equipment for cleaning cores with steam. A standard hot water bath was modified with a hardware screen and an insulated box equipped with a condenser coil at the top.

Core plugs can be obtained with a cutting drill using liquid nitrogen for the lubricant. The growth of ice crystals in the pores can cause mechanical disturbance of the grain contacts; therefore, frozen cores must be cut at a very slow rate of drill penetration. The core plugs can be encapsulated in a plastic or steel tube with resin painted on the outside to provide a tight fit, or heat-shrink tubing may be used for a jacket.

5. Some crude oils contain polar organic compounds that render rocks oil-wet and react with atmospheric oxygen to form large molecules, appearing in the oil as a flocculent material. Filtration for their removal is not possible because precipitation of more oxygen-linked compounds occurs due to material exposure to more atmospheric oxygen. If the crude oil behaves in this manner, a sample will have to be obtained at the well under a blanket of nitrogen (displace air from the container with nitrogen and then introduce the oil, at the well, using a piece of tubing that reaches almost to the bottom of the container). All laboratory tests with the crude oil will have to be conducted under a blanket of nitrogen. Any contact with atmospheric oxygen will immediately cause formation of the flocculent precipitates.

SYNTHETIC BRINE SOLUTIONS

Synthetic brines are used to ensure the purity of the solution and the availability of a constant supply. Consequently, reagent-grade salts, which have been dried in a vacuum oven at 150°C or a nonvacuum oven at 250°C to constant mass should be used. The water should be deionized or triple distilled and free of dissolved carbon dioxide. The presence of atmospheric carbon dioxide in the purified water can be inferred by determining the resistivity of the water. Water in equilibrium with atmospheric carbon dioxide has a resistivity of about 15,000 ohm-m, whereas pure water, which is free of carbon dioxide, has a resistivity of about 100,000 ohm-m.

The synthetic brines should contain calcium chloride to provide divalent cations as well as monovalent sodium from sodium chloride. The ratio of calcium chloride to sodium chloride should generally be about 1–5, respectively.

If the brine solution is to be used over a long period of time (several months), a small amount of mercuric chloride (about 10 ppm) should be added to the solution to prevent the growth of microbes in the brine solution.

The most accurate method for preparing the solution is to accurately weigh each of the components. Select an electric balance that is capable of obtaining the mass of the final solution. Place the empty container on the balance, obtain the mass of the salts on a separate balance, transfer the salts to the solution container, and observe the mass of the container and the salts. Add some water to the container and shake to completely dissolve the salts, replace the container on the balance, and add water to the desired level and obtain its mass.

$$\text{ppm} = \frac{\text{grams solute}}{(\text{grams solvent} \times 10^6)}$$

If the solution is prepared by adding a measured volume of water to a given mass of salt, calculation of the parts per million (ppm) must be made using the density of the water at the temperature at which the volume of water was measured:

$$\text{ppm} = \frac{\text{grams solute}}{[\text{volume solvent}(\text{cm}^3) \times \text{density of solvent}(\text{g/cm}^3) \times 10^6]}$$

DEADWEIGHT TESTER

The deadweight tester is used to calibrate and adjust bourdon-tube-type pressure gauges. The deadweight tester essentially consists of a set of weights and a piston. A known force exerted by the weights on the piston is transmitted to the gauge through a column of oil. The gauge pressure readings are plotted as a function of the true pressure exerted by the deadweights to obtain the correction to be applied to the gauge. If the gauge pressure is parallel to a line that represents the true deadweight pressure (45°-line on linear graph paper), the gauge can be mechanically adjusted by resetting the position of the measuring arm (the hand). If the gauge pressure deviates (plus or minus) monotonically from the true pressure, the gauge can be corrected by adjusting the length of the connecting arm between the bourdon tube and the supporting pin (the pin supporting the hand).

PROCEDURE

1. Make sure that oil fills the pipe that leads to the gauge connection. Attach the pressure gauge to the deadweight tester at the port provided and tighten with a wrench.
2. Make sure that oil completely fills the tube that holds the deadweight piston before inserting the piston. Turn the screw to properly support the weight-piston in the cylinder. The weight-piston should turn freely.
3. Place known weights on the weight-piston and rotate them freely as measurements of the pressure readings of the gauge are recorded beside the weight being used. Tap the pressure gauge lightly with a finger during the calibration test to ensure that parts of the gauge mechanism are not sticking because of friction.
4. After reaching the limit of the gauge, continue the calibration procedure by removing weights and recording the results.
5. Prepare a chart of the true pressure; note any deviation from the exact pressure exerted by the weights and the gauge pressure. If the chart indicates that the gauge can be adjusted, make the necessary adjustments as detailed and recalibrate the gauge to ensure that it is operating correctly. If the deviation is such that mechanical adjustments of the gauge will not correct the problem, then prepare a calibration chart to be used with that gauge to obtain correct pressures.

REFERENCES

Donaldson EC, Ewall N, Singh B. Characteristics of capillary pressure curves. J Pet Sci Eng 1991;6:249–61.

Donaldson EC, Kendall RF, Baker BA, Manning FS. Surface area measurement of geologic materials. J Pet Sci Eng 1975;15(2):111–16. April.

Worthington AE, Hedges JH, Pallatt N. SCA guidelines for sample preparation and porosity measurement of electrical resistivity samples, Part I. The Log Analyst 1990;31(1):20–8 January–February.

Lerner DB, Dacy JM, Raible CJ, Rathmell JJ, Swanson G, Walls JD. SCA guidelines for sample preparation and porosity measurement of electrical resistivity samples, Part II. The Log Analyst 1990;31(2):57–63.

Index

C

D

E

H

J

K

L

M

N

O

P

Q

R

S

T

X

Y

Z